Trigonometry

Right Triangle Trigonometry

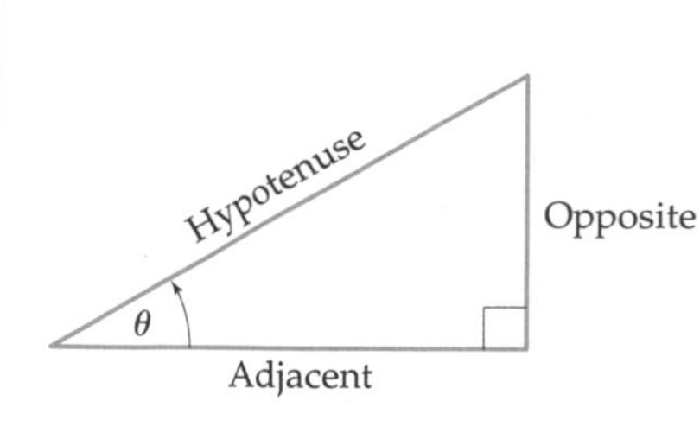

$$\sin\theta = \frac{\text{Opposite}}{\text{Hypotenuse}}$$

$$\cos\theta = \frac{\text{Adjacent}}{\text{Hypotenuse}}$$

$$\tan\theta = \frac{\text{Opposite}}{\text{Adjacent}}$$

Special Right Triangles

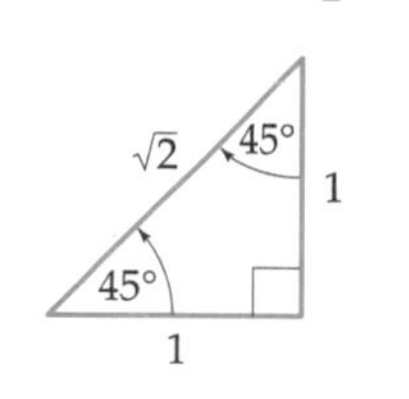

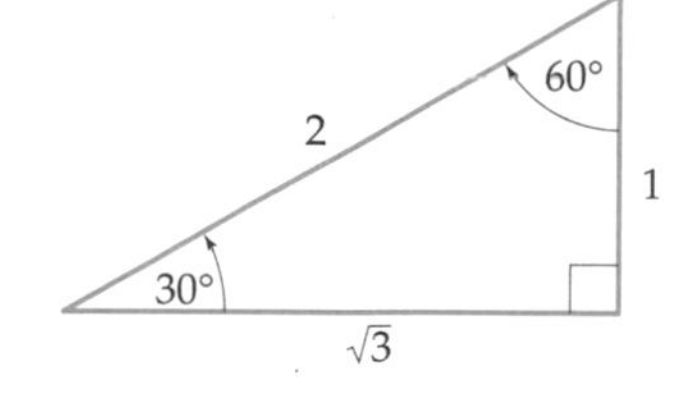

Special Values

θ Degrees	θ Radians	$\sin\theta$	$\cos\theta$	$\tan\theta$
0°	0	0	1	0
30°	$\frac{\pi}{6}$	$\frac{1}{2}$	$\frac{\sqrt{3}}{2}$	$\frac{\sqrt{3}}{3}$
45°	$\frac{\pi}{4}$	$\frac{\sqrt{2}}{2}$	$\frac{\sqrt{2}}{2}$	1
60°	$\frac{\pi}{3}$	$\frac{\sqrt{3}}{2}$	$\frac{1}{2}$	$\sqrt{3}$
90°	$\frac{\pi}{2}$	1	0	undefined

Law of Cosines

$$a^2 = b^2 + c^2 - 2bc\cos A$$

$$b^2 = a^2 + c^2 - 2ac\cos B$$

$$c^2 = a^2 + b^2 - 2ab\cos C$$

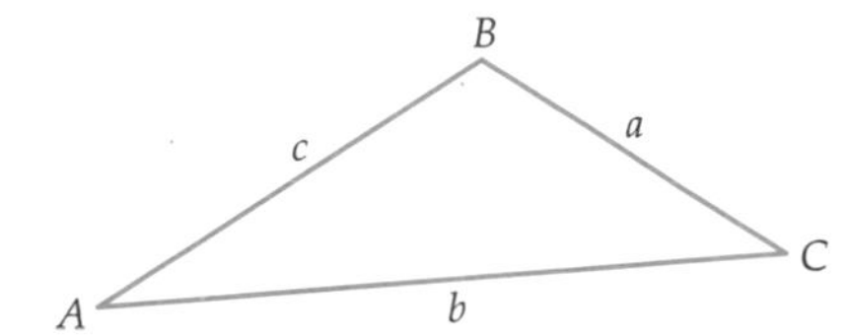

Law of Sines

$$\frac{a}{\sin A} = \frac{b}{\sin B} = \frac{c}{\sin C}$$

Triangle Area Formulas

Heron's Formula: Area $= \sqrt{s(s-a)(s-b)(s-c)}$, where $s = \frac{1}{2}(a+b+c)$

Area $= \frac{1}{2}ab\sin C$

Unit Circle Description

If t is a real number and P is the point where the terminal side of an angle of t radians in standard position meets the unit circle, then

$\cos t = x$-coordinate of P $\qquad$ $\sin t = y$-coordinate of P

$$\tan t = \frac{\sin t}{\cos t} \qquad \csc t = \frac{1}{\sin t} \qquad \sec t = \frac{1}{\cos t} \qquad \cot t = \frac{\cos t}{\sin t}$$

Point-in-the-Plane Description

For any real number t and point (x, y) on the terminal side of an angle of t radians in standard position:

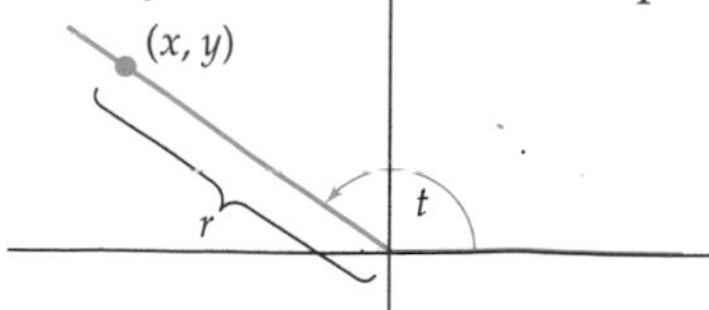

$$\sin t = \frac{y}{r} \qquad \cos t = \frac{x}{r} \qquad \tan t = \frac{y}{x} \quad (x \neq 0)$$

$$\csc t = \frac{r}{y} \quad (y \neq 0) \qquad \sec t = \frac{r}{x} \quad (x \neq 0) \qquad \cot t = \frac{x}{y} \quad (y \neq 0)$$

Periodic Graphs

If $A \neq 0$ and $b > 0$, then each of $f(t) = A\sin(bt + c)$ and $g(t) = A\cos(bt + c)$ has

Amplitude $|A|$, $\qquad$ Period $2\pi/b$, $\qquad$ Phase shift $-c/b$.

Jenny Garcia —— Salvatore 08

Paul M-F - 12-13

Contemporary Trigonometry

A Graphing Approach

THOMAS W. HUNGERFORD
Cleveland State University
Saint Louis University

THOMSON
BROOKS/COLE™

Australia • Canada • Mexico • Singapore • Spain
United Kingdom • United States

Contemporary Trigonometry
A Graphing Approach
Thomas W. Hungerford

Sponsoring Editor: *John-Paul Ramin*
Development Editor: *Leslie Lahr*
Assistant Editor: *Katherine Brayton*
Editorial Assistant: *Leata Holloway*
Project Manager, Editorial Production: *Janet Hill*
Technology Project Manager: *Earl Perry*
Senior Marketing Manager: *Karin Sandberg*
Marketing Assistant: *Erin Mitchell*
Managing Marketing Communications Project Manager: *Bryan Vann*
Print/Media Buyer: *Barbara Britton*
Permissions Editor: *Kiely Sisk*
Production Service: *Hearthside Publishing Services/Laura Horowitz*
Text Designer: *Terri Wright*
Art Editor: *Hearthside Publishing Services*
Photo Researcher: *Terri Wright*
Illustrator: *Hearthside Publishing Services*
Cover Designer: *Terri Wright*
Cover Printer: *Phoenix Color Corp*
Compositor: *Better Graphics, Inc.*
Printer: *R.R. Donnelley/Willard*

Printed in the United States of America

1 2 3 4 5 6 7 07 06 05

Library of Congress Control Number: 2005920288

Student Edition: ISBN 0-534-46638-9

Instructor's Edition: ISBN 0-534-46639-7

Thomson Brooks/Cole
10 Davis Drive
Belmont, CA 94002
USA

Asia
Thomson Learning
5 Shenton Way #01-01
UIC Building
Singapore 068808

Australia/New Zealand
Thomson Learning
102 Dodds Street
Southbank, Victoria 3006
Australia

Canada
Nelson
1120 Birchmount Road
Toronto, Ontario M1K 5G4
Canada

Europe/Middle East/Africa
Thomson Learning
High Holborn House
50/51 Bedford Row
London WC1R 4LR
United Kingdom

Latin America
Thomson Learning
Seneca, 53
Colonia Polanco
11560 Mexico D.F.
Mexico

Spain/Portugal
Paraninfo
Calle/Magallanes, 25
28015 Madrid, Spain

To the memory of my teacher and friend
Charles Conway
St. Louis University High School

CONTENTS

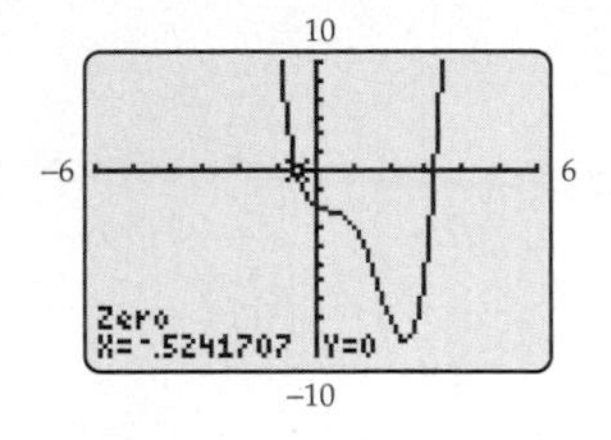

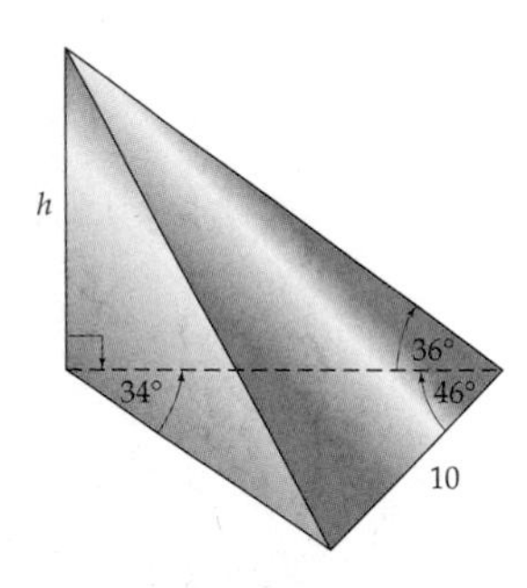

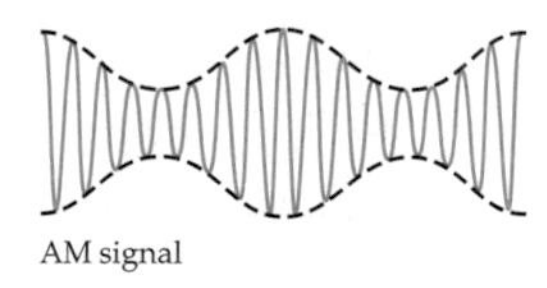
AM signal

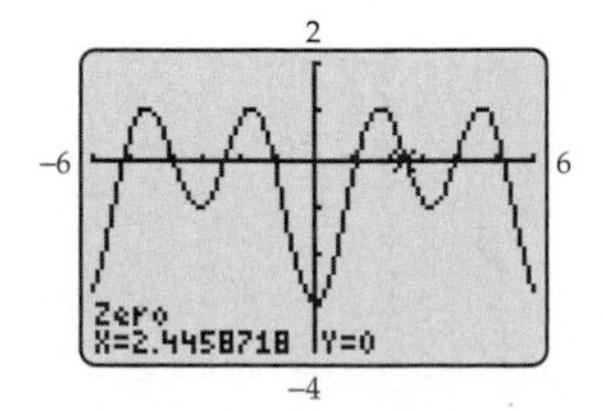

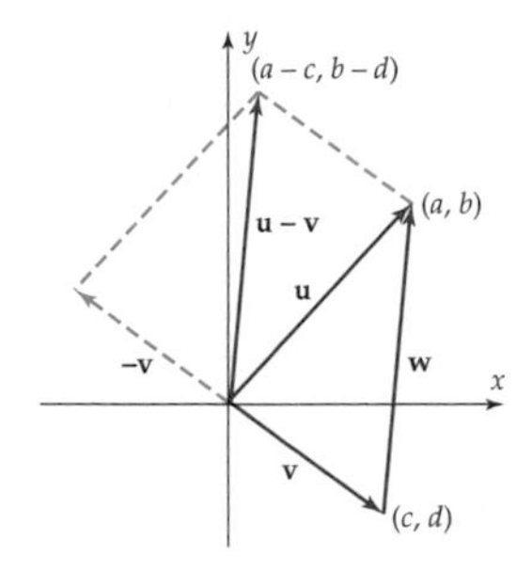
y
(a − c, b − d)
u − v
(a, b)
u
−v
w
x
v
(c, d)

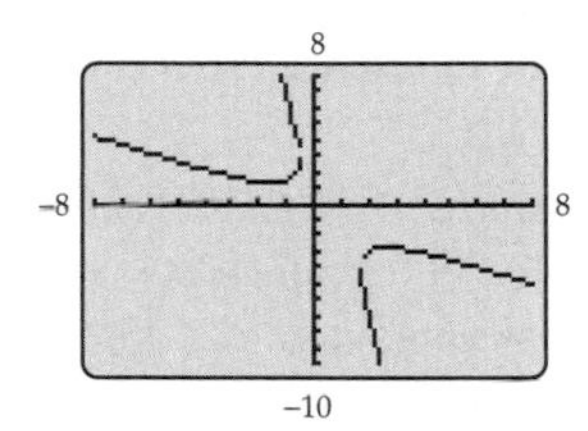
8
−8
8
−10

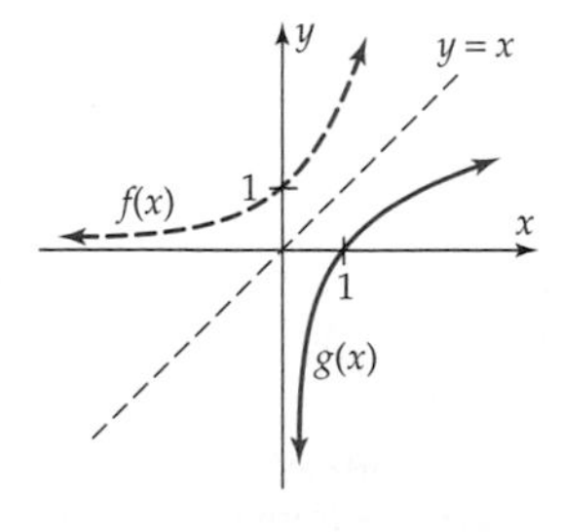
y
y = x
f(x)
1
x
1
g(x)

PREFACE

This book is designed to provide an introduction to trigonometry and its applications for students who have had two years of high school algebra or a college algebra course. The key prerequisites (algebra, functions, technology, and graphs) are reviewed in the (optional) Chapter 0.

As in other books in this series,* mathematical concepts are presented in an informal manner that stresses meaningful motivation, careful explanations, and numerous examples, with an ongoing focus on real-world problem solving. Technology is integrated into the text and students are expected to use it to participate actively in exploring various topics from algebraic, graphical, and numerical perspectives.

Mathematical and Pedagogical Features

Concepts are introduced informally, with an emphasis on understanding over formal proof (although proofs are included when appropriate). In particular, triangle trigonometry is introduced first, with trigonometric functions of a real variable considered later. However, instructors who wish to reverse this order may easily do so by using two alternate sections, as explained in "To the Instructor' on page x.

The helpful pedagogical features used in other books of this series are retained here, including:

Cautions These boxes alert students to common misconceptions and mistakes.

Exercises There are a wide variety of exercises, proceeding from routine drill to those requiring some thought, including graph interpretation and applied problems. Some exercise sets include *Thinkers* that challenge students to "think outside the box" (many of these are not difficult—just different).

Real-Data Applications A significant number of exercises and examples are based on real data. They provide a partial answer to the perennial question: "What's this stuff good for?".

Chapter Openers Each chapter begins with a brief example of an application of the mathematics treated in that chapter, together with a reference to appropriate exercises. The opener also lists the sections of the chapters and their interdependence, which makes it easy for instructors to rearrange the order to suit the needs of a particular class.

Chapter Reviews Each chapter concludes with a list of important concepts (referenced by section and page number), a summary of important facts and formulas, and a set of review questions.

Discovery Projects At the end of each chapter, there is an investigative problem (suitable for small group work) that enables students to apply some of the mathematics in the chapter to solve a real-world problem.

Geometry Review Appendix Frequently used facts about angles, right triangles, congruent and similar triangles, and straight lines are summarized, with examples and exercises.

Technology (particularly, the graphing calculator) is used as an important tool for effectively developing a fuller understanding of the underlying mathematics. It

**Contemporary College Algebra*, 2nd Edition, *Contemporary College Algebra and Trigonometry*, 2nd Edition, and *Contemporary Precalculus*, 4th Edition is integrated into the presentation, not just an optional add-on. Several features are designed to assist students to get the most out of the technological tools available to them.

Calculator Investigations These precede some exercise sets and encourage students to become familiar with the capabilities and limitations of graphing calculators.

Technology Tips Tips in the margin provide information and assistance with carrying out various procedures on specific calculators.

Graphing Explorations These enable students to participate actively in the development of various mathematical topics, by completing a discussion or by exploring appropriate examples.

For the Instructor

INSTRUCTOR'S EDITION

This special version of the complete student text contains a Resource Integration Guide and a complete set of answers printed in the back of the text.
ISBN: 0-534-46639-7

TEST BANK

The *Test Bank* includes six tests per chapter as well as three final exams. The tests are made up of a combination of multiple-choice, free-response, true/false, and fill-in-the-blank questions. **ISBN: 0-534-46646-X**

COMPLETE SOLUTIONS MANUAL

The *Complete Solutions Manual* provides worked out solutions to all of the problems in the text. **ISBN: 0-534-46642-7**

TEXT-SPECIFIC VIDEOTAPES

These text-specific videotape sets, available at no charge to qualified adopters of the text, feature 10- to 20-minute problem-solving lessons that cover each section of every chapter. **ISBN: 0-534-46645-1**

iLrn INSTRUCTOR VERSION

With a balance of efficiency and high-performance, simplicity and versatility, iLrn gives you the power to transform the teaching and learning experience. The iLrn Instructor Version is made up of two components, iLrn Testing and iLrn Tutorial. iLrn Testing is a revolutionary, internet-ready, text-specific testing suite that allows instructors to customize exams and track student progress in an accessible, browser-based format. iLrn offers full algorithmic generation of problems and free-response mathematics. iLrn Tutorial is a text-specific, interactive tutorial software program that is delivered via the web (at http://ilrn.com) and is offered in both student and instructor versions. Like iLrn Testing, it is browser-based, making it an intuitive mathematical guide even for students with little technological proficiency. So sophisticated, it's simple, iLrn Tutorial allows students to work with real math notation in real time, providing instant analysis and feedback. The tracking program built into the instructor version of the software enables instructors to carefully monitor student progress. The complete integration of the testing, tutorial, and course management components simplifies your routine tasks. Results flow automatically to your gradebook and you can easily communicate with individuals, sections, or entire courses.

WEBTUTOR TOOLBOX ON WEBCT AND BLACKBOARD

ISBN: 0-534-27488-9 WebCT, ISBN: 0-534-27489-7 Blackboard

Preloaded with content and available free via PIN code when packaged with this text, WebTutor ToolBox for WebCT pairs all the content of this text's rich Book Companion website with all the sophisticated course management functionality of a WebCT product. You can assign materials (including online quizzes) and have the results flow automatically to your gradebook. ToolBox is ready to use as soon as you log on—or, you can customize its preloaded content by uploading images and other resources, adding web links, or creating your own practice materials. Students only have access to student resources on the website. Instructors can enter a PIN code for access to password-protected Instructor Resources.

For the Student

STUDENT SOLUTIONS MANUAL

The *Student Solutions Manual* provides worked-out solutions to the odd-numbered problems in the text, as well as practice tests and summaries of key formulas and concepts. **ISBN: 0-534-46644-3**

BROOKS/COLE MATHEMATICS WEBSITE

http://mathematics.brookscole.com

The Book Companion Website offers a range of learning resources that can be used to help your students get the most from their class. Chapter-specific quizzes and tests allow students to assess their understanding—and receive immediate feedback. Students also have access to *Web Explorations* and an *Online Graphing Calculator Manual* that feature additional explorations of concepts.

iLrn TUTORIAL STUDENT VERSION

A complete content mastery resource offered online, iLrn Tutorial allows students to work with real math notation in real time, with unlimited practice problems, instant analysis and feedback, and streaming video to illustrate key concepts. And live, online, text-specific tutorial help is just a click away with vMentor, accessed seamlessly through iLrn Tutorial. The iLrn Tutorial access card located inside the cover of the text provides a PIN code for students to utilize when logging into iLrn.

INTERACTIVE VIDEO SKILLBUILDER CD-ROM

Think of it as portable office hours. The Interactive Video SkillBuilder CD-ROM contains more than eight hours of video instruction. The problems worked during each video lesson are shown next to the viewing screen so that students can try working them before watching the solution. To help students evaluate their progress, each section contains a 10-question web quiz (the results of which can be emailed to the instructor) and each chapter contains a chapter test, with the answer to each problem on each test. This CD-ROM also includes MathCue tutorial and quizzing software, featuring a SkillBuilder that presents problems to solve and evaluates answers with step-by-step explanations; a Quiz function that enables students to generate quiz problems keyed to problem types from each section of the book; a Chapter Test that provides many problems keyed to problem types from each chapter; and a Solution Finder that allows students to enter their own basic problems and receive step-by-step help as if they were working with a tutor.

Acknowledgments

This book would not have been possible, without the diligent work of

Phil Embree, William Woods University

who checked examples, exercises, and answers for accuracy, wrote several Discovery Projects, and provided helpful suggestions on other aspects of the text. I am also grateful to

Margaret Donlan, University of Delaware and
Sudhir Goel, Valdosta State University

who reviewed the text for accuracy and saved me from a number of embarrassments. Similar fine work was done by

Fred Safier, City College of San Francisco

who prepared the various solution manuals for the text and suggested a number of improvements. I also want to thank Lee Windsperger, a student at St. Louis University, who assisted me with manuscript preparation, as well as those who have prepared various supplements for the text.

Nancy Matthews, University of Oklahoma [Test Bank]
Eric Howe [Resource Integration Guide]

It is a pleasure to acknowledge the Brooks/Cole staff members who have been instrumental in producing this book.

John-Paul Ramin, Sponsoring Editor
Katherine Brayton, Assistant Editor
Leata Holloway, Editorial Assistant
Karin Sandberg, Senior Marketing Manager
Earl Perry, Technology Project Manager
Janet Hill, Senior Project Manager, Editorial Production
Vernon Boes, Senior Art Director
Bryan Vann, Marketing Communications Project Manager

My warmest thanks go to

Leslie Lahr, Developmental Editor,

who took on the difficult task of assembling the final manuscript and did her usual excellent work. Her enthusiasm, energy, and good common sense have made an otherwise difficult job much easier than I had any right to expect.

I also want to thank the outside production staff.

Terri Wright of Terri Wright Design

Special accolades are due to

Laura Horowitz of Hearthside Publishing Services

who coordinated the production of the book with tact, kindness, and finesse.

I especially want to thank my former colleagues at Cleveland State University, where this book was begun, for their friendship and support over the years. Thanks also go to Mike May, S.J. and my other new colleagues at Saint Louis University, where the book was finished, for the warm welcome they have given me.

Finally, I thank my wife Mary Alice, once again, for her support, understanding, and love, without which I would not long survive.

Thomas W. Hungerford

TO THE INSTRUCTOR

This book contains more than enough material for a single semester. By using the chart on the facing page (and the similar ones at the beginning of each chapter that show the interdependence of sections within the chapter) you can easily design a course to fit the needs of your students and the constraints of time. When planning your syllabus, three items are worth noting:

Prerequisites Chapter 0 (Preliminaries) is a prerequisite for the entire book. Students coming directly from a college algebra course, however, can probably omit (or skim) most or all of it.

Special Topics Sections with this label are usually related to the immediately preceding section and are not prerequisites for other sections of the text. I am inclined to omit most of them (particularly in a short course), but I know that a number of people consider some of them to be essential material. So feel free to include as many or as few as you wish.

Trigonometry Triangle trigonometry is introduced in Chapter 1, with trigonometric functions of a real variable (the "unit circle approach") in Chapter 2. Those who wish may reverse this order, as follows: Begin with Chapter 2, using Alternate Section 2.2 in place of Section 2.2. Chapter 1 may then be covered at any time, using Alternate Section 1.1 in place of Section 1.1.

This text assumes the use of technology. You and your students will need either a graphing calculator or a computer with appropriate software. All discussions of calculators in the text apply (with obvious modifications) to computer software. You should be aware of the following facts.

Minimal Requirements Current calculator models that meet the minimal requirements include TI-82 and above, HP-38 and above, and Casio 9850 and above (including FX 2.0).

Technology Tips The Tips in the margin provide general information and advice, as well as listing the proper menus or keys needed to carry out procedures on specific calculators. Unless noted otherwise:

Tips for	also apply to
TI-83 Plus	TI-83, TI-82, TI-84
TI-86	TI-85
TI-89	TI-92
Casio 9850	Casio 9970
HP-39+	HP-38, HP-39

You may want to explain to your students that the icon in the text indicates examples and exercises that are discussed in the Interactive Video SkillBuilder CD-ROM that accompanies the text.

INTERDEPENDENCE OF CHAPTERS

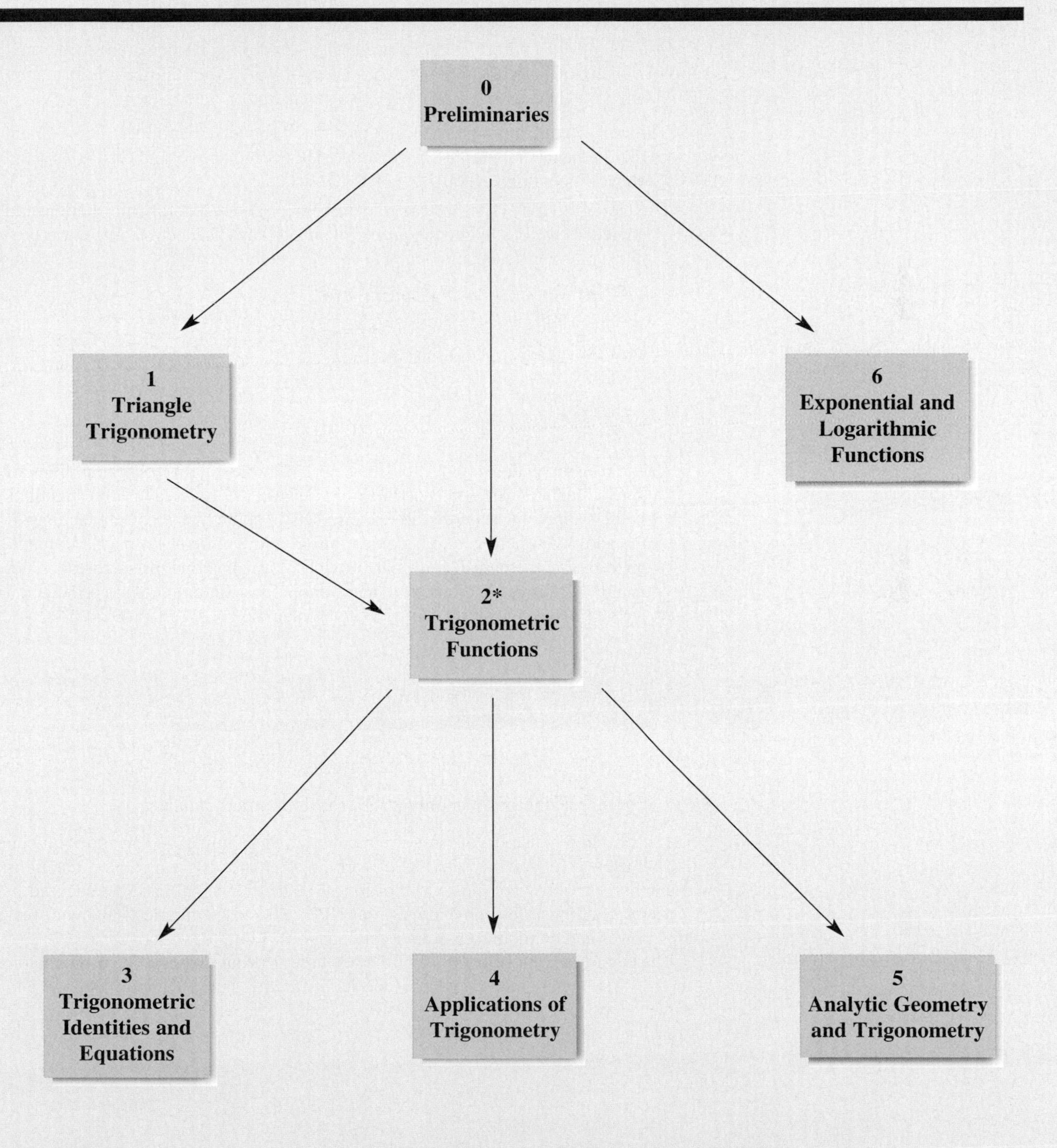

*Chapter 2 may be read before Chapter 1, as explained on the preceding page.

TO THE STUDENT

This text assumes the use of technology. You will need either a graphing calculator or a computer with appropriate software. All discussions of calculators in the text apply (with obvious modifications) to computer software. The following information should help you choose and effectively use a calculator.

Minimal Requirements Current calculator models that meet the minimal requirements include TI-82 and above, HP-38 and above, and Casio 9850 and above (including FX 2.0).

Technology Tips The Tips in the margin provide general information and advice, as well as listing the proper menus or keys needed to carry out procedures on specific calculators. Unless noted otherwise:

Tips for	also apply to
TI-83 Plus	TI-83, TI-82, TI-84 Plus
TI-86	TI-85
TI-89	TI-92
Casio 9850	Casio 9970
HP-39+	HP-38, HP-39

Calculator Investigations You may not be aware of the full capabilities of your calculator (or some of its limitations). The Calculator Investigations (which appear just before the exercise sets in some of the earlier sections of the book) will help you to become familiar with your calculator and to maximize the mathematical power it provides. Even if your instructor does not assign these investigations, you may want to look through them to be sure you are getting the most you can from your calculator.

Getting the Most Out of This Course

With all this talk about calculators, don't lose sight of this crucial fact:

Technology is only a *tool* for doing mathematics.

You can't build a house if you only use a hammer. A hammer is great for pounding nails, but useless for sawing boards. Similarly, a calculator is great for computations and graphing, but it is not the right tool for every mathematical task. To succeed in this course, you must develop and use your algebraic and geometric skills, your reasoning power and common sense, and you must be willing to work.

The key to success is to use all of the resources at your disposal: your instructor, your fellow students, your calculator (and its instruction manual), and this book and the Interactive Video SkillBuilder CD-ROM that accompanies it. Here are some tips for making the most of these resources.

Ask Questions Remember the words of Hillel:

The bashful do not learn.

There is no such thing as a "dumb question" (assuming, of course, that you have attended class and read the text). Your instructor will welcome questions that arise from a serious effort on your part.

Read the Book Not just the homework exercises, but the rest of the text as well. There is no way your instructor can possibly cover the essential topics, clarify ambiguities, explain the fine points, and answer all your questions during class time. You simply will not develop the level of understanding you need to succeed in this course and in calculus unless you read the text fully and carefully.

Be an Interactive Reader You can't read a math book the way you read a novel or history book. You need pencil, paper, and your calculator at hand to work out the statements you don't understand and to make notes of things to ask your fellow students and/or your instructor.

Do the Graphing Explorations When you come to a box labeled "Graphing Exploration," use your calculator as directed to complete the discussion. Typically, this will involve graphing one or more equations and answering some questions about the graphs. Doing these explorations as they arise will improve your understanding and clarify issues that might otherwise cause difficulties.

Do Your Homework Remember that

Mathematics is not a spectator sport.

You can't expect to learn mathematics without doing mathematics, any more than you can learn to swim without getting wet. Like swimming or dancing or reading or any other skill, mathematics takes practice. Homework assignments are where you get the practice that is essential for passing this course.

CHAPTER 0

Preliminaries

Stockphoto.com/Michael Howell

On a clear day, can you see forever?

If you are at the top of the Sears Tower in Chicago, how far can you see? In earlier centuries, the lookout on a sailing ship was posted atop the highest mast because he could see farther from there than from the deck. How much farther? These questions, and similar ones, can be answered (at least approximately) by using basic algebra and geometry. See Exercise 101 on page 13.

Chapter Outline

Interdependence of Sections

Each section in this chapter depends on the preceding one.

This chapter reviews the essential facts about equations, functions, graphing, and technology that are needed in the remainder of the book. Students who have studied these topics (in a college algebra course, for example) may be able to omit (or at least skim) this material.

0.1 The Real Numbers and Equations

You have been using **real numbers** most of your life. They include the **natural numbers** (or **positive integers**): 1, 2, 3, 4, . . . and the **integers:**

$$\ldots, -5, -4, -3, -2, -1, 0, 1, 2, 3, 4, 5, \ldots$$

A real number is said to be a **rational number** if it can be expressed as a fraction $\frac{r}{s}$, with r and s integers and $s \neq 0$; for instance,

$$\frac{1}{2}, \qquad -.983 = -\frac{983}{1000}, \qquad 47 = \frac{47}{1}, \qquad 8\tfrac{3}{5} = \frac{43}{5}.$$

Alternatively, rational numbers may be described as numbers that can be expressed as terminating decimals, such as $.25 = \frac{1}{4}$, or as nonterminating repeating decimals, in which a single digit or a block of digits repeats forever, such as

$$\frac{5}{3} = 1.66666\cdots \qquad \text{or} \qquad \frac{58}{333} = .174174174\cdots.$$

A real number that cannot be expressed as a fraction with integer numerator and denominator is called an **irrational number.** Alternatively, an irrational number is one that can be expressed as a nonterminating, nonrepeating decimal (no block of digits repeats forever). For example, the number π, which is used to calculate the area of a circle, is irrational.[†]

*Section 0.6 is first used in Sections 2.4 and 2.5. It may be postponed until needed.

[†]This fact is difficult to prove. In the past you may have used $22/7$ as π; a calculator might display π as 3.141592654. However, these numbers are just *approximations* of π (close, but not quite *equal* to π).

The real numbers are often represented geometrically as points on a **number line,** as in Figure 0–1. We shall assume that there is exactly one point on the line for every real number (and vice versa) and use phrases such as "the point 3.6" or "a number on the line." This mental identification of real numbers and points on the line is often extremely helpful.

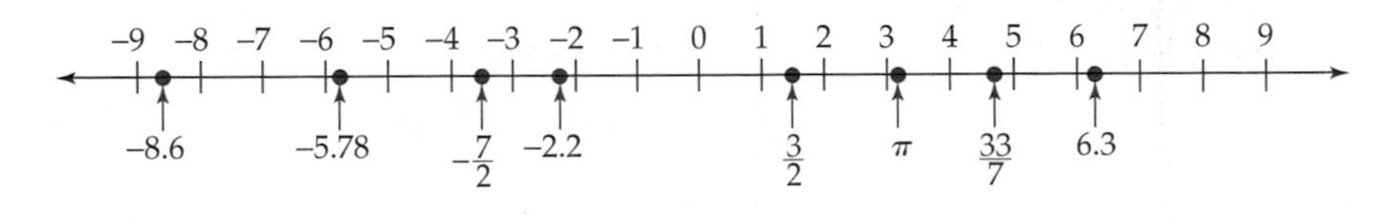

Figure 0–1

ORDER

The statement $c < d$, which is read ***"c* is less than *d*,"** and the statement $d > c$ (read ***"d* is greater than *c*"**) mean exactly the same thing:

c lies to the *left* of d on the number line.

For example, Figure 0–1 shows that $-5.78 < -2.2$ and $4 > \pi$.

The statement $c \leq d$, which is read ***"c* is less than or equal to *d*,"** means

Either c is less than d or c is equal to d.

Only one part of an "either . . . or" statement needs to be true for the entire statement to be true. So the statement $5 \leq 10$ is true because $5 < 10$, and the statement $5 \leq 5$ is true because $5 = 5$. The statement $d \geq c$ (read ***"d* is greater than or equal to *c*"**) means exactly the same thing as $c \leq d$.

The statement $b < c < d$ means

$$b < c \qquad \text{and simultaneously} \qquad c < d.$$

For example, $3 < x < 7$ means that x is a number that is strictly between 3 and 7 on the number line (greater than 3 and less than 7). Similarly, $b \leq c < d$ means

$$b \leq c \qquad \text{and simultaneously} \qquad c < d,$$

and so on.

TECHNOLOGY TIP

To enter a negative number, such as −5, on most calculators, you must use the negation key: (−) 5. If you use the subtraction key on such calculators and enter − 5, the display will read

ANS − 5

which tells the calculator to subtract 5 from the previous answer.

NEGATIVE NUMBERS AND NEGATIVES OF NUMBERS

The **positive numbers** are those to the right of 0 on the number line, that is,

All numbers c with $c > 0$.

The **negative numbers** are those to the left of 0, that is,

All numbers c with $c < 0$.

The **nonnegative numbers** are the numbers c with $c \geq 0$.

The word "negative" has a second meaning in mathematics. The **negative *of a number c*** is the number $-c$. For example, the negative of 5 is -5, and the negative of -3 is $-(-3) = 3$. Thus the negative of a negative number is a positive number. Zero is its own negative, since $-0 = 0$. In summary,

Negatives

> The negative of the number c is $-c$.
>
> If c is a positive number, then $-c$ is a negative number.
>
> If c is a negative number, then $-c$ is a positive number.

SQUARE ROOTS

The **square root** of a nonnegative real number d is defined as the nonnegative number whose square is d and is denoted $\sqrt{d}$. For instance,

$$\sqrt{25} = 5 \qquad \text{because} \qquad 5^2 = 25.$$

In the past you may have said that $\sqrt{25} = \pm 5$, since $(-5)^2$ is also 25. It is preferable, however, to have a single unambiguous meaning for the symbol $\sqrt{25}$. So in the real number system, the term "square root" and the radical symbol $\sqrt{\ }$ always denote a *nonnegative* number. To express -5 in terms of radicals, we write $-5 = -\sqrt{25}$.

Although $-\sqrt{25}$ is a real number, the expression $\sqrt{-25}$ is *not defined* in the real numbers because there is no real number whose square is -25. In fact, since the square of every real number is nonnegative,

No negative number has a square root in the real numbers.

Some square roots can be found (or verified) by hand, such as

$$\sqrt{225} = 15 \qquad \text{and} \qquad \sqrt{1.21} = 1.1.$$

Usually, however, a calculator is needed to obtain rational *approximations* of roots. For instance, we know that $\sqrt{87}$ is between 9 and 10 because $9^2 = 81$ and $10^2 = 100$. A calculator shows that $\sqrt{87} \approx 9.327379$.*

TECHNOLOGY TIP

To compute an expression such as $\sqrt{7^2 + 51}$ on a calculator, you must use parentheses:

$$\sqrt{\ }\,(7^2 + 51).$$

Without the parentheses, the calculator will compute

$$\sqrt{7^2} + 51 = 7 + 51 = 58$$

instead of the correct answer:

$$\begin{aligned}\sqrt{7^2 + 51} &= \sqrt{49 + 51}\\ &= \sqrt{100}\\ &= 10.\end{aligned}$$

ABSOLUTE VALUE

On an informal level, most students think of absolute value like this:

> The absolute value of a nonnegative number is the number itself.
>
> The absolute value of a negative number is found by "erasing the minus sign."

If $|c|$ denotes the absolute value of c, then, for example, $|5| = 5$ and $|-4| = 4$.

This informal approach is inadequate, however, for finding the absolute value of a number such as $\pi - 6$. It doesn't make sense to "erase the minus sign" here. So we must develop a more precise definition. The statement $|5| = 5$ suggests that the absolute value of a positive number ought to be the number itself. For negative numbers, such as -4, note that $|-4| = 4 = -(-4)$, that is, the absolute value of the negative number -4 is the *negative* of -4. These facts are the basis of the formal definition.

*$\approx$ means "approximately equal."

Absolute Value

The **absolute value** of a real number c is denoted $|c|$ and is defined as follows.

$$\text{If } c \geq 0, \text{ then } |c| = c.$$

$$\text{If } c < 0, \text{ then } |c| = -c.$$

EXAMPLE 1

(a) $|3.5| = 3.5$ and $|-7/2| = -(-7/2) = 7/2$.

(b) To find $|\pi - 6|$, note that $\pi \approx 3.14$, so $\pi - 6 < 0$. Hence, $|\pi - 6|$ is defined to be the *negative* of $\pi - 6$, that is,

$$|\pi - 6| = -(\pi - 6) = -\pi + 6.$$

(c) $|5 - \sqrt{2}| = 5 - \sqrt{2}$ because $5 - \sqrt{2} \geq 0$. ■

TECHNOLOGY TIP

To find $|9 - 3\pi|$ on a calculator, key in

ABS(9 − 3π).

The ABS key is located in this menu/submenu:

TI: MATH/NUM

Casio: OPTN/NUM

HP-39+: Keyboard

When c is a positive number, then $\sqrt{c^2} = c$, but when c is negative, this is *false*. For example, if $c = -3$, then

$$\sqrt{c^2} = \sqrt{(-3)^2} = \sqrt{9} = 3 \qquad (\textit{not } -3),$$

so $\sqrt{c^2} \neq c$. In this case, however, $|c| = |-3| = 3$, so $\sqrt{c^2} = |c|$. The same thing is true for any negative number c. It is also true for positive numbers (since $|c| = c$ when c is positive). In other words,

Square Roots of Squares

For every real number c,

$$\sqrt{c^2} = |c|.$$

DISTANCE ON THE NUMBER LINE

Observe that the distance from -5 to 3 on the number line is 8 units.

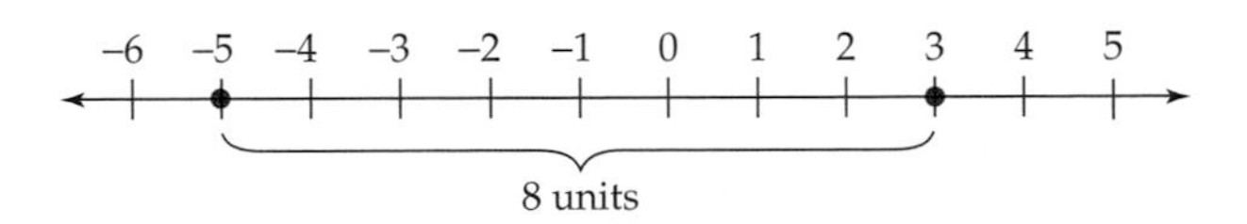

Figure 0–2

This distance can be expressed in terms of absolute value by noting that

$$|(-5) - 3| = 8.$$

That is, the distance is the *absolute value of the difference* of the two numbers. Furthermore, the order in which you take the difference doesn't matter; $|3 - (-5)|$ is also 8. This reflects the geometric fact that the distance from -5 to 3 is the same as the distance from 3 to -5. The same thing is true in the general case.

Distance on the Number Line

The distance between c and d on the number line is the number

$$|c - d| = |d - c|.$$

EXAMPLE 2

The distance from 4.2 to 9 is $|4.2 - 9| = |-4.8| = 4.8$, and the distance from 6 to $\sqrt{2}$ is $|6 - \sqrt{2}|$. ■

When $d = 0$, the distance formula shows that $|c - 0| = |c|$. Hence,

Distance to Zero

$|c|$ is the distance between c and 0 on the number line.

EQUATIONS

A **solution** of an equation such as

$$4x + 3 = 9 \qquad \text{or} \qquad x^2 + 8x + 3 = 0$$

is a number that when substituted for the variable x produces a true statement.* For example, 5 is a solution of $3x + 2 = 17$ because $3 \cdot 5 + 2 = 17$ is a true statement. To **solve an equation** means to find all its solutions.

An equation is called an **identity** if every real number for which all terms of the equation are defined is a solution. For instance,

$$x^2 - 4 = (x + 2)(x - 2)$$

is an identity because every real number is a solution.

Two equations are said to be **equivalent** if they have the same solutions. For example, the equations

$$3x + 2 = 17, \qquad 3x = 15, \qquad \text{and} \qquad x = 5$$

are all equivalent, since 5 is the only solution of each one.

We assume that you are experienced in solving linear equations, such as

$$4x - 2 = 5 \qquad \text{or} \qquad 3x + 7 = 6 - 12x$$

A **quadratic equation** is one that can be written in the form

$$ax^2 + bx + c = 0$$

for some constants a, b, c, with $a \neq 0$. Quadratic equations can sometimes be solved by factoring.

*Any letter may be used for the variable.

EXAMPLE 3

To solve $3x^2 + x = 10$, we first rearrange the terms to make one side 0 and then factor:

Subtract 10 from each side: $\quad 3x^2 + x - 10 = 0$

Factor left side: $\quad (3x - 5)(x + 2) = 0.$

CAUTION

To guard against mistakes check your solutions by substituting each one in the *original* equation to make sure it really *is* a solution.

If a product of real numbers is 0, then at least one of the factors must be 0. So this equation is equivalent to

$$3x - 5 = 0 \quad \text{or} \quad x + 2 = 0$$
$$3x = 5 \qquad\qquad x = -2$$
$$x = 5/3$$

Therefore the solutions are $5/3$ and -2. ■

When factoring doesn't work, you can always find the solutions of a quadratic equation by using the following fact.

The Quadratic Formula

The solutions of the quadratic equation $ax^2 + bx + c = 0$ are

$$x = \frac{-b \pm \sqrt{b^2 - 4ac}}{2a}.$$

You should memorize the quadratic formula.

EXAMPLE 4

Solve $x^2 + 3 = -8x$.

SOLUTION Rewrite the equation as $x^2 + 8x + 3 = 0$, and apply the quadratic formula with $a = 1$, $b = 8$, and $c = 3$:

$$x = \frac{-b \pm \sqrt{b^2 - 4ac}}{2a} = \frac{-8 \pm \sqrt{8^2 - 4 \cdot 1 \cdot 3}}{2 \cdot 1}$$
$$= \frac{-8 \pm \sqrt{52}}{2} = \frac{-8 \pm \sqrt{4 \cdot 13}}{2}$$
$$= \frac{-8 \pm 2\sqrt{13}}{2} = -4 \pm \sqrt{13}.$$

The equation has two distinct real solutions: $-4 + \sqrt{13}$ and $-4 - \sqrt{13}$. ■

EXAMPLE 5

Solve $x^2 - 194x + 9409 = 0$.

SOLUTION Use a calculator and the quadratic formula with $a = 1, b = -194$, and $c = 9409$:

$$x = \frac{-b \pm \sqrt{b^2 - 4ac}}{2a} = \frac{-(-194) \pm \sqrt{(-194)^2 - 4 \cdot 1 \cdot 9409}}{2 \cdot 1}$$

$$= \frac{194 \pm \sqrt{37636 - 37636}}{2} = \frac{194 \pm 0}{2} = 97.$$

Thus, 97 is the only solution of the equation. ■

EXAMPLE 6

Solve $2x^2 + x + 3 = 0$.

SOLUTION Using the quadratic formula with $a = 2, b = 1$, and $c = 3$:

$$x = \frac{-b \pm \sqrt{b^2 - 4ac}}{2a} = \frac{-1 \pm \sqrt{1^2 - 4 \cdot 2 \cdot 3}}{2 \cdot 2} = \frac{-1 \pm \sqrt{1 - 24}}{4}$$

$$= \frac{-1 \pm \sqrt{-23}}{4}.$$

Since $\sqrt{-23}$ is not a real number, this equation has *no real solutions* (that is, no solutions in the real number system). ■

EXAMPLE 7

If an object is thrown upward, dropped, or thrown downward and travels in a straight line subject only to gravity (with wind resistance ignored), the height h of the object above the ground (in feet) after t seconds is given by

$$h = -16t^2 + v_0 t + h_0,$$

where h_0 is the height of the object when $t = 0$ and v_0 is the initial velocity at time $t = 0$. The value of v_0 is taken as positive if the object moves upward and negative if it moves downward. If a baseball is thrown down from the top of a 640-foot-high building with an initial velocity of 52 feet per second, how long does it take to reach the ground?

SOLUTION In this case, v_0 is -52 and h_0 is 640, so that the height equation is

$$h = -16t^2 - 52t + 640.$$

The object is on the ground when $h = 0$, so we must solve the equation

$$0 = -16t^2 - 52t + 640.$$

Using the quadratic formula and a calculator, we see that

$$t = \frac{-(-52) \pm \sqrt{(-52)^2 - 4(-16)(640)}}{2(-16)} = \frac{52 \pm \sqrt{43{,}664}}{-32} \approx \begin{cases} -8.15 \\ \text{or} \\ 4.90. \end{cases}$$

Only the positive answer makes sense in this case. So it takes about 4.9 seconds for the baseball to reach the ground. ■

The quadratic formula and a calculator can be used to solve quadratic equations. Experiment with your calculator to find the most efficient sequence of keystrokes for doing this.

EXAMPLE 8

TECHNOLOGY TIP
Most calculators have built-in polynomial equation solvers that will solve quadratic and other polynomial equations. See Calculator Investigation 4 at the end of this section.

Use a calculator to solve $3.2x^2 + 15.93x - 7.1 = 0$.

First compute $\sqrt{b^2 - 4ac} = \sqrt{15.93^2 - 4(3.2)(-7.1)}$ and store the result in memory D. Then the solutions of the equation are given by

$$x = \frac{-b \pm \sqrt{b^2 - 4ac}}{2a} = \frac{-15.93 \pm D}{2(3.2)}.$$

They are easily approximated on a calculator:

$$\frac{-15.93 + D}{2(3.2)} \approx .411658467 \qquad \text{and} \qquad \frac{-15.93 - D}{2(3.2)} \approx -5.389783467$$

Remember that these answers are *approximations*, so they might not check exactly when substituted in the original equation. ■

CALCULATOR INVESTIGATIONS 0.1

1. EDIT AND REPLAY

Consider the equation $y = x^3 + 6x^2 - 5$.

(a) Find the value of y when $x = -7$ by keying in

$$(*) \qquad (-7)^3 + 6(-7)^2 - 5$$

and pressing ENTER.*

(b) To find the value of y when $x = 9$, without retyping the entire calculation, use the **replay** feature as follows:

Calculator	Procedure
TI	Press 2nd ENTER.
Casio	Press left or right arrow key *or* press AC and the up arrow key.
HP-39+	Use up arrow key to shade the previous computation; then press COPY.

Your screen now returns to the previous calculation (*). Use the left/right arrow keys and the DEL(ete) key to move through the equation and replace −7 by 9. Press ENTER, and the result of the new computation is displayed.

(c) Use the replay feature again to find the value of y when $x = 108$. On some calculators, you might need to use the INS(ert) key when replacing 9 by 108 to avoid typing over some of the computation.

(d) Press the replay keys repeatedly [on Casio, press AC and then press the up arrow repeatedly]. The calculator will display all the preceding computations, in reverse order. Go back to the first one (*) and compute it again.

*Here and throughout the book, Casio users should read "EXE" in place of "ENTER."

2. MATHEMATICAL OPERATIONS

(a) Key in each of the following, and explain why your answers are different. [See the Technology Tip on page 5.]

ABS(−) 9 + 2 ENTER and ABS ((−) 9 + 2) ENTER

(b) Find INT or FLOOR in the NUM or REAL submenu of the MATH or OPTN menu. Find out what this command does when you follow it by a number and ENTER. Try all kinds of numbers (positive, negative, decimals, fractions). Answers on Casio will sometimes differ from those on other calculators.

3. SYMBOLIC CALCULATIONS

(a) To store the number 2 in memory A of a calculator, type

2 STO ▶ A ENTER [TI-85/86] or

2 STO ▶ ALPHA A ENTER [TI-83+/89, HP-39+] or

2 → ALPHA A EXE [Casio].

If you now key in ALPHA A ENTER, what does the calculator display?

(b) In a similar fashion, store the number 5 in memory B and −10 in memory C. Then, using the ALPHA keys, display this expression on the screen: B + C/A. If you press ENTER, what happens? Explain what the calculator is doing.

(c) Experiment with other expressions, such as $B^2 - 4AC$.

4. POLYNOMIAL EQUATION SOLVERS

If you have one of the calculators mentioned below, use its polynomial equation solver to solve the equation $3.2x^2 + 15.93x - 7.1 = 0$. How do the answers you obtain compare with those in Example 8?

	Solver Name and Location	*Special Syntax (if needed)*
TI-86	POLY on Keyboard	For "order," enter the degree of the polynomial.
TI-89	SOLVE in ALGEBRA menu	SOLVE $(3.2x^2 + 15.93x - 7.1 = 0, x)$
Casio	EQUATION in main menu	
HP-39+	POLYROOT in POLYNOM submenu of MATH menu	POLYROOT([3.2, 15.93, −7.1])

Note: TI-83+ and TI-89 users can download a free polynomial solver similar to the one on TI-86 [http://education.ti.com/apps].

Use the same technique to solve the following equations.

(a) $x^2 + 2x - 4 = 0$

(b) $x^2 + 5x + 2 = 0$

EXERCISES 0.1

1. Draw a number line and mark the location of each of these numbers: 0, -7, $8/3$, 10, -1, -4.75, $1/2$, -5, and 2.25.
2. (a) Use the π key on your calculator to find a decimal approximation of π.
   (b) Use your calculator to determine which of the following rational numbers is the best approximation of the irrational number π.

$$\frac{22}{7}, \quad \frac{355}{113}, \quad \frac{103{,}993}{33{,}102}, \quad \frac{2{,}508{,}429{,}787}{798{,}458{,}000}.$$

If your calculator says that one of these numbers equals π, it's lying. All you can conclude is that the number agrees with π for as many decimal places as your calculator can handle (usually 12–14).

In Exercises 3–14, express the given statement in symbols.

3. 12 is greater than 9.
4. 5 is less than 7.
5. -4 is greater than -8.
6. -17 is less than 14.
7. π is less than 100.
8. x is nonnegative.
9. y is less than or equal to 7.5.
10. z is greater than or equal to -4.
11. t is positive.
12. d is not greater than 2.
13. c is at most 3.
14. z is at least -17.

In Exercises 15–20, fill the blank with $<$, $=$, or $>$ so that the resulting statement is true.

15. -6 ____ -2
16. 5 ____ -3
17. $3/4$ ____ .75
18. 3.1 ____ π
19. $1/3$ ____ .33.
20. 2 ____ $\sqrt{2}$

*The consumer price index for urban consumers (CPI-U) measures the cost of consumer goods and services such as food, housing, transportation, medical costs, etc. The table shows the yearly percentage increase in the CPI-U over a decade.**

Year	Percentage change
1993	3.0
1994	2.6
1995	2.8
1996	3.0
1997	2.3
1998	1.6
1999	2.2
2000	3.4
2001	2.8
2002	1.6

In Exercises 21–25, let p denote the yearly percentage increase in the CPI-U. Find the number *of years in this period in which p satisfied the given inequality.*

21. $p \geq 2.8$
22. $p < 2.6$
23. $p > 2.3$
24. $p \leq 3.0$
25. $p > 3.4$
26. Galileo discovered that the period of a pendulum depends only on the length of the pendulum and the acceleration of gravity. The period T of a pendulum (in seconds) is

$$T = 2\pi\sqrt{\frac{l}{g}}$$

where l is the length of the pendulum in feet and $g \approx 32.2 \text{ ft/sec}^2$ is the acceleration due to gravity. Find the period of a pendulum whose length is 4 ft.

In Exercises 27–28, use a calculator and list the given numbers in order from smallest to largest.

27. $\frac{189}{37}, \quad \frac{4587}{691}, \quad \sqrt{47}, \quad 6.735, \quad \sqrt{27}, \quad \frac{2040}{523}$
28. $\frac{385}{177}, \quad \sqrt{10}, \quad \frac{187}{63}, \quad \pi, \quad \sqrt{\sqrt{85}}, \quad 2.9884$

In Exercises 29–34, fill the blank so as to produce two equivalent statements. For example, the arithmetic statement "a is negative" is equivalent to the geometric statement "the point a lies to the left of the point 0*."*

Arithmetic Statement	**Geometric Statement**
29. ____________	a lies c units to the right of b.
30. ____________	a lies between b and c.
31. $a - b > 0$	____________

*U.S. Bureau of Labor Statistics.

Arithmetic Statement	**Geometric Statement**
32. a is positive.	______________
33. ______________	a lies to the left of b.
34. $a \geq b$	______________

In Exercises 35–42, simplify and write the given number without using absolute values.

35. $3 - |2 - 5|$

36. $-2 - |-2|$

37. $|6 - 4| + |-3 - 5|$

38. $|-6| - |6|$

39. $|(-13^2)|$

40. $-|-5|^2$

41. $|\pi - \sqrt{2}|$

42. $|\sqrt{2} - 2|$

In Exercises 43–48, fill the blank with <, =, or > so that the resulting statement is true.

43. $|-2|$ ____ $|-5|$

44. 5 ____ $|-2|$

45. $|3|$ ____ $-|4|$

46. $|-3|$ ____ 0

47. -7 ____ $|-1|$

48. $-|-4|$ ____ 0

In Exercises 49–54, find the distance between the given numbers.

49. -3 and 4

50. 7 and 107

51. -7 and $15/2$

52. $-3/4$ and -10

53. π and 3

54. π and -3

55. Ann rode her bike down the Blue Ridge Parkway, beginning at Linville Falls (mile post 316.3) and ending at Mount Pisgah (mile post 408.6). Use absolute value notation to describe how far Ann rode.

56. A broker predicts that over the next six months, the price p of a particular stock will not vary from its current price of \$25.75 by more than \$4. Express this prediction as an inequality.

57. According to data from the Center for Science in the Public Interest, the healthy weight range for a person depends on the person's height. For example,

Height	**Healthy Weight Range (lb)**
5 ft 8 in.	143 ± 21
6 ft 0 in.	163 ± 26

Express each of these ranges as an absolute value inequality in which x is the weight of the person.

*The wind-chill factor, shown in the table, calculates how a given temperature feels to your skin when the wind is taken into account. For example, the table shows that a temperature of 20° in a 40 mph wind feels like −1°.**

	Wind (mph)								
Temperature (°F)	**Calm**	**5**	**10**	**15**	**20**	**25**	**30**	**35**	**40**
	40	36	34	32	30	29	28	28	27
	30	25	21	19	17	16	15	14	13
	20	13	9	6	4	3	1	0	−1
	10	1	−4	−7	−9	−11	−12	−14	−15
	0	−11	−16	−19	−22	−24	−26	−27	−29
	−10	−22	−28	−32	−35	−37	−39	−41	−43
	−20	−34	−41	−45	−48	−51	−53	−55	−57
	−30	−46	−53	−58	−61	−64	−67	−69	−71
	−40	−57	−66	−71	−74	−78	−80	−82	−84

In Exercises 58–60, find the absolute value of the difference of the two given wind-chill factors. For example, the difference between the wind-chill at 30° with a 15 mph wind and one at −10° with a 10 mph wind is $|19-(-28)| = 47°$ or, equivalently, $|-28 - 19| = 47°$.

58. 10° with a 25 mph wind and 20° with a 20 mph wind

59. 30° with a 10 mph wind and 10° with a 30 mph wind

60. −30° with a 5 mph wind and 0° with a 10 mph wind

61. The use of digital devices (cell phones, DVD players, PCs, etc.) continues to grow. The graph shows the approximate number of digital devices (in billions) in use worldwide from 2000 to 2004.†

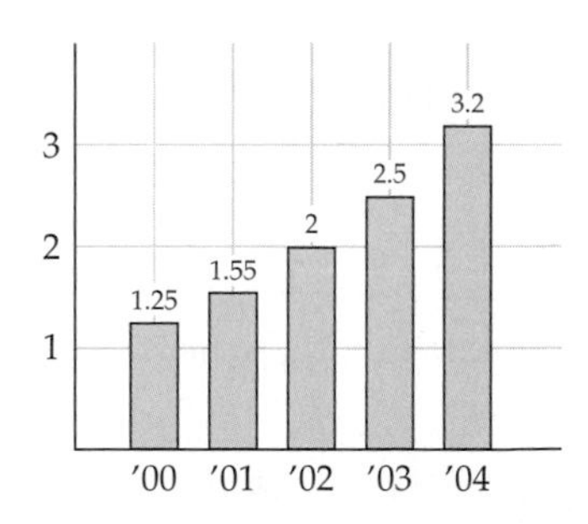

In what years was the following statement true

$$|x - 2{,}000{,}000{,}000| \geq 500{,}000{,}000,$$

where x is the number of digital devices in use in that year?

62. At Statewide Insurance, each department's expenses are reviewed monthly. A department can fail to pass the budget variance test in a category if either (i) the absolute

*Table from the Joint Action Group for Temperature Indices, 2001

†Data, estimates, and projections from *IDC*

value of the difference between actual expenses and the budget is more than $500 or (ii) the absolute value of the difference between the actual expenses and the budget is more than 5% of the budgeted amount. Which of the following items fail the budget variance test? Explain your answers.

Item	Budgeted Expense ($)	Actual Expense ($)
Wages	220,750	221,239
Overtime	10,500	11,018
Shipping and Postage	530	589

In Exercises 63–74, solve the equation by factoring.

63. $x^2 - 8x + 15 = 0$

64. $x^2 + 5x + 6 = 0$

65. $x^2 - 5x = 14$

66. $x^2 + x = 20$

67. $2y^2 + 5y - 3 = 0$

68. $3t^2 - t - 2 = 0$

69. $4t^2 + 9t + 2 = 0$

70. $9t^2 + 2 = 11t$

71. $3u^2 + u = 4$

72. $5x^2 + 26x = -5$

73. $12x^2 + 13x = 4$

74. $18x^2 = 23x + 6$

In Exercises 75–84, use the quadratic formula to solve the equation.

75. $x^2 - 4x + 1 = 0$

76. $x^2 - 2x - 1 = 0$

77. $x^2 + 6x + 7 = 0$

78. $x^2 + 4x - 3 = 0$

79. $x^2 + 6 = 2x$

80. $x^2 + 11 = 6x$

81. $4x^2 - 4x = 7$

82. $4x^2 - 4x = 11$

83. $4x^2 - 8x + 1 = 0$

84. $2t^2 + 4t + 1 = 0$

In Exercises 85–94, solve the equation by any method.

85. $x^2 + 9x + 18 = 0$

86. $3t^2 - 11t - 20 = 0$

87. $4x(x + 1) = 1$

88. $25y^2 = 20y + 1$

89. $2x^2 = 7x + 15$

90. $2x^2 = 6x + 3$

91. $t^2 + 4t + 13 = 0$

92. $5x^2 + 2x = -2$

93. $\frac{7x^2}{3} = \frac{2x}{3} - 1$

94. $25x + \frac{4}{x} = 20$

In Exercises 95 and 96, use a calculator and the quadratic formula to find approximate solutions of the equation.

95. $4.42x^2 - 10.14x + 3.79 = 0$

96. $8.06x^2 + 25.8726x - 25.047256 = 0$

97. The atmospheric pressure a (in pounds per square foot) at height h thousand feet above sea level is approximately

$$a = .8315h^2 - 73.93h + 2116.1.$$

(a) Find the atmospheric pressure at sea level and at the top of Mount Everest, the tallest mountain in the world (29,035 feet*). [Remember that h is measured in thousands.]

(b) The atmospheric pressure at the top of Mount Rainier is 1223.43 pounds per square foot. How high is Mount Rainier?

98. Data from the U.S. Department of Health and Human Services indicates that the cumulative number N of reported cases of AIDS in the United States in year x can be approximated by the equation

$$N = 3362.1x^2 - 17{,}270.3x + 24{,}043,$$

where $x = 0$ corresponds to 1980. In what year did the total reach 550,000?

99. According to data from the U.S. Census Bureau, the population P of Cleveland, Ohio (in thousands) in year x can be approximated by $P = .08x^2 - 13.08x + 927$, where $x = 0$ corresponds to 1950. In what year in the past was the population about 804,200?

100. According to data from the National Highway Traffic Safety Administration, the driver fatality rate D in automobile accidents per 1000 licensed drivers every 100 million miles can be approximated by the equation

$$D = .0031x^2 - .291x + 7.1,$$

where x is the age of the driver.

(a) For what ages is the driver fatality rate approximately 1 death per 1000 drivers every 100 million miles?

(b) For what ages is the rate three times greater than in part (a).

101. If you are located h feet aove the ground, then because of the curvature of the earth, the maximum distance you can see is approximately d miles, where

$$d = \sqrt{1.5h + \frac{3.587h^2}{10^8}}.$$

How far can you see from

(a) the 500-foot high Smith Tower in Seattle?

(b) the 1454-foot high Sears Tower in Chicago?

(c) How high must you be above the ground to see for 25 miles? [*Hint:* Square both sides of the formula and substitute the appropriate value for d. Only one solution of the resulting equation makes sense in the context of the problem.]

102. Suppose you are k miles (not feet) above the ground. The radius of the earth is approximately 3960 miles, and the point where your line of sight meets the earth is perpen-

*Based on measurements in 1999 by climbers sponsored by the Boston Museum of Science and the National Geographic Society, using satellite-based technology.

dicular to the radius of the earth at that point, as shown in the figure.

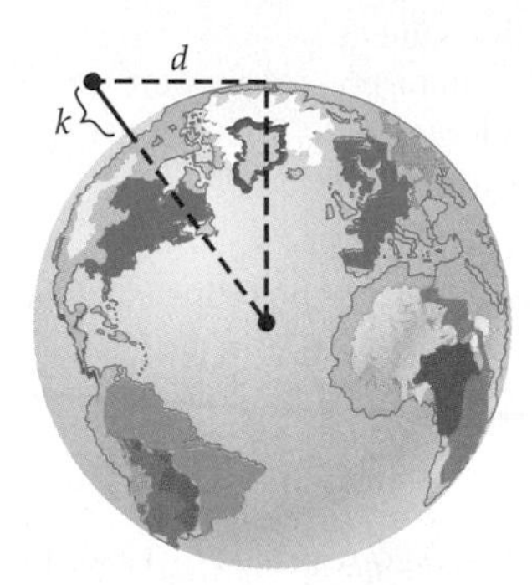

(a) Use the Pythagorean Theorem (see the Geometry Review Appendix) to show that

$$d = \sqrt{(3960 + k)^2 - 3960^2}$$

(b) Show that the equation in part (a) simplifies to $d = \sqrt{7920k + k^2}$.

(c) If you are h feet above the ground, then you are $h/5280$ miles high. Why? Use this fact and the equation in part (b) to obtain the formula used in Exercise 101.

103. A 13-foot-long ladder leans on a wall, as shown in the figure. The bottom of the ladder is 5 feet from the wall. If the bottom is pulled out 3 feet farther from the wall, how far does the top of the ladder move down the wall? [*Hint:* Draw pictures of the right triangle formed by the ladder, the ground, and the wall before and after the ladder is moved. In each case, use the Pythagorean Theorem to find the distance from the top of the ladder to the ground.]

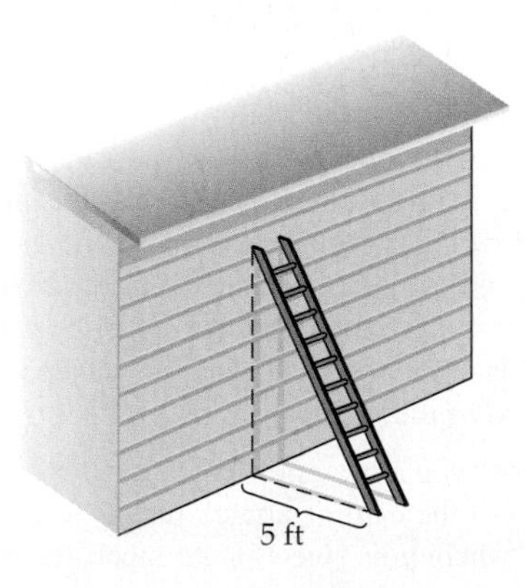

104. A 15-foot-long pole leans against a wall. The bottom is 9 feet from the wall. How much farther should the bottom be pulled away from the wall so that the top moves the same amount down the wall?

105. A concrete walk of uniform width is to be built around a circular pool, as shown in the figure. The radius of the pool is 12 meters and enough concrete is available to cover 52π square meters. If all the concrete is to be used, how wide should the walk be?

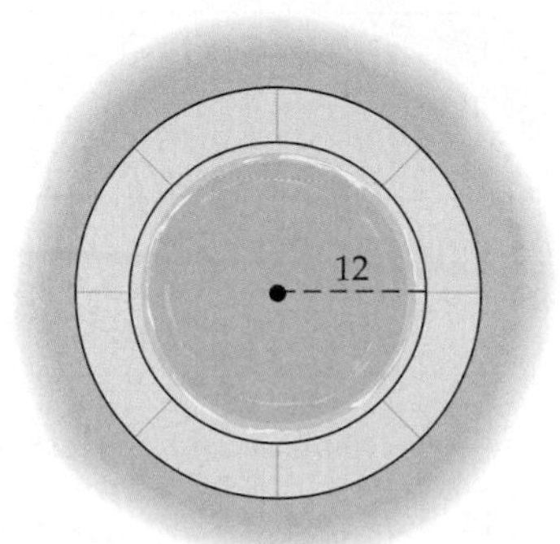

106. A decorator has 92 one-foot-square tiles that will be laid around the edges of a 12-by-15-foot room. A rectangular rug that is 3 feet longer than it is wide is to be placed over the center area where there are no tiles. To the nearest quarter foot, find the dimensions of the smallest rug that will cover the untiled part of the floor. [Assume that all the tiles are used and that none of them are split.]

In Exercises 107–110, use the height equation in Example 7. Note that an object that is dropped (rather than thrown downward) has initial velocity $v_0 = 0$.

107. How long does it take a baseball to reach the ground if it is dropped from the top of a 640-foot-high building? Compare with Example 7.

108. You are standing on a cliff that is 200 feet high. How long will it take a rock to reach the ground if

(a) you drop it?

(b) you throw it downward at an initial velocity of 40 feet per second?

(c) How far does the rock fall in 2 seconds if you throw it downward with an initial velocity of 40 feet per second?

109. A rocket is fired straight up from ground level with an initial velocity of 800 feet per second.

(a) How long does it take the rocket to rise 3200 feet?

(b) When will the rocket hit the ground?

110. A rocket loaded with fireworks is to be shot vertically upward from ground level with an initial velocity of 200 feet per second. When the rocket reaches a height of 400 feet on its upward trip the fireworks will be detonated. How many seconds after liftoff will this take place?

Thinkers

111. Explain why the statement $|a| + |b| + |c| > 0$ is algebraic shorthand for "at least one of the numbers *a, b, c,* is different from zero."

112. Find an algebraic shorthand version of the statement "the numbers *a, b, c,* are all different from zero."

0.2 The Coordinate Plane

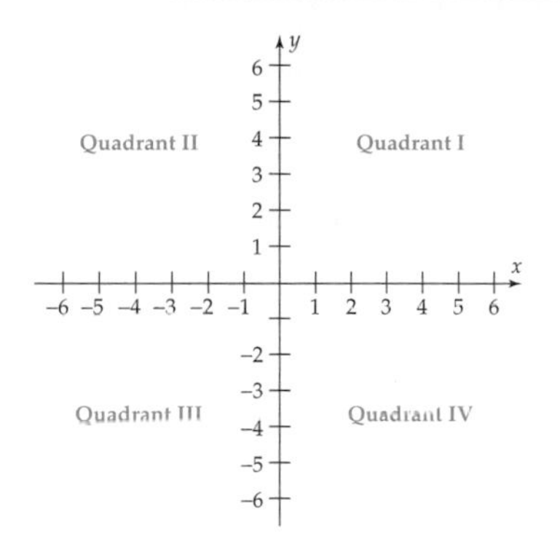

Figure 0–3

Just as real numbers are identified with points on the number line, ordered *pairs* of real numbers can be identified with points in the plane. To do this, draw two number lines in the plane, one horizontal (usually called the **x-axis**) and one vertical (usually called the **y-axis**), as in Figure 0–3.* The point where the axes intersect is the **origin.** The axes divide the plane into four regions, called **quadrants** (see Figure 0–3).

Let P be a point in the plane. If the vertical line through P intersects the x-axis at the number c and the horizontal line through P intersects the y-axis at d, then we say that P has **coordinates** (c, d).

EXAMPLE 1 

Plot the points $(4, -3)$, $(-5, 2)$, $(2, -5)$, $(0, 13/3)$, $(2, 2)$, and $(-4, -1.2)$.

SOLUTION You can think of the coordinates of a point as directions for locating it. To find $(4, -3)$, for instance, start at the origin and move 4 units to the right along the x-axis (positive direction), then move 3 units downward (negative direction), as shown in Figure 0–4. To find $(-5, 2)$, start at the origin and move 5 units to the left along the x-axis (negative direction), then move 2 units upward (positive direction). The other points are plotted similarly, as shown in Figure 0–4. ■

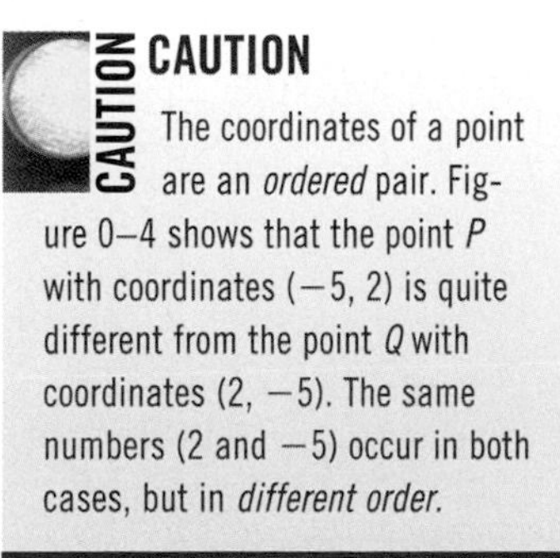

CAUTION

The coordinates of a point are an *ordered* pair. Figure 0–4 shows that the point P with coordinates $(-5, 2)$ is quite different from the point Q with coordinates $(2, -5)$. The same numbers (2 and -5) occur in both cases, but in *different order.*

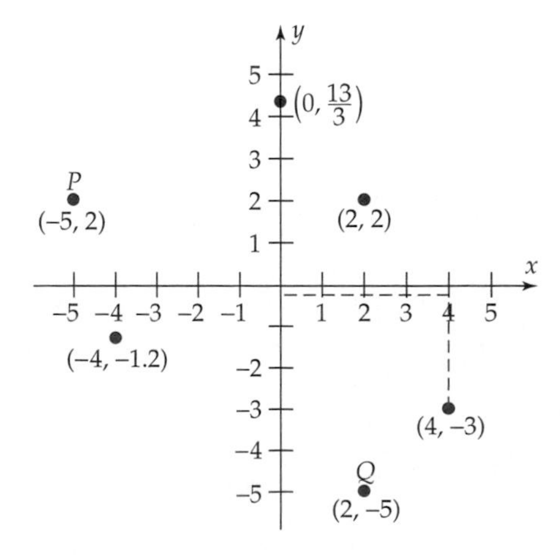

Figure 0–4

*Any letters may be used in place of x and y to label the axes.

THE DISTANCE FORMULA

We shall often identify a point with its coordinates and refer, for example, to the point (2, 3). When dealing with several points simultaneously, it is customary to label the coordinates of the first point (x_1, y_1), the second point (x_2, y_2), the third point (x_3, y_3), and so on.* Once the plane is coordinatized, it is easy to compute the distance between any two points:

The Distance Formula

> The distance between points (x_1, y_1) and (x_2, y_2) is
> $$\sqrt{(x_1 - x_2)^2 + (y_1 - y_2)^2}.$$

Before proving the distance formula, we shall see how it is used.

EXAMPLE 2

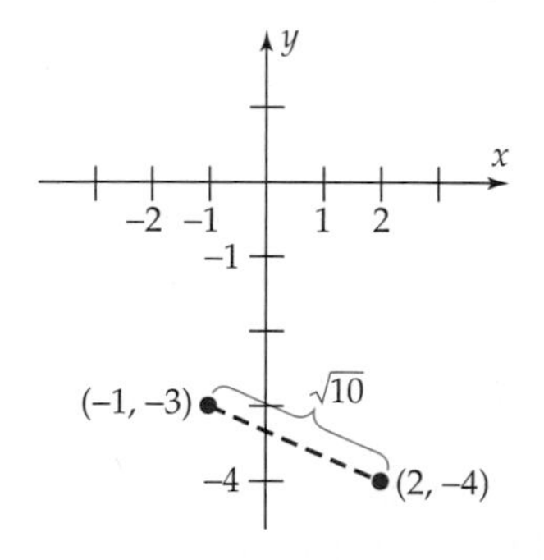

Figure 0–5

To find the distance between the points $(-1, -3)$ and $(2, -4)$ in Figure 0–5, substitute $(-1, -3)$ for (x_1, y_1) and $(2, -4)$ for (x_2, y_2) in the distance formula:

$$\begin{aligned}
&\text{Distance formula:} && \text{Distance} = \sqrt{(x_1 - x_2)^2 + (y_1 - y_2)^2} \\
&\text{Substitute:} && = \sqrt{(-1-2)^2 + (-3-(-4))^2} \\
&\text{Simplify:} && = \sqrt{(-3)^2 + (-3+4)^2} \\
& && = \sqrt{9+1} = \sqrt{10}.
\end{aligned}$$

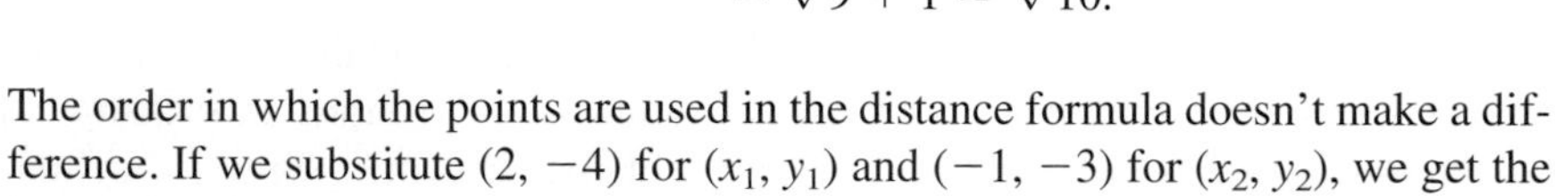

The order in which the points are used in the distance formula doesn't make a difference. If we substitute $(2, -4)$ for (x_1, y_1) and $(-1, -3)$ for (x_2, y_2), we get the same answer:

$$\sqrt{[2-(-1)]^2 + [-4-(-3)]^2} = \sqrt{3^2 + (-1)^2} = \sqrt{10}. \quad ■$$

Proof of the Distance Formula Figure 0–6 shows typical points P and Q in the plane. We must find length d of line segment PQ.

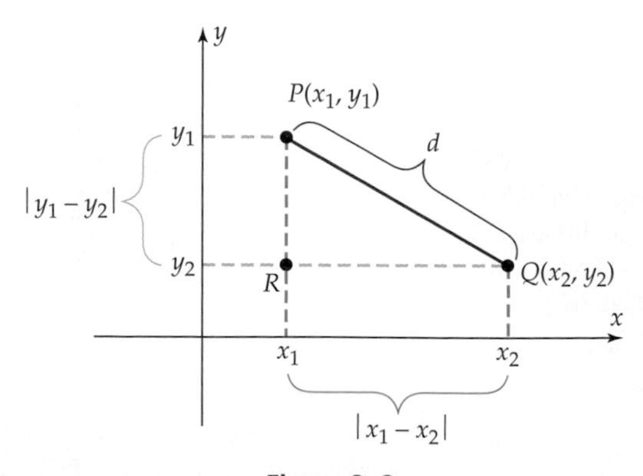

Figure 0–6

---

*"x_1" is read "x-one" or "x-sub-one"; it is a *single symbol* denoting the first coordinate of the first point, just as c denotes the first coordinate of (c, d). Analogous remarks apply to y_1, x_2, and so on.

As shown in Figure 0–6, the length of RQ is the same as the distance from x_1 to x_2 on the x-axis (number line), namely, $|x_1 - x_2|$. Similarly, the length of PR is the same as the distance from y_1 to y_2 on the y-axis, namely, $|y_1 - y_2|$. According to the Pythagorean Theorem* the length d of PQ is given by

$$(\text{Length } PQ)^2 = (\text{length } RQ)^2 + (\text{length } PR)^2$$

$$d^2 = |x_1 - x_2|^2 + |y_1 - y_2|^2.$$

Since $|c|^2 = |c| \cdot |c| = |c^2| = c^2$ (because $c^2 \geq 0$), this equation becomes

$$d^2 = (x_1 - x_2)^2 + (y_1 - y_2)^2.$$

Since the length d is nonnegative, we must have

$$d = \sqrt{(x_1 - x_2)^2 + (y_1 - y_2)^2}. \quad \blacksquare$$

The distance formula can be used to prove the following useful fact (see Exercise 64).

The Midpoint Formula

> The midpoint of the line segment from (x_1, y_1) to (x_2, y_2) is
>
> $$\left(\frac{x_1 + x_2}{2}, \frac{y_1 + y_2}{2}\right).$$

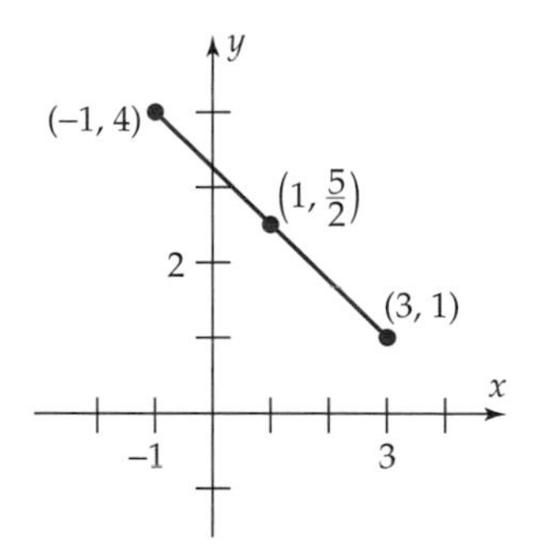

Figure 0–7

EXAMPLE 3

To find the midpoint of the segment joining $(-1, 4)$ and $(3, 1)$, use the formula in the box with $x_1 = -1$, $y_1 = 4$, $x_2 = 3$, and $y_2 = 1$. The midpoint is

$$\left(\frac{x_1 + x_2}{2}, \frac{y_1 + y_2}{2}\right) = \left(\frac{-1 + 3}{2}, \frac{4 + 1}{2}\right) = \left(1, \frac{5}{2}\right)$$

as shown in Figure 0–7. ■

GRAPHS

A **graph** is a set of points in the plane. Many graphs arise from equations, as follows. A **solution** of an equation in variables x and y is a pair of numbers such that the substitution of the first number for x and the second for y produces a true statement. For instance, $(3, -2)$ is a solution of $5x + 7y = 1$ because

$$5 \cdot 3 + 7(-2) = 1,$$

and $(-2, 3)$ is *not* a solution because $5(-2) + 7 \cdot 3 \neq 1$. The **graph of an equation** in two variables is the set of points in the plane whose coordinates are solutions of the equation. Thus the graph is a *geometric picture of the solutions.*

*See the Geometry Review Appendix.

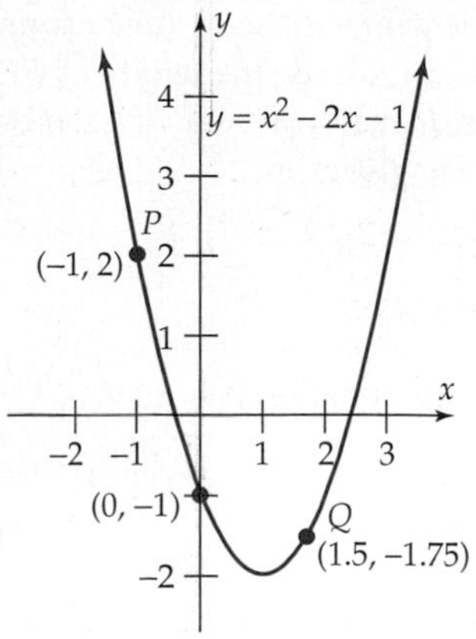

Figure 0–8

EXAMPLE 4

The graph of $y = x^2 - 2x - 1$ is shown in Figure 0–8. Verify that the coordinates of P and Q actually are solutions of the equation.

SOLUTION For the point $(-1, 2)$, substitute -1 for x in the equation:

$$y = x^2 - 2x - 1$$

$$y = (-1)^2 - 2(-1) - 1 = 1 + 2 - 1 = 2.$$

Therefore, $(-1, 2)$ is a solution. Similarly, substituting $x = 1.5$ shows that

$$y = x^2 - 2x - 1 = 1.5^2 - 2(1.5) - 1 = 2.25 - 3 - 1 = -1.75.$$

Hence, $(1.5, -1.75)$ is also a solution. ■

A graph may intersect the x- or y-axis at one or more points. The x-coordinate of a point where the graph intersects the x-axis is called an ***x*-intercept** of the graph. Similarly, the y-coordinate of a point where the graph intersects the y-axis is called a ***y*-intercept** of the graph.

EXAMPLE 5

Find the x- and y-intercepts of the graph of $y = x^2 - 2x - 1$ in Figure 0–8.

SOLUTION The points where the graph intersects the x-axis have 0 as their y-coordinate (see Figure 0–8). We can find their x-coordinates by setting $y = 0$ and solving the resulting equation,

$$x^2 - 2x - 1 = 0.$$

By the quadratic formula

$$x = \frac{-(-2) \pm \sqrt{(-2)^2 - 4 \cdot 1 \cdot (-1)}}{2 \cdot 1} = \frac{2 \pm \sqrt{8}}{2} \approx \begin{cases} -.4142 \\ 2.4142. \end{cases}$$

So the x-intercepts are approximately $-.4142$ and 2.4142.

In this case, you can read the y-intercept from the graph; it is -1. Because points on the y-axis have 0 as their x-coordinate, the y-intercept can be found algebraically by setting $x = 0$ in the equation and solving for y. ■

The process in Example 5 can be summarized as follows.

x- and *y*-Intercepts

> To find the x-intercepts of the graph of an equation, set $y = 0$ and solve for x.
>
> To find the y-intercepts, set $x = 0$ and solve for y.

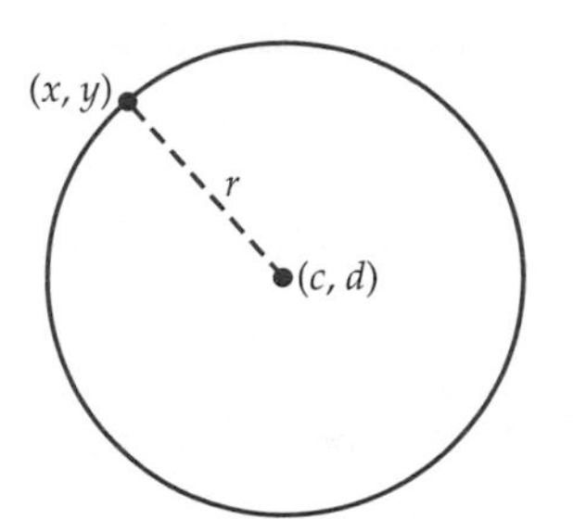

Figure 0–9

CIRCLES

If (c, d) is a point in the plane and r a positive number, then the **circle with center (c, d) and radius r** consists of all points (x, y) that lie r units from (c, d), as shown in Figure 0–9. According to the distance formula, the statement that "the distance from (x, y) to (c, d) is r units" is equivalent to:

$$\sqrt{(x-c)^2 + (y-d)^2} = r$$

Squaring both sides shows that (x, y) satisfies this equation:

$$(x-c)^2 + (y-d)^2 = r^2$$

Reversing the procedure shows that any solution (x, y) of this equation is a point on the circle. Therefore

Circle Equation

> The circle with center (c, d) and radius r is the graph of
> $$(x-c)^2 + (y-d)^2 = r^2.$$

We say that $(x-c)^2 + (y-d)^2 = r^2$ is the **equation of the circle** with center (c, d) and radius r.

EXAMPLE 6

Identify the graph of the equation $(x-4)^2 + (y-2)^2 = 9$.

SOLUTION Since $9 = 3^2$, we can write the equation as

$$(x-4)^2 + (y-2)^2 = 3^2.$$

Now the equation is of the form shown in the box above, with $c = 4$, $d = 2$ and $r = 3$. So the graph is a circle with center (4, 2) and radius 3, as shown in Figure 0–10. ■

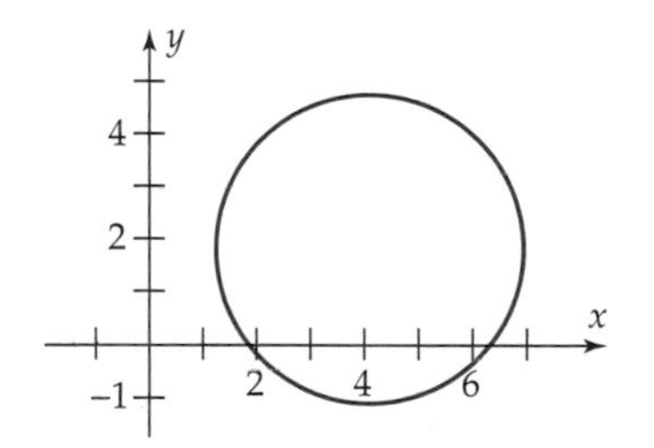

Figure 0–10

EXAMPLE 7

Find the equation of the circle with center $(-3, 2)$ and radius 2 and sketch its graph.

SOLUTION Here the center is $(c, d) = (-3, 2)$ and the radius is $r = 2$, so the equation of the circle is:

$$(x-c)^2 + (y-d)^2 = r^2$$
$$[x-(-3)]^2 + (y-2)^2 = 2^2$$
$$(x+3)^2 + (y-2)^2 = 4.$$

Its graph is in Figure 0–11. ■

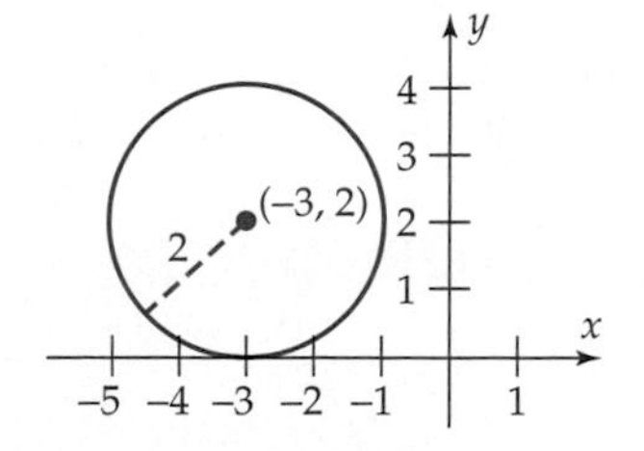

Figure 0–11

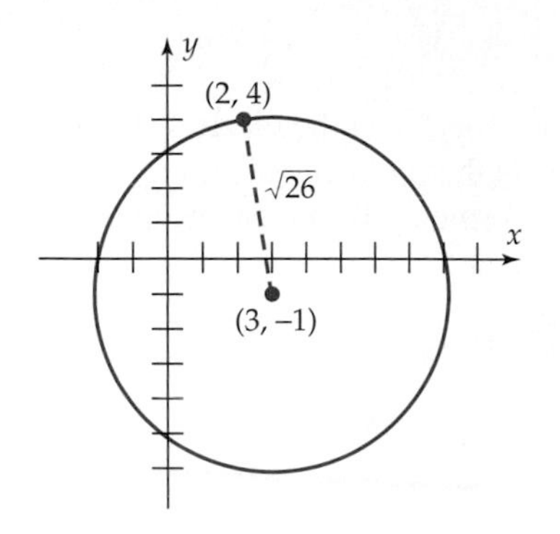

Figure 0–12

EXAMPLE 8

Find the equation of the circle with center (3, −1) that passes through (2, 4) and sketch its graph.

SOLUTION We must first find the radius. Since (2, 4) is on the circle, the radius is the distance from (2, 4) to (3, −1) as shown in Figure 0–12, namely,

$$\sqrt{(2-3)^2 + (4-(-1))^2} = \sqrt{1+25} = \sqrt{26}.$$

The equation of the circle with center at (3, −1) and radius $\sqrt{26}$ is

$$(x-3)^2 + (y-(-1))^2 = (\sqrt{26})^2$$
$$(x-3)^2 + (y+1)^2 = 26$$
$$x^2 - 6x + 9 + y^2 + 2y + 1 = 26$$
$$x^2 + y^2 - 6x + 2y - 16 = 0. \quad ■$$

When the center of a circle of radius r is at the origin (0, 0), its equation takes a simpler form.

Circle at the Origin

> The circle with center (0, 0) and radius r is the graph of
> $$x^2 + y^2 = r^2.$$

Proof Substitute $c = 0$ and $d = 0$ in the equation for the circle with center (c, d) and radius r.

$$(x-c)^2 + (y-d)^2 = r^2$$
$$(x-0)^2 + (y-0)^2 = r^2$$
$$x^2 + y^2 = r^2. \quad ■$$

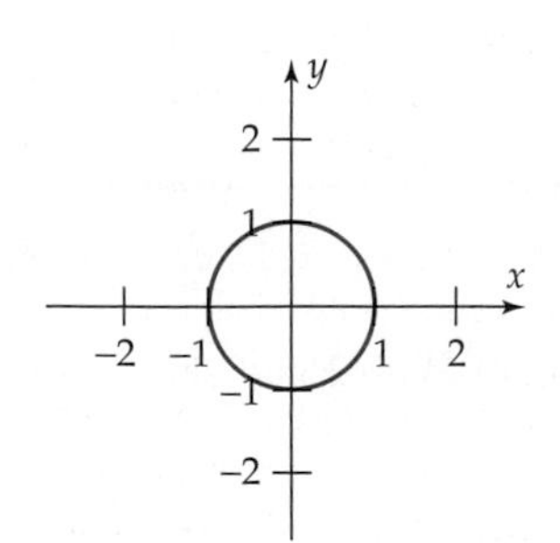

Figure 0–13

EXAMPLE 9

Letting $r = 1$ shows that the graph of $x^2 + y^2 = 1$ is the circle of radius 1 centered at the origin, as shown in Figure 0–13. This circle is called the **unit circle.** ■

EXERCISES 0.2

1. Find the coordinates of points *A*–*I*.

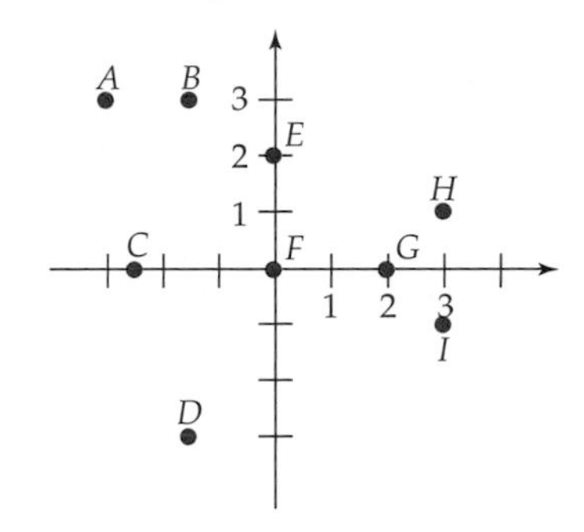

In Exercises 2–5, find the coordinates of the point P.

2. *P* lies 4 units to the left of the *y*-axis and 5 units below the *x*-axis.
3. *P* lies 3 units above the *x*-axis and on the same vertical line as $(-6, 7)$.
4. *P* lies 2 units below the *x*-axis, and its *x*-coordinate is three times its *y*-coordinate.
5. *P* lies 4 units to the right of the *y*-axis, and its *y*-coordinate is half its *x*-coordinate.
6. In what quadrant(s) does a point lie if the product of its coordinates is
   (a) positive? (b) negative?
7. (a) Plot the points $(3, 2)$, $(4, -1)$, $(-2, 3)$, and $(-5, -4)$.
   (b) Change the sign of the *y*-coordinate in each of the points in part (a), and plot these new points.
   (c) Explain how the points (a, b) and $(a, -b)$ are related graphically. [*Hint:* What are their relative positions with respect to the *x*-axis?]
8. (a) Plot the points $(5, 3)$, $(4, -2)$, $(-1, 4)$, and $(-3, -5)$.
   (b) Change the sign of the *x*-coordinate in each of the points in part (a), and plot these new points.
   (c) Explain how the points (a, b) and $(-a, b)$ are related graphically. [*Hint:* What are their relative positions with respect to the *y*-axis?]

In Exercises 9–16, find the distance between the two points and the midpoint of the segment joining them.

9. $(-3, 5)$, $(2, -7)$
10. $(2, 4)$, $(1, 5)$
11. $(1, -5)$, $(2, -1)$
12. $(-2, 3)$, $(-3, 2)$
13. $(\sqrt{2}, 1)$, $(\sqrt{3}, 2)$
14. $(-1, \sqrt{5})$, $(\sqrt{2}, -\sqrt{3})$
15. (a, b), (b, a)
16. (s, t), $(0, 0)$
17. Find the perimeter of the shaded area in the figure.

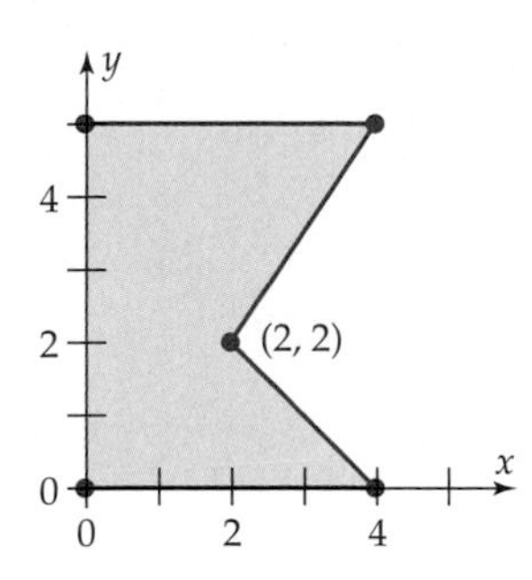

18. What is the perimeter of the triangle with vertices $(1, 1)$, $(5, 4)$, and $(-2, 5)$?

In Exercises 19–20, show that the three points are the vertices of a right triangle, and state the length of the hypotenuse. [You may assume that a triangle with sides of lengths a, b, c is a right triangle with hypotenuse c provided that $a^2 + b^2 = c^2$.]

19. $(0, 0)$, $(1, 1)$, $(2, -2)$
20. $(3, -2)$, $(0, 4)$, $(-2, 3)$
21. Sales of organically grown foods in the United States were \$7.8 billion in 2000 and are projected to be \$16 billion in 2004.*
   (a) Represent this data graphically by two points.
   (b) Find the midpoint of the line segment joining these two points.
   (c) How might this midpoint be intepreted? What assumptions, if any, are needed to make this interpretation?
22. The figure on the next page shows Wrigley Field (home of the Chicago Cubs) placed on a coordinate plane, with home plate at the origin and the right-field foul line along the *x*-axis.
   (a) A ball is hit to the right fielder, who is 5 feet from the wall and 5 feet from the right-field foul line, as indicated in the figure. If the fielder throws the ball to second base, how far does he throw the ball? [*Hint:* What are the coordinates of the fielder and of second base?]
   (b) If the left fielder is at the point $(50, 325)$, how far is he from first base?
   (c) How far is the left fielder in part (b) from the right fielder, who has now moved to the point $(280, 20)$?

*www.datamonitor.com

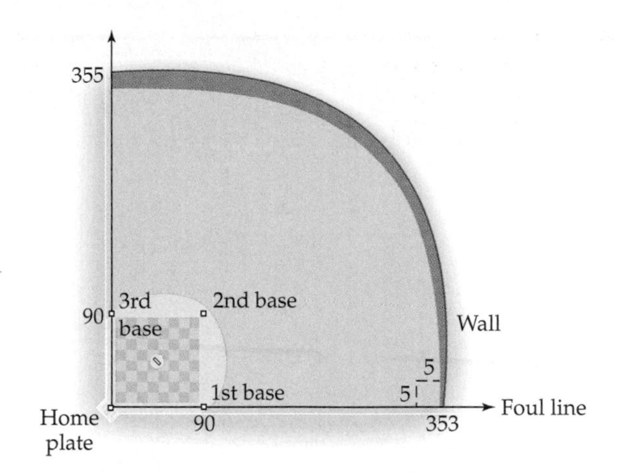

23. A standard football field is 100 yards long and $53\frac{1}{3}$ yards wide. The quarterback, who is standing on the 10-yard line, 20 yards from the left sideline, throws the ball to a receiver who is on the 45-yard line, 5 yards from the right sideline, as shown in the figure.
   (a) How long was the pass? [*Hint:* Place the field in the first quadrant of the coordinate plane, with the left sideline on the y-axis and the goal line on the x-axis. What are the coordinates of the quarterback and the receiver?]
   (b) A player is standing halfway between the quarterback and the receiver. What are his coordinates?

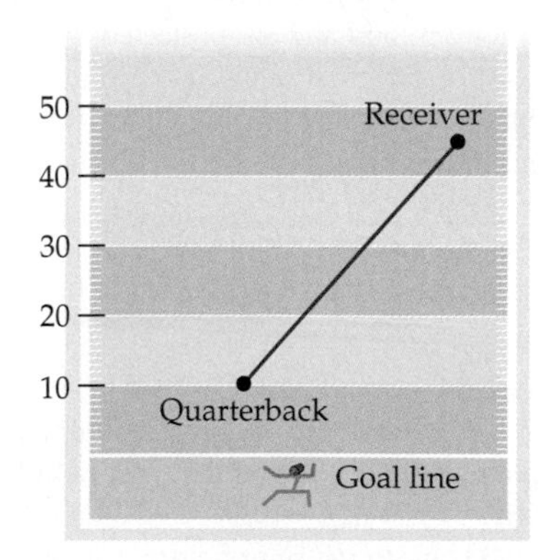

24. How far is the quarterback in Exercise 23 from a player who is on the 50-yard line, halfway between the sidelines?

In Exercises 25–30, determine whether the point is on the graph of the given equation.

25. $(1, -2)$; $3x - y - 5 = 0$
26. $(2, -1)$; $x^2 + y^2 - 6x + 8y = -15$
27. $(6, 2)$; $3y + x = 12$
28. $(1, -2)$; $3x + y = 12$
29. $(3, 4)$; $(x - 2)^2 + (y + 5)^2 = 4$
30. $(1, -1)$; $\frac{x^2}{2} + \frac{y^2}{3} = 1$

In Exercises 31–34, find the equation of the circle with given center and radius r.

31. $(-3, 4)$; $r = 2$
32. $(-2, -1)$; $r = 3$
33. $(0, 0)$; $r = \sqrt{2}$
34. $(5, -2)$; $r = 1$

In Exercises 35–38, sketch the graph of the equation. Label the x- and y-intercepts.

35. $(x - 5)^2 + (y + 2)^2 = 5$
36. $(x + 6)^2 + y^2 = 4$
37. $(x + 1)^2 + (y - 3)^2 = 9$
38. $(x - 2)^2 + (y - 4)^2 = 1$

In Exercises 39–46, find the equation of the circle.

39. Center $(2, 2)$; passes through the origin.
40. Center $(-1, -3)$; passes through $(-4, -2)$.
41. Center $(1, 2)$; intersects x-axis at -1 and 3.
42. Center $(3, 1)$; diameter 2.
43. Center $(-5, 4)$; tangent (touching at one point) to the x-axis.
44. Center $(2, -6)$; tangent to the y-axis.
45. Endpoints of diameter are $(3, 3)$ and $(1, -1)$.
46. Endpoints of diameter are $(-3, 5)$ and $(7, -5)$.
47. One diagonal of a square has endpoints $(-3, 1)$ and $(2, -4)$. Find the endpoints of the other diagonal.
48. Find the vertices of all possible squares with this property: Two of the vertices are $(2, 1)$ and $(2, 5)$. [*Hint:* There are three such squares.]
49. Find a number x such that $(0, 0)$, $(3, 2)$, and $(x, 0)$ are the vertices of an isosceles triangle, neither of whose two equal sides lie on the x-axis.
50. Do Exercise 49 if one of the two equal sides lies on the positive x-axis.
51. The graph, which is based on data from the Actuarial Society of South Africa and assumes no changes in current behavior, shows the projected new cases of AIDS in South Africa (in millions) in coming years ($x = 0$ corresponds to 2000).
   (a) Estimate the number of new cases in 2010.
   (b) Estimate the year in which the largest number of new cases will occur. About how many new cases will there be in that year?
   (c) In what years will the number of new cases be below 7,000,000?

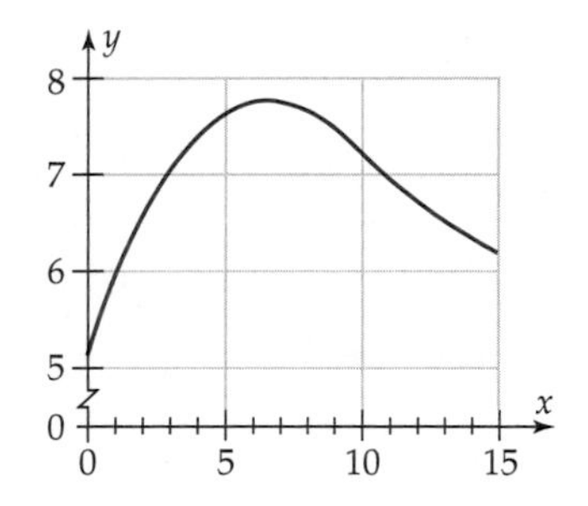

52. The graph shows the total number of alcohol-related car crashes in Ohio at a particular time of the day for the years 1991–2000.* Time is measured in hours after midnight. During what periods is the number of crashes

(a) below 5000?

(b) above 15,000?

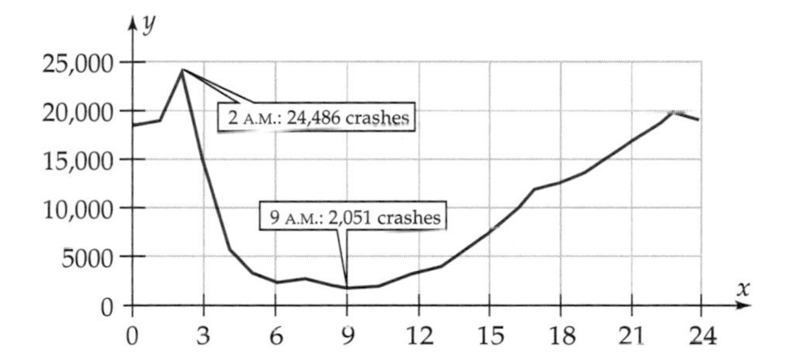

53. Many companies are changing their traditional employee pension plans to so-called cash balance plans. The graph shows pension accrual by age for two hypothetical plans.†

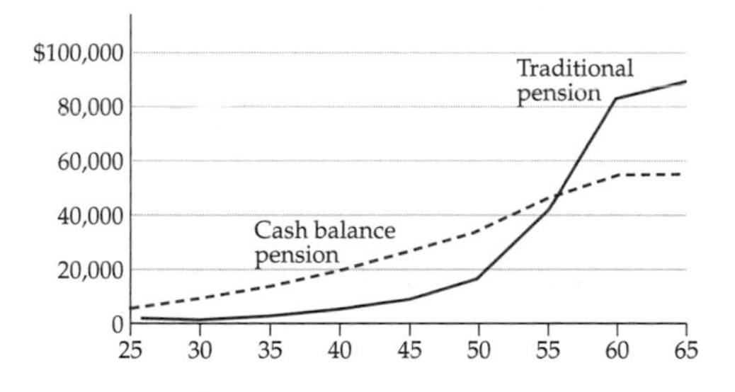

(a) Assuming that you can take your accrued pension benefits in cash when you leave the company before retirement, for what age group is the cash-balance plan better?

(b) At what age is the accrued amount the same for either type of pension plan?

(c) If you remain with the company until retirement, how much better off are you with a traditional instead of cash-balance plan?

54. In an ongoing consumer confidence survey, respondents are asked two questions: Are jobs plentiful? Are jobs hard to get? The graph shows the percentage of people answering "yes" to each question over the years.‡

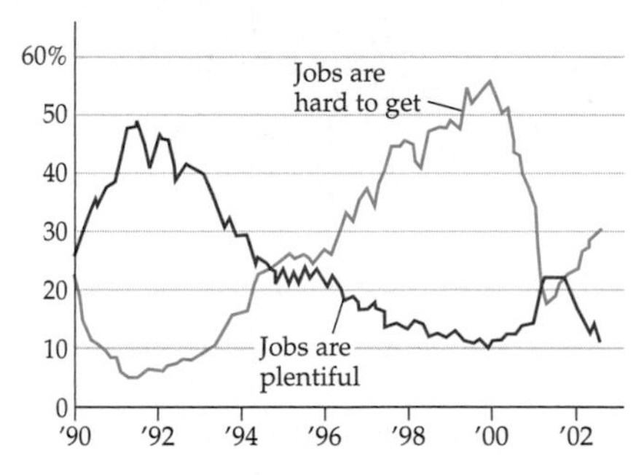

(a) In what year did the most people feel that jobs were plentiful? In that year, approximately what percentage of people felt that jobs were hard to get?

(b) In what year did the most people feel that jobs were hard to get? In that year approximately what percentage of people felt that jobs were plentiful?

(c) In what years was the percentage of those who thought jobs were plentiful the same as the percentage of those that thought jobs were hard to get?

In Exercises 55–58, determine which of graphs A, B, C best describes the given situation.

55. You have a job that pays a fixed salary for the week. The graph shows your salary.

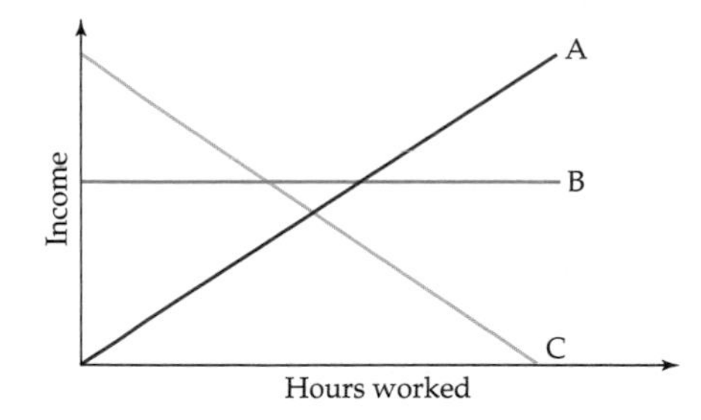

56. You have a job that pays an hourly wage. The graph shows your salary.

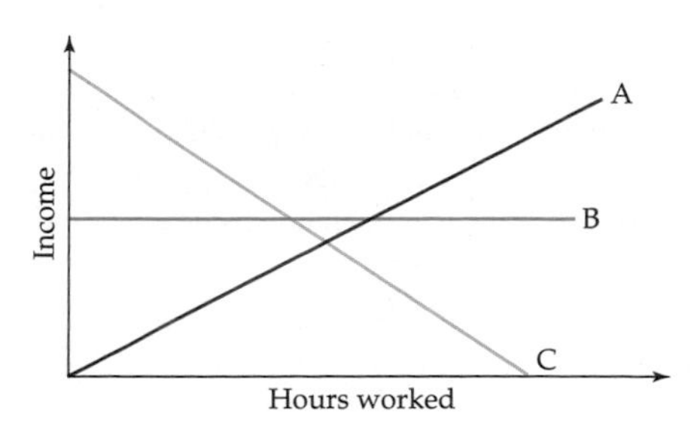

*The Cleveland *Plain Dealer*

†New York Times, March 9, 2003

‡New York Times, March 1, 2003

57. You take a ride on a Ferris wheel. The graph shows your distance from the ground.*

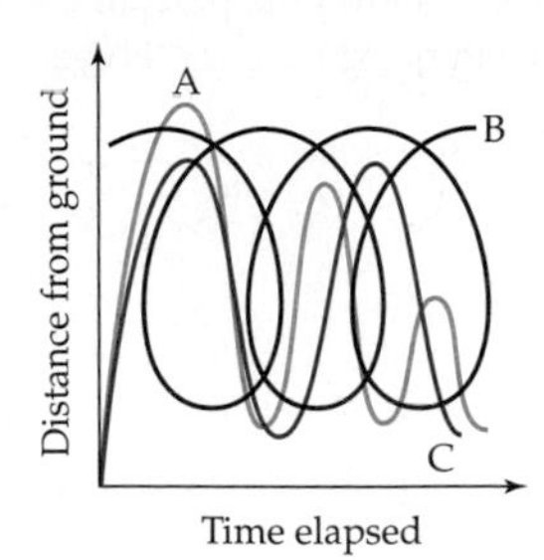

58. Alison's wading pool is filled with a hose by her big sister Emily, and Alison plays in the pool. When they are finished, Emily empties the pool. The graph shows the water level of the pool.

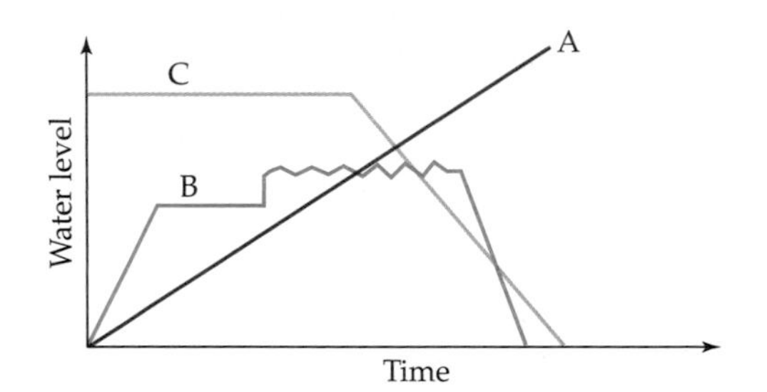

59. Show that the midpoint M of the hypotenuse of a right triangle is equidistant from the vertices of the triangle. [*Hint:* Place the triangle in the first quadrant of the plane, with right angle at the origin so that the situation looks like the figure.]

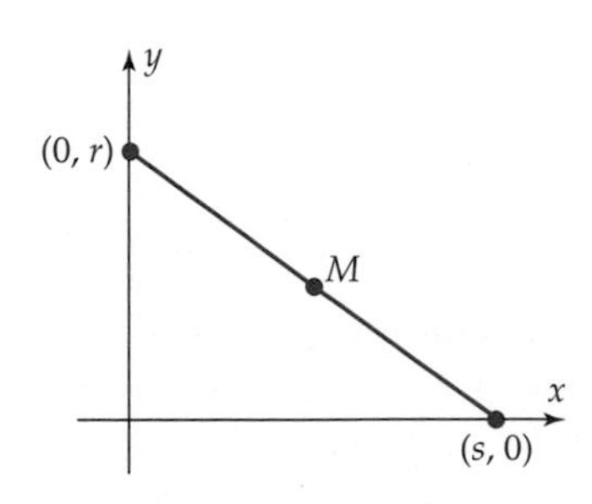

60. Show that the diagonals of a parallelogram bisect each other. [*Hint:* Place the parallelogram in the first quadrant with a vertex at the origin and one side along the x-axis so that the situation looks like the figure.]

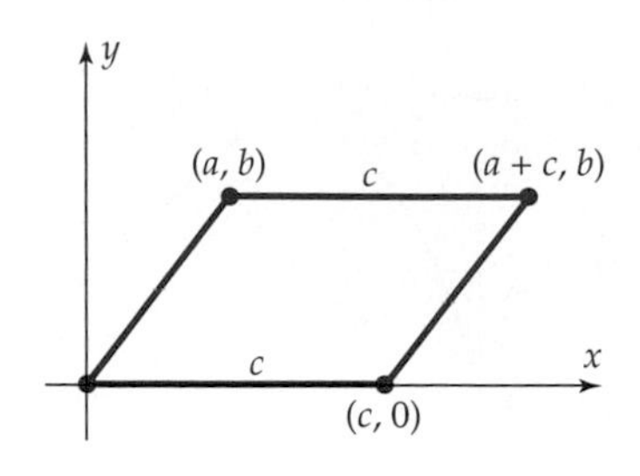

Thinkers

61. For each nonzero real number k, the graph of $(x - k)^2 + y^2 = k^2$ is a circle. Describe all possible such circles.

62. Suppose every point in the coordinate plane is moved 5 units straight up.

(a) To what point does each of these points go: $(0, -5)$, $(2, 2)$, $(5, 0)$, $(5, 5)$, $(4, 1)$?
(b) Which points go to each of the points in part (a)?
(c) To what point does (a, b) go?
(d) To what point does $(a, b - 5)$ go?
(e) What point goes to $(-4a, b)$?
(f) What points go to themselves?

63. Let (c, d) be any point in the plane with $c \neq 0$. Prove that (c, d) and $(-c, -d)$ lie on the same straight line through the origin, on opposite sides of the origin, the same distance from the origin. [*Hint:* Find the midpoint of the line segment joining (c, d) and $(-c, -d)$.]

64. *Proof of the Midpoint Formula* Let P and Q be the points (x_1, y_1) and (x_2, y_2), respectively, and let M be the point with coordinates

$$\left(\frac{x_1 + x_2}{2}, \frac{y_1 + y_2}{2}\right).$$

Use the distance formula to compute the following:

(a) The distance d from P to Q;
(b) The distance d_1 from M to P;
(c) The distance d_2 from M to Q.
(d) Verify that $d_1 = d_2$.
(e) Show that $d_1 + d_2 = d$. [*Hint:* Verify that $d_1 = \frac{1}{2}d$ and $d_2 = \frac{1}{2}d$.]
(f) Explain why parts (d) and (e) show that M is the midpoint of PQ.

---

*_Mathematics Teacher,_ Vol. 95, No. 9, December 2002.

0.3 Graphs, Technology, and Equations

The traditional method of graphing an equation "by hand" is as follows: Construct a table of values with a reasonable number of entries, plot the corresponding points, and use whatever algebraic or other information is available to make an educated guess about the rest.

EXAMPLE 1

The graph of $y = x^2$ consists of all points (x, x^2), where x is a real number. You can easily construct a table of values and plot the corresponding points, as in Figure 0–14. These points suggest that the graph looks like the one in Figure 0–15, which is obtained by connecting the plotted points and extending the graph upward. ■

x	$y = x^2$
−2.5	6.25
−2	4
−1.5	2.25
−1	1
−.5	.25
0	0
.5	.25
1	1
1.5	2.25
2	4
2.5	6.25

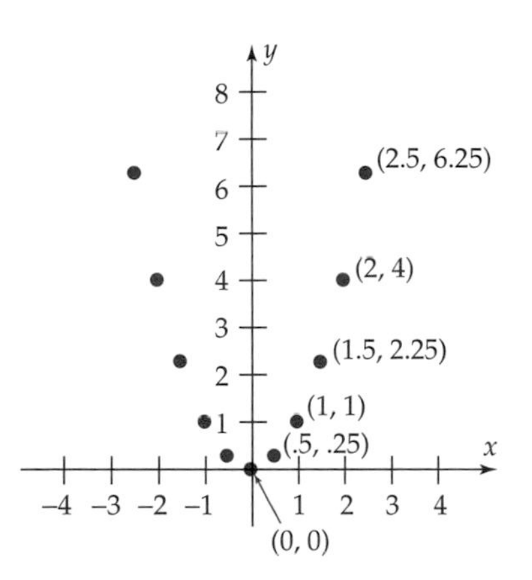

Figure 0–14

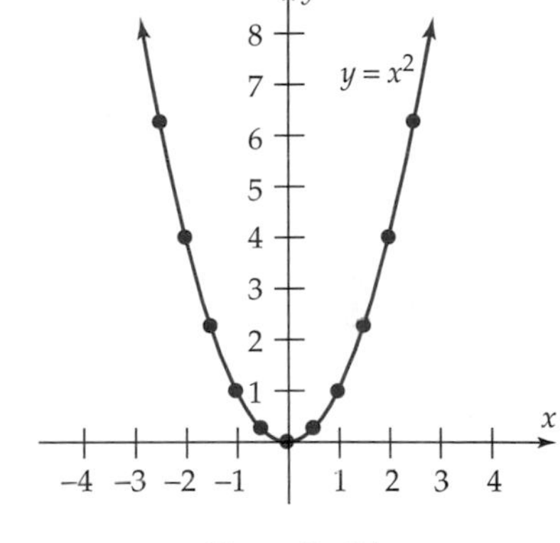

Figure 0–15

GRAPHING WITH TECHNOLOGY

A graphing calculator or computer graphing program graphs in the same way you would graph by hand, but with much greater speed: It plots 95 or more points and simultaneously connects them with line segments. These graphs are generally quite accurate, but no technology is perfect. Algebra and geometry may be needed to interpret misleading (or incorrect) screen images. The next example illustrates the basics of graphing with technology.

EXAMPLE 2

Graph the equation $2x^3 - 8x - 2y + 4 = 0$

(a) using a calculator;

(b) using a computer graphing program.

SOLUTION (a) For calculator graphing, proceed as follows:

Step 1 *Set the **viewing window***—the portion of the coordinate plane that will appear on the screen. Since we don't know yet where the graph lies, we'll try the window with $-10 \le x \le 10$ and $-10 \le y \le 10$. We can change it later if necessary. Press the WINDOW (or RANGE or V-WINDOW or PLOT SET-UP) key,* and enter the appropriate numbers, as in Figure 0–16 (which shows a TI-83+; other calculators are similar). Then the calculator will display the portion of the plane shown in Figure 0–17.

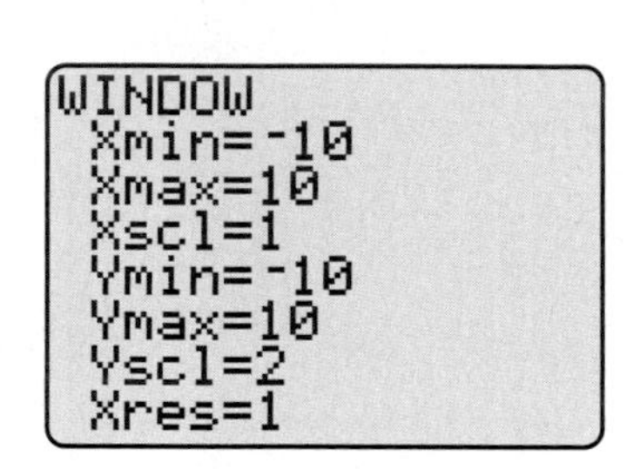

Figure 0–16

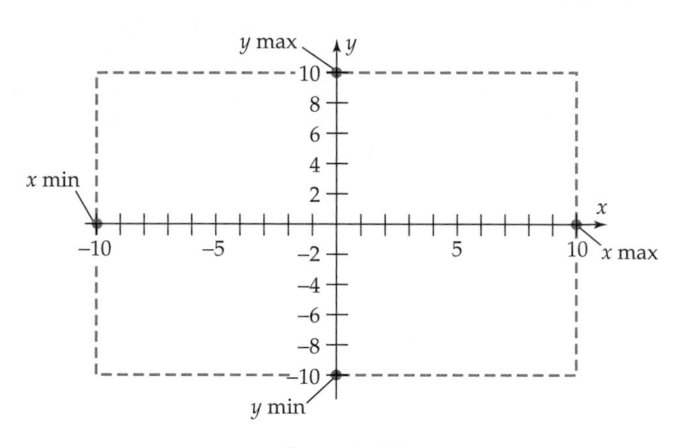

Figure 0–17

Setting Xscl = 1 in Figure 0–16 puts the tick marks 1 unit apart on the x-axis; similarly, Yscl = 2 puts the tick marks 2 units apart on the y-axis.†

Step 2 *Enter the equation in the calculator.* To do this, you must first solve the equation for y:

$$2x^3 - 8x - 2y + 4 = 0$$

Rearrange terms: $$-2y = -2x^3 + 8x - 4$$

Divide by -2: $$y = x^3 - 4x + 2.$$

Now call up the **equation memory** by pressing Y= on TI (or SYMB on HP-39+ or GRAPH (main menu) on Casio). Use the

*On TI-86, press GRAPH first, then WINDOW will appear as a menu choice.

†Xscl is labeled "X scale" on Casio and "Xtick" on HP-39+. Some calculators do not have an Xres setting. On those that do, it should normally be set at 1 (or at "detail" on HP-39+).

TECHNOLOGY TIP

When entering an equation for graphing, use the "variable" key rather than the ALPHA *X* key. It has a label such as *X*, T, *θ* or *X*, T, *θ*, *n* or *X*, *θ*, T or x-VAR. On TI-89, however, use the *x* key on the keyboard.

Technology Tip in the margin to enter the equation, as shown in Figure 0–18.

Step 3 *Graph the equation.* Press GRAPH (or PLOT or DRAW) to obtain Figure 0–19.

Figure 0–18

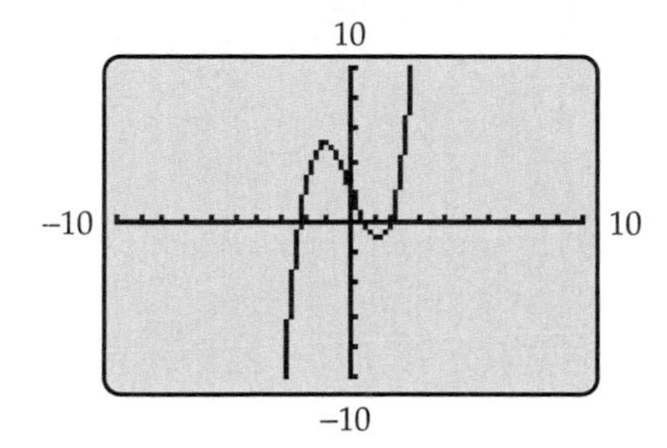

Figure 0–19

Because of the limited resolution of a calculator screen, the graph appears to consist of short adjacent line segments rather than a smooth unbroken curve.

Step 4 *If necessary, adjust the viewing window for a better view.* The graph in Figure 0–19 is scrunched up in the middle of the screen. So we change the viewing window (Figure 0–20) and press GRAPH again to obtain Figure 0–21.

TECHNOLOGY TIP

On TI-83+/86 you may get an error message when you press GRAPH if one of Plot 1, Plot 2, or Plot 3, at the top of the Y= menu (see Figure 0–18) is shaded. In this case, move the cursor over the shading and press ENTER to remove it.

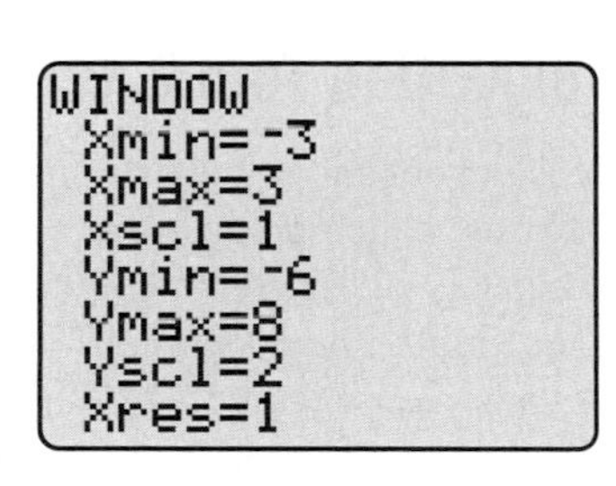

Figure 0–20

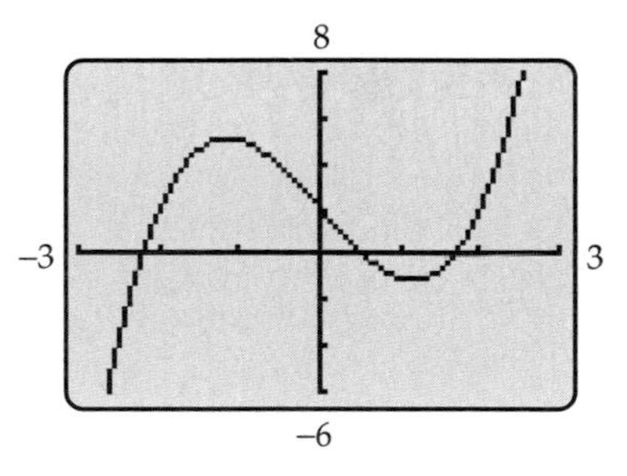

Figure 0–21

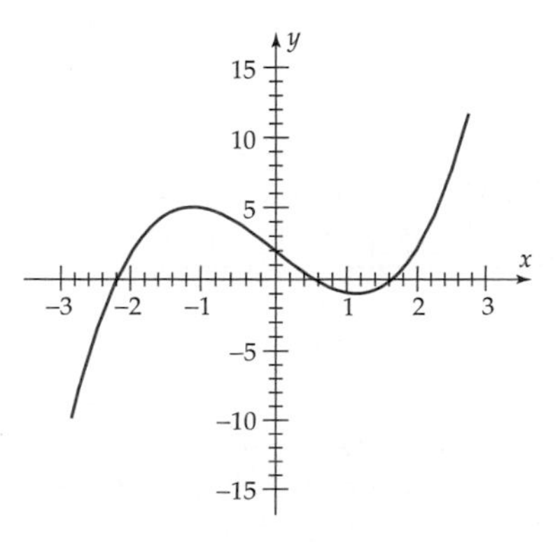

Figure 0–22

(b) A typical computer graphing program requires the same information as does a calculator, namely, the equation to be graphed and the viewing window in which to graph it. The procedures for entering this information vary widely, so check your instruction manual or help key index. For example, the following command in Maple produces Figure 0–22, in which the range of *y*-values was chosen automatically by Maple.

```
plot(x^3 -4*x + 2, x = -3..3);
```

To duplicate Figure 0–21, we specify the range of *y*-values and the number of tick marks on each axis:

```
plot(x^3 -4*x + 2, x = -3..3, y = -6..8,
xtickmarks = 6, ytickmarks = 6);
```

The result is Figure 0–23.

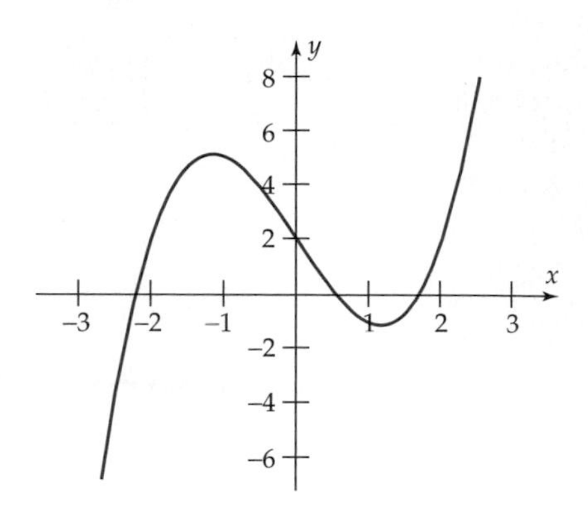

Figure 0–23

As is typical of computer-generated graphs, this one appears smooth and connected, as it should. ■

TECHNOLOGY TIP

An equation is "on" if its equal sign is shaded or if there is a check mark next to it. Only equations that are "on" are graphed when you press GRAPH.

To turn an equation "on" or "off" on TI-83+, move the cursor over the equal sign and press ENTER. On other calculators, move the cursor to the equation and press SELECT or CHECK.

NOTE An equation stays in a calculator's equation memory until you delete it. When several equations are in the memory, you must turn "on" those you want graphed and turn "off" those you don't want graphed, as explained in the Technology Tip in the margin.

GRAPHING TOOLS: THE TRACE FEATURE

A calculator obtains a graph by plotting points and simultaneously connecting them. To see which points were actually plotted, press TRACE, and a flashing cursor appears on the graph. Use the left and right arrow keys to move the cursor along the graph. The coordinates of the point the cursor is on appear at the bottom of the screen. Figure 0–24 illustrates this for the graph of $y = x^3 - 4x + 2$.

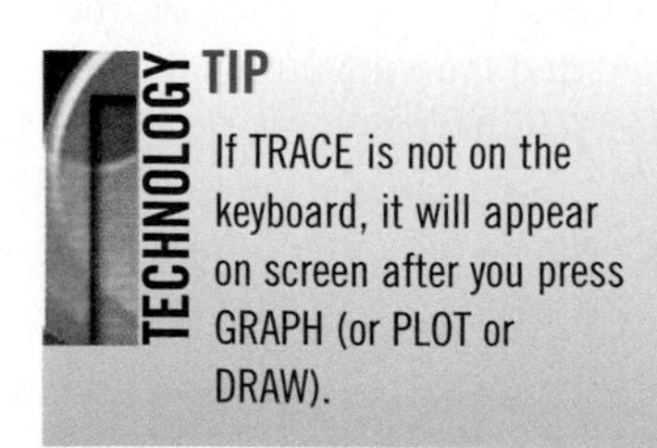

TIP

If TRACE is not on the keyboard, it will appear on screen after you press GRAPH (or PLOT or DRAW).

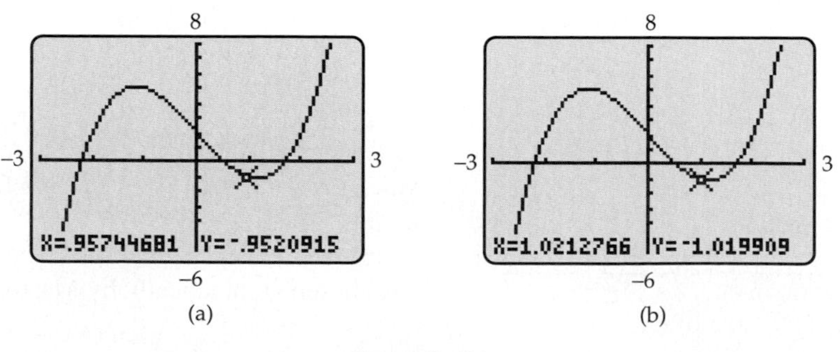

Figure 0–24

The trace cursor displays only the points that the calculator actually plotted. For instance, $(1, -1)$ is on the graph of $y = x^3 - 4x + 2$, as you can easily verify, but was not one of the points the calculator plotted to produce Figure 0–24. So the trace lands on two nearby points that *were* plotted, but skips $(1, -1)$.

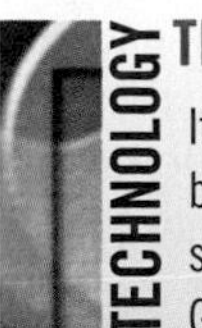

TECHNOLOGY TIP

If ZOOM is not on the keyboard, it will appear on screen after you press GRAPH (or PLOT or DRAW).

To set the zoom factors, look for FACT, ZFACT, or (SET) FACTORS in the ZOOM menu (or its MEMORY submenu).

GRAPHING TOOLS: ZOOM IN/OUT

The ZOOM-OUT and ZOOM-IN keys on the ZOOM menu make it easy to increase or decrease the size of the viewing window with a single keystroke. Figure 0–25 shows the graph of $y = x^3 - 2x^2 + 1$ in two windows. Window (b) was obtained from window (a) by zooming in at the origin by a factor of 4.

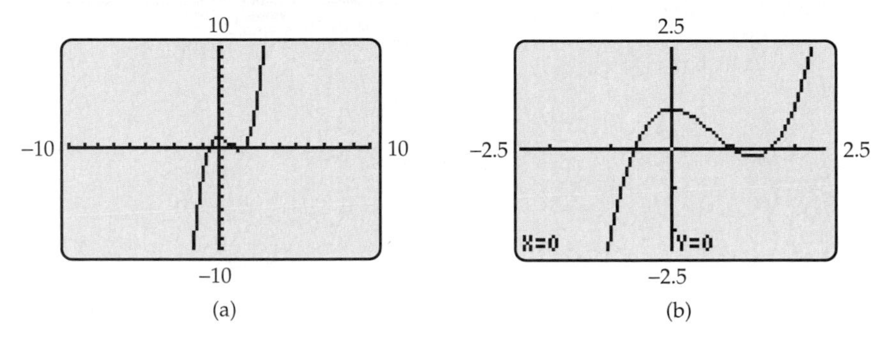

Figure 0–25

The calculator automatically changes the range of x- and y-values when zooming, but does not change the *X*scl or *Y*scl settings. This may cause occasional viewing problems.

NOTE

The heading GRAPHING EXPLORATION indicates that you are to use your calculator or computer as directed to complete the discussion.

GRAPHING EXPLORATION

Graph $y = x^3 - 2x^2 + 1$ in the same window as Figure 0–25(a). Then zoom *out* from the origin by a factor of 4. Can you read the tick marks on the axes? Change the settings to $X\text{scl} = 5$ and $Y\text{scl} = 5$, and regraph. Can you read them now?

GRAPHING TOOLS: SPECIAL VIEWING WINDOWS

The ZOOM menu also has keys that enable you to obtain frequently used viewing windows in a single keystroke, including most or all of the following:

1. The **standard viewing window** (also labeled **Zstandard** or **Zstd** or **Default**) has $-10 \le x \le 10$ and $-10 \le y \le 10$.
2. A **decimal window** (also labeled **Zdecimal** or **Zdecm** or **Init**) is one in which the horizontal distance between two adjacent pixels on the screen is .1. The size of the decimal window depends on the width of your calculator screen.

GRAPHING EXPLORATION

Graph $y = x^4 - 2x^2 - 1$ in the standard window. Use the TRACE key, and watch the values of the x-coordinates. Now regraph by using DECIMAL in the ZOOM menu. Then use the TRACE key again. How do the x-coordinates change at each step? Finally, look in WINDOW to find the dimensions of your decimal window.

TECHNOLOGY TIP

For an approximately square window, the y-axis should be 2/3 as long as the x-axis on TI-83+ and 3/5 as long on TI-86. It should be half as long on TI-89, Casio, and HP-39+.

On calculators other than TI-86, the decimal window is a square window.

3. In a **square window,** a one-unit segment on the x-axis has the same length on the screen as a one-unit segment on the y-axis. Because calculator screens are wider than they are high, the y-axis in a square window must be shorter than the x-axis (see the Technology Tip in the margin). The next example illustrates the usefulness of square windows.

EXAMPLE 3

Graph the circle $x^2 + y^2 = 9$ on a calculator.

SOLUTION First, we solve the equation for y:

$$y^2 = 9 - x^2$$

$$y = \sqrt{9 - x^2} \quad \text{or} \quad y = -\sqrt{9 - x^2}.$$

Graphing both of these equations on the same screen will produce the graph of the circle. In the standard viewing window (Figure 0–26), however, the graph doesn't look like a circle.

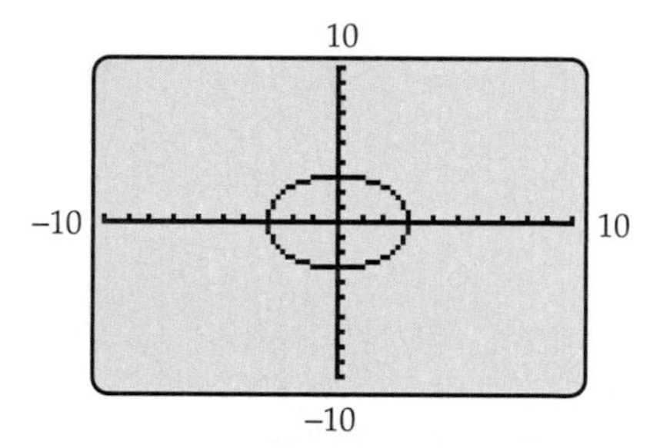

Figure 0–26

The reason is that this is *not* a square window (the 20 one-unit segments on the y-axis occupy less space than the 20 one-unit segments on the x-axis, which means that the ones on the x-axis are a tad longer than the ones on the y-axis). So we press SQUARE (or ZSQUARE or ZOOMSQR) on the ZOOM menu and obtain Figure 0–27. The size of the x-axis has been changed to produce a square window in which the circle looks round, as it should.* ■

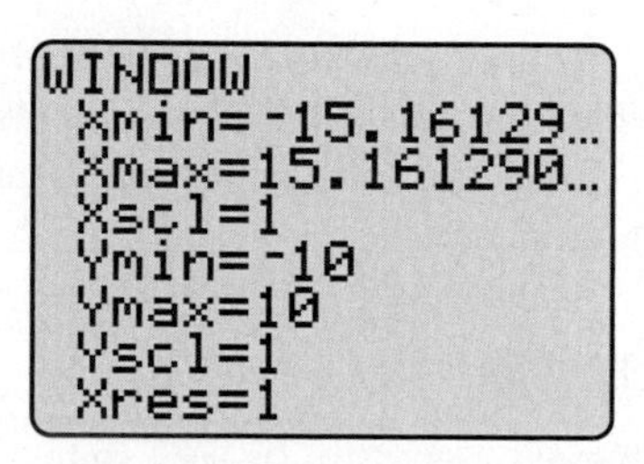

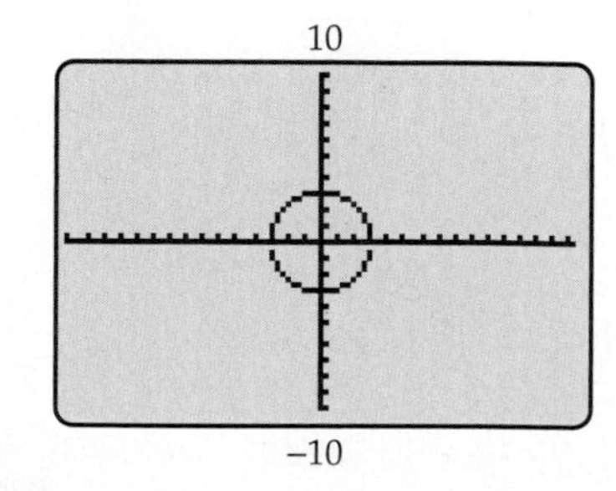

Figure 0–27

*On some calculators, the SQUARE key changes the y-axis length instead of the x-axis. The result in either case is a square window.

Although any convenient viewing window is usually OK, you should use a square window when you want circles to look round and perpendicular lines to look perpendicular.

GRAPHING EXPLORATION

The lines $y = .5x$ and $y = -2x + 2$ are perpendicular (why?). Graph them in the standard viewing window. Do they look perpendicular? Now graph them in a square window. Do they look perpendicular?

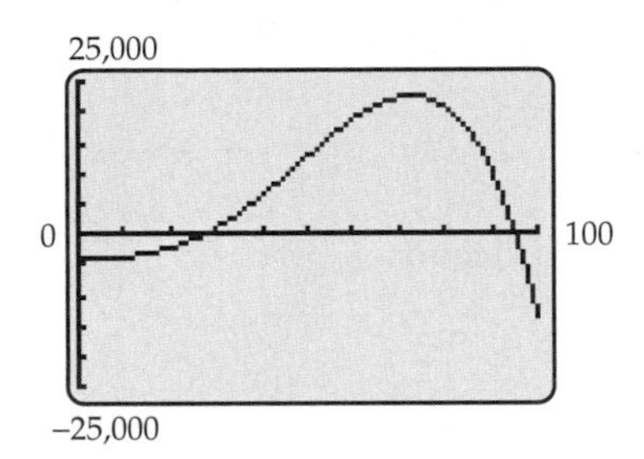

Figure 0–28

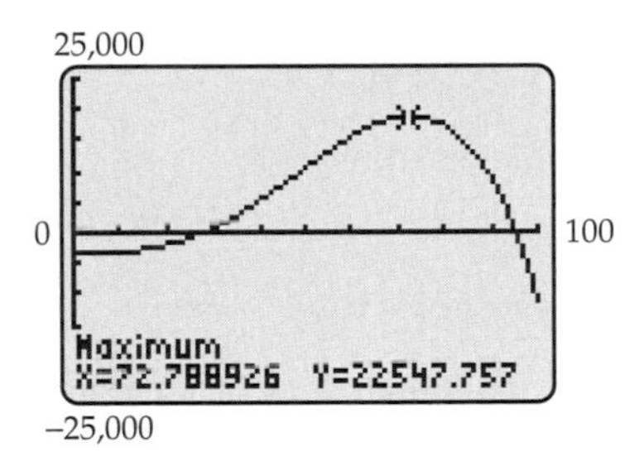

Figure 0–29

TECHNOLOGY TIP

The graphical maximum finder is in the following menu/submenu:

TI-83+: CALC

TI-86/89: GRAPH/MATH

Casio 9850: GRAPH/G-SOLV

Casio FX2.0: GRPH-TBL/G-SLV

HP-39+: PLOT/FNC

It is labeled MAXIMUM, MAX, FMAX or EXTREMUM.

On some TI calculators, you must first select a left (or lower) bound, meaning an *x*-value to the left of the highest point, and a right (or upper) bound, meaning an *x*-value to its right, and make an initial guess. On other calculators, you may have to move the cursor near the point you are seeking.

GRAPHING TOOLS: THE MAXIMUM/MINIMUM FINDER

Many graphs have peaks and valleys (for instance, see Figure 0–25 on page 29 or Figure 0–28). A calculator's maximum finder or minimum finder can locate these points with a high degree of accuracy, as illustrated in the next example.

EXAMPLE 4

The Cortopassi Computer Company can produce a maximum of 100,000 computers a year. Their annual profit is given by

$$y = -.003x^4 + .3x^3 - x^2 + 5x - 4000,$$

where y is the profit (in thousands of dollars) from selling x thousand computers. Use graphical methods to estimate how many computers should be sold to make the largest possible profit.

SOLUTION We first choose a viewing window. The number x of computers is nonnegative, and no more than 100,000 can be produced, so that $0 \le x \le 100$ (because x is measured in thousands). The profit y may be positive or negative (the company could lose money). So we try a window with $-25{,}000 \le y \le 25{,}000$ and obtain Figure 0–28. For each point on the graph,

The x-coordinate is the number of thousands of computers produced;

The y-coordinate is the profit (in thousands) on that number of computers.

The largest possible profit occurs at the point with the largest y-coordinate, that is, the highest point in the window. The maximum finder on a TI-83+ (see the Technology Tip in the margin) produced Figure 0–29. Since x and y are measured in thousands, we see that making about 72,789 computers results in a maximum profit of about \$22,547,757. ■

GRAPHING EXPLORATION

Graph $y = .3x^3 + .8x^2 - 2x - 1$ in the window with $-5 \le x \le 5$ and $-5 \le y \le 5$. Use your maximum finder to approximate the coordinates of the highest point to the left of the y-axis. Then use your minimum finder (in the same menu) to approximate the coordinates of the *lowest* point to the right of the y-axis. How do these answers compare with the ones you get by using the trace feature?

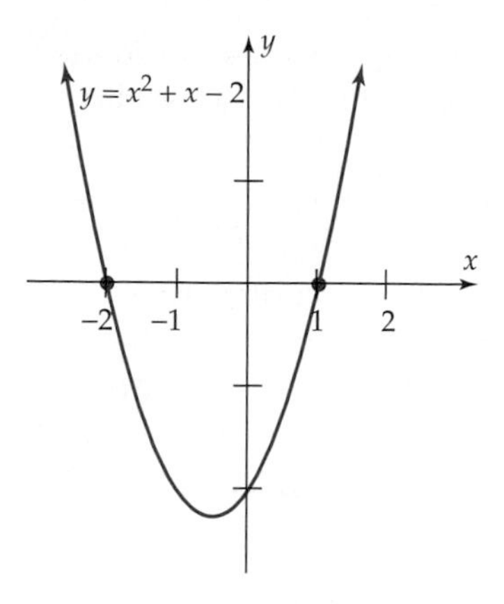

Figure 0–30

SOLVING EQUATIONS GRAPHICALLY

The graph of $y = x^2 + x - 2$ in Figure 0–30 has x-intercepts at -2 and 1, which means that the graph intersects the x-axis at the points $(-2, 0)$ and $(1, 0)$. To say that $(1, 0)$ is on the graph means that $x = 1$ and $y = 0$ satisfy the equation:

$$y = x^2 + x - 2$$
$$0 = 1^2 + 1 - 2.$$

In other words, the x-intercept 1 is a *solution* of the equation

$$x^2 + x - 2 = 0.$$

Similarly, the other x-intercept, -2, is also a solution of $x^2 + x - 2 = 0$ because $(-2)^2 + (-2) - 2 = 0$. The same argument works in the general case.

Solutions and Intercepts

> The real solutions of a one-variable equation of the form
>
> $$\text{expression in } x = 0$$
>
> are the x-intercepts of the graph of the two-variable equation
>
> $$y = \text{expression in } x.$$

Therefore, to solve an equation, you need only find the x-intercepts of its graph, as illustrated in the next example.

EXAMPLE 5 

Solve the equation $x^4 - 4x^3 + 3x^2 - x = 2$ graphically.

SOLUTION We first rewrite the equation so that one side is 0.

$$x^4 - 4x^3 + 3x^2 - x - 2 = 0.$$

As we have just seen, the solutions of this equation are the x-intercepts of the graph of

$$y = x^4 - 4x^3 + 3x^2 - x - 2.$$

These x-intercepts may be found by graphing the equation and using the graphical root finder, as in Figure 0–31.* (See the Technology Tip in the margin.)

TECHNOLOGY TIP

The graphical root finder is labeled ROOT or ZERO or X-INCPT and is in this menu/submenu:

TI-83+: CALC
TI-86/89: GRAPH/MATH
Casio 9850: GRAPH/G-SOLV
Casio FX2.0: GRPH-TBL/G-SLV
HP-39+: PLOT/FCN

On most TI calculators, you must select a left and a right bound (x-values to the left and right of the intercept) and make an initial guess.

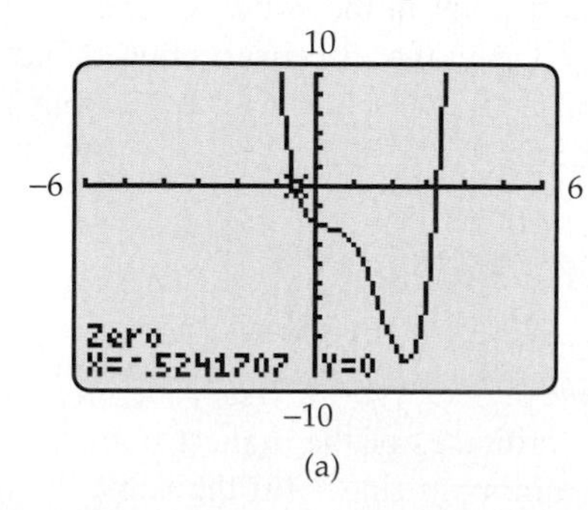

(a)

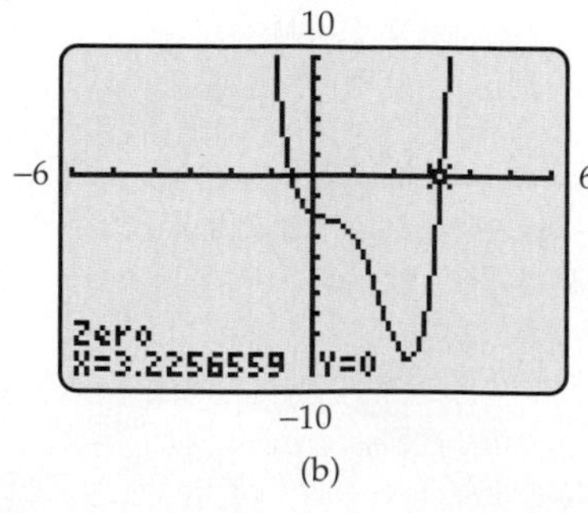

(b)

Figure 0–31

*Unless stated otherwise in the examples and exercises of this section, you may assume that the given viewing window includes all x-intercepts of the graph.

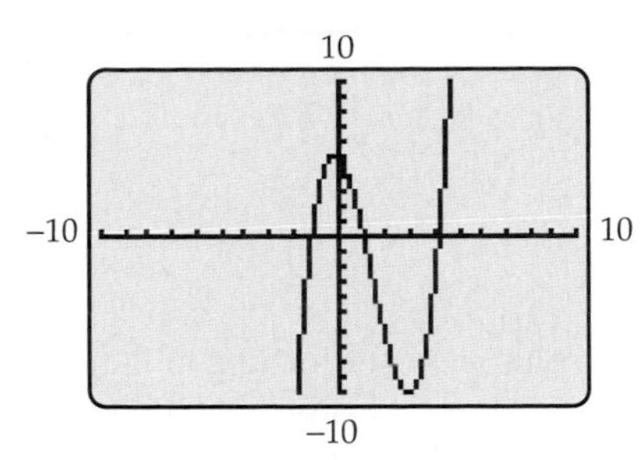

Figure 0–32

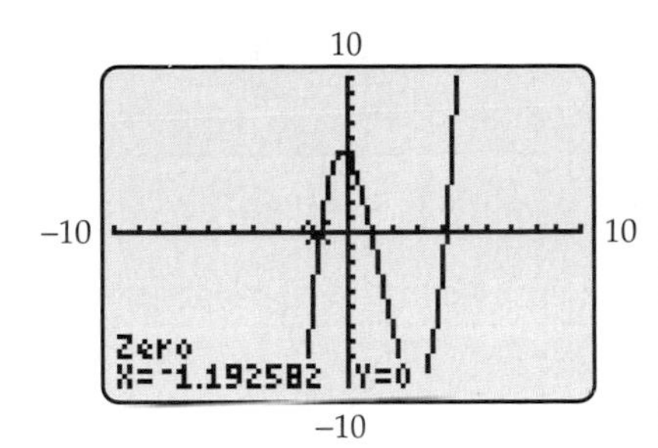

Figure 0–33

Therefore, the approximate solutions of the original equation are $x \approx -.5242$ and $x \approx 3.2257$. ■

EXAMPLE 6

Solve: $x^3 - 4x^2 - 2x + 5 = 0$.

SOLUTION We graph $y = x^3 - 4x^2 - 2x + 5$ (Figure 0–32) and see that it has three x-intercepts. So the equation has three real solutions. A root finder shows that one solutions is $x \approx -1.1926$ (Figure 0–33). ■

GRAPHING EXPLORATION

Use your graphical root finder to find the other two solutions (x-intercepts) of the equation in Example 6.

THE INTERSECTION METHOD

The next examples illustrate an alternative graphical method, which is sometimes more convenient for solving equations.

EXAMPLE 7

Solve: $|x^2 - 4x - 3| = x^3 + x - 6$.

SOLUTION Let $y_1 = |x^2 - 4x - 3|$ and $y_2 = x^3 + x - 6$, and graph both equations on the same screen (Figure 0–34). Consider the point where the two graphs intersect. Since it is on the graph of y_1, its second coordinate is $|x^2 - 4x - 3|$, and since it is also on the graph of y_2, its second coordinate is $x^3 + x - 6$. So for this number x, we must have $|x^2 - 4x - 3| = x^3 + x - 6$. In other words, *the x-coordinate of the intersection point is the solution of the equation.*

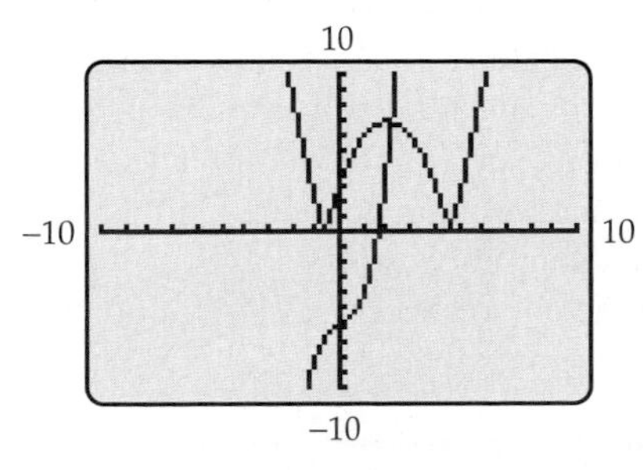

Figure 0–34

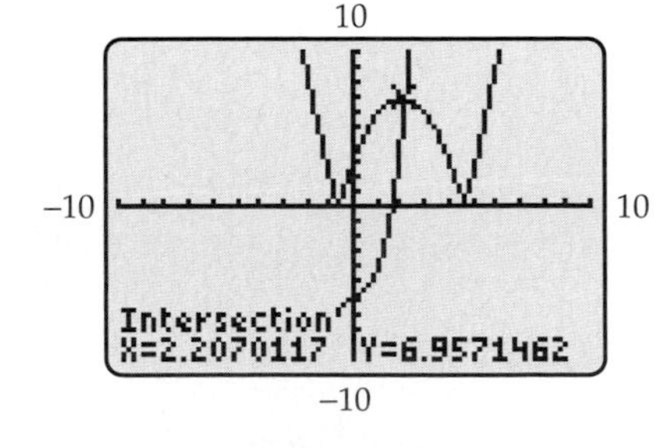

Figure 0–35

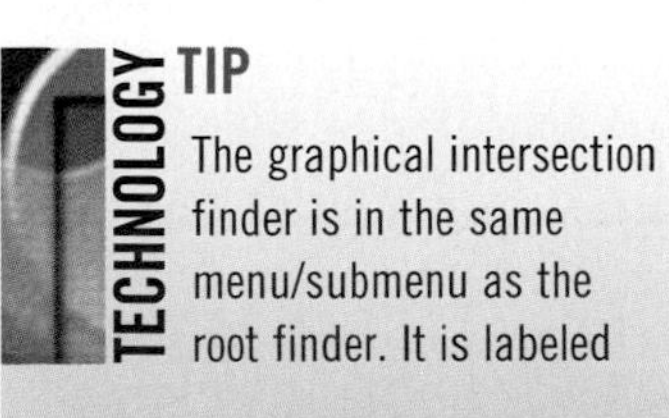

TECHNOLOGY TIP The graphical intersection finder is in the same menu/submenu as the root finder. It is labeled INTERSECTION, INTERSECT, ISECT, ISCT, or INTSECT.

This coordinate can be approximated by using a graphical intersection finder (see the Technology Tip in the margin), as shown in Figure 0–35. Therefore, the solution of the original equation is $x \approx 2.207$ ■

EXAMPLE 8

Solve: $x^2 - 2x - 6 = \sqrt{2x + 7}$.

SOLUTION We graph $y_1 = x^2 - 2x - 6$ and $y_2 = \sqrt{2x + 7}$ on the same screen. Figure 0–36 shows that there are two solutions (intersection points). According to an intersection finder (Figure 0–37), the positive solution is $x \approx 4.3094$. ■

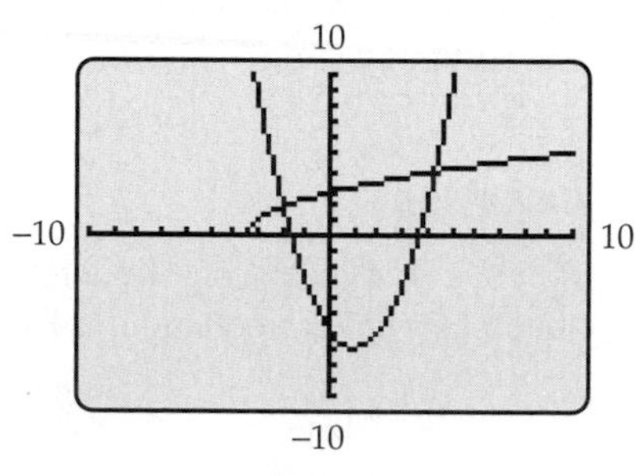

Figure 0–36

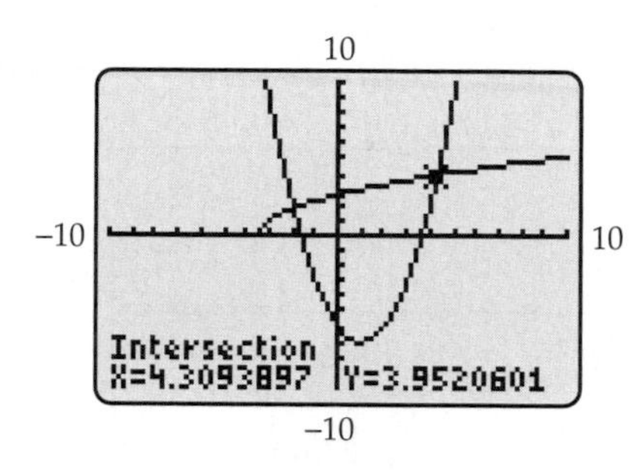

Figure 0–37

EXPLORATION

Use a graphical intersection finder to approximate the negative solution of the equation in Example 8.

NUMERICAL METHODS

In addition to graphical tools for solving equations, calculators and computer algebra systems have **equation solvers** that can find or approximate the solutions of most equations. On most calculators, the solver finds one solution at a time. You must enter the equation and an initial guess and possibly an interval in which to search. A few calculators (such as TI-89) and most computer systems have **one-step equation solvers** that will find or approximate all the solutions in a single step. Check you calculator instruction manual or your computer's help menu for directions on using these solvers.

Several calculators and many computer algebra systems also have **polynomial solvers,** one-step solvers designed specifically for polynomial equations; you need only enter the degree and coefficients of the polynomial. A few of these (such as Casio 9850) are limited to equations of degree 2 and 3. Directions for using polynomial solvers on calculators are in Calculator Investigation 4 at the end of Section 0.1. Computer users should check the help menu for the proper syntax.

EXAMPLE 9

When asked to search the interval $-10 \le x \le 10$ with an initial guess of 3, the equation solver in the MATH menu of a TI-83+ found one solution of

$$4x^5 - 12x^3 + 8x - 1 = 0,$$

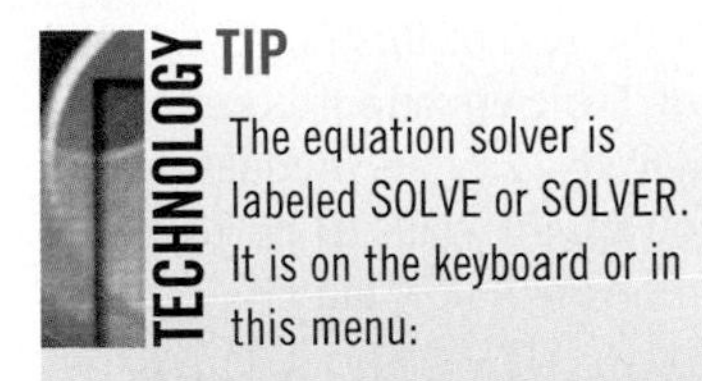

TIP

The equation solver is labeled SOLVE or SOLVER. It is on the keyboard or in this menu:

TI-83+: MATH

TI-89: ALGEBRA

HP-39+: APLET

Casio: EQUA (Main Menu)

The syntax varies, so check your instruction manual.

namely, $x \approx .128138691376$ (Figure 0–38). The POLY solver on a TI-86 produced that solution and four others (Figure 0–39) as did the downloadable POLY solver for TI-83+.

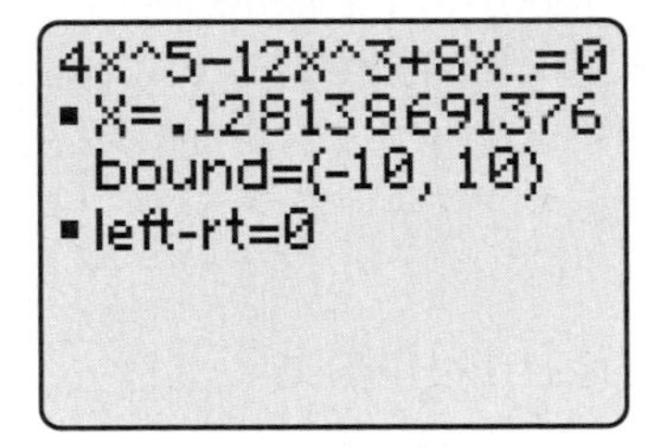

Figure 0–38

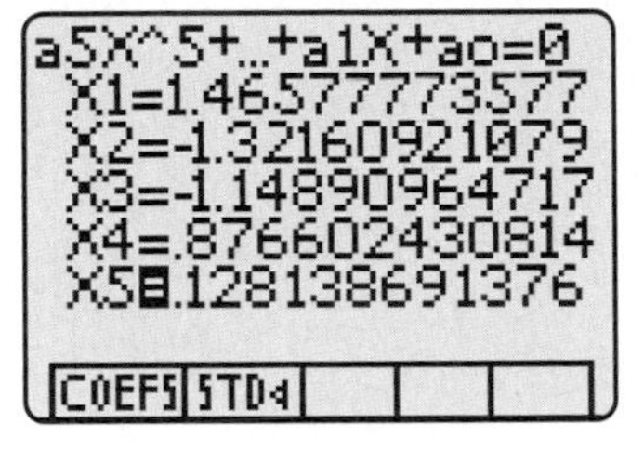

Figure 0–39

On Maple, the command

fsolve(4*x^5 − 12*x^3 + 8*x − 1 = 0, x);

also produced all five solutions of the equation. ■

CHOOSING A SOLUTION METHOD

We have seen that equations can be solved by algebraic, graphical, and numerical methods. Each method has both advantages and disadvantages, as summarized in the table.

Solution Method	Advantages	Possible Disadvantages
Algebraic	Produces exact solutions. Easiest method for most linear and quadratic equations.	May be difficult or impossible to use with complicated equations.
Graphical Root Finder or Intersection Finder	Works well for a large variety of equations. Gives visual picture of the location of the solutions.	Solutions may be approximations. Finding a useable viewing window may take a lot of time.
Numerical Equation Solvers		Solutions may be approximations.
Polynomial solver	Fast and easy.	Works only for polynomial equations.
One-step solver	Fast and easy.	May miss some solutions or be unable to solve certain equations.
Other solvers		May require a considerable amount of work to find the particular solution you want

The choice of solution method is up to you. In the rest of this book, we normally use algebraic means for solving linear and quadratic equations because this is often the fastest and most accurate method. Naturally, any such equation can also be solved graphically or numerically (and you might want to do that as a check against errors). Graphical and numerical methods will normally be used for more complicated equations.

CALCULATOR

INVESTIGATIONS 0.3

1. TICK MARKS

(a) Set Xscl = 1 so that adjacent tick marks on the x-axis are one unit apart. Find the largest range of x values such that the tick marks on the x-axis are clearly distinguishable and appear to be equally spaced.

(b) Do part (a) with y in place of x.

2. VIEWING WINDOWS

Look in the ZOOM menu to find out how many built-in viewing windows your calculator has. Take a look at each one.

3. MAXIMUM/MINIMUM FINDERS

Use your minimum finder to approximate the x-coordinates of the lowest point on the graph of $y = x^3 - 2x + 5$ in the window with $0 \le x \le 5$ and $-3 \le y \le 8$. The correct answer is

$$x = \sqrt{\frac{2}{3}} \approx .816496580928.$$

How good is your approximation?

4. SQUARE WINDOWS

Find a square viewing window on your calculator that has $-10 \le x \le 10$.

5. DOT GRAPHING MODE

To see which points your calculator actually graphs (without any connecting line segments), change the graphing mode. Select DOT (or DRAW DOT or PLOT) in the appropriate menu/submenu:

TI-83+: MODE	TI-89: Y= / STYLE
TI-86: GRAPH/FORMAT	Casio: SETUP/DRAWTYPE

On HP-39+, uncheck "connect" on the second page of the PLOT SETUP menu.

Now graph $y = .5x^3 - 2x^2 + 1$ in the standard window. Try some other equations as well.

EXERCISES 0.3

Exercises 1–4 are representations of calculator screens. State the approximate coordinates of the points P and Q.

1.

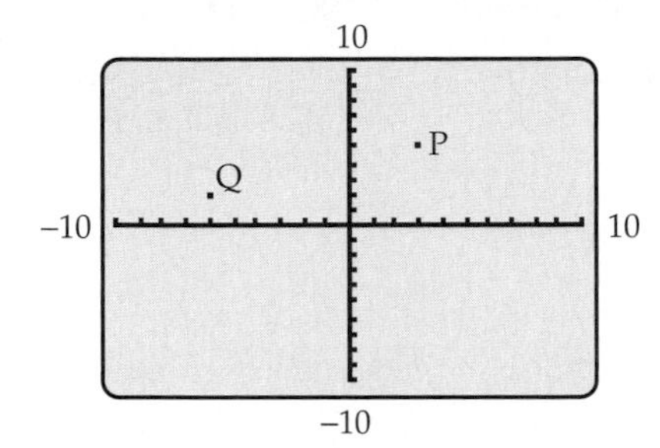

2.

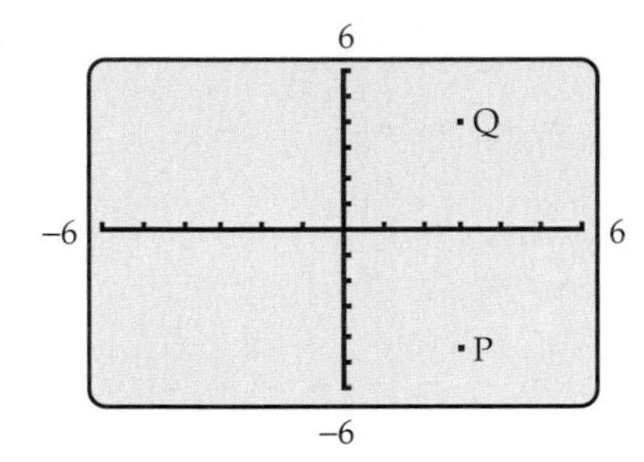

3.

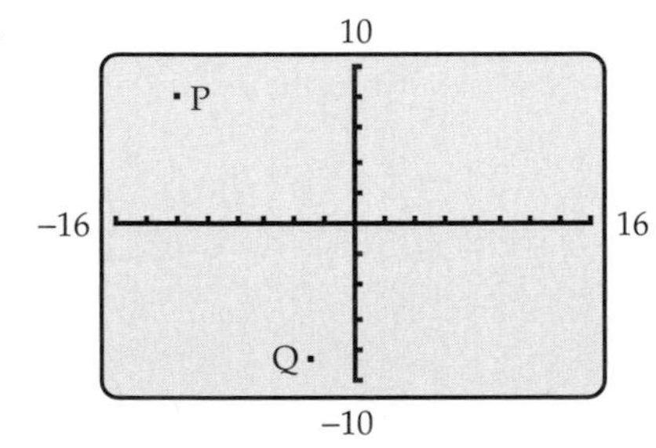

4.

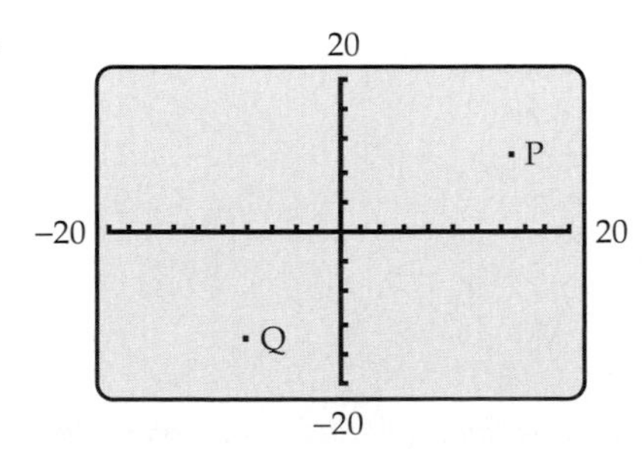

In Exercises 5–10, graph the equation by hand by plotting no more than six points and filling in the rest of the graph as best you can. Then use the calculator to graph the equation and compare the results.

5. $y = |x - 2|$ **6.** $y = \sqrt{x + 5}$

7. $y = x^2 - x$ **8.** $y = x^2 + x + 1$

9. $y = x^3 + 1$ **10.** $y = \frac{1}{x}$

In Exercises 11–16, find the graph of the equation in the standard window.

11. $3 + y = .5x$ **12.** $y - 2x = 4$

13. $y = x^2 - 5x + 2$ **14.** $y = .3x^2 + x - 4$

15. $y = .2x^3 + .1x^2 - 4x + 1$

16. $y = .2x^4 - .2x^3 - 2x^2 - 2x + 5$

In Exercises 17–22, determine which of the following viewing windows gives the best view of the graph of the given equation.

(a) $-10 \le x \le 10; \quad -10 \le y \le 10$
(b) $-5 \le x \le 25; \quad 0 \le y \le 20$
(c) $-10 \le x \le 10; \quad -100 \le y \le 100$
(d) $-20 \le x \le 15; \quad -60 \le y \le 250$
(e) None of a, b, c, d gives a complete graph.

17. $y = 18x - 3x^2$ **18.** $y = 4x^2 + 80x + 350$

19. $y = \frac{1}{3}x^3 - 25x + 100$ **20.** $y = x^4 + x - 5$

21. $y = x^2 + 50x + 625$ **22.** $y = .01(x - 15)^4$

23. A toy rocket is shot straight up from ground level and then falls back to earth; wind resistance is negligible. Use your calculator to determine which of the following equations has a graph whose portion above the x-axis provides the most plausible model of the path of the rocket.
(a) $y = .1(x - 3)^3 - .1x^2 + 5$
(b) $y = -x^4 + 16x^3 - 88x^2 + 192x$
(c) $y = -16x^2 + 117x$
(d) $y = .16x^2 - 3.2x + 16$
(e) $y = -(.1x - 3)^6 + 600$

24. Monthly profits at DayGlo Tee Shirt Company appear to be given by the equation

$$y = -.00027(x - 15{,}000)^2 + 60{,}000,$$

where x is the number of shirts sold that month and y is the profit. DayGlo's maximum production capacity is 15,000 shirts per month.

(a) If you plan to graph the profit equation, what range of x values should you use? [*Hint:* You can't make a negative number of shirts.]

(b) The president of DayGlo wants to motivate the sales force (who are all in the profit-sharing plan) so he asks you to prepare a graph that shows DayGlo's profits increasing *dramatically* as sales increase. Using the profit equation and the x range from part (a), what viewing window is suitable?

(c) The City Council is talking about imposing more taxes. The president asks you to prepare a graph showing that DayGlo's profits are essentially flat. Using the profit equation and the x range from part (a), what viewing window is suitable?

25. (a) Graph $y = x^3 - 2x^2 + x - 2$ in the standard window.
(b) Use the trace feature to show that the portion of the graph with $0 \le x \le 1.5$ is not actually horizontal. [*Hint:* All the points on a horizontal segment must have the same y-coordinate. (Why?)]
(c) Find a viewing window that clearly shows that the graph is not horizontal when $0 \le x \le 1.5$.

26. (a) Graph $y = \dfrac{1}{x^2 + 1}$ in the standard window.
(b) Does the graph appear to stop abruptly part-way along the x-axis? Use the trace feature to explain why this happens. [*Hint:* In this viewing window, each pixel represents a rectangle that is approximately .32 units high.]
(c) Find a viewing window with $-10 \le x \le 10$, which shows a complete graph that does not fade into the x-axis.

In Exercises 27–36, use the techniques of Example 3 to graph the equation in a suitable square viewing window.

27. $x^2 + y^2 = 16$

28. $y^2 = x + 2$

29. $3x^2 + 2y^2 = 48$

30. $25(x - 5)^2 + 36(y + 4)^2 = 900$

31. $(x - 4)^2 + (y + 2)^2 = 25$

32. $9x^2 + 4y^2 = 36$

33. $4x^2 - 9y^2 = 36$

34. $9y^2 - x^2 = 9$

35. $9x^2 + 5y^2 = 45$

36. $x = y^2 - 2$

In Exercises 37–38, use zoom-in or a maximum/minimum finder to determine the highest and lowest point on the graph in the given window.

37. $y = .4x^3 - 3x^2 + 4x + 3$
$(0 \le x \le 5 \text{ and } -5 \le y \le 5)$

38. $y = .07x^5 - .3x^3 + 1.5x^2 - 2$
$(-3 \le x \le 2 \text{ and } -6 \le y \le 6)$

In Exercises 39–42, find an appropriate viewing window for the graph of the equation (which may take some experimentation) and use a maximum/minimum finder to answer the question.

39. The population y of New Orleans (in thousands) in year x of the 20th century is approximated by

$$y = .000046685x^4 - .0108x^3 + .7194x^2 - 9.2426x + 305 \qquad (0 \le x \le 100),$$

where $x = 0$ corresponds to 1900. According to this model, in what year was the population largest?

40. The NASDAQ stock index between March 2000 and March 2002 is roughly approximated by

$$y = 11.19x^2 - 470x + 6358 \qquad (3 \le x \le 27),$$

where y is the index in month x, with $x = 3$ corresponding to March 2000. Determine the months during this period when the NASDAQ reached its high and its low. What were the approximate values of the NASDAQ at these times?

41. The number y of subscribers to direct broadcast satellite TV (in millions) in year x can be approximated by the equation

$$y = -.0255x^3 + .57x^2 - .65x - 10.24 \qquad (8 \le x \le 15),$$

where $x = 8$ corresponds to 1998.* In what year was the number of subscribers the largest?

42. The following equation gives the approximate number of thefts at Cleveland businesses each hour of the day:

$$y = .00357x^4 - .3135x^3 + 6.87x^2 - 38.3x + 118.4 \qquad (0 \le x \le 23)$$

where x is measured in hours after midnight.
(a) About how many thefts occurred around 5 A.M.?
(b) When did the largest number of thefts occur?

In Exercises 43–48, determine graphically the number *of solutions of the equation, but don't solve the equation. You may need a viewing window other than the standard one to find all the x-intercepts.*

43. $x^5 + 5 = 3x^4 + x$

44. $x^3 + 5 = 3x^2 + 24x$

45. $x^7 - 10x^5 + 15x + 10 = 0$

46. $x^5 + 36x + 25 = 13x^3$

47. $x^4 + 500x^2 - 8000x = 16x^3 - 32{,}000$

48. $6x^5 + 80x^3 + 45x^2 + 30 = 45x^4 + 86x$

In Exercises 49–56, use graphical approximation (a root finder or an intersection finder) to find a solution of the equation in the given interval.

49. $x^3 + 4x^2 + 10x + 15 = 0; \quad -3 < x < -2$

50. $x^3 + 9 = 3x^2 + 6x; \quad 1 < x < 2$

51. $x^4 + x - 3 = 0; \quad x > 0$

---

*Based on data and projections in *Newsweek*, December 23, 2002.

52. $x^5 + 5 = 3x^4 + x; \quad x < 1$

53. $\sqrt{x^4 + x^3 - x - 3} = 0; \quad x > 0$

54. $\sqrt{8x^4 - 14x^3 - 9x^2 + 11x - 1} = 0; \quad x < 0$

55. $\dfrac{2x^5 - 10x + 5}{x^3 + x^2 - 12x} = 0; \quad x < 0$

56. $\dfrac{3x^5 - 15x + 5}{x^7 - 8x^5 + 2x^2 - 5} = 0; \quad x > 1$

In Exercises 57–70, use algebraic, graphical, or numerical methods to find all real solutions of the equation, approximating when necessary.

57. $2x^3 - 4x^2 + x - 3 = 0$

58. $6x^3 - 5x^2 + 3x - 2 = 0$

59. $x^5 - 6x + 6 = 0$

60. $x^3 - 3x^2 + x - 1 = 0$

61. $10x^5 - 3x^2 + x - 6 = 0$

62. $\frac{1}{4}x^4 - x - 4 = 0$

63. $2x - \frac{1}{2}x^2 - \frac{1}{12}x^4 - 0$

64. $\frac{1}{4}x^4 + \frac{1}{3}x^2 + 3x - 1 = 0$

65. $\dfrac{5x}{x^2 + 1} - 2x + 3 = 0$

66. $\dfrac{2x}{x + 5} = 1$

67. $|x^2 - 4| = 3x^2 - 2x + 1$

68. $|x^3 + 2| = 5 + x - x^2$

69. $\sqrt{x^2 + 3} = \sqrt{x - 2} + 5$

70. $\sqrt{x^3 + 2} = \sqrt{x + 5} + 4$

71. According to data from the U.S. Department of Education, the average cost y of tuition and fees at four-year public colleges and universities in year x is approximated by

$$y = .325x^3 - 7.624x^2 + 220.276x + 2034,$$

where $x = 0$ corresponds to 1990. If this model continues to be accurate, in what year will tuition and fees reach \$5000?

72. According to data from the U.S. Department of Health and Human Services, the cumulative number y of AIDS cases (in thousands) as of year x is approximated by

$$y = .1006x^4 - 2.581x^3 + 19.282x^2 + 23.731x + 188 \quad (0 \le x \le 11),$$

where $x = 0$ corresponds to 1990.

(a) When did the cumulative number of cases reach 750,000?

(b) If this model remains accurate after 2001, in what year does the cumulative number of cases reach one million?

0.4 Functions

To understand the origin of the concept of function it may help to consider some "real-life" situations in which one numerical quantity depends on, corresponds to, or determines another.

EXAMPLE 1

The amount of income tax you pay depends on the amount of your income. The way in which the income determines the tax is given by the tax law. ■

EXAMPLE 2

The graph in Figure 0–40 on the next page shows the temperatures in Cleveland, Ohio, on April 11, 2001, as recorded by the U.S. Weather Bureau at Hopkins Airport. The graph indicates the temperature that corresponds to each given time. ■

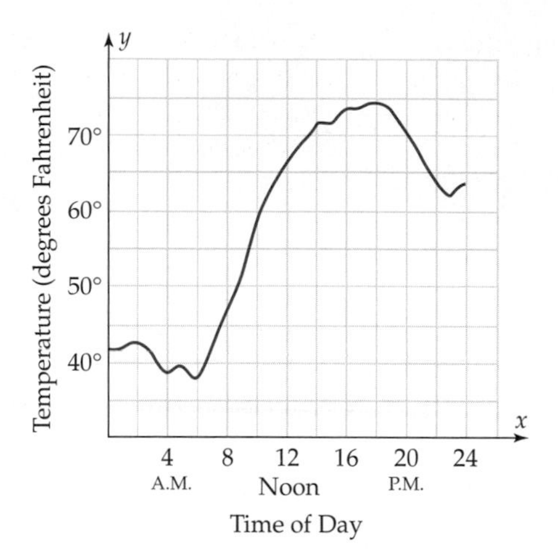

Figure 0–40

EXAMPLE 3

Suppose a rock is dropped straight down from a high place. Physics tells us that the distance traveled by the rock in t seconds is $16t^2$ feet. So the distance depends on the time. ■

These examples share several common features. Each involves two sets of numbers, which we can think of as inputs and outputs. In each case, there is a rule by which each input determines an output, as summarized here.

	Set of Inputs	Set of Outputs	Rule
Example 1	All incomes	All tax amounts	The tax law
Example 2	Hours since midnight	Temperatures during the day	Time/temperature graph
Example 3	Seconds elapsed after dropping the rock	Distance rock travels	Distance $= 16t^2$

Each of these examples may be mentally represented by an idealized calculator that has a single operation key: A number is entered [*input*], the rule key is pushed [*rule*], and an answer is displayed [*output*]. The formal definition of function incorporates these common features (input/rule/output), with a slight change in terminology.

Functions

A **function** consists of

A set of inputs (called the **domain**);

A **rule** by which each input determines one and only one output;

A set of outputs (called the **range**).

The phrase "one and only one" in the definition of the rule of a function may need some clarification. In Example 2, for each time of day (input), there is one and only one temperature (output). But it is quite possible to have the same temperature (output) at different times (inputs). In general,

> **For each input, the rule of a function determines exactly one output. But different inputs may produce the same output.**

Although real-world situations, such as Examples 1–3, are the motivation for functions, much of the emphasis in mathematics courses is on the functions themselves, independent of possible interpretations in specific situations.

FUNCTIONS DEFINED BY EQUATIONS

Equations in two variables are *not* the same things as functions. However, many equations can be used to define functions.

EXAMPLE 4

The equation $4x - 2y^3 + 5 = 0$ can be solved uniquely for y:

$$2y^3 = 4x + 5$$

$$y^3 = 2x + \frac{5}{2}$$

$$y = \sqrt[3]{2x + \frac{5}{2}}.$$

If a number is substituted for x in this equation, then exactly one value of y is produced. So we can define a function whose domain is the set of all real numbers and whose rule is

The input x produces the output $\sqrt[3]{2x + 5/2}$.

In this situation, we say that the equation defines **y as a function of x.**

The original equation can also be solved for x:

$$4x = 2y^3 - 5$$

$$x = \frac{2y^3 - 5}{4}.$$

Now if a number is substituted for y, exactly one value of x is produced. So we can think of y as the input and the corresponding x as the output and say that the equation defines **x as a function of y.** ■

EXAMPLE 5

If you solve the equation

$$y^2 - x + 1 = 0$$

for y, you obtain

$$y^2 = x - 1$$
$$y = \pm\sqrt{x - 1}.$$

This equation does *not* define y as a function of x because, for example, the input $x = 5$ produces two outputs: $y = \pm 2$. ■

EXAMPLE 6

The equation $y = x^3 - 2x + 3$ defines y as a function of x. Use the table feature of a calculator* to find the outputs for each of the following inputs:

(a) $-3, -2, -1, 0, 1, 2, 3, 4, 5$ (b) $-5, -11, 8, 7.2, -.44$

TECHNOLOGY TIP

To find the table setup screen, look for TBLSET (or RANG or NUM SETUP) on the keyboard or in the TABLE menu.

The increment is called ΔTBL on TI (and PITCH or NUM-STEP on others).

The table type is called INDPNT on TI, and NUMTYPE on HP-39+.

SOLUTION

(a) To use the table feature, we first enter $y = x^3 - 2x + 3$ in the equation memory, say, as y_1. Then we call up the setup screen (see the Technology Tip in the margin and Figure 0–41) and enter the *starting number* (−3), the *increment* (the amount the input changes for each subsequent entry, which is 1 here), and the *table type* (AUTO, which means the calculator will compute all the outputs at once).* Then press TABLE to obtain the table in Figure 0–42. To find values that don't appear on the screen in Figure 0–43, use the up and down arrow keys to scroll through the table.

(b) With an apparently random list of inputs, as here, we change the table type to ASK (or USER or BUILD YOUR OWN).† Then key in each value of x, and hit ENTER. This produces the table one line at a time, as in Figure 0–43. ■

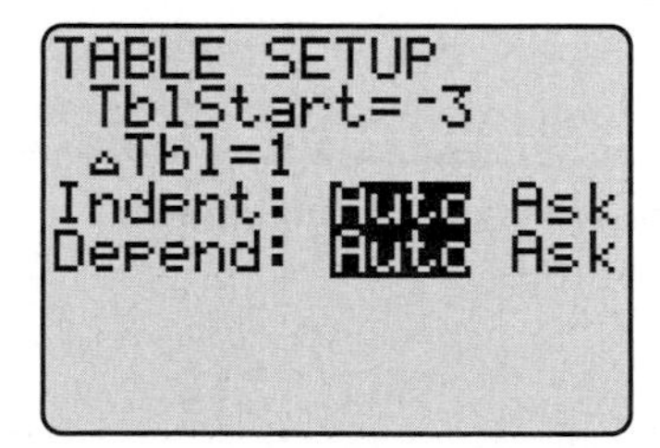

Figure 0–41

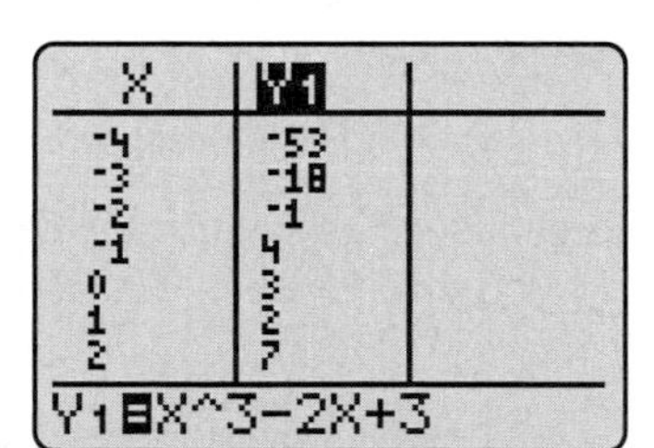

Figure 0–42

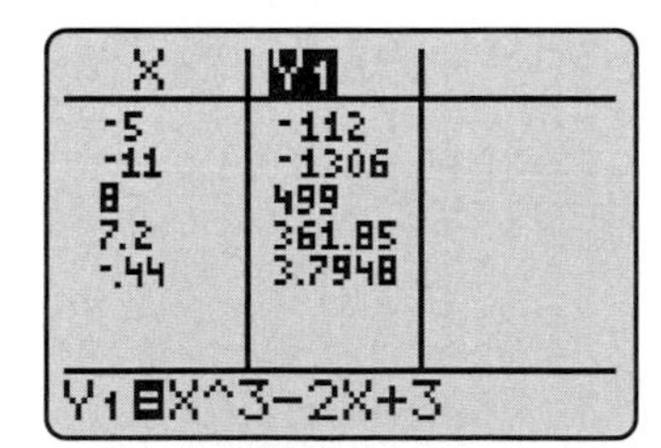

Figure 0–43

CALCULATOR EXPLORATION

Construct a table of values for the function in Example 6 that shows the outputs for these inputs: 2, 2.4, 2.8, 3.2, 3.6, and 4. What is the increment here?

*TI-85 does not have a built-in table feature, but an alternative is available. (See Calculator Investigation 3 at the end of this section.)

†Casio users should follow the directions in Calculator Investigation 2.

FUNCTIONAL NOTATION

Functional notation is a convenient shorthand language that facilitates the analysis of mathematical problems involving functions. Suppose a function is given. Denote the function by f and let x denote a number in the domain. Then

$f(x)$ denotes the output produced by input x.

For example, $f(6)$ is the output produced by the input 6. The sentence

"y is the output produced by input x according to the rule of the function f"

is abbreviated

$$y = f(x),$$

which is read "y equals f of x." The output $f(x)$ is sometimes called the **value** of the function f at x.

In actual practice, functions are seldom presented in the style of domain, rule, range, as they have been here. Usually, you will be given a phrase such as "the function $f(x) = \sqrt{x^2 + 1}$." This should be understood as a set of directions:

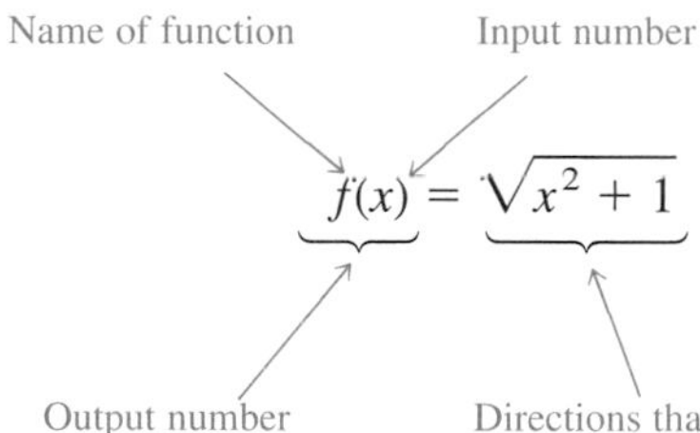

TECHNOLOGY TIP

Functional notation can be used on calculators other than TI-85 and some Casio models. If the function is y_1 in the function memory, then

$y_1(5)$ ENTER

gives the value of the function at $x = 5$.

You can key in y_1 from the keyboard on TI-86/89 and HP-39+. On TI-83+, y_1 is in the FUNCTION submenu of the Y-VARS menu.

Warning: Keying in $y_1(5)$ on TI-85 or Casio will produce an answer, but it usually will be wrong.

For example, to find $f(3)$, the output of the function f for input 3, simply replace x by 3 in the formula:

$$f(x) = \sqrt{x^2 + 1}$$
$$f(3) = \sqrt{3^2 + 1} = \sqrt{10}.$$

Similarly, replacing x by -5 and 0 shows that

$$f(-5) = \sqrt{(-5)^2 + 1} = \sqrt{26} \quad \text{and} \quad f(0) = \sqrt{0^2 + 1} = 1.$$

EXAMPLE 7 

The expression

$$h(x) = \frac{x^2 + 5}{x - 1}$$

defines the function h whose rule is

For input x, the output is the number $\dfrac{x^2 + 5}{x - 1}$.

Find each of the following:

$$h(\sqrt{3}), \quad h(-2), \quad h(-a), \quad h(r^2 + 3), \quad h(\sqrt{c} + 2).$$

SOLUTION To find $h(\sqrt{3})$ and $h(-2)$, replace x by $\sqrt{3}$ and -2, respectively, in the rule of h:

$$h(\sqrt{3}) = \frac{(\sqrt{3})^2 + 5}{\sqrt{3} - 1} = \frac{8}{\sqrt{3} - 1} \quad \text{and} \quad h(-2) = \frac{(-2)^2 + 5}{-2 - 1} = -3.$$

TECHNOLOGY TIP

One way to evaluate a function $f(x)$ is to enter its rule as an equation $y = f(x)$ in the equation memory and use TABLE or (on TI-85/86) EVAL; see Example 6 or Calculator Investigation 3 at the end of this section.

The value of the function h at any quantity, such as $-a$, $r^2 + 3$, etc., can be found by using the same procedure: *Replace x in the formula for $h(x)$ by that quantity:*

$$h(-a) = \frac{(-a)^2 + 5}{-a - 1} = \frac{a^2 + 5}{-a - 1}$$

$$h(r^2 + 3) = \frac{(r^2 + 3)^2 + 5}{(r^2 + 3) - 1} = \frac{r^4 + 6r^2 + 9 + 5}{r^2 + 2} = \frac{r^4 + 6r^2 + 14}{r^2 + 2}$$

$$h(\sqrt{c + 2}) = \frac{(\sqrt{c + 2})^2 + 5}{\sqrt{c + 2} - 1} = \frac{c + 2 + 5}{\sqrt{c + 2} - 1} = \frac{c + 7}{\sqrt{c + 2} - 1}. \quad ■$$

As the preceding examples illustrate, functional notation is a specialized shorthand language. Treating it as ordinary algebraic notation can lead to mistakes.

EXAMPLE 8

CAUTION

Common Mistakes with Functional Notation

Each of the following statements may be FALSE:

1. $f(a + b) = f(a) + f(b)$
2. $f(a - b) = f(a) - f(b)$
3. $f(ab) = f(a)f(b)$
4. $f(ab) = af(b)$
5. $f(ab) = f(a)b$

Here are examples of three of the errors listed in the Caution box.

1. If $f(x) = x^2$, then

$$f(3 + 2) = f(5) = 5^2 = 25.$$

But

$$f(3) + f(2) = 3^2 + 2^2 = 9 + 4 = 13.$$

So $f(3 + 2) \neq f(3) + f(2)$.

3. If $f(x) = x + 7$, then

$$f(3 \cdot 4) = f(12) = 12 + 7 = 19.$$

But

$$f(3)f(4) = (3 + 7)(4 + 7) = 10 \cdot 11 = 110.$$

So $f(3 \cdot 4) \neq f(3)f(4)$.

5. If $f(x) = x^2 + 1$, then

$$f(2 \cdot 3) = (2 \cdot 3)^2 + 1 = 36 + 1 = 37.$$

But

$$f(2) \cdot 3 = (2^2 + 1)3 = 5 \cdot 3 = 15.$$

So $f(2 \cdot 3) \neq f(2) \cdot 3$. ■

DOMAINS

When the rule of a function is given by a formula, as in Examples 7 and 8, its domain (set of inputs) is determined by the following convention.

Domain Convention

> Unless specific information to the contrary is given, the domain of a function f includes every real number (input) for which the rule of the function produces a real number as output.

Thus, the domain of a polynomial function such as $f(x) = x^3 - 4x + 1$ is the set of all real numbers, since $f(x)$ is defined for every value of x. In cases in which applying the rule of a function leads to division by zero or to the square root of a negative number, however, the domain may not consist of all real numbers.

EXAMPLE 9

Find the domain of the function given by

(a) $k(x) = \dfrac{x^2 - 6x}{x - 1}$ (b) $f(u) = \sqrt{u + 2}$

SOLUTION

(a) When $x = 1$, the denominator of $\dfrac{x^2 - 6x}{x - 1}$ is 0, and the fraction is not defined. When $x \neq 1$, however, the denominator is nonzero, and the fraction *is* defined. Therefore, the domain of the function k consists of all real numbers *except* 1.

(b) Since negative numbers do not have real square roots, $\sqrt{u + 2}$ is a real number only when $u + 2 \geq 0$, that is, when $u \geq -2$. Therefore, the domain of f consists of all real numbers greater than or equal to -2. ■

OPERATIONS ON FUNCTIONS

If f and g are functions, then their **sum** is the function h defined by the rule

$$h(x) = f(x) + g(x).$$

For example, if $f(x) = 3x^2 + x$ and $g(x) = 4x - 2$, then

$$h(x) = f(x) + g(x) = (3x^2 + x) + (4x - 2) = 3x^2 + 5x - 2.$$

Instead of using a different letter h for the sum function, we shall usually denote it by $f + g$. Thus, the sum $f + g$ is defined by the rule

$$\mathbf{(f + g)(x) = f(x) + g(x).}$$

This rule is *not* just a formal manipulation of symbols. If x is a number, then so are $f(x)$ and $g(x)$. The plus sign in $f(x) + g(x)$ is addition of *numbers,* and the result is a number. But the plus sign in $f + g$ is addition of *functions,* and the result is a new function.

The **difference** $f - g$ is the function defined by the rule

$$\mathbf{(f - g)(x) = f(x) - g(x).}$$

The domain of the sum and difference functions is the set of all real numbers that are in both the domain of f and the domain of g.

EXAMPLE 10

If $f(x) = \sqrt{9 - x^2}$ and $g(x) = \sqrt{x - 2}$, find the rules of the functions $f + g$ and $f - g$ and their domains.

SOLUTION We have

$$(f + g)(x) = f(x) + g(x) = \sqrt{9 - x^2} + \sqrt{x - 2};$$
$$(f - g)(x) = f(x) - g(x) = \sqrt{9 - x^2} - \sqrt{x - 2}.$$

The domain of f consists of all x such that $9 - x^2 \geq 0$ (so that the square root will be defined), that is, all x with $-3 \leq x \leq 3$. Similarly, the domain of g consists of all x such that $x \geq 2$. The domain of $f + g$ and $f - g$ consists of all real numbers in both the domain of f and the domain of g, namely, all x such that $2 \leq x \leq 3$. ■

TECHNOLOGY TIP

If you have two functions entered in the equation memory as y_1 and y_2, you can graph their sum by entering $y_1 + y_2$ as y_3 in the equation memory and graphing y_3. Differences, products, and quotients are graphed similarly. To find the correct keys for y_1 and y_2, see the Tip on page 43.

The **product** and **quotient** of functions f and g are the functions defined by the rules

$$(fg)(x) = f(x)g(x) \qquad \text{and} \qquad \left(\frac{f}{g}\right)(x) = \frac{f(x)}{g(x)}.$$

The domain of fg consists of all real numbers in both the domain of f and the domain of g. The domain of f/g consists of all real numbers x in both the domain of f and the domain of g such that $g(x) \neq 0$.

EXAMPLE 11

If $f(x) = \sqrt{3x}$ and $g(x) = x^2 - 1$, find the rules of the functions fg and f/g and their domains.

SOLUTION The rules are

$$(fg)(x) = f(x)g(x) = \sqrt{3x}\,(x^2 - 1) = \sqrt{3x}\,x^2 - \sqrt{3x};$$
$$\left(\frac{f}{g}\right)(x) = \frac{f(x)}{g(x)} = \frac{\sqrt{3x}}{x^2 - 1}.$$

The domain of fg consists of all numbers x in both the domain of f (all nonnegative real numbers) and the domain of g (all real numbers), that is, all $x \geq 0$. The domain of f/g consists of all these x for which $g(x) \neq 0$, that is, all nonnegative real numbers *except* $x = 1$. ■

If c is a real number and f is a function, then the product of f and the constant function $g(x) = c$ is usually denoted cf. For example, if the function

$$f(x) = x^3 - x + 2,$$

and $c = 5$, then $5f$ is the function given by

$$(5f)(x) = 5 \cdot f(x) = 5(x^3 - x + 2) = 5x^3 - 5x + 10.$$

CALCULATOR

INVESTIGATIONS 0.4

1. FUNCTION EVALUATION

The equation $y = x^3 - 4x + 1$ defines y as a function of x. On calculators other than HP-39+, the function may be evaluated as follows.

(a) Enter this equation as y_1 in the equation memory, and return to the home screen. To evaluate y at $x = 5$, store 5 in memory X*, where X* denotes the "variable" key that is used to enter the equation y_1 in the function memory. Then choose y_1, which is found in the following menu/submenu:

TI-83+: VARS/Y-VARS

Casio: VARS/GRPH [choose Y and type in 1]

TI-86/89: Type in y_1 on the keyboard.

Now press ENTER and the calculator displays the value of y at $x = 5$.

(b) Evaluate y at the following values of x: 3.5, 2, −1, −4.7.

(c) Enter the equation $y = x^2 + x - 3$ as y_2 in the equation memory.

Evaluate this function at $x = 7$ and $x = -5.5$.

2. USER-CONTROLLED TABLES ON CASIO

To construct the table in Example 6(b), proceed as follows.

(a) Enter the equation $y = x^3 - 2x + 3$ in the equation memory. This can be done by selecting either TABLE or GRAPH in the MAIN menu.

(b) Return to the MAIN menu and select LIST. Enter the numbers at which you want to evaluate this function [in this case, −5, −11, 8, 7.2, −.44] as List 1.

(c) Return to the MAIN menu and select TABLE. Then press SET-UP [that is, 2nd MENU] and select LIST as the Variable; on the LIST menu, choose List 1. Press EXIT and then press TABL to produce the table.

(d) Use the up/down arrow key to scroll through the table. If you change an entry in the X column, the corresponding y_1 value will automatically change. Find the value of y_1 at $x = 11$ and $x = -6$.

3. FUNCTION EVALUATION ON TI-85/86

The equation $y = x^2 - 3.7x + 4.5$ defines y as a function of x.

(a) Enter this equation as y_1 in the equation memory, and return to the home screen by pressing 2nd QUIT. To determine the value of y when $x = 29$, key in EVAL 29 ENTER (you will find EVAL in the MISC submenu of the MATH menu). Use this method to find the value of y when $x = .45$ and 611.

(b) Enter two more equations in the equation memory: $y_2 = x^3 - 6$ and $y_3 = 5x + 2$. Now when you key in EVAL 29 ENTER, the value of all three functions at $x = 29$ will be displayed as a list, which you can scroll through by using the arrow keys.

(c) In the equation memory, turn off y_2, and return to the home screen. If you use EVAL now, what happens?

4. CUSTOM MENU ON TI-85/86

If you press CUSTOM, the calculator displays a menu that may be blank. You may fill in any entries you want, so that frequently used operations (such as EVAL and FRAC) may be accessed quickly from the CUSTOM menu, rather than searching for them in varius menus and submenus. Check your instruction manual for details.

EXERCISES 0.4

In Exercises 1–4, determine whether or not the given table could possibly be a table of values of a function. Give reasons for your answer.

1.

Input	−2	0	3	1	−5
Output	2	3	−2.5	2	14

2.

Input	−5	3	0	−3	5
Output	7	3	0	5	−3

3.

Input	−5	1	3	−5	7
Output	0	2	4	6	8

4.

Input	1	−1	2	−2	3
Output	1	−2	±5	−6	8

In Exercises 5–12, determine whether the equation defines y as a function of x or defines x as a function of y.

5. $y = 3x^2 - 12$ **6.** $y = 2x^4 + 3x^2 - 2$

7. $y^2 = 4x + 1$ **8.** $5x - 4y^4 + 64 = 0$

9. $3x + 2y = 12$ **10.** $y - 4x^3 - 14 = 0$

11. $x^2 + y^2 = 9$ **12.** $y^2 - 3x^4 + 8 = 0$

In Exercises 13–16, each equation defines y as a function of x. Create a table that shows the values of the function for the given values of x.

13. $y = x^2 + x - 4$; $x = -2, -1.5, -1, \ldots, 3, 3.5, 4$.

14. $y = x^3 - x^2 + 4x + 1$; $x = 3, 3.1, 3.2, \ldots, 3.9, 4$

15. $y = \sqrt{4 - x^2}$; $x = -2, -1.2, -.04, .04, 1.2, 2$

16. $y = |x^2 - 5|$; $x = -8, -6, \ldots, 8, 10, 12$

17. The amount of postage required to mail a first-class letter is determined by its weight. In this situation, is weight a function of postage? Or vice versa? Or both?

18. Could the following statement ever be the rule of a function?

For input x, the output is the number whose square is x.

Why or why not? If there is a function with this rule, what is its domain and range?

19. (a) Use the following chart to make two tables of values (one for an average man and one for an average woman) in which the inputs are the number of drinks per hour and the outputs are the corresponding blood alcohol contents.*

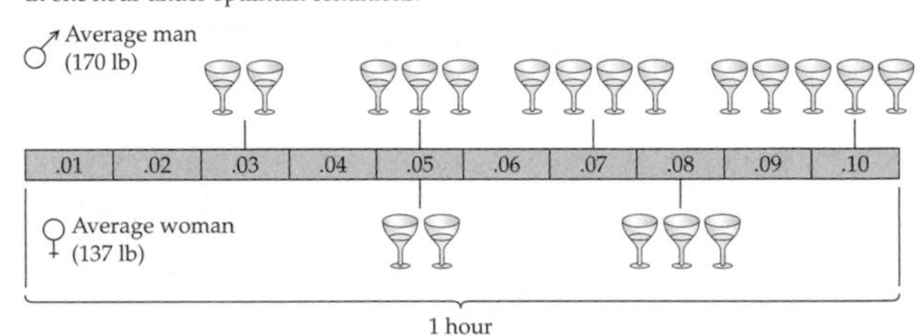

(b) Does each of these tables define a function? If so, what are the domain and range of each function? [Remember that you can have part of a drink.]

20. The table relates a large-framed woman's height to the weight at which the woman should live longest.† In this situation, is weight a function of height? Is height a function of weight? Jusify your answer.

Height (in shoes)	Weight (in pounds, in indoor clothing)
4′10″	118–131
4′11″	120–134
5′0″	122–137
5′1″	125–140
5′2″	128–143
5′3″	131–147
5′4″	134–151
5′5″	137–155
5′6″	140–159
5′7″	143–163
5′8″	146–167
5′9″	149–170
5′10″	152–173
5′11″	155–176
6′0″	158–179

*National Highway Traffic Safety Administration. Art by AP/Amy Kranz.

†Metropolitan Life Insurance Company.

21. Find an equation that expresses the area A of a circle as a function of its
(a) radius r (b) diameter d

22. Find an equation that expresses the area of a square as a function of its
(a) side x (b) diagonal d

23. A box with a square base of side x is four times higher than it is wide. Express the volume V of the box as a function of x.

24. The surface area of a cylindrical can of radius r and height h is $2\pi r^2 + 2\pi rh$. If the can is twice as high as the diameter of its top, express its surface area S as a function of r.

In Exercises 25 and 26, find the indicated values of the function by hand and *by using the table feature of a calculator (or the EVAL key on TI-85/86). If your answers do not agree with each other or with those at the back of the book, you are either making algebraic mistakes or incorrectly entering the function in the equation memory.*

25. $f(x) = \dfrac{x-3}{x^2+4}$

(a) $f(-1)$ (b) $f(0)$ (c) $f(1)$ (d) $f(2)$ (e) $f(3)$

26. $g(x) = \sqrt{x+4} - 2$

(a) $g(-2)$ (b) $g(0)$ (c) $g(4)$ (d) $g(5)$ (e) $g(12)$

Exercises 27–48 refer to these three functions:

$$f(x) = \sqrt{x+3} - x + 1 \qquad g(t) = t^2 - 1$$

$$h(x) = x^2 + \frac{1}{x} + 2$$

In each case, find the indicated value of the function.

27. $f(0)$ **28.** $f(1)$

29. $f(\sqrt{2})$ **30.** $f(\sqrt{2} - 1)$

31. $f(-2)$ **32.** $f(-3/2)$

33. $h(3)$ **34.** $h(-4)$

35. $h(3/2)$ **36.** $h(\pi + 1)$

37. $h(a + k)$ **38.** $h(-x)$

39. $h(2 - x)$ **40.** $h(x - 3)$

41. $g(3)$ **42.** $g(-2)$

43. $g(0)$ **44.** $g(x)$

45. $g(s + 1)$ **46.** $g(1 - r)$

47. $g(-t)$ **48.** $g(t + h)$

49. If $f(x) = x^3 + cx^2 + 4x - 1$ for some constant c and $f(1) = 2$, find c. [*Hint:* Use the rule of f to compute $f(1)$.]

50. If $g(x) = \sqrt{cx - 4}$ and $g(8) = 6$, find c.

51. If $f(x) = \dfrac{dx - 5}{x - 3}$ and $f(4) = 3$, find d.

52. If $g(x) = \dfrac{dx + 1}{x + 2}$ and $g(3) = .2$, find d.

53. In each part, compute $f(a)$, $f(b)$, and $f(a + b)$ and determine whether the statement "$f(a + b) = f(a) + f(b)$" is true or false for the given function.
(a) $f(x) = x^2$ (b) $f(x) = 3x$

54. In each part, compute $g(a)$, $g(b)$, and $g(ab)$ and determine whether the statment "$g(ab) = g(a) \cdot g(b)$" is true or false for the given function.
(a) $g(x) = x^3$ (b) $g(x) = 5x$

In Exercises 55–68, determine the domain of the function according to the usual convention.

55. $f(x) = x^2$ **56.** $g(x) = \dfrac{1}{x^2} + 2$

57. $h(t) = |t| - 1$ **58.** $k(u) = \sqrt{u}$

59. $k(x) = |x| + \sqrt{x} - 1$ **60.** $h(x) = \sqrt{(x+1)^2}$

61. $g(u) = \dfrac{|u|}{u}$ **62.** $h(x) = \dfrac{\sqrt{x-1}}{x^2 - 1}$

63. $g(y) = [-y]$ **64.** $f(t) = \sqrt{-t}$

65. $g(u) = \dfrac{u^2 + 1}{u^2 - u - 6}$ **66.** $f(t) = \sqrt{4 - t^2}$

67. $f(x) = -\sqrt{9 - (x - 9)^2}$

68. $f(x) = \sqrt{-x} + \dfrac{2}{x + 1}$

69. Give an example of two different functions f and g that have all of the following properties:

$$f(-1) = 1 = g(-1) \quad \text{and} \quad f(0) = 0 = g(0)$$
$$\text{and} \quad f(1) = 1 = g(1).$$

70. Give an example of a function g with the property that $g(x) = g(-x)$ for every real number x.

In Exercises 71–74, the rule of a function f is given. Write an algebraic formula for f(x).

71. Double the input, subtract 5, and take the square root of the result.

72. Square the input, multiply by 3, and subtract the result from 8.

73. Cube the input, add 6, and divide the result by 5.

74. Take the square root of the input, add 7, divide the result by 8, and add this result to the original input.

75. Data and projections from the International Trade Administration indicate that the number of foreign tourists visiting the United States was 43.3 million in 1995, 50.9 million in 2000, and 56.3 million in 2005. This data can be modeled by the following functions (where $x = 5$ corresponds to 1995 and $g(x)$ is in millions):

$$g(x) = 1.1x + 39$$

$$g(x) = -.005x^2 + 1.1x + 44$$

$$g(x) = .07x^3 - 2x^2 + 20x - 14.8$$

(a) Make a table or otherwise evaluate each function at the values of x corresponding to 1995, 2000, and 2005. Determine which function provides the best model for the given data.

(b) Use the function determined in part (a) to estimate the number of foreign tourists in 2006.

76. According to the National Center for Health Statistics, the number of marriages in the United States was 1,523,000 in 1960, 2,443,000 in 1990, and 2,327,000 in 2001. Here are three functions that model this data (where $x = 0$ corresponds to 1960 and $f(x)$ is in thousands):

$$f(x) = 15.9x + 1822$$

$$f(x) = -1.2x^2 + 65.8x + 1600$$

$$f(x) = .02x^3 - 2.4x^2 + 84.2x + 1526.$$

(a) Make a table or otherwise evaluate each function at the values of x corresponding to 1960, 1990, and 2001. Determine which function provides the best model for the given data.

(b) Use the function determined in part (a) to estimate the number of marriages in 2000.

77. Jack and Jill are salespeople in the suit department of a clothing store. Jack is paid \$200 per week plus \$5 for each suit he sells, whereas Jill is paid \$10 for every suit she sells.

(a) Let $f(x)$ denote Jack's weekly income, and let $g(x)$ denote Jill's weekly income from selling x suits. Find the rules of the functions f and g.

(b) Use algebra or a table to find: $f(20)$ and $g(20)$; $f(35)$ and $g(35)$; $f(50)$ and $g(50)$.

(c) If Jack sells 50 suits a week, how many must Jill sell to have the same income as Jack?

78. A rectangular region of 6000 sq ft is to be fenced in on three sides with fencing costing \$3.75 per foot and on the fourth side with fencing costing \$2.00 per foot. Express the cost of the fence as a function of the length x of the fourth side.

79. Suppose that the width and height of the box in the figure are equal and that the sum of the length and the girth is 108 (the maximum size allowed by the post office).

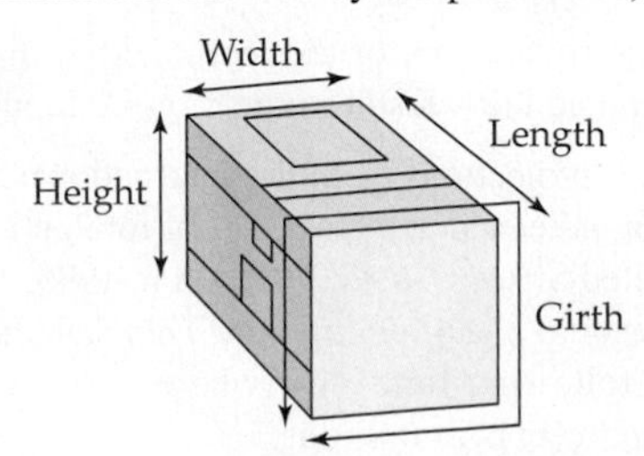

(a) Express the length y as a function of the width x. [*Hint:* Use the girth.]

(b) Express the volume V of the box as a function of the width x. [*Hint:* Find a formula for the volume and use part (a).]

80. A person who needs crutches can determine the correct length as follows: a 50-inch-tall person needs a 38-inch-long crutch. For each additional inch in the person's height, add .72 inch to the crutch length.

(a) If a person is y inches taller than 50 inches, write an expression for the proper crutch length.

(b) Write the rule of a function f such that $f(x)$ is the proper crutch length (in inches) for a person who is x inches tall. [*Hint:* Replace y in your answer to part (a) with an expression in x. How are x and y related?]

In Exercises 81–84, find $(f+g)(x)$, $(f-g)(x)$, and $(g-f)(x)$.

81. $f(x) = -3x + 2, \quad g(x) = x^3$

82. $f(x) = x^2 + 2, \quad g(x) = -4x + 7$

83. $f(x) = 1/x, \quad g(x) = x^2 + 2x - 5$

84. $f(x) = \sqrt{x}, \quad g(x) = x^2 + 1$

In Exercises 85–88, find $(fg)(x)$, $(f/g)(x)$, and $(g/f)(x)$.

85. $f(x) = -3x + 2, \quad g(x) = x^3$

86. $f(x) = 4x^2 + x^4, \quad g(x) = \sqrt{x^2 + 4}$

87. $f(x) = x^2 - 3, \quad g(x) = \sqrt{x - 3}$

88. $f(x) = \sqrt{x^2 - 1}, \quad g(x) = \sqrt{x - 1}$

In Exercises 89–92, find the domains of fg and f/g.

89. $f(x) = x^2 + 1, \quad g(x) = 1/x$

90. $f(x) = x + 2. \quad g(x) = \dfrac{1}{x+2}$

91. $f(x) = \sqrt{4 - x^2}, \quad g(x) = \sqrt{3x + 4}$

92. $f(x) = 3x^2 + x^4 + 2, \quad g(x) = 4x - 3$

In Exercises 93–96, find the indicated values, where

$$g(t) = t^2 - t \quad \textit{and} \quad f(x) = 1 + x.$$

93. $g(f(0))$

94. $f(g(3))$

95. $g(f(2) + 3)$

96. $f(2g(1))$

0.5 Graphs of Functions

The graph of a function f is the graph of the equation $y = f(x)$. Hence,

> **The graph of the function f consists of the points $(x, f(x))$ for every number x in the domain of f.**

When the rule of a function is given by an algebraic formula, the graph is easily obtained with technology. So the emphasis here will be on the properties of graphs and the relationship between the rule of a function and its graph—can you tell the shape of the graph just by looking at the rule?

EXAMPLE 1 Linear Functions

The graph of a function of the form

$$f(x) = mx + b \qquad \text{(with } m \text{ and } b \text{ constants)},$$

that is, the graph of the equation $y = mx + b$ is a straight line, with y-intercept b.* The number m is the **slope** of the line. The line moves upward to the right when m is positive, and downward to the right when m is negative. The line is horizontal when $m = 0$. Some typical linear functions are graphed in Figure 0–44. On a calculator, these graphs will usually look a bit bumpy and jagged. ■

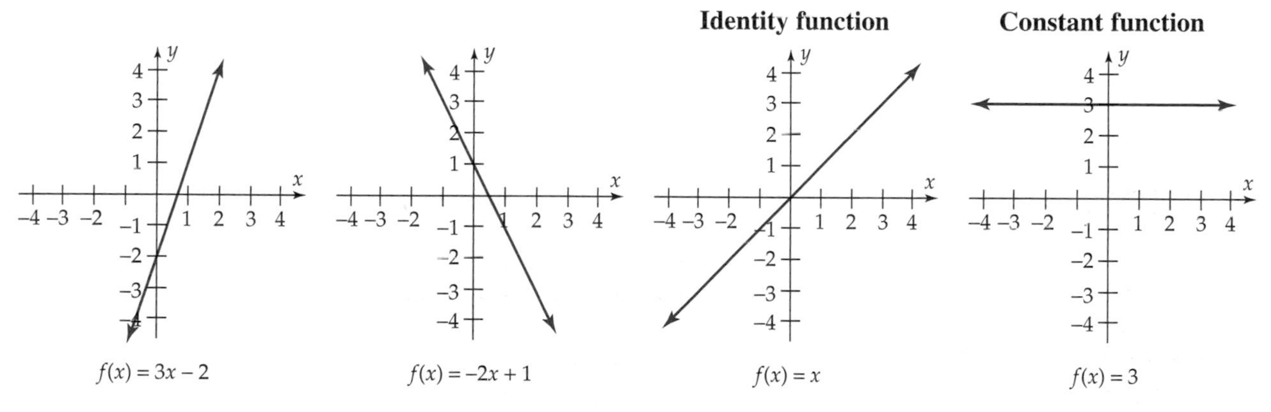

Figure 0–44

EXAMPLE 2 The Absolute Value Function

Recall the definition of absolute value

$$|x| = \begin{cases} x & \text{if } x \geq 0 \\ -x & \text{if } x < 0 \end{cases}.$$

Consequently, the graph of the absolute value function $f(x) = |x|$ can be obtained either by drawing the part of the line $y = x$ to the right of the origin and the part of the line $y = -x$ to the left of the origin (Figure 0–45 on the next page) or by graphing $y =$ ABS x on a calculator (Figure 0–46). ■

*Properties of lines are discussed in the Geometry Review Appendix.

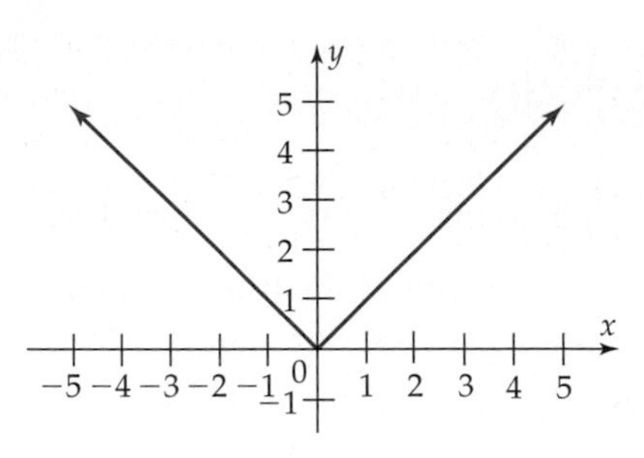

Figure 0–45

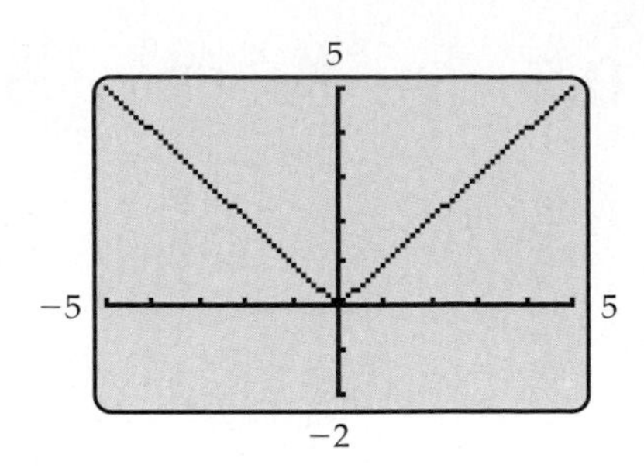

Figure 0–46

LOCAL MAXIMA AND MINIMA

The graph of a function may include some peaks and valleys (Figure 0–47). A peak is not necessarily the highest point on the graph, but it is the highest point in its neighborhood. Similarly, a valley is the lowest point in the neighborhood but not necessarily the lowest point on the graph.

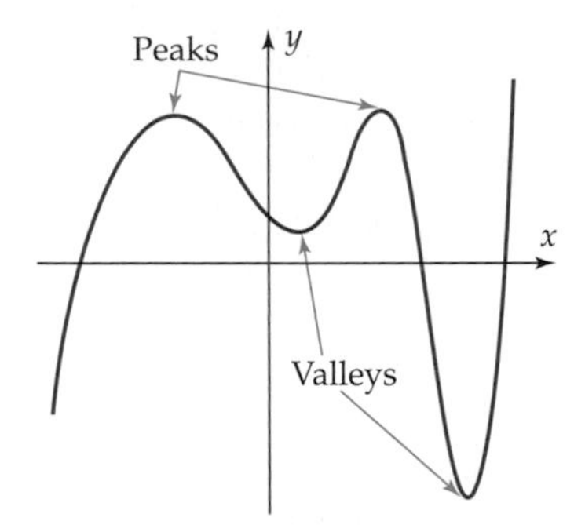

Figure 0–47

More formally, we say that a function f has a **local maximum** at $x = c$ if the graph of f has a peak at the point $(c, f(c))$. This means that all nearby points $(x, f(x))$ have smaller y-coordinates, that is,

$$f(x) \leq f(c) \qquad \text{for all } x \text{ near } c.$$

Similarly, a function has a **local minimum** at $x = d$ provided that

$$f(x) \geq f(d) \qquad \text{for all } x \text{ near } d.$$

In other words, the graph of f has a valley at $(d, f(d))$, since all nearby points $(x, f(x))$ have larger y-coordinates.

Calculus is usually needed to find the exact location of local maxima and minima (the plural forms of maximum and minimum). However, they can be accurately approximated by the maximum finder or minimum finder of a calculator.

EXAMPLE 3

The graph of $f(x) = x^3 - 1.8x^2 + x + 1$ in Figure 0–48 does not appear to have any local maxima or minima. However, if you use the trace feature to move along the flat segment to the right of the y-axis, you find that the y-coordinates increase, then decrease, then increase (try it!). To see what's really going on, we change viewing windows (Figure 0–49) and see that the function actually has a local maximum and a local minimum (Figure 0–50). The calculator's minimum finder shows that the local minimum occurs when $x \approx .7633$. ■

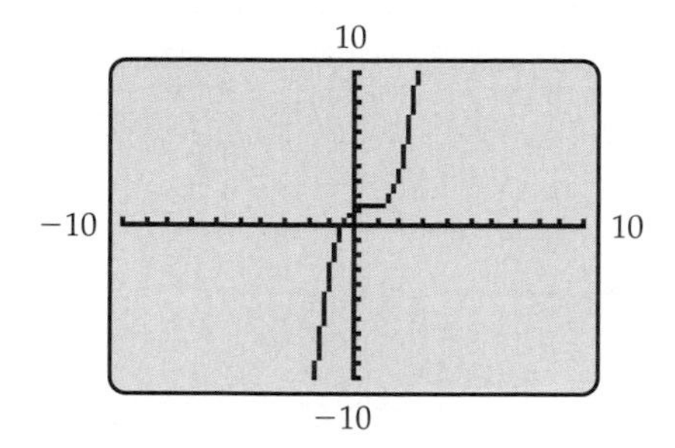

Figure 0–48

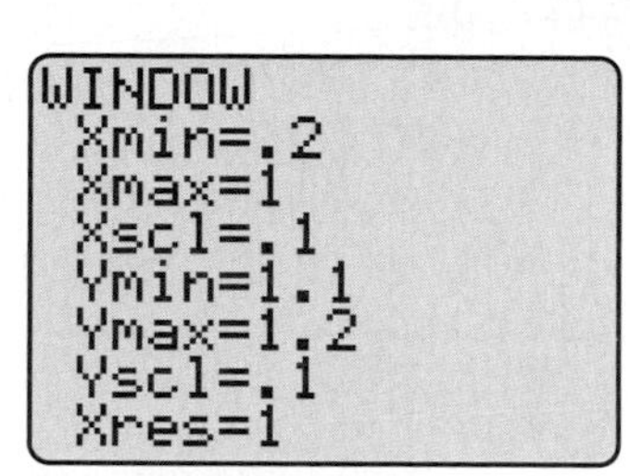

Figure 0–49

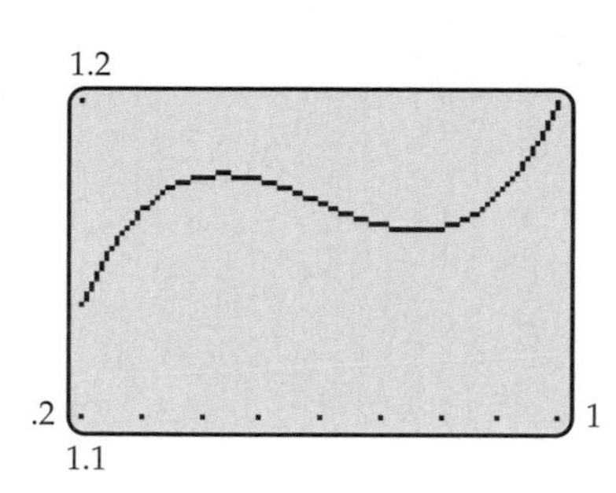

Figure 0–50

GRAPHING **EXPLORATION**

Graph the function in Example 3 in the viewing window of Figure 0–49. Use the maximum finder to approximate the location of the local maximum.

EXAMPLE 4

U.S. military troop levels (in millions) during 1980–2003 can be approximated by the function

$$g(x) = (2.063 \times 10^{-5})x^4 - (5.9 \times 10^{-4})x^3 - (4.88 \times 10^{-4})x^2 + .047x + 2.052,$$

where $x = 0$ corresponds to 1980. During this period, when were troop levels at their highest?

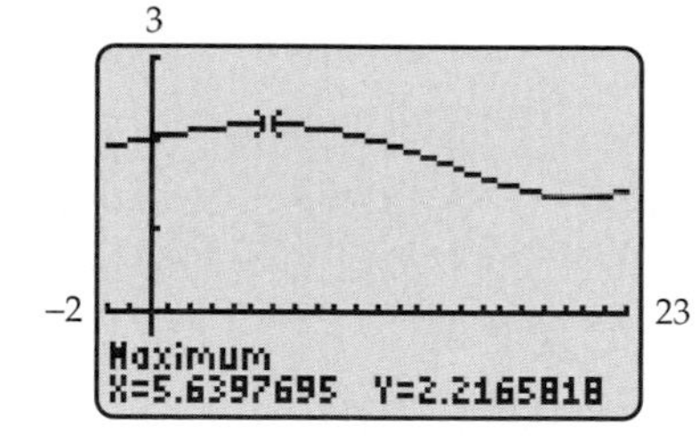

Figure 0–51

SOLUTION We graph the function and use the maximum finder to determine that the local maximum occurs when $x \approx 5.640$ and $g(5.640) \approx 2.217$, as shown in Figure 0–51. Therefore, troop levels reached their highest level (about 2,217,000) in late 1985. ■

INCREASING AND DECREASING FUNCTIONS

A function is said to be **increasing on an interval** if its graph always rises as you move from left to right over the interval. It is **decreasing on an interval** if its graph always falls as you move from left to right over the interval. A function is said to be **constant on an interval** if its graph is horizontal over the interval.

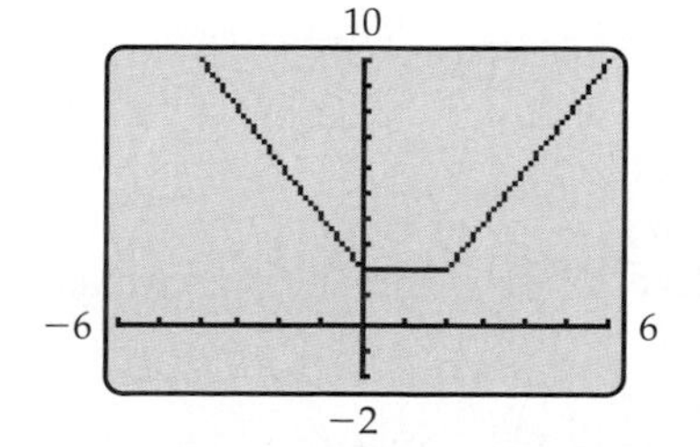

Figure 0–52

EXAMPLE 5

Figure 0–52 suggests that $f(x) = |x| + |x - 2|$ is decreasing when $x < 0$, increasing when $x > 2$, and constant between $x = 0$ and $x = 2$. You can confirm that the function is actually constant between 0 and 2 by using the trace feature to move along the graph there (the y-coordinates remain the same, as they should on a horizontal segment). For an algebraic proof that f is constant between 0 and 2, see Exercise 8. ■

CAUTION

A horizontal segment on a calculator graph does not always mean that the function is constant there. There may be **hidden behavior,** as was the case in Example 3. When in doubt, use the trace feature to see whether the y-coordinates remain constant as you move along the "horizontal" segment, or change the viewing window.

EXAMPLE 6

On what (approximate) intervals is the function $g(x) = .5x^3 - 3x$ increasing or decreasing?

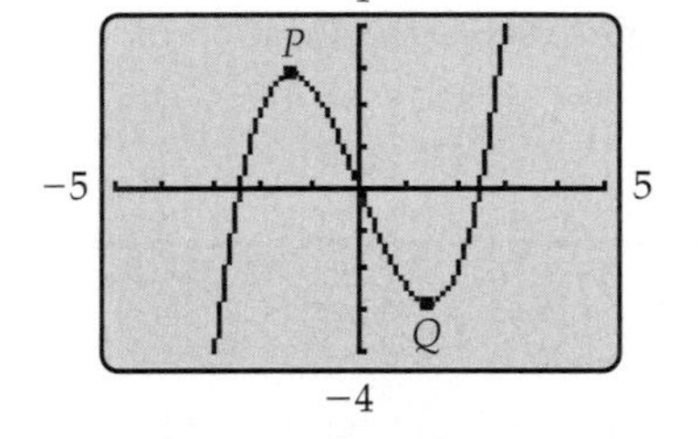

Figure 0–53

SOLUTION The graph of g in Figure 0–53 shows that g has a local maximum at P and a local minimum at Q. The maximum and minimum finders show that the approximate coordinates of P and Q are

$$P = (-1.4142, 2.8284) \qquad \text{and} \qquad Q = (1.4142, -2.8284).$$

Therefore, f is increasing when $x < -1.4142$ and when $x > 1.4142$. It is decreasing when $-1.4142 < x < 1.4142$. ■

THE VERTICAL LINE TEST

The following fact, which distinguishes graphs of functions from other graphs, can also be used to interpret some calculator-generated graphs.

Vertical Line Test

The graph of a function $y = f(x)$ has this property:

No vertical line intersects the graph more than once.

Conversely, any graph with this property is the graph of a function.

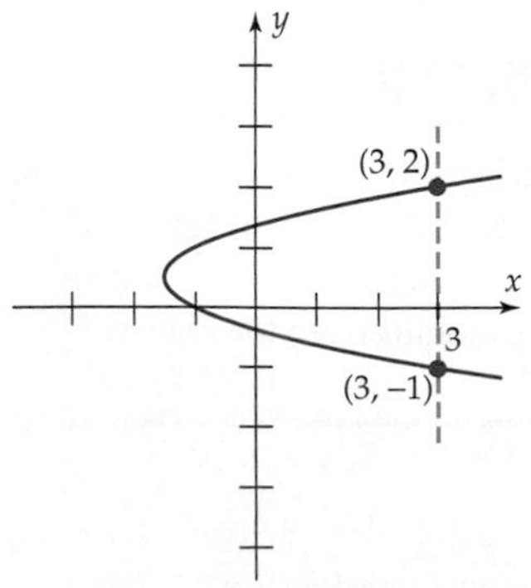

Figure 0–54

To see why this is true, consider Figure 0–54, in which the graph intersects the vertical line at two points. If this were the graph of a function f, then we would have $f(3) = 2$ [because (3, 2) is on the graph] *and* $f(3) = -1$ [because (3, −1) is on the graph]. This means that the input 3 produces two different outputs, which is impossible for a function. Therefore, Figure 0–54 is not the graph of a function. A similar argument works in the general case.

Care must be used when applying the Vertical Line Test to a calculator graph, as the following example illustrates.

EXAMPLE 7

The rule $g(x) = x^{15} + 2$ does give a function (each input produces exactly one output), but its graph in Figure 0–55 appears to be a vertical line near $x = 1$. To see why this is not actually true, change the viewing window. Figure 0–56 shows that the graph is not vertical between $x = 1$ and $x = 1.2$. This detail is lost in Figure 0–55 because all the x values in Figure 0–56 occupy only a single pixel width in Figure 0–55. ■

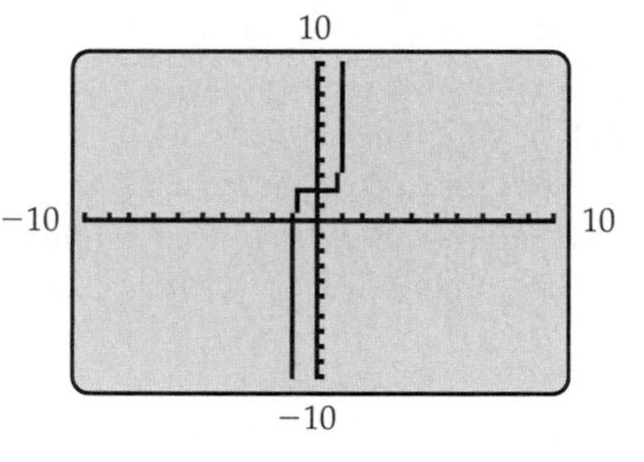

Figure 0–55

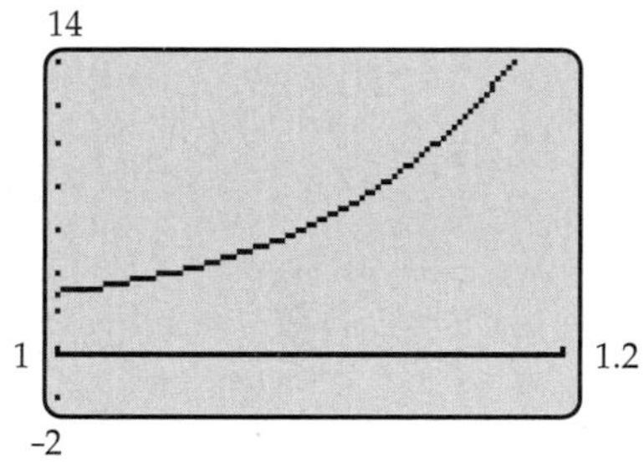

Figure 0–56

GRAPHING **EXPLORATION**

Find a viewing window that shows that the graph of the function g in Example 7 is not actually vertical near $x = -1$.

PARAMETRIC GRAPHING*

Most of the functions we have seen up to now can be described by equations in which y is a function of x and, hence, are easily graphed on a calculator. When you must graph an equation that expresses x as a function of y, a different calculator graphing technique is needed.

EXAMPLE 8

Graph $x = y^3 - 3y^2 - 4y + 7$.

SOLUTION Let t be any real number. If $y = t$, then

$$x = y^3 - 3y^2 - 4y + 7 = t^3 - 3t^2 - 4t + 7.$$

Thus, the graph consists of all points (x, y) such that

$$x = t^3 - 3t^2 - 4t + 7 \qquad \text{and} \qquad y = t \quad (t \text{ any real number}).$$

*This material is used only in Section 2.2, Chapter 5, and occasional exercises. It may be postponed until needed.

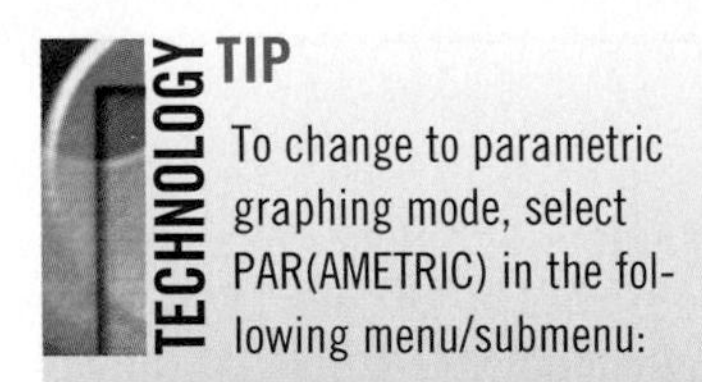

TECHNOLOGY TIP To change to parametric graphing mode, select PAR(AMETRIC) in the following menu/submenu:

TI: MODE

Casio: GRAPH/TYPE

HP-39+: APLET

When written in this form, the equation can be graphed as follows.

Change the graphing mode to **parametric mode** (see the Technology Tip in the margin), and enter the equations

$$x_{1t} = t^3 - 3t^2 - 4t + 7$$

$$y_{1t} = t$$

in the equation memory.* Next set the viewing window so that

$$-6 \le t \le 6, \qquad -10 \le x \le 10, \qquad -6 \le y \le 6,$$

as partially shown in Figure 0–57 (scroll down to see the rest). We use the same range for t and y here because $y = t$. Note that we must also set "t-step" (or "t pitch"), which determines how much t changes each time a point is plotted. A t-step between .05 and .15 usually produces a relatively smooth graph in a reasonable amount of time. Finally, pressing GRAPH (or PLOT or DRAW) produces the graph in Figure 0–58. ■

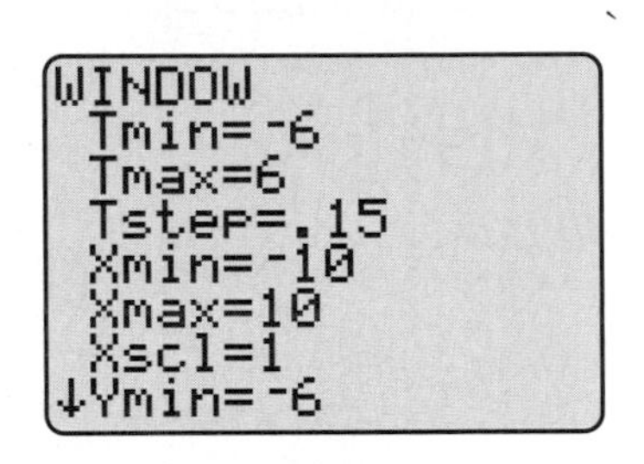

Figure 0–57

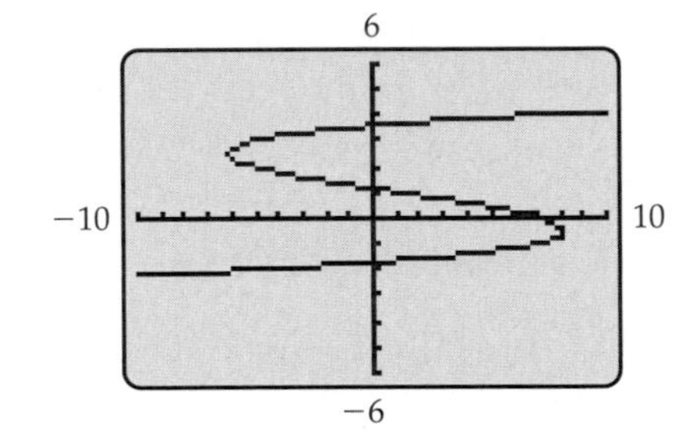

Figure 0–58

TECHNOLOGY TIP If you have trouble finding appropriate ranges for t, x, and y, it might help to use the TABLE feature to display a table of t-x-y values produced by the parametric equations.

As is illustrated in Example 8, the underlying idea of **parametric graphing** is to express both x and y as functions of a third variable t. The equations that define x and y are called **parametric equations,** and the variable t is called the **parameter.** Example 8 illustrates just one of the many applications of parametric graphing. It can also be used to graph curves that are not graphs of a single equation in x and y.

EXAMPLE 9

Graph the curve given by

$$x = t^2 - t - 1 \qquad \text{and} \qquad y = t^3 - 4t - 6 \qquad (-2 \le t \le 3).$$

SOLUTION Using the standard viewing window, we obtain the graph in Figure 0–59. Note that the graph crosses over itself at one point and that it does not extend forever to the left and right but has endpoints.

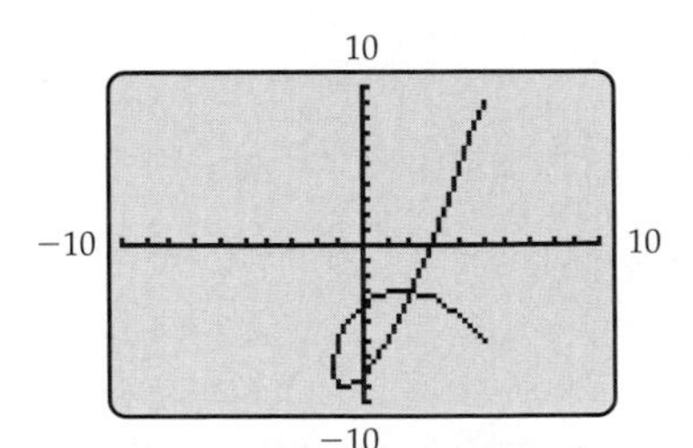

Figure 0–59

*On some calculators, x_{1t} is denoted x_{t1} or $x_1(T)$, and similarly for y_{1t}.

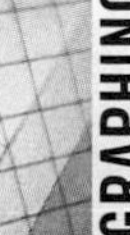

EXPLORATION

Graph these same parametric equations, but set the range of t values so that $-4 \le t \le 4$. What happens to the graph? Now change the range of t values so that $-10 \le t \le 10$. Find a viewing window large enough to show the entire graph, including endpoints. ■

Any function of the form $y = f(x)$ can be expressed in terms of parametric equations and graphed that way. For instance, to graph $f(x) = x^2 + 1$, let $x = t$ and $y = f(t) = t^2 + 1$.

✓ EXERCISES 0.5

In Exercises 1–4, state whether or not the graph is the graph of a function. If it is, find $f(3)$.

1.

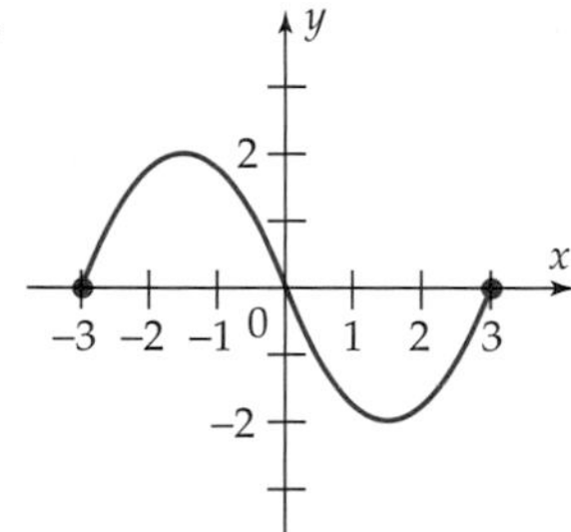

2.

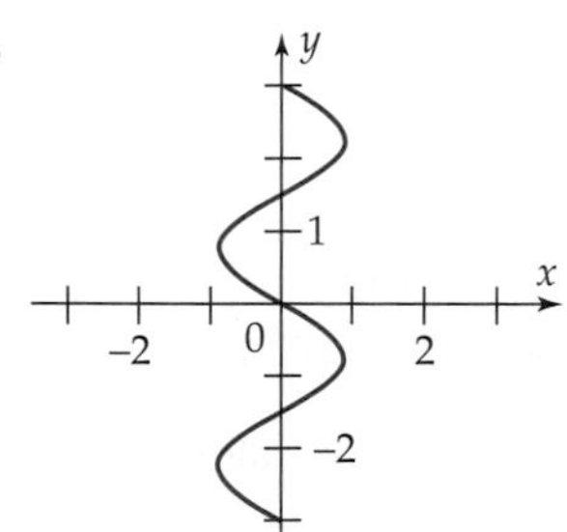

3.

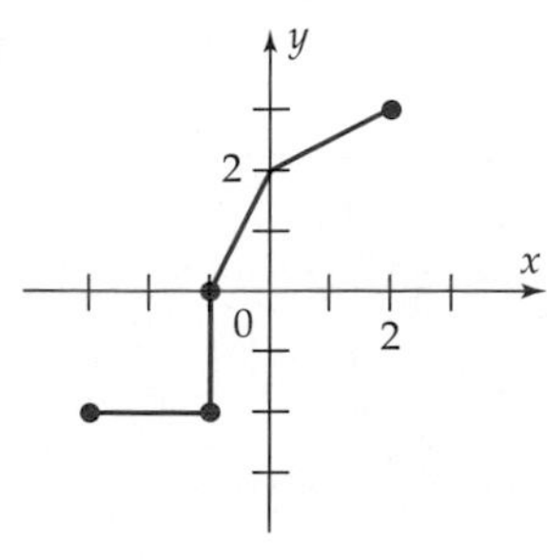

4.

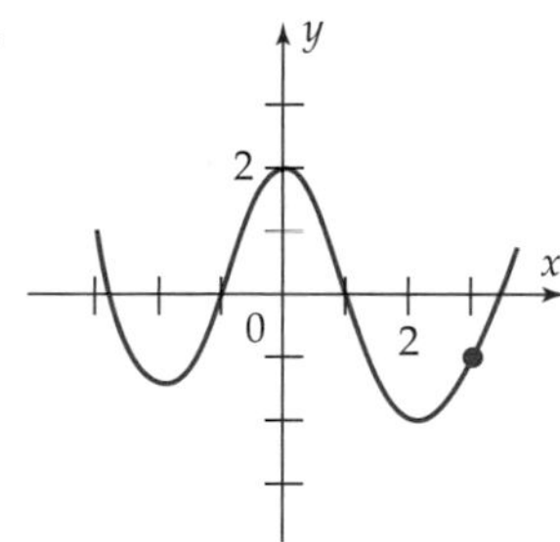

In Exercises 5–7, graph the function.

5. $f(x) = |x| + 2$

6. $g(x) = |x| - 4$

7. $g(x) = |x + 3|$.

8. Show that the function $f(x) = |x| + |x - 2|$ is constant when $0 \le x \le 2$. [*Hint:* Use the definition of absolute value (see Example 2) to compute $f(x)$ when $0 \le x \le 2$.]

In Exercises 9–14, find the approximate location of all local maxima and minima of the function.

9. $f(x) = x^3 - x$

10. $g(t) = -\sqrt{16 - t^2}$

11. $h(x) = \dfrac{x}{x^2 + 1}$

12. $k(x) = x^3 - 3x + 1$

13. $f(x) = x^3 - 1.8x^2 + x + 2$

14. $g(x) = 2x^3 + x^2 + 1$

In Exercises 15–16, find the approximate intervals on which the function whose graph is shown is increasing and those on which it is decreasing.

15.

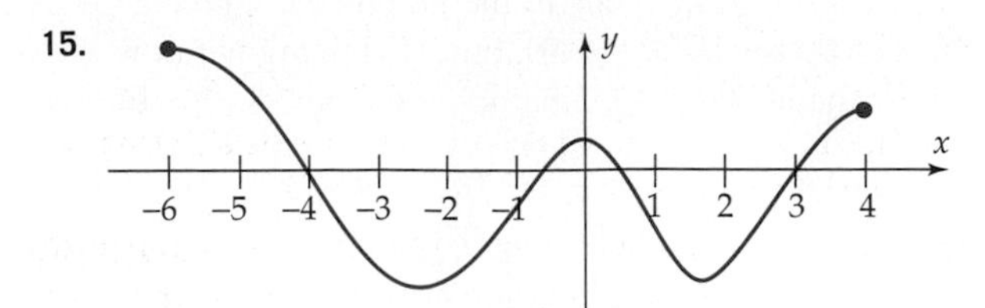

16.

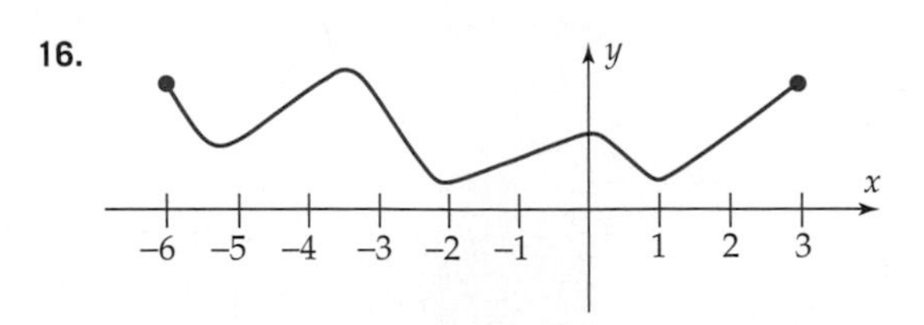

In Exercises 17–22, find the approximate intervals on which the function is increasing, those on which it is decreasing, and those on which it is constant.

17. $f(x) = |x - 1| - |x + 1|$

18. $g(x) = |x - 1| + |x + 2|$

19. $f(x) = -x^3 - 8x^2 + 8x + 5$

20. $f(x) = x^4 - .7x^3 - .6x^2 + 1$

21. $g(x) = .2x^4 - x^3 + x^2 - 2$

22. $g(x) = x^4 + x^3 - 4x^2 + x - 1$

In Exercises 23–24, sketch the graph of a function f that satisfies all of the given conditions. The function whose graph you sketch need not be given by an algebraic formula.

23. (i) The domain of f consists of all real numbers x with $-2 \le x \le 4$.
(ii) The range of f consists of all real numbers y with $-5 \le y \le 6$.
(iii) $f(-1) = f(3)$
(iv) $f\left(\frac{1}{2}\right) = 0$

24. (i) $f(-1) = 2$
(ii) $f(x) \ge 2$ when $-1 \le x < 1/2$
(iii) $f(x)$ starts decreasing when $x = 1$.
(iv) $f(0) = 3$ and $f(3) = 3$
(v) $f(x)$ starts increasing when $x = 5$.

25. *Find the dimensions of the rectangle with perimeter 100 inches and largest possible area, as follows.

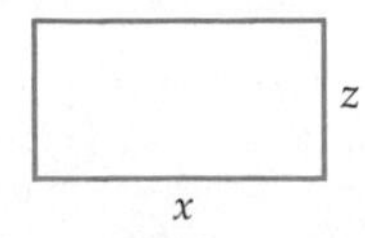

(a) Use the figure to write an equation in x and z that expresses the fact that the perimeter of the rectangle is 100.
(b) The area A of the rectangle is given by $A = xz$ (why?). Write an equation that expresses A as a function of x. [*Hint:* Solve the equation in part (a) for z, and substitute the result in the area equation.]
(c) Graph the function in part (b), and find the value of x that produces the largest possible value of A. What is z in this case?

26. Find the dimensions of the rectangle with area 240 square inches and smallest possible perimeter, as follows.
(a) Using the figure for Exercise 25, write an equation for the perimeter P of the rectangle in terms of x and z.
(b) Write an equation in x and z that expresses the fact that the area of the rectangle is 240.
(c) Write an equation that expresses P as a function of x. [*Hint:* Solve the equation in part (b) for z, and substitute the result in the equation of part (a).]
(d) Graph the function in part (c), and find the value of x that produces the smallest possible value of P. What is z in this case?

27. Find the dimensions of a box with a square base that has a volume of 867 cubic inches and the smallest possible surface area, as follows.

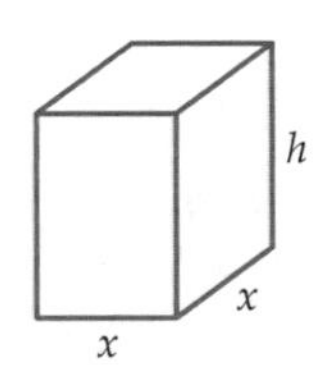

(a) Write an equation for the surface area S of the box in terms of x and h. [Be sure to include all four sides, the top, and the bottom of the box.]
(b) Write an equation in x and h that expresses the fact that the volume of the box is 867.
(c) Write an equation that expresses S as a function of x. [*Hint:* Solve the equation in part (b) for h, and substitute the result in the equation of part (a).]
(d) Graph the function in part (c), and find the value of x that produces the smallest possible value of S. What is h in this case?

28. Find the radius r and height h of a cylindrical can with a surface area of 60 square inches and the largest possible volume, as follows.

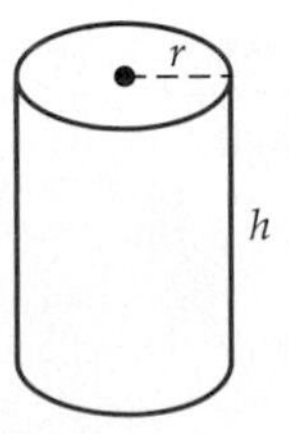

* **C** and ● indicate examples, exercises, and sections that are relevant to calculus.

(a) Write an equation for the volume V of the can in terms of r and h.

(b) Write an equation in r and h that expresses the fact that the surface area of the can is 60. [*Hint:* Think of cutting the top and bottom off the can; then cut the side of the can lengthwise and roll it out flat; it's now a rectangle. The surface area is the area of the top and bottom plus the area of this rectangle. The length of the rectangle is the same as the circumference of the original can (why?).]

(c) Write an equation that expresses V as a function of r. [*Hint:* Solve the equation in part (b) for h, and substitute the result in the equation of part (a).]

(d) Graph the function in part (c), and find the value of r that produces the largest possible value of V. What is h in this case?

29. Match each of the following functions with the graph that best fits the situation.

(a) The phases of the moon as a function of time;

(b) The demand for a product as a function of its price;

(c) The height of a ball thrown from the top of a building as a function of time;

(d) The distance a woman runs at constant speed as a function of time;

(e) The temperature of an oven turned on and set to 350° as a function of time.

(i)

(ii)

(iii)

(iv)

(v)

In Exercises 30–32, sketch a plausible graph of the given function. Label the axes and specify a reasonable domain and range.

30. The distance from the top of your head to the ground as you jump on a trampoline is a function of time.

31. The amount you spend on gas each week is a function of the number of gallons you put in your car.

32. The temperature of an oven that is turned on, set to 350°, and 45 minutes later turned off is a function of time.

In Exercises 33–34, the graph of a function f is shown. Find and label the given points on the graph.

33. (a) $(k, f(k))$

(b) $(-k, f(-k))$

(c) $(k, -f(k))$

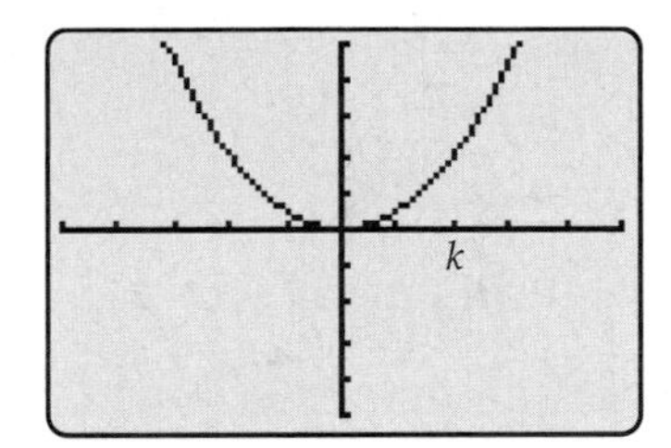

34. (a) $(k, f(k))$

(b) $(k, .5f(k))$

(c) $(.5k, f(.5k))$

(d) $(2k, f(2k))$

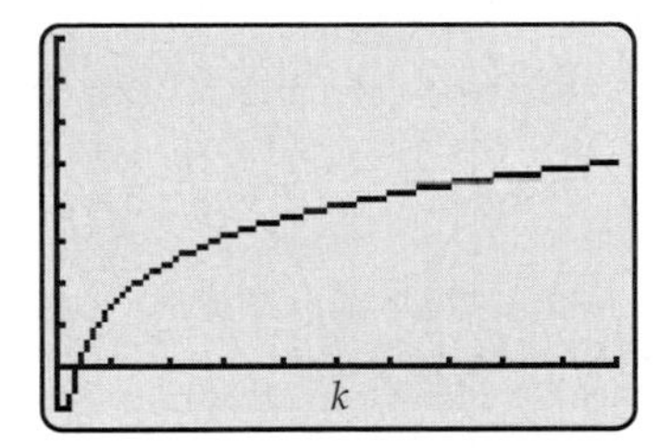

35. The graph of the function f, whose rule is $f(x)$ = average interest rate on a 30-year fixed-rate mortgage in year x, is shown in the figure.* Use it to answer these questions (reasonable approximations are OK).

(a) $f(1980) = ?$ (b) $f(2000) = ?$ (c) $f(2003) = ?$

(d) In what year between 1990 and 1996 were rates the lowest? The highest?

(e) During what three-year period were rates changing the fastest? How do you determine this from the graph?

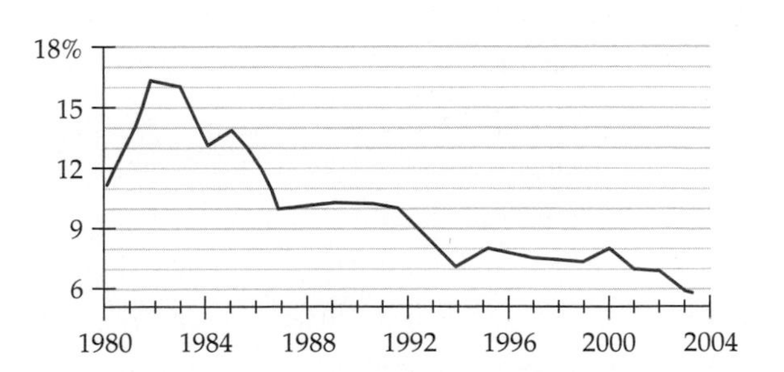

*Federal Home Mortgage Corporation.

36. The figure shows the graph of the function g whose rule is $g(x)$ = the Pentagon's budget in year x (in billions of 2001 dollars, adjusted for inflation), with $x = 0$ corresponding to 1975.* Use it to answer the following questions.

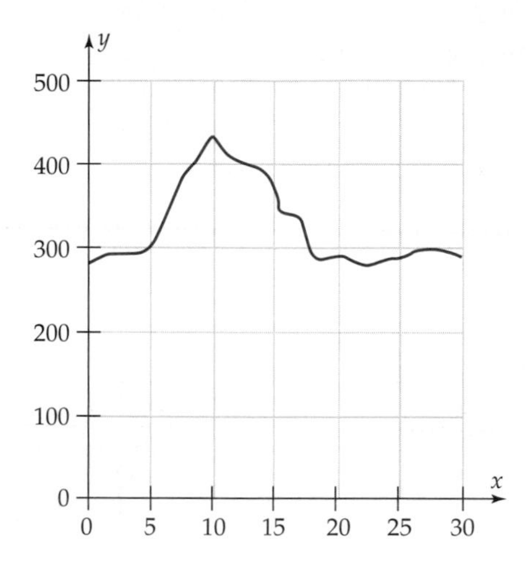

(a) Over what intervals was this function approximately constant?
(b) Over what intervals was this function decreasing?
(c) When did the Pentagon budget reach a local maximum?

Exercises 37–46 deal with the function g whose entire graph is shown in the figure.

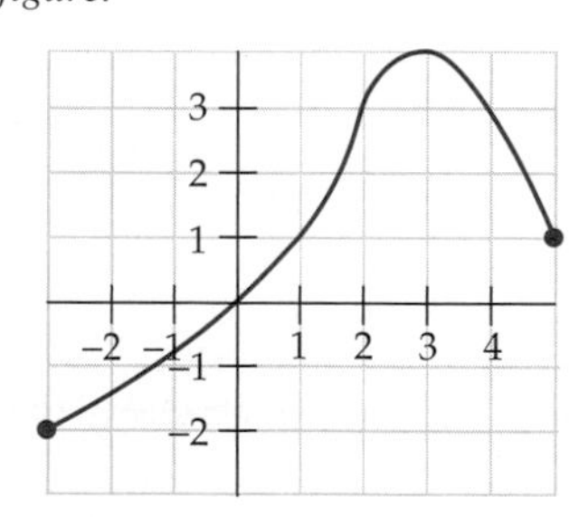

37. What is the domain of g?

38. What is the range of g?

39. If $t = 1.5$, then $g(2t) = ?$

40. If $t = 1.5$, then $2g(t) = ?$

41. If $y = 2$, then $g(y + 1.5) = ?$

42. If $y = 2$, then $g(y) + g(1.5) = ?$

43. If $y = 2$, then $g(y) + 1.5 = ?$

44. For what values of x is $g(x) < 0$?

45. For what values of z is $g(z) = 1$?

46. For what values of z is $g(z) = -1$?

For Exercises 47–50, use the Cleveland temperature graph from Example 2 of Section 0.4, which is reproduced below. Let T(x) denote the temperature at time x hours after midnight.

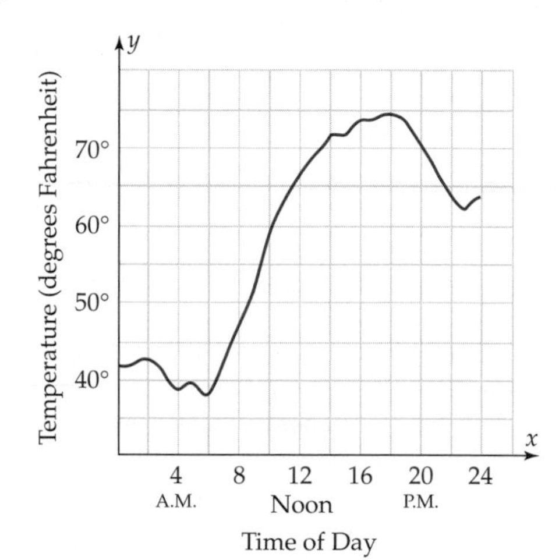

47. Find $T(4)$ and $T(7 + 11)$.

48. Is $T(8)$ larger than, equal to, or less than $T(16)$?

49. Find an eight-hour period in which $T(x) > 65°$ for all x in this period. [There are many possible correct answers.]

50. Determine whether the following statements are true or false.

(a) $T(4 \cdot 3) = T(4) \cdot T(3)$
(b) $T(4 \cdot 3) = 4 \cdot T(3)$
(c) $T(4 + 14) = T(4) + T(14)$

In Exercises 51–52, express the length h as a function of x.

51.

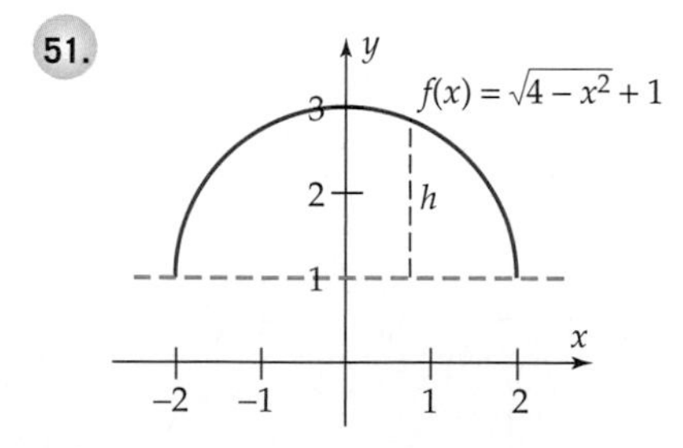

52.

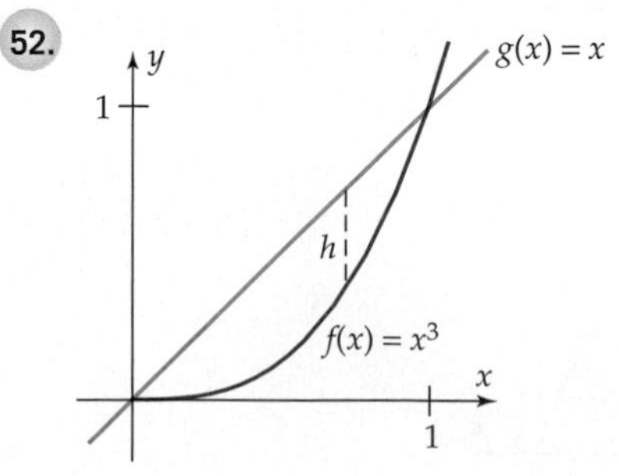

*Department of Defense; figures for 2002–2005 are estimates.

In Exercises 53–58, use parametric graphing. Find a viewing window that shows a complete graph of the equation.

53. $x = y^3 + 5y^2 - 4y - 5$

54. $\sqrt[3]{y^2 - y - 1} - x + 2 = 0$

55. $xy^2 + xy + x = y^3 - 2y^2 + 4$
[*Hint:* First solve for x.]

56. $2y = xy^2 + 180x$

57. $x - \sqrt{y} + y^2 + 8 = 0$

58. $y^2 - x - \sqrt{y + 5} + 4 = 0$

In Exercises 59–64, find a viewing window that shows a complete graph of the curve determined by the parametric equations.

59. $x = 3t^2 - 5$ and $y = t^2$ $(-4 \le t \le 4)$

60. The Zorro curve: $x = .1t^3 - .2t^2 - 2t + 4$ and $y = 1 - t$ $(-5 \le t \le 6)$

61. $x = t^2 - 3t + 2$ and $y = 8 - t^3$ $(-4 \le t \le 4)$

62. $x = t^2 - 6t$ and $y = \sqrt{t + 7}$ $(-5 \le t \le 9)$

63. $x = 1 - t^2$ and $y = t^3 - t - 1$ $(-4 \le t \le 4)$

64. $x = t^2 - t - 1$ and $y = 1 - t - t^2$

Thinkers

65. A jogger begins her daily run from her home. The graph shows her distance from home at time t minutes. The graph shows, for example, that she ran at a slow but steady pace for 10 minutes, then increased her pace for 5 minutes, all the time moving farther from home. Describe the rest of her run.

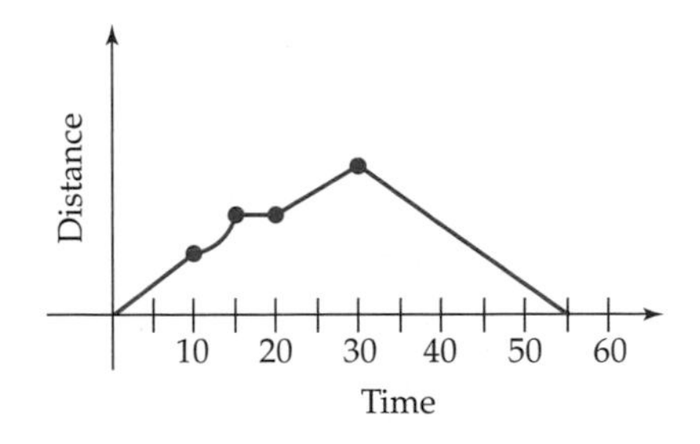

66. The graph shows the speed (in mph) at which a driver is going at time t minutes. Describe his journey.

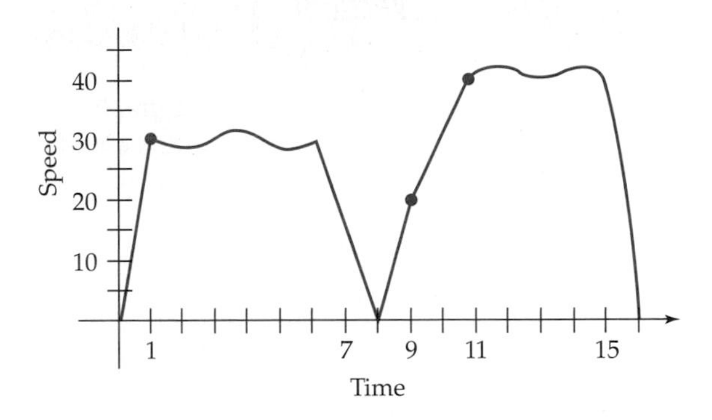

67. Graph the curve given by

$$x = (t^2 - 1)(t^2 - 4)(t + 5) + t + 3$$

$$y = (t^2 - 1)(t^2 - 4)(t^3 + 4) + t - 1$$

$$(-2.5 \le t \le 2.5)$$

How many times does this curve cross itself?

68. Use parametric equations to describe a curve that crosses itself more times than the curve in Exercise 67. [Many correct answers are possible.]

0.6 Graphs and Transformations

If the rule of a function is algebraically changed, so as to produce a new function, then the graph of the new function can sometimes be obtained from the graph of the original function by a simple geometric transformation, as we now see.

One way to change the rule of a function is to add or subtract a positive constant. For example, from $f(x) = x^2$, we obtain the new functions

$$g(x) = x^2 + 5 \qquad \text{and} \qquad h(x) = x^2 - 7.$$

The graphs of g and h are related to the graph of $f(x) = x^2$ as follows.

Vertical Shifts

Let f be a function and c a positive constant.

The graph of $g(x) = f(x) + c$ is the graph of f shifted c units upward.

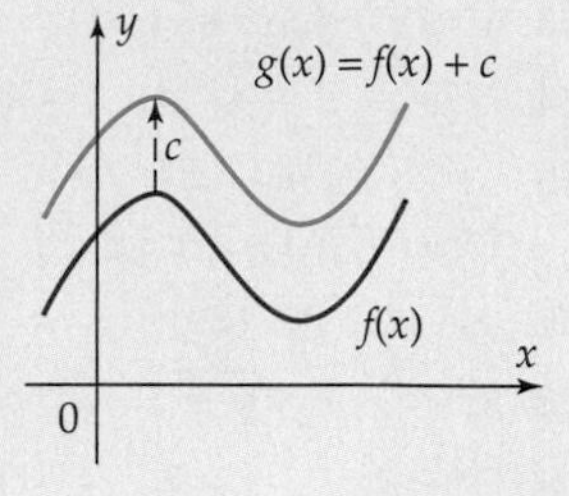

The graph of $h(x) = f(x) - c$ is the graph of f shifted c units downward.

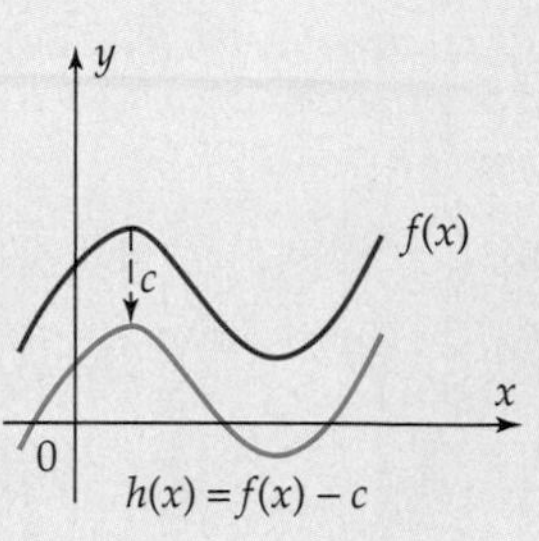

GRAPHING EXPLORATION

Illustrate the facts presented in the preceding box by graphing the functions

$$f(x) = x^2, \qquad g(x) = x^2 + 5, \qquad \text{and} \qquad h(x) = x^2 - 7$$

on the same screen, using the standard window.

EXAMPLE 1 

A calculator was used to obtain the graph of $f(x) = .04x^3 - x - 3$ in Figure 0–60. The graph of

$$h(x) = f(x) - 4 = (.04x^3 - x - 3) - 4 = .04x^3 - x - 7$$

is the graph of f shifted 4 units downward, as shown in Figure 0–61.

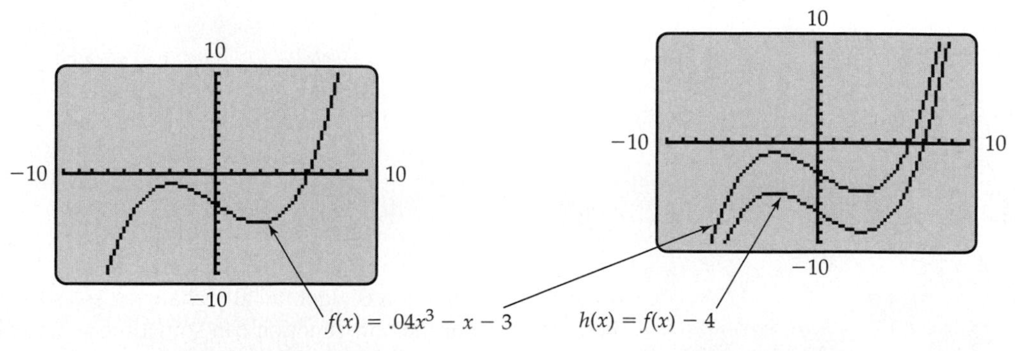

Figure 0–60 Figure 0–61

Although it may appear that the graph of h is closer to the graph of f at the outer edges of Figure 0–61 than in the center, this is an optical illusion. The *vertical* distance between the graphs is always 4 units.

GRAPHING EXPLORATION

Use the trace feature of your calculator as follows to confirm that the vertical distance is always 4:*

Move the cursor to any point on the graph of f, and note its coordinates.

Use the down arrow to drop the cursor to the graph of h, and note the coordinates of the cursor in its new position.

The x-coordinates will be the same in both cases, and the new y-coordinate will be 4 less than the original y-coordinate. ■

HORIZONTAL SHIFTS

Another way to change the rule of a function $f(x)$ is to replace the variable x by $x + c$ or $x - d$ for some constants c and d. For example, from $f(x) = 2x^3$, we obtain

$$g(x) = 2(x + 6)^3 \qquad \text{and} \qquad h(x) = 2(x - 8)^3.$$

The graphs of g and h are related to the graph of $f(x) = 2x^3$ as follows.

Horizontal Shifts

Let f be a function and c a positive constant.

The graph of $g(x) = f(x + c)$ is the graph of f shifted horizontally c units to the left.

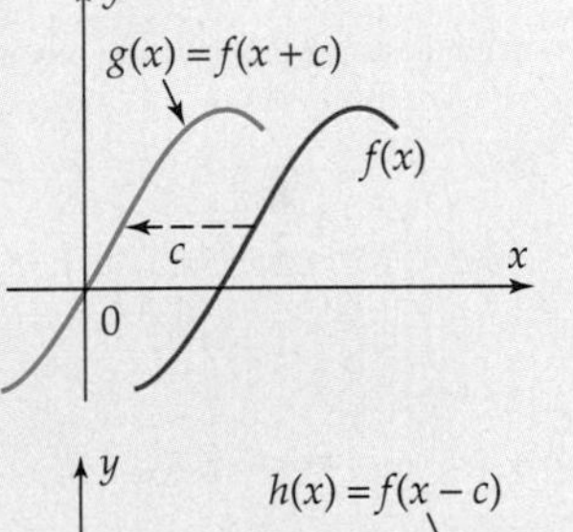

The graph of $h(x) = f(x - c)$ is the graph of f shifted horizontally c units to the right.

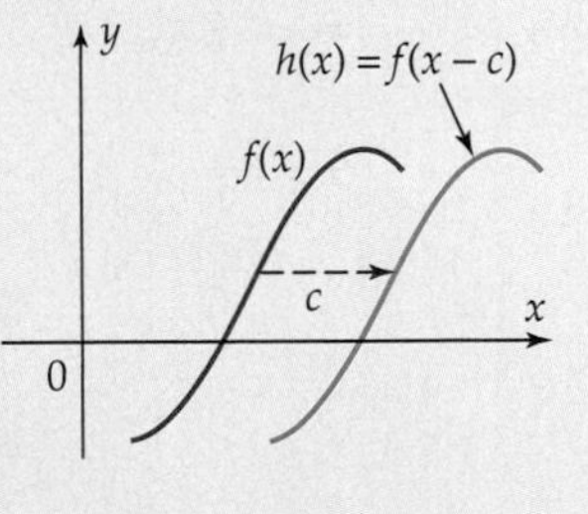

GRAPHING EXPLORATION

Illustrate the facts in the preceding box by graphing the functions

$$f(x) = 2x^3, \qquad g(x) = 2(x + 6)^3, \qquad \text{and} \qquad h(x) = 2(x - 8)^3$$

on the same screen, using the standard window.

---

*The trace cursor can be moved vertically from graph to graph by using the up and down arrows.

TECHNOLOGY TIP

If the function f of Example 2 is entered as $y_1 = x^2 - 7$, then the functions g and h can be entered as $y_2 = y_1(x + 5)$ and $y_3 = y_1(x - 4)$ on calculators other than TI-85 and Casio.

EXAMPLE 2

In some cases, shifting the graph of a function f horizontally may produce a graph that overlaps the graph of f. For instance, a complete graph of $f(x) = x^2 - 7$ is shown in red in Figure 0–62. The graph of

$$g(x) = f(x + 5) = (x + 5)^2 - 7 = x^2 + 10x + 25 - 7 = x^2 + 10x + 18$$

is the graph of f shifted 5 units to the left, and the graph of

$$h(x) = f(x - 4) = (x - 4)^2 - 7 = x^2 - 8x + 16 - 7 = x^2 - 8x + 9$$

is the graph of f shifted 4 units to the right, as shown in Figure 0–62. ■

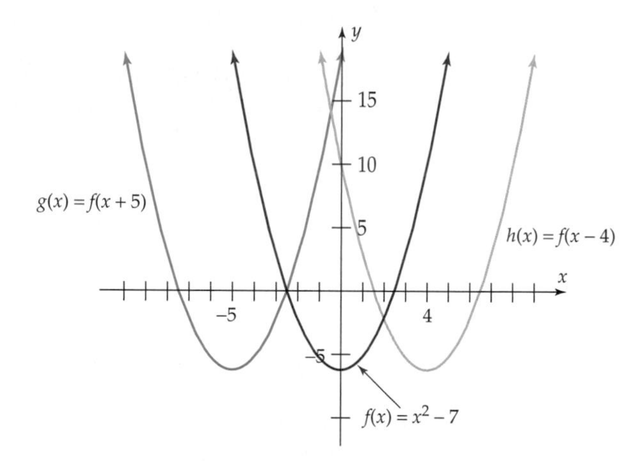

Figure 0–62

EXPANSIONS AND CONTRACTIONS

The following example illustrates the effect that multiplying the rule of a function by a positive constant has on the graph of the function.

EXAMPLE 3

Graph the functions

$$f(x) = x^2 - 4 \qquad \text{and} \qquad g(x) = 3f(x) = 3(x^2 - 4),$$

and explain how the graphs are related to one another.

SOLUTION The table of values in Figure 0–63 shows that the y-coordinates on the graph of $Y_2 = g(x)$ are always 3 times the y-coordinates of the corresponding points on the graph of $Y_1 = f(x)$. To translate this into visual terms, imagine that the graph of f is nailed to the x-axis at its intercepts (± 2). The graph of g is then obtained by "stretching" the graph of f away from the x-axis (with the nails holding the x-intercepts in place) by a factor of 3, as shown in Figure 0–64. ■

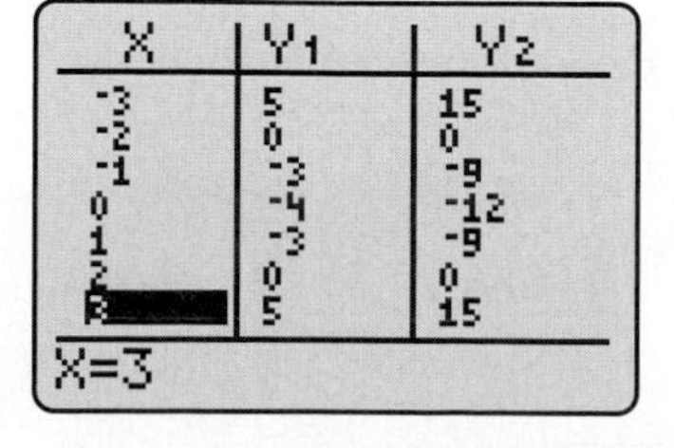

X	Y1	Y2
-3	5	15
-2	0	0
-1	-3	-9
0	-4	-12
1	-3	-9
2	0	0
3	5	15

X=3

Figure 0–63

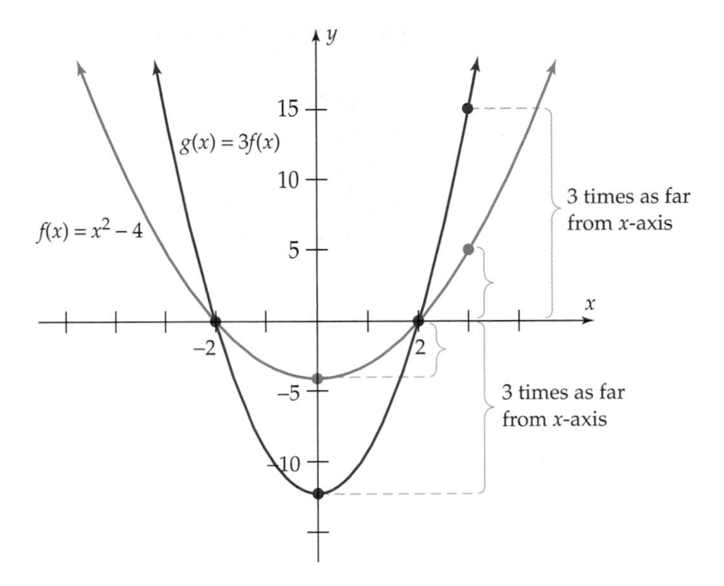

Figure 0–64

GRAPHING EXPLORATION

In the viewing window with $-5 \leq x \leq 5$ and $-5 \leq y \leq 10$, graph these functions on the same screen:

$$f(x) = x^2 - 4 \qquad h(x) = \frac{1}{4}(x^2 - 4).$$

Your screen should suggest that the graph of h is the graph of f "shrunk" vertically toward the x-axis by a factor of $1/4$.

Analogous facts are true in the general case.

Expansions and Contractions

If $c > 1$, then the graph of $g(x) = cf(x)$ is the graph of f stretched vertically away from the x-axis by a factor of c.

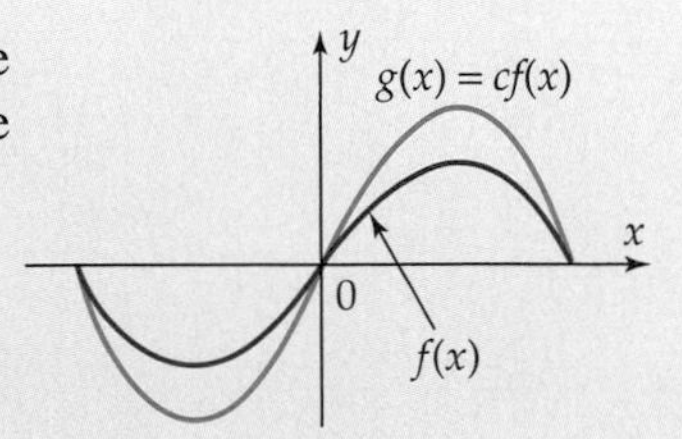

If $0 < c < 1$, then the graph of $h(x) = cf(x)$ is the graph of f shrunk vertically toward the x-axis by a factor of c.

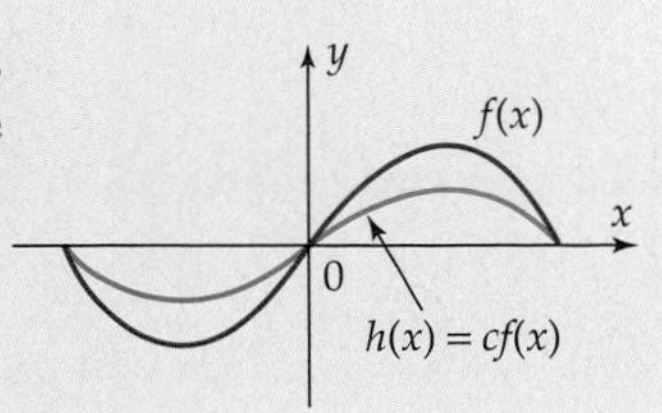

REFLECTIONS

GRAPHING EXPLORATION

In the standard viewing window, graph these functions on the same screen.

$$f(x) = .04x^3 - x \qquad g(x) = -f(x) = -(.04x^3 - x).$$

By moving your trace cursor from graph to graph, verify that for every point on the graph of f, there is a point on the graph of g with the same first coordinate that is on the opposite side of the x-axis, the same distance from the x-axis.

This Exploration shows that the graph of g is the mirror image (reflection) of the graph of f, with the x-axis being the (two-way) mirror. The same thing is true in the general case.

Reflections

Let f be a function. The graph of $g(x) = -f(x)$ is the graph of f reflected in the x-axis.

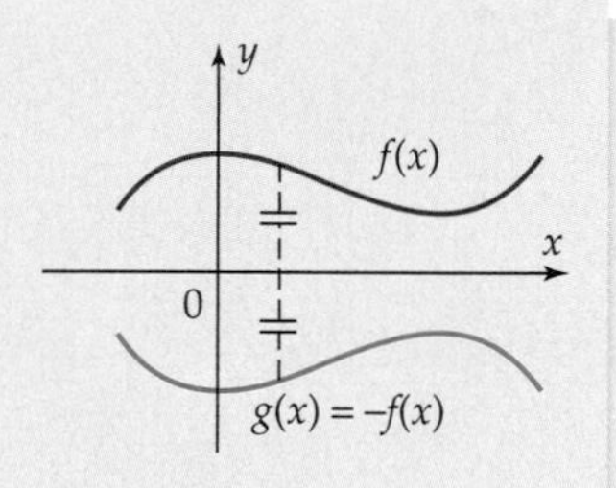

EXAMPLE 4 

If $f(x) = x^2 - 3$, then the graph of

$$g(x) = -f(x) = -(x^2 - 3)$$

is the reflection of the graph of f in the x-axis, as shown in Figure 0–65. ■

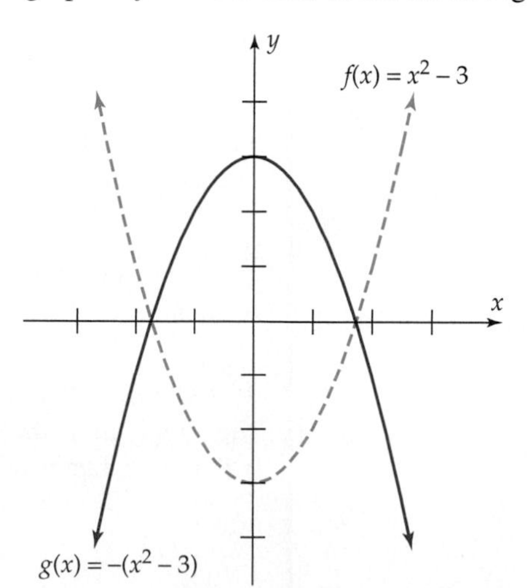

Figure 0–65

GRAPHING

EXPLORATION

In the standard viewing window, graph these functions on the same screen:

$$f(x) = \sqrt{5x + 10} \quad \text{and} \quad h(x) = f(-x) = \sqrt{5(-x) + 10}.$$

Reflect carefully: How are the two graphs related to the y-axis? Now graph these two functions on the same screen:

$$f(x) = x^2 + 3x - 3$$

$$h(x) = f(-x) = (-x)^2 + 3(-x) - 3 = x^2 - 3x - 3.$$

Are the graphs of f and h related in the same way as the first pair?

This Exploration shows that the graph of h in each case is the mirror image (reflection) of the graph of f, with the y-axis as the mirror. The same thing is true in the general case.

Reflections

Let f be a function. The graph of $h(x) = f(-x)$ is the graph of f reflected in the y-axis.

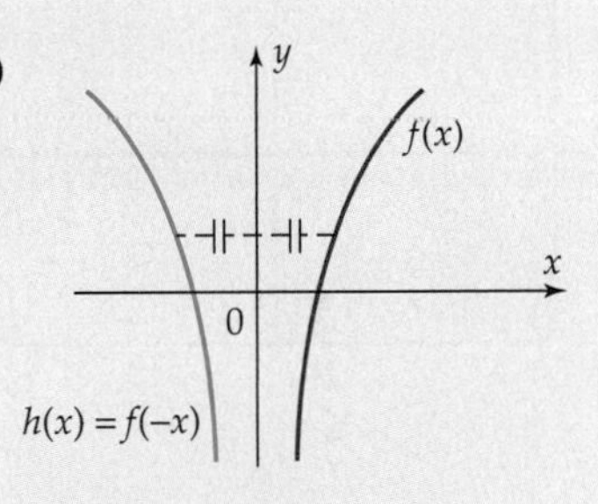

Other algebraic operations and their graphical effects are considered in Exercises 43–50.

EXERCISES 0.6

First, graph these functions (on separate screens):

$$f(x) = x^2, \qquad g(x) = x^3, \qquad h(x) = \sqrt{x}.$$

Use the graphs of f, g, and h, the information in this section, and no *additional graphing to match each function in Exercises 1–8 with its graph, which is one of those shown here.*

A.

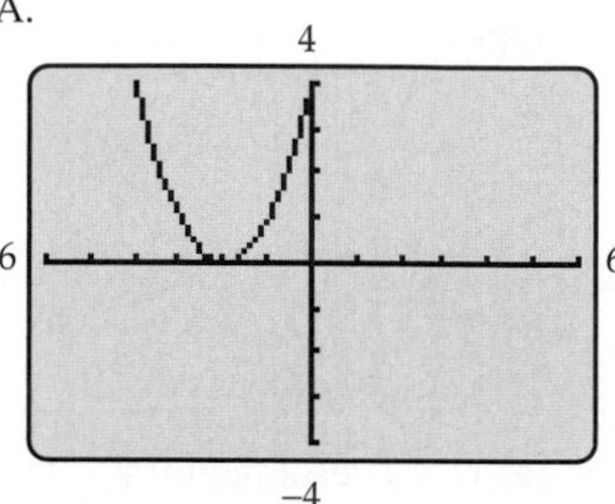

B.

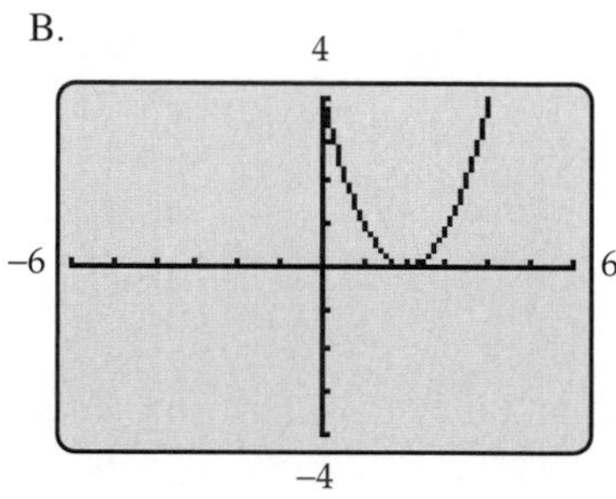

C.

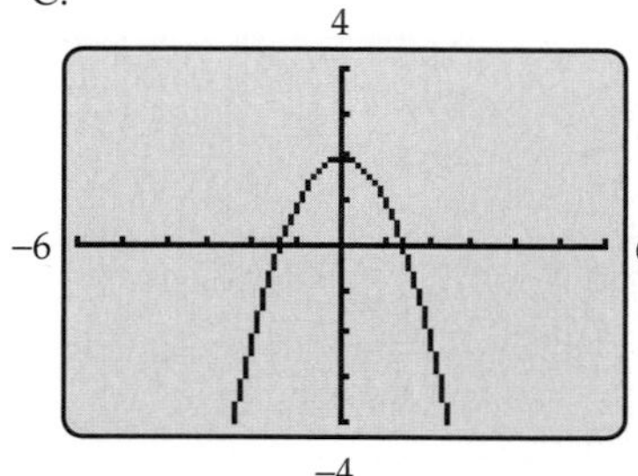

D.

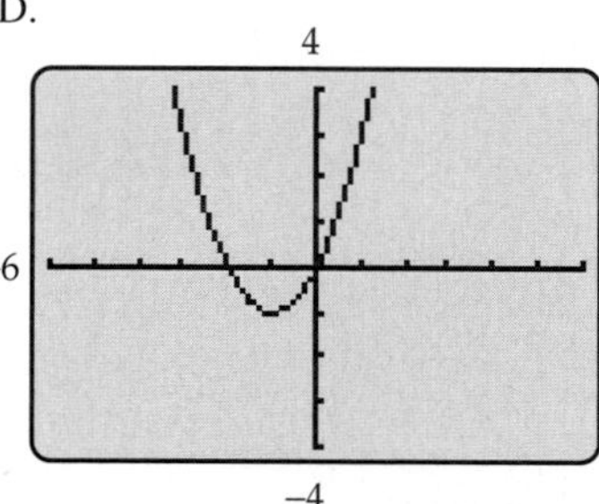

E.

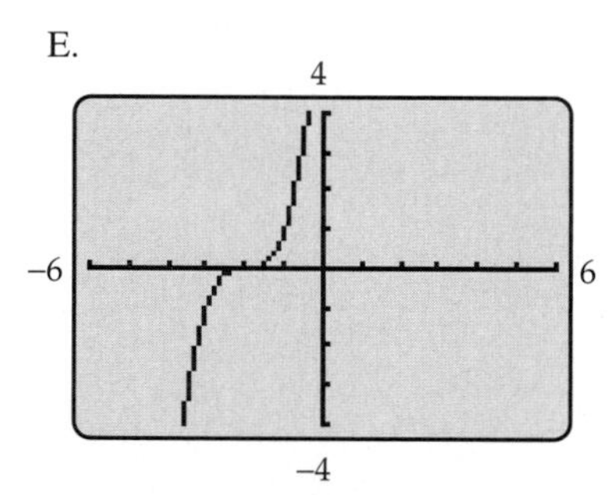

F.

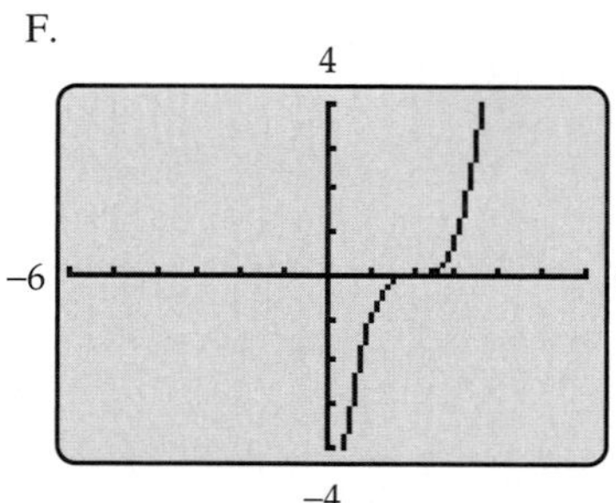

G.

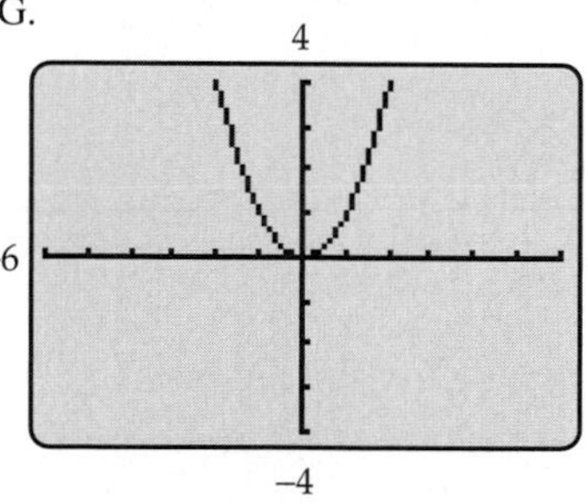

H.

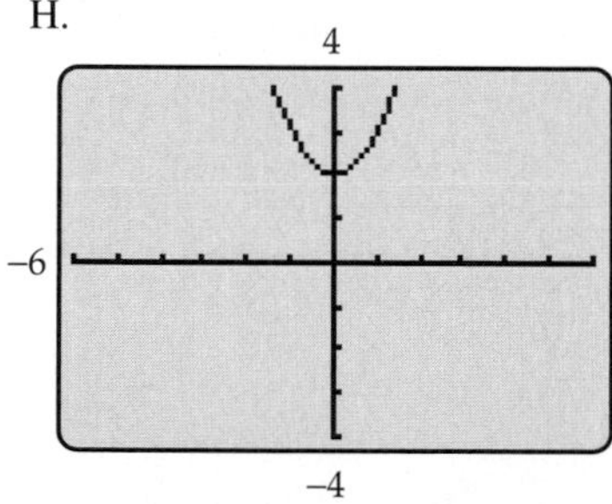

I.

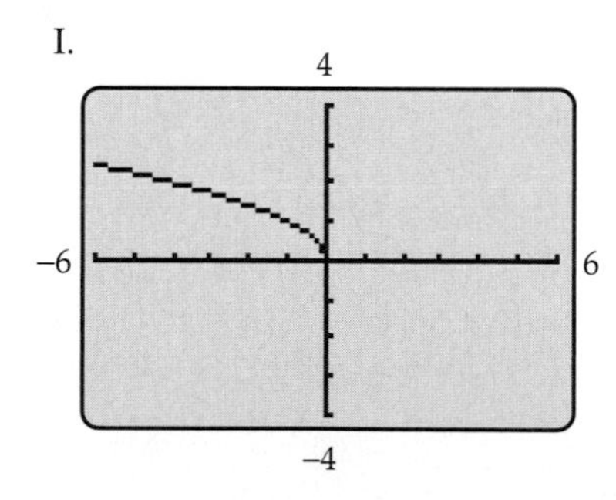

J.

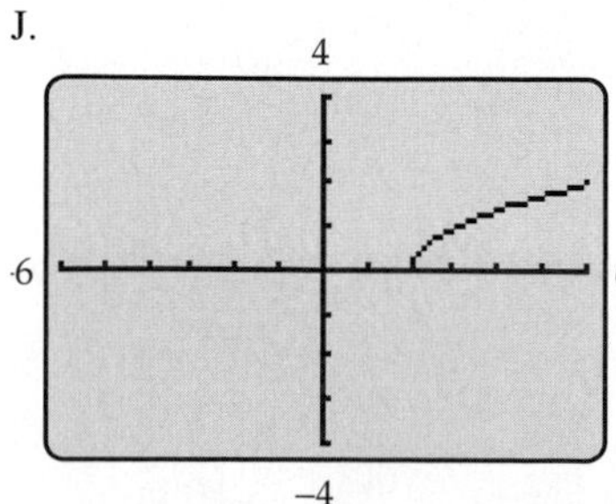

K.

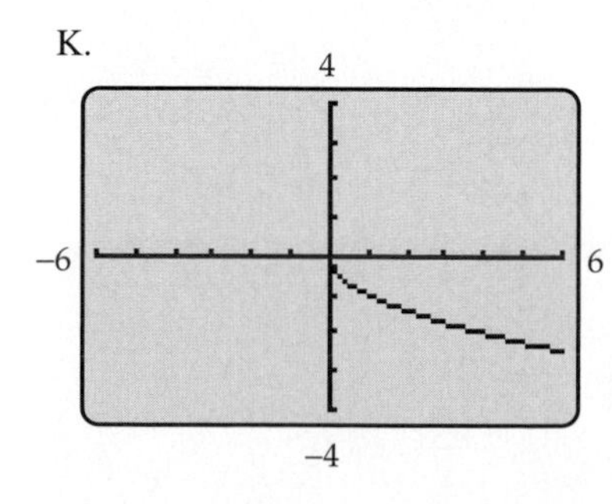

L.

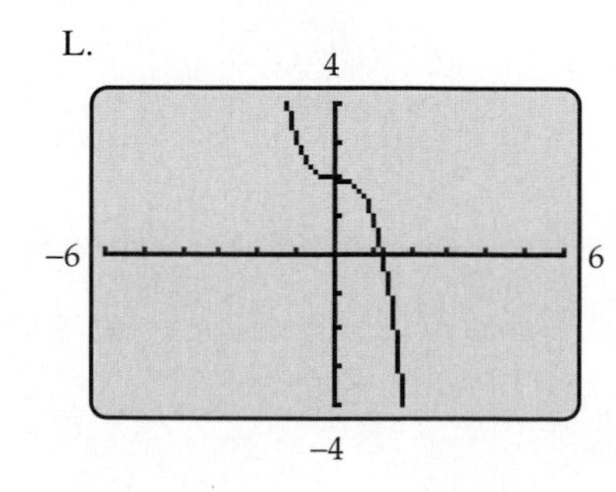

1. $f(x) = x^2 + 2$
2. $f(x) = \sqrt{x - 2}$
3. $g(x) = (x - 2)^3$
4. $g(x) = (x - 2)^2$
5. $f(x) = -\sqrt{x}$
6. $f(x) = (x + 1)^2 - 1$
7. $g(x) = -x^2 + 2$
8. $g(x) = -x^3 + 2$

In Exercises 9–12, use Figure 0–45 on page 52 and information from this section (but not a calculator) to sketch the graph of the function.

9. $f(x) = |x - 2|$
10. $(g)|x| = |x| - 2$
11. $g(x) = -|x|$
12. $f(x) = |x + 2| - 2$

In Exercises 13–16, find a single viewing window that shows complete graphs of the functions f, g, and h.

13. $f(x) = .25x^3 - 9x + 5;\quad g(x) = f(x) + 15;$
$h(x) = f(x) - 20$
14. $f(x) = \sqrt{x^2 - 9} - 5;\quad g(x) = 3f(x);$
$h(x) = .5f(x)$
15. $f(x) = |x^2 - 5|;\quad g(x) = f(x + 8);$
$h(x) = f(x - 6)$
16. $f(x) = .125x^3 - .25x^2 - 1.5x + 5;$
$g(x) = f(x) - 5;\quad h(x) = 5 - f(x)$

In Exercises 17 and 18, find complete graphs of the functions f and g in the same viewing window.

17. $f(x) = \dfrac{4 - 5x^2}{x^2 + 1};\quad g(x) = -f(x)$
18. $f(x) = x^4 - 4x^3 + 2x^2 + 3;\quad g(x) = f(-x)$

In Exercises 19–22, describe a sequence of transformations that will transform the graph of the function f into the graph of the function g.

19. $f(x) = x^2 + x;\quad g(x) = (x - 3)^2 + (x - 3) + 2$
20. $f(x) = x^2 + 5;\quad g(x) = (x + 2)^2 + 10$
21. $f(x) = \sqrt{x^3 + 5};\quad g(x) = -\frac{1}{2}\sqrt{x^3 + 5} - 6$
22. $f(x) = \sqrt{x^4 + x^2 + 1};\quad g(x) = 10 - \sqrt{4x^4 + 4x^2 + 4}$

In Exercises 23–26, write the rule of a function g whose graph can be obtained from the graph of the function f by performing the transformations in the order given.

23. $f(x) = x^2 + 2$; shift the graph horizontally 5 units to the left and then vertically upward 4 units.
24. $f(x) = x^2 - x + 1$; reflect the graph in the x-axis, then shift it vertically upward 3 units.
25. $f(x) = \sqrt{x}$; shift the graph horizontally 6 units to the right, stretch it away from the x-axis by a factor of 2, and shift it vertically downward 3 units.
26. $f(x) = \sqrt{-x}$; shift the graph horizontally 3 units to the left, then reflect it in the x-axis, and shrink it toward the x-axis by a factor of $1/2$.

C

In Exercises 27–28, the **difference quotient** *of a function f is the quantity*

$$\frac{f(x + h) - f(x)}{h},$$

where h is a nonzero constant.

27. Let $f(x) = x^2 + 3x$, and let $g(x) = f(x) + 2$.
    (a) Write the rule of $g(x)$.
    (b) Find the difference quotients of $f(x)$ and $g(x)$. How are they related?
28. Let $f(x) = x^2 + 5$, and let $g(x) = f(x - 1)$.
    (a) Write the rule of $g(x)$ and simplify.
    (b) Find the difference quotients of $f(x)$ and $g(x)$.
    (c) Let $d(x)$ denote the difference quotient of $f(x)$. Show that the difference quotient of $g(x)$ is $d(x - 1)$.

In Exercises 29–31, use the graph of the function f in the figure to sketch the graph of the function g.

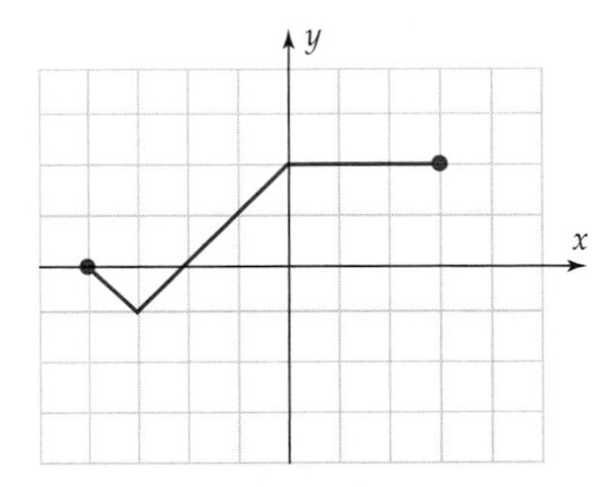

29. $g(x) = f(x) + 3$
30. $g(x) = f(x) - 1$
31. $g(x) = 3f(x)$

32. The graph shows the approximate annual net revenue of Intel (in billions of dollars) over a ten year period, with $x = 0$ corresponding to 1993.* Sketch the revenue graph corresponding to each of the following conditions.
(a) Revenue was \$5 billion smaller each year.
(b) Revenue was 25% larger each year.

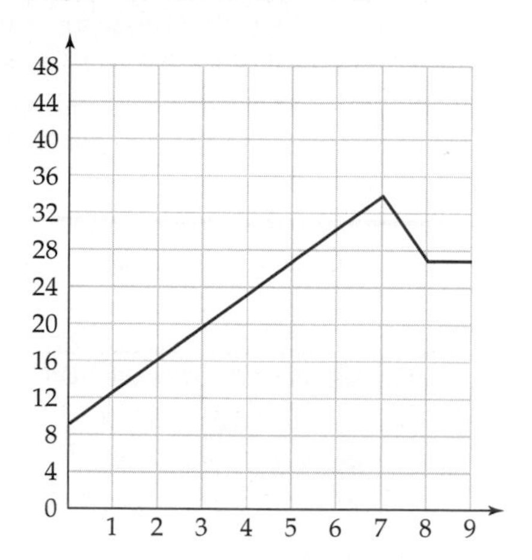

In Exercises 33–36, use the graph of the function f in the figure to sketch the graph of the function h.

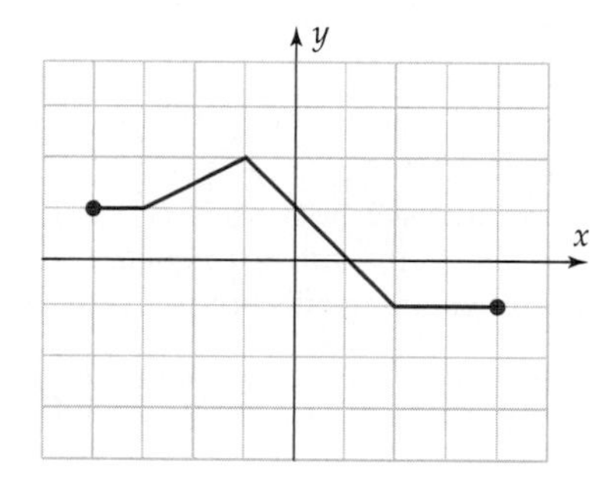

33. $h(x) = -f(x)$ **34.** $h(x) = -4f(x)$

35. $h(x) = f(-x)$ **36.** $h(x) = f(-x) + 2$

In Exercises 37–40, use the graph of the function f in the figure to sketch the graph of the function g.

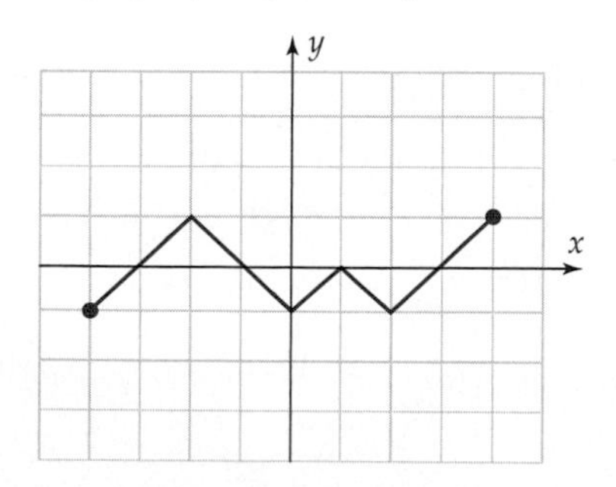

37. $g(x) = f(x + 3)$ **38.** $g(x) = f(x - 2)$

39. $g(x) = f(x - 2) + 3$ **40.** $g(x) = 2 - f(x)$

41. Graph $f(x) = -|x - 3| - |x - 17| + 20$ in the window with $0 \le x \le 20$ and $-2 \le y \le 12$. Think of the x-axis as a table and the graph as a side view of a fast-food carton placed upside down on the table (the flat part of the graph is the bottom of the carton). Find the rule of a function g whose graph (in this viewing window) looks like another fast-food carton, which has been placed right side up on top of the first one.

42. A factory has a linear cost function $c(x) = ax + b$, where b represents fixed costs and a represents the variable costs (labor and materials) of making one item, both in thousands of dollars. If property taxes (part of the fixed costs) are increased by \$28,000 per year, what effect does this have on the graph of the cost function?

In Exercises 43–45, assume $f(x) = (.2x)^6 - 4$. Use the standard viewing window to graph the functions f and g on the same screen.

43. $g(x) = f(2x)$ **44.** $g(x) = f(3x)$ **45.** $g(x) = f(4x)$

46. On the basis of the results of Exercises 43–45, describe the transformation that transforms the graph of a function $f(x)$ into the graph of the function $f(cx)$, where c is a constant with $c > 1$. [*Hint:* How are the two graphs related to the y-axis? Stretch your mind.]

In Exercises 47–49, assume $f(x) = x^2 - 3$. Use the standard viewing window to graph the functions f and g on the same screen.

47. $g(x) = f\left(\frac{1}{2}x\right)$ **48.** $g(x) = f\left(\frac{1}{3}x\right)$

49. $g(x) = f\left(\frac{1}{4}x\right)$

50. On the basis of the results of Exercises 47–49, describe the transformation that transforms the graph of a function $f(x)$ into the graph of the function $f(cx)$, where c is a constant with $0 < c < 1$. [*Hint:* How are the two graphs related to the y-axis?]

*Based on data in the 2002 Annual Report of Intel Corporation

Chapter 0 Review

IMPORTANT

CONCEPTS

Section 0.1

Real numbers, integers, rationals, irrationals 2
Number line 3
Order ($<, \leq, >, \geq$) 3
Negatives 3
Square roots 4
Absolute value 5
Distance on the number line 6
Quadratic equations 6
Quadratic formula 7

Section 0.2

Coordinate plane, x-axis, y-axis, quadrants 15
Distance formula 16
Midpoint formula 17
Graph of an equation 17
x- and y-intercepts 18
Circle, center, radius 19
Equation of the circle 19
Unit circle 20

Section 0.3

Graphing by hand 25
Viewing window 26
Equation memory 26
Trace 28
Zoom-in/out 29
Standard viewing window 29
Decimal window 29
Square window 30
Maximum/minimum finder 31
Solving equations graphically 32
Graphical root finder 32
Graphical intersection finder 33
Equation solvers 34
Solution methods 35

Section 0.4

Function 40
Domain 40
Range 40
Rule 40
Functions defined by equations 41
Functional notation 43
Common mistakes with functional notation 44
Domain convention 45
Sums, differences, products, and quotients of functions 45–46

Section 0.5

Graphs of functions 51
Local maxima and minima 52
Increasing and decreasing functions 53
Vertical Line Test 54
Parametric graphing 55

Section 0.6

Vertical shifts 62
Horizontal shifts 63
Expansions and contractions 65
Reflections in the axes 66–67

IMPORTANT

FACTS & FORMULAS

- *Quadratic Formula:* If $a \neq 0$, then the solutions of $ax^2 + bx + c = 0$ are
$$x = \frac{-b \pm \sqrt{b^2 - 4ac}}{2a}.$$

- *Distance Formula:* The distance from (x_1, y_1) to (x_2, y_2) is
$$\sqrt{(x_1 - x_2)^2 + (y_1 - y_2)^2}.$$

- *Midpoint Formula:* The midpoint of the line segment from (x_1, y_1) to (x_2, y_2) is
$$\left(\frac{x_1 + x_2}{2}, \frac{y_1 + y_2}{2}\right).$$

- The equation of the circle with center (c, d) and radius r is
$$(x - c)^2 + (y - d)^2 = r^2.$$

- The equation of the unit circle is
$$x^2 + y^2 = 1.$$

REVIEW QUESTIONS

1. Fill the blanks with one of the symbols $<$, $=$, or $>$ so that the resulting statement is true.
   (a) 142 ____ $|-51|$ (b) $\sqrt{2}$ ____ $|-2|$
   (c) -1000 ____ $\frac{1}{10}$ (d) $|-2|$ ____ $-|6|$
   (e) $|u - v|$ ____ $|v - u|$ where u and v are fixed real numbers.
2. List two real numbers that are *not* rational numbers.
3. Express in symbols:
   (a) y is negative, but greater than -10;
   (b) x is nonnegative and not greater than 10.
4. Express in symbols:
   (a) $c - 7$ is nonnegative;
   (b) .6 is greater than $|5x - 2|$.
5. Express in symbols:
   (a) x is less than 3 units from -7 on the number line;
   (b) y is farther from 0 than x is from 3 on the number line.
6. Simplify: $|b^2 - 2b + 1|$
7. (a) $|\pi - 7| =$ ____
   (b) $|\sqrt{23} - \sqrt{3}| =$ ____
8. If c and d are real numbers with $c \neq d$ what are the possible values of $\frac{c - d}{|c - d|}$?
9. A number x is one of the five numbers A, B, C, D, E on the number line shown in the figure.

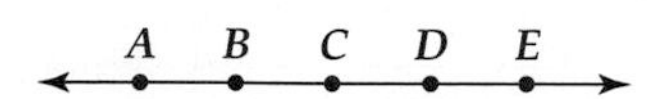

   If x satisfies *both* $A \leq x < D$ *and* $B < x \leq E$, then
   (a) $x = A$ (b) $x = B$
   (c) $x = C$ (d) $x = D$
   (e) $x = E$
10. Which one of these statements is *always* true for any real numbers x, y?
   (a) $|2x| = 2x$ (b) $\sqrt{x^2} = x$
   (c) $|x - y| = x - y$ (d) $|x - y| = |y - x|$
   (e) $|x - y| = |x| + |y|$
11. Solve for x: $3x^2 - 2x + 5 = 0$
12. Solve for x: $5x^2 + 6x = 7$
13. Solve for y: $3y^2 - 2y = 5$
14. The population P (in thousands) of St. Louis, Missouri, can be approximated by
$$P = .11x^2 - 15.95x + 864,$$
where $x = 0$ corresponds to 1950. When did the population first drop below 450,000?
15. Find the distance from $(1, -2)$ to $(4, 5)$.
16. Find the distance from $(3/2, 4)$ to $(3, 5/2)$.
17. Find the distance from (c, d) to $(c - d, c + d)$.
18. Find the midpoint of the line segment from $(-4, 7)$ to $(9, 5)$.
19. Find the midpoint of the line segment from (c, d) to $(2d - c, c + d)$.
20. Find the equation of the circle with center $(-3, 4)$ that passes through the origin.
21. (a) If $(1, 1)$ is on a circle with center $(2, -3)$, what is the radius of the circle?
   (b) Find the equation of the circle in part (a).
22. Sketch the graph of $3x^2 + 3y^2 = 12$.
23. Sketch the graph of $(x - 5)^2 + y^2 - 9 = 0$.
24. Find the equation of the circle of radius 4 whose center is the midpoint of the line segment joining $(3, 5)$ and $(-5, -1)$.
25. Which of statements (a)–(d) are descriptions of the circle with center $(0, -2)$ and radius 5?
   (a) The set of points (x, y) that satisfy $|x| + |y + 2| = 5$.
   (b) The set of all points whose distance from $(0, -2)$ is 5.
   (c) The set of all points (x, y) such that $x^2 + (y + 2)^2 = 5$.
   (d) The set of all points (x, y) such that $\sqrt{x^2 + (y + 2)^5} = 5$.
26. If the equation of a circle is $3x^2 + 3(y - 2)^2 = 12$, which of the following statements is true?
   (a) The circle has diameter 3.
   (b) The center of the circle is $(2, 0)$.
   (c) The point $(0, 0)$ is on the circle.
   (d) The circle has radius $\sqrt{12}$.
   (e) The point $(1, 1)$ is on the circle.

In Questions 27–32,
(a) Determine which of the viewing windows a–e shows a "complete graph" of the equation, that is, a graph that shows all the important features (peaks, valleys, intercepts, etc.) and suggests the general shape of the portions of the graph that aren't in the viewing window.
(b) For each viewing window that does not show a complete graph, explain why.

REVIEW QUESTIONS 

(c) *Find a viewing window that gives a "better" complete graph than windows a–e (meaning that the window is small enough to show as much detail as possible, yet large enough to show a complete graph).*

(a) Standard viewing window
(b) $-10 \le x \le 10, -200 \le y \le 200$
(c) $-20 \le x \le 20, -500 \le y \le 500$
(d) $-50 \le x \le 50, -50 \le y \le 50$
(e) $-1000 \le x \le 1000, -1000 \le y \le 1000$

27. $y = .2x^3 - .8x^2 - 2.2x + 6$

28. $y = x^3 - 11x^2 - 25x + 275$

29. $y = x^4 - 7x^3 - 48x^2 + 180x + 200$

30. $y = x^3 - 6x^2 - 4x + 24$

31. $y = .03x^5 - 3x^3 + 69.12x$

32. $y = .00000002x^6 - .0000014x^5 - .00017x^4 + .0107x^3 + .2568x^2 - 12.096x$

33. According to data from the U.S. Department of Labor, the number y of unemployed people in the labor force (in millions) from 1988 to 2003 can be approximated by the equation

$$y = (5.55 \times 10^{-4})x^4 - .0208x^3 + .1393x^2 + 1.8466x - 9.23,$$

where $x = 8$ corresponds to 1988.
(a) In what year between 1988 and 2003 was unemployment the lowest?
(b) In what year between 1988 and 2000 was unemployment the highest?
(c) In what year between 2000 and 2003 was unemployment the highest?

34. Data and projections from the U.S. National Center for Educational Statistics show that the number y of children enrolled in public elementary schools (in millions) between 1965 and 2005 can be approximated by the equation

$$y = (-4.28 \times 10^{-5})x^4 + .00356x^3 - .0819x^2 + .4135x + 31.08,$$

where $x = 0$ corresponds to 1965. According to this model, when was enrollment the lowest? The highest?

In Questions 35–40, find a solution of the equation that lies in the given interval.

35. $x^4 + x^3 - 10x^2 = 8x + 16; \quad (x > 0)$

36. $2x^4 + x^3 - 2x^2 + 6x + 2 = 0; \quad (x < -1)$

37. $\dfrac{x^3 + 2x^2 - 3x + 4}{x^2 + 2x - 15} = 0; \quad (x > -10)$

38. $\dfrac{3x^4 + x^3 - 6x^2 - 2x}{x^5 + x^3 + 2} = 0; \quad (x \ge 0)$

39. $\sqrt{x^3 + 2x^2 - 3x - 5} = 0; \quad (x \ge 0)$

40. $\sqrt{1 + 2x - 3x^2 + 4x^3 - x^4} = 0; \quad (-5 < x < 5)$

41. According to data from the U.S. Department of Justice, the total number of prisoners y in state and federal prisons (in thousands) can be approximated by the equation

$$y = .074x^4 - 1.91x^3 + 13.75x^2 + 37.94x + 738.613,$$

where $x = 0$ corresponds to 1990. In what year did the prison population first reach 1,300,000?

42. Since 1995, the amount spent on federal student assistance (such as Pell grants and other programs) can be approximated by

$$y = .1366x^3 - 3.3107x^2 + 27.434x - 35.924,$$

where $x = 5$ corresponds to 1995 and y is in billions of dollars. In what year did student assistance reach \$50 billion?

In Questions 43–50, find all real solutions of the equation.

43. $x^4 - 2x^2 - 15 = 0$

44. $x^4 - x^2 = 6$

45. $x^6 + 7x^3 - 8 = 0$

46. $3y^7 - 3y^5 - 15y^3 = 0$

47. $\sqrt{x - 1} = 2 - x$

48. $\sqrt[3]{1 - t^2} = -2$

49. $\sqrt{x + 1} + \sqrt{x - 1} = 1$

50. $\sqrt{3x - 1} + \sqrt{x} = 2$

51. Let f be the function given by the rule $f(x) = 7 - 2x$. Complete this table.

x	0	1	2	-4	t	k	$b - 1$	$1 - b$	$6 - 2u$
$f(x)$	7								

52. What is the domain of the function g given by

$$g(t) = \frac{\sqrt{t - 2}}{t - 3}?$$

53. In each case, give a *specific* example of a function and numbers a, b to show that the given statement may be *false*.
(a) $f(a + b) = f(a) + f(b)$ (b) $f(ab) = f(a)f(b)$

54. If $f(x) = |3 - x|\sqrt{x - 3} + 7$, then

$$f(7) - f(4) = \_\_\_\_ .$$

REVIEW QUESTIONS 

55. What is the domain of the function given by

$$g(r) = \sqrt{r - 4} + \sqrt{r} - 2?$$

56. What is the domain of the function $f(x) = \sqrt{-x + 2}$?

57. If $h(x) = x^2 - 3x$, then $h(t + 2) =$ ____.

58. If $f(x) = 2x^3 + x + 1$, then $f(x/2) =$ ____.

59. The radius of an oil spill (in meters) is 50 times the square root of the time t (in hours).
 (a) Write the rule of a function f that gives the radius of the spill at time t.
 (b) Write the rule of a function g that gives the area of the spill at time t.
 (c) What are the radius and area of the spill after 9 hours?
 (d) When will the spill have an area of 100,000 square meters?

60. Which of the following are graphs of functions of x?

(a)

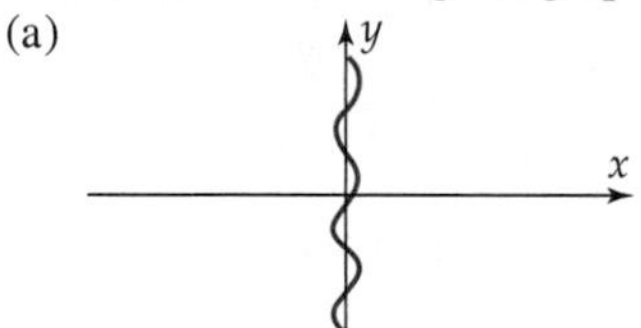

(b)

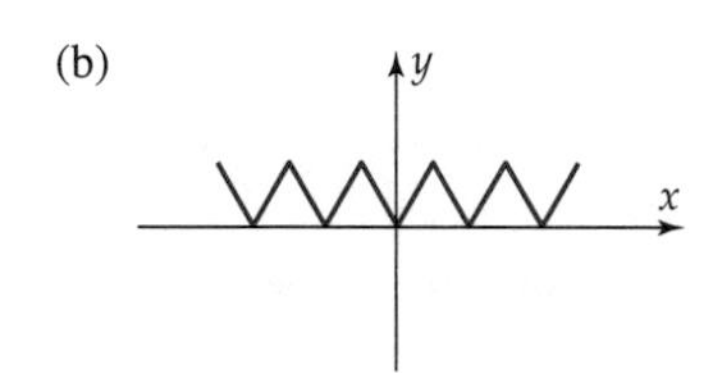

In Questions 61–64, determine the local maxima and minima of the function, the intervals on which the function is increasing, and the intervals on which it is decreasing.

61. $g(x) = \sqrt{x^2 + x + 1}$

62. $f(x) = 2x^3 - 5x^2 + 4x - 3$

63. $g(x) = x^3 + 8x^2 + 4x - 3$

64. $f(x) = .5x^4 + 2x^3 - 6x^2 - 16x + 2$

65. Sketch the graph of $g(x) = 5 + \dfrac{4}{x - 5}$.

Use the graph of the function f in the figure to answer Questions 66–70.

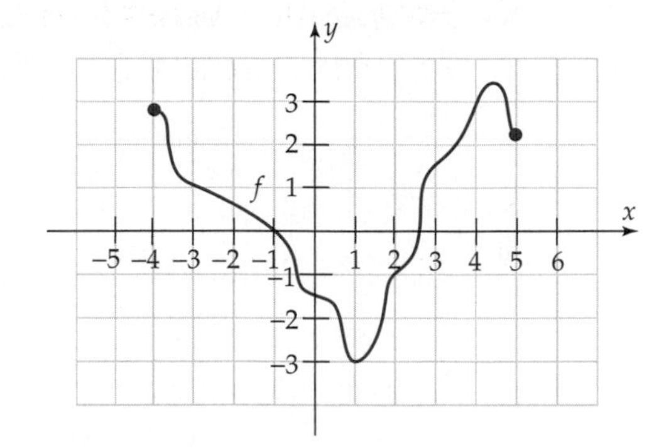

66. $f(-3) =$ ____.

67. $f(2 + 2) =$ ____.

68. $f(-1) + f(1) =$ ____.

69. True or false: $2f(2) = f(4)$.

70. True or false: $3f(2) = -f(4)$.

In Questions 71 and 72, sketch the graph of the curve given by the parametric equations.

71. $x = t^2 - 4$ and $y = 2t + 1 \quad (-3 \le t \le 3)$

72. $x = t^3 + 3t^2 - 1$ and $y = t^2 + 1 \quad (-3 \le t \le 2)$

In Questions 73–76, list the transformations, in the order in which they should be performed on the graph of $g(x) = x^2$, so as to produce a complete graph of the function f.

73. $f(x) = (x - 2)^2$

74. $f(x) = .25x^2 + 2$

75. $f(x) = -(x + 4)^2 - 5$

76. $f(x) = -3(x - 7)^2 + 2$

REVIEW QUESTIONS

77. The graph of a function f is shown in the figure. On the same coordinate plane, carefully draw the graphs of the functions g and h whose rules are:

$$g(x) = -f(x) \quad \text{and} \quad h(x) = 1 - f(x)$$

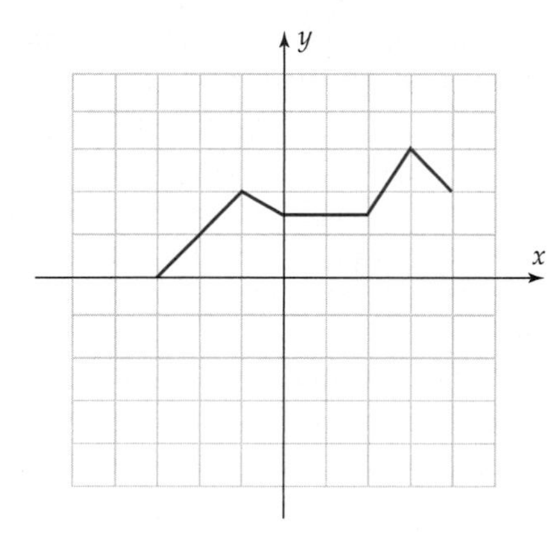

78. The figure shows the graph of a function f. If g is the function given by $g(x) = f(x + 2)$, then which of these statements about the graph of g is true?

(a) It does not cross the x-axis.
(b) It does not cross the y-axis.
(c) It crosses the y-axis at $y = 4$.
(d) It crosses the y-axis at the origin.
(e) It crosses the x-axis at $x = -3$.

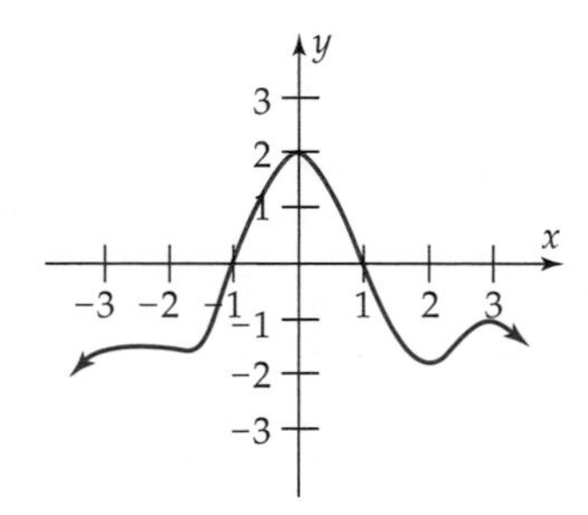

79. If $f(x) = 3x + 2$ and $g(x) = x^3 + 1$, find:
(a) $(f + g)(-1)$ (b) $(f - g)(2)$ (c) $(fg)(0)$

80. If $f(x) = \dfrac{1}{x - 1}$ and $g(x) = \sqrt{x^2 + 5}$, find:
(a) $(f/g)(2)$ (b) $(g/f)(x)$
(c) $(fg)(c + 1)$ $(c \neq 0)$

DISCOVERY PROJECT 0 Feedback—Good and Bad

Tom Sobolik/Black Star Publishing/PictureQuest

Whether it's for a concert or a worship service, a sporting event or a graduation ceremony, a shopping mall or a lecture hall, audio engineers are concerned to avoid audio feedback. If the system is not set up correctly, then sound from the speakers reaches the microphone with enough clarity that it is fed back into the audio system and amplified. This feedback cycle repeats again and again. Each time the amplified sound follows very quickly behind the previous one so that the audience hears an unpleasant hum that rapidly becomes a loud screech. This is feedback we would prefer to avoid.

Other types of feedback are considerably more pleasing. Consider, for example, an investment of x dollars that earns 4% interest compounded annually. Then the amount in the account (principal plus interest) after one year is $1.04x$. Assume the initial deposit is left in the account without adding deposits or withdrawing funds. If \$1000 is invested, then the balance (rounded to the nearest penny) is

Beginning	1000
End of year 1	$1.04 \cdot 1000 = 1040$
End of year 2	$1.04 \cdot 1040 = 1081.60$
End of year 3	$1.04 \cdot 1081.60 = 1124.86$

and so on. You can easily carry out this process on a calculator as follows: key in 1000 and press ENTER; then key in "× 1.04" and press ENTER repeatedly. Each time you press ENTER, the calculator computes as shown in Figure 1. The result is an ever-increasing bank balance.

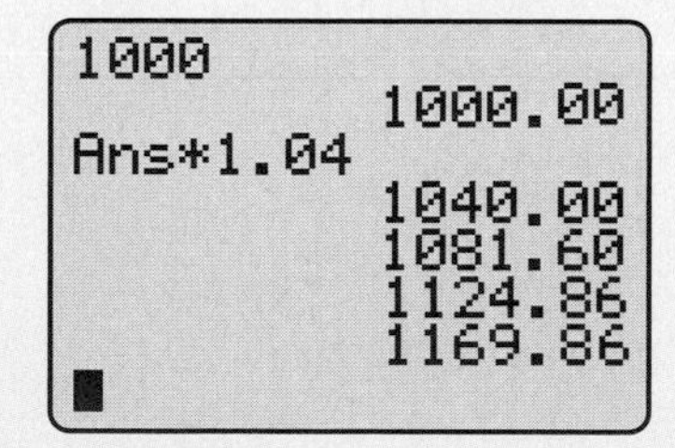

1169.86
1216.65
1265.32
1315.93
1368.57
1423.31
1480.24

Figure 1

In each of the preceding examples, the output of the process was fed back as input and the process repeated to generate the result—unpleasant in the first case and quite pleasant in the second. Now we strip away the specifics and look at the process in more general terms. Suppose we have a function f and x is in its domain. We find $f(x)$ and feed that value back to the function to find

$f(f(x))$. Then feed that value back to find $f(f(f(x)))$. The sequence of values obtained by continuing this process

$$f(x), f(f(x)), f(f(f(x))), f(f(f(f(x)))), \ldots$$

is called the **orbit** of x and the feedback process is called **iteration.**

Technology makes it easy to compute orbits. For example, let $f(x) = \sqrt{x}$ and enter f as Y_1 in your calculator. Key in 8 and press ENTER. Then key in "STO➡ X : Y_1" and press ENTER repeatedly.* Each time you press ENTER, the calculator stores the answer from the preceding calculation as X and evaluates Y_1 at X, as shown in Figure 2.

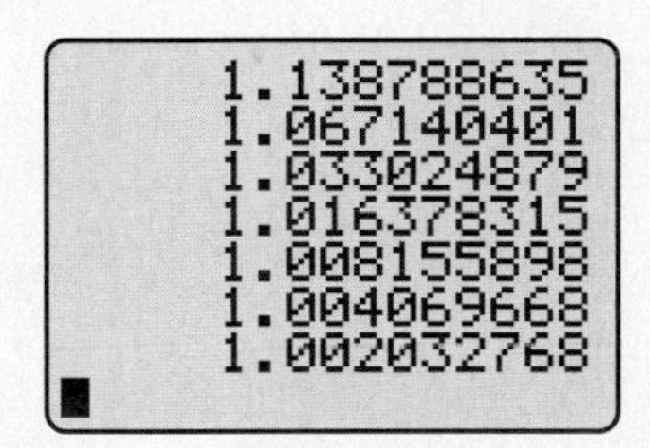

Figure 2

Thus, the (approximate) orbit of 8 is

$$\{8, 2.828, 1.682, 1.297, 1.139, 1.067, 1.033, \ldots\}.$$

Note that the terms of the orbit are getting closer and closer to 1.

1. Find the orbit of .2 and two other numbers of your choice (except 0 and 1). Do the terms of these orbits get closer and closer to 1?

For the function $f(x) = \sqrt{x}$ we can see that $f(1) = 1$, so that iteration produces the orbit $\{1, 1, 1, \ldots\}$; we say that 1 is a **fixed point** for this function. Another fixed point is zero because $f(0) = 0$. A fixed point a is called an **attracting fixed point** if nearby numbers have orbits that approach a or a **repelling fixed point** if nearby numbers have orbits that move farther away from a. The following exercises illustrate these ideas for the function $f(x) = \sqrt{x}$.

2. Confirm numerically that 1 is an attracting fixed point by computing the orbits of .5, 2, and 11.†

---

*The colon (:) is on the TI and HP-39+ keyboards and in the Casio PRGM menu (press SHIFT VARS).

†Your calculator may tell you that the terms of the orbits eventually are equal to 1, but this is due to round-off error.

3. Explain mathematically why 1 is an attracting fixed point. [*Hint:* You may assume that if $u < v$, then $\sqrt{u} < \sqrt{v}$. Is $\sqrt{x}$ larger or smaller than x when $x < 1$? When $x > 1$?]
4. Determine whether zero is an attracting fixed point or a repelling fixed point.
5. Do you think this function has any other fixed points? Why or why not?

Generalizations are interesting in their own right but you may be wondering whether these ideas have any relevance to the real world. So let's look at another application. Ecologists want to model the growth of the population of some animal in an ecosystem. To simplify notation, let's let the variable x represent a fraction of the theoretical maximum population, so that $0 \leq x \leq 1$. [If the population were the maximum the ecosystem could sustain, then $x = 1$ and if there are none of these animals in the ecosystem, then $x = 0$. We're interested in the usual situation where x is somewhere in between.] With a growth rate r, the function $f(x) = rx$ seems to fill the bill.* But logistics, the limitation of resources (food, nesting locations, etc.), demands an adjustment. The **logistic function** $f(x) = rx(1 - x)$ gives a more realistic model of the growth, which is nearly exponential when x is small (and $1 - x$ is nearly 0). It also reflects the influence of logistics in that $(1 - x)$ approaches 0 as x approaches 1.

Do these exercises for the logistic function $f(x) = rx(1 - x)$ and the given value of r.

6. Use $r = 2$ and different initial values (such as .25, .8, and .1) to confirm that $f(x) = 2x(1 - x)$ has fixed point .5.
7. Find the fixed point (to the nearest hundredth) when $r =$

   (a) 2.5 (b) 2.8 (c) 3
8. Verify that for $r = 3.3$, the orbit of any point x (with $0 < x < 1$) seems to (eventually) jump back and forth between *two* values. Somewhere between 3.2 and 3.3 a **bifurcation** has occurred. Instead of a single fixed point we now have a periodic oscillation, with **period 2.** Find the two values.
9. Verify that periodic oscillation also occurs when $r = 3.4$. Is the period still 2?
10. Verify that $r = 3.5$ eventually oscillates among *four* values (that is, the period is $4 = 2^2$). Another bifurcation has occurred. Find the four values.

*Notice the similarity to the 4% interest example, where the "balance function" was $f(x) = 1.04x$. Notice also, that if r is less than 1, the population declines to extinction.

11. Verify that for $r = 3.54$ the period is still four but at $r = 3.56$ there are *eight* values (period is $8 = 2^3$). Find them.

12. Is it possible to determine exactly where bifurcations occur? When r is very large, do you think that the system breaks down to "chaos" or that there are values of r that have period 2^n for large values of n?

13. If you are able to consult a naturalist, determine whether some animal populations follow yearly patterns consistent with the logistic model.

CHAPTER 1

Triangle Trigonometry

Chris Alan Wilton/Getty Images

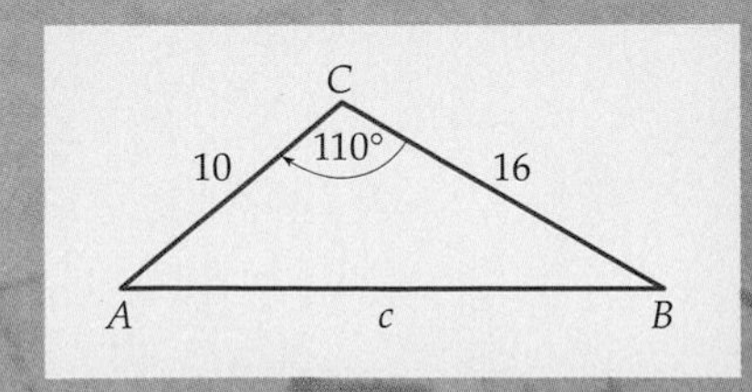

Where are we?

Navigators at sea must determine their location. Surveyors need to determine the height of a mountain or the width of a canyon when direct measurement isn't feasible. A fighter plane's computer must set the course of a missile so that it will hit a moving target. These and many similar problems can be solved by using triangle trigonometry. See Exercise 59 on page 115 and Exercise 33 on page 123.

Chapter Outline

Interdependence of Sections

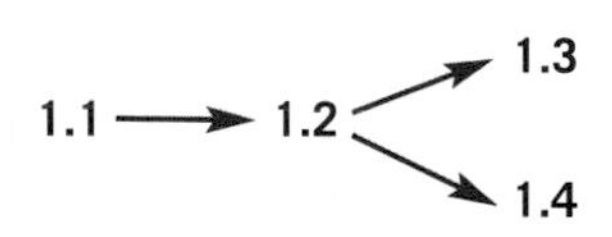

Roadmap

Instructors who want to introduce trigonometric functions of real numbers before doing triangle trigonometry should cover Chapter 2 before Chapter 1, following the roadmap at the beginning of Chapter 2.

Trigonometry was first used by the ancient Greeks for astronomical calculations that involved triangles. In fact, *trigonometry* means "triangle measurement." Later, trigonometry was used to solve problems in navigation and surveying. This kind of trigonometry is still used by engineers and scientists to deal with a wide variety of situations, some of which are considered here.

1.1 Trigonometric Functions of Acute Angles

NOTE
If you have read Chapter 2, begin this chapter with Alternate Section 1.1 on page 90.

Before reading this section, it might be a good idea to read the Geometry Review Appendix, which presents the basic facts about angles and triangles that frequently are used here. In particular, recall that a **right triangle** is one that contains a **right angle,** that is, an angle of 90°. An **acute angle** is an angle whose measure is less than 90°. Consider the right triangles in Figure 1–1, each of which has an acute angle of θ degrees.

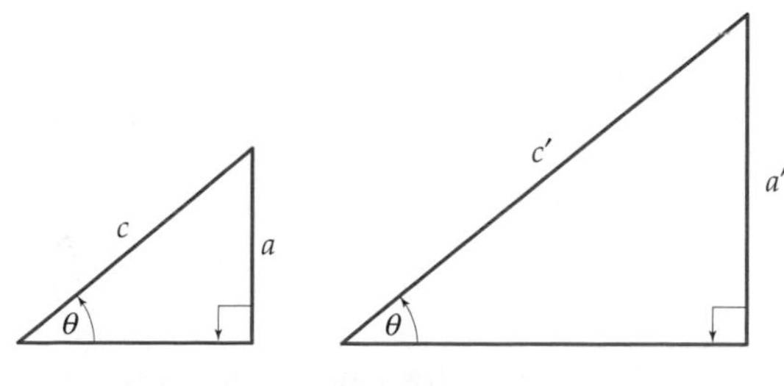

Figure 1–1

Since the sum of the angles of any triangle is 180°, we see that the third angle in each of these triangles has the same measure, namely, $180° - 90° - \theta$. Thus, the two triangles have equal corresponding angles and, therefore, are similar.

Consequently, by the Note after the Ratios Theorem of the Geometry Review Appendix, we know that the ratio of corresponding sides is the same, that is,

$$\frac{a}{c} = \frac{a'}{c'}.$$

Each of these fractions is the ratio

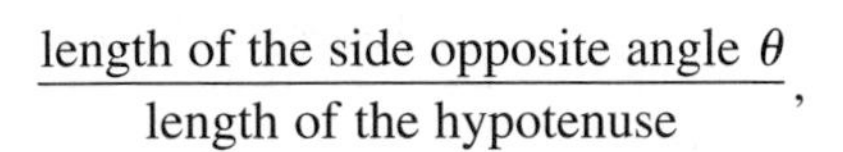

$$\frac{\text{length of the side opposite angle } \theta}{\text{length of the hypotenuse}},$$

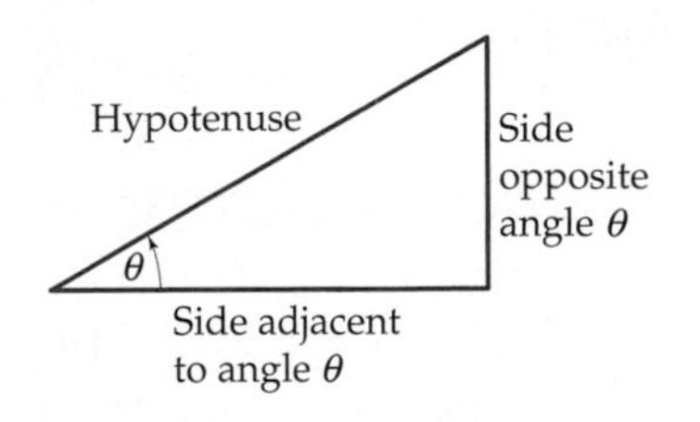

Figure 1–2

as indicated in Figure 1–2. Consequently, *this ratio depends only on the angle* θ and *not* on the size of the triangle. Similar remarks apply to the ratios of other sides of the triangle in Figure 1–2 and make it possible to define three new functions. For each function, the input is an acute angle θ, and the corresponding output is a ratio of sides in any right triangle containing angle θ, as summarized here.

Trigonometric Functions of Acute Angles

Name of Function	Abbreviation	Rule of Function
sine	sin	$\sin\theta = \dfrac{\text{length of side opposite angle } \theta}{\text{length of hypotenuse}}$
cosine	cos	$\cos\theta = \dfrac{\text{length of side adjacent to angle } \theta}{\text{length of hypotenuse}}$
tangent	tan	$\tan\theta = \dfrac{\text{length of side opposite angle } \theta}{\text{length of side adjacent to angle } \theta}$

EXAMPLE 1

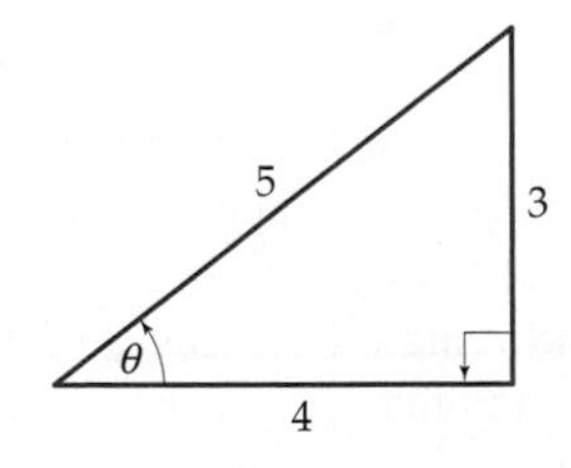

Figure 1–3

Evaluate $\sin\theta$, $\cos\theta$, and $\tan\theta$ for the angle θ shown in Figure 1–3.

SOLUTION The side opposite angle θ has length 3, and the side adjacent to angle θ has length 4. Hence,

$$\sin\theta = \frac{\text{length of side opposite angle } \theta}{\text{length of hypotenuse}} = \frac{3}{5}$$

$$\cos\theta = \frac{\text{length of side adjacent to angle } \theta}{\text{length of hypotenuse}} = \frac{4}{5}$$

$$\tan\theta = \frac{\text{length of side opposite angle } \theta}{\text{length of side adjacent to angle } \theta} = \frac{3}{4}. \quad \blacksquare$$

Now that you have the idea, we shall abbreviate the ratio

$$\frac{\text{length of the side adjacent to angle } \theta}{\text{length of the hypotenuse}} \quad \text{as} \quad \frac{\text{adjacent}}{\text{hypotenuse}}$$

and similarly for the others.

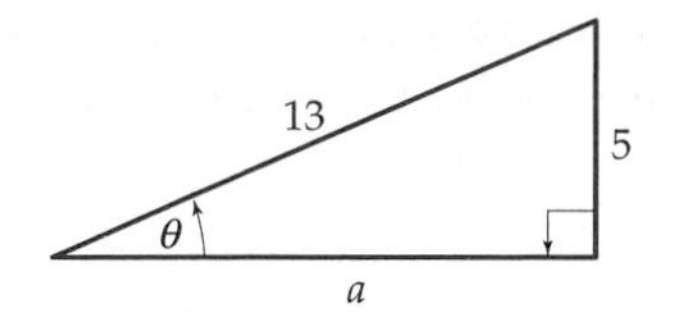

Figure 1–4

TECHNOLOGY TIP

Throughout this chapter, be sure your calculator is set for DEGREE mode. Use the MODE(S) menu on TI and HP and the SETUP menu on Casio.

EXAMPLE 2

Evaluate $\sin\theta$, $\cos\theta$, and $\tan\theta$ when θ is the angle shown in Figure 1–4.

SOLUTION First, we find the length of the third side a by using the Pythagorean Theorem.

$$a^2 + 5^2 = 13^2$$

Multiply out terms: $$a^2 + 25 = 169$$

Subtract 25 from both sides: $$a^2 = 144$$

Take square roots on both sides.* $$a = \sqrt{144} = 12$$

Now we can calculate the values of the trigonometric functions.

$$\sin\theta = \frac{\text{opposite}}{\text{hypotenuse}} = \frac{5}{13} \qquad \cos\theta = \frac{\text{adjacent}}{\text{hypotenuse}} = \frac{12}{13}$$

$$\tan\theta = \frac{\text{opposite}}{\text{adjacent}} = \frac{5}{12} \qquad ■$$

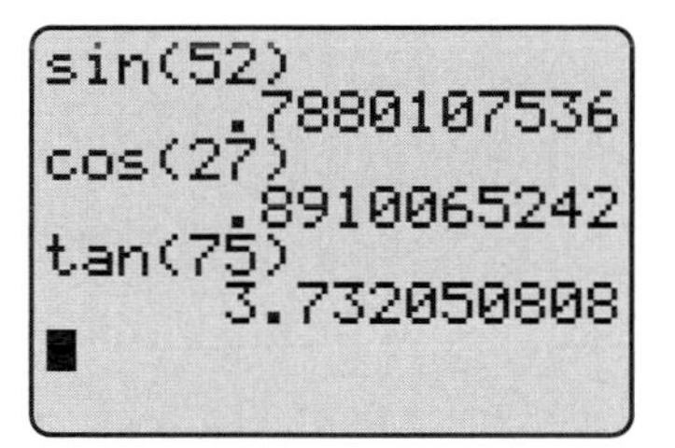

Figure 1–5

Unless you are given an appropriate triangle, whose sides are known (or can be computed) as in Example 2, it may be difficult to find the exact values of the trigonometric functions at an angle θ. Fortunately, however, your calculator can provide good approximations, as illustrated in Figure 1–5.

In a few cases, however, we can find the exact values of sine, cosine, and tangent. You should memorize the values found in Examples 3 and 4.

EXAMPLE 3

Evaluate $\sin\theta$, $\cos\theta$, $\tan\theta$ when (a) $\theta = 30°$, (b) $\theta = 60°$.

SOLUTION

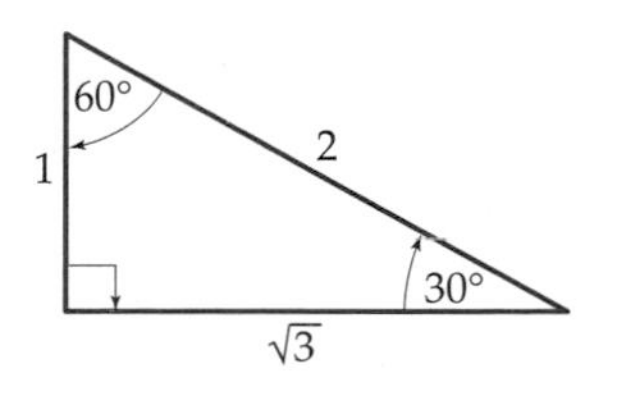

Figure 1–6

(a) Consider a 30°-60°-90° triangle whose hypotenuse has length 2. As is explained in Example 3 of the Geometry Review Appendix, the side opposite the 30° angle must have length 1 (half the hypotenuse) and the side adjacent to this angle must have length $\sqrt{3}$, as shown in Figure 1–6. According to the right triangle description,

$$\sin 30° = \frac{\text{opposite}}{\text{hypotenuse}} = \frac{1}{2}, \qquad \cos 30° = \frac{\text{adjacent}}{\text{hypotenuse}} = \frac{\sqrt{3}}{2},$$

$$\tan 30° = \frac{\text{opposite}}{\text{adjacent}} = \frac{1}{\sqrt{3}} = \frac{\sqrt{3}}{3}.$$

*The equation $a^2 = 144$ has two solutions, 12 and -12, but only the positive one applies here since a is the side of a triangle.

(b) The same triangle can be used to evaluate the trigonometric functions at 60°. In this case, the opposite side has length $\sqrt{3}$ and the adjacent side has length 1. Therefore,

$$\sin 60° = \frac{\text{opposite}}{\text{hypotenuse}} = \frac{\sqrt{3}}{2}, \qquad \cos 60° = \frac{\text{adjacent}}{\text{hypotenuse}} = \frac{1}{2},$$

$$\tan 60° = \frac{\text{opposite}}{\text{adjacent}} = \frac{\sqrt{3}}{1} = \sqrt{3}. \quad ■$$

EXAMPLE 4

Evaluate sin 45°, cos 45°, and tan 45°.

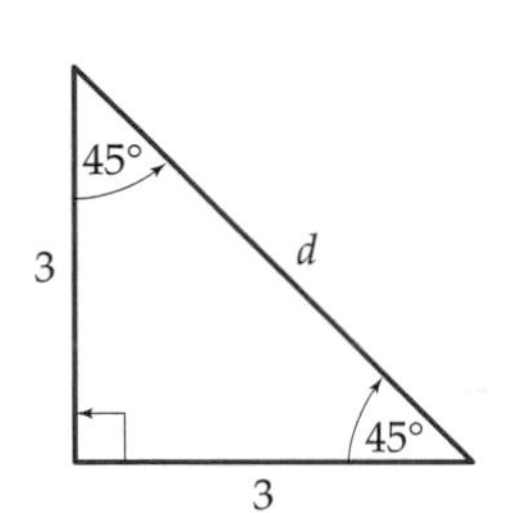

Figure 1–7

SOLUTION Consider a 45°-45°-90° triangle whose legs each have length 3 (Figure 1–7). According to the Pythagorean Theorem, the hypotenuse d satisfies

$$d^2 = 3^2 + 3^2 = 18,$$

so

$$d = \sqrt{18} = \sqrt{9 \cdot 2} = \sqrt{9}\,\sqrt{2} = 3\sqrt{2}.$$

Therefore,

$$\sin 45° = \frac{\text{opposite}}{\text{hypotenuse}} = \frac{3}{3\sqrt{2}} = \frac{1}{\sqrt{2}} = \frac{\sqrt{2}}{2}$$

$$\cos 45° = \frac{\text{adjacent}}{\text{hypotenuse}} = \frac{3}{3\sqrt{2}} = \frac{1}{\sqrt{2}} = \frac{\sqrt{2}}{2}$$

$$\tan 45° = \frac{\text{opposite}}{\text{adjacent}} = \frac{3}{3} = 1. \quad ■$$

FINDING SIDES AND ANGLES OF TRIANGLES

Up to now, we have started with a triangle and found the values of the trigonometric functions at various angles. This process can be reversed: By using the values of the trigonometric functions, we can find the sides or angles of a right triangle.

EXAMPLE 5

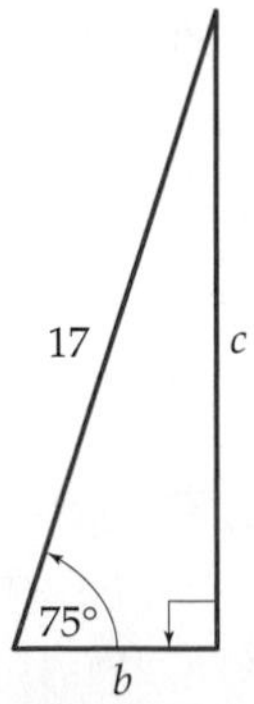

Find the lengths of sides b and c in the right triangle shown in Figure 1–8.

Figure 1–8

SOLUTION Since the side c is opposite the 75° angle and the hypotenuse is 17, we have

$$\sin 75° = \frac{\text{opposite}}{\text{hypotenuse}} = \frac{c}{17}.$$

We can solve this equation for c.

$$\frac{c}{17} = \sin 75^\circ$$

Multiply both sides by 17: $$c = 17 \sin 75^\circ$$

Use a calculator to evaluate sin 75°: $$c \approx 17(.9659) \approx 16.42$$

Side b can now be found by the Pythagorean Theorem or by using the fact that

$$\cos 75^\circ = \frac{\text{adjacent}}{\text{hypotenuse}} = \frac{b}{17}.$$

Solving this equation and using a calculator shows that

$$\frac{b}{17} = \cos 75^\circ$$

$$b = 17 \cos 75^\circ \approx 17(.2588) \approx 4.40. \quad \blacksquare$$

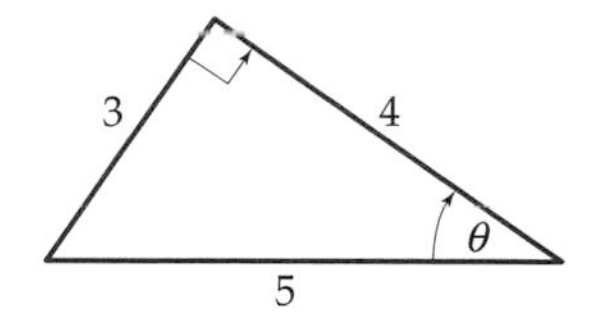

Figure 1–9

EXAMPLE 6

Find the degree measure of the angle θ in Figure 1–9.

SOLUTION We first note that

$$\cos \theta = \frac{\text{adjacent}}{\text{hypotenuse}} = \frac{4}{5} = .8.$$

Before calculators were available, θ was found by using a table of cosine values, as follows: Look through the column of cosine values for the closest one to .8, then look in the first column for the corresponding value of θ. You can do the same thing by having your calculator generate a table for $y_1 = \cos x$, as in Figure 1–10. The closest entry to .8 in the cosine (y_1) column is .79968, which corresponds to an angle of 36.9°. Hence, $\theta \approx 36.9^\circ$.

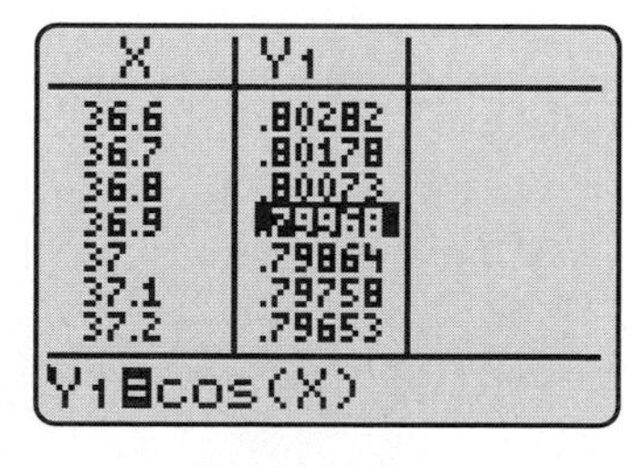

Figure 1–10

Figure 1–11

> **NOTE**
> In this chapter, we shall use the COS^{-1} key, and the analogous keys SIN^{-1} and TAN^{-1}, as they are used in Example 6: as a way to find an angle θ in a triangle, when sin θ or cos θ or tan θ is known. The other uses of these keys are discussed in Section 3.4, which deals with the inverse functions of sine, cosine, and tangent.

A faster and more accurate method of finding θ is to use the COS^{-1} key on your calculator (labeled ACOS on some models). When you key in COS^{-1} .8, as in Figure 1–11, the calculator produces an acute angle whose cosine is .8, namely, $\theta \approx 36.8699^\circ$. Thus, the COS^{-1} key provides the electronic equivalent of searching the cosine table, without actually having to construct the table. $\blacksquare$

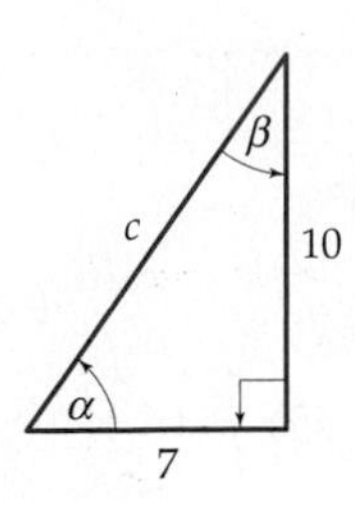

Figure 1–12

Figure 1–13

EXAMPLE 7

Without using the Pythagorean Theorem, find angles α and β and side c of the triangle in Figure 1–12.

SOLUTION We first use the fact that

$$\tan \alpha = \frac{\text{opposite}}{\text{adjacent}} = \frac{10}{7}.$$

The TAN^{-1} key on a calculator (Figure 1–13) shows that $\alpha \approx 55.01°$. Since the sum of the angles of a triangle is 180°, we have

$$\alpha + \beta + 90° = 180°$$

$$55.01 + \beta + 90° \approx 180°$$

$$\beta \approx 180° - 90° - 55.01° = 34.99°.$$

Next, we note that

$$\sin \alpha = \frac{\text{opposite}}{\text{hypotenuse}} = \frac{10}{c}$$

$$\sin 55.01° = \frac{10}{c}.$$

Multiplying both sides of this equation by c shows that

$$c \sin 55.01° = 10$$

$$c = \frac{10}{\sin 55.01°}.$$

A calculator shows that $c \approx 12.21$ (Figure 1–14). ■

Figure 1–14

EXERCISES 1.1

All decimal approximations should be rounded off to one decimal place at the end *of the computation.*

In Exercises 1–6, find $\sin\theta$, $\cos\theta$, $\tan\theta$.

1.

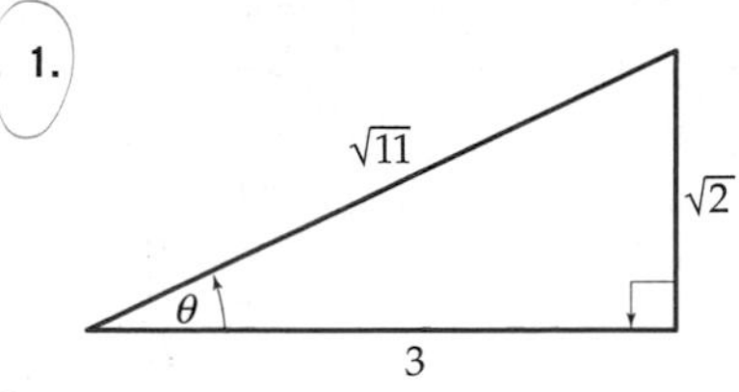

2.

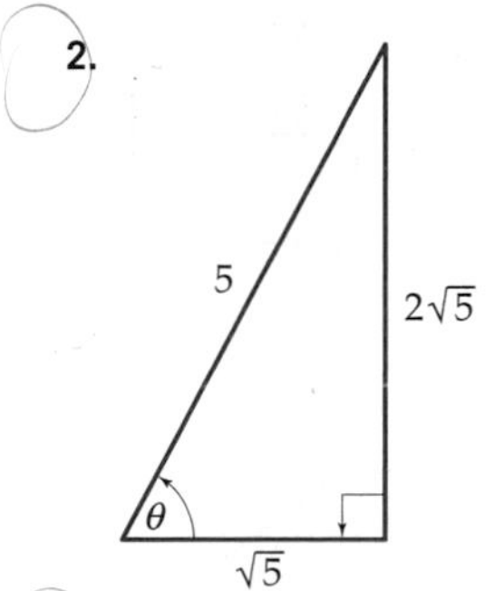

3.

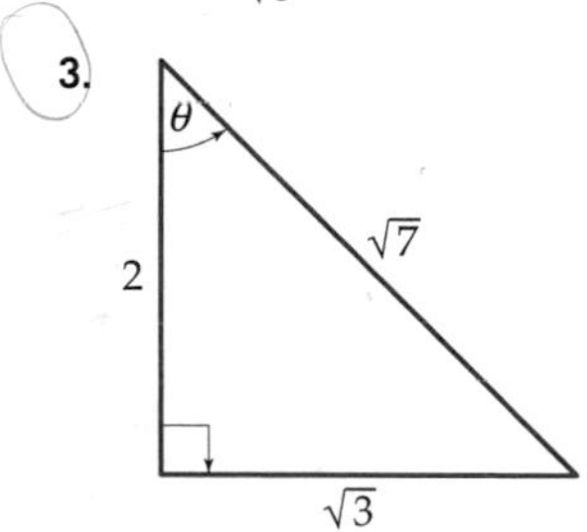

4.

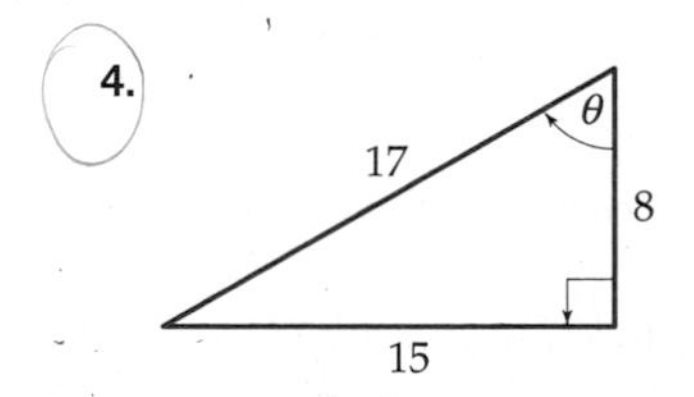

5.

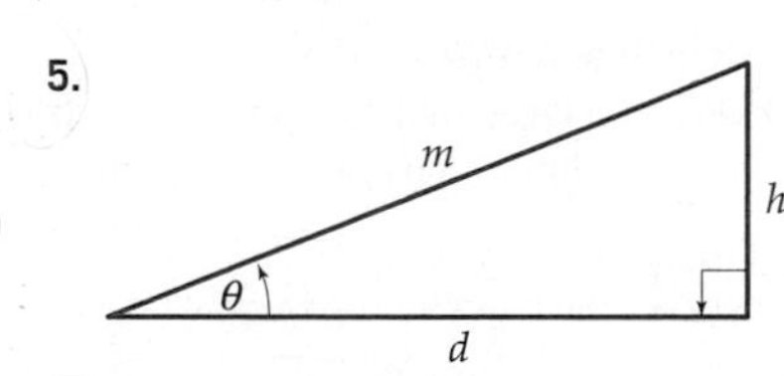

6.

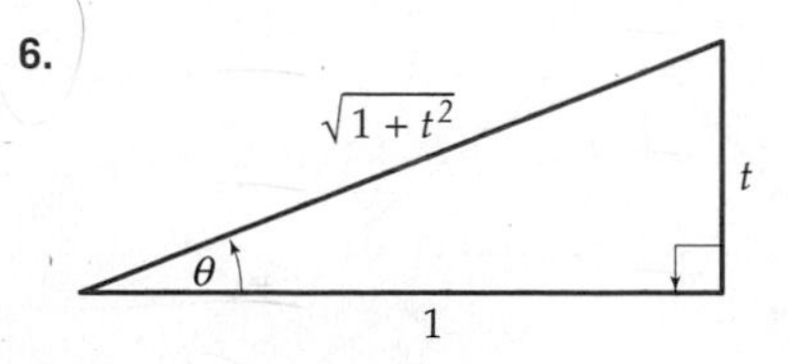

In Exercises 7–12, find side c of the right triangle in the figure under the given conditions.

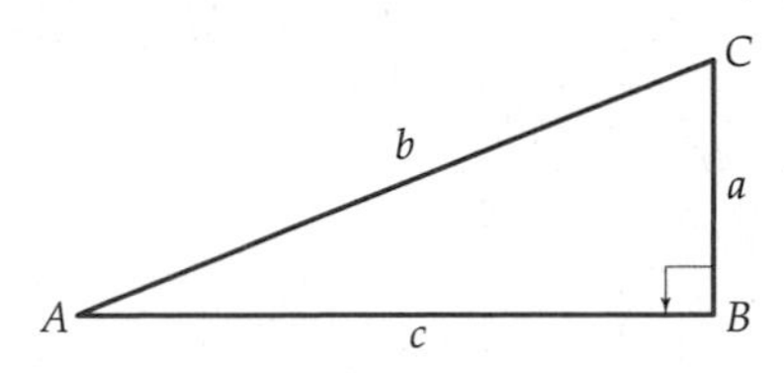

7. $\cos A = 12/13$ and $b = 39$

8. $\sin C = 3/4$ and $b = 12$

9. $\tan A = 5/12$ and $a = 15$

10. $\cos A = 1/2$ and $b = 8$

11. $\tan A = 1/6$ and $a = 1.4$

12. $\sin C = 2/3$ and $b = 4.5$

In Exercises 13–18, find the length h of the side of the right triangle, without using a calculator.

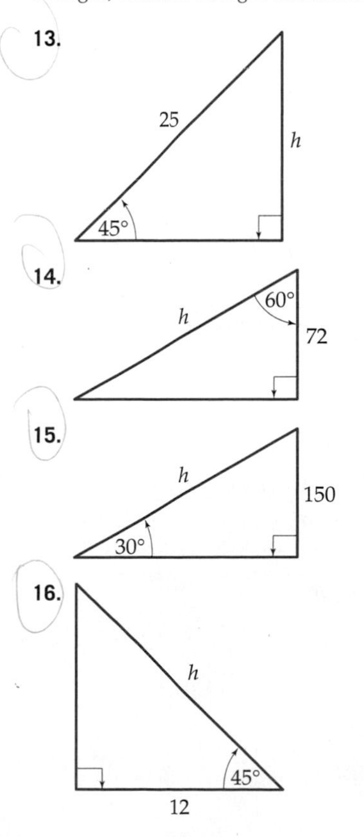

17.

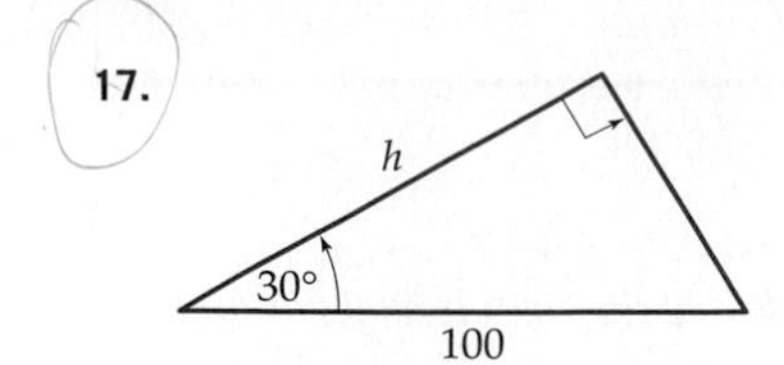

18.

In Exercises 19–22, find the required side without using a calculator.

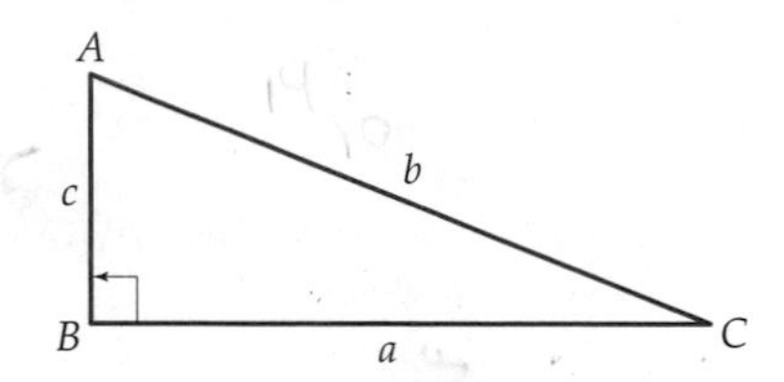

19. $a = 4$ and angle A measures $60°$; find c.

20. $c = 5$ and angle A measures $60°$; find a.

21. $c = 10$ and angle A measures $30°$; find a.

22. $a = 12$ and angle A measures $30°$; find c.

In Exercises 23–28, use the figure for Exercises 19–22. [See Example 5.]

23. If $b = 10$ and $\measuredangle\, C = 50°$, find a and c.

24. If $c = 12$ and $\measuredangle\, C = 37°$, find a and b.

25. If $a = 6$ and $\measuredangle\, A = 14°$, find b and c.

26. If $a = 8$ and $\measuredangle\, A = 40°$, find b and c.

27. If $c = 5$ and $\measuredangle\, A = 65°$, find a and b.

28. If $c = 4$ and $\measuredangle\, C = 28°$, find a and b.

In Exercises 29–32, find angle θ.

29.

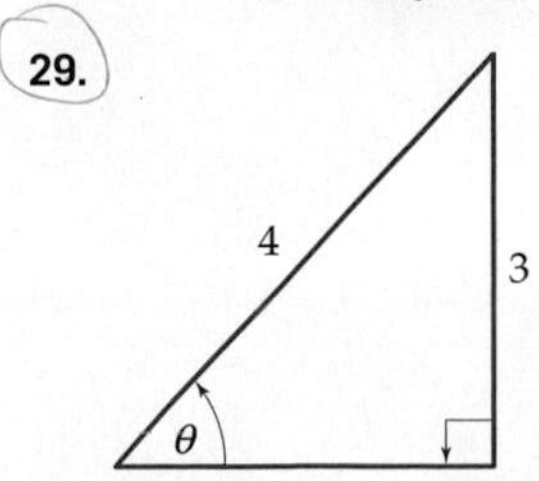

30.

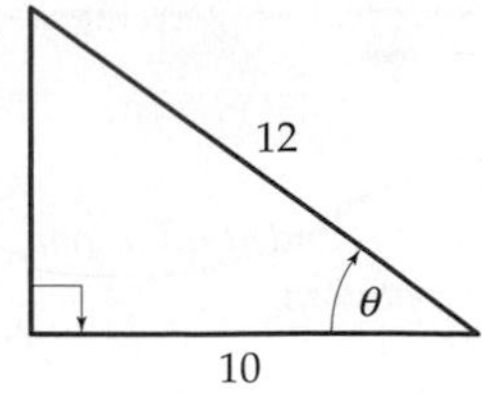

31.

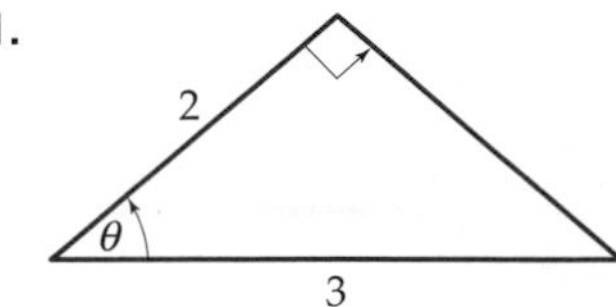

32.

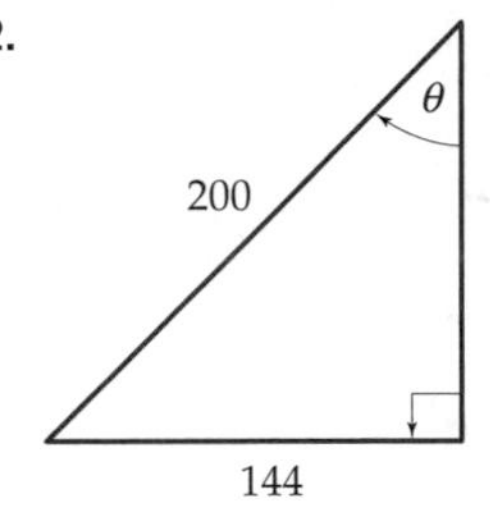

In Exercises 33–38, use the figure for Exercises 19–22 to find angles A and C under the given conditions. [See Example 7.]

33. $a = 4$ and $c = 6$

34. $b = 14$ and $c = 5$

35. $a = 7$ and $b = 10$

36. $a = 5$ and $c = 3$

37. $b = 18$ and $c = 12$

38. $a = 4$ and $b = 9$

39. Let θ be an acute angle with sides a and b in a triangle, as in the figure below.
(a) Find the area of the triangle (in terms of h and a).
(b) Find $\sin \theta$.
(c) Use part (b) to show that $h = b \sin \theta$.
(d) Use parts (a) and (c) to show that the area A of a triangle in which an acute angle θ has sides a and b is

$$A = \frac{1}{2}ab \sin \theta.$$

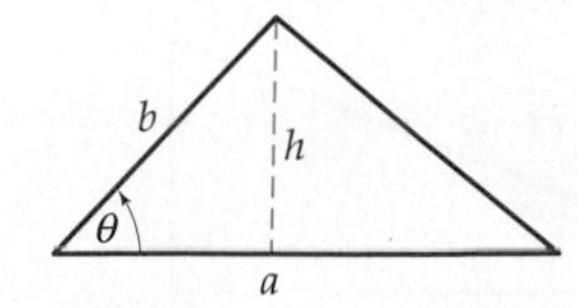

In Exercises 40–44, use the result of Exercise 39 to find the area of the given triangle.

40.

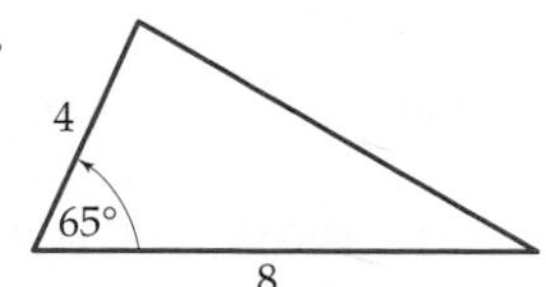

41.

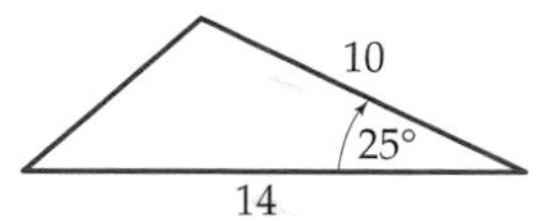

42.

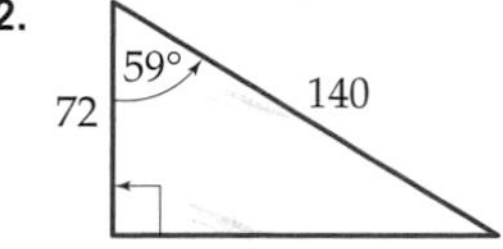

43.

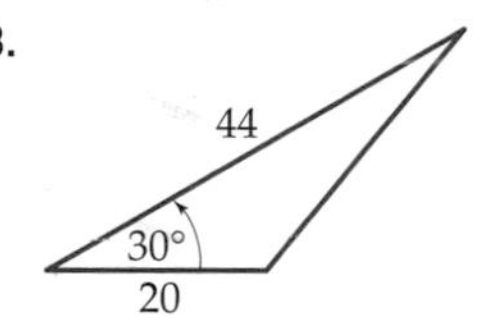

44.

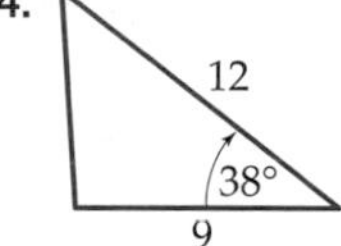

Exercises 45–52, develop some interconnections among the trigonometric functions. Some of these will be explored in greater detail in Chapter 2.

45. Let θ be any acute angle, as shown in the figure.

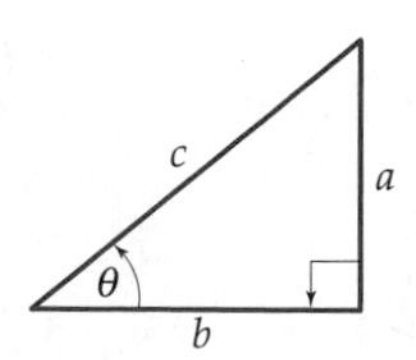

(a) Express c in terms of a and b. [*Hint:* Think Pythagorean.]
(b) Find $\sin\theta$ and $\cos\theta$.
(c) Use parts (a) and (b) to show that for any acute angle θ,

$$(\sin\theta)^2 + (\cos\theta)^2 = 1.$$

This equation is called the *Pythagorean identity* for sine and cosine.

46. Use the figure for Exercise 45 to show that for any acute angle θ,

$$\tan\theta = \frac{\sin\theta}{\cos\theta}.$$

In Exercises 47–50, use the results of Exercises 45–46 to find the exact values of sin θ, cos θ, and tan θ under the given circumstances.

47. $\sin\theta = 3/5$

48. $\sin\theta = 7/12$

49. $\cos\theta = 3/4$

50. $\cos\theta = 3/7$

51. Let θ and α be acute angles of a right triangle, as shown in the figure.

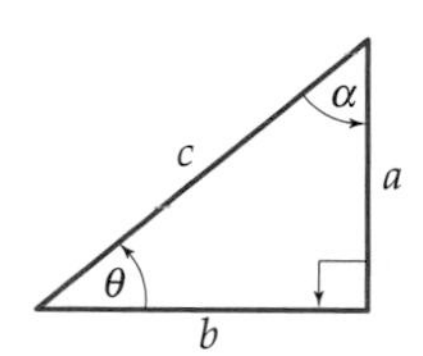

(a) Find $\sin\theta$ and $\cos\alpha$.
(b) Explain why $\theta + \alpha = 90°$.
(c) Use parts (a) and (b) to conclude that for any acute angle θ,

$$\cos(90° - \theta) = \sin\theta.$$

52. (a) Using the figure for Exercise 51, find $\cos\theta$ and $\sin\alpha$.
(b) Use part (a) and part (b) of Exercise 51 to show that for any acute angle θ,

$$\sin(90° - \theta) = \cos\theta.$$

This equation and the one in Exercise 51(c) are called *cofunction identities.*

1.1 *ALTERNATE* Trigonometric Functions of Angles

> **NOTE**
> If you have not read Chapter 2, omit this section. If you have read Chapter 2, use this section in place of Section 1.1.

Trigonometric functions are defined in Chapter 2 as functions whose domains consist of real numbers. In the classical approach, however, the domains of the trigonometric functions consist of *angles*. In other words, instead of starting with a number t and then moving to an angle of t radians, we start directly with the angle, as summarized here.

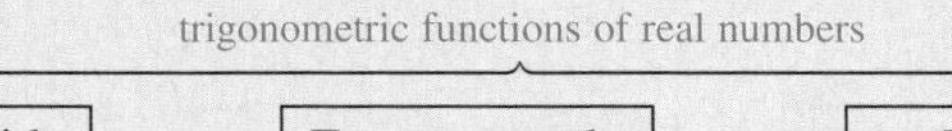

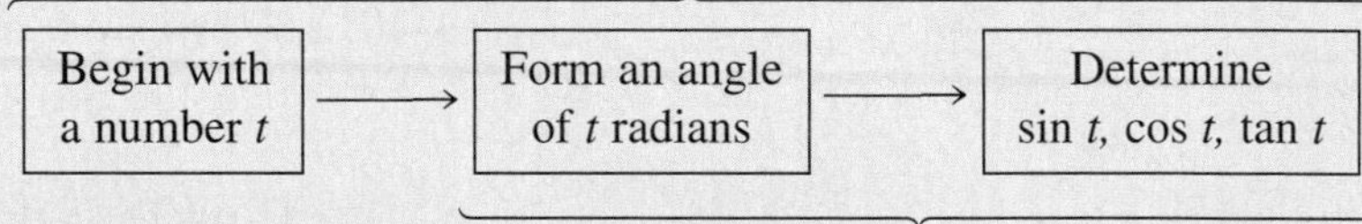

In this chapter (and hereafter, whenever convenient), we shall take this classical approach and begin with angles. From there on, everything is essentially the same. The *values* of the trigonometric functions are still *numbers* and are obtained as before. For example, the point-in-the-plane description now reads as follows.

Point-in-the-Plane Description

Let θ be an angle in standard position and let (x, y) be any point (except the origin) on the terminal side of θ. Let $r = \sqrt{x^2 + y^2}$. Then, the values of the six trigonometric functions of the angle θ are given by

$$\sin\theta = \frac{y}{r} \qquad \cos\theta = \frac{x}{r} \qquad \tan\theta = \frac{y}{x}$$

$$\csc\theta = \frac{r}{y} \qquad \sec\theta = \frac{r}{x} \qquad \cot\theta = \frac{x}{y}.$$

EXAMPLE A

Evaluate the six trigonometric functions at the angle θ shown in Figure 1–15.

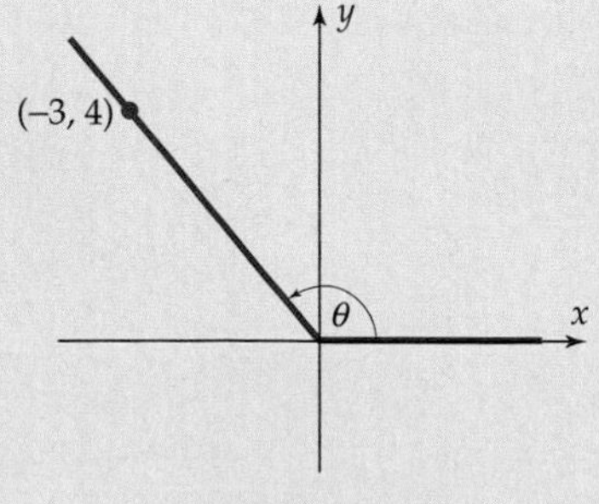

Figure 1–15

SOLUTION We use $(-3, 4)$ as the point (x, y), so

$$r = \sqrt{x^2 + y^2} = \sqrt{9 + 16} = \sqrt{25} = 5.$$

Thus,

$$\sin\theta = \frac{y}{r} = \frac{4}{5} \qquad \cos\theta = \frac{x}{r} = \frac{-3}{5} \qquad \tan\theta = \frac{y}{x} = \frac{4}{-3}$$

$$\csc\theta = \frac{r}{y} = \frac{5}{4} \qquad \sec\theta = \frac{r}{x} = \frac{5}{-3} \qquad \cot\theta = \frac{x}{y} = \frac{-3}{4}.$$

■

DEGREES AND RADIANS

Angles can be measured in either degrees or radians. If radian measure is used (as is the case in Chapter 2), then everything is the same as before. For example, sin 30 denotes the sine of an angle of 30 radians.

But when angles are measured in degrees (as will be done in the rest of this chapter), new notation is needed. To denote the value of the sine function at an angle of 30 *degrees,* we write

$$\sin 30^\circ \qquad \text{[note the degree symbol]}$$

The degree symbol here is essential for avoiding error.

EXAMPLE B

Since an angle of 30° is the same as one of $\pi/6$ radians, sin 30° is the same number as $\sin \pi/6$. Hence,

$$\sin 30^\circ = \sin \pi/6 = 1/2.$$

This is *different* from sin 30 (the sine of an angle of 30 *radians*); a calculator in radian mode shows that $\sin 30 = -.988$. ■

RIGHT TRIANGLE DESCRIPTION OF TRIGONOMETRIC FUNCTIONS

For angles between 0° and 90°, the trigonometric functions may be evaluated by using right triangles as follows. Suppose θ is an angle in a right triangle. Place the triangle so that angle θ is in standard position, with the hypotenuse as its terminal side, as shown in Figure 1–16.

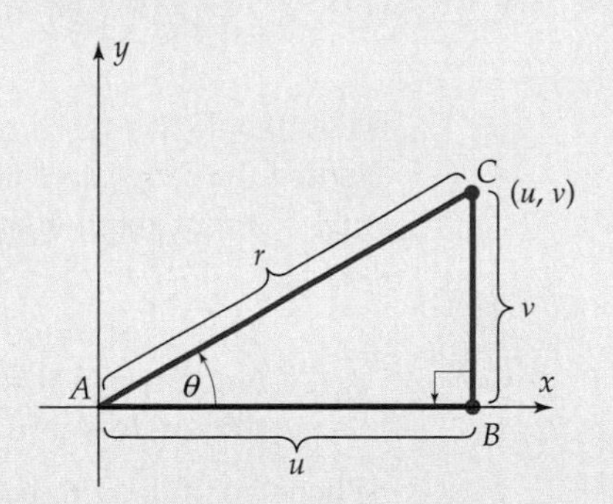

Figure 1–16

Denote the length of the side *AB* (the one *adjacent* to angle θ) by u and the length of side *BC* (the one *opposite* angle θ) by v. Then the coordinates of C are (u, v). Let r be the length of the *hypotenuse AC* (the distance from (u, v) to the origin). Then the point-in-the-plane description shows that

$$\sin \theta = \frac{v}{r} = \frac{\text{length of opposite side}}{\text{length of hypotenuse}} \qquad \cos \theta = \frac{u}{r} = \frac{\text{length of adjacent side}}{\text{length of hypotenuse}}$$

$$\tan \theta = \frac{v}{u} = \frac{\text{length of opposite side}}{\text{length of adjacent side}},$$

and similarly for the other trigonometric functions. These facts can be summarized as follows.

Right Triangle Description

Consider a right triangle containing an acute angle θ.

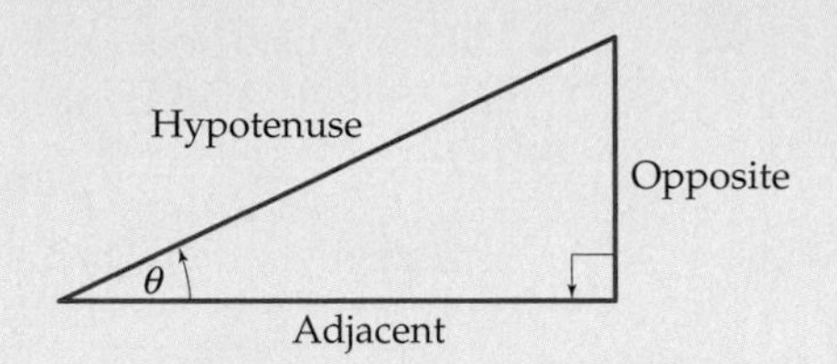

The values of the six trigonometric functions of the angle θ are given by

$$\sin\theta = \frac{\text{opposite}}{\text{hypotenuse}} \qquad \cos\theta = \frac{\text{adjacent}}{\text{hypotenuse}} \qquad \tan\theta = \frac{\text{opposite}}{\text{adjacent}}$$

$$\csc\theta = \frac{\text{hypotenuse}}{\text{opposite}} \qquad \sec\theta = \frac{\text{hypotenuse}}{\text{adjacent}} \qquad \cot\theta = \frac{\text{adjacent}}{\text{opposite}}.$$

To finish the discussion of trigonometric functions of angles, you should now go to page 82 of Section 1.1 and begin reading at Example 1.

1.2 Applications of Right Triangle Trigonometry

Many applications of trigonometry involve **"solving a triangle."** This means finding the lengths of all three sides and the measures of all three angles when only some of these quantities are given. Solving right triangles depends on this fact.

> **The right angle description of a trigonometric function (such as $\sin\theta$ = opposite/hypotenuse) relates three quantities: the angle θ and two sides of the right triangle.**

When two of these three quantities are known, then the third can always be found.

EXAMPLE 1

Solve the right triangle in Figure 1–17.

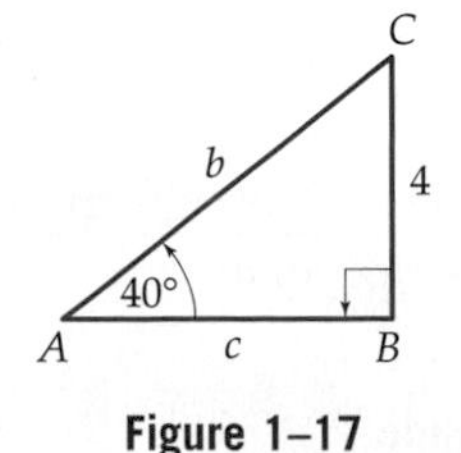

Figure 1–17

SOLUTION We must find the measure of $\measuredangle C$ and the lengths of sides b and c. Since the sum of the angles of a triangle is 180°, we have

$$40^\circ + 90^\circ + \measuredangle C = 180^\circ$$

$$\measuredangle C = 180^\circ - 40^\circ - 90^\circ = 50^\circ.$$

Furthermore, Figure 1–17 shows that

$$\sin A = \frac{4}{b}$$

Since $A = 40°$: $$\sin 40° = \frac{4}{b}$$

Multiply both sides by b: $$b \sin 40° = 4$$

Divide both sides by sin 40°: $$b = \frac{4}{\sin 40°} \approx 6.22.$$

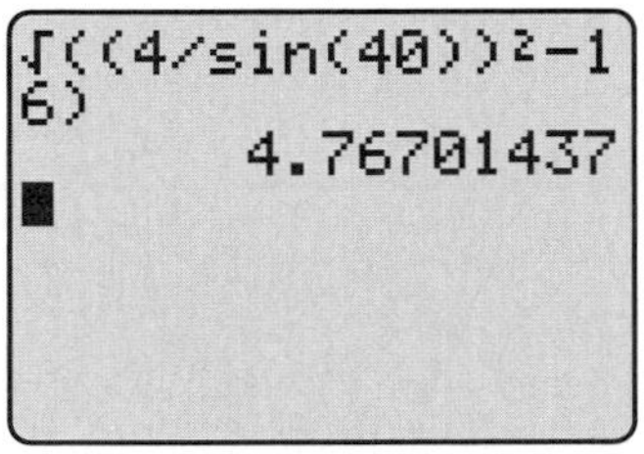

```
√((4/sin(40))²-1
6)
      4.76701437
```

Figure 1–18

Now side c can be found by using the Pythagorean Theorem.

$$4^2 + c^2 = b^2$$

$$c^2 = b^2 - 16$$

$$c = \sqrt{b^2 - 16} = \sqrt{\left(\frac{4}{\sin 40°}\right)^2 - 16}$$

A calculator shows that $c \approx 4.77$ (Figure 1–18). ■

CALCULATOR EXPLORATION

In Example 1, use the approximation $b \approx 6.22$ and the Pythagorean Theorem to find c. Is your answer the same as the length of c found in Figure 1–18? Why not? The moral here is: Don't use approximations in intermediate steps if you can avoid it. However, rounding your final answer is usually appropriate.

APPLICATIONS

The following examples illustrate a variety of practical applications of triangle trigonometry.

EXAMPLE 2

Bob and Pat Burger see a tree on the river's edge directly opposite them. They walk along the riverbank for 120 feet and note that the angle formed by their path and a line to the tree measures 70°, as indicated in Figure 1–19. How wide is the river?

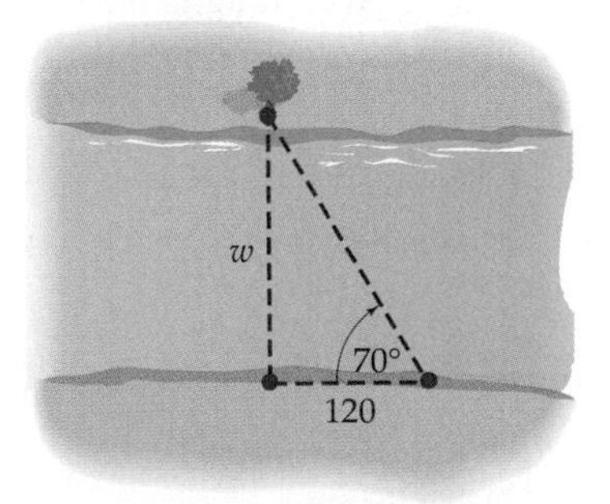

Figure 1–19

SOLUTION Using the right triangle whose one leg is the width w of the river and whose other leg is the 120-foot path on this side of the river, we see that

$$\tan 70^\circ = \frac{\text{opposite}}{\text{adjacent}}$$

$$\tan 70^\circ = \frac{w}{120}$$

Multiply both sides by 120: $$w = 120 \tan 70^\circ \approx 329.7.$$

So the river is about 330 feet wide. ■

EXAMPLE 3

A plane takes off, making an angle of 18° with the ground. After the plane travels three miles along this flight path, how high (in feet) is it above the ground?

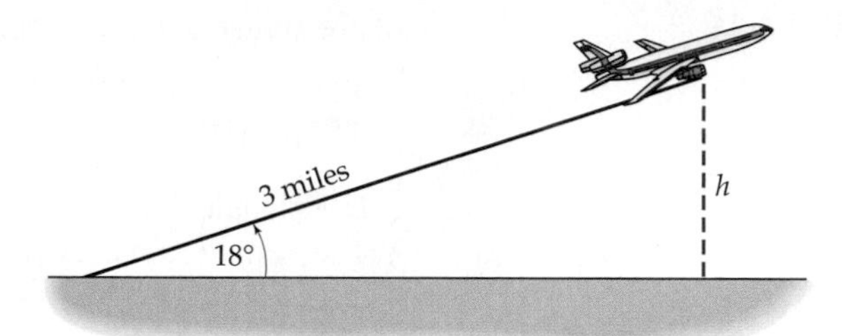

Figure 1–20

SOLUTION Figure 1–20 shows that

$$\sin 18^\circ = \frac{\text{opposite}}{\text{hypotenuse}} = \frac{h}{3}$$

Multiply both sides by 3: $$h = 3 \sin 18^\circ \approx .92705 \text{ miles.}$$

Since there are 5280 feet in a mile, the height of the plane in feet is

$$h = .92705 \cdot 5280 \approx 4894.8 \text{ feet.}$$ ■

EXAMPLE 4 

According to the safety sticker on an extension ladder, the distance from the foot of the ladder to the base of the wall on which it leans should be one-fourth of the length of the ladder. If the ladder is in this position, what angle does it make with the ground?

SOLUTION Let c be the distance from the foot of the ladder to the base of the wall. Then the ladder's length is $4c$, as shown in Figure 1–21. If θ is the angle the ladder makes with the ground, then

$$\cos \theta = \frac{\text{adjacent}}{\text{hypotenuse}} = \frac{c}{4c} = \frac{1}{4}.$$

Using the COS^{-1} key, we find that $\theta \approx 75.52^\circ$, as shown in Figure 1–22. ■

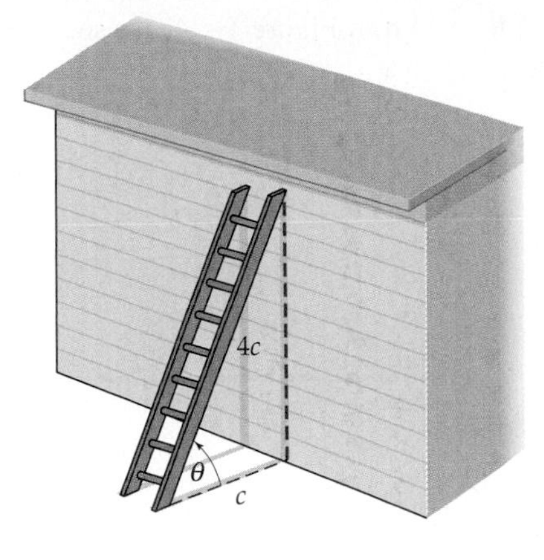

Figure 1–21

```
cos⁻¹(1/4)
        75.52248781
```

Figure 1–22

In many practical applications, one uses the angle between the horizontal and some other line (for instance, the line of sight from an observer to a distant object). This angle is called the **angle of elevation** or the **angle of depression,** depending on whether the line is above or below the horizontal, as shown in Figure 1–23.

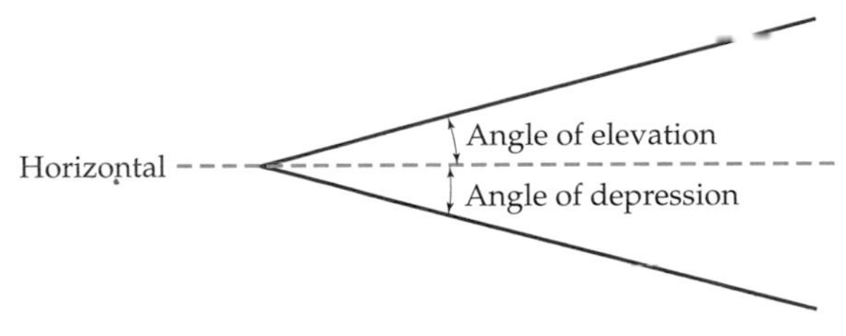

Figure 1–23

EXAMPLE 5 

A surveyor stands on one edge of a ravine. By using the method in Example 2, she determines that the ravine is 125 feet wide. She then determines that the angle of depression from the edge where she is standing to a point on the bottom of the ravine is 57.5°, as shown in Figure 1–24 (which is not to scale). How deep is the ravine?

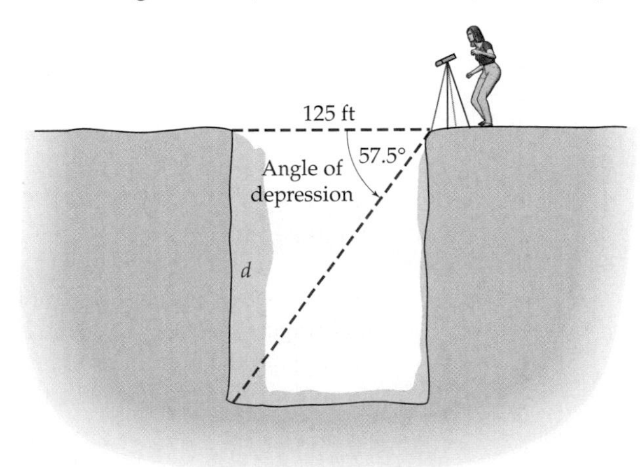

Figure 1–24

SOLUTION From Figure 1–24, we see that

$$\tan 57.5° = \frac{\text{opposite}}{\text{adjacent}} = \frac{d}{125}$$

Multiply both sides by 125: $$d = 125 \tan 57.5°$$

$$d \approx 196.21 \text{ feet} \quad ■$$

EXAMPLE 6

A wire is to be stretched from the top of a 10-meter-high building to a point on the ground. From the top of the building, the angle of depression to the ground point is 22°. How long must the wire be?

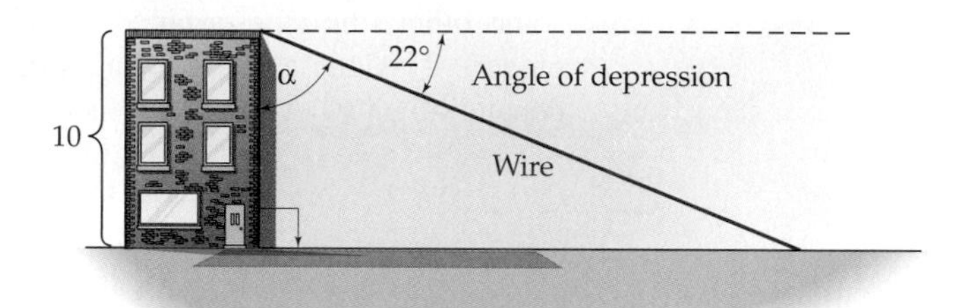

Figure 1–25

SOLUTION Figure 1–25 shows that the sum of the angle of depression and the angle α is 90°. Hence, α measures $90° - 22° = 68°$. We know the length of the side of the triangle adjacent to the angle α and must find the hypotenuse w (the length of the wire). Using the cosine function, we see that

$$\cos 68° = \frac{\text{adjacent}}{\text{hypotenuse}} = \frac{10}{w}$$

Multiply both sides by w: $$(\cos 68°)w = 10$$

Divide both sides by cos 68°: $$w = \frac{10}{\cos 68°} \approx 26.7 \text{ meters} \quad ■$$

EXAMPLE 7

A large American flag flies from a pole on top of the Terminal Tower in Cleveland (Figure 1–26). At a point 203 feet from the base of the tower, the angle of elevation to the bottom of the flag pole is 74°, and the angle of elevation to the top of the pole is 75.285°. To the nearest foot, how long is the flagpole?

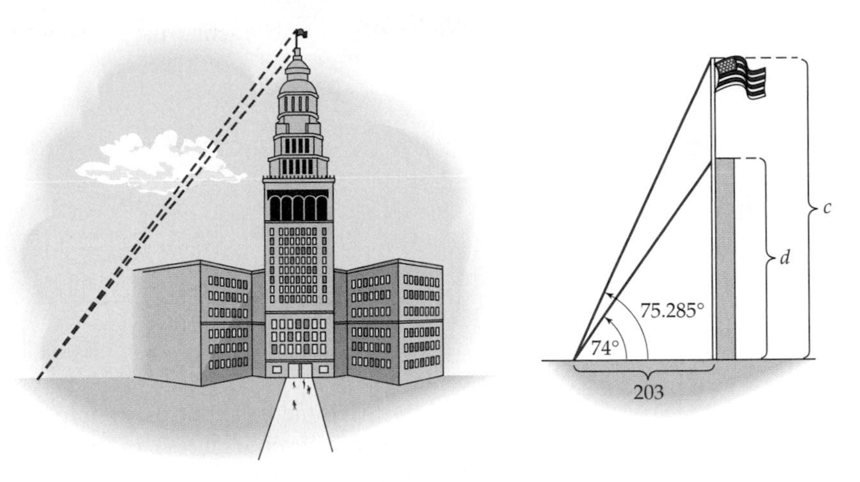

Figure 1–26 **Figure 1–27**

SOLUTION By abstracting the given information, we see that there are two right triangles, as shown in Figure 1–27 (which is not to scale). The length of the flagpole is $c - d$. We can use the two triangles to find c and d.

Figure 1–28

Larger Triangle	**Smaller Triangle**
$\frac{c}{203} = \frac{\text{opposite}}{\text{adjacent}} = \tan 75.285°$	$\frac{d}{203} = \frac{\text{opposite}}{\text{adjacent}} = \tan 74°$
$c = 203 \tan 75.285° \approx 773$	$d = 203 \tan 74° \approx 708$

As shown in Figure 1–28, the length of the flagpole is

$$c - d \approx 773 - 708 = 65 \text{ feet.}$$ ■

EXAMPLE 8 

Phil Embree stands on the edge of one bank of a canal and observes a lamp post on the edge of the other bank of the canal. His eye level is 152 centimeters above the ground (approximately 5 feet). The angle of elevation from eye level to the top of the lamp post is 12°, and the angle of depression from eye level to the bottom of the lamp post is 7°, as shown in Figure 1–29. How wide is the canal? How high is the lamp post?

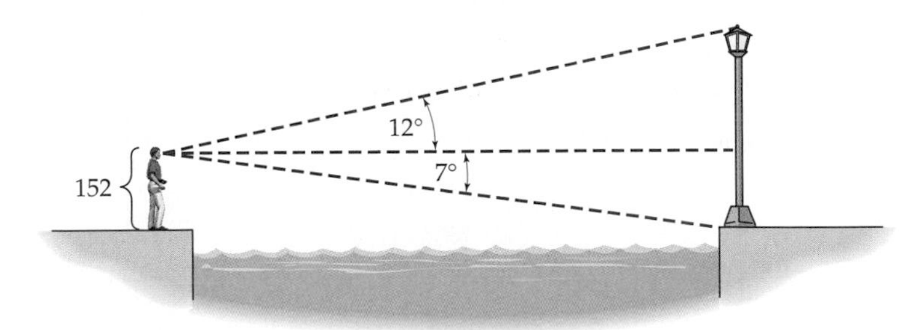

Figure 1–29

SOLUTION Abstracting the essential information, we obtain the diagram in Figure 1–30.

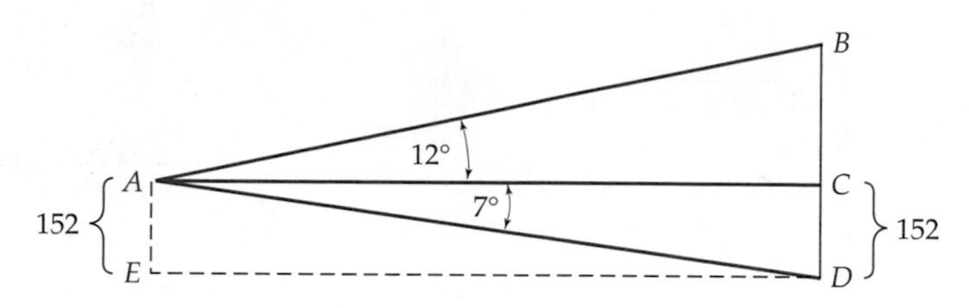

Figure 1–30

We must find the height of the lamp post BD and the width of the canal AC (or ED). The eye level height AE of the observer is 152 centimeters. Since AC and ED are parallel, CD also has length 152 centimeters. In right triangle ACD, we know the angle of 7° and the side CD opposite it. We must find the adjacent side AC. The tangent function is needed.

$$\tan 7^\circ = \frac{\text{opposite}}{\text{adjacent}} = \frac{152}{AC} \quad \text{or, equivalently,} \quad AC = \frac{152}{\tan 7^\circ}$$

$$AC = \frac{152}{\tan 7^\circ} \approx 1237.94 \text{ centimeters.}$$

So the canal is approximately 12.3794 meters* wide (about 40.6 feet). Now using right triangle ACB, we see that

$$\tan 12^\circ = \frac{\text{opposite}}{\text{adjacent}} = \frac{BC}{AC} \approx \frac{BC}{1237.94}$$

or, equivalently,

$$BC \approx 1237.94(\tan 12^\circ) \approx 263.13 \text{ centimeters.}$$

Therefore, the height of the lamp post BD is $BC + CD \approx 263.13 + 152 = 415.13$ centimeters or, equivalently, 4.1513 meters. ■

*Remember, 100 centimeters = 1 meter.

EXERCISES 1.2

In solving triangles here, all decimal approximations should be rounded to one decimal place at the end of the computation.

In Exercises 1–4, solve the right triangle.

1.

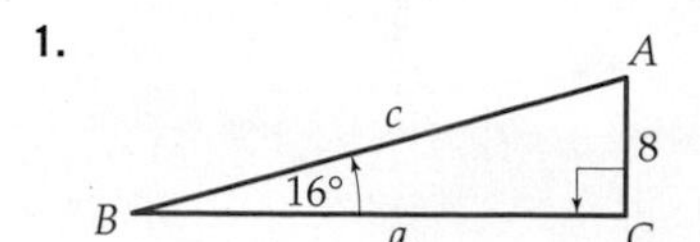

2.

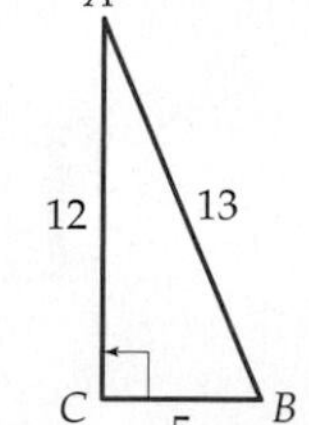

3.

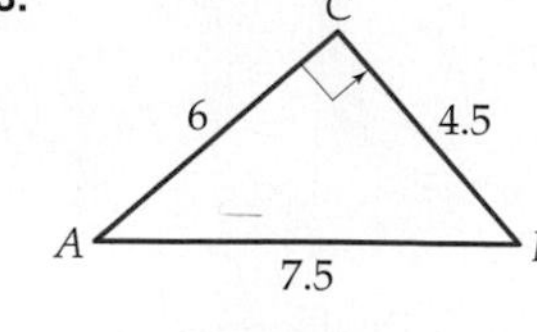

4.

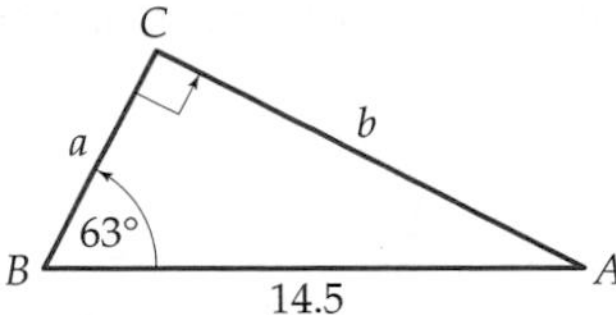

In Exercises 5–12, solve right triangle ABC under the given conditions.

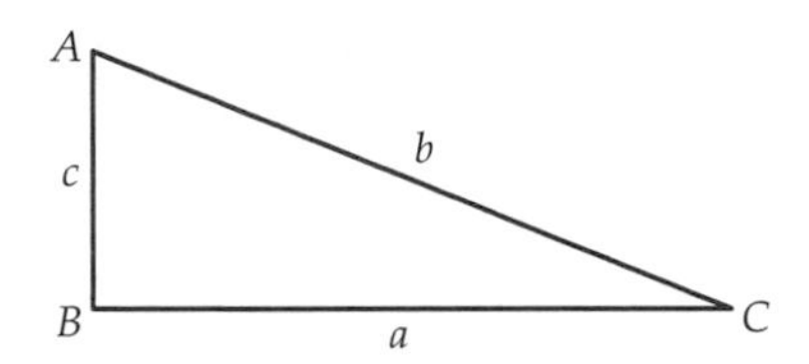

5. $a = 5$ and $\measuredangle A = 30°$ **6.** $b = 10$ and $\measuredangle C = 45°$

7. $b = 3.5$ and $\measuredangle A = 72°$ **8.** $a = 4.2$ and $\measuredangle C = 33°$

9. $a = 2.5$ and $c = 1.4$ **10.** $b = 3.7$ and $c = 2.2$

11. $a = 6$ and $b = 10$ **12.** $a = 12$ and $c = 4$

13. What is the width of the river in Example 2 if the angle to the tree is 40° (and all the other information is the same)?

14. Suppose the plane in Example 3 takes off at an angle of 5° and travels along this path for one mile. How high (in feet) is the plane above the ground?

15. Suppose you have a 24-foot-long ladder and you ignore the safety advice in Example 4 by placing the foot of the ladder 9 feet from the base of the wall. What angle does the ladder make with the ground?

16. The surveyor in Example 5 stands at the edge of another ravine, which is known to be 115 feet wide. She notes that the angle of depression from the edge she is standing on to the bottom of the oposite side is 64.3°. How deep is this ravine?

17. How long a wire is needed in Example 6 if the angle of depression is 25.8°?

18. Suppose that the flagpole on the Terminal Tower (Example 7) has been replaced. Now from a point 240 feet from the base of the tower, the angle of elevation to the bottom of the flagpole is 71.3°, and the angle of elevation to the top of the pole is 72.9°. To the nearest foot, how long is the new flagpole?

19. A 20-foot-long ladder leans on a wall of a building. The foot of the ladder makes an angle of 50° with the ground. How far above the ground does the top of the ladder touch the wall?

20. A pilot flying at an altitude of 14,500 feet notes that his angle of depression to the control tower of a nearby airport is 15°. If the plane continues flying at this altitude toward the control tower, how far must it travel before it is directly over the tower?

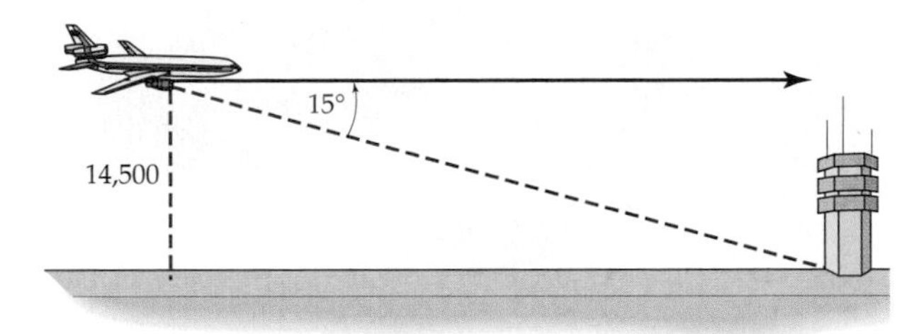

21. A straight road leads from an ocean beach into the nearby hills. The road has a constant upward grade of 3°. After taking this road for one mile, how high above sea level (in feet) are you?

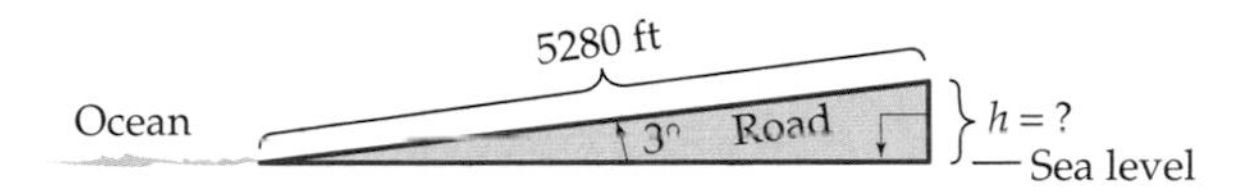

22. A wire from the top of a TV tower to the ground makes an angle of 49.5° with the ground and touches ground 225 feet from the base of the tower. How high is the tower?

23. A powerful searchlight projects a beam of light vertically upward so that it shines on the bottom of a cloud. A clinometer, 600 feet from the searchlight, measures the angle θ, as shown in the figure. If θ measures 80°, how high is the cloud?

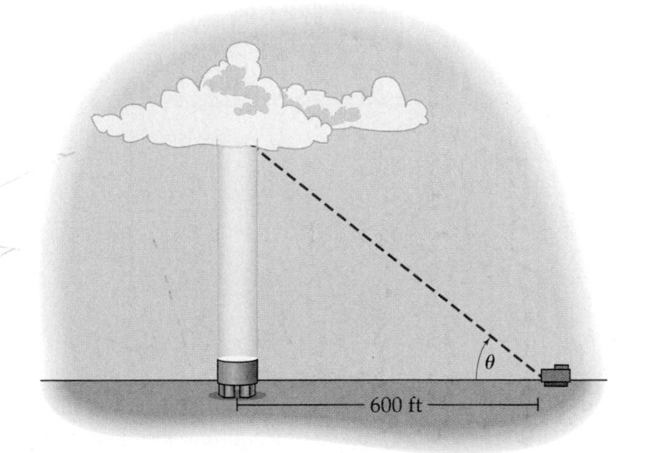

24. The Seattle Space Needle casts a 225-foot-long shadow. If the angle of elevation from the tip of the shadow to the top of the Space Needle is 69.6°, how high is the Space Needle?

25. Find the distance across the pond (from B to C) if AC is 110 feet and angle A measures 38°.

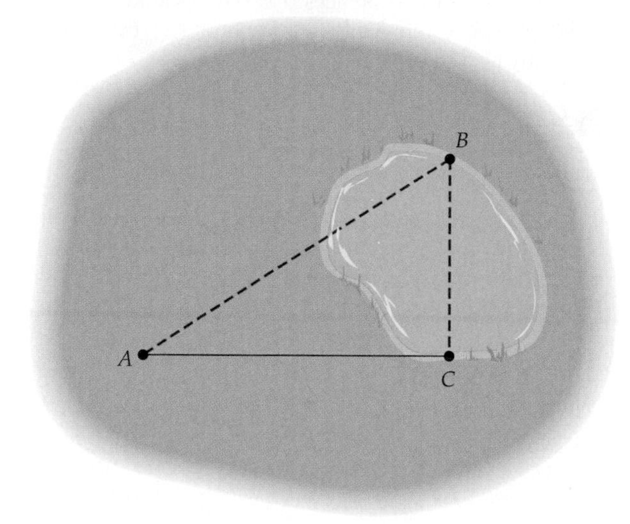

26. From the top of a 130-foot-high lighthouse, the angle of depression to a boat in Lake Erie is 2.5°. How far is the boat from the lighthouse?

27. Batman is on the edge of a 200-foot-deep chasm and wants to jump to the other side. A tree on the edge of the chasm is directly across from him. He walks 20 feet to his right and notes that the angle to the tree is 54°. His jet belt enables him to jump a maximum of 24 feet. How wide is the chasm, and is it safe for Batman to jump?

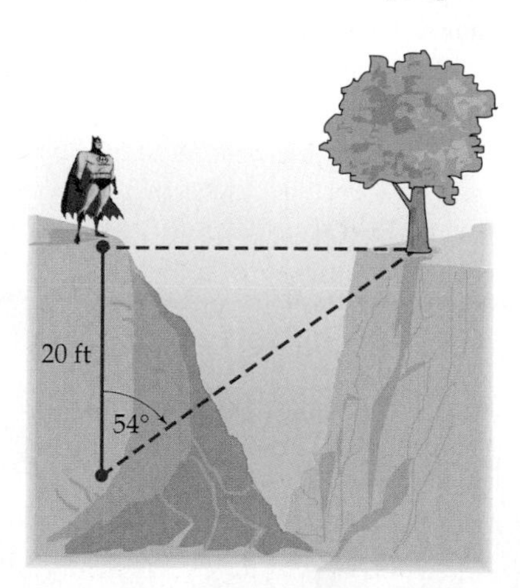

28. Alice is flying a kite. Her hand is three feet above ground level and is holding the end of a 300-foot-long kite string, which makes an angle of 57° with the horizontal. How high is the kite above the ground?

29. If you stand upright on a mountainside that makes a 62° angle with the horizontal and stretch your arm straight out at shoulder height, you may be able to touch the mountain (as shown in the figure). Can a person with an arm reach of 27 inches, whose shoulder is five feet above the ground, touch the mountain?

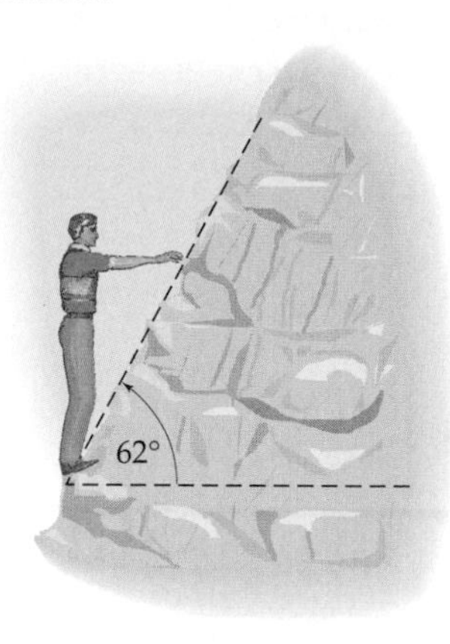

30. A swimming pool is three feet deep in the shallow end. The bottom of the pool has a steady downward drop of 12°. If the pool is 50 feet long, how deep is it at the deep end?

31. It is claimed that the Ohio Turnpike never has an uphill grade of more than 3°. How long must a straight uphill segment of the road be to allow a vertical rise of 450 feet?

32. A buoy in the ocean is observed from the top of a 40-meter-high radar tower on shore. The angle of depression from the top of the tower to the base of the buoy is 6.5°. How far is the buoy from the base of the radar tower?

33. Consider a 16-foot-long drawbridge on a medieval castle, as shown in the figure. The royal army is engaged in ignominious retreat. The king would like to raise the end of the drawbridge 8 feet off the ground so that Sir Rodney can jump onto the drawbridge and scramble into the castle while the enemy's cavalry are held at bay. Through how much of an angle must the drawbridge be raised for the end of it to be 8 feet off the ground?

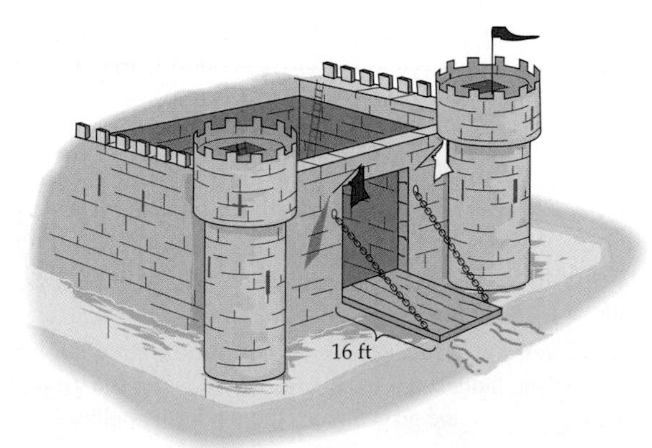

34. A 150-foot-long ramp connects a ground-level parking lot with the entrance of a building. If the entrance is 8 feet above the ground, what angle does the ramp make with the ground?

35. A plane flies a straight course. On the ground directly below the flight path, observers two miles apart spot the plane at the same time. The plane's angle of elevation is 46° from one observation point and 71° from the other. How high is the plane?

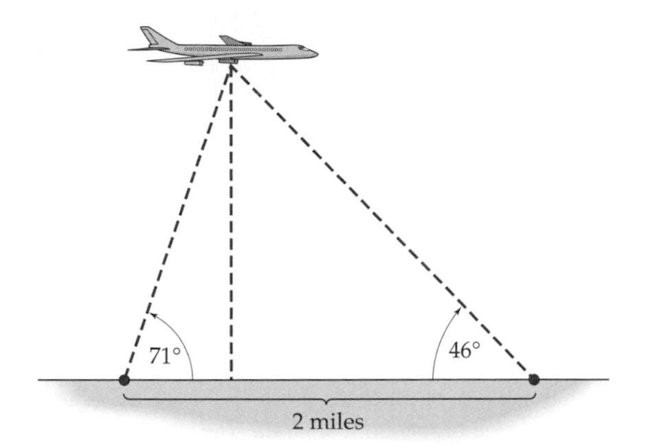

36. A man stands 12 feet from a statue. The angle of elevation from eye level to the top of the statue is 30°, and the angle of depression to the base of the statue is 15°. How tall is the statue?

37. Two boats lie on a straight line with the base of a lighthouse. From the top of the lighthouse (21 meters above water level), it is observed that the angle of depression of the nearer boat is 53° and the angle of depression of the farther boat is 27°. How far apart are the boats?

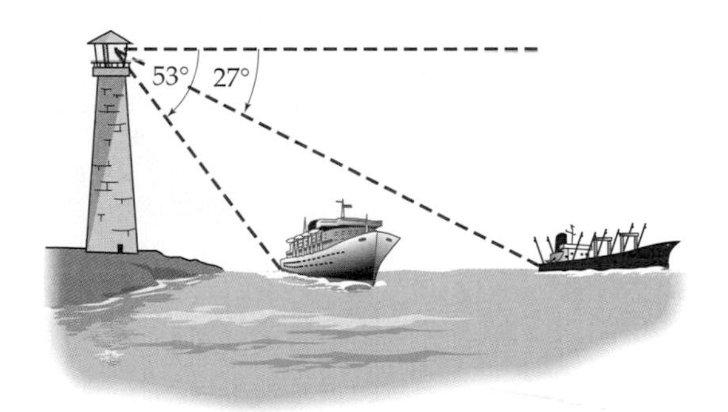

38. A rocket shoots straight up from the launchpad. Five seconds after liftoff, an observer two miles away notes that the rocket's angle of elevation is 3.5°. Four seconds later, the angle of elevation is 41°. How far did the rocket rise during those four seconds?

39. From a 35-meter-high window, the angle of depression to the top of a nearby streetlight is 55°. The angle of depression to the base of the streetlight is 57.8°. How high is the streetlight?

40. A plane takes off at an angle of 6° traveling at the rate of 200 feet/second. If it continues on this flight path at the same speed, how many minutes will it take to reach an altitude of 8000 feet?

41. A car on a straight road passes under a bridge. Two seconds later an observer on the bridge, 20 feet above the road, notes that the angle of depression to the car is 7.4°. How fast (in miles per hour) is the car traveling? [*Note:* 60 mph is equivalent to 88 feet/second.]

42. A plane passes directly over your head at an altitude of 500 feet. Two seconds later, you observe that its angle of elevation is 42°. How far did the plane travel during those two seconds?

43. A woman 5.5 feet tall stands 10 feet from a streetlight and casts a 4-foot-long shadow. How tall is the streetlight? What is angle θ?

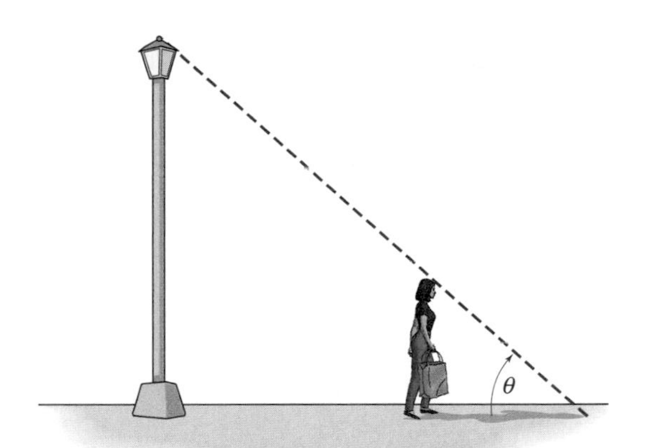

44. One plane flies straight east at an altitude of 31,000 feet. A second plane is flying west at an altitude of 14,000 feet on a course that lies directly below that of the first plane and directly above the straight road from Thomasville to Johnsburg. As the first plane passes over Thomasville, the second is passing over Johnsburg. At that instant, both planes spot a beacon next to the road between Thomasville to Johnsburg. The angle of depression from the first plane to the beacon is 61°, and the angle of depression from the second plane to the beacon is 34°. How far is Thomasville from Johnsburg?

45. A schematic diagram of a pedestrian overpass is shown in the figure. If you walk on the overpass from one end to the other, how far have you walked?

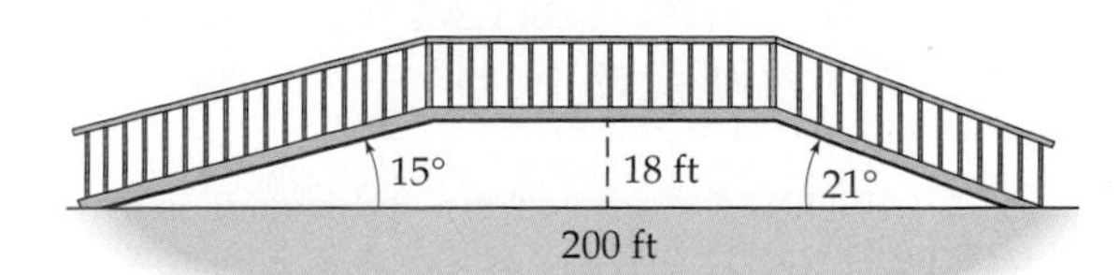

46. A 5-inch-high plastic beverage glass has a 2.5-inch-diameter base. Its sides slope outward at a 4° angle as shown. What is the diameter of the top of the glass?

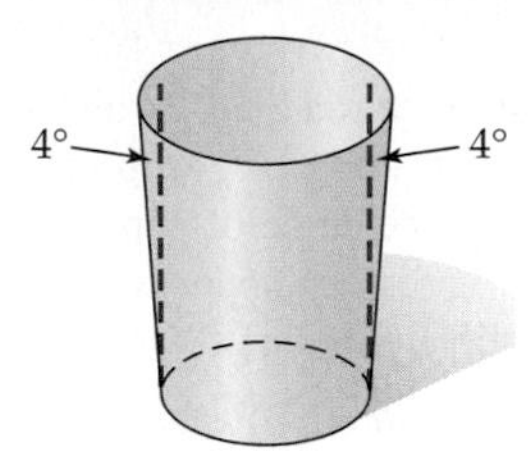

47. In aerial navigation, directions are given in degrees clockwise from north. Thus, east is 90°, south is 180°, and so on, as shown in the figure. A plane travels from an airport for 200 miles in the direction 300°. How far west of the airport is the plane then?

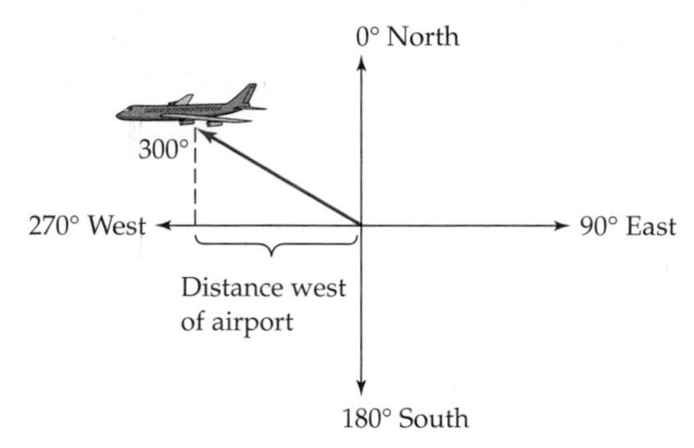

48. A plane travels at a constant 300 mph in the direction 65° (see Exercise 47).

(a) How far east of its starting point is the plane after half an hour?

(b) How far north of its starting point is the plane after 2 hours and 24 minutes?

49. A closed 60-foot-long drawbridge is 24 feet above water level. When open, the bridge makes an angle of 33° with the horizontal.

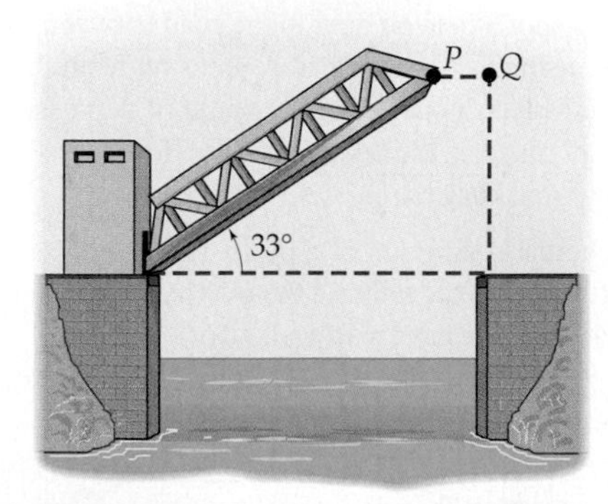

(a) How high is the tip P of the open bridge above the water?

(b) When the bridge is open, what is the distance from P to Q?

Thinkers

50. A gutter is to be made from a strip of metal 24 inches wide by bending up the sides to form a trapezoid.

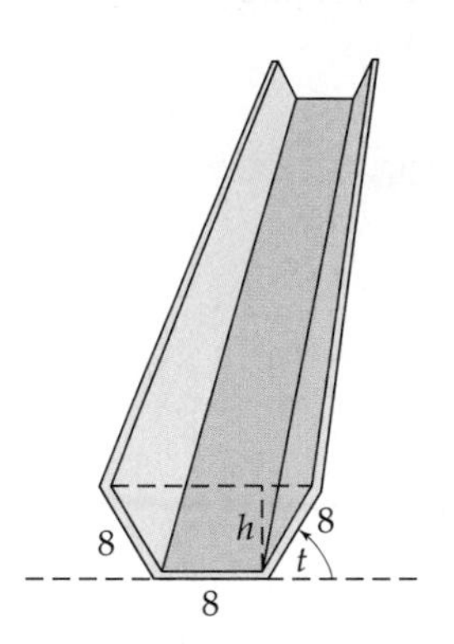

(a) Express the area of the cross section of the gutter as a function of the angle t. [*Hint:* The area of a trapezoid with bases b and b' and height h is $h(b + b')/2$.]

(b) For what value of t will this area be as large as possible?

51. The cross section of a tunnel is a semicircle with radius 10 meters. The interior walls of the tunnel form a rectangle.

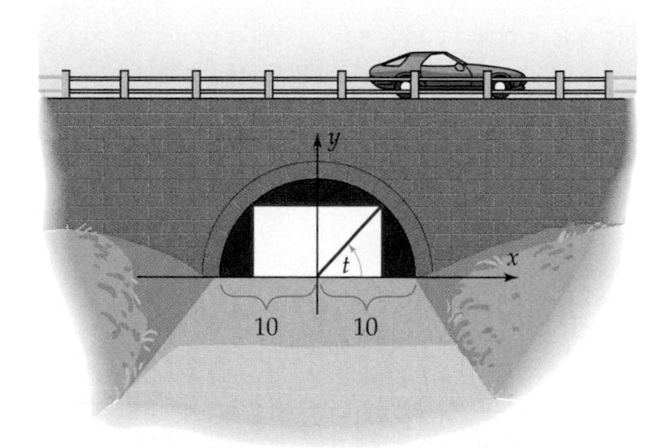

(a) Express the area of the rectangular cross section of the tunnel opening as a function of angle t.

(b) For what value of t is the cross-sectional area of the tunnel opening as large as possible? What are the dimensions of the tunnel opening in this case?

52. A 50-foot-high flagpole stands on top of a building. From a point on the ground, the angle of elevation of the top of the pole is 43°, and the angle of elevation of the bottom of the pole is 40°. How high is the building?

53. Two points on level ground are 500 meters apart. The angles of elevation from these points to the top of a nearby hill are 52° and 67°, respectively. The two points and the ground level point directly below the top of the hill lie on a straight line. How high is the hill?

1.3 The Law of Cosines

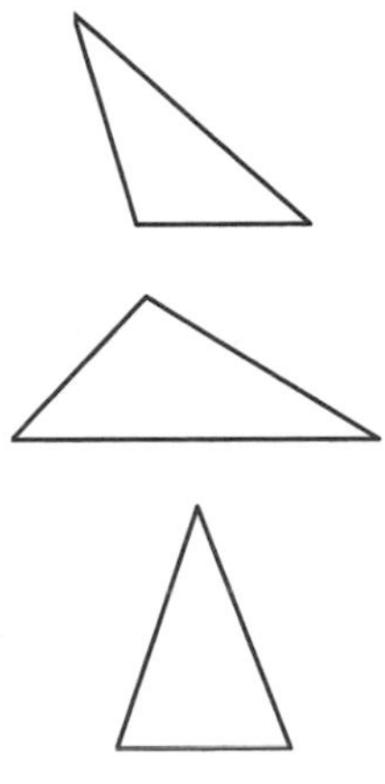

Figure 1–31

We have seen that right triangle trigonometry has many useful applications. The next step is to learn how to solve other types of triangles, such as those in Figure 1–31. To do this, we must find a description of the trigonometric functions that applies to all angles, not just to acute angles in a right triangle. First, we introduce some terminology.

An angle in the coordinate plane is said to be in **standard position** if its vertex is at the origin and one of its sides (which we call the **initial side**) is on the positive x-axis, as shown in Figure 1–32. The other side of the angle will be called its **terminal side.**

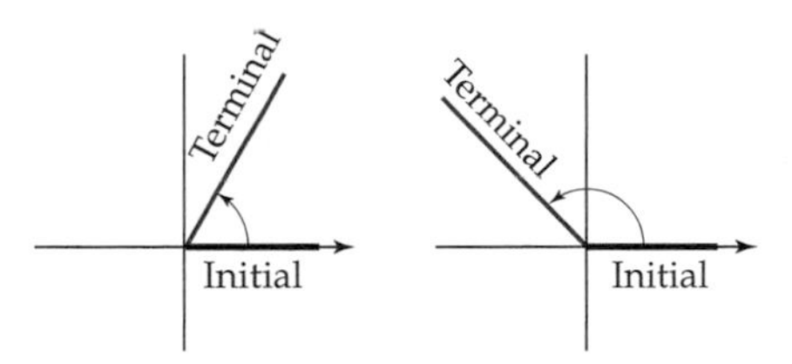

Figure 1–32

Next, we develop a new description of the trigonometric functions. Let θ be an acute angle in standard position. Choose any point (x, y) on the terminal side of θ (except the origin) and consider the right triangle with vertices $(0, 0)$, $(x, 0)$, and (x, y) shown in Figure 1–33.

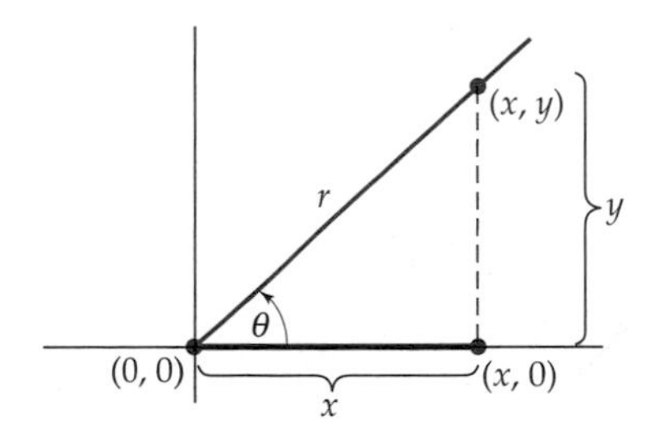

Figure 1–33

The legs of this triangle have lengths x and y, respectively. The distance formula shows that the hypotenuse has length

$$\sqrt{(x-0)^2 + (y-0)^2} = \sqrt{x^2 + y^2},$$

which we denote by r. The triangle shows that

$$\sin \theta = \frac{y}{r} \qquad \cos \theta = \frac{x}{r} \qquad \tan \theta = \frac{y}{x}.$$

We now have a description of the trigonometric functions in terms of the coordinate plane rather than right triangles. Furthermore, this description makes sense for *any* angle. Consequently, we make the following definition.

Point-in-the-Plane Description

> Let θ be an angle in standard position and (x, y) any point (except the origin) on its terminal side. Then the trigonometric functions are defined by these rules:
>
> $$\sin\theta = \frac{y}{r}, \qquad \cos\theta = \frac{x}{r}, \qquad \tan\theta = \frac{y}{x} \quad (x \neq 0),$$
>
> where $r = \sqrt{x^2 + y^2}$ is the distance from (x, y) to the origin.

The discussion preceding the box shows that when θ is an acute angle, these definitions produce the same numbers for $\sin\theta$, $\cos\theta$, and $\tan\theta$ as does the right triangle definition on page 82. It can be shown that the values of the trigonometric functions of θ are independent of the point that is chosen on the terminal side (just as the definition for acute angles is independent of the size of the right triangle).

EXAMPLE 1 

Find $\sin\theta$, $\cos\theta$, and $\tan\theta$ for the angle θ shown in Figure 1–34.

SOLUTION Since $(-5, 7)$ is on the terminal side of θ, we apply the definitions in the preceding box with $(x, y) = (-5, 7)$ and

$$r = \sqrt{x^2 + y^2} = \sqrt{(-5)^2 + 7^2} = \sqrt{74}.$$

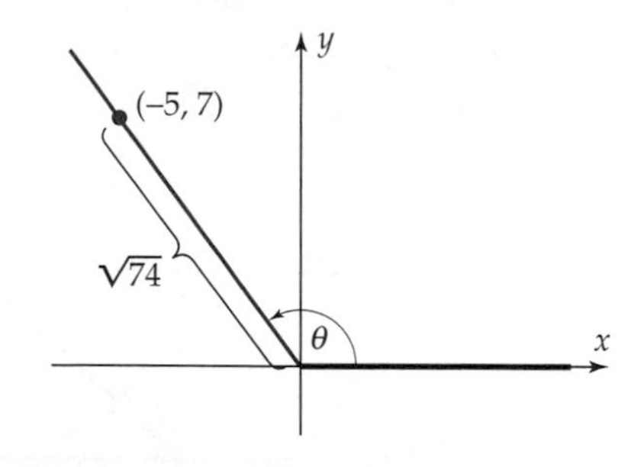

Figure 1–34

$$\sin\theta = \frac{y}{r} = \frac{7}{\sqrt{74}}$$

$$\cos\theta = \frac{x}{r} = \frac{-5}{\sqrt{74}}$$

$$\tan\theta = \frac{y}{x} = \frac{7}{-5} = -\frac{7}{5}$$

■

```
sin(125)
     .8191520443
cos(178)
     -.999390827
tan(98)
     -7.115369722
```

Figure 1–35

For most angles, we use a calculator to approximate the values of the trigonometric functions, as in Figure 1–35. But there are a few angles for which we can compute the exact values of the trigonometric functions.

EXAMPLE 2

Find the exact values of the trigonometric functions at $\theta = 90°$.

SOLUTION We use the point (0, 1) on the terminal side of an angle of 90° in standard position (Figure 1–36). In this case, $r = \sqrt{x^2 + y^2} = \sqrt{0^2 + 1^2} = 1$, so

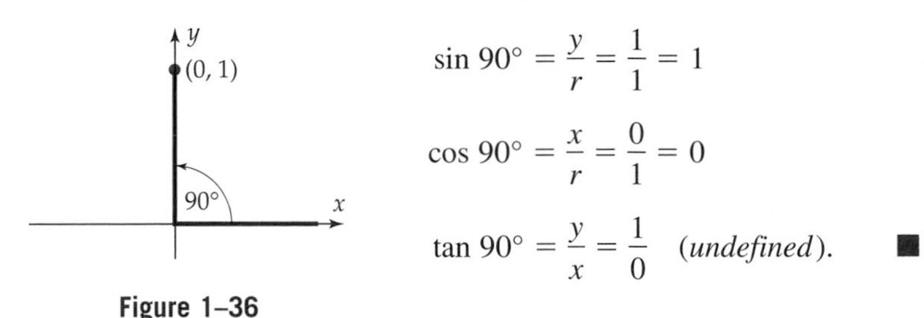

Figure 1–36

$$\sin 90° = \frac{y}{r} = \frac{1}{1} = 1$$

$$\cos 90° = \frac{x}{r} = \frac{0}{1} = 0$$

$$\tan 90° = \frac{y}{x} = \frac{1}{0} \quad (\textit{undefined}).$$ ■

The last part of Example 2 shows that the domain of the tangent function excludes 90°, whereas the domains of sine and cosine include all angles.

EXAMPLE 3

Find the exact values of sin 135°, cos 135°, and tan 135°.

SOLUTION Construct an angle of 135° in standard position and let P be the point on the terminal side that is 1 unit from the origin (Figure 1–37). Draw a vertical line from P to the x-axis, forming a right triangle with hypotenuse 1 and two angles of 45°. Each side of this triangle has length $\sqrt{2}/2$, as explained in Example 2 of the Geometry Review Appendix. Therefore, the coordinates of P are $(-\sqrt{2}/2, \sqrt{2}/2)$, and we have

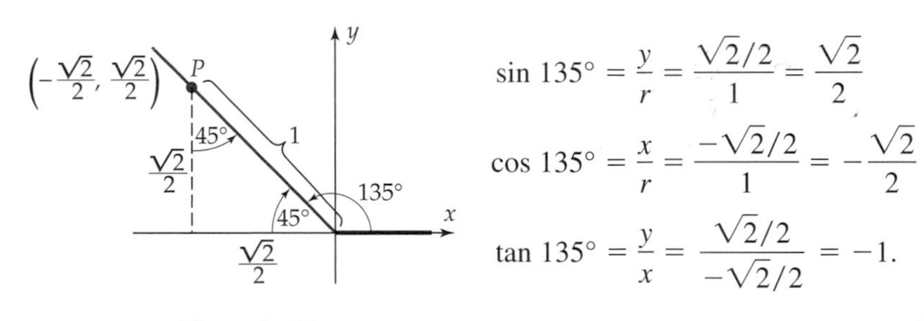

Figure 1–37

$$\sin 135° = \frac{y}{r} = \frac{\sqrt{2}/2}{1} = \frac{\sqrt{2}}{2}$$

$$\cos 135° = \frac{x}{r} = \frac{-\sqrt{2}/2}{1} = -\frac{\sqrt{2}}{2}$$

$$\tan 135° = \frac{y}{x} = \frac{\sqrt{2}/2}{-\sqrt{2}/2} = -1.$$ ■

THE LAW OF COSINES

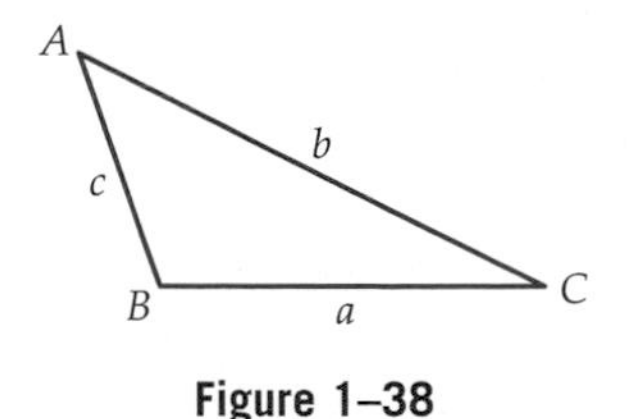

Figure 1–38

We now consider the solution of *oblique* triangles (ones that don't contain a right angle). We shall use **standard notation** for triangles: Each vertex is labeled with a capital letter, and the length of the side opposite that vertex is denoted by the same letter in lower case, as shown in Figure 1–38. The letter A will also be used to label the *angle* at vertex A and similarly for B and C. So we shall make statements such as $A = 37°$ or $\cos B = .326$.

The first fact needed to solve oblique triangles is the Law of Cosines, whose proof is given at the end of this section.

Law of Cosines

In any triangle ABC, with sides of lengths a, b, c, as in Figure 1–38,

$$a^2 = b^2 + c^2 - 2bc \cos A$$
$$b^2 = a^2 + c^2 - 2ac \cos B$$
$$c^2 = a^2 + b^2 - 2ab \cos C.$$

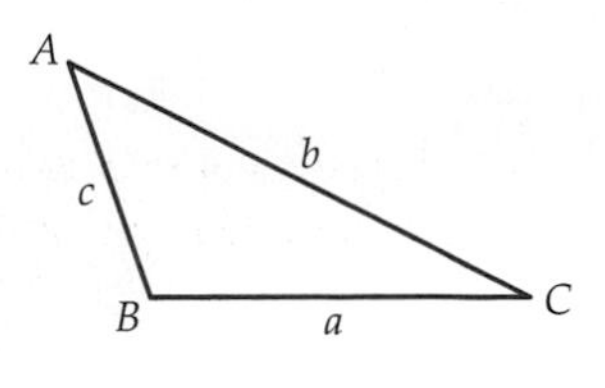

Figure 1–38

You need only memorize one of these equations, since each of them provides essentially the same information: a description of one side of a triangle in terms of the angle opposite it and the other two sides.

> NOTE. When C is a right angle, then c is the hypotenuse, and
>
> $$\cos C = \cos 90° = 0,$$
>
> so the third equation in the Law of Cosines becomes the Pythagorean Theorem:
>
> $$c^2 = a^2 + b^2.$$

Sometimes it is more convenient to use the Law of Cosines in a slightly different form. To do this, we solve the first equation in the Law of Cosines for $\cos A$.

$$a^2 = b^2 + c^2 - 2bc \cos A$$

Add $2bc \cos A$ to both sides: $$2bc \cos A + a^2 = b^2 + c^2$$

Subtract a^2 from both sides: $$2bc \cos A = b^2 + c^2 - a^2$$

Divide both sides by $2bc$: $$\cos A = \frac{b^2 + c^2 - a^2}{2bc}$$

So we have the following result.

Law of Cosines: Alternate Form

In any triangle ABC, with sides of lengths a, b, c, as in Figure 1–38,

$$\cos A = \frac{b^2 + c^2 - a^2}{2bc}.$$

The other two equations can be similarly rewritten. In this form, the Law of Cosines provides a description of each angle of a triangle in terms of the three sides. Consequently, the Law of Cosines can be used to solve triangles in these cases:

1. Two sides and the angle between them are known (SAS).
2. Three sides are known (SSS).

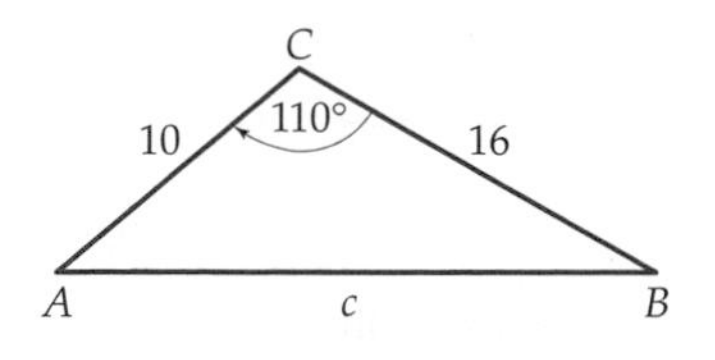

Figure 1–39

EXAMPLE 4  SAS

Solve triangle ABC in Figure 1–39.

SOLUTION We have $a = 16$, $b = 10$, and $C = 110°$. The right side of the third equation in the Law of Cosines involves only these known quantities. Hence,

$$\begin{aligned} c^2 &= a^2 + b^2 - 2ab \cos C \\ &= 16^2 + 10^2 - 2 \cdot 16 \cdot 10 \cos 110° \\ &\approx 256 + 100 - 320(-.342) \approx 465.4.^* \end{aligned}$$

Therefore, $c \approx \sqrt{465.4} \approx 21.6$. Now use the alternate form of the Law of Cosines.

$$\cos A = \frac{b^2 + c^2 - a^2}{2bc} \approx \frac{10^2 + (21.6)^2 - 16^2}{2 \cdot 10 \cdot 21.6} \approx .7172.$$

The COS^{-1} key (in degree mode) produces an angle with cosine .7172: $A \approx 44.2°$. Hence, $B \approx 180° - (44.2° + 110°) = 25.8°$. ■

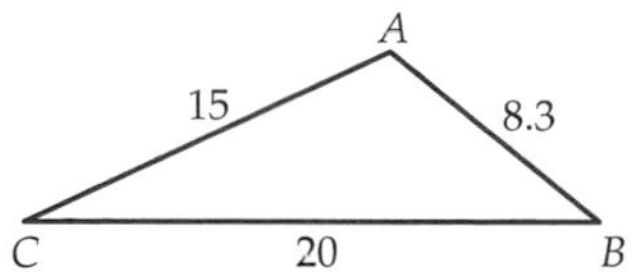

Figure 1–40

EXAMPLE 5 SSS

Find the angles of triangle ABC in Figure 1–40.

SOLUTION In this case, $a = 20$, $b = 15$, and $c = 8.3$. By the alternate form of the Law of Cosines,

$$\cos A = \frac{b^2 + c^2 - a^2}{2bc} = \frac{15^2 + 8.3^2 - 20^2}{2 \cdot 15 \cdot 8.3} = \frac{-106.11}{249} \approx -.4261.$$

The COS^{-1} key shows that $A \approx 115.2°$. Similarly, the alternate form of the Law of Cosines yields

$$\begin{aligned} \cos B &\approx \frac{a^2 + c^2 - b^2}{2ac} = \frac{20^2 + 8.3^2 - 15^2}{2 \cdot 20 \cdot 8.3} = \frac{243.89}{332} \approx .7346 \\ B &\approx 42.7°. \end{aligned}$$

Therefore, $C \approx 180° - (115.2° + 42.7°) = 180° - 157.9° = 22.1°$. ■

*Throughout this chapter, all decimals are printed in rounded-off form for reading convenience, but no rounding is done in the actual computation until the final answer is obtained.

APPLICATIONS

EXAMPLE 6

Two trains leave a station on different tracks. The tracks make an angle of 125° with the station as vertex. The first train travels at an average speed of 100 kilometers per hour, and the second travels at an average speed of 65 kilometers per hour. How far apart are the trains after two hours?

SOLUTION The first train A traveling at 100 kilometers per hour for two hours goes a distance of $100 \times 2 = 200$ kilometers. The second train B travels a distance of $65 \times 2 = 130$ kilometers. So we have the situation shown in Figure 1–41.

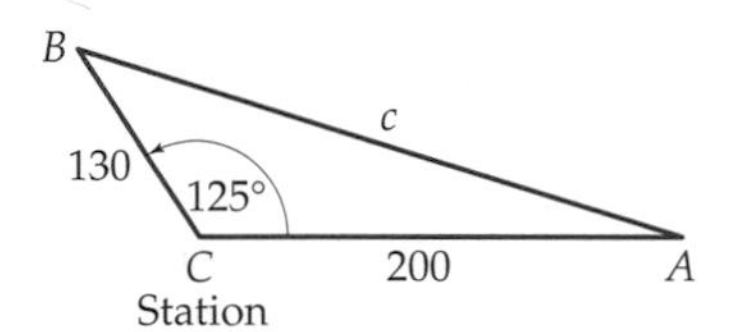

Figure 1–41

By the Law of Cosines,

$$\begin{aligned} c^2 &= a^2 + b^2 - 2ab\cos C \\ &= 130^2 + 200^2 - 2 \cdot 130 \cdot 200 \cos 125° \\ &= 56{,}900 - 52{,}000 \cos 125° \approx 86{,}725.97 \\ c &\approx \sqrt{86{,}725.97} = 294.5 \text{ kilometers.} \end{aligned}$$

The trains are 294.5 kilometers apart after two hours. ■

EXAMPLE 7

A small powerboat leaves Chicago and sails 35 miles due east on Lake Michigan. It then changes course 59° northward, heading for Grand Haven, Michigan, as shown in Figure 1–42. After traveling 60 miles on this course, how far is the boat from Chicago?

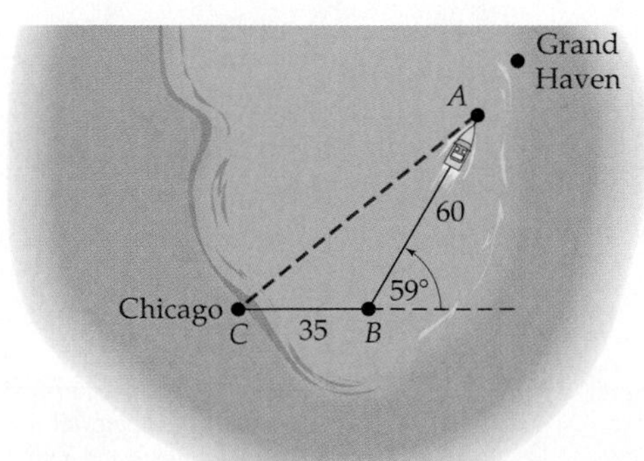

Figure 1–42

SOLUTION Figure 1–42 shows that

$$B + 59° = 180°$$

$$B = 180° - 59° = 121°.$$

We must find the side of the triangle opposite angle B. By the Law of Cosines,

$$b^2 = a^2 + c^2 - 2ac \cos B$$

$$b^2 = 35^2 + 60^2 - 2 \cdot 35 \cdot 60 \cos 121°$$

$$b^2 \approx 6988.1599$$

$$b \approx \sqrt{6988.1599} \approx 83.5952.$$

So the boat is about 83.6 miles from Chicago. ■

EXAMPLE 8

A sculpture is being placed in front of a new office building. The sculpture consists of two steel beams of lengths 10 and 12 feet, respectively, and a 15.2-foot cable, as shown in Figure 1–43. If the 10-foot beam makes an angle of 50° with the ground, what angle should the 12-foot beam make with the ground?

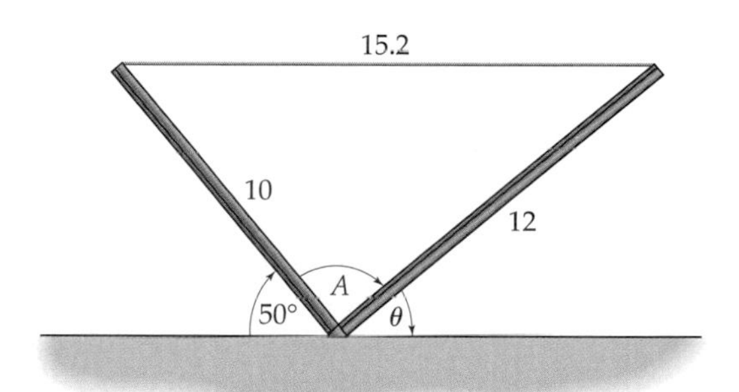

Figure 1–43

SOLUTION We must find the measure of angle θ. As you can see in Figure 1–43,

$$50° + A + \theta = 180°.$$

We first find the measure of angle A and then solve this equation for θ. The triangle formed by the beams and cable has sides of lengths 10, 12, and 15.2, with angle A opposite the 15.2-foot side. By the alternate form of the Law of Cosines,

$$\cos A = \frac{10^2 + 12^2 - 15.2^2}{2 \cdot 10 \cdot 12} = .054.$$

Figure 1–44

Hence, the measure of angle A is about 86.9° (Figure 1–44). Therefore,

$$50° + A + \theta = 180°$$

$$50° + 86.9° + \theta = 180°$$

$$\theta = 180° - 50° - 86.9° = 43.1°.$$ ■

EXAMPLE 9

A 100-foot-tall antenna tower is to be placed on a hillside that makes an angle of 12° with the horizontal. It is to be anchored by two cables from the top of the tower to points 85 feet uphill and 95 feet downhill from the base. How much cable is needed?

SOLUTION The situation is shown in Figure 1–45.

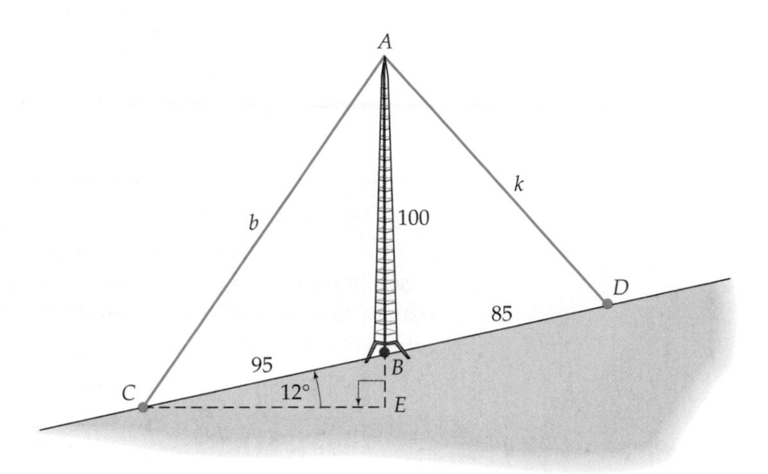

Figure 1–45

In triangle BEC, angle E is a right angle, and by hypothesis, angle C measures 12°. Since the sum of the angles of a triangle is 180°, we must have

$$\measuredangle CBE = 180° - (90° + 12°) = 78°.$$

As shown in the figure, the sum of angles CBE and CBA is a straight angle (180°). Hence,

$$\measuredangle CBA = 180° - 78° = 102°.$$

Applying the Law of Cosines to triangle ABC, we have

$$\begin{aligned} b^2 &= 95^2 + 100^2 - 2 \cdot 95 \cdot 100 \cos 102° \\ &= 9025 + 10{,}000 - 19{,}000 \cos 102° \\ &\approx 22{,}975.32. \end{aligned}$$

Therefore, the length of the downhill cable is $b \approx \sqrt{22{,}975.32} \approx 151.58$ feet.

To find the length of the uphill cable, note that the sum of angles CBA and DBA is a straight angle, so

$$\measuredangle DBA = 180° - \measuredangle CBA = 180° - 102° = 78°.$$

Applying the Law of Cosines to triangle DBA, we have

$$\begin{aligned} k^2 &= 85^2 + 100^2 - 2 \cdot 85 \cdot 100 \cos 78° \\ &= 7225 + 10{,}000 - 17{,}000 \cos 78° \approx 13{,}690.50. \end{aligned}$$

Hence, the length of the uphill cable is $k = \sqrt{13{,}690.50} \approx 117.01$ feet. ■

PROOF OF THE LAW OF COSINES

Given triangle ABC, position it on a coordinate plane so that angle A is in standard position with initial side c and terminal side b. Depending on the size of angle A, there are two possibilities, as shown in Figure 1–46.

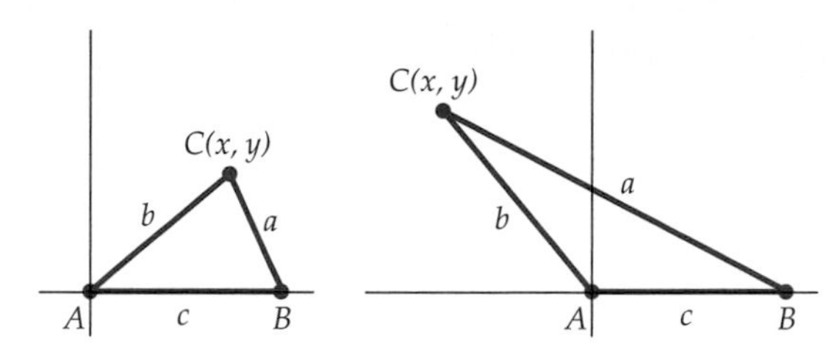

Figure 1–46

The coordinates of B are $(c, 0)$. Let (x, y) be the coordinates of C. Now C is a point on the terminal side of angle A, and the distance from C to the origin A is obviously b. Therefore, according to the point-in-the-plane description of sine and cosine, we have

$$\frac{x}{b} = \cos A \qquad \text{or, equivalently,} \qquad x = b\cos A$$

$$\frac{y}{b} = \sin A \qquad \text{or, equivalently,} \qquad y = b\sin A.$$

Using the distance formula on the coordinates of B and C, we have

$$\begin{aligned} a &= \text{distance from } C \text{ to } B \\ &= \sqrt{(x-c)^2 + (y-0)^2} = \sqrt{(b\cos A - c)^2 + (b\sin A - 0)^2}. \end{aligned}$$

Squaring both sides of this last equation and simplifying, using the Pythagorean identity (Exercise 64), yields

$$\begin{aligned} a^2 &= (b\cos A - c)^2 + (b\sin A)^2 \\ a^2 &= b^2\cos^2 A - 2bc\cos A + c^2 + b^2\sin^2 A \\ a^2 &= b^2(\sin^2 A + \cos^2 A) + c^2 - 2bc\cos A \\ a^2 &= b^2 + c^2 - 2bc\cos A. \end{aligned}$$

This proves the first equation in the Law of Cosines. Similar arguments beginning with angle B or C in standard position prove the other two equations.

EXERCISES 1.3

Standard notation for triangle ABC is used throughout. When solving triangles here, all decimal approximations should be rounded to one decimal place at the end of the computation.

In Exercises 1–4, find $\sin\theta$, $\cos\theta$, *and* $\tan\theta$.

1.

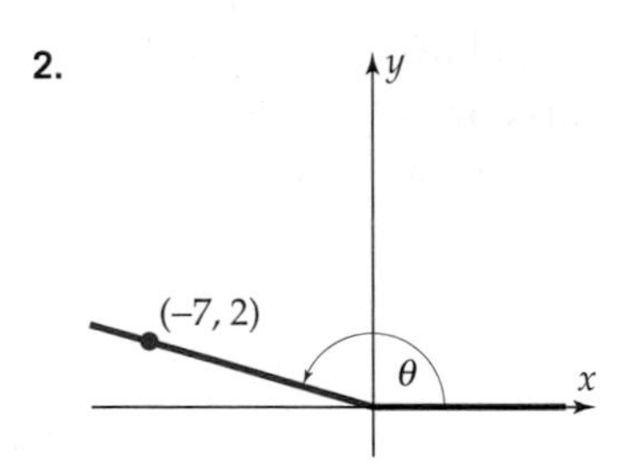

2.

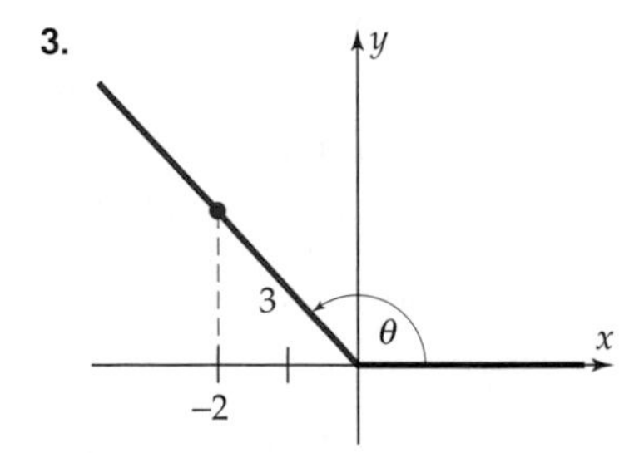

3.

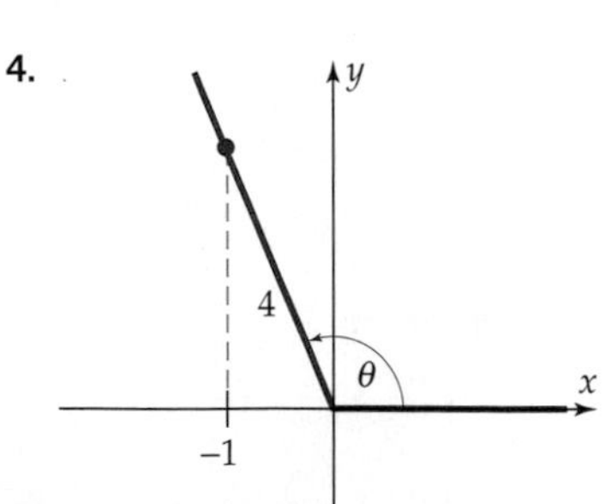

4.

In Exercises 5–10, find $\sin\theta$, $\cos\theta$, *and* $\tan\theta$, *where* θ *is an angle in standard position whose terminal side passes through the given point.*

5. $(2, 7)$ **6.** $(-3, 2)$ **7.** $(-5, 6)$

8. $(\sqrt{3}, 10)$ **9.** $(\sqrt{2}, \sqrt{3})$ **10.** $(-\pi, 2)$

In Exercises 11–14, use a calculator to find the required number.

11. $\sin 147°$ **12.** $\cos 100°$

13. $\tan 110°$ **14.** $\sin 160°$

In Exercises 15–18, assume that an angle of θ *degrees in standard position (with* $0° < \theta < 180°$*) has its terminal side on the straight line whose equation is given. Use the equation to find a point on the terminal side of* θ *and then find* $\sin\theta$, $\cos\theta$, *and* $\tan\theta$.

15. $4x - 2y = 0$ **16.** $3x - 4y = 0$

17. $2x + 5y = 0$ **18.** $4x + 10y = 0$

19. Find the exact values of $\sin 150°$, $\cos 150°$, and $\tan 150°$. [*Hint:* Use Example 3 in the Geometry Review Appendix to determine the coordinates of P in the figure.]

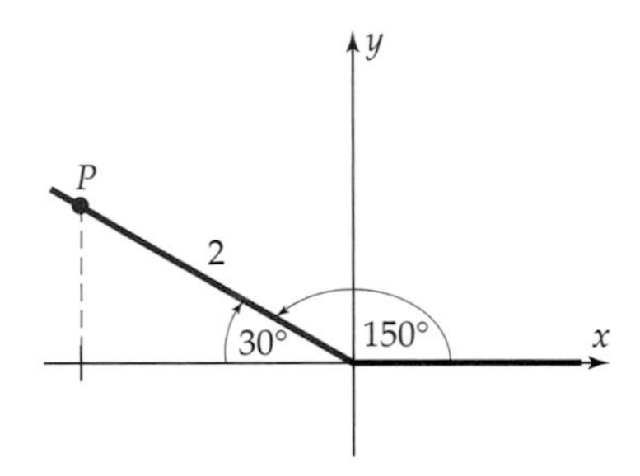

20. Find the exact values of $\sin 120°$, $\cos 120°$, and $\tan 120°$. [*Hint:* Use Example 4 in the Geometry Review Appendix to determine the coordinates of P in the figure.]

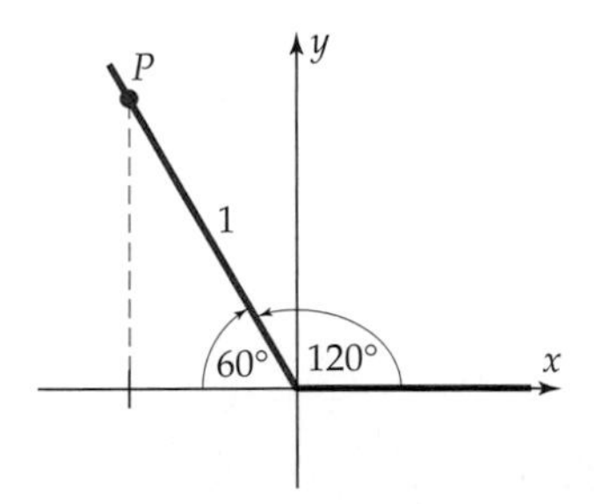

In Exercises 21–36, solve the triangle ABC under the given conditions.

21. $A = 20°, b = 10, c = 7$

22. $B = 40°, a = 12, c = 20$

23. $C = 118°, a = 6, b = 10$

24. $C = 52.5°, a = 6.5, b = 9$

25. $A = 140°, b = 12, c = 14$

26. $B = 25.4°, a = 6.8, c = 10.5$

27. $C = 78.6°, a = 12.1, b = 20.3$

28. $A = 118.2°, b = 16.5, c = 10.7$

29. $a = 7, b = 3, c = 5$

30. $a = 8, b = 5, c = 10$

31. $a = 16, b = 20, c = 32$

32. $a = 5.3, b = 7.2, c = 10$

33. $a = 7.2, b = 6.5, c = 11$

34. $a = 6.8, b = 12.4, c = 15.1$

35. $a = 12, b = 16.5, c = 21.3$

36. $a = 5.7, b = 20.4, c = 16.8$

37. Find the angles of the triangle whose vertices are $(0, 0)$, $(5, -2)$, $(1, -4)$.

38. Find the angles of the triangle whose vertices are $(-3, 4)$, $(6, 1)$, $(2, -1)$.

39. In Example 6, suppose that the angle between the two tracks is 112° and that the average speeds are 90 kilometers per hour for the first train and 55 kilometers per hour for the second train. How far apart are the trains after two hours and 45 minutes?

40. Suppose that the boat in Example 7 goes 25 miles due east and then changes course 56° northward. After traveling 50 miles on this course, how far is the boat from Chicago?

41. The sculptor builds a smaller version of the sculpture in Example 8, in which the beams are six feet and nine feet long, respectively, and the cable is 10.4 feet long. If the six-foot cable makes an angle of 40° with the ground, what angle does the nine-foot beam make with the ground?

42. Suppose that the tower in Example 9 is 175 feet high and that the cable on the downhill side is 120 feet from the base of the tower. How long is that cable?

43. An engineer wants to measure the width *CD* of a sinkhole. So she places a stake *B* and determines the measurements shown in the figure. How wide is the sinkhole?

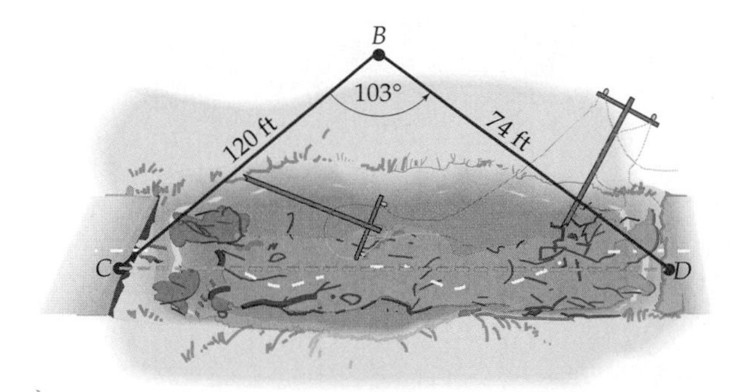

44. A straight tunnel is to be dug through a hill. Two people stand on opposite sides of the hill where the tunnel entrances are to be located. Both can see a stake located 530 meters from the first person and 755 meters from the second. The angle determined by the two people and the stake (vertex) is 77°. How long must the tunnel be?

45. The pitcher's mound on a standard baseball diamond (which is actually a square) is 60.5 feet from home plate (see the figure). How far is the pitcher's mound from first base?

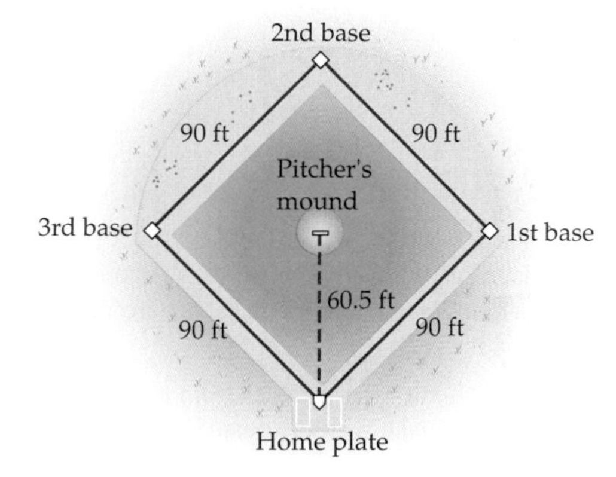

46. At Wrigley Field in Chicago, the straight-line distance from home plate over second base to the center field wall is 400 feet. How far is it from first base to the same point at the center field wall? [*Hint:* Adapt and extend the figure from Exercise 45.]

47. A stake is located 10.8 feet from the end of a closed gate that is 8 feet long. The gate swings open, and its end hits the stake. Through what angle did the gate swing?

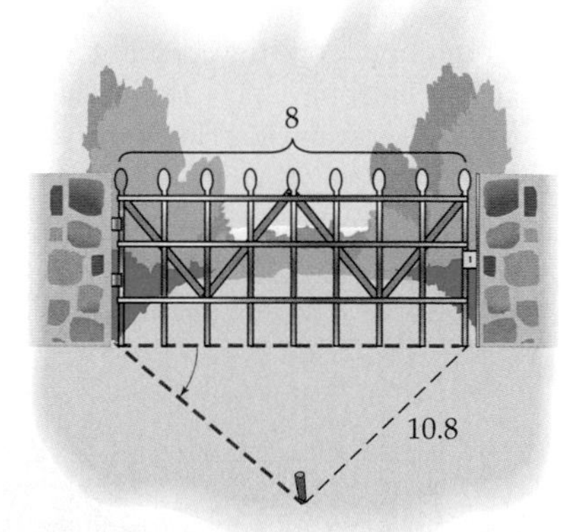

48. The distance from Chicago to St. Louis is 440 kilometers, that from St. Louis to Atlanta 795 kilometers, and that from Atlanta to Chicago 950 kilometers. What are the angles in the triangle with these three cities as vertices?

49. A satellite is placed in an orbit such that the satellite remains stationary 24,000 miles over a fixed point on the surface of the earth. The angle *CES*, where *C* is Cape Canaveral, *E* is the center of the earth, and *S* is the satellite, measures 60°. Assuming that the radius of the earth is 3960 miles, how far is the satellite from Cape Canaveral?

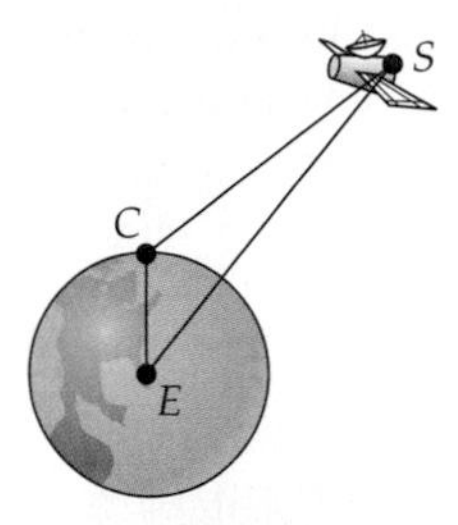

50. One plane flies west from Cleveland at 350 mph. A second plane leaves Cleveland at the same time and flies southeast at 200 mph. How far apart are the planes after 1 hour and 36 minutes?

51. A weight is hung by two cables from a beam. What angles do the cables make with the beam?

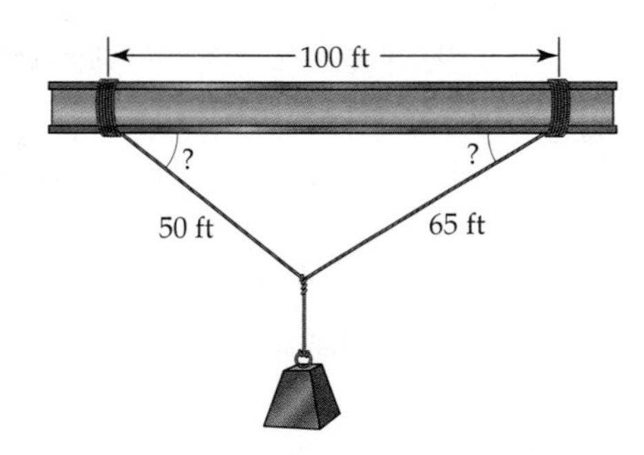

52. Two ships leave port, one traveling in a straight course at 22 mph and the other traveling a straight course at 31 mph. Their courses diverge by 38°. How far apart are they after 3 hours?

53. A boat runs in a straight line for 3 kilometers, then makes a 45° turn and goes for another 6 kilometers (see the figure). How far is the boat from its starting point?

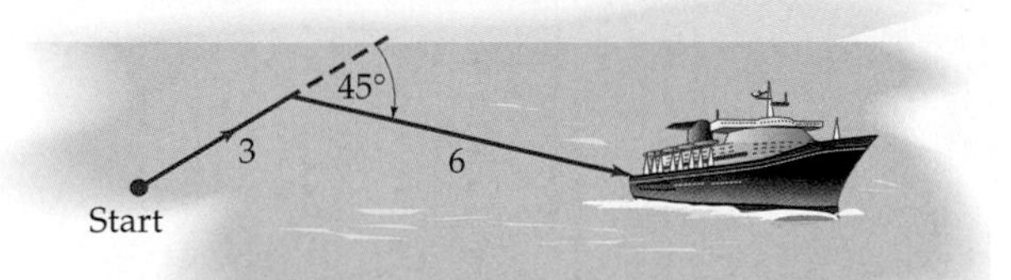

54. A plane flies in a straight line at 400 mph for 1 hour and 12 minutes. It makes a 15° turn and flies at 375 mph for 2 hours and 27 minutes. How far is it from its starting point?

55. Two planes at the same altitude approach an airport. One plane is 16 miles from the control tower and the other is 22 miles from the tower. The angle determined by the planes and the tower, with the tower as vertex, is 11°. How far apart are the planes?

56. One diagonal of a parallelogram is 6 centimeters long, and the other is 13 centimeters long. They form an angle of 42° with each other. How long are the sides of the parallelogram? [*Hint:* The diagonals of a parallelogram bisect each other.]

57. A 400-foot-high tower stands on level ground, anchored by two cables on the west side. The end of the cable closest to the tower makes an angle of 70° with the horizontal. The two cable ends are 100 feet apart, as shown in the figure. How long are the cables?

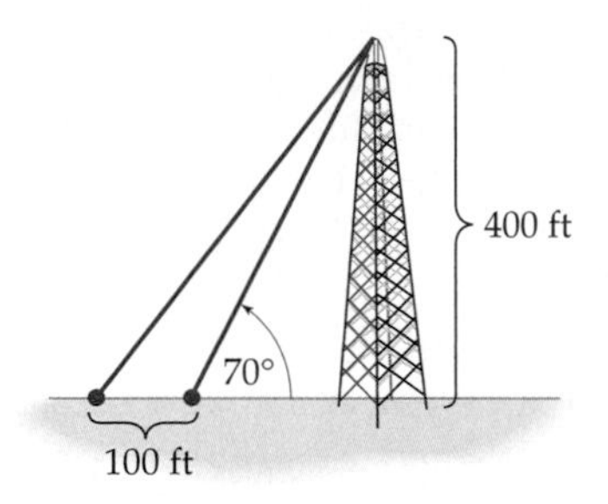

58. A surveyor has determined the distance and angles in the figure. He wants you to find the straight-line distance from *A* to *B*.

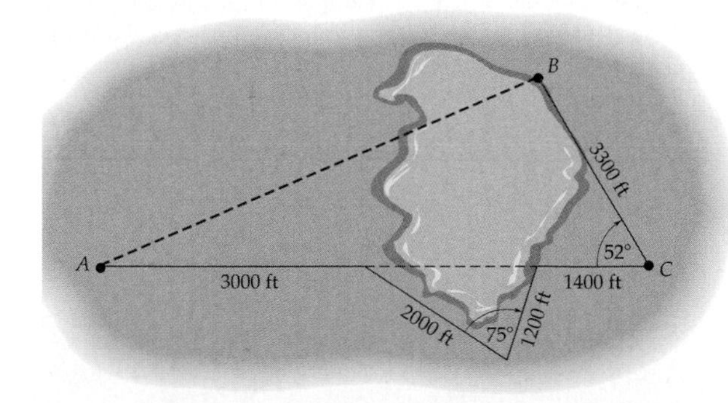

59. A ship is traveling at 18 mph from Corsica to Barcelona, a distance of 350 miles. To avoid bad weather, the ship leaves Corsica on a route 22° south of the direct route (see the figure). After seven hours, the bad weather has been bypassed. Through what angle should the ship now turn to head directly to Barcelona?

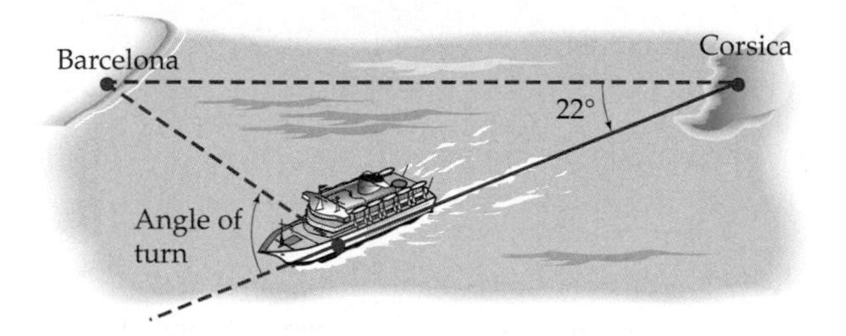

60. A plane leaves South Bend for Buffalo, 400 miles away, intending to fly a straight course in the direction 70° (aerial navigation is explained in Exercise 47 on page 102). After flying 180 miles, the pilot realizes that an error has been made and that he has actually been flying in the direction 55°.

(a) At that time, how far is the plane from Buffalo?

(b) In what direction should the plane now go to reach Buffalo?

61. Assume that the earth is a sphere of radius 3960 miles. A satellite travels in a circular orbit around the earth, 900 miles above the equator, making one full orbit every six hours. If it passes directly over a tracking station at 2 P.M., what is the distance from the satellite to the tracking station at 2:05 P.M.?

62. A parallelogram has diagonals of lengths 12 and 15 inches that intersect at an angle of 63.7°. How long are the sides of the parallelogram? [See the hint for Exercise 56.]

Thinkers

63. Let $P = (x, y)$ be a point (other than the origin) on the terminal side of an angle θ in standard position, and let r be the distance from P to the origin O. Let $Q = (u, v)$ be another point (not the origin) on the terminal side of θ, and let s be the distance from Q to O.

(a) Show that the value of $\sin\theta$ computed by using P is the same as that computed by using Q. [*Hint:* Consider the figure, and note that triangles POS and QOR are similar.]

(b) Do part (a) with $\cos\theta$ in place of $\sin\theta$.

(c) Do part (a) with $\tan\theta$ in place of $\sin\theta$.

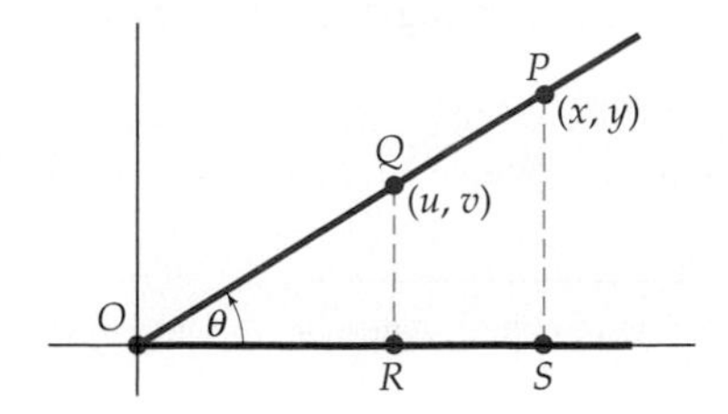

64. Let θ be an angle in standard position, and let (x, y) a point on its terminal side, as shown in the figure. Let r be the distance from (x, y) to the origin. Then by definition,

$$\sin\theta = \frac{y}{r} \text{ and } \cos\theta = \frac{x}{r}.$$

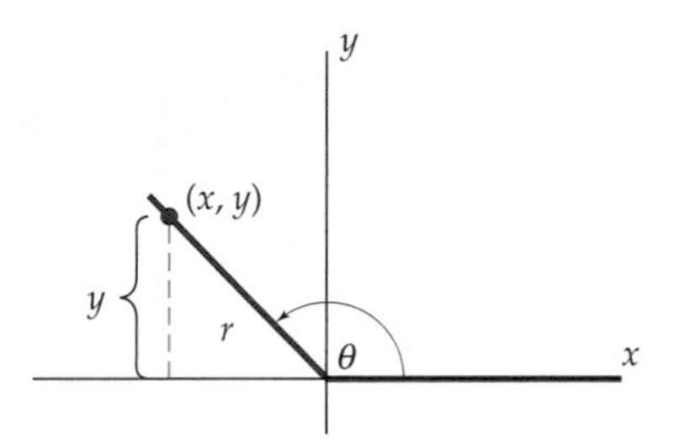

(a) Use the Pythagorean Theorem to show that

$$x^2 + y^2 = r^2.$$

(b) Use part (a) and the definition of $\sin\theta$ and $\cos\theta$ to show that

$$(\sin\theta)^2 + (\cos\theta)^2 = 1.$$

This identity, which was proved for acute angles in Exercise 45 of Section 9.1, is called the *Pythagorean identity.*

65. Assuming that the circles in the figure are mutually tangent, find the lengths of the sides and the measures of the angles in triangle ABC.

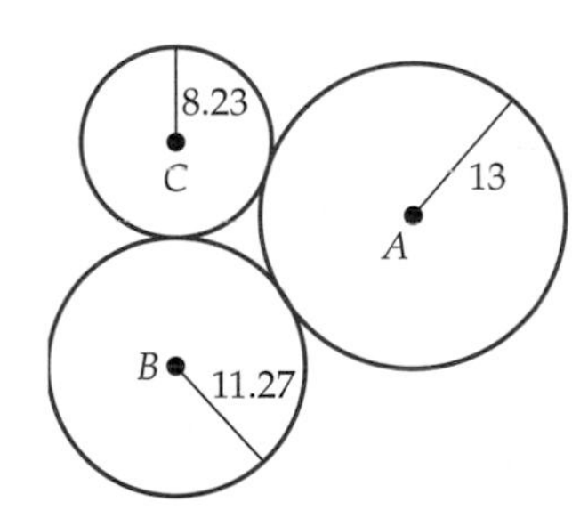

66. Assuming that the circles in the figure are mutually tangent, find the lengths of the sides and the measures of the angles in triangle ABC.

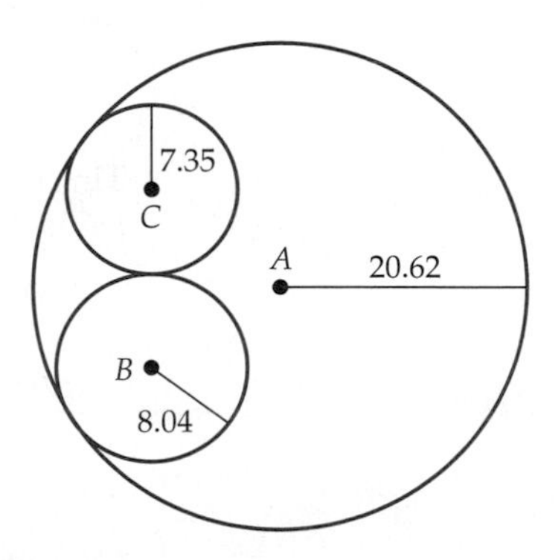

1.4 The Law of Sines

To solve oblique triangles in cases in which the Law of Cosines cannot be used, we need the following fact.

Law of Sines

> In any triangle ABC (in standard notation),
>
> $$\frac{a}{\sin A} = \frac{b}{\sin B} = \frac{c}{\sin C}.^{*}$$

Proof Position triangle ABC on a coordinate plane so that angle C is in standard position, with initial side b and terminal side a, as shown in Figure 1–47.

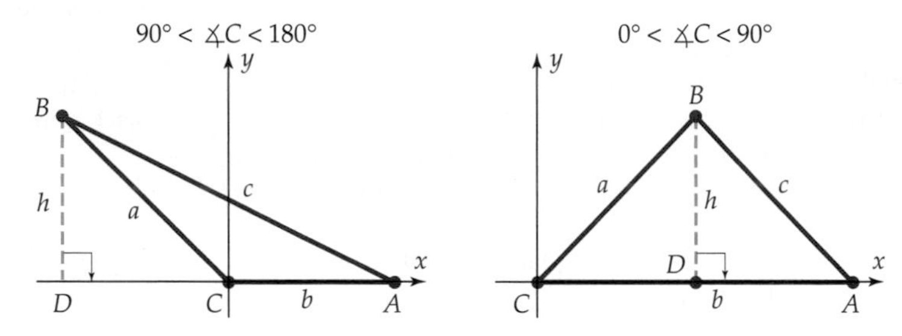

Figure 1–47

In each case, we can compute $\sin C$ by using the point B on the terminal side of angle C. The second coordinate of B is h, and the distance from B to the origin is a. Therefore, by the point-in-the-plane description of sine,

$$\sin C = \frac{h}{a} \qquad \text{or, equivalently,} \qquad h = a \sin C.$$

In each case, right triangle ADB shows that

$$\sin A = \frac{\text{opposite}}{\text{hypotenuse}} = \frac{h}{c} \qquad \text{or, equivalently,} \qquad h = c \sin A.$$

Combining this with the fact that $h = a \sin C$, we have

$$a \sin C = c \sin A.$$

Since angles in a triangle are nonzero, $\sin A \neq 0$ and $\sin C \neq 0$. Dividing both sides of the last equation by $\sin A \sin C$ yields

$$\frac{a}{\sin A} = \frac{c}{\sin C}.$$

This proves one equation in the Law of Sines. Similar arguments beginning with angle A or B in standard position prove the other equations. ■

The Law of Sines can be used to solve triangles in these cases.

1. Two angles and one side are known (AAS).
2. Two sides and the angle opposite one of them are known (SSA).

---

*An equality of the form $u = v = w$ is shorthand for the statement $u = v$ and $v = w$ and $w = u$.

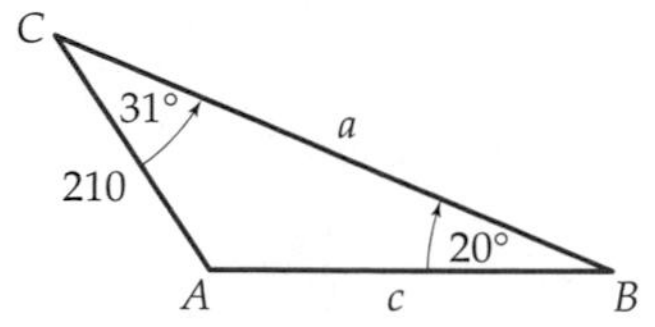

Figure 1–48

EXAMPLE 1 AAS

If $B = 20°$, $C = 31°$, and $b = 210$ in Figure 1–48, find the other angles and sides.

SOLUTION Since the sum of the angles of a triangle is 180°,

$$A = 180° - (20° + 31°) = 180° - 51° = 129°.$$

To find side a, we observe that we know three of the four quantities in one of the equations given by the Law of Sines.

$$\frac{a}{\sin A} = \frac{b}{\sin B}$$

$$\frac{a}{\sin 129°} = \frac{210}{\sin 20°}.$$

Multiplying both sides by sin 129°, we obtain

$$a = \frac{210(\sin 129°)}{\sin 20°} \approx 477.2.$$

Side c is found similarly. Beginning with an equation from the Law of Sines involving c and three known quantities, we have

$$\frac{c}{\sin C} = \frac{b}{\sin B}$$

$$\frac{c}{\sin 31°} = \frac{210}{\sin 20°}$$

$$c = \frac{210 \sin 31°}{\sin 20°} \approx 316.2. \quad ■$$

In the AAS case, there is exactly one triangle that satisfies the given data.* But when two sides of a triangle and the angle opposite one of them are known (SSA), there may be one, two, or no triangles that satisfy the given data (the **ambiguous case**). To see why this can happen, suppose sides a and b and angle A are given. Place angle A in standard position with terminal side b, as shown on the next page. If angle A is less than 90°, then there are four possibilities for side a:

---

*Once you know two angles, you know all three (their sum must be 180°). Hence, you know two angles and the included side. Any two triangles satisfying these conditions will be congruent by ASA in the Congruent Triangles Theorem in the Geometry Review Appendix.

(i) Side a is too short to reach the third side: *no solution.*

(ii) Side a just reaches the third side and is perpendicular to it: *one solution.*

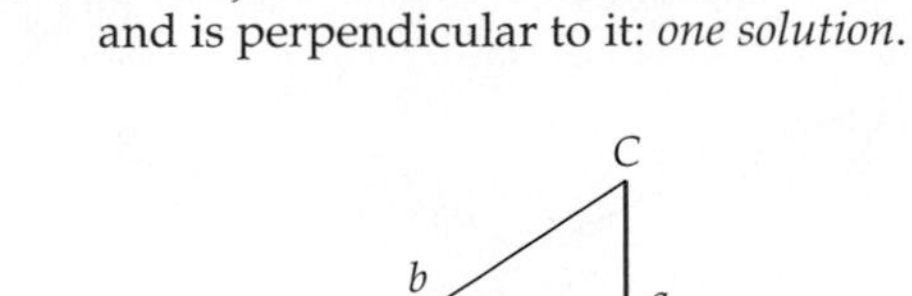

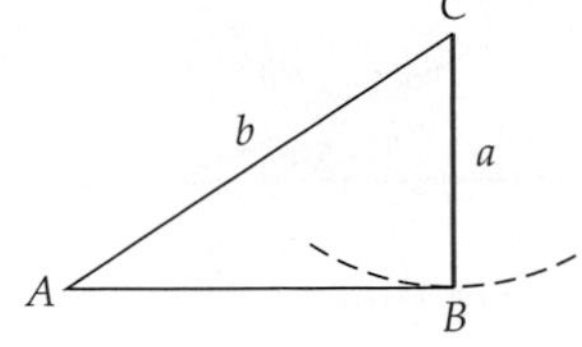

(iii) An arc of radius a meets the third side at two points to the right of A: *two solutions.*

(iv) $a \geq b$, so that an arc of radius a meets the third side at just one point to the right of A: *one solution.*

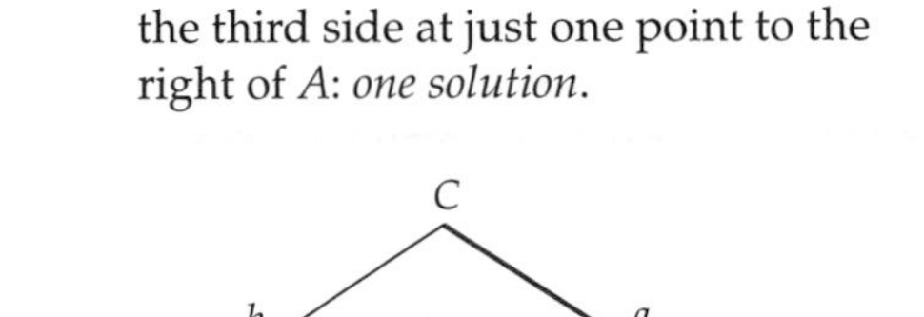

When angle A is more than 90°, then there are only two possibilities:

(i) $a \leq b$, so that side a is too short to reach the third side: *no solution.*

(ii) $a > b$, so that an arc of radius a meets the third side at just one point to the right of A: one *solution.*

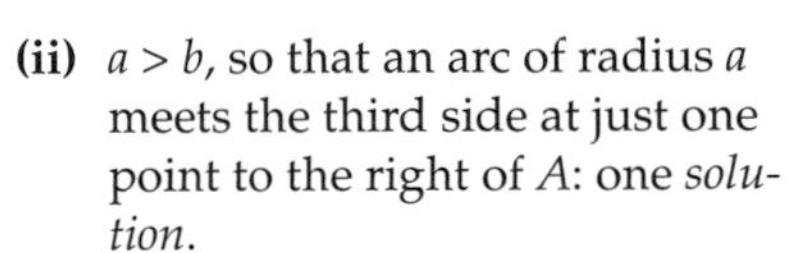

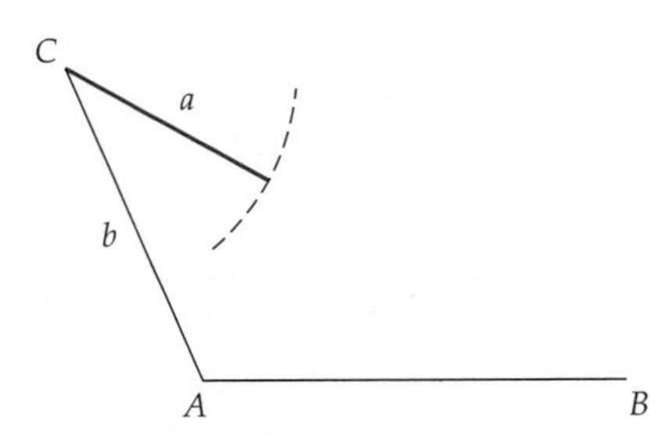

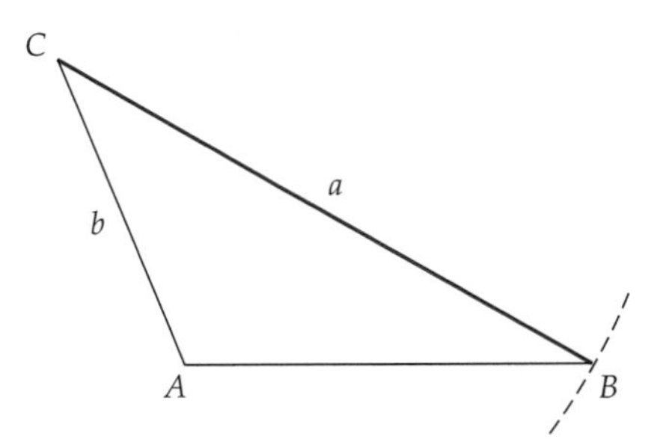

To deal with the case of two solutions, we need the following identity.

Supplementary Angle Identity

If $0° \leq \theta \leq 90°$, then

$$\sin \theta = \sin(180° - \theta).$$

Proof Place the angle $180° - \theta$ in standard position, and choose a point D on its terminal side. Let r be the distance from D to the origin. The situation looks like Figure 1–49.

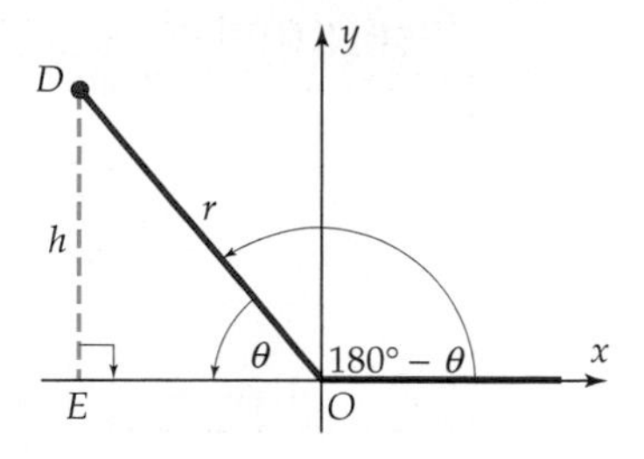

Figure 1–49

Since h is the second coordinate of D, we have $\sin(180° - \theta) = h/r$. Right triangle OED shows that

$$\sin \theta = \frac{\text{opposite}}{\text{hypotenuse}} = \frac{h}{r} = \sin(180° - \theta).$$

EXAMPLE 2 SSA

Given triangle ABC with $a = 6$, $b = 7$, and $A = 65°$, find angle B.

SOLUTION We use an equation from the Law of Sines involving the known quantities.

$$\frac{b}{\sin B} = \frac{a}{\sin A}$$

$$\frac{7}{\sin B} = \frac{6}{\sin 65°}$$

$$\sin B = \frac{7 \sin 65°}{6} \approx 1.06.$$

There is no angle B whose sine is greater than 1. Therefore, there is no triangle satisfying the given data. ■

EXAMPLE 3

An airplane A takes off from carrier B and flies in a straight line for 12 kilometers. At that instant, an observer on destroyer C, located 5 kilometers from the carrier, notes that the angle determined by the carrier, the destroyer (vertex), and the plane is 37°. How far is the plane from the destroyer?

SOLUTION The given data provide Figure 1–50.

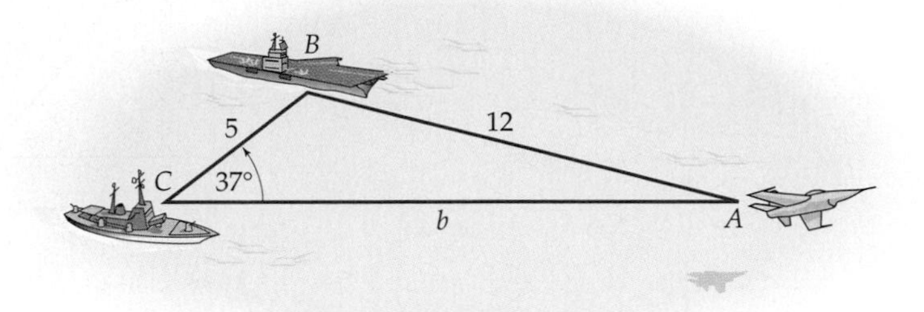

Figure 1–50

We must find side b. To do this, we first use the Law of Sines to find angle A.

$$\frac{a}{\sin A} = \frac{c}{\sin C}$$

$$\frac{5}{\sin A} = \frac{12}{\sin 37°}$$

$$\sin A = \frac{5 \sin 37°}{12} \approx .2508.$$

The SIN^{-1} key on a calculator in degree mode shows that 14.5° is an angle whose sine is .2508. The supplementary angle identity shows that $180° - 14.5° = 165.5°$ is also an angle with sine .2508. But if $A = 165.5°$ and $C = 37°$, the sum of angles A, B, C would be greater than 180°. Since this is impossible, $A = 14.5°$ is the only solution here. Therefore,

$$B = 180° - (37° + 14.5°) = 180° - 51.5° = 128.5°.$$

Using the Law of Sines again, we have

$$\frac{b}{\sin B} = \frac{c}{\sin C}$$

$$\frac{b}{\sin 128.5°} = \frac{12}{\sin 37°}$$

$$b = \frac{12 \sin 128.5°}{\sin 37°} \approx 15.6.$$

Thus, the plane is approximately 15.6 kilometers from the destroyer. ■

EXAMPLE 4 **SSA**

Solve triangle ABC when $a = 7.5$, $b = 12$, and $A = 35°$.

SOLUTION The Law of Sines shows that

$$\frac{b}{\sin B} = \frac{a}{\sin A}$$

$$\frac{12}{\sin B} = \frac{7.5}{\sin 35^\circ}$$

$$\sin B = \frac{12 \sin 35^\circ}{7.5} \approx .9177.$$

The SIN^{-1} key shows that 66.6° is a solution of $\sin B = .9177$. Therefore, $180^\circ - 66.6^\circ = 113.4^\circ$ is also a solution of $\sin B = .9177$ by the supplementary angle identity. In each case the sum of angles A and B is less than 180°, so there are two triangles ABC satisfying the given data, as shown in Figure 1–51.

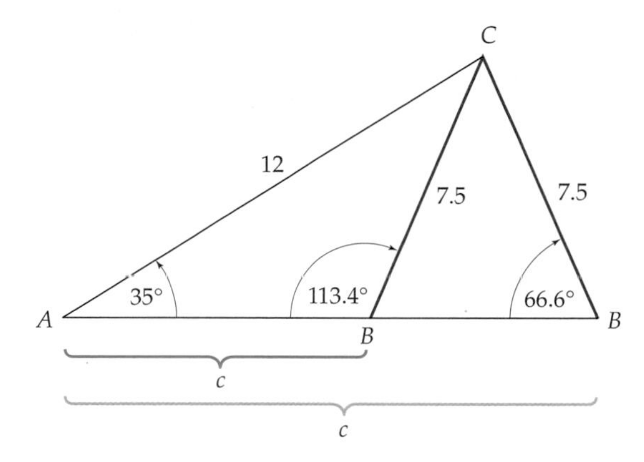

Figure 1–51

Case 1. If $B = 66.6^\circ$, then $C = 180^\circ - (35^\circ + 66.6^\circ) = 78.4^\circ$. By the Law of Sines,

$$\frac{c}{\sin C} = \frac{a}{\sin A}$$

$$\frac{c}{\sin 78.4^\circ} = \frac{7.5}{\sin 35^\circ}$$

$$c = \frac{7.5 \sin 78.4^\circ}{\sin 35^\circ} \approx \frac{7.5(.9796)}{.5736} \approx 12.8.$$

Case 2. If $B = 113.4^\circ$, then $C = 180^\circ - (35^\circ + 113.4^\circ) = 31.6^\circ$. Consequently,

$$\frac{c}{\sin C} = \frac{a}{\sin A}$$

$$\frac{c}{\sin 31.6^\circ} = \frac{7.5}{\sin 35^\circ}$$

$$c = \frac{7.5 \sin 31.6^\circ}{\sin 35^\circ} \approx \frac{7.5(.5240)}{.5736} \approx 6.9.$$ ■

EXAMPLE 5

A plane flying in a straight line passes directly over point A on the ground and later directly over point B, which is three miles from A. A few minutes after the plane passes over B, the angle of elevation from A to the plane is 43° and the angle of elevation from B to the plane is 67°. How high is the plane at that moment?

SOLUTION If C represents the plane when the angles of elevation are noted, then the situation is represented in Figure 1–52. We must find the length of h.

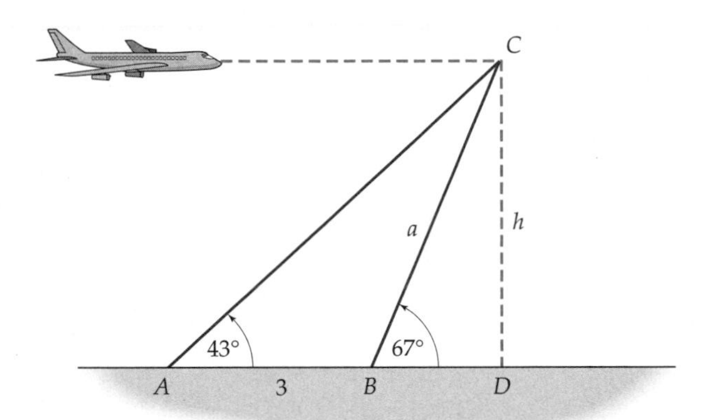

Figure 1–52

Note that angle ABC measures $180° - 67° = 113°$; hence,

$$\measuredangle BCA = 180° - (43° + 113°) = 24°.$$

Use the Law of Sines to find side a of triangle ABC.

$$\frac{a}{\sin 43°} = \frac{3}{\sin 24°}$$

$$a = \frac{3 \sin 43°}{\sin 24°} \approx 5.03.$$

Now in the right triangle CBD, we have

$$\sin 67° = \frac{\text{opposite}}{\text{hypotenuse}} = \frac{h}{a} \approx \frac{h}{5.03}.$$

Therefore, $h \approx 5.03 \sin 67° \approx 4.63$ miles. ■

EXERCISES 1.4

Standard notation for triangle ABC is used throughout. Use a calculator and round off your answers to one decimal place at the end of the computation.

In Exercises 1–8, solve triangle ABC under the given conditions.

1. $A = 48°, B = 22°, a = 5$
2. $B = 33°, C = 46°, b = 4$
3. $A = 116°, C = 50°, a = 8$
4. $A = 105°, B = 27°, b = 10$
5. $B = 44°, C = 48°, b = 12$
6. $A = 67°, C = 28°, a = 9$
7. $A = 102.3°, B = 36.2°, a = 16$
8. $B = 97.5°, C = 42.5°, b = 7$

In Exercises 9–32, solve the triangle. The Law of Cosines might be needed in Exercises 19–32.

9. $b = 15, c = 25, B = 47°$
10. $b = 30, c = 50, C = 60°$
11. $a = 12, b = 5, B = 20°$
12. $b = 12.5, c = 20.1, B = 37.3°$
13. $a = 5, c = 12, A = 102°$
14. $a = 9, b = 14, B = 95°$
15. $b = 11, c = 10, C = 56°$
16. $a = 12.4, c = 6.2, A = 72°$
17. $A = 41°, B = 67°, a = 10.5$
18. $a = 30, b = 40, A = 30°$
19. $b = 4, c = 10, A = 75°$
20. $a = 50, c = 80, C = 45°$
21. $a = 6, b = 12, c = 16$
22. $B = 20.67°, C = 34°, b = 185$
23. $a = 16.5, b = 18.2, C = 47°$
24. $a = 21, c = 15.8, B = 71°$
25. $b = 17.2, c = 12.4, B = 62.5°$
26. $b = 24.1, c = 10.5, C = 26.3°$
27. $a = 10.1, b = 18.2, A = 50.7°$
28. $b = 14.6, c = 7.8, B = 40.4°$
29. $b = 12.2, c = 20, A = 65°$
30. $a = 44, c = 84, C = 42.2°$
31. $A = 19°, B = 35°, a = 110$
32. $b = 15.4, c = 19.3, A = 42°$
33. A surveyor marks points A and B 200 meters apart on one bank of a river. She sights a point C on the opposite bank and determines the angles shown in the figure. What is the distance from A to C?

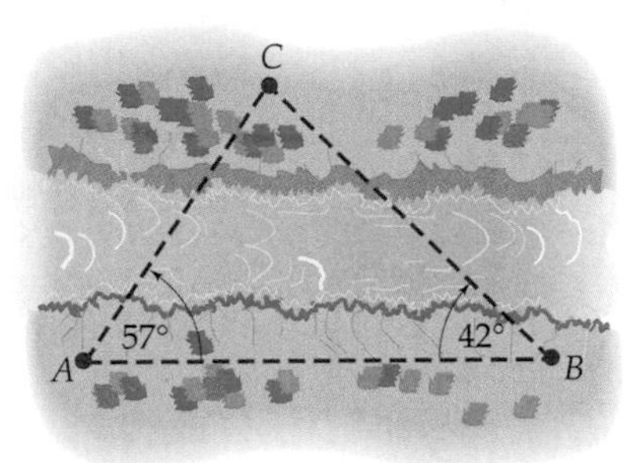

34. A forest fire is spotted from two fire towers. The triangle determined by the two towers and the fire has angles of 28° and 37° at the tower vertices. If the towers are 3000 meters apart, which one is closer to the fire?
35. A string of lights is to be placed over one end of a pond (from A to B in the figure). If angle A measures 49°, angle B measures 128°, and BC is 144 meters long, what is the minimum possible length for the string of lights?

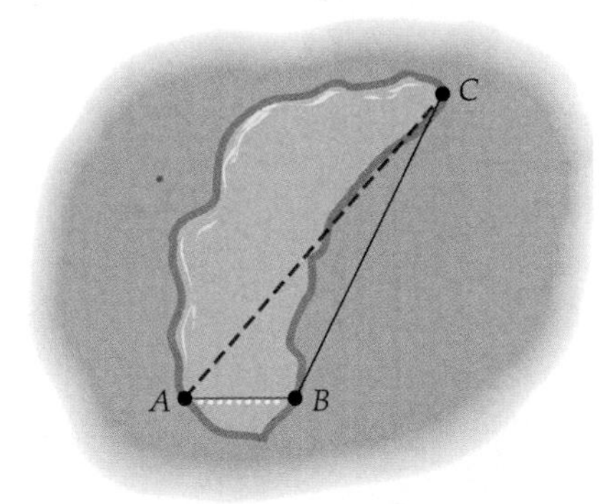

36. A fence post is located 50 feet from one corner of a building and 40 feet from the adjacent corner. Fences are put up between the post and the building corners to form a triangular garden area. The 40-foot fence makes a 58° angle with the building. How long is the building wall?

37. Two straight roads meet at an angle of 40° in Harville, one leading to Eastview and the other to Wellston. Eastview is 18 kilometers from Harville and 20 kilometers from Wellston. What is the distance from Harville to Wellston?

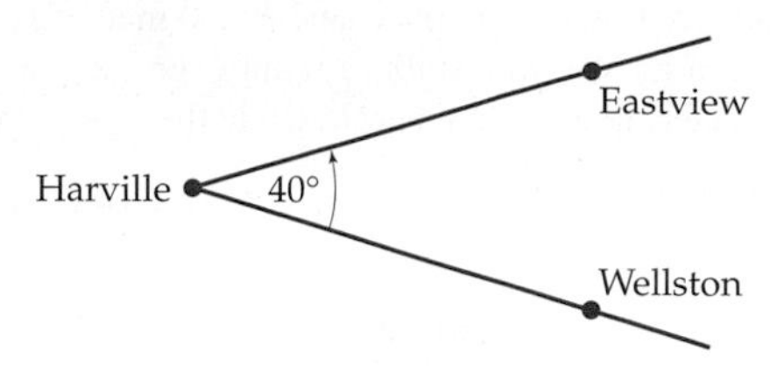

38. A triangular piece of land has two sides that are 80 feet and 64 feet long, respectively. The 80-foot side makes an angle of 28° with the third side. An advertising firm wants to know whether a 30-foot long sign can be placed along the third side. What would you tell them?

39. Al visited the Leaning Tower of Pisa and observed that the tower's shadow was 40 meters long and that the angle of elevation from the tip of the shadow to the top of the tower was 57°. The tower is now 54 meters tall (measured from the ground to the top along the center line of the tower). Approximate the angle α that the center line of the tower makes with the vertical.

40. A pole tilts at an angle 9° from the vertical, away from the sun, and casts a shadow 24 feet long. The angle of elevation from the end of the pole's shadow to the top of the pole is 53°. How long is the pole?

41. A side view of a bus shelter is shown in the figure. The brace d makes an angle of 37.25° with the back and an angle of 34.85° with the top of the shelter. How long is this brace?

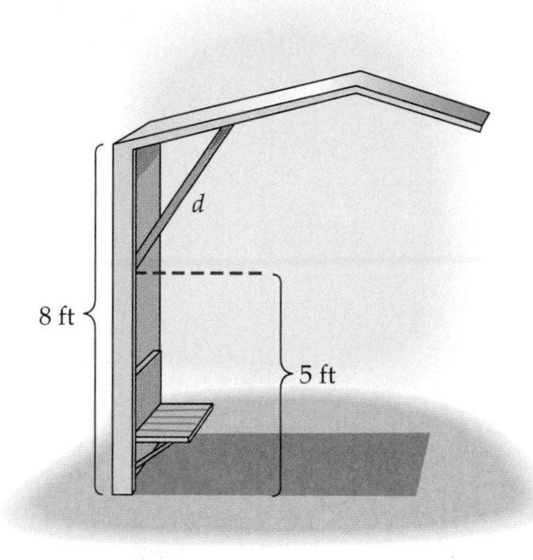

42. A straight path makes an angle of 6° with the horizontal. A statue at the higher end of the path casts a 6.5-meter-long shadow straight down the path. The angle of elevation from the end of the shadow to the top of the statue is 32°. How tall is the statue?

43. A vertical statue 6.3 meters high stands on top of a hill. At a point on the side of the hill 35 meters from the statue's base, the angle between the hillside and a line from the top of the statue is 10°. What angle does the side of the hill make with the horizontal?

44. Each of two observers 400 feet apart measures the angle of elevation to the top of a tree that sits on the straight line between them. These angles are 51° and 65°, respectively. How tall is the tree? How far is the base of its trunk from each observer?

45. From the top of the 800-foot-tall Cartalk Tower, Tom sees a plane; the angle of elevation is 67°. At the same instant, Ray, who is on the ground, one mile from the building, notes that his angle of elevation to the plane is 81° and that his angle of elevation to the top of Cartalk Tower is 8.6°. Assuming that Tom and Ray and the airplane are in a plane perpendicular to the ground, how high is the airplane?

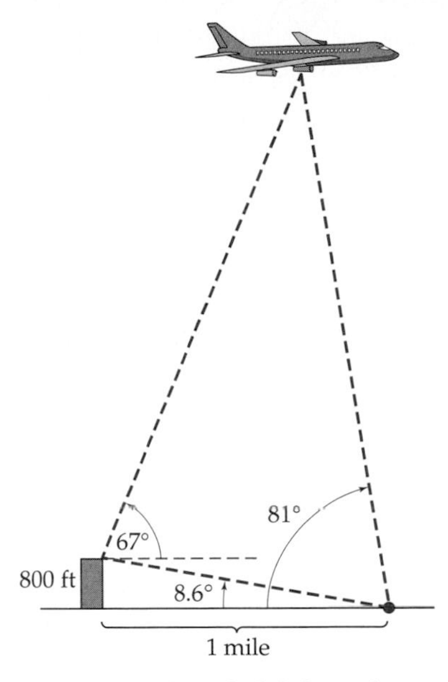

46. A plane flies in a direction of 105° from airport A. After a time, it turns and proceeds in a direction of 267°. Finally, it lands at airport B, 120 miles directly south of airport A. How far has the plane traveled? [*Note:* Aerial navigation directions are explained in Exercise 47 on page 102.]

47. Charlie is afraid of water; he can't swim and refuses to get in a boat. However, he must measure the width of a river for his geography class. He has a long tape measure but no way to measure angles. While pondering what to do, he paces along the side of the river using the five paths joining points A, B, C, and D. If he can't determine the width of the river, he will flunk the course.

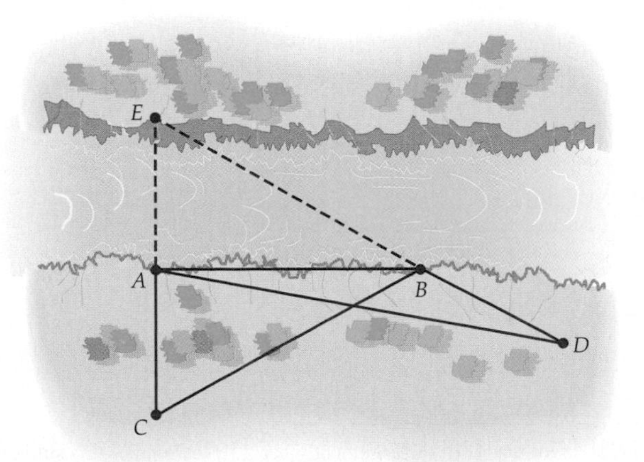

(a) Save Charlie from disaster by explaining how he can determine the width AE simply by measuring the lengths AB, AC, AD, BC, and BD and using trigonometry.

(b) Charlie determines that $AB = 75$ feet, $AC = 25$ feet, $AD = 90$ feet, $BC = 80$ feet, and $BD = 22$ feet. How wide is the river between A and E?

48. A plane flies in a direction of 85° from Chicago. It then turns and flies in the direction of 200° for 150 miles. It is then 195 miles from its starting point. How far did the plane fly in the direction of 85°? (See the note in Exercise 46.)

49. A hinged crane makes an angle of 50° with the ground. A malfunction causes the lock on the hinge to fail and the top part of the crane swings down. How far from the base of the crane does the top hit the ground?

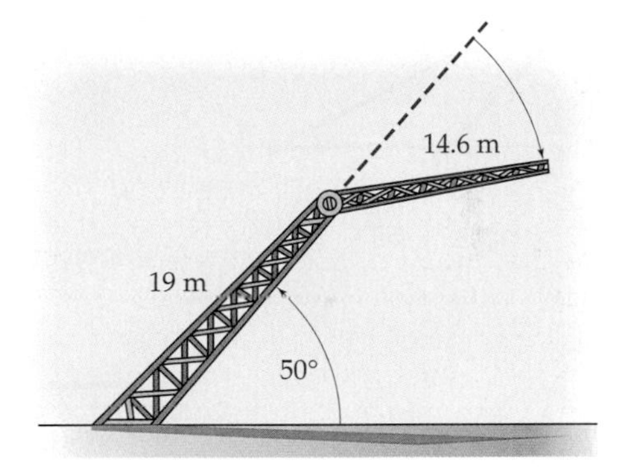

1.4.A *SPECIAL TOPICS* The Area of a Triangle

The proof of the Law of Sines enables us to prove a useful fact.

Area of a Triangle

The area of a triangle containing an angle C with sides of lengths a and b is

$$\frac{1}{2}ab \sin C.$$

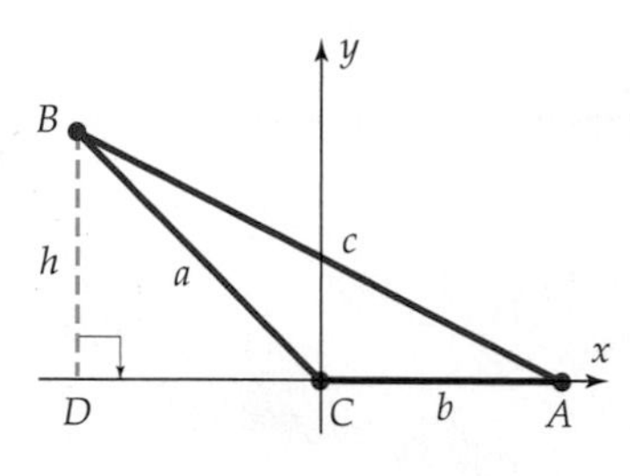

Figure 1–53

Proof Place the vertex of angle C at the origin, with side b on the positive x-axis (Figure 1–53).* Then b is the base and h is the altitude of the triangle so that

$$\text{area of triangle } ABC = \frac{1}{2} \times \text{base} \times \text{altitude} = \frac{1}{2} \cdot b \cdot h.$$

The proof of the Law of Sines on page 116 shows that $h = a \sin C$. Therefore,

$$\text{area of triangle } ABC = \frac{1}{2} \cdot b \cdot h = \frac{1}{2} \cdot b \cdot a \sin C = \frac{1}{2}ab \sin C.$$ ■

EXAMPLE 1 

Find the area of the triangle shown in Figure 1–54.

SOLUTION

$$\frac{1}{2} \cdot 8 \cdot 13 \sin 130° \approx 39.83 \text{ square centimeters.}$$ ■

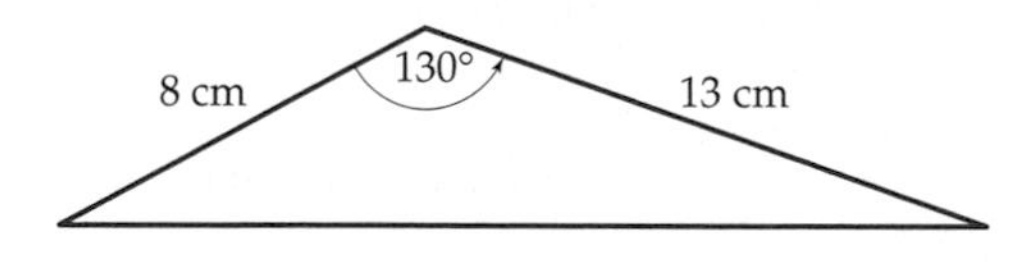

Figure 1–54

Here is a useful formula for the area of a triangle in terms of its sides.

Heron's Formula

The area of a triangle with sides a, b, c is

$$\sqrt{s(s-a)(s-b)(s-c)},$$

where $s = \frac{1}{2}(a + b + c)$.

*Figure 1–53 is the case when C is larger than 90°; the argument when C is less than 90° is similar.

Proof The preceding area formula and the Pythagorean identity,

$$\sin^2 C = 1 - \cos^2 C = (1 + \cos C)(1 - \cos C),$$

show that the area of triangle ABC (standard notation) is

$$\begin{aligned}\frac{1}{2}ab \sin C &= \sqrt{\left(\frac{1}{2}ab \sin C\right)^2} = \sqrt{\frac{1}{4}a^2b^2 \sin^2 C} \\ &= \sqrt{\frac{1}{4}a^2b^2(1 - \cos^2 C)} \\ &= \sqrt{\frac{1}{2}ab(1 + \cos C)\frac{1}{2}ab(1 - \cos C)}.\end{aligned}$$

Exercise 22 uses the Law of Cosines to show that

$$\begin{aligned}\frac{1}{2}ab(1 + \cos C) &= \frac{(a + b)^2 - c^2}{4} = \frac{(a + b) + c}{2} \cdot \frac{(a + b) - c}{2} \\ &= s(s - c)\end{aligned}$$

and

$$\begin{aligned}\frac{1}{2}ab(1 - \cos C) &= \frac{c^2 - (a - b)^2}{4} = \frac{c - (a - b)}{2} \cdot \frac{c + (a - b)}{2} \\ &= (s - a)(s - b).\end{aligned}$$

Combining these facts completes the proof.

$$\begin{aligned}\text{Area} = \frac{1}{2}ab \sin C &= \sqrt{\frac{1}{2}ab(1 + \cos C)\frac{1}{2}ab(1 - \cos C)} \\ &= \sqrt{s(s - a)(s - b)(s - c)}. \quad \blacksquare\end{aligned}$$

EXAMPLE 2 

Find the area of the triangle whose sides have lengths 7, 9, and 12.

SOLUTION Apply Heron's Formula with $a = 7$, $b = 9$, $c = 12$, and

$$s = \frac{1}{2}(a + b + c) = \frac{1}{2}(7 + 9 + 12) = 14.$$

The area is

$$\begin{aligned}\sqrt{s(s - a)(s - b)(s - c)} &= \sqrt{14(14 - 7)(14 - 9)(14 - 12)} \\ &= \sqrt{980} \approx 31.3 \text{ square units.} \quad \blacksquare\end{aligned}$$

EXERCISES 1.4.A

In Exercises 1–8, find the area of triangle ABC (standard notation) under the given conditions.

1. $a = 4, b = 8, C = 27°$
2. $b = 10, c = 14, A = 36°$
3. $c = 7, a = 10, B = 68°$
4. $a = 9, b = 13, C = 75°$
5. $a = 11, b = 15, c = 18$
6. $a = 4, b = 12, c = 14$
7. $a = 7, b = 9, c = 11$
8. $a = 17, b = 27, c = 40$

In Exercises 9–11, find the area of the triangle with the given vertices.

9. $(0, 0), (2, -5), (-3, 1)$
10. $(-4, 2), (5, 7), (3, 0)$

In Exercises 11 and 12, find the area of the polygonal region. [Hint: Divide the region into triangles.]

11.

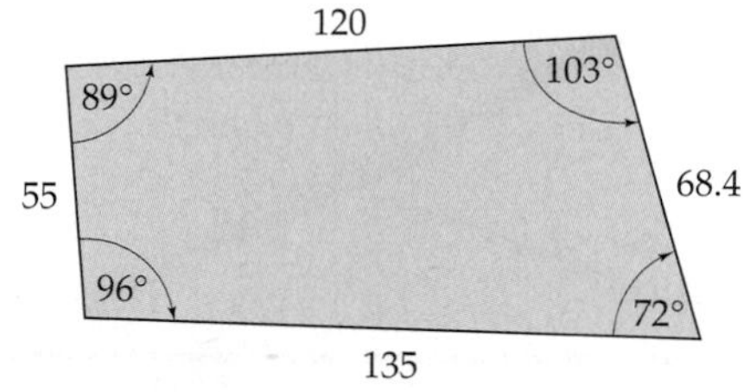

12.

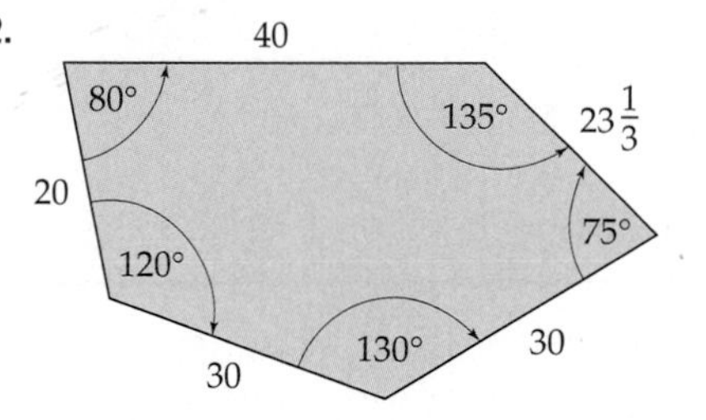

13. A triangular banner is hung from a window along the side of a building. The edges that touch the window are 20 and 24 feet long respectively. The third side is parallel to the ground. The angle between the 20-foot side and the third side is 44°. What is the area of the banner?

14. A triangular lot has sides of length 120 feet and 160 feet. The angle between these sides is 42°. Adjacent to this lot is a rectangular lot whose longest side has length 200 feet and whose shortest side is the same length as the shortest side of the triangular lot. What is the total area of both lots?

15. If a gallon of paint covers 400 square feet, how many gallons are needed to paint a triangular deck with sides of lengths 65 feet, 72 feet, and 88 feet?

16. Find the volume of the prism in the figure. The volume is given by the formula $V = \frac{1}{3}Bh$, where B is the area of the base and h is the height.

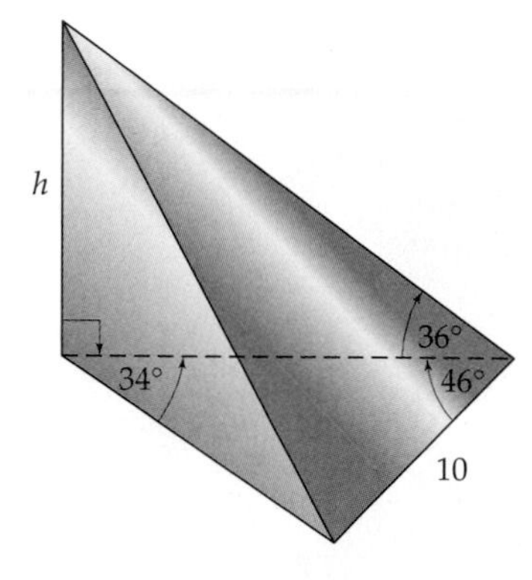

17. A rigid plastic triangle ABC rests on three vertical rods, as shown in the figure. What is its area?

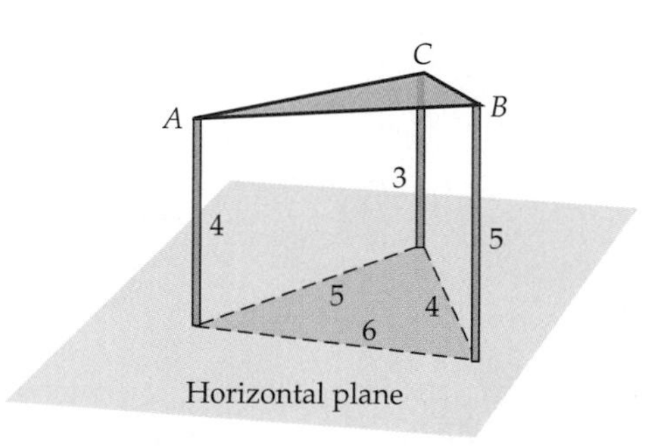

18. Prove that the area of triangle ABC (standard notation) is given by $\dfrac{a^2 \sin B \sin C}{2 \sin A}$.

19. A *regular polygon* has n equal sides and n equal angles formed by the sides. For example, a regular polygon of three sides is an equilateral triangle and a regular polygon of 4 sides is a square. If a regular polygon of n sides is inscribed in a circle of radius r, then Exercise 20 shows that its area is

$$\frac{1}{2}nr^2 \sin\left(\frac{360}{n}\right)^\circ.$$

Find the area of
(a) A square inscribed in a circle of radius 5.
(b) A regular hexagon (6 sides) inscribed in a circle of radius 2.

20. Prove the area formula in Exercise 19 as follows.

(a) Draw lines from each vertex of the polygon to the center of the circle, as shown in the figure for $n = 6$.

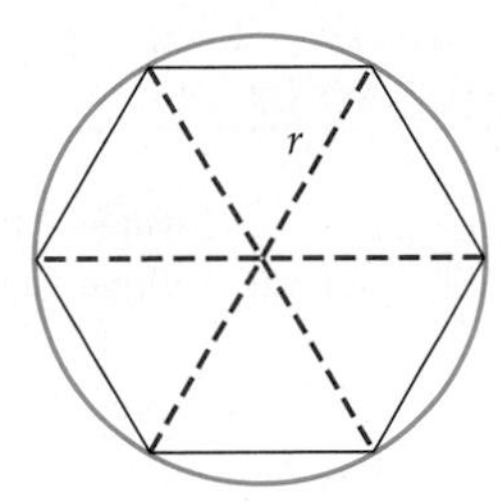

(b) Explain why all the triangles are congruent and hence, all the angles at the center of the circle have equal measure. [*Hint:* SSS.]

(c) Explain why the measure of each center angle is $360°/n$.

(d) Find the area of each triangle. [*Hint:* See the first box on page 126.]

(e) The area of the polygon is the sum of the areas of these n triangles. Use this fact and part (d) to obtain the area formula in Exercise 19.

21. What is the area of a triangle whose sides have lengths 12, 20, and 36? (If your answer turns out strangely, try drawing a picture.)

22. Complete the proof of Heron's Formula as follows. Let $s = \frac{1}{2}(a + b + c)$.

(a) Show that

$$\frac{1}{2}ab(1 + \cos C) = \frac{(a+b)^2 - c^2}{4} = \frac{(a+b)+c}{2} \cdot \frac{(a+b)-c}{2} = s(s-c).$$

[*Hint:* Use the Law of Cosines to express $\cos C$ in terms of a, b, c; then simplify.]

(b) Show that

$$\frac{1}{2}ab(1 - \cos C) = \frac{c^2 - (a-b)^2}{4} = \frac{c-(a-b)}{2} \cdot \frac{c+(a-b)}{2} = (s-a)(s-b).$$

Chapter 1 Review

IMPORTANT CONCEPTS

IMPORTANT FACTS & FORMULAS

- *Right Triangle Description:* In a right triangle containing an acute angle θ,

$$\sin\theta = \frac{\text{opposite}}{\text{hypotenuse}}$$

$$\cos\theta = \frac{\text{adjacent}}{\text{hypotenuse}}$$

$$\tan\theta = \frac{\text{opposite}}{\text{adjacent}}.$$

- *Point-in-the-Plane Description:* If (x, y) is a point (other than the origin) on the terminal side of angle θ in standard position and r is the distance from (x, y) to the origin, then

$$\sin\theta = \frac{y}{r} \qquad \cos\theta = \frac{x}{r} \qquad \tan\theta = \frac{y}{x}.$$

- *Law of Cosines:* $a^2 = b^2 + c^2 - 2bc\cos A$
- *Law of Cosines—Alternate Form:* $\cos A = \dfrac{b^2 + c^2 - a^2}{2bc}$
- *Law of Sines:* $\dfrac{a}{\sin A} = \dfrac{b}{\sin B} = \dfrac{c}{\sin C}$
- *Supplementary Angle Identity:* $\sin D = \sin(180° - D)$
- Area of triangle $ABC = \dfrac{ab\sin C}{2}$
- *Heron's Formula:*

$$\text{Area of triangle } ABC = \sqrt{s(s-a)(s-b)(s-c)}$$

where $s = \dfrac{1}{2}(a + b + c)$.

REVIEW QUESTIONS

Note: Standard notation is used for triangles.

In Questions 1 and 2, find $\sin\theta$, $\cos\theta$, *and* $\tan\theta$.

1.

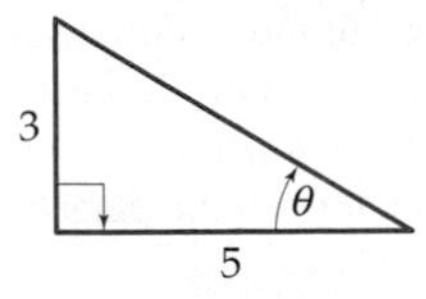

2.

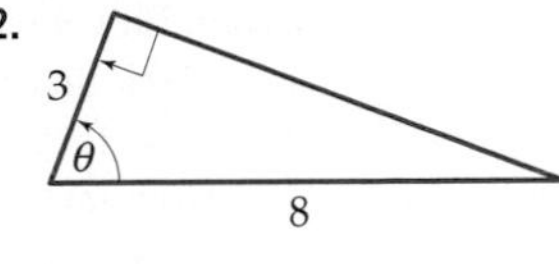

3. Which of the following statements about the angle θ is true?

(a) $\sin\theta = 3/4$
(b) $\cos\theta = 5/4$
(c) $\tan\theta = 3/5$
(d) $\sin\theta = 4/5$
(e) $\sin\theta = 4/3$

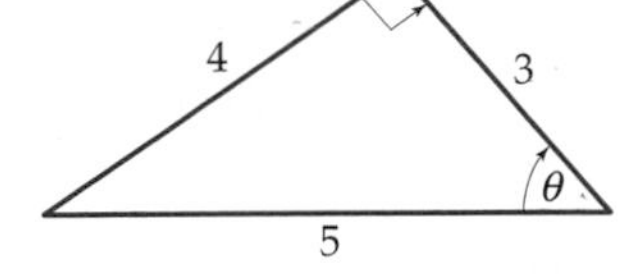

4. Suppose θ is an acute angle. Consider the right triangle with sides as shown in the figure. Then

(a) $x = 1$
(b) $x = 2$
(c) $x = 4$
(d) $x = 2(\cos\theta + \sin\theta)$
(e) none of the above

[*Hint*: Exercise 45(c) of Section 1.1]

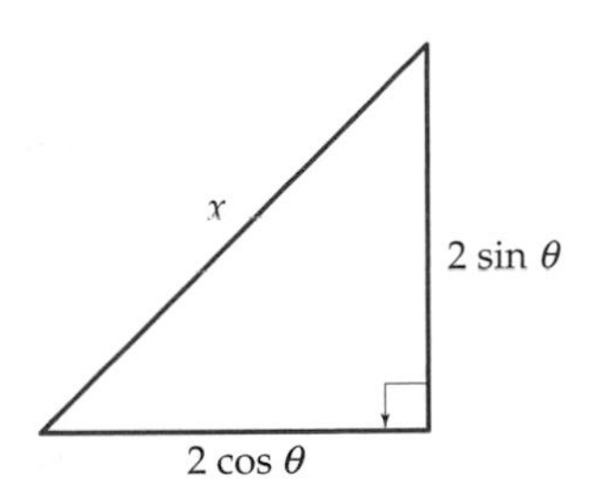

5. If θ is an angle, denote the number $\dfrac{1}{\cos\theta}$ by $\sec\theta$. Use the right triangle in the figure to find $\sec\theta$.

(a) $\sec\theta = 7/4$
(b) $\sec\theta = 4/\sqrt{65}$
(c) $\sec\theta = 7/\sqrt{65}$
(d) $\sec\theta = \sqrt{65}/7$
(e) $\sec\theta = \sqrt{65}/4$

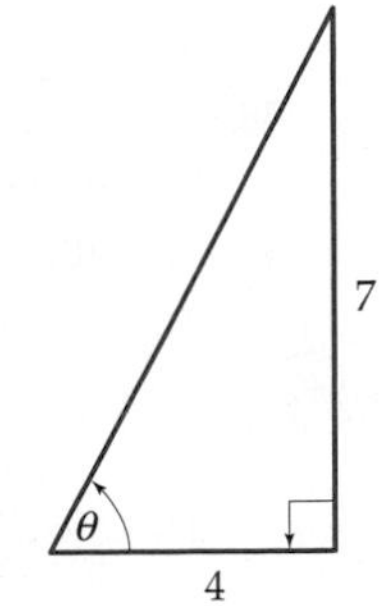

6. Find the length of side h in the triangle, where angle A measures 40° and the distance from C to A is 25.

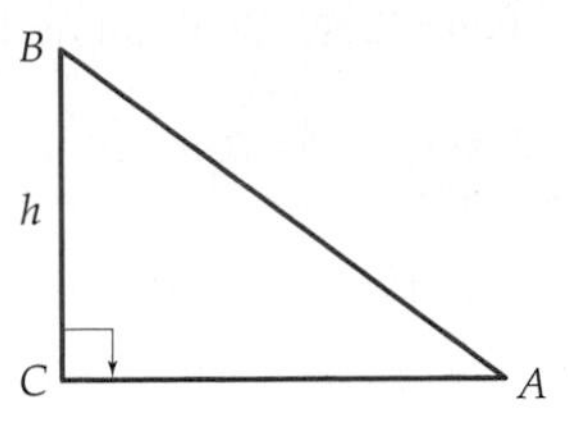

In Questions 7–10, angle B is a right angle. Solve triangle ABC.

7. $a = 12$, $c = 13$

8. $A = 40°$, $b = 10$

9. $C = 35°$, $a = 12$

10. $A = 56°$, $a = 11$

11. From a point on level ground 145 feet from the base of a tower, the angle of elevation to the top of the tower is 57.3.°. How high is the tower?

12. A pilot in a plane at an altitude of 22,000 feet observes that the angle of depression to a nearby airport is 26°. How many *miles* is the airport from a point on the ground directly below the plane?

13. A road rises 140 feet per horizontal mile. What angle does the road make with the horizontal?

14. A lighthouse keeper 100 feet above the water sees a boat sailing in a straight line directly toward her. As she watches, the angle of depression to the boat changes from 25° to 40°. How far has the boat traveled during this time?

In Exercises 15–17, find $\sin\theta$, $\cos\theta$, *and* $\tan\theta$, *where* θ *is an angle in standard position whose terminal side passes through the given point.*

15. $(-5, 5)$

16. $(-4, 12)$

17. $(6, 10)$

18. The terminal side of angle θ in standard position (with $0° < \theta < 180°$) lies on the line with equation $3x + 2y = 0$. Find $\sin\theta$, $\cos\theta$, and $\tan\theta$.

In Questions 19–22, use the Law of Cosines to solve triangle ABC.

19. $a = 12$, $b = 10$, $c = 15$

20. $a = 7.5$, $b = 3.2$, $c = 6.4$

21. $a = 10$, $c = 14$, $B = 130°$

22. $a = 7$, $b = 8.6$, $C = 72.4°$

23. Two trains depart simultaneously from the same station. The angle between the two tracks on which they leave is 120°. One train travels at an average speed of 45 mph and the other at 70 mph. How far apart are the trains after 3 hours?

REVIEW QUESTIONS 

24. A 40-foot-high flagpole sits on the side of a hill. The hillside makes a 17° angle with the horizontal. How long is a wire that runs from the top of the pole to a point 72 feet downhill from the base of the pole?

In Questions 25–30, use the Law of Sines to solve triangle ABC.

25. $B = 124°$, $C = 31°$, $c = 3.5$
26. $A = 96°$, $B = 44°$, $b = 12$
27. $a = 75$, $c = 84$, $C = 62°$
28. $a = 5$, $c = 2.5$, $C = 30°$
29. $a = 3.5$, $b = 4$, $A = 60°$
30. $a = 3.8$, $c = 2.8$, $C = 41°$
31. Find the area of triangle ABC if $b = 24$, $c = 15$, and $A = 55°$.
32. Find the area of triangle ABC if $a = 10$, $c = 14$, and $B = 75°$.
33. A boat travels for 8 kilometers in a straight line from the dock. It is then sighted from a lighthouse that is 6.5 kilometers from the dock. The angle determined by the dock, the lighthouse (vertex), and the boat is 25°. How far is the boat from the lighthouse?
34. A pole tilts 12° from the vertical, away from the sun, and casts a 34-foot-long shadow on level ground. The angle of elevation from the end of the shadow to the top of the pole is 64°. How long is the pole?

In Questions 35–38, solve triangle ABC.

35. $A = 48°$, $B = 57°$, $b = 47$
36. $A = 67°$, $c = 125$, $a = 100$
37. $a = 5$, $c = 8$, $B = 76°$
38. $a = 90$, $b = 70$, $c = 40$
39. Two surveyors, Joe and Alice, are 240 meters apart on a riverbank. Each sights a flagpole on the opposite bank. The angle from the pole to Joe (vertex) to Alice is 63°. The angle from the pole to Alice (vertex) to Joe is 54°. How far are Joe and Alice from the pole?
40. A surveyor stakes out points A and B on opposite sides of a building. Point C on the side of the building is 300 feet from A and 440 feet from B. Angle ACB measures 38°. What is the distance from A to B?
41. A woman on the top of a 448-foot-high building spots a small plane. As she views the plane, its angle of elevation is 62°. At the same instant, a man at the ground-level entrance to the building sees the plane and notes that its angle of elevation is 65°.
    (a) How far is the woman from the plane?
    (b) How far is the man from the plane?
    (c) How high is the plane?
42. A straight road slopes at an angle of 10° with the horizontal. When the angle of elevation of the sun (from horizontal) is 62.5°, a telephone pole at the side of the road casts a 15-foot shadow downhill, parallel to the road. How high is the telephone pole?
43. Find angle ABC.

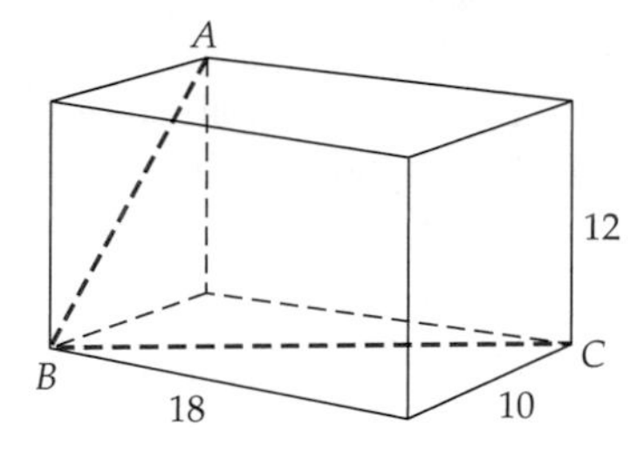

44. Use the Law of Sines to prove Engelsohn's equations: For any triangle ABC (standard notation),

$$\frac{a+b}{c} = \frac{\sin A + \sin B}{\sin C}$$

and

$$\frac{a-b}{c} = \frac{\sin A - \sin B}{\sin C}.$$

In Questions 45–48, find the area of triangle ABC under the given conditions.

45. There is an angle of 30°, the sides of which have lengths 5 and 8.
46. There is an angle of 40°, the sides of which have lengths 3 and 12.
47. The sides have lengths 7, 11, and 14.
48. The sides have lengths 4, 8, and 10.

DISCOVERY PROJECT 1 Life on a Sphere

Russell Illig/Getty Images

Although we experience the surface of the earth as a level surface, it is not. The earth, of course, is a sphere whose radius is about 6370 kilometers. When you stand on a sphere, strange things happen, especially when the surface is particularly smooth.

Granted, the horizon is often interfered with by hills, forests, buildings, and such. However, if you are out in an area of flat land like that found in western Kansas or northern Ontario, the hills are very slight, and the ground is very much like a sphere. The same is true of large lakes and oceans.

Generally speaking, you are not as interested in seeing the horizon as you are in seeing objects that the horizon might hide. On the ocean, for instance, you might want to see another ship. On land, you might be looking for a particular building or vehicle.

1. Suppose that a person is standing on a highway and a 3.5-meter-high truck passes by. Assume also that this person has eyes that are 1.55 meters off the ground. If the road is flat and straight, how far away is the truck when it disappears below the horizon?

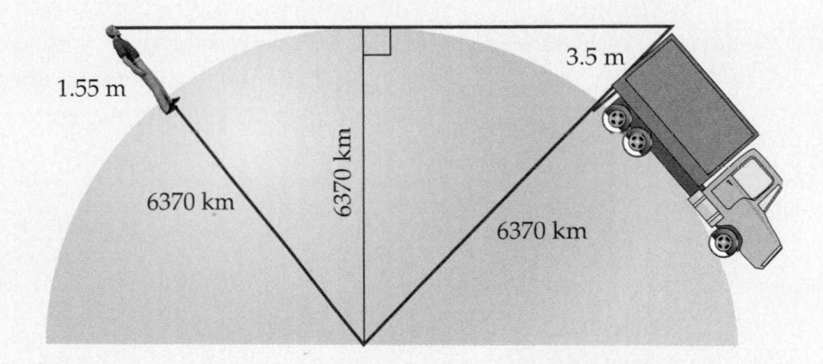

(Project continues)

2. Engineers building large bridges must also account for the curvature of the earth. In the figure, a bridge with towers 200 meters tall and 900 meters apart (straight-line distance) at the base is constructed. How much farther apart are the tops of the towers than the bases?

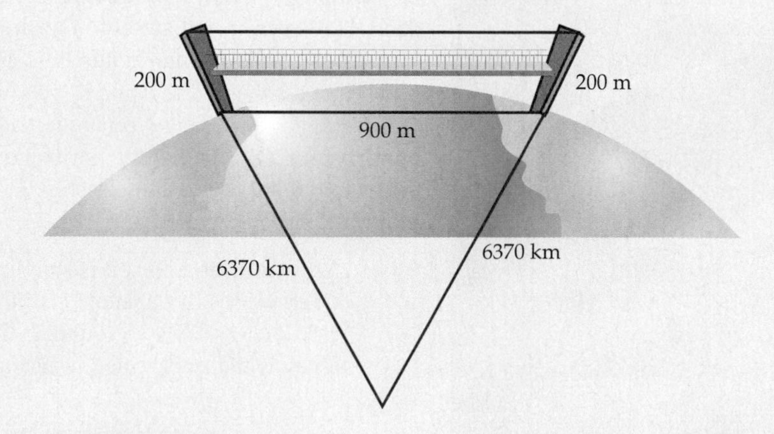

3. How much difference does it make if the distance between the towers is measured along the curve of the earth?

CHAPTER 2

Trigonometric Functions

Don Smetzer/Getty Images

AM signal

FM signal

Don't touch that dial!

Radio stations transmit by sending out a signal in the form of an electromagnetic wave that can be described by a trigonometric function. The shape of this signal is modified by the sounds being transmitted. AM radio signals are modified by varying the "height," or amplitude, of the waves, whereas FM signals are modified by varying the frequency of the waves. See Exercise 47 on page 193.

Chapter Outline

Interdependence of Sections

2.1 → 2.2 → 2.3 → 2.4 → 2.5
2.4 → 2.6*

Roadmap

Instructors who cover Chapter 2 before Chapter 1 should replace Section 2.2 with Alternate Section 2.2. In Chapter 1, they should replace Section 1.1 with Alternate Section 1.1.

The ancient Greeks developed trigonometry for measuring angles and sides of triangles to solve problems in astronomy, navigation, and surveying.[†] But with the invention of calculus in the seventeenth century and the subsequent explosion of knowledge in the physical sciences, a different viewpoint toward trigonometry arose.

Whereas the ancients dealt only with *angles*, the classical trigonometric concepts of sine and cosine are now considered as *functions* with domain the set of all *real numbers*. The advantage of this switch in viewpoint is that almost any phenomenon involving rotation or vibration can be described in terms of trigonometric functions, including light rays, sound waves, electron orbitals, planetary orbits, radio transmission, vibrating strings, pendulums, and many more.

The presentation of trigonometry here reflects this modern viewpoint. Nevertheless, angles still play an important role in defining the trigonometric functions, so the chapter begins with them.

2.1 Angles and Their Measurement

In Chapter 1, we considered angles as static figures, consisting of two fixed line segments (the sides of a triangle) that begin at the same point. In this chapter, we take a different viewpoint and think of an **angle** as being formed dynamically by *rotating* a line segment around its endpoint (the **vertex**), as shown in Figure 2–1. The rotation may be clockwise or counterclockwise as indicated by the arrow. The position of the line segment at the beginning is the **initial side,** and its final position is the **terminal side** of the angle.

---

*Parts of Section 2.6 may be covered much earlier; see the Roadmap at the beginning of Section 2.2.

[†]In fact, *trigonometry* means "triangle measurement."

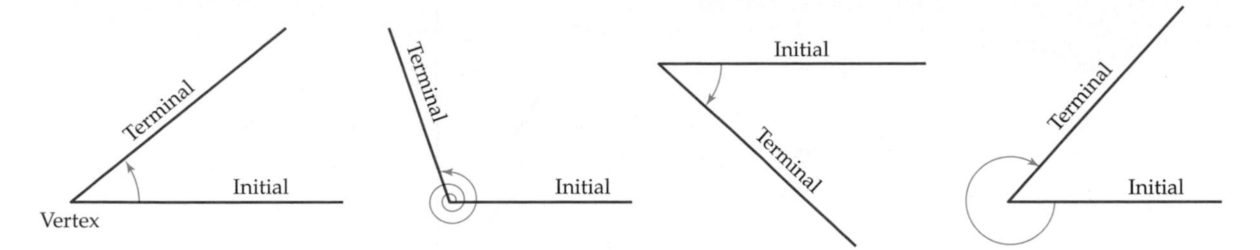

Figure 2–1

Figure 2–2 shows that different angles (that is, angles obtained by different rotations) may have the same initial and terminal side.* Such angles are said to be **coterminal.**

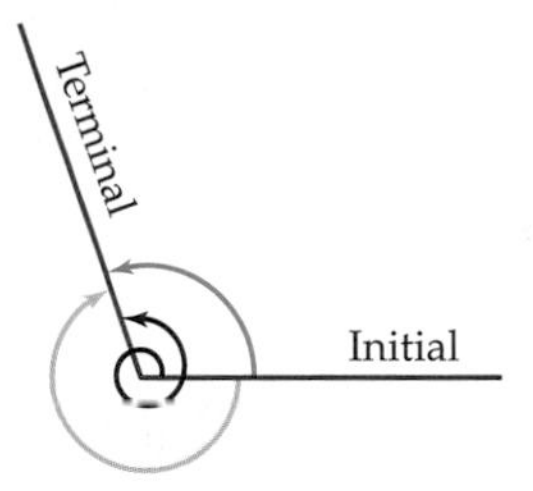

Figure 2–2

An angle in the coordinate plane is said to be in **standard position** if its vertex is at the origin and its initial side on the positive x-axis, as in Figure 2–3. When measuring angles in standard position, we use positive numbers for angles obtained by counterclockwise rotation (**positive angles**) and negative numbers for ones obtained by clockwise rotation (**negative angles**).

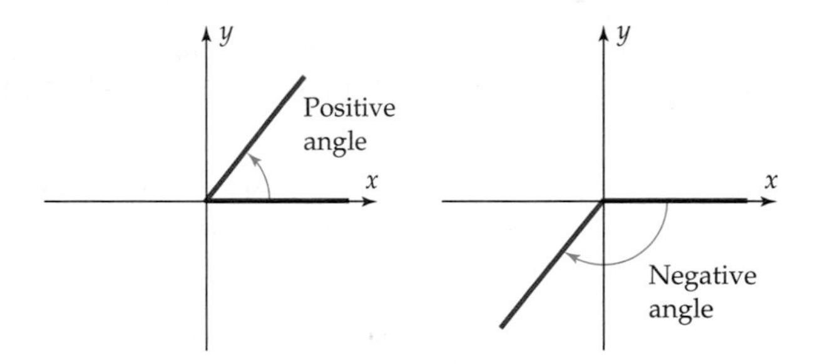

Figure 2–3

The classical unit for angle measurement is the **degree** (in symbols, °). You should be familiar with the positive angles in standard position shown in Figure 2–4 on the next page. Note that a 360° angle corresponds to one full revolution and thus is coterminal with an angle of 0°.

*They are *not* the same angle, however. For instance, both $\frac{1}{2}$ turn and $1\frac{1}{2}$ turns put a circular faucet handle in the same position, but the water flow is quite different.

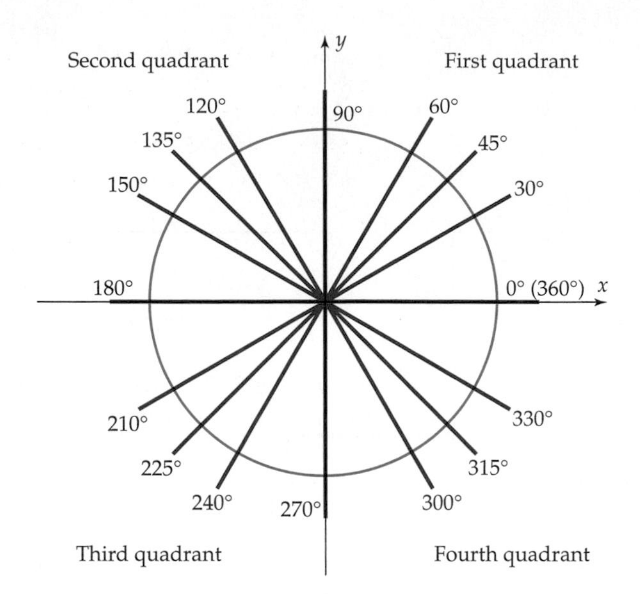

Figure 2–4

RADIAN MEASURE

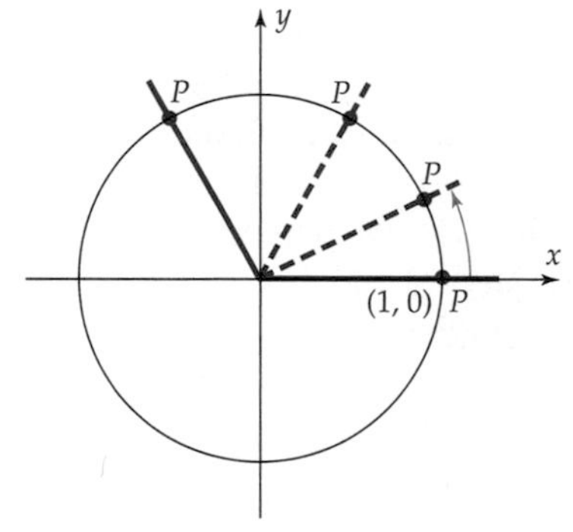

Figure 2–5

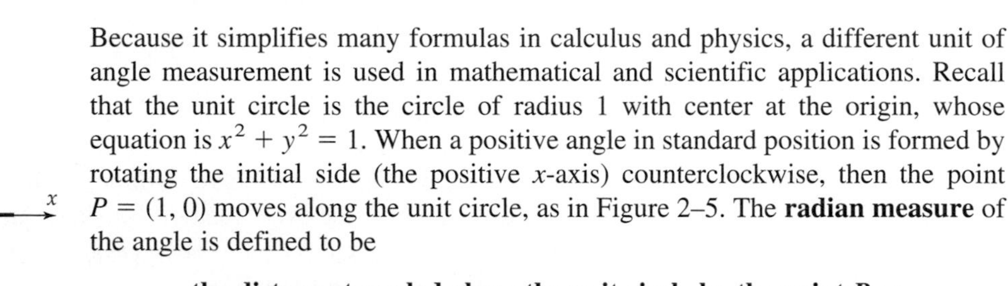

Because it simplifies many formulas in calculus and physics, a different unit of angle measurement is used in mathematical and scientific applications. Recall that the unit circle is the circle of radius 1 with center at the origin, whose equation is $x^2 + y^2 = 1$. When a positive angle in standard position is formed by rotating the initial side (the positive x-axis) counterclockwise, then the point $P = (1, 0)$ moves along the unit circle, as in Figure 2–5. The **radian measure** of the angle is defined to be

> **the distance traveled along the unit circle by the point P as it moves from its starting position on the initial side to its final position on the terminal side of the angle.**

The radian measure of a negative angle in standard position is found in the same way, except that you move clockwise along the unit circle. Figure 2–6 shows angles of 3.75, 7, and -2 radians, respectively.

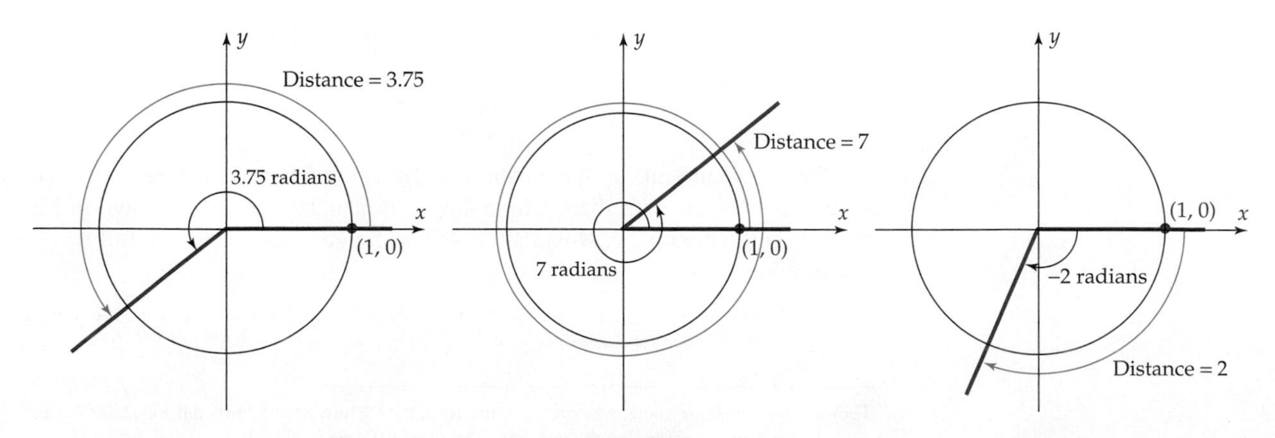

Figure 2–6

To become comfortable with radian measure, think of the terminal side of the angle revolving around the origin: When it makes one full revolution, it produces an angle of 2π radians (because the circumference of the unit circle is 2π). When it makes half a revolution, it forms an angle whose radian measure is $1/2$ of 2π, that is, π radians, and so on, as illustrated in Figure 2–7 and the table below.

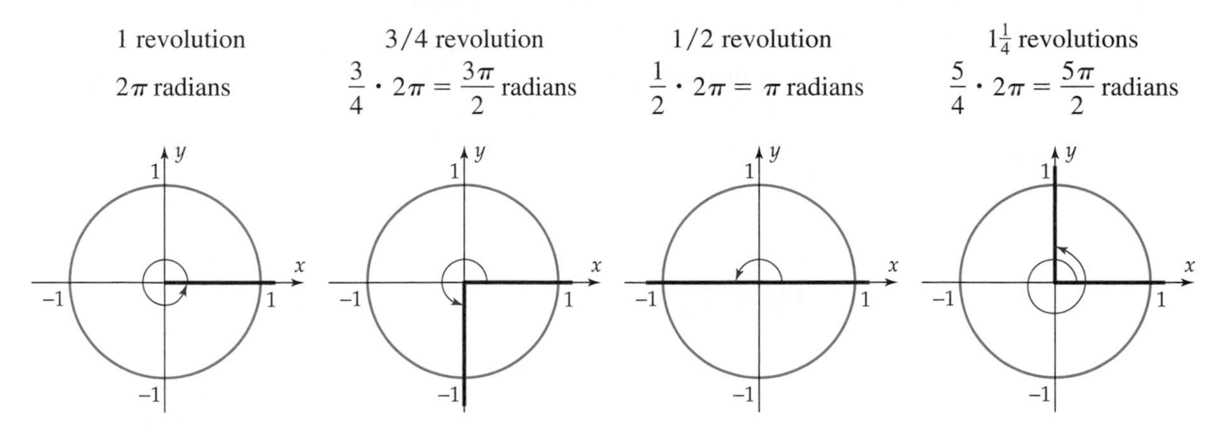

Figure 2–7

Terminal Side	Radian Measure of Angle	Equivalent Degree Measure
1 revolution	2π	360°
$\frac{7}{8}$ revolution	$\frac{7}{8} \cdot 2\pi = \frac{7\pi}{4}$	$\frac{7}{8} \cdot 360 = 315°$
$\frac{3}{4}$ revolution	$\frac{3}{4} \cdot 2\pi = \frac{3\pi}{2}$	$\frac{3}{4} \cdot 360 = 270°$
$\frac{2}{3}$ revolution	$\frac{2}{3} \cdot 2\pi = \frac{4\pi}{3}$	$\frac{2}{3} \cdot 360 = 240°$
$\frac{1}{2}$ revolution	$\frac{1}{2} \cdot 2\pi = \pi$	$\frac{1}{2} \cdot 360 = 180°$
$\frac{1}{3}$ revolution	$\frac{1}{3} \cdot 2\pi = \frac{2\pi}{3}$	$\frac{1}{3} \cdot 360 = 120°$
$\frac{1}{4}$ revolution	$\frac{1}{4} \cdot 2\pi = \frac{\pi}{2}$	$\frac{1}{4} \cdot 360 = 90°$
$\frac{1}{6}$ revolution	$\frac{1}{6} \cdot 2\pi = \frac{\pi}{3}$	$\frac{1}{6} \cdot 360 = 60°$
$\frac{1}{8}$ revolution	$\frac{1}{8} \cdot 2\pi = \frac{\pi}{4}$	$\frac{1}{8} \cdot 360 = 45°$
$\frac{1}{12}$ revolution	$\frac{1}{12} \cdot 2\pi = \frac{\pi}{6}$	$\frac{1}{12} \cdot 360 = 30°$

Although equivalent degree measures are given in the table, you should learn to "think radian" as much as possible rather than mentally translating from radians to degrees.

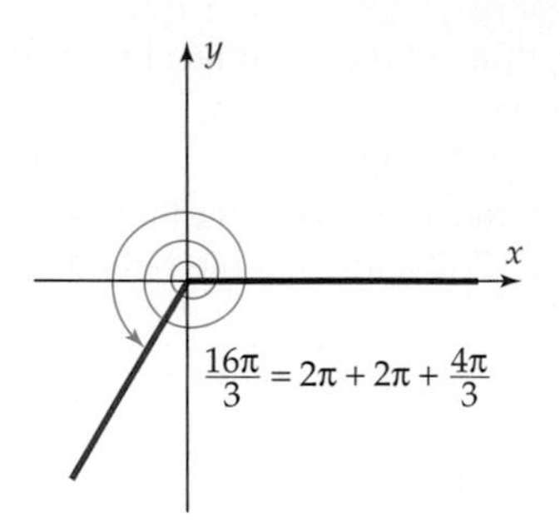

Figure 2–8

EXAMPLE 1

To construct an angle of $16\pi/3$ radians in standard position, note that

$$\frac{16\pi}{3} = \frac{12\pi}{3} + \frac{4\pi}{3} = 4\pi + \frac{4\pi}{3} = 2\pi + 2\pi + \frac{4\pi}{3}.$$

So the terminal side must be rotated counterclockwise through two complete revolutions (each full-circle revolution is 2π radians) and then rotated an additional $2/3$ of a revolution (since $4\pi/3$ is $2/3$ of a complete revolution of 2π radians), as shown in Figure 2–8. ■

EXAMPLE 2

Since $-5\pi/4 = -\pi - \pi/4$, an angle of $-5\pi/4$ radians in standard position is obtained by rotating the terminal side *clockwise* for half a revolution (π radians) plus an additional $1/8$ of a revolution (since $\pi/4$ is $1/8$ of a full-circle revolution of 2π radians), as shown in Figure 2–9. ■

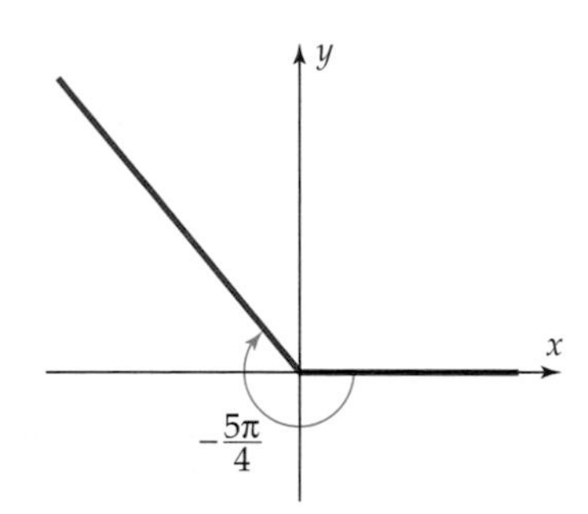

Figure 2–9

Consider an angle of t radians in standard position (Figure 2–10). Since 2π radians corresponds to a full revolution of the terminal side, this angle has the same terminal side as an angle of $t + 2\pi$ radians or $t - 2\pi$ radians or $t + 4\pi$ radians.

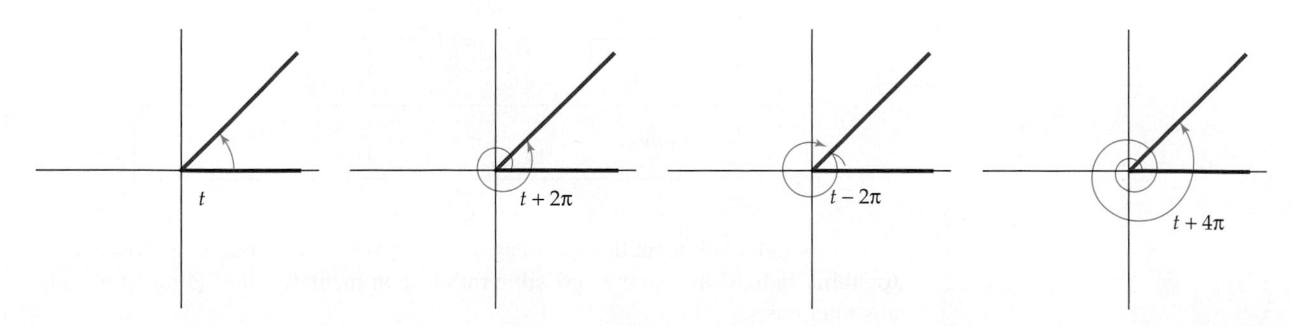

Figure 2–10

The same thing is true in general.

Coterminal Angles

> Increasing or decreasing the radian measure of an angle by an integer multiple of 2π results in a coterminal angle.

EXAMPLE 3

Find angles in standard position that are coterminal with an angle of

(a) $23\pi/5$ radians (b) $-\pi/12$ radians.

SOLUTION

(a) We can subtract 2π to obtain a coterminal angle whose measure is

$$\frac{23\pi}{5} - 2\pi = \frac{23\pi}{5} - \frac{10\pi}{5} = \frac{13\pi}{5} \text{ radians,}$$

or we can subtract 4π to obtain a coterminal angle of measure

$$\frac{23\pi}{5} - 4\pi = \frac{3\pi}{5} \text{ radians.}$$

Subtracting 6π produces a coterminal angle of

$$\frac{23\pi}{5} - 6\pi = -\frac{7\pi}{5} \text{ radians.}$$

(b) An angle of $\frac{-\pi}{12}$ radians is coterminal with an angle of

$$-\frac{\pi}{12} + 2\pi = \frac{23\pi}{12} \text{ radians}$$

and with an angle of

$$-\frac{\pi}{12} - 2\pi = -\frac{25\pi}{12} \text{ radians.}$$ ■

RADIAN/DEGREE CONVERSION

Although we shall generally work with radians, it may occasionally be necessary to convert from radian to degree measure or vice versa. The key to doing this is the fact that

(*) $$\boldsymbol{\pi \text{ radians} = 180^\circ.}$$

Dividing both sides of (*) by π shows that

$$1 \text{ radian} = \frac{180}{\pi} \text{ degrees} \approx 57.3^\circ,$$

and dividing both sides of (*) by 180 shows that

$$1^\circ = \frac{\pi}{180} \text{ radians} \approx .0175 \text{ radians.}$$

Consequently, we have the following rules.

Radian/Degree Conversion

> To convert radians to degrees, multiply by $\frac{180}{\pi}$.
>
> To convert degrees to radians, multiply by $\frac{\pi}{180}$.

EXAMPLE 4

(a) To find the degree measure of an angle of 2.4 radians, multiply by $\frac{180}{\pi}$:

$$(2.4)\left(\frac{180}{\pi}\right) = \frac{432}{\pi} \approx 137.51°.$$

(b) An angle of $-.3$ radians has degree measure

$$(-.3)\left(\frac{180}{\pi}\right) = \frac{-54}{\pi} \approx -17.19°. \quad ■$$

Angle conversions from radian to degree measure or vice versa can be done directly on a calculator. Since procedures vary from one model to another, check your instruction manual.

EXERCISES 2.1

In Exercises 1–5, find the radian measure of the angle in standard position formed by rotating the terminal side the given amount.

1. 1/9 of a circle **2.** 1/24 of a circle

3. 1/18 of a circle **4.** 1/72 of a circle

5. 1/36 of a circle

6. State the radian measure of *every* standard position angle in the figure.*

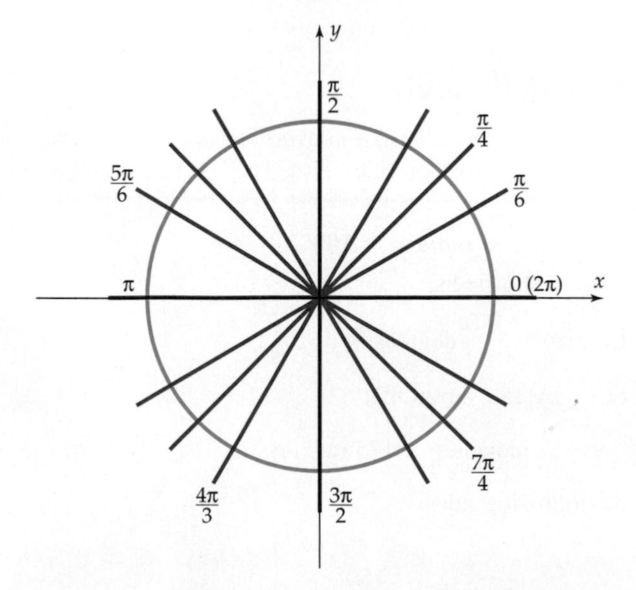

*This is the same diagram that appears in Figure 2–4 on page 138, showing positive angles in standard position.

In Exercises 7–10, find the radian measure of four angles in standard position that are coterminal with the angle in standard position whose measure is given.

7. $\pi/4$ **8.** $7\pi/5$

9. $-\pi/6$ **10.** $-9\pi/7$

In Exercises 11–18, find the radian measure of an angle in standard position that has measure between 0 and 2π and is coterminal with the angle in standard position whose measure is given.

11. $-\pi/3$ **12.** $-3\pi/4$ **13.** $19\pi/4$

14. $16\pi/3$ **15.** $-7\pi/5$ **16.** $45\pi/8$

17. 7 **18.** 18.5

In Exercises 19–30, convert the given degree measure to radians.

19. 6° **20.** −10° **21.** −12° **22.** 36°

23. 75° **24.** −105° **25.** 135° **26.** −165°

27. −225° **28.** 252° **29.** 930° **30.** −585°

In Exercises 31–42, convert the given radian measure to degrees.

31. $\pi/5$ **32.** $-\pi/6$ **33.** $-\pi/10$

34. $2\pi/5$ **35.** $3\pi/4$ **36.** $-5\pi/3$

37. $\pi/45$ **38.** $-\pi/60$ **39.** $-5\pi/12$

40. $7\pi/15$ **41.** $27\pi/5$ **42.** $-41\pi/6$

In Exercises 43–48, determine the positive radian measure of the angle that the second hand of a clock traces out in the given time.

43. 40 seconds

44. 50 seconds

45. 35 seconds

46. 2 minutes and 15 seconds.

47. 3 minutes and 25 seconds

48. 1 minute and 55 seconds

In Exercises 49–56, a wheel is rotating around its axle. Find the angle (in radians) through which the wheel turns in the given time when it rotates at the given number of revolutions per minute (rpm). Assume that $t > 0$ and $k > 0$.

49. 3.5 minutes, 1 rpm

50. t minutes, 1 rpm

51. 1 minute, 2 rpm

52. 3.5 minutes, 2 rpm

53. 4.25 minutes, 5 rpm

54. t minutes, 5 rpm

55. 1 minute, k rpm

56. t minutes, k rpm

2.1.A *SPECIAL TOPICS* Arc Length and Angular Speed

Consider a circle of radius r, with center at the origin, and an angle of θ radians in standard position, as shown in Figure 2–11. As you can see, the sides of the angle of θ radians determine an arc length s along the circle. We say that the **central angle** of θ radians **subtends an arc** of length s on the circle. It can be shown that the ratio of the arc length s to the circumference of the entire circle (namely, $2\pi r$) is the same as the ratio of the angle of θ radians to the full-circle angle of 2π radians; that is,

$$\frac{s}{2\pi r} = \frac{\theta}{2\pi}.$$

Solving this equation for s, we obtain the following fact.

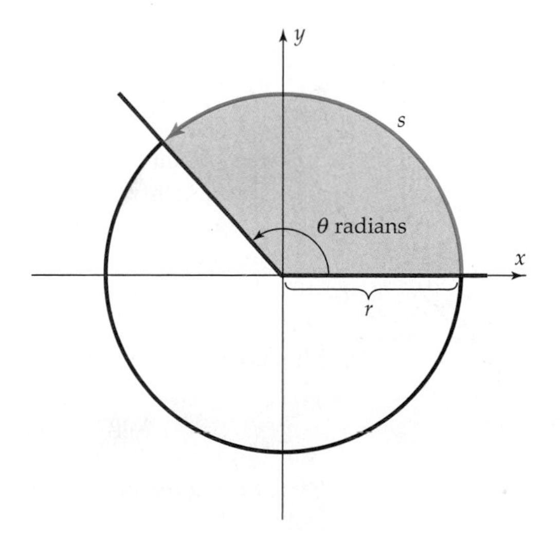

Figure 2–11

Arc Length

> A central angle of θ radians in a circle of radius r subtends an arc of length
> $$s = \theta r,$$
> that is, the length s of the arc is the product of the radian measure of the angle and the radius of the circle.

EXAMPLE 1

The second hand on a large clock is 6 inches long. How far does the tip of the second hand move in 15 seconds?

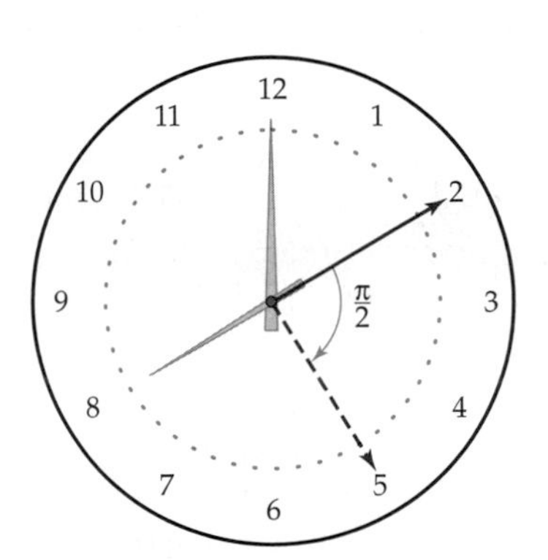

Figure 2–12

SOLUTION The second hand makes a full revolution every 60 seconds, that is, it moves through an angle of 2π radians. During a 15-second interval, it will make $\frac{15}{60} = \frac{1}{4}$ of a revolution, moving through an angle of $\pi/2$ radians (Figure 2–12). If we think of the second hand as the radius of a circle, then during a 15-second interval, its tip travels along the arc subtended by an angle of $\pi/2$ radians. Therefore, the distance (arc length) traveled by the tip of the second hand is

$$s = \theta r = \frac{\pi}{2} \cdot 6 = 3\pi \approx 9.425 \text{ inches.} \qquad \blacksquare$$

EXAMPLE 2

Suppose the circle in Figure 2–11 has radius 6 and the arc length s is 14.64. We can find the measure of the central angle that subtends the arc s by solving $s = \theta r$ for θ.

$$\theta = \frac{s}{r} = \frac{14.64}{6} = 2.44 \text{ radians.} \qquad \blacksquare$$

ANGULAR SPEED

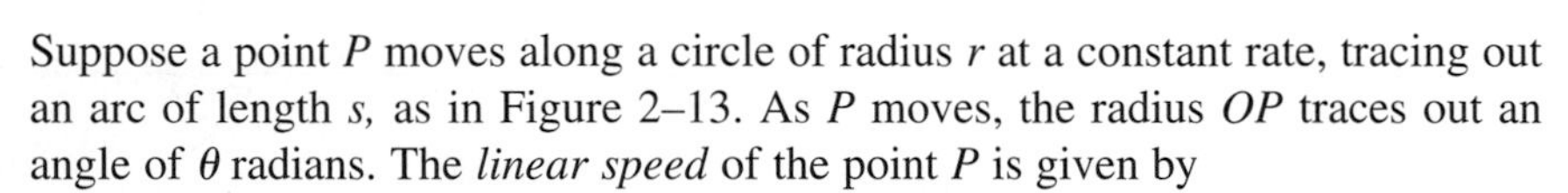

Suppose a point P moves along a circle of radius r at a constant rate, tracing out an arc of length s, as in Figure 2–13. As P moves, the radius OP traces out an angle of θ radians. The *linear speed* of the point P is given by

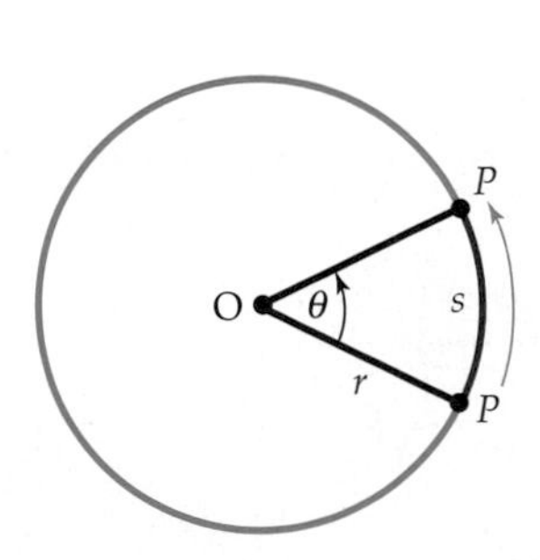

Figure 2–13

$$\textbf{Linear speed} = \frac{\text{Distance traveled by } P}{\text{Times elapsed}} = \frac{s}{t} = \frac{\theta r}{t}.$$

Similarly, the *angular speed* of P is defined by

$$\textbf{Angular speed} = \frac{\text{Angle traced out by } OP}{\text{Time elapsed}} = \frac{\theta}{t}.$$

By rewriting the linear speed equation, we obtain the relationship between linear and angular speed.

$$\frac{\theta r}{t} = r \cdot \frac{\theta}{t}$$

$$\textbf{Linear speed} = r \cdot \textbf{(Angular speed).}$$

EXAMPLE 3

A merry-go-round makes eight revolutions per minute.

(a) What is the angular speed of the merry-go-round in radians per minute?

(b) How fast is a horse 12 feet from the center traveling?

(c) How fast is a horse 4 feet from the center traveling?

SOLUTION

(a) Each revolution of the merry-go-round corresponds to a central angle of 2π radians, so it travels through an angle of $8 \cdot 2\pi = 16\pi$ radians in one minute. Therefore, its

$$\text{Angular speed} = \frac{\theta}{t} = \frac{16\pi}{1} = 16\pi \text{ radians per minute.}$$

(b) The horse that is 12 feet from the center travels along a circle of radius 12 and travels through an angle of 16π radians in 1 minute. Therefore, its

$$\text{Linear speed} = \frac{\theta r}{t} = \frac{16\pi \cdot 12}{1} = 192\pi \text{ feet per minute,}$$

which is approximately 6.9 miles per hour.

(c) For the horse that is 4 feet from the center, $r = 4$, $\theta = 16\pi$, and $t = 1$. Hence,

$$\text{Linear speed} = \frac{\theta r}{t} = \frac{16\pi \cdot 4}{1} = 64\pi \text{ feet per minute (about 2.28 mph).}$$

■

EXAMPLE 4

A bicycle, each of whose wheels is 26 inches in diameter, travels at a constant 14 mph.

(a) What is the angular speed of one of its wheels (in radians per hour)?

(b) How many revolutions per minute does each wheel make?

SOLUTION Within each computation, we must be careful to use comparable units.

(a) Since the wheel size is given in inches, we also express the speed in terms of inches. Recall that there are 5280 feet in a mile, which means that

$$1 \text{ mile} = 5280 \cdot 12 = 63{,}360 \text{ inches.}$$

Thus, the bike's linear speed is

$$14 \text{ miles per hour} = 14(63{,}360) \text{ inches per hour.}$$

As we saw above, linear speed $= r \cdot$ (angular speed), so

$$\text{Angular speed} = \frac{\text{Linear speed}}{r} = \frac{14(63{,}360)}{13} \approx 68{,}233.85 \text{ radians per hour.}$$

(b) Each revolution corresponds to an angle of 2π radians, so the number of revolutions per hour is $\dfrac{68{,}233.85}{2\pi}$. Dividing this number by 60 gives the revolutions per minute.

$$\frac{1}{60}\left(\frac{68{,}233.85}{2\pi}\right) \approx 181 \text{ revolutions per minute.} \quad \blacksquare$$

EXERCISES 2.1.A

1. The second hand on a clock is six centimeters long. How far does its tip travel in 40 seconds? (See Exercise 43 in Section 2.1.)
2. The second hand on a clock is five centimeters long. How far does its tip travel in 2 minutes and 15 seconds? (See Exercise 46 in Section 2.1.)
3. If the radius of the circle in the figure is 20 centimeters and the length of arc s is 85 centimeters, what is the radian measure of the angle θ?

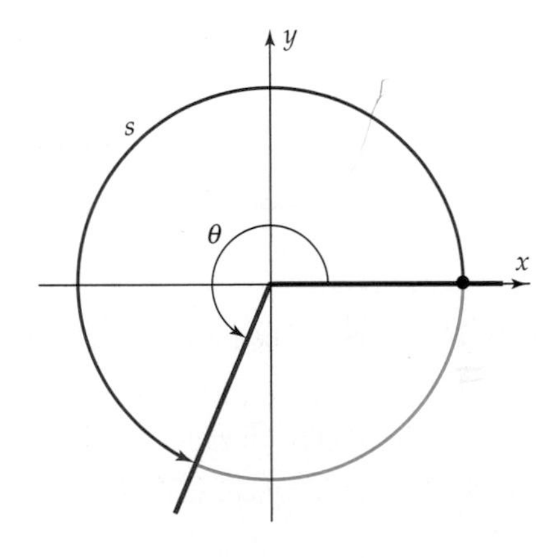

4. Find the radian measure of the angle θ in the preceding figure if the *diameter* of the circle is 150 centimeters and s has length 360 centimeters.

In Exercises 5–8, assume that a wheel on a car has radius 36 centimeters. Find the angle (in radians) that the wheel turns while the car travels the given distance.

5. 2 meters (= 200 centimeters)
6. 5 meters
7. 720 meters
8. 1 kilometer (= 1000 meters)

In Exercises 9–12, find the length of the circular arc subtended by the central angle whose radian measure is given. Assume the circle has diameter 10.

9. 1 radian
10. 2 radians
11. 1.75 radians
12. 2.2 radians

*In Exercises 13–16, the latitudes of a pair of cities are given. Assume that one city is directly south of the other and that the earth is a perfect sphere of radius 4000 miles. Find the distance between the two cities. [The **latitude** of a point P on the earth is the degree measure of the angle θ between the point and the plane of the equator (with the center of the earth being the vertex), as shown in the figure. Remember that angles are measured in radians in the arc length formula.]*

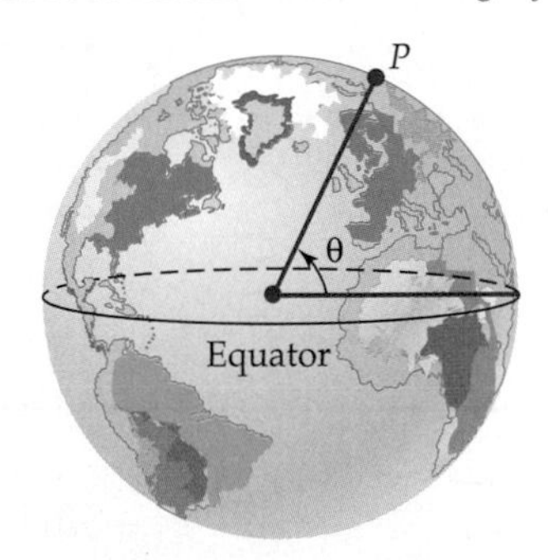

13. The North Pole (latitude 90° north) and Springfield, Illinois (latitude 40° north).
14. San Antonio, Texas (latitude 29.5° north), and Mexico City, Mexico (latitude 20° north).
15. Cleveland, Ohio (latitude 41.5° north), and Tampa, Florida (latitude 28° north).
16. Copenhagen, Denmark (latitude 54.3° north), and Rome, Italy (latitude 42° north).

17. One end of a rope is attached to a winch (circular drum) of radius 2 feet and the other to a steel beam on the ground. When the winch is rotated, the rope wraps around the drum and pulls the object upward (see the figure). Through what angle must the winch be rotated to raise the beam 6 feet above the ground?

18. A spy plane on a practice run over the Midwest takes a picture that shows Cleveland, Ohio, on the eastern horizon and St. Louis, Missouri, on the western horizon (the figure is not to scale).

(a) The arc of the earth between St. Louis and Cleveland is 520 miles long. Assume that the radius of the earth is 3950 miles. Use the arc length formula to find the radian measure of angle θ.

(b) If you have read Section 1.1, use trigonometry to find the height of the plane when the picture was taken. [*Hint:* The sight lines from the plane to the horizons are tangent to the earth and a tangent line to a circle is perpendicular to the radius at that point. Note that $\alpha = \theta/2$ and θ is in radians.]

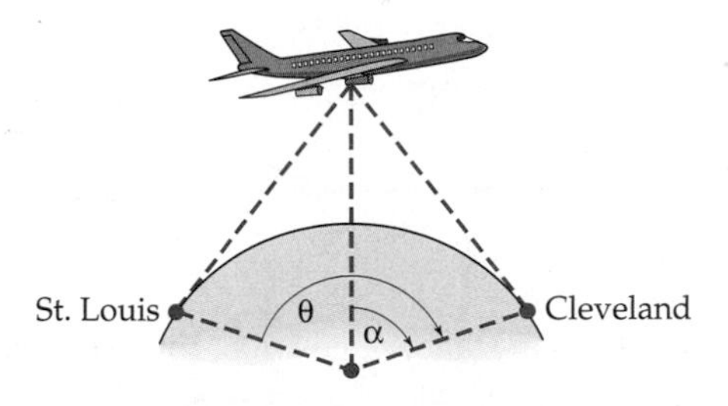

19. A circular gear rotates at the rate of 200 revolutions per minute (rpm).

(a) What is the angular speed of the gear in radians per minute?

(b) What is the linear speed of a point on the gear 2 inches from the center in inches per minute and in feet per minute?

20. A circular saw blade has an angular speed of 15,000 radians per minute.

(a) How many revolutions per minute does the saw make?

(b) How long will it take the saw to make 6000 revolutions?

21. A bicycle has wheels that are 26 inches in diameter. If the bike is traveling at 18 mph, what is the angular speed of each wheel?

22. A wheel in a large machine is 2.8 feet in diameter and rotates at 1200 rpm.

(a) What is the angular speed of the wheel?

(b) How fast is a point on the circumference of the wheel traveling in feet per minute? In miles per hour?

23. A riding lawn mower has wheels that are 15 inches in diameter. If the wheels are making 2.5 revolutions per second

(a) What is the angular speed of a wheel?

(b) How fast is the lawn mower traveling in miles per hour?

24. A weight is attached to the end of a four-foot-long leather cord. A boy swings the cord in a circular motion over his head. If the weight makes 12 revolutions every 10 seconds, what is its linear speed?

25. A merry-go-round horse is traveling at 10 feet per second, and the merry-go-round is making 6 revolutions per minute. How far is the horse from the center of the merry-go-round?

26. The pedal sprocket of a bicycle has radius 4.5 inches, and the rear wheel sprocket has radius 1.5 inches (see figure). If the rear wheel has a radius of 13.5 inches and the cyclist is pedaling at the rate of 80 rpm, how fast is the bicycle traveling in feet per minute? In miles per hour?

2.2 The Sine, Cosine, and Tangent Functions

NOTE

If you have not read Chapter 1, use Alternate Section 2.2 on page 157 in place of this section.

Trigonometric functions of any angle were defined in Section 1.3, using a point on the terminal side of the angle. According to that definition, the domains of the sine, cosine, and tangent functions consist of *angles.* We now define trigonometric functions whose domains consist of *real numbers.* The basic idea is quite simple: If t is a real number, then

sin t is defined to be the sine of an angle of t radians.

cos t is defined to be the cosine of an angle of t radians.

tan t is defined to be the tangent of an angle of t radians.

Instructors who wish to cover all six trigonometric functions simultaneously should incorporate Section 2.6 into Sections 2.2–2.4, as follows.

Subsection of Section 2.6	Cover at the end of
Definitions; Alternate Descriptions	Section 2.2
Algebra and Identities	Section 2.3
Graphs	Section 2.4

In other words, instead of starting with an angle as in Chapter 1, we now start with a number t and then move to an angle of t radians, as summarized here:

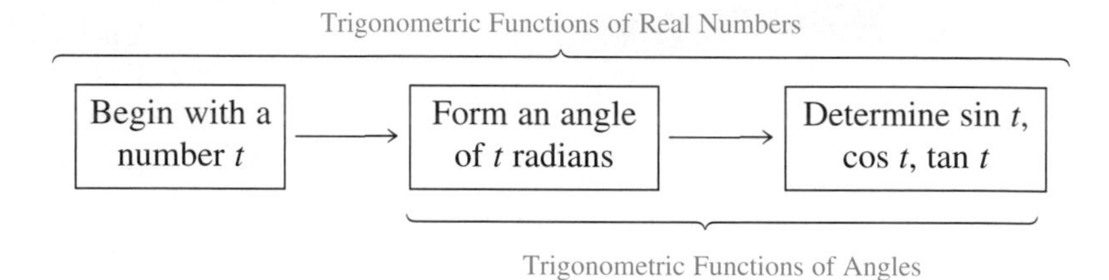

Adapting the definition of Section 1.3 to this new viewpoint, we have the following.

Point-in-the-Plane Description

Let t be a real number. Let (x, y) be any point (except the origin) on the terminal side of an angle of t radians in standard position. Then

$$\sin t = \frac{y}{r}, \qquad \cos t = \frac{x}{r}, \qquad \tan t = \frac{y}{x},$$

where $r = \sqrt{x^2 + y^2}$ is the distance from (x, y) to the origin.

EXAMPLE 1

Figure 2–14 shows an angle of t radians in standard position. Find $\sin t$, $\cos t$, and $\tan t$.

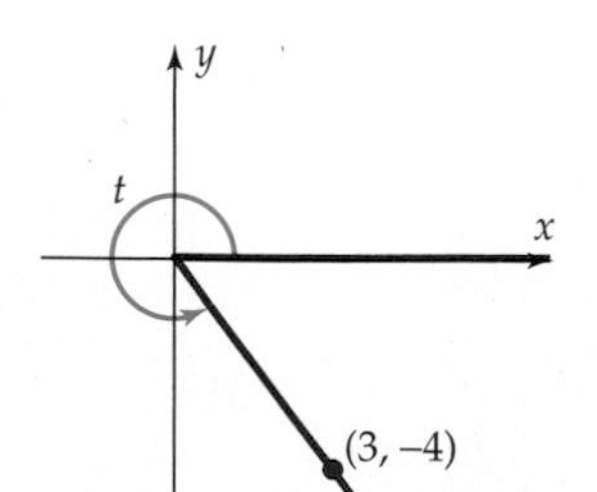

Figure 2–14

SOLUTION Apply the facts in the box above with $(x, y) = (3, -4)$ and

$$r = \sqrt{x^2 + y^2} = \sqrt{3^2 + (-4)^2} = \sqrt{25} = 5.$$

Then we have

$$\sin t = \frac{y}{r} = \frac{-4}{5} = -\frac{4}{5}, \qquad \cos t = \frac{x}{r} = \frac{3}{5}, \qquad \tan t = \frac{y}{x} = \frac{-4}{3} = -\frac{4}{3}.$$ ■

EXAMPLE 2

The terminal side of a first-quadrant angle of t radians in standard position lies on the line with equation $2x - 3y = 0$. Evaluate the three trigonometric functions at t.

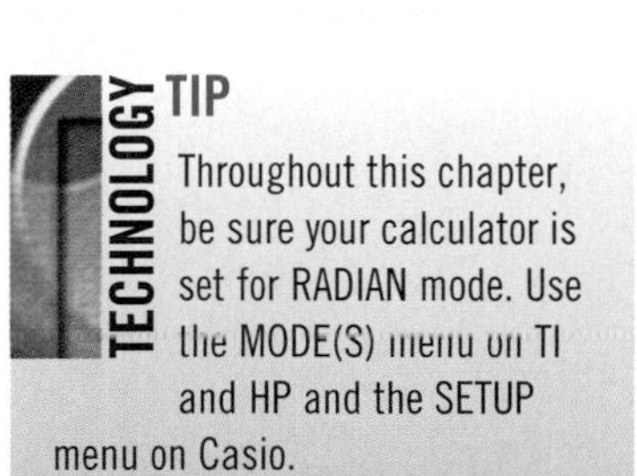

Figure 2–15

SOLUTION Verify that the point (3, 2) satisfies the equation and hence lies on the terminal side of the angle (Figure 2–15). Now we have $(x, y) = (3, 2)$ and $r = \sqrt{x^2 + y^2} = \sqrt{3^2 + 2^2} = \sqrt{13}$. Therefore,

$$\sin t = \frac{y}{r} = \frac{2}{\sqrt{13}}, \qquad \cos t = \frac{x}{r} = \frac{3}{\sqrt{13}}, \qquad \tan t = \frac{y}{x} = \frac{2}{3}. \quad \blacksquare$$

In most cases, evaluating trigonometric functions is not as simple as it was in the preceding examples. Usually, you must use a calculator (in radian mode) to approximate the values of these functions, as illustrated in Figure 2–16.

TECHNOLOGY TIP

Throughout this chapter, be sure your calculator is set for RADIAN mode. Use the MODE(S) menu on TI and HP and the SETUP menu on Casio.

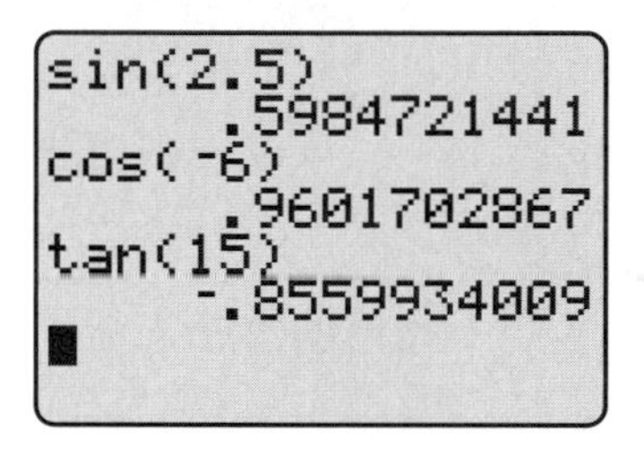

Figure 2–16

EXAMPLE 3

When a baseball is hit by a bat, the horizontal distance d traveled by the ball is approximated by

$$d = \frac{v^2 \sin t \cos t}{16},$$

where the ball leaves the bat at an angle of t radians and has initial velocity v feet per second, as shown in Figure 2–17.

(a) How far does the ball travel when the initial velocity is 90 feet per second and $t = .7$?

(b) If the initial velocity is 105 feet per second and $t = 1$, how far does the ball travel?

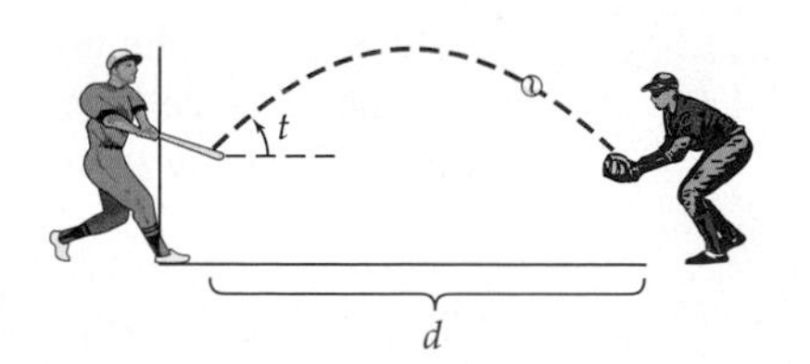

Figure 2–17

SOLUTION

(a) Let $v = 90$ and $t = .7$ in the formula for d. Then a calculator (in radian mode) shows that

$$d = \frac{v^2 \sin t \cos t}{16} = \frac{90^2 \sin .7 \cos .7}{16} \approx 249.44 \text{ feet.}$$

(b) Now let $v = 105$ and $t = 1$. Then

$$d = \frac{v^2 \sin t \cos t}{16} = \frac{105^2 \sin 1 \cos 1}{16} \approx 313.28. \quad \blacksquare$$

Although a calculator approximation is usually necessary, the trigonometric functions can be evaluated exactly for a few numbers by using our knowledge of special angles.

EXAMPLE 4

Find the exact value of

(a) $\sin \pi/6$ (b) $\cos \pi/6$ (c) $\tan 3\pi/4$

SOLUTION

(a) An angle of $\pi/6$ radians is the same as an angle of 30° and from Example 3(a) of Section 1.1, we know that $\sin 30° = 1/2$. Therefore,

$$\sin \pi/6 = \sin 30° = \frac{1}{2}.$$

(b) In Example 3(a) of Section 1.1, we also saw that $\cos 30° = \sqrt{3}/2$. Hence,

$$\cos \pi/6 = \cos 30° = \frac{\sqrt{3}}{2}.$$

(c) An angle of $3\pi/4$ is the same as an angle of 135°, so from Example 3 of Section 1.3, we now have

$$\tan 3\pi/4 = \tan 135° = -1. \quad \blacksquare$$

THE UNIT CIRCLE DESCRIPTION

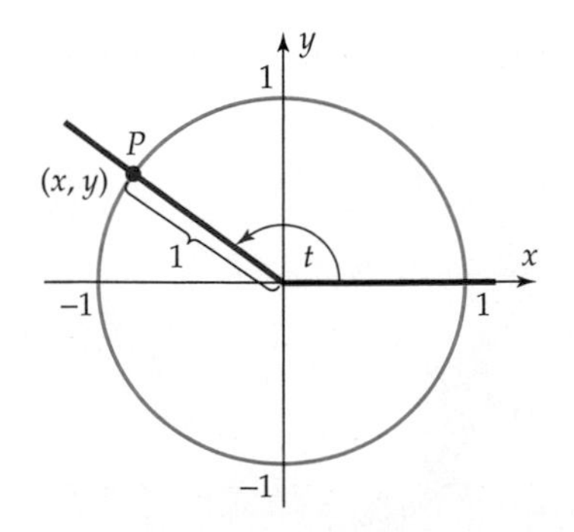

Figure 2–18

We now develop a description of sine, cosine, and tangent that is based on the **unit circle,** which is the circle of radius 1 with center at the origin, whose equation is $x^2 + y^2 = 1$. Let t be any real number and construct an angle of t radians in standard position. Let $P = (x, y)$ be the point where the terminal side of this angle intersects the unit circle, as shown in Figure 2–18.

The distance from (x, y) to the origin is 1 unit because the radius of the unit circle is 1. Using the point (x, y) and $r = 1$, we see that

$$\cos t = \frac{x}{r} = \frac{x}{1} = x \quad \text{and} \quad \sin t = \frac{y}{r} = \frac{y}{1} = y.$$

In other words, we have the following.

Unit Circle Description

> If P is the point where the terminal side of an angle of t radians in standard position meets the unit circle, then
>
> $$P \text{ has coordinates } (\cos t, \sin t).$$

This description is often used as a definition of sine and cosine. To get a better feel for it, do the following Graphing Exploration.

GRAPHING EXPLORATION

With your calculator in radian mode and parametric graphing mode, set the range values as follows:

$$0 \le t \le 2\pi, \qquad -1.8 \le x \le 1.8, \qquad -1.2 \le y \le 1.2^{*}$$

Then graph the curve given by the parametric equations.

$$x = \cos t \qquad \text{and} \qquad y = \sin t.$$

The graph is the unit circle. Use the trace to move around the circle. At each point, the screen will display three numbers: the values of t, x, and y. For each t, the cursor is on the point where the terminal side of an angle of t radians meets the unit circle, so the corresponding x is the number $\cos t$ and the corresponding y is the number $\sin t$.

Suppose an angle of t radians in standard position intersects the unit circle at the point (x, y), as in Figure 2–18. From our earlier definition, $\tan t = y/x$, and by the unit circle description, we have $x = \cos t$ and $y = \sin t$. Therefore, we have another description of the tangent function:

$$\tan t = \frac{\sin t}{\cos t}.$$

EXAMPLE 5 

Use the unit circle description and the preceding equation to find the exact values of $\sin t$, $\cos t$, and $\tan t$ when

(a) $t = \pi$ (b) $t = \pi/2$.

*Parametric graphing is explained at the end of Section 0.5. The window settings here give a square viewing window on calculators with a screen measuring approximately 95 by 63 pixels (such as TI-83+); hence, the unit circle will look like a circle. For wider screens, adjust the x range settings to obtain a square window.

SOLUTION

(a) Construct an angle of π radians, as in Figure 2–19. Its terminal side lies on the negative x-axis and intersects the unit circle at $P = (-1, 0)$. Hence,

$$\sin \pi = y\text{-coordinate of } P = 0$$

$$\cos \pi = x\text{-coordinate of } P = -1$$

$$\tan \pi = \frac{\sin \pi}{\cos \pi} = \frac{0}{-1} = 0.$$

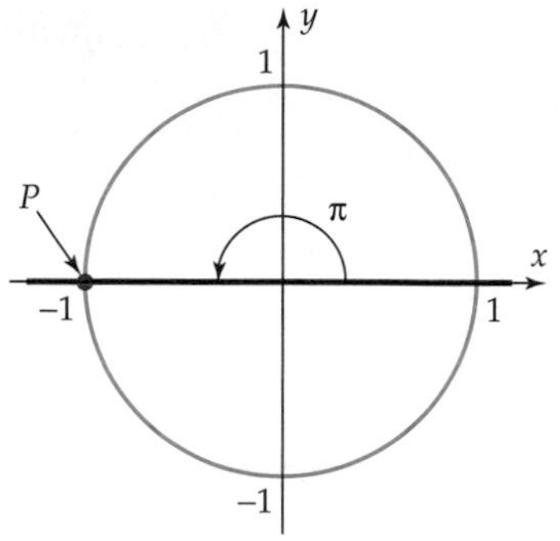

Figure 2–19

(b) An angle of $\pi/2$ radians (Figure 2–20) has its terminal side on the positive y-axis and intersects the unit circle at $P = (0, 1)$.

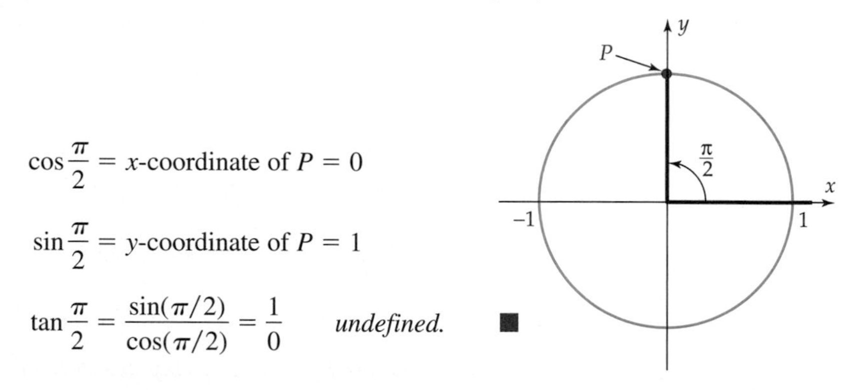

$$\cos \frac{\pi}{2} = x\text{-coordinate of } P = 0$$

$$\sin \frac{\pi}{2} = y\text{-coordinate of } P = 1$$

$$\tan \frac{\pi}{2} = \frac{\sin(\pi/2)}{\cos(\pi/2)} = \frac{1}{0} \quad \textit{undefined.}$$ ■

Figure 2–20

The definitions of sine and cosine show that $\sin t$ and $\cos t$ are defined for every real number t. Example 5(b), however, shows that $\tan t$ is *not* defined when the x-coordinate of the point P is 0. This occurs when P has coordinates $(0, 1)$ or $(0, -1)$, that is, when $t = \pm\pi/2, \pm 3\pi/2, \pm 5\pi/2$, etc. Consequently, the domain (set of inputs) of each trigonometric function is as follows.

Function	Domain
$f(t) = \sin t$	All real numbers
$g(t) = \cos t$	All real numbers
$h(t) = \tan t$	All real numbers except $\pm\pi/2, \pm 3\pi/2, \pm 5\pi/2, \ldots$

The coordinates of every point on the unit circle lie between -1 and 1 (see Figure 2–20 above). Since the point P in the unit circle description has coordinates $(\cos t, \sin t)$, we have this fact about the range (set of outputs) of sine and cosine.

Range of Sine and Cosine

For every real number t

$$-1 \leq \sin t \leq 1$$

and

$$-1 \leq \cos t \leq 1.$$

As we shall see in Section 2.4,

The range of the tangent function consists of all real numbers.

You can confirm this fact by doing the following Exploration.

CALCULATOR EXPLORATION

Use the table feature to evaluate $\tan\left(\frac{\pi}{2} + x\right)$ when $x = .01, .001, .0001$, and so on. What does this suggest about the outputs of the tangent function? Now evaluate when $x = -.01, -.001, -.0001,$

SPECIAL VALUES

The trigonometric functions can be evaluated exactly at $t = \pi/3$, $t = \pi/4$, $t = \pi/6$, and any integer multiples of these numbers by using the following facts (which are explained in the Geometry Review Appendix):

A right triangle with hypotenuse 1 and angles of $\pi/6$ and $\pi/3$ radians has sides of lengths $1/2$ (opposite the angle of $\pi/6$) and $\sqrt{3}/2$ (opposite the angle of $\pi/3$).

A right triangle with hypotenuse 1 and two angles of $\pi/4$ radians has two sides of length $\sqrt{2}/2$.

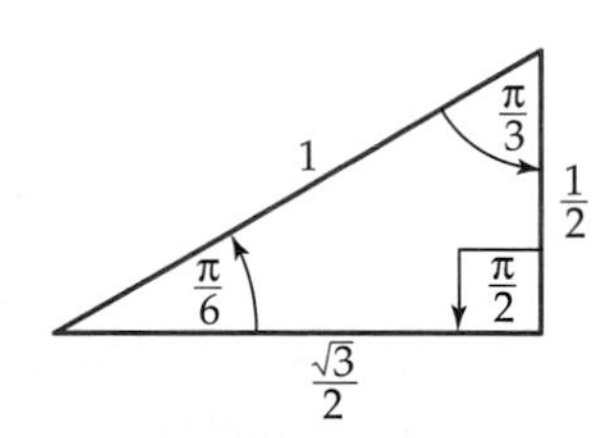

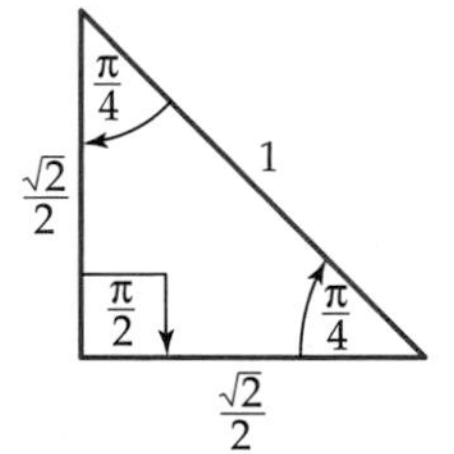

Figure 2–21

EXAMPLE 6

Find the exact values of the three trigonometric functions at $t = -5\pi/4$.

SOLUTION Construct an angle of $-5\pi/4$ radians in standard position and let P be the point where the terminal side intersects the unit circle, as in Figure 2–22 on the next page.

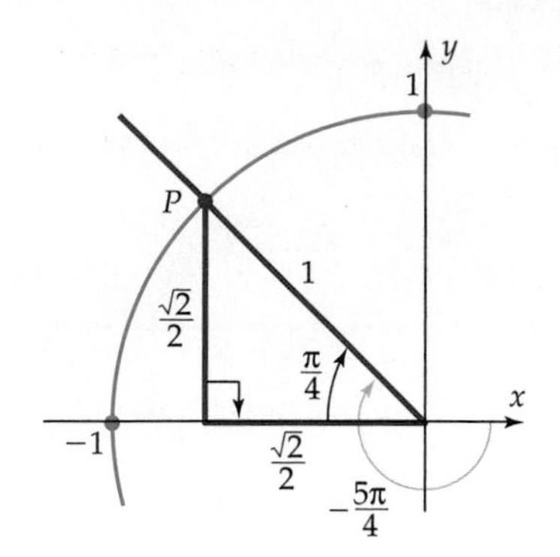

Figure 2–22

Draw a vertical line from P to the x-axis, forming a right triangle with hypotenuse 1 and two angles of $\pi/4$ radians (45°), as shown in Figure 2–22. Each side of this triangle has length $\sqrt{2}/2$, as explained above. Therefore, the coordinates of P are $(-\sqrt{2}/2, \sqrt{2}/2)$, so

$$\sin \frac{-5\pi}{4} = y\text{-coordinate of } P = \frac{\sqrt{2}}{2}$$

$$\cos \frac{-5\pi}{4} = x\text{-coordinate of } P = -\frac{\sqrt{2}}{2}$$

$$\tan \frac{-5\pi}{4} = \frac{\sin t}{\cos t} = \frac{\sqrt{2}/2}{} = -1. \quad ■$$

EXAMPLE 7

Evaluate the trigonometric functions exactly at $t = 11\pi/3$.

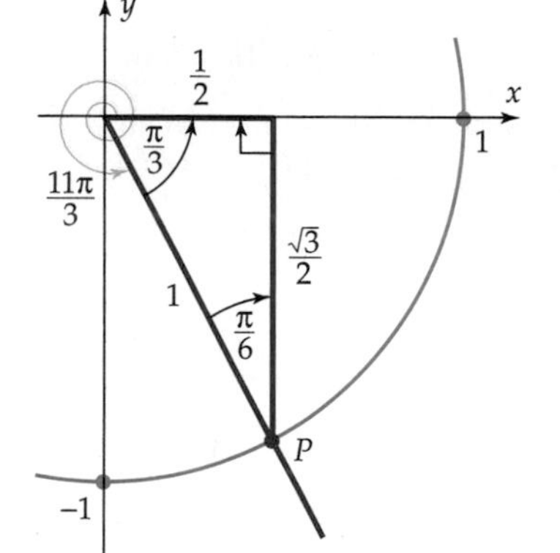

Figure 2–23

SOLUTION Construct an angle of $11\pi/3$ radians in standard position and draw a vertical line from the x-axis to the point P where the terminal side of the angle meets the unit circle, as shown in Figure 2–23. The right triangle formed in this way has hypotenuse 1 and angles of $\pi/6$ and $\pi/3$ radians. The sides of the triangle must have lengths $1/2$ and $\sqrt{3}/2$, as explained above. So the coordinates of P are $(1/2, -\sqrt{3}/2)$, and

$$\sin t = y\text{-coordinate of } P = -\frac{\sqrt{3}}{2}$$

$$\cos t = x\text{-coordinate of } P = \frac{1}{2}$$

$$\tan t = \frac{\sin t}{\cos t} = \frac{-\sqrt{3}/2}{1/2} = -\sqrt{3}. \quad ■$$

✓ EXERCISES 2.2

Note: Unless stated otherwise, all angles are in standard position.

In Exercises 1–6, find sin t, cos t, and tan t, when the terminal side of an angle of t radians in standard position passes through the given point.

1. $(2, 7)$ **2.** $(-3, 2)$ **3.** $(-5, -6)$

4. $(4, -3)$ **5.** $(\sqrt{3}, -10)$ **6.** $(-\pi, 2)$

In Exercises 7–16, use the definition (not a calculator) to find the function value.

7. $\sin(3\pi/2)$ **8.** $\sin(-\pi)$ **9.** $\cos(3\pi/2)$

10. $\cos(-\pi/2)$ **11.** $\tan(4\pi)$ **12.** $\tan(-\pi)$

13. $\cos(-3\pi/2)$ **14.** $\sin(9\pi/2)$ **15.** $\cos(-11\pi/2)$

16. $\tan(-13\pi)$

In Exercises 17–20, assume that the terminal side of an angle of t radians passes through the given point on the unit circle. Find sin t, cos t, tan t.

17. $(-2/\sqrt{5}, 1/\sqrt{5})$ **18.** $(1/\sqrt{10}, -3/\sqrt{10})$

19. $(-3/5, -4/5)$ **20.** $(.6, -.8)$

In Exercises 21–35, find the exact value of the sine, cosine, and tangent of the number, without using a calculator.

21. $\pi/3$ **22.** $2\pi/3$ **23.** $\pi/4$

24. $5\pi/4$ **25.** $3\pi/4$ **26.** $-7\pi/3$

27. $5\pi/6$ **28.** 3π **29.** $-23\pi/6$

30. $11\pi/6$ **31.** $-19\pi/3$ **32.** $-10\pi/3$

33. $-15\pi/4$ **34.** $-25\pi/4$ **35.** $-17\pi/2$

36. Fill the blanks in the following table. Write each entry as a fraction with denominator 2 and with a radical in the numerator. For example,

$$\sin\frac{\pi}{2} = 1 = \frac{\sqrt{4}}{2}.$$

Some students find the resulting pattern an easy way to remember these functional values.

t	0	$\pi/6$	$\pi/4$	$\pi/3$	$\pi/2$
sin t					
cos t					

In Exercises 37–42, write the expression as a single real number. Do not use decimal approximations. [Hint: *Exercises 21–27 might be helpful.*]

37. $\sin(\pi/3)\cos(\pi) + \sin(\pi)\cos(\pi/3)$

38. $\sin(\pi/6)\cos(\pi/2) - \cos(\pi/6)\sin(\pi/2)$

39. $\cos(\pi/2)\cos(\pi/4) - \sin(\pi/2)\sin(\pi/4)$

40. $\cos(2\pi/3)\cos\pi + \sin(2\pi/3)\sin\pi$

41. $\sin(3\pi/4)\cos(5\pi/6) - \cos(3\pi/4)\sin(5\pi/6)$

42. $\sin(-7\pi/3)\cos(5\pi/4) + \cos(-7\pi/3)\sin(5\pi/4)$

In Exercises 43–46, use a calculator in radian mode.

43. When a plane flies faster than the speed of sound, the sound waves it generates trail the plane in a cone shape, as shown in the figure. When the bottom part of the cone hits the ground, you hear a sonic boom. The equation that describes this situation is

$$\sin\left(\frac{t}{2}\right) = \frac{w}{p},$$

where t is the radian measure of the angle of the cone, w is the speed of the sound wave, p is the speed of the plane, and $p > w$.

(a) Find the speed of the sound wave when the plane flies at 1200 mph and $t = .8$.

(b) Find the speed of the plane if the sound wave travels at 500 mph and $t = .7$.

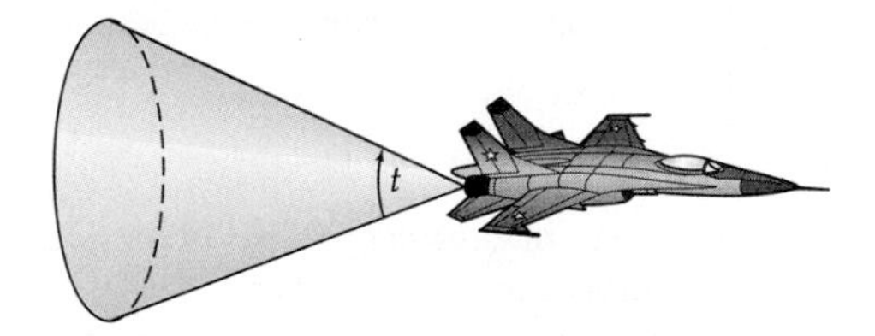

44. Suppose the batter in Example 3 hits the ball with an initial velocity of 100 feet per second.

(a) Complete this table.

t	.5	.6	.7	.8	.9
d					

(b) By experimentation, find the value of t (to two decimal places) that produces the longest distance.

(c) If $t = 1.6$, what is d? Explain your answer.

45. The average daily temperature in St. Louis, Missouri (in degrees Fahrenheit), is approximated by the function

$$T(x) = 24.6\sin(.522x - 2.1) + 56.3 \qquad (1 \le x < 13),$$

where $x = 1$ corresponds to January 1, $x = 2$ to February 1, etc.*

(a) Complete this table.

Date	Average Temperature
Jan. 1	
Mar. 1	
May 1	
July 1	
Sept. 1	
Nov. 1	

(b) Make a table that shows the average temperature every third day in June, beginning on June 1. [Assume that three days = 1/10 of a month.]

*Based on data for 1971–2000 from the National Climatic Data Center.

46. A regular polygon has n equal sides and n equal angles formed by the sides. For example, a regular polygon of three sides is an equilateral triangle, and a regular polygon of four sides is a square. If a regular polygon of n sides is circumscribed around a circle of radius r, as shown in the figure for $n = 4$ and $n = 5$, then the area of the polygon is given by

$$A = nr^2 \tan\left(\frac{\pi}{n}\right).$$

(a) Find the area of a regular polygon of 12 sides circumscribed around a circle of radius 5.

(b) Complete the following table for a regular polygon of n sides circumscribed around the unit circle (which, as you recall, has radius 1).

n	5	50	500	5000	10,000
Area					

(c) As n gets larger and larger, what number does the area get very close to? [*Hint:* What is the area of the unit circle?]

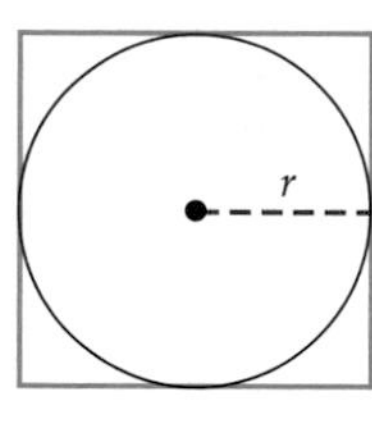

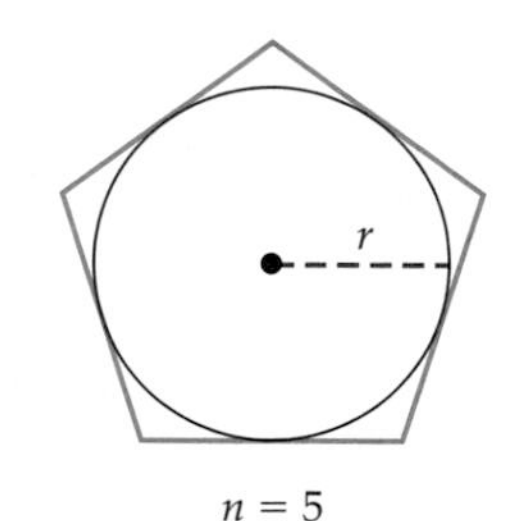

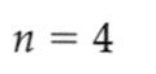

$n = 4$ $n = 5$

In Exercises 47–50, assume that the terminal side of an angle of t radians in standard position lies in the given quadrant on the given straight line. Find sin t, cos t, tan t. [Hint: *Find a point on the terminal side of the angle.*]

47. Quadrant IV; line with equation $y = -3x$.

48. Quadrant III; line with equation $2y - 4x = 0$.

49. Quadrant IV; line through $(-3, 5)$ and $(-9, 15)$.

50. Quadrant III; line through the origin parallel to

$$7x - 2y = -6.$$

51. Quadrant II; line through the origin parallel to

$$2y + x = 6.$$

52. Quadrant I; line through the origin perpendicular to

$$3y + x = 6.$$

53. The values of $\sin t$, $\cos t$, and $\tan t$ are determined by the point (x, y) where the terminal side of an angle of t radians in standard position intersects the unit circle. The coordinates x and y are positive or negative, depending on what quadrant (x, y) lies in. For instance, in the second quadrant x is negative and y is positive, so that $\cos t$ (which is x by definition) is negative. Fill the blanks in this chart with the appropriate sign (+ or −).

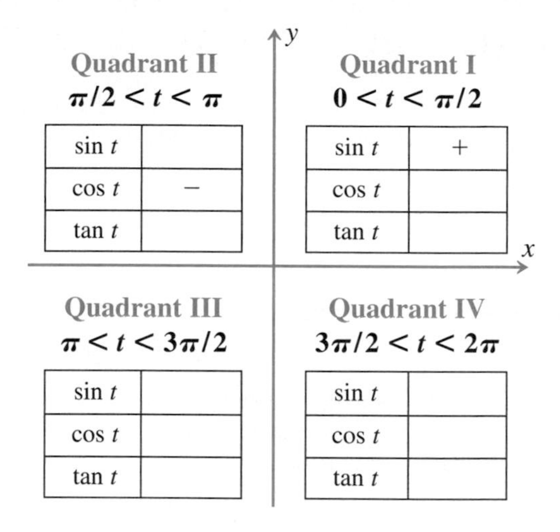

54. (a) Find two numbers c and d such that

$$\sin(c + d) \neq \sin c + \sin d.$$

(b) Find two numbers c and d such that

$$\cos(c + d) \neq \cos c + \cos d.$$

In Exercises 55–60, draw a rough sketch to determine if the given number is positive.

55. $\sin 1$ [*Hint:* The terminal side of an angle of 1 radian lies in the first quadrant (why?), so any point on it will have a positive y-coordinate.]

56. $\cos 2$ **57.** $\tan 3$ **58.** $(\cos 2)(\sin 2)$

59. $\tan 1.5$ **60.** $\cos 3 + \sin 3$

In Exercises 61–66, find all the solutions of the equation.

61. $\sin t = 1$ **62.** $\cos t = -1$ **63.** $\tan t = 0$

64. $\sin t = -1$ **65.** $|\sin t| = 1$ **66.** $|\cos t| = 1$

Thinkers

67. Using only the definition and no calculator, determine which number is larger: $\sin(\cos 0)$ or $\cos(\sin 0)$.

68. With your calculator in radian mode and function graphing mode, graph the following functions on the same screen, using the viewing window with $0 \le x \le 2\pi$ and $-3 \le y \le 3$: $f(x) = \cos(x^3)$ and $g(x) = (\cos x)^3$. Are the graphs the same? What do you conclude about the statement $\cos(x^3) = (\cos x)^3$?

69. Figure R is a diagram of a merry-go-round that includes horses A through F. The distance from the center P to A is 1 unit and the distance from P to D is 5 units. Define six functions as follows:

$A(t)$ = vertical distance from horse A to the x-axis at time t minutes;

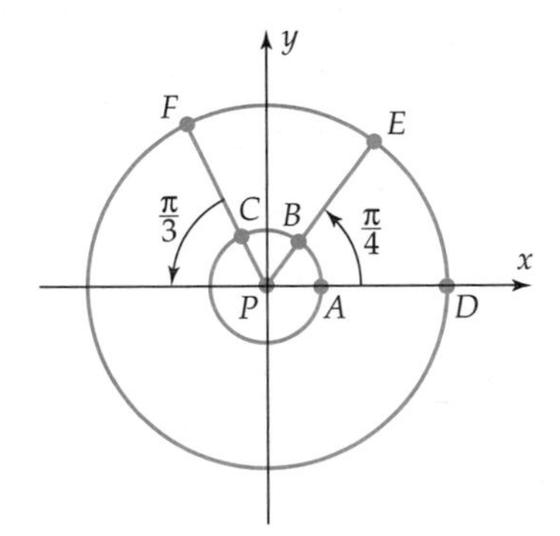

Figure R

and similarly for $B(t)$, $C(t)$, $D(t)$, $E(t)$, $F(t)$. The merry-go-round rotates counterclockwise at a rate of 1 revolution per minute, and the horses are in the positions shown in Figure R at the starting time $t = 0$. As the merry-go-round rotates, the horses move around the circles shown in Figure R.

(a) Show that $B(t) = A(t + 1/8)$ for every t.
(b) In a similar manner, express $C(t)$ in terms of the function $A(t)$.
(c) Express $E(t)$ and $F(t)$ in terms of the function $D(t)$.
(d) Explain why Figure S is valid and use it and similar triangles to express $D(t)$ in terms of $A(t)$.
(e) In a similar manner, express $E(t)$ and $F(t)$ in terms of $A(t)$.
(f) Show that $A(t) = \sin(2\pi t)$ for every t. [*Hint:* Exercises 49–56 in Section 2.1 may be helpful.]
(g) Use parts (a), (b), and (f) to express $B(t)$ and $C(t)$ in terms of the sine function.
(h) Use parts (d), (e), and (f) to express $D(t)$, $E(t)$, and $F(t)$ in terms of the sine function.

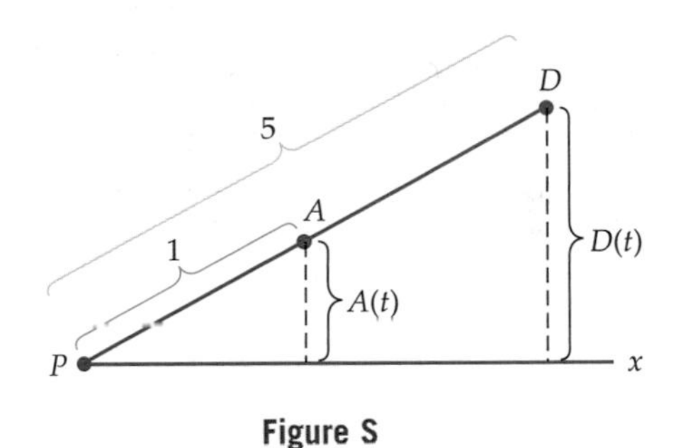

Figure S

2.2 *ALTERNATE* The Sine, Cosine, and Tangent Functions

> **NOTE**
> If you have read Chapter 1, omit this section. If you have not read Chapter 1, use this section in place of Section 2.2.

Unlike most of the functions seen thus far, the definitions of the sine and cosine functions do not involve algebraic formulas. Instead, these functions are defined geometrically, using the unit circle.* Recall that the unit circle is the circle of radius 1 with center at the origin, whose equation is $x^2 + y^2 = 1$.

Both the sine and cosine functions have the set of all real numbers as domain. Their rules are given by the following three-step geometric process.

1. Given a real number t, construct an angle of t radians in standard position.
2. Find the coordinates of the point P where the terminal side of this angle meets the unit circle $x^2 + y^2 = 1$, say, $P = (a, b)$, as shown on the next page.

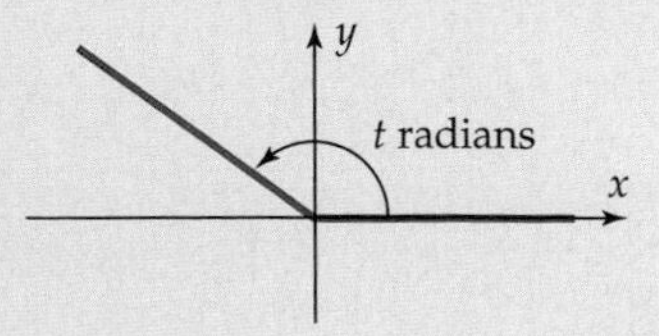

*If you have previously studied the trigonometry of triangles, the definition given here may not look familiar. If this is the case, just concentrate on this definition and don't worry about relating it to any definition you remember from the past. The connection between this definition and the trigonometry of triangles is explained in Alternate Section 1.1.

Roadmap

Instructors who wish to cover all six trigonometric functions simultaneously should incorporate Section 2.6 into Sections 2.2–2.4, as follows.

Subsection of Section 2.6	Cover at the end of
Definitions; Alternate Descriptions	Section 2.2
Algebra and Identities	Section 2.3
Graphs	Section 2.4

3. The value of the **cosine function** at t (denoted $\cos t$) is the x-coordinate of P:

$$\cos t = a.$$

The value of the **sine function** at t (denoted $\sin t$) is the y-coordinate of P:

$$\sin t = b.$$

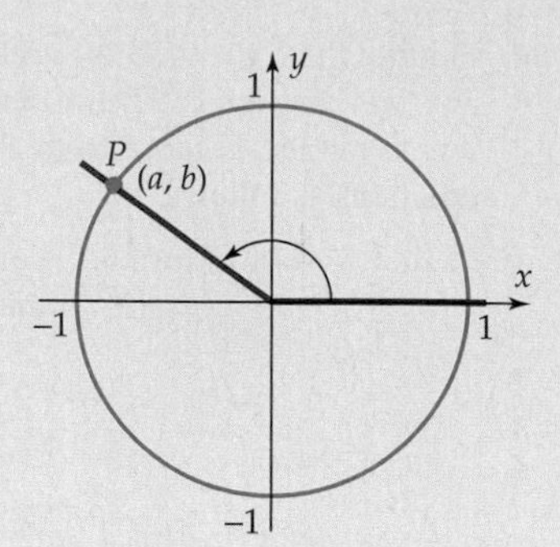

In other words, we have the following.

Unit Circle Description

If P is the point where the terminal side of an angle of t radians in standard position meets the unit circle, then

$$P \text{ has coordinates } (\cos t, \sin t).$$

GRAPHING EXPLORATION

With your calculator in radian mode and parametric graphing mode, set the range values as follows:

$$0 \le t \le 2\pi, \qquad -1.8 \le x \le 1.8, \qquad -1.2 \le y \le 1.2^*$$

Then graph the curve given by the parametric equations

$$x = \cos t \qquad \text{and} \qquad y = \sin t.$$

The graph is the unit circle. Use the trace to move around the circle. At each point, the screen will display three numbers: the values of t, x, and y. For each t, the cursor is on the point where the terminal side of an angle of t radians meets the unit circle, so the corresponding x is the number $\cos t$ and the corresponding y is the number $\sin t$.

TECHNOLOGY TIP

Throughout this chapter, be sure your calculator is set for RADIAN mode. Use the MODE(S) menu on TI and HP and the SETUP menu on Casio.

The **tangent function** is defined as the quotient of the sine and cosine functions. Its value at the number t, denoted $\tan t$, is given by

$$\tan t = \frac{\sin t}{\cos t}$$

Every calculator has SIN, COS, and TAN keys for evaluating the sine, cosine, and tangent functions. For instance, a calculator (in radian mode) gives these approximations (Figure 2–24).

```
sin(2.5)
     .5984721441
cos(-6)
     .9601702867
tan(15)
    -.8559934009
```

Figure 2–24

Nevertheless, you should know how to use the definition to find exact values of these functions when possible, as illustrated in Examples 1–4.

*Parametric graphing is explained at the end of Section 0.5. The window settings here give a square viewing window on calculators with a screen measuring approximately 95 by 63 pixels (such as TI-83+); hence, the unit circle will look like a circle. For wider screens, adjust the x range settings to obtain a square window.

EXAMPLE 1

Evaluate the three trigonometric functions at

(a) $t = \pi$ (b) $t = \pi/2$.

SOLUTION

(a) Construct an angle of π radians, as in Figure 2–25. Its terminal side lies on the negative x-axis and intersects the unit circle at $P = (-1, 0)$. Hence,

$$\sin \pi = y\text{-coordinate of } P = 0$$

$$\cos \pi = x\text{-coordinate of } P = -1$$

$$\tan \pi = \frac{\sin \pi}{\cos \pi} = \frac{0}{-1} = 0.$$

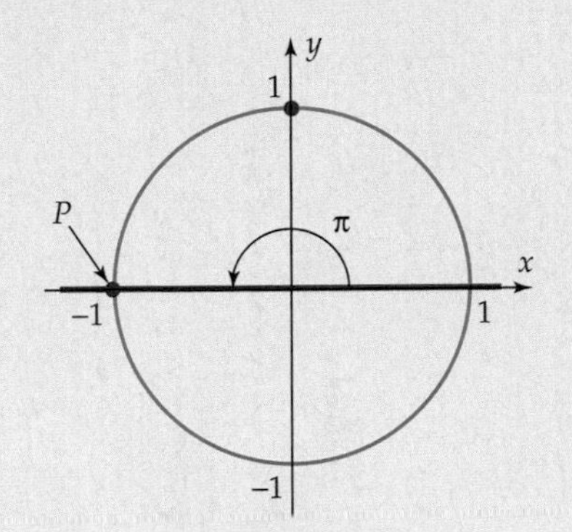

Figure 2–25

(b) An angle of $\pi/2$ radians (Figure 2–26) has its terminal side on the positive y-axis and intersects the unit circle at $P = (0, 1)$.

$$\cos \frac{\pi}{2} = x\text{-coordinate of } P = 0$$

$$\sin \frac{\pi}{2} = y\text{-coordinate of } P = 1$$

$$\tan \frac{\pi}{2} = \frac{\sin(\pi/2)}{\cos(\pi/2)} = \frac{1}{0} \quad \textit{undefined.}$$ ■

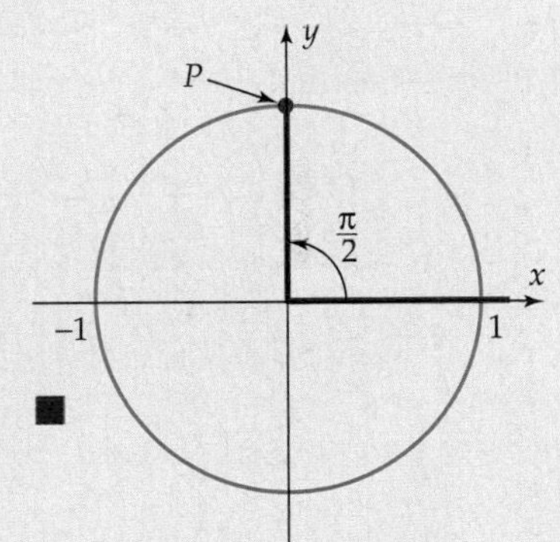

Figure 2–26

The definition of sine and cosine shows that $\sin t$ and $\cos t$ are defined for every real number t. Example 1(b), however, shows that $\tan t$ is *not* defined when the x-coordinate of the point P is 0. This occurs when P has coordinates $(0, 1)$ or $(0, -1)$, that is when $t = \pm\pi/2, \pm 3\pi/2, \pm 5\pi/2$, etc. Consequently, the domains of the trigonometric functions are as follows.

Function	Domain
$f(t) = \sin t$	All real numbers
$g(t) = \cos t$	All real numbers
$h(t) = \tan t$	All real numbers except $\pm\pi/2, \pm 3\pi/2, \pm 5\pi/2, \ldots$

The coordinates of every point on the unit circle lie between -1 and 1 (see Figure 2–26 on page 159). Since the point P in the unit circle description has coordinates $(\cos t, \sin t)$, we have this fact about the range (set of outputs) of sine and cosine.

Range of Sine and Cosine

For every real number t,

$$-1 \leq \sin t \leq 1$$

and

$$-1 \leq \cos t \leq 1.$$

As we shall see in Section 2.4,

The range of the tangent function consists of all real numbers.

You can confirm this fact by doing the following Exploration.

CALCULATOR EXPLORATION

Use the table feature to evaluate $\tan\left(\frac{\pi}{2} + x\right)$ when $x = .01, .001, .0001$, and so on. What does this suggest about the outputs of the tangent function? Now evaluate when $x = -.01, -.001, -.0001$,

SPECIAL VALUES

The trigonometric functions can be evaluated exactly at $t = \pi/3$, $t = \pi/4$, $t = \pi/6$, and any integer multiples of these numbers by using the following facts (which are explained in the Geometry Review Appendix):

A right triangle with hypotenuse 1 and angles of $\pi/6$ and $\pi/3$ radians has sides of lengths $1/2$ (opposite the angle of $\pi/6$) and $\sqrt{3}/2$ (opposite the angle of $\pi/3$).

A right triangle with hypotenuse 1 and two angles of $\pi/4$ radians has two sides of length $\sqrt{2}/2$.

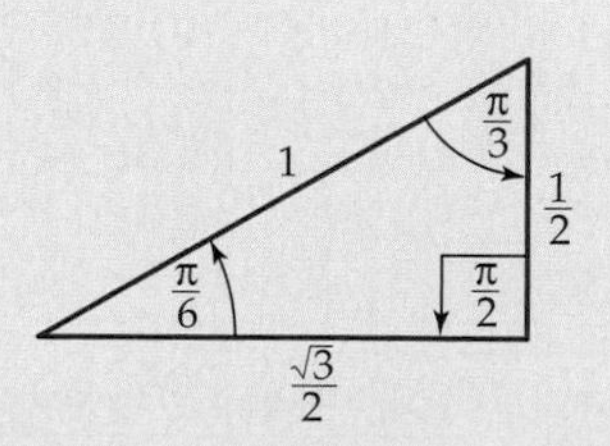

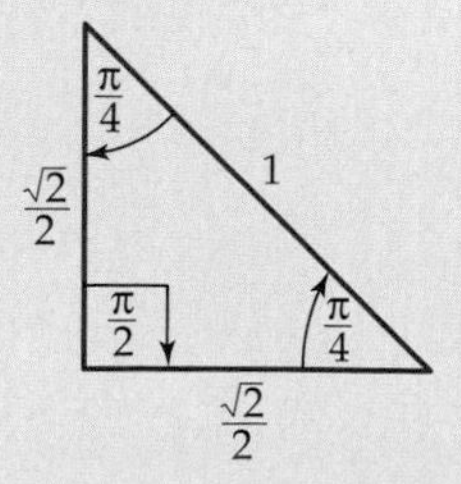

Figure 2–27

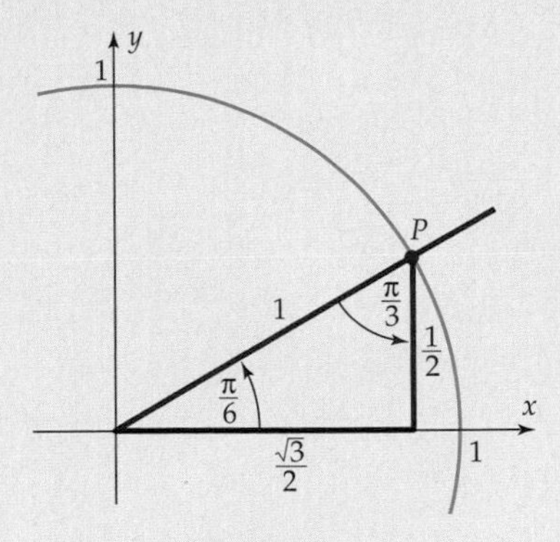

Figure 2–28

EXAMPLE 2

Evaluate the three trigonometric functions at $t = \pi/6$.

SOLUTION Construct an angle of $\pi/6$ radians in standard position and let P be the point where its terminal side intersects the unit circle (Figure 2–28). Draw a vertical line from P to the x-axis, forming a right triangle with hypotenuse 1, angles of $\pi/6$ and $\pi/3$ radians, and sides of lengths $1/2$ and $\sqrt{3}/2$, as described above. Then P has coordinates $(\sqrt{3}/2, 1/2)$, and by definition,

$$\sin\frac{\pi}{6} = y\text{-coordinate of } P = \frac{1}{2}$$

$$\cos\frac{\pi}{6} = x\text{-coordinate of } P = \frac{\sqrt{3}}{2}$$

$$\tan\frac{\pi}{6} = \frac{\sin(\pi/6)}{\cos(\pi/6)} = \frac{1/2}{\sqrt{3}/2} = \frac{1}{\sqrt{3}} = \frac{\sqrt{3}}{3}. \quad \blacksquare$$

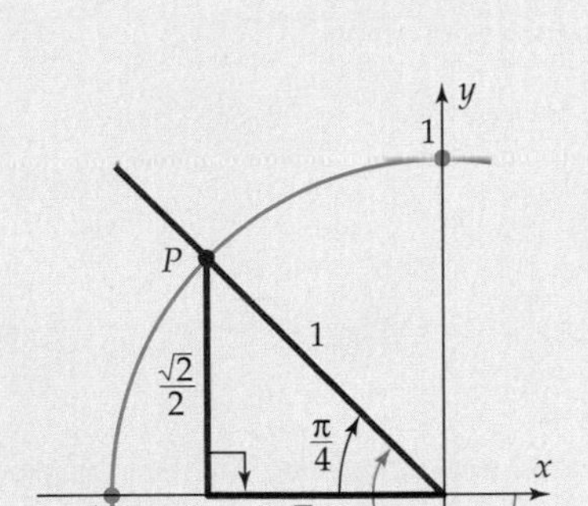

Figure 2–29

EXAMPLE 3

Evaluate the trigonometric functions at $-5\pi/4$.

SOLUTION Construct an angle of $-5\pi/4$ radians in standard position and let P be the point where the terminal side intersects the unit circle. Draw a vertical line from P to the x-axis, forming a right triangle with hypotenuse 1 and two angles of $\pi/4$ radians, as shown in Figure 2–29.

Each side of this triangle has length $\sqrt{2}/2$, as explained above. Therefore, the coordinates of P are $(-\sqrt{2}/2, \sqrt{2}/2)$, so

$$\sin\frac{-5\pi}{4} = y\text{-coordinate of } P = \frac{\sqrt{2}}{2}$$

$$\cos\frac{-5\pi}{4} = x\text{-coordinate of } P = -\frac{\sqrt{2}}{2}$$

$$\tan\frac{-5\pi}{4} = \frac{\sin t}{\cos t} = \frac{\sqrt{2}/2}{-\sqrt{2}/2} = -1. \quad \blacksquare$$

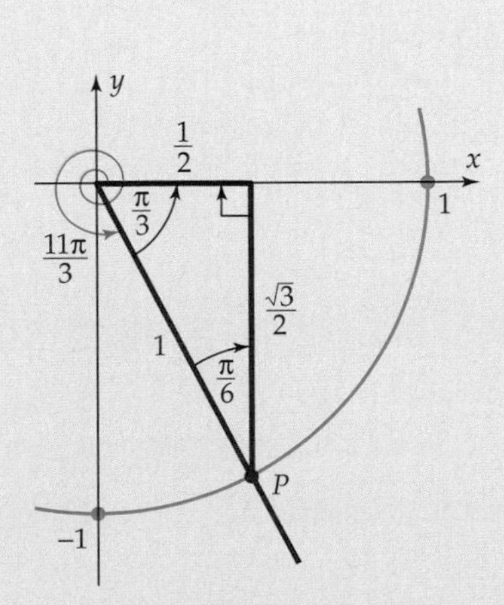

Figure 2–30

EXAMPLE 4

Evaluate the trigonometric functions at $11\pi/3$.

SOLUTION Construct an angle of $11\pi/3$ radians in standard position and draw a vertical line from the x-axis to the point P where the terminal side of the angle meets the unit circle, as shown in Figure 2–30. The right triangle formed in this way has hypotenuse 1 and angles of $\pi/6$ and $\pi/3$ radians. The sides of the triangle must have lengths $1/2$ and $\sqrt{3}/2$, so the coordinates of P are $(1/2, -\sqrt{3}/2)$, and

$$\sin t = y\text{-coordinate of } P = -\frac{\sqrt{3}}{2}$$

$$\cos t = x\text{-coordinate of } P = \frac{1}{2}$$

$$\tan t = \frac{\sin t}{\cos t} = \frac{-\sqrt{3}/2}{1/2} = -\sqrt{3}. \quad \blacksquare$$

POINT-IN-THE-PLANE DESCRIPTION OF TRIGONOMETRIC FUNCTIONS

In evaluating $\sin t$, $\cos t$ and $\tan t$, from the definition, we use the point where the unit circle intersects the terminal side of an angle of t radians in standard position. Here is an alternative method of evaluating the trigonometric functions that uses *any* point on the terminal side of the angle (except the origin).

Point-in-the-Plane Description

Let t be a real number. Let (x, y) be any point (except the origin) on the terminal side of an angle of t radians in standard position. Then,

$$\sin t = \frac{y}{r}, \qquad \cos t = \frac{x}{r}, \qquad \tan t = \frac{y}{x},$$

where $r = \sqrt{x^2 + y^2}$ is the distance from (x, y) to the origin.

Proof Let Q be the point on the terminal side of the standard position angle of t radians and let P be the point where the terminal side meets the unit circle, as in Figure 2–31. The definition of sine and cosine shows that P has coordinates $(\cos t, \sin t)$. The distance formula shows that the segment OQ has length $\sqrt{x^2 + y^2}$, which we denote by r.

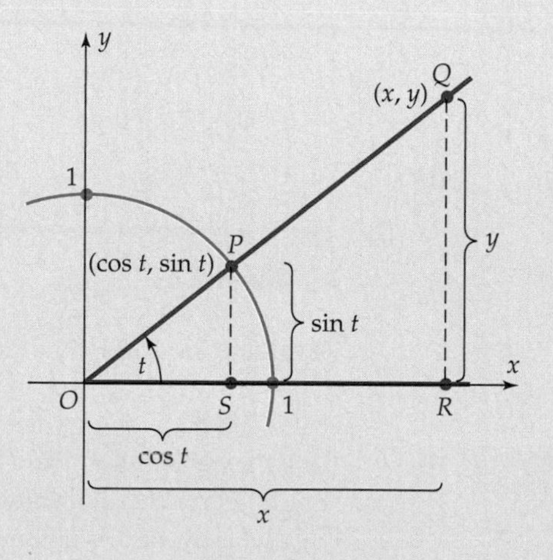

Figure 2–31

Both triangles QOR and POS are right triangles containing an angle of t radians. Therefore, these triangles are *similar.** Consequently,

$$\frac{\text{length } OP}{\text{length } OQ} = \frac{\text{length } PS}{\text{length } QR} \quad \text{and} \quad \frac{\text{length } OP}{\text{length } OQ} = \frac{\text{length } OS}{\text{length } OR}.$$

Figure 2–31 shows what each of these lengths is. Hence,

$$\frac{1}{r} = \frac{\sin t}{y} \qquad \text{and} \qquad \frac{1}{r} = \frac{\cos t}{x}$$

$$r \sin t = y \qquad\qquad r \cos t = x$$

$$\sin t = \frac{y}{r} \qquad\qquad \cos t = \frac{x}{r}$$

Similar arguments work when the terminal side is not in the first quadrant. In every case, $\tan t = \dfrac{\sin t}{\cos t} = \dfrac{y/r}{x/r} = \dfrac{y}{x}$. This completes the proof of the statements in the box. ■

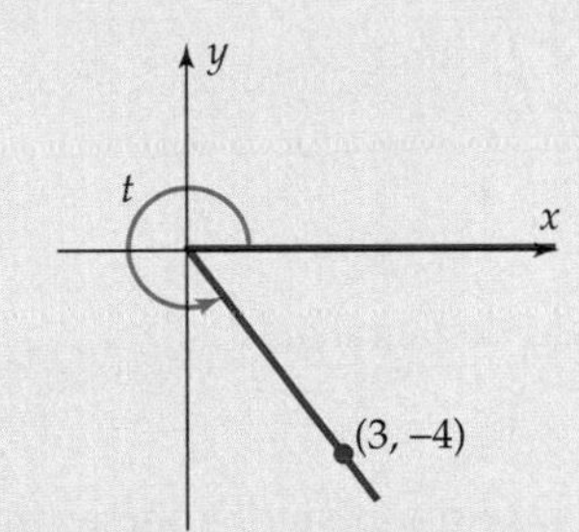

Figure 2–32

EXAMPLE 5

Figure 2–32 shows an angle of t radians in standard position. Evaluate the three trigonometric functions at t.

SOLUTION Apply the facts in the box above with $(x, y) = (3, -4)$ and

$$r = \sqrt{x^2 + y^2} = \sqrt{3^2 + (-4)^2} = \sqrt{25} = 5.$$

Then we have

$$\sin t = \frac{y}{r} = \frac{-4}{5} = -\frac{4}{5}, \qquad \cos t = \frac{x}{r} = \frac{3}{5}, \qquad \tan t = \frac{y}{x} = \frac{-4}{3} = -\frac{4}{3}. \quad ■$$

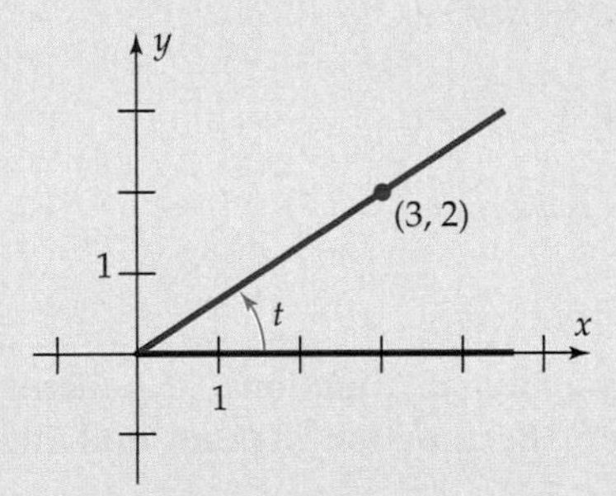

Figure 2–33

EXAMPLE 6

The terminal side of a first-quadrant angle of t radians in standard position lies on the line with equation $2x - 3y = 0$. Evaluate the three trigonometric functions at t.

SOLUTION Verify that the point (3, 2) satisfies the equation and hence lies on the terminal side of the angle (Figure 2–33). Now we have

$$(x, y) = (3, 2) \qquad \text{and} \qquad r = \sqrt{x^2 + y^2} = \sqrt{3^2 + 2^2} = \sqrt{13}.$$

Therefore,

$$\sin t = \frac{y}{r} = \frac{2}{\sqrt{13}}, \qquad \cos t = \frac{x}{r} = \frac{3}{\sqrt{13}}, \qquad \tan t = \frac{y}{x} = \frac{2}{3}. \quad ■$$

---

*See the Geometry Review Appendix for the basic facts about similar triangles.

Use the exercises for Section 2.2 on page 154.

2.3 Algebra and Identities

In the previous section, we concentrated on evaluating the trigonometric functions. In this section, the emphasis is on the algebra of such functions. In dealing with trigonometric functions, two notational conventions are normally observed.

1. **Parentheses are omitted whenever no confusion can result.**

   For example,

   $$\sin(t) \quad \text{is written} \quad \sin t$$
   $$-(\cos(5t)) \quad \text{is written} \quad -\cos 5t$$
   $$4(\tan t) \quad \text{is written} \quad 4 \tan t.$$

   On the other hand, parentheses *are* needed to distinguish

   $$\cos(t + 3) \qquad \text{from} \qquad \cos t + 3,$$

   because the first one says, "Add 3 to t and take the cosine of the result," but the second one says, "Take the cosine of t and add 3 to the result." When $t = 5$, for example, these are different numbers, as shown in Figure 2–34.

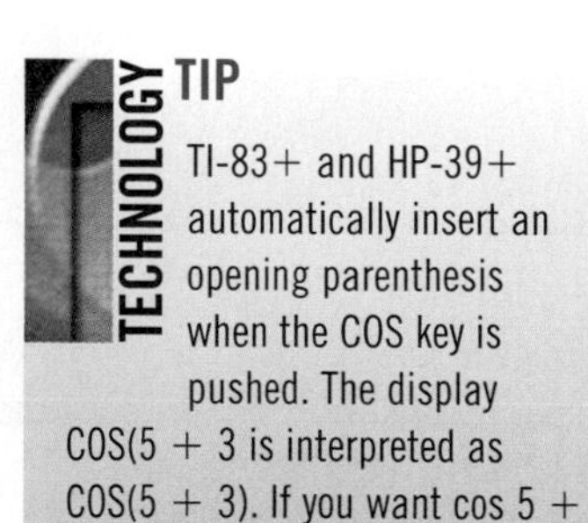

TECHNOLOGY TIP

TI-83+ and HP-39+ automatically insert an opening parenthesis when the COS key is pushed. The display COS(5 + 3 is interpreted as COS(5 + 3). If you want cos 5 + 3, you must insert a parenthesis after the 5: COS(5) + 3.

Figure 2–34

2. **When dealing with powers of trigonometric functions, exponents (other than −1) are written between the function symbol and the variable.**

   For example,

   $$(\cos t)^3 \quad \text{is written} \quad \cos^3 t$$
   $$(\sin t)^4(\tan 7t)^2 \quad \text{is written} \quad \sin^4 t \tan^2 7t.$$

   Furthermore,

   $$\sin t^3 \quad \text{means} \quad \sin(t^3) \quad [\textit{not } (\sin t)^3 \quad \text{or} \quad \sin^3 t]$$

For instance, when $t = 4$, we have Figure 2–35.

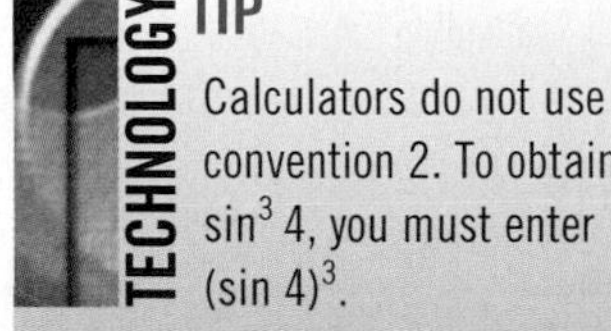

TECHNOLOGY TIP

Calculators do not use convention 2. To obtain $\sin^3 4$, you must enter $(\sin 4)^3$.

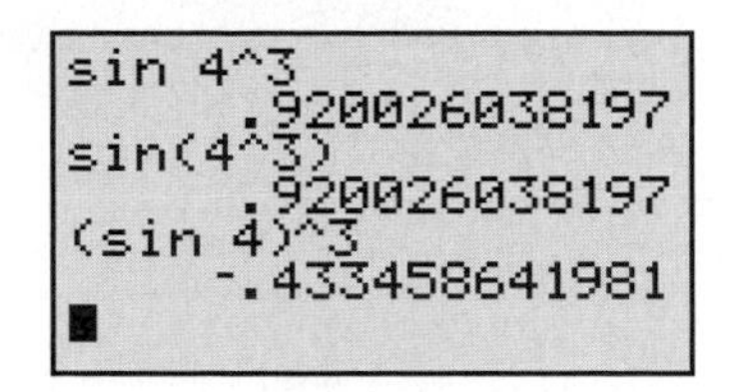

Figure 2–35

CAUTION

Convention 2 is not used when the exponent is -1. For example, $\sin^{-1} t$ does *not* mean $(\sin t)^{-1}$ or $1/(\sin t)$; it has an entirely different meaning that will be discussed in Section 3.4. Similar remarks apply to $\cos^{-1} t$ and $\tan^{-1} t$.

Except for these two conventions and the Caution in the margin, the algebra of trigonometric functions is just like the algebra of other functions. They may be added, subtracted, multiplied, composed, etc.

EXAMPLE 1

If $f(t) = \sin^2 t + \tan t$ and $g(t) = \tan^3 t + 5$, then the product function fg is given by the rule

$$(fg)(t) = f(t)g(t) = (\sin^2 t + \tan t)(\tan^3 t + 5)$$
$$= \sin^2 t \tan^3 t + 5 \sin^2 t + \tan^4 t + 5 \tan t. \quad \blacksquare$$

EXAMPLE 2

Factor $2 \cos^2 t - 5 \cos t - 3$.

SOLUTION You can do this directly, but it may be easier to understand if you make a substitution. Let $u = \cos t$; then

$$\begin{aligned} 2 \cos^2 t - 5 \cos t - 3 &= 2(\cos t)^2 - 5 \cos t - 3 \\ &= 2u^2 - 5u - 3 \\ &= (2u + 1)(u - 3) \\ &= (2 \cos t + 1)(\cos t - 3). \quad \blacksquare \end{aligned}$$

EXAMPLE 3

If $f(t) = \cos^2 t - 9$ and $g(t) = \cos t + 3$, then the quotient function f/g is given by the rule

$$\left(\frac{f}{g}\right)(t) = \frac{f(t)}{g(t)} = \frac{\cos^2 t - 9}{\cos t + 3} = \frac{(\cos t + 3)(\cos t - 3)}{\cos t + 3} = \cos t - 3. \quad \blacksquare$$

CAUTION

You are dealing with *functional notation* here, so the symbol sin *t* is a *single entity,* as are cos *t* and tan *t*. Don't try some nonsensical "canceling" operation, such as

$$\frac{\sin t}{\cos t} = \frac{\sin}{\cos} \quad \text{or} \quad \frac{\cos t^2}{\cos t} = \frac{\cos t}{\cos} = t.$$

EXAMPLE 4 

If $f(t) = \sin t$ and $g(t) = t^2 + 3$, then the composite function $g \circ f$ is given by the rule

$$(g \circ f)(t) = g(f(t)) = g(\sin t) = \sin^2 t + 3.$$

The composite function $f \circ g$ is given by the rule

$$(f \circ g)(t) = f(g(t)) = f(t^2 + 3) = \sin(t^2 + 3).$$

The parentheses are absolutely necessary here because $\sin(t^2 + 3)$ is *not* the same function as $\sin t^2 + 3$. For instance, a calculator in radian mode shows that for $t = 5$,

$$\sin(5^2 + 3) = \sin(25 + 3) = \sin 28 \approx .2709,$$

whereas

$$\sin 5^2 + 3 = \sin 25 + 3 \approx (-.1324) + 3 = 2.8676.$$ ■

THE PYTHAGOREAN IDENTITY

Trigonometric functions have numerous interrelationships that are usually expressed as *identities*. An **identity** is an equation that is true for all values of the variable for which every term of the equation is defined. Here is one of the most important trigonometric identities.

Pythagorean Identity

> For every real number t,
>
> $$\sin^2 t + \cos^2 t = 1.$$

Proof For each real number t, the point P, where the terminal side of an angle of t radians intersects the unit circle, has coordinates $(\cos t, \sin t)$, as in Figure 2–36. Since P lies on the unit circle, its coordinates must satisfy the equation of the unit circle: $x^2 + y^2 = 1$, that is, $\cos^2 t + \sin^2 t = 1$. ■

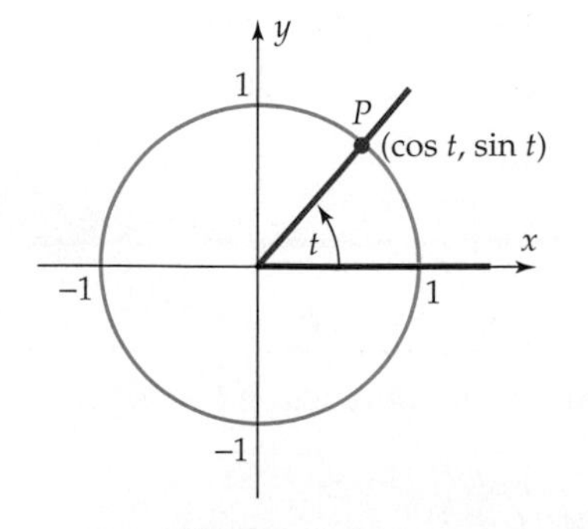

Figure 2–36

EXPLORATION

Recall that the graph of $y = 1$ is a horizontal line through (0, 1). Verify the Pythagorean identity by graphing the equation

$$y = (\sin x)^2 + (\cos x)^2$$

in the window with $-10 \le x \le 10$ and $-3 \le y \le 3$ and using the trace feature.

EXAMPLE 5

If $\pi/2 < t < \pi$ and $\sin t = 2/3$, find $\cos t$ and $\tan t$.

SOLUTION By the Pythagorean identity,

$$\cos^2 t = 1 - \sin^2 t = 1 - \left(\frac{2}{3}\right)^2 = 1 - \frac{4}{9} = \frac{5}{9}.$$

So there are two possibilities:

$$\cos t = \sqrt{5/9} = \sqrt{5}/3 \qquad \text{or} \qquad \cos t = -\sqrt{5/9} = -\sqrt{5}/3.$$

Since $\pi/2 < t < \pi$, $\cos t$ is negative (see Exercise 53 on page 156). Therefore, $\cos t = -\sqrt{5}/3$, and

$$\tan t = \frac{\sin t}{\cos t} = \frac{2/3}{-\sqrt{5}/3} = \frac{-2}{\sqrt{5}} = \frac{-2\sqrt{5}}{5}. \qquad \blacksquare$$

EXAMPLE 6

The Pythagorean identity is valid for *any* number t. For instance, if $t = 3k + 7$, then $\sin^2(3k + 7) + \cos^2(3k + 7) = 1$. ■

EXAMPLE 7

Simplify: $\tan^2 t \cos^2 t + \cos^2 t$.

SOLUTION By the definition of tangent and the Pythagorean identity,

$$\begin{aligned}\tan^2 t \cos^2 t + \cos^2 t &= \frac{\sin^2 t}{\cos^2 t}\cos^2 t + \cos^2 t \\ &= \sin^2 t + \cos^2 t = 1. \qquad \blacksquare\end{aligned}$$

PERIODICITY IDENTITIES

Let t be any real number and construct two angles in standard position of measure t and $t + 2\pi$ radians, respectively, as shown in Figure 2–37. As we saw in Section 2.1, both of these angles have the same terminal side. Therefore, the point P where the terminal side meets the unit circle is the *same* in both cases.

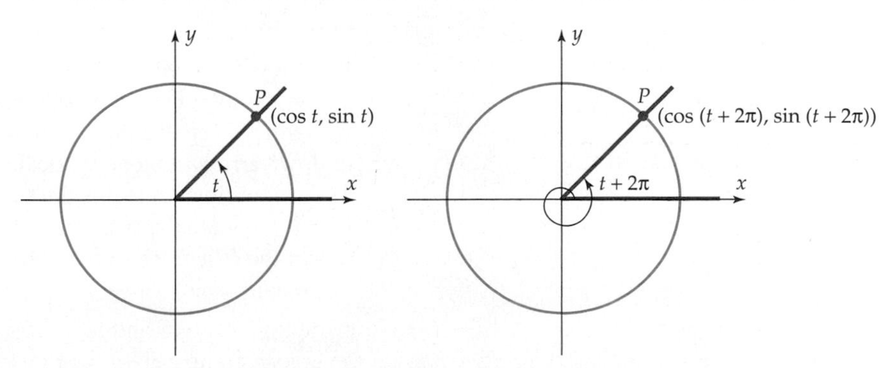

Figure 2–37

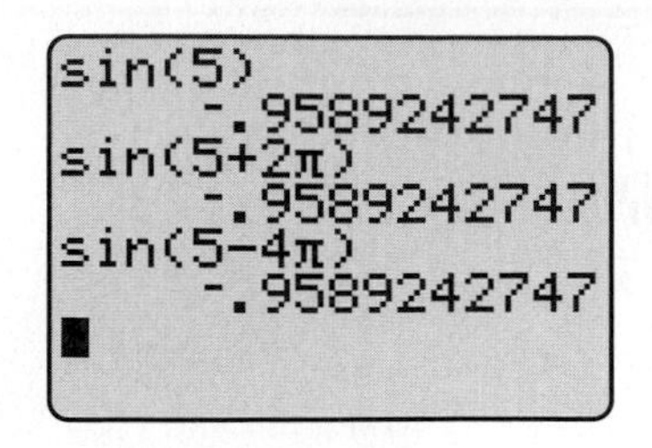

Figure 2–38

Therefore, the coordinates of P are the same, that is,

$$\sin t = \sin(t + 2\pi) \qquad \text{and} \qquad \cos t = \cos(t + 2\pi).$$

Furthermore, since an angle of t radians has the same terminal side as angles of radian measure $t \pm 2\pi$, $t \pm 4\pi$, $t \pm 6\pi$, and so forth, the same argument shows that

$$\sin t = \sin(t \pm 2\pi) = \sin(t \pm 4\pi) = \sin(t \pm 6\pi) = \cdots$$

$$\cos t = \cos(t \pm 2\pi) = \cos(t \pm 4\pi) = \cos(t \pm 6\pi) = \cdots,$$

as illustrated (for $t = 5$) in Figure 2–38.

There is a special name for functions that repeat their values at regular intervals. A function f is said to be **periodic** if there is a positive constant k such that $f(t) = f(t + k)$ for every number t in the domain of f. There will be more than one constant k with this property; the smallest one is called the **period** of the function f. We have just seen that sine and cosine are periodic with $k = 2\pi$. Exercises 65 and 66 show that 2π is the smallest such positive constant k. Therefore, we have the following.

Period of Sine and Cosine

The sine and cosine functions are periodic with period 2π: For every real number t,

$$\sin t = \sin(t \pm 2\pi) \qquad \text{and} \qquad \cos t = \cos(t \pm 2\pi).$$

EXAMPLE 8

Find $\sin \dfrac{13\pi}{6}$.

SOLUTION The periodicity identity shows that

$$\sin\frac{13\pi}{6} = \sin\left(\frac{\pi}{6} + \frac{12\pi}{6}\right) = \sin\left(\frac{\pi}{6} + 2\pi\right) = \sin\frac{\pi}{6} = \frac{1}{2}. \qquad \blacksquare$$

The tangent function is also periodic (see Exercise 36), but its period is π rather than 2π, that is,

$$\mathbf{tan}(t + \pi) = \mathbf{tan}\, t \quad \textbf{for every real number } t,$$

as we shall see in Section 2.4.

NEGATIVE ANGLE IDENTITIES

GRAPHING EXPLORATION

(a) In a viewing window with $-2\pi \le x \le 2\pi$, graph $y_1 = \sin x$ and $y_2 = \sin(-x)$ on the same screen. Use trace to move along $y_1 = \sin x$. Stop at a point and note its y-coordinate. Use the up or down arrow to move vertically to the graph of $y_2 = \sin(-x)$. The x-coordinate remains the same, but the y-coordinate is different. How are the two y-coordinates related? Is one the negative of the other? Repeat the procedure for other points. Are the results the same?

(b) Now graph $y_1 = \cos x$ and $y_2 = \cos(-x)$ on the same screen. How do the graphs compare?

(c) Repeat part (a) for $y_1 = \tan x$ and $y_2 = \tan(-x)$. Are the results similar to those for sine?

The preceding Graphing Exploration suggests the truth of the following statement.

Negative Angle Identities

For every real number t,

$$\sin(-t) = -\sin t$$
$$\cos(-t) = \cos t$$
$$\tan(-t) = -\tan t.$$

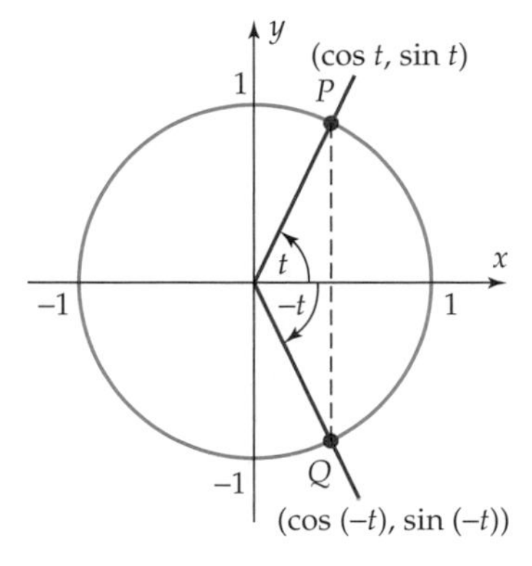

Figure 2–39

Proof Consider angles of t radians and $-t$ radians in standard position, as in Figure 2–39. By the definition of sine and cosine, P has coordinates $(\cos t, \sin t)$, and Q has coordinates $(\cos(-t), \sin(-t))$. As Figure 2–39 suggests, P and Q lie on the same vertical line. Therefore, they have the same first coordinate, that is, $\cos(-t) = \cos t$. As the figure also suggests, P and Q lie at equal distances from the x-axis.* So the y-coordinate of Q must be the negative of the y-coordinate of P, that is, $\sin(-t) = -\sin t$. Finally, by the definition of the tangent function and the two identities just proved, we have

$$\tan(-t) = \frac{\sin(-t)}{\cos(-t)} = \frac{-\sin t}{\cos t} = -\frac{\sin t}{\cos t} = -\tan t. \quad \blacksquare$$

EXAMPLE 9

Evaluate the trigonometric functions at $t = -\frac{\pi}{6}$.

SOLUTION In Example 4 of Section 2.2 (or Example 2 of Alternate Section 2.2), we saw that

$$\sin\frac{\pi}{6} = \frac{1}{2} \quad \text{and} \quad \cos\frac{\pi}{6} = \frac{\sqrt{3}}{2}.$$

Using the negative angle identities, we have

$$\sin\left(-\frac{\pi}{6}\right) = -\sin\frac{\pi}{6} = -\frac{1}{2} \quad \text{and} \quad \cos\left(-\frac{\pi}{6}\right) = \cos\frac{\pi}{6} = \frac{\sqrt{3}}{2},$$

Hence,

$$\tan\left(-\frac{\pi}{6}\right) = -\tan\frac{\pi}{6} = -\frac{\sin\frac{\pi}{6}}{\cos\frac{\pi}{6}} = -\frac{\frac{1}{2}}{\frac{\sqrt{3}}{2}} = -\frac{1}{\sqrt{3}} = -\frac{\sqrt{3}}{3}. \quad \blacksquare$$

EXAMPLE 10

To simplify $(1 + \sin t)(1 + \sin(-t))$, we use the negative angle identity and the Pythagorean identity.

$$\begin{aligned}(1 + \sin t)(1 + \sin(-t)) &= (1 + \sin t)(1 - \sin t)\\ &= 1 - \sin^2 t\\ &= \cos^2 t. \quad \blacksquare\end{aligned}$$

*These facts can be proved by using congruent triangles. (See Exercise 13 in the Geometry Review Appendix.)

EXERCISES 2.3

In Exercises 1–4, find the rule of the product function fg.

1. $f(t) = 3 \sin t; \quad g(t) = \sin t + 2 \cos t$
2. $f(t) = 5 \tan t; \quad g(t) = \tan^3 t - 1$
3. $f(t) = 3 \sin^2 t; \quad g(t) = \sin t + \tan t$
4. $f(t) = \sin 2t + \cos^4 t; \quad g(t) = \cos 2t + \cos^2 t$

In Exercises 5–14, factor the given expression.

5. $\cos^2 t - 4$
6. $25 - \tan^2 t$
7. $\sin^2 t - \cos^2 t$
8. $\sin^3 t - \sin t$
9. $\tan^2 t + 6 \tan t + 9$
10. $\cos^2 t - \cos t - 2$
11. $6 \sin^2 t - \sin t - 1$
12. $\tan t \cos t + \cos^2 t$
13. $\cos^4 t + 4 \cos^2 t - 5$
14. $3 \tan^2 t + 5 \tan t - 2$

In Exercises 15–18, find the rules of the composite functions $f \circ g$ and $g \circ f$. (The notation is explained in Example 4.)

15. $f(t) = \cos t; \quad g(t) = 2t + 4$
16. $f(t) = \sin t + 2; \quad g(t) = t^2$
17. $f(t) = \tan(t + 3); \quad g(t) = t^2 - 1$
18. $f(t) = \cos^2(t - 2); \quad g(t) = 5t + 2$

In Exercises 19–24, determine if it is possible for a number t to satisfy the given conditions. [Hint: *Think Pythagorean.*]

19. $\sin t = 5/13$ and $\cos t = 12/13$
20. $\sin t = -2$ and $\cos t = 1$
21. $\sin t = -1$ and $\cos t = 1$
22. $\sin t = 1/\sqrt{2}$ and $\cos t = -1/\sqrt{2}$
23. $\sin t = 1$ and $\tan t = 1$
24. $\cos t = 8/17$ and $\tan t = 15/8$

In Exercises 25–28, use the Pythagorean identity to find sin t.

25. $\cos t = -.5$ and $\pi < t < 3\pi/2$
26. $\cos t = -3/\sqrt{10}$ and $\pi/2 < t < \pi$
27. $\cos t = 1/2$ and $0 < t < \pi/2$
28. $\cos t = 2/\sqrt{5}$ and $3\pi/2 < t < 2\pi$

In Exercises 29–35, assume that $\sin t = 3/5$ and $0 < t < \pi/2$. Use identities in the text to find the number.

29. $\sin(-t)$
30. $\sin(t + 10\pi)$
31. $\sin(2\pi - t)$
32. $\cos t$
33. $\tan t$
34. $\cos(-t)$
35. $\tan(2\pi - t)$
36. (a) Show that $\tan(t + 2\pi) = \tan t$ for every t in the domain of $\tan t$. [*Hint:* Use the definition of tangent and some identities proved in the text.]
    (b) Verify that it appears true that $\tan(x + \pi) = \tan x$ for every t in the domain by using your calculator's table feature to make a table of values for $y_1 = \tan(x + \pi)$ and $y_2 = \tan x$.

In Exercises 37–42, assume that

$$\cos t = -2/5 \quad \textit{and} \quad \pi < t < 3\pi/2.$$

Use identities to find the number.

37. $\sin t$
38. $\tan t$
39. $\cos(2\pi - t)$
40. $\cos(-t)$
41. $\sin(4\pi + t)$
42. $\tan(4\pi - t)$

In Exercises 43–46, assume that

$$\sin(\pi/8) = \frac{\sqrt{2 - \sqrt{2}}}{2}$$

and use identities to find the exact functional value.

43. $\cos(\pi/8)$
44. $\tan(\pi/8)$
45. $\sin(17\pi/8)$
46. $\tan(-15\pi/8)$

In Exercises 47–58, use algebra and identities in the text to simplify the expression. Assume all denominators are nonzero.

47. $(\sin t + \cos t)(\sin t - \cos t)$

48. $(\sin t - \cos t)^2$

49. $\tan t \cos t$

50. $(\sin t)/(\tan t)$

51. $\sqrt{\sin^3 t \cos t}\sqrt{\cos t}$

52. $(\tan t + 2)(\tan t - 3) - (6 - \tan t) + 2 \tan t$

53. $\left(\dfrac{4\cos^2 t}{\sin^2 t}\right)\left(\dfrac{\sin t}{4\cos t}\right)^2$

54. $\dfrac{5\cos t}{\sin^2 t} \cdot \dfrac{\sin^2 t - \sin t \cos t}{\sin^2 t - \cos^2 t}$

55. $\dfrac{\cos^2 t + 4\cos t + 4}{\cos t + 2}$

56. $\dfrac{\sin^2 t - 2\sin t + 1}{\sin t - 1}$

57. $\dfrac{1}{\cos t} - \sin t \tan t$

58. $\dfrac{1 - \tan^2 t}{1 + \tan^2 t} + 2\sin^2 t$

*In Exercises 59–63, show that the given function is periodic with period less than 2π. [*Hint: *Find a positive number k with $k < 2\pi$ such that $f(t + k) = f(t)$ for every t in the domain of f.]*

59. $f(t) = \sin 2t$

60. $f(t) = \cos 3t$

61. $f(t) = \sin 4t$

62. $f(t) = \cos(3\pi t/2)$

63. $f(t) = \tan 2t$

64. If you know what even and odd functions are, fill the blanks with "even" or "odd" so that the resulting statement is true. Then prove the statement by using an appropriate identity.
(a) $f(t) = \sin t$ is an ____ function.
(b) $g(t) = \cos t$ is an ____ function.
(c) $h(t) = \tan t$ is an ____ function.
(d) $f(t) = t \sin t$ is an ____ function.
(e) $g(t) = t + \tan t$ is an ____ function.

65. Here is a proof that the cosine function has period 2π. We saw in the text that $\cos(t + 2\pi) = \cos t$ for every t. We must show that there is no positive number smaller than 2π with this property. Do this as follows:
(a) Find all numbers k such that $0 < k < 2\pi$ and $\cos k = 1$. [*Hint:* Draw a picture and use the definition of the cosine function.]
(b) Suppose k is a number such that $\cos(t + k) = \cos t$ for every number t. Show that $\cos k = 1$. [*Hint:* Consider $t = 0$.]
(c) Use parts (a) and (b) to show that there is no positive number k less than 2π with the property that $\cos(t + k) = \cos t$ for *every* number t. Therefore, $k = 2\pi$ is the smallest such number, and the cosine function has period 2π.

66. Here is proof that the sine function has period 2π. We saw in the text that $\sin(t + 2\pi) = \sin t$ for every t. We must show that there is no positive number smaller than 2π with this property. Do this as follows:
(a) Find a number t such that $\sin(t + \pi) \neq \sin t$.
(b) Find all numbers k such that $0 < k < 2\pi$ and $\sin k = 0$. [*Hint:* Draw a picture and use the definition of the sine function.]
(c) Suppose k is a number such that $\sin(t + k) = \sin t$ for every number t. Show that $\sin k = 0$. [*Hint:* Consider $t = 0$.]
(d) Use parts (a)–(c) to show that there is no positive number k less than 2π with the property that $\sin(t + k) = \sin t$ for *every* number t. Therefore, $k = 2\pi$ is the smallest such number, and the sine function has period 2π.

2.4 Basic Graphs

Although a graphing calculator will quickly sketch the graphs of the sine, cosine, and tangent functions, it will not give you much insight into why these graphs have the shapes they do and why these shapes are important. So the emphasis here is on the connection between the definitions of these functions and their graphs.

If P is the point where the unit circle meets the terminal side of an angle of t radians, then the y-coordinate of P is the number $\sin t$. We can get a rough sketch of the graph of $f(t) = \sin t$ by watching the y-coordinate of P.

As t Increases	The Point P Moves	The y-Coordinate of $P (= \sin t)$	Rough Sketch of the Graph
from 0 to $\frac{\pi}{2}$	from (1, 0) to (0, 1) 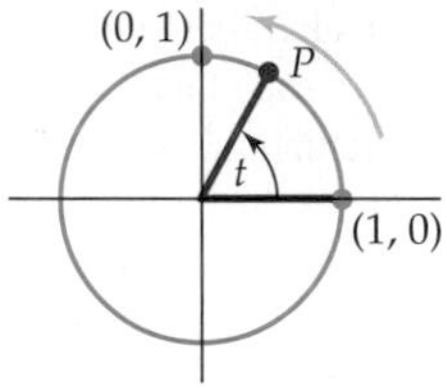	increases from 0 to 1	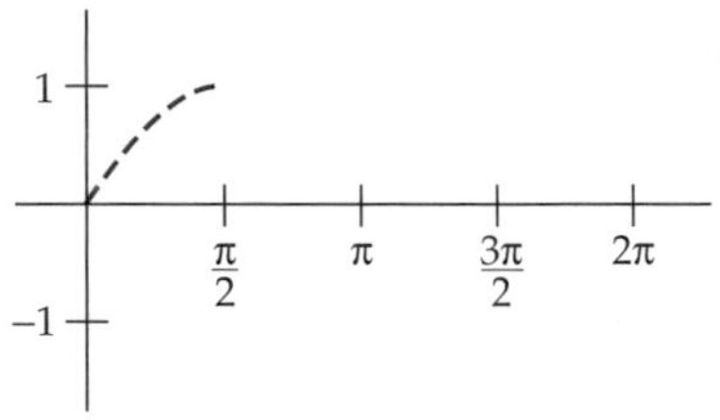
from $\frac{\pi}{2}$ to π	from (0, 1) to (−1, 0) 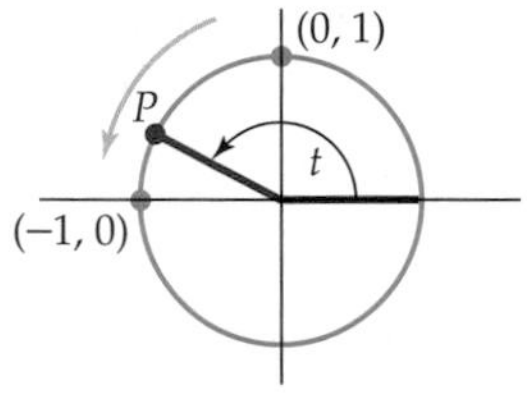	decreases from 1 to 0	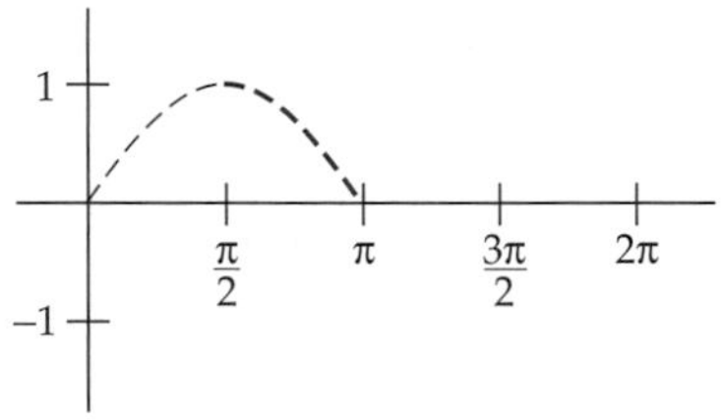
from π to $\frac{3\pi}{2}$	from (−1, 0) to (0, −1) 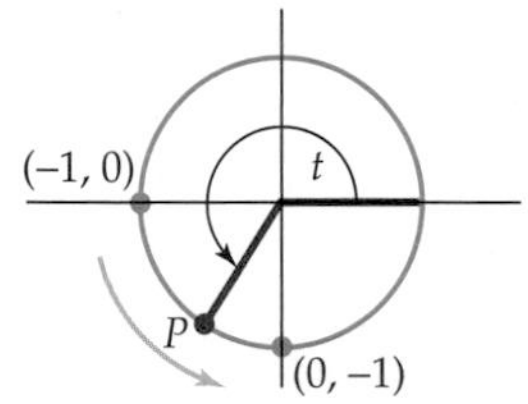	decreases from 0 to −1	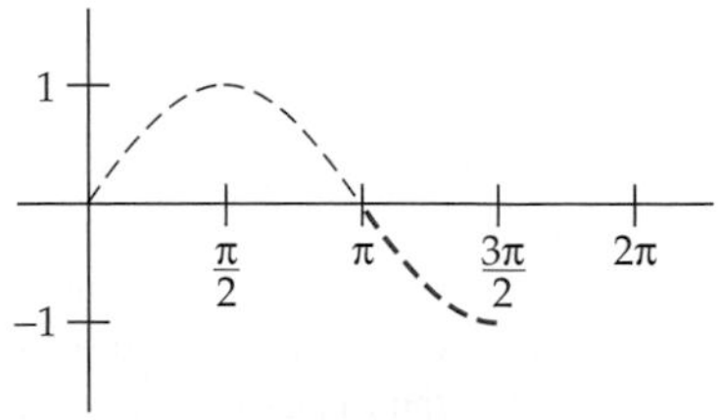
from $\frac{3\pi}{2}$ to 2π	from (0, −1) to (1, 0) 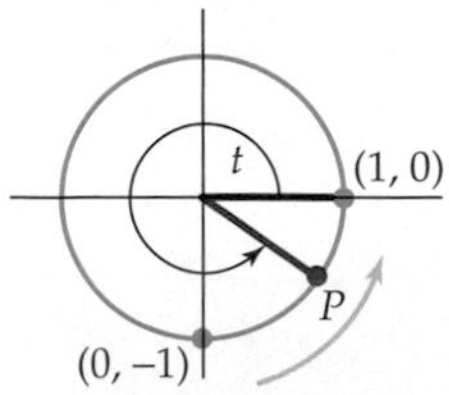	increases from −1 to 0	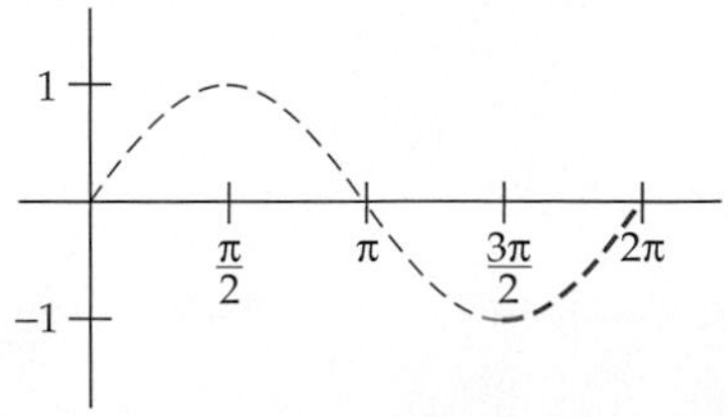

GRAPHING EXPLORATION

Your calculator can provide a dynamic simulation of this process. Put it in parametric graphing mode, and set the range values as follows:

$$0 \leq t \leq 6.28 \qquad -1 \leq x \leq 6.28 \qquad -2.5 \leq y \leq 2.5.$$

On the same screen, graph the two functions given by

$$x_1 = \cos t, \quad y_1 = \sin t \qquad \text{and} \qquad x_2 = t, \quad y_2 = \sin t.$$

Using the trace feature, move the cursor along the first graph (the unit circle). Stop at a point on the circle, and note the value of t and the y-coordinate of the point. Then switch the trace to the second graph (the sine function) by using the up or down cursor arrows. The value of t remains the same. What are the x- and y-coordinates of the new point? How does the y-coordinate of the new point compare with the y-coordinate of the original point on the unit circle?

NOTE
Parametric graphing is explained at the end of Section 0.5.

To complete the graph of the sine function, note that as t goes from 2π to 4π, the point P on the unit circle *retraces* the path it took from 0 to 2π, so *the same wave shape will repeat* on the graph. The same thing happens when t goes from 4π to 6π, or from -2π to 0, and so on. This repetition of the same pattern is simply the graphical expression of the fact that the sine function has period 2π: For any number t, the points

$$(t, \sin t) \qquad \text{and} \qquad (t + 2\pi, \sin(t + 2\pi))$$

on the graph have the same second coordinate.

A graphing calculator or some point plotting with an ordinary calculator now produces the graph of $f(t) = \sin t$ (Figure 2–40).

TECHNOLOGY TIP

Calculators have built-in windows for trigonometric functions, in which the x-axis tick marks are at intervals of $\pi/2$.
Choose TRIG or ZTRIG in this menu:

TI: ZOOM

HP-39+: VIEWS

Casio: V-WINDOW

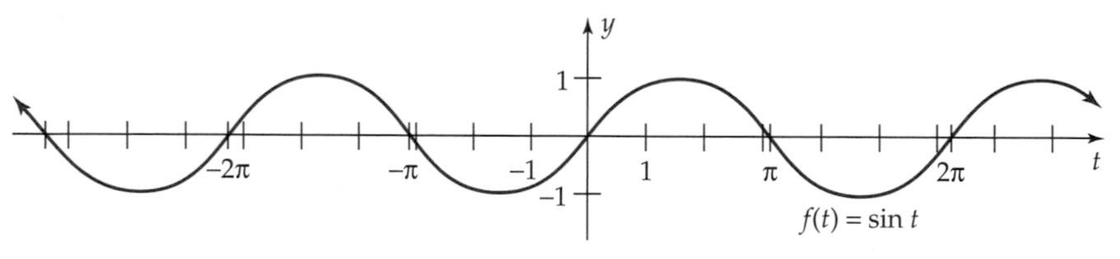

Figure 2–40

NOTE Throughout this chapter, we use t as the variable for trigonometric functions to avoid any confusion with the x's and y's that are part of the definition of these functions. For calculator graphing in "function mode," however, you must use x as the variable: $f(x) = \sin x$, $g(x) = \cos x$, etc.

The graph of the sine function and the techniques of Section 0.6 can be used to graph other trigonometric functions.

EXAMPLE 1

The graph of $h(t) = 3 \sin t$ is the graph of $f(t) = \sin t$ stretched away from the horizontal axis by a factor of 3, as shown in Figure 2–41. ■

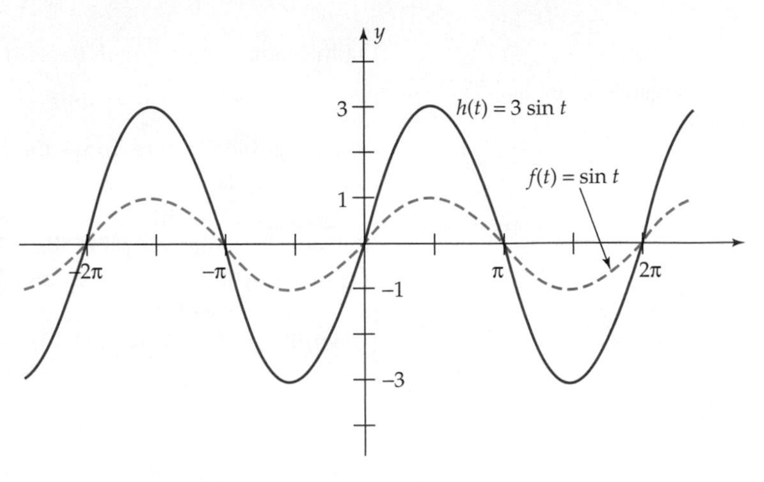

Figure 2–41

EXAMPLE 2

The graph of $k(t) = -\frac{1}{2} \sin t$ is the graph of $f(t) = \sin t$ shrunk by a factor of $1/2$ toward the horizontal axis and then reflected in the horizontal axis, as shown in Figure 2–42. ■

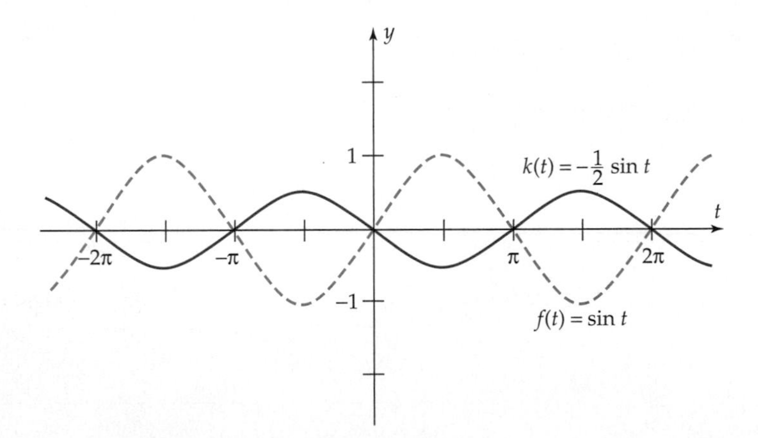

Figure 2–42

GRAPH OF THE COSINE FUNCTION

To obtain the graph of $g(t) = \cos t$, we follow the same procedure as with sine, except that we now watch the x-coordinate of P (which is $\cos t$).

As t Increases	The Point P Moves	The x-coordinate of P $(= \cos t)$	Rough Sketch of the Graph
from 0 to $\frac{\pi}{2}$	from (1, 0) to (0, 1)	decreases from 1 to 0	
from $\frac{\pi}{2}$ to π	from (0, 1) to (−1, 0)	decreases from 0 to −1	
from π to $\frac{3\pi}{2}$	from (−1, 0) to (0, −1)	increases from −1 to 0	
from $\frac{3\pi}{2}$ to 2π	from (0, −1) to (1, 0)	increases from 0 to 1	

As t takes larger values, P begins to retrace its path around the unit circle, so the graph of $g(t) = \cos t$ repeats the same wave pattern, and similarly for negative values of t. So the graph of $\cos t$ looks like Figure 2–43.

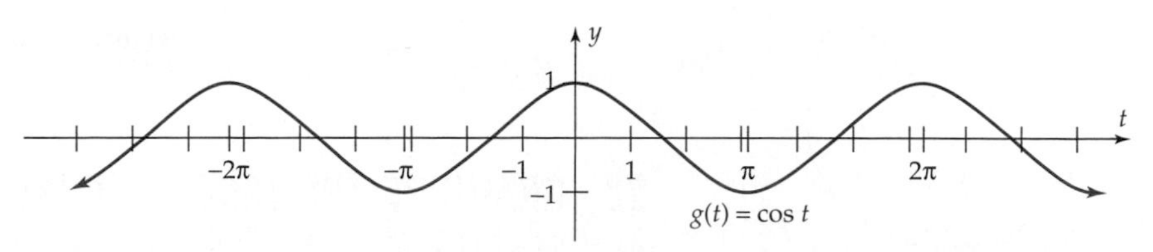

Figure 2–43

For a dynamic simulation of the cosine graphing process described above, see Exercise 51.

The techniques of Section 0.6 can be used to graph variations of the cosine function.

EXAMPLE 3 

The graph of $h(t) = 4\cos t$ is the graph of $g(t) = \cos t$ stretched away from the horizontal axis by a factor of 4, as shown in Figure 2–44. ■

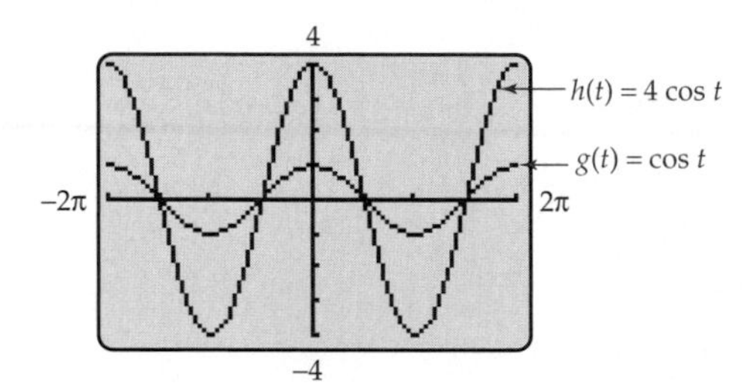

Figure 2–44

EXAMPLE 4

The graph of $k(t) = -2\cos t + 3$ is the graph of $g(t) = \cos t$ stretched away from the horizontal axis by a factor of 2, reflected in the horizontal axis, and shifted vertically 3 units upward as shown in Figure 2–45. ■

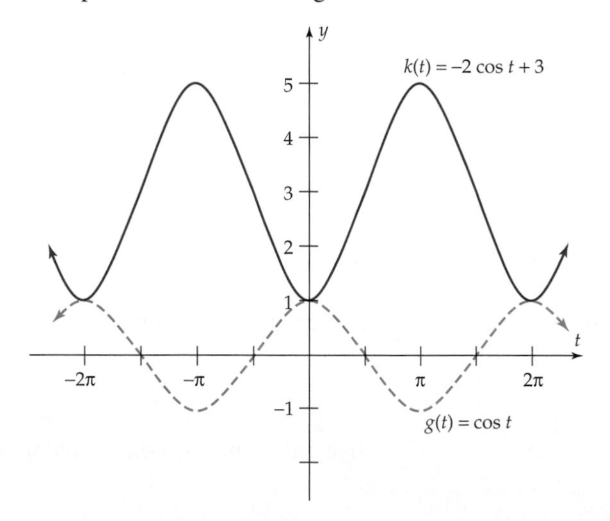

Figure 2–45

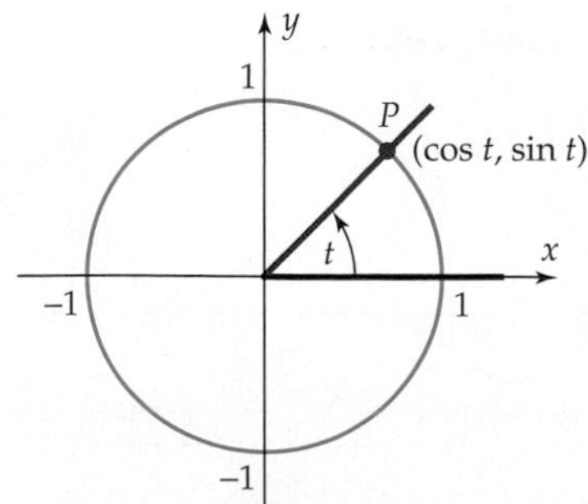

Figure 2–46

GRAPH OF THE TANGENT FUNCTION

To determine the shape of the graph of $h(t) = \tan t$, we use an interesting connection between the tangent function and straight lines. As shown in Figure 2–46, the point P where the terminal side of an angle of t radians in standard position meets the unit circle has coordinates (cos t, sin t). We can use this point and the point (0, 0) to compute the *slope* of the terminal side.*

$$\text{slope} = \frac{\sin t - 0}{\cos t - 0} = \frac{\sin t}{\cos t} = \tan t$$

Therefore, we have the following.

Slope and Tangent

> The slope of the terminal side of an angle of t radians in standard position is the number tan t.

The graph of $h(x) = \tan t$ can now be sketched by watching the slope of the terminal side of an angle of t radians, as t takes different values. Recall that the more steeply a line rises from left to right, the larger its slope. Similarly, lines that fall from left to right have negative slopes that increase in absolute value as the line falls more steeply.

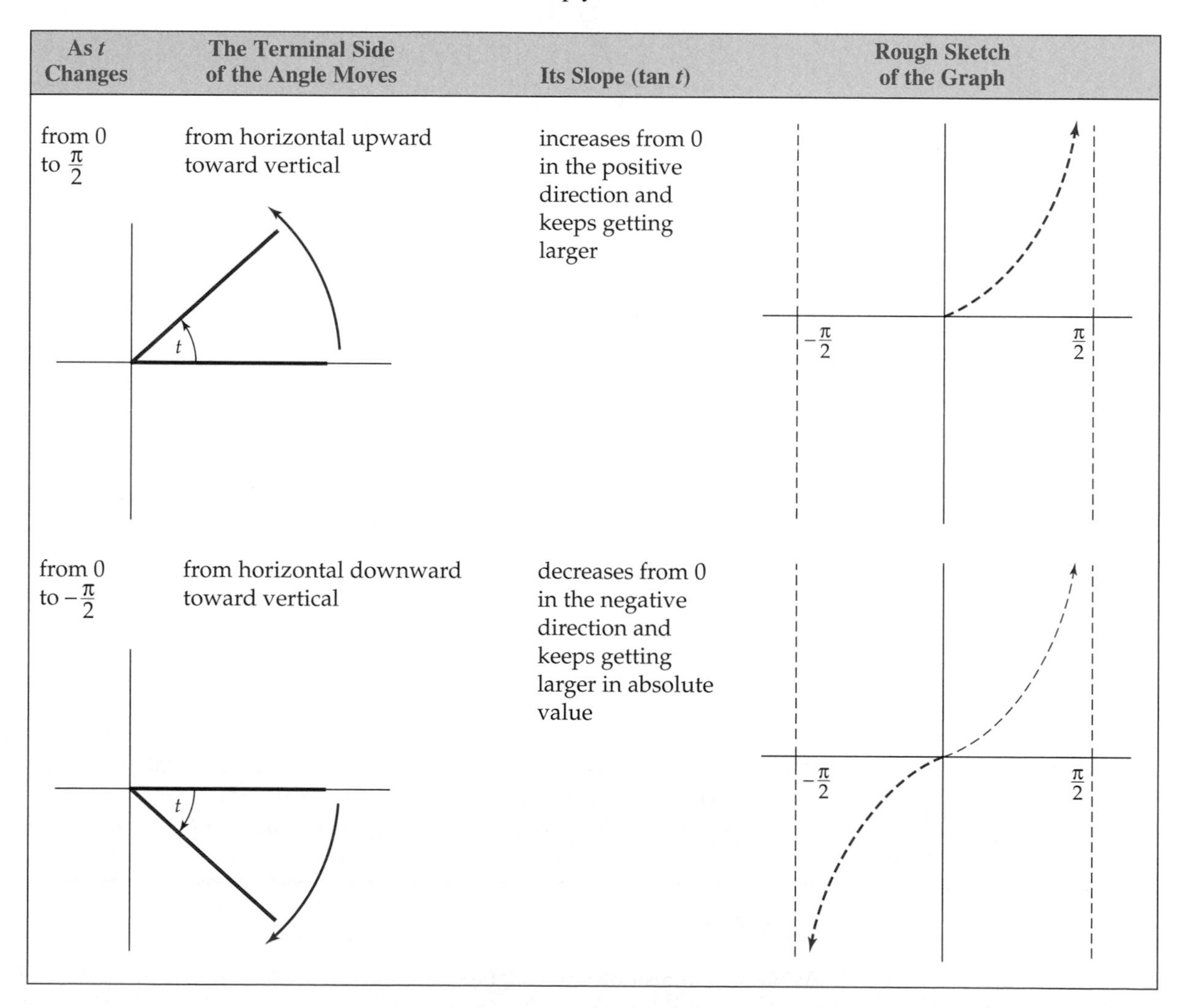

As t Changes	The Terminal Side of the Angle Moves	Its Slope (tan t)	Rough Sketch of the Graph
from 0 to $\frac{\pi}{2}$	from horizontal upward toward vertical	increases from 0 in the positive direction and keeps getting larger	
from 0 to $-\frac{\pi}{2}$	from horizontal downward toward vertical	decreases from 0 in the negative direction and keeps getting larger in absolute value	

* The slope of a line is discussed in the Geometry Review Appendix.

When $t = \pm\pi/2$, the terminal side of the angle is vertical, and hence its slope is not defined. This corresponds to the fact that the tangent function is not defined when $t = \pm\pi/2$. The vertical lines through $\pm\pi/2$ are vertical asymptotes of the graph: It gets closer and closer to these lines but never touches them.

As t goes from $\pi/2$ to $3\pi/2$, the terminal side goes from almost vertical with negative slope to horizontal to almost vertical with positive slope (draw a picture), exactly as it does between $-\pi/2$ and $\pi/2$. So the graph repeats the same pattern. The same thing happens between $3\pi/2$ and $5\pi/2$, between $-3\pi/2$ and $-\pi/2$, etc. Therefore, the entire graph looks like Figure 2–47.

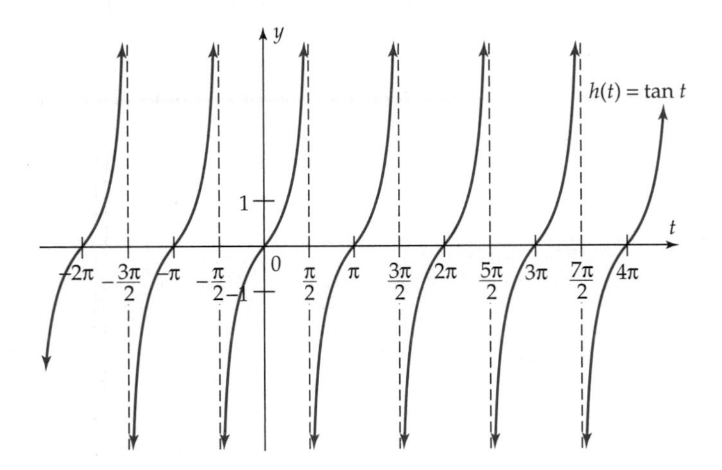

Figure 2–47

Because calculators sometimes do not graph accurately across vertical asymptotes, the graph may look slightly different on a calculator screen (with vertical line segments where the asymptotes should be).

The graph of the tangent function repeats the same pattern at intervals of length π. This means that the tangent function repeats its values at intervals of π.

Period of Tangent

> The tangent function is periodic with period π: For every real number t in its domain,
>
> $$\tan(t \pm \pi) = \tan t.$$

Near each of the vertical asymptotes, the graph of the tangent function extends upward and downward forever. This means that $\tan t$ takes all possible real number values. Hence, we have proved the statement made in Section 2.2: *The range (set of outputs) of the tangent function is the set of all real numbers.*

EXAMPLE 5

As we saw in Section 0.6, the graph of

$$k(t) = \tan\left(t - \frac{\pi}{2}\right)$$

is the graph of $h(t) = \tan t$ shifted horizontally $\pi/2$ units to the right (Figure 2–48). ■

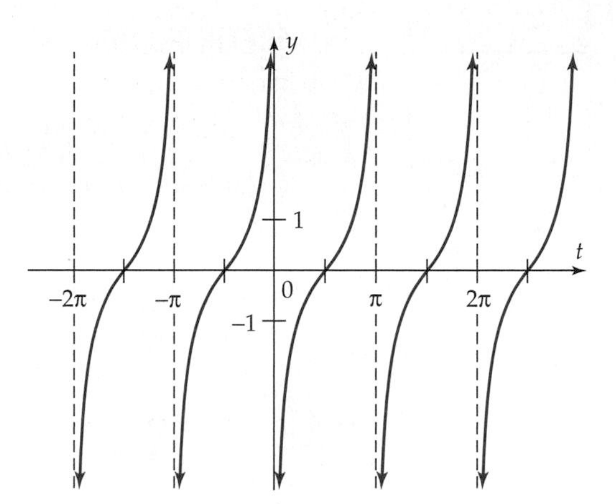

Figure 2–48

GRAPHS AND IDENTITIES

Graphing calculators can be used to identify equations that could possibly be identities. A calculator cannot *prove* that such an equation is an identity; but it can provide evidence that it *might* be one. On the other hand, a calculator *can* prove that a particular equation is not an identity.

EXAMPLE 6

Which of the following equations could possibly be an identity?

(a) $\cos\left(\frac{\pi}{2} + t\right) = \sin t$ (b) $\cos\left(\frac{\pi}{2} - t\right) = \sin t$

SOLUTION

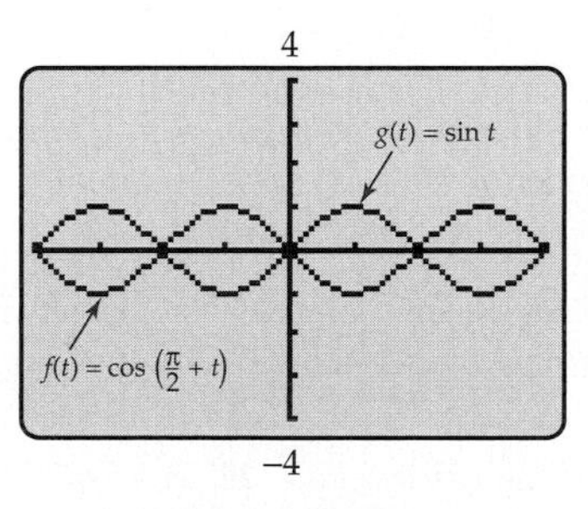

Figure 2–49

(a) Consider the functions $f(t) = \cos\left(\frac{\pi}{2} + t\right)$ and $g(t) = \sin t$, whose rules are given by the two sides of the equation

$$\cos\left(\frac{\pi}{2} + t\right) = \sin t.$$

If this equation is an identity, then $f(t) = g(t)$ for every real number t, and hence, f and g have the same graph. But the graphs of f and g (Figure 2–49) are obviously different. Therefore, this equation is *not* an identity.

(b) We can test this equation in the same manner. The graph of the left side, that is, the graph of

$$h(t) = \cos\left(\frac{\pi}{2} - t\right),$$

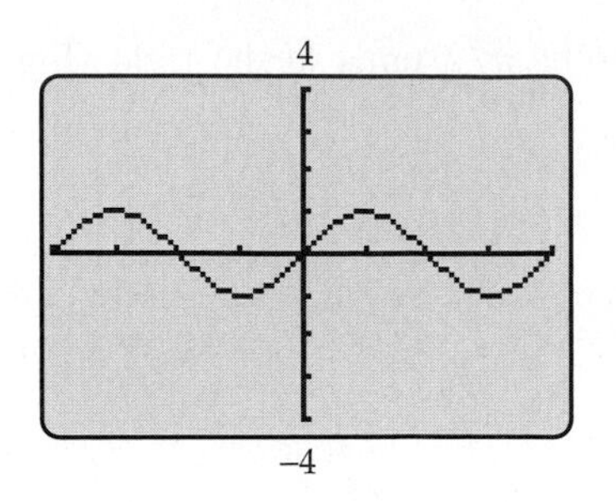

Figure 2–50

in Figure 2–50 appears to be the same as the graph of $g(t) = \sin t$ when $-2\pi \le t \le 2\pi$ (Figure 2–49). So let's explore further.

GRAPHING EXPLORATION

Graph $h(t) = \cos\left(\frac{\pi}{2} - t\right)$ and $g(t) = \sin t$ on the same screen and use the trace feature to confirm that the graphs appear to be identical.

The fact that the graphs appear to be identical means that the two functions have the same value at every number t that the calculator computed in making the graphs (at least 95 numbers). This evidence strongly suggests that the equation $\cos\left(\frac{\pi}{2} - t\right) = \sin t$ is an identity but does not prove it. All we can say at this point is that the equation could possibly be an identity. ■

CAUTION

Do not assume that two graphs that look the same on a calculator screen actually are the same. Depending on the viewing window, two graphs that are actually quite different may appear to be identical. See Exercises 43–50 for some examples.

EXERCISES 2.4

In Exercises 1–6, use the graphs of the sine and cosine functions to find all the solutions of the equation.

1. $\sin t = 0$ **2.** $\cos t = 0$ **3.** $\sin t = 1$

4. $\sin t = -1$ **5.** $\cos t = -1$ **6.** $\cos t = 1$

In Exercises 7–10, find tan t, where the terminal side of an angle of t radians lies on the given line.

7. $y = 11x$ **8.** $y = 1.5x$ **9.** $y = 1.4x$

10. $y = .32x$

In Exercises 11–22, list the transformations needed to change the graph of $f(t)$ into the graph of $g(t)$. [See Section 0.6.]

11. $f(t) = \sin t$; $g(t) = \sin t + 3$

12. $f(t) = \cos t$; $g(t) = \cos t - 2$

13. $f(t) = \cos t$; $g(t) = -\cos t$

14. $f(x) = \sin t$; $g(t) = -3 \sin t$

15. $f(t) = \tan t$; $g(t) = \tan t + 5$

16. $f(t) = \tan t$; $g(t) = -\tan t$

17. $f(t) = \cos t$; $g(t) = 3 \cos t$

18. $f(t) = \sin t$; $g(t) = -2 \sin t$

19. $f(t) = \sin t$; $g(t) = 3 \sin t + 2$

20. $f(t) = \cos t$; $g(t) = 5 \cos t + 3$

21. $f(t) = \sin t$; $g(t) = \sin(t - 2)$

22. $f(t) = \cos t$; $g(t) = 3 \cos(t + 2) - 3$

In Exercises 23–30, use the graphs of the trigonometric functions to determine the number *of solutions of the equation between 0 and 2π.*

23. $\sin t = 3/5$ [*Hint:* How many points on the graph of $f(t) = \sin t$ between $t = 0$ and $t = 2\pi$ have second coordinate $3/5$?]

24. $\cos t = -1/4$ **25.** $\tan t = 4$ **26.** $\cos t = 2/3$

27. $\sin t = -1/2$

28. $\sin t = k$, where k is a nonzero constant such that $-1 < k < 1$.

29. $\cos t = k$, where k is a constant such that $-1 < k < 1$.

30. $\tan t = k$, where k is any constant.

In Exercises 31–42, use graphs to determine whether the equation could possibly be an identity or definitely is not an identity.

31. $\sin(-t) = -\sin t$

32. $\cos(-t) = \cos t$

33. $\sin^2 t + \cos^2 t = 1$

34. $\sin(t + \pi) = -\sin t$

35. $\sin t = \cos(t - \pi/2)$

36. $\sin^2 t - \tan^2 t = -(\sin^2 t)(\tan^2 t)$

37. $\dfrac{\sin t}{1 + \cos t} = \tan t$

38. $\dfrac{\cos t}{1 - \sin t} = \dfrac{1}{\cos t} + \tan t$

39. $\cos\left(\dfrac{\pi}{2} + t\right) = -\sin t$

40. $\sin\left(\dfrac{\pi}{2} + t\right) = -\cos t$

41. $(1 + \tan t)^2 = \dfrac{1}{\cos t}$

42. $(\cos^2 t - 1)(\tan^2 t + 1) = -\tan^2 t$

Thinkers

Exercises 43–46 explore various ways in which a calculator can produce inaccurate graphs of trigonometric functions. These exercises also provide examples of two functions, with different graphs, whose graphs appear identical in certain viewing windows.

43. Choose a viewing window with $-3 \leq y \leq 3$ and $0 \leq x \leq k$, where k is chosen as follows.

Width of Screen	k
95 pixels (TI-83+)	188π
127 pixels (TI-86, Casio)	252π
131 pixels (HP-39+)	260π
159 pixels (TI-89)	316π

(a) Graph $y = \cos x$ and the constant function $y = 1$ on the same screen. Do the graphs look identical? Are the functions the same?

(b) Use the trace feature to move the cursor along the graph of $y = \cos x$, starting at $x = 0$. For what values of x did the calculator plot points? [*Hint:* $2\pi \approx 6.28$.] Use this information to explain why the two graphs look identical.

44. Using the viewing window in Exercise 43, graph $y = \tan x + 2$ and $y = 2$ on the same screen. Explain why the graphs look identical even though the functions are not the same.

45. The graph of $g(x) = \cos x$ is a series of repeated waves (see Figure 2–43). A full wave (from the peak, down to the trough, and up to the peak again) starts at $x = 0$ and finishes at $x = 2\pi$.

(a) How many full waves will the graph make between $x = 0$ and $x = 502.65$ ($\approx 80 \cdot 2\pi$)?

(b) Graph $g(t) = \cos t$ in a viewing window with $0 \leq t \leq 502.65$. How many full waves are shown on the graph? Is your answer the same as in part (a)? What's going on?

46. Find a viewing window in which the graphs of $y = \cos x$ and $y = .54$ appear identical. [*Hint:* See the chart in Exercise 43 and note that $\cos 1 \approx .54$.]

Exercises 47–50 provide further examples of functions with different graphs, whose graphs appear identical in certain viewing windows.

47. Approximating trigonometric functions by polynomials. For each odd positive integer n, let f_n be the function whose rule is

$$f_n(t) = t - \frac{t^3}{3!} + \frac{t^5}{5!} - \frac{t^7}{7!} + \cdots - \frac{t^n}{n!}.$$

Since the signs alternate, the sign of the last term might be $+$ instead of $-$, depending on what n is. Recall that $n!$ is the product of all integers from 1 to n; for instance, $5! = 1 \cdot 2 \cdot 3 \cdot 4 \cdot 5 = 120$.

(a) Graph $f_7(t)$ and $g(t) = \sin t$ on the same screen in a viewing window with $-2\pi \leq t \leq 2\pi$. For what values of t does f_7 appear to be a good approximation of g?

(b) What is the smallest value of n for which the graphs of f_n and g appear to coincide in this window? In this case, determine how accurate the approximation is by finding $f_n(2)$ and $g(2)$.

48. For each even positive integer n, let f_n be the function whose rule is

$$f_n(t) = 1 - \frac{t^2}{2!} + \frac{t^4}{4!} - \frac{t^6}{6!} + \frac{t^8}{8!} - \cdots + \frac{t^n}{n!}.$$

(The sign of the last term may be $-$ instead of $+$, depending on what n is.)

(a) In a viewing window with $-2\pi \leq t \leq 2\pi$, graph f_6, f_{10}, and f_{12}.

(b) Find a value of n for which the graph of f_n appears to coincide (in this window) with the graph of a well-known trigonometric function. What is the function?

49. Find a rational function whose graph appears to coincide with the graph of $h(t) = \tan t$ when

$$-2\pi \leq t \leq 2\pi.$$

[*Hint:* See Exercises 47 and 48.]

50. Find a periodic function whose graph consists of "square waves." [*Hint:* Consider the sum

$$\sin \pi t + \frac{1}{3}\sin 3\pi t + \frac{1}{5}\sin 5\pi t + \frac{1}{7}\sin 7\pi t + \cdots.]$$

51. With your calculator in parametric graphing mode and the range values

$$0 \le t \le 6.28 \qquad -1 \le x \le 6.28 \qquad -2.5 \le y \le 2.5,$$

graph the following two functions on the same screen:

$$x_1 = \cos t, \quad y_1 = \sin t \qquad \text{and} \qquad x_2 = t, \quad y_2 = \cos t.$$

Using the trace feature, move the cursor along the first graph (the unit circle). Stop at a point on the circle, note the value of t and the x-coordinate of the point. Then switch the trace to the second graph (the cosine function) by using the up or down cursor arrows. The value of t remains the same. How does the y-coordinate of the new point compare with the x-coordinate of the original point on the unit circle? Explain what's going on.

52. If you know what even and odd functions are, answer these questions.

(a) Judging from their graphs, which of the functions $f(t) = \sin t$, $g(t) = \cos t$, and $h(t) = \tan t$ appear to be even functions? Which appear to be odd functions?

(b) Confirm your answers in part (a) algebraically by using appropriate identities from Section 2.3.

2.5 Periodic Graphs and Simple Harmonic Motion

We now analyze functions whose rule is of the form

$$f(t) = A\sin(bt + c) \qquad \text{or} \qquad g(t) = A\cos(bt + c),$$

where A, b, and c are constants. Many periodic phenomena can be modeled by such functions, as we shall see below.

PERIOD

The functions $f(t) = \sin t$ and $g(t) = \cos t$ have period 2π, so each of their graphs makes one full wave between 0 and 2π. The sine wave begins on the horizontal axis, rises to height 1, falls to -1, and returns to the axis (Figure 2–51). The cosine wave between 0 and 2π begins at height 1, falls to -1, and rises to height 1 again (Figure 2–52).

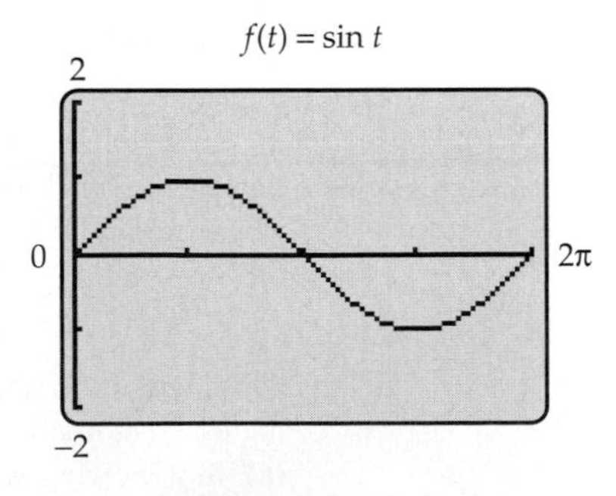

Figure 2–51

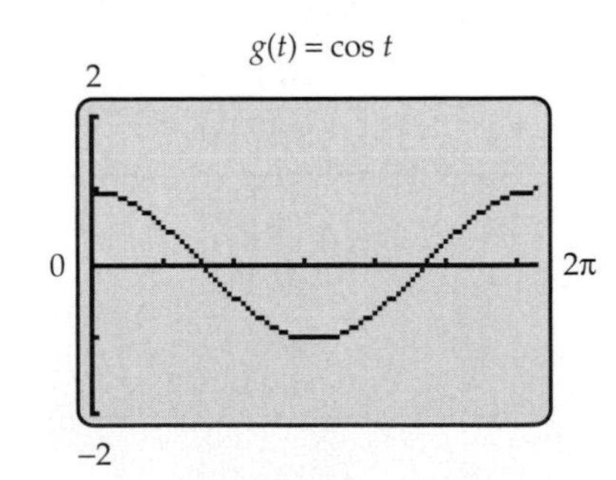

Figure 2–52

GRAPHING EXPLORATION

Graph the following functions, one at a time, in a viewing window with $0 \le t \le 2\pi$. Determine the number of complete waves in each graph and the period of the function (the length of one wave).

$$f(t) = \sin 2t, \qquad g(t) = \cos 3t, \qquad h(x) = \sin 4t, \qquad k(x) = \cos 5t.$$

This exploration suggests the following.

Period

> If $b > 0$, then the graph of either
>
> $$f(t) = \sin bt \qquad \text{or} \qquad g(t) = \cos bt$$
>
> makes b complete waves between 0 and 2π. Hence, each function has period $2\pi/b$.

Although we arrived at this statement by generalizing from several graphs, it can also be explained algebraically.

EXAMPLE 1

The graph of $g(t) = \cos t$ makes one complete wave as t takes values from 0 to 2π. Similarly, the graph of $k(t) = \cos 3t$ will complete one wave as the quantity $3t$ takes values from 0 to 2π. However,

$$3t - 0 \text{ when } t - 0 \qquad \text{and} \qquad 3t = 2\pi \text{ when } t = 2\pi/3.$$

So the graph of $k(t) = \cos 3t$ makes one complete wave between $t = 0$ and $t = 2\pi/3$, as shown in Figure 2–53, and hence k has period $2\pi/3$. Similarly, the graph makes a complete wave from $t = 2\pi/3$ to $t = 4\pi/3$ and another one from $t = 4\pi/3$ to $t = 2\pi$, as shown in Figure 2–53. ■

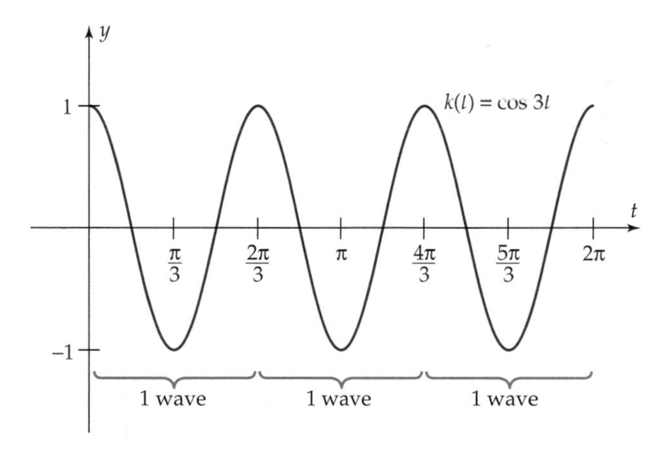

Figure 2–53

EXAMPLE 2

According to the box above, the function $f(t) = \sin \frac{1}{2}t$ has period $\dfrac{2\pi}{1/2} = 4\pi$. Its graph makes *half* a wave from $t = 0$ to $t = 2\pi$ (just as $\sin t$ does from $t = 0$ to $t = \pi$) and the other half of the wave from $t = 2\pi$ to $t = 4\pi$, as shown in Figure 2–54. ■

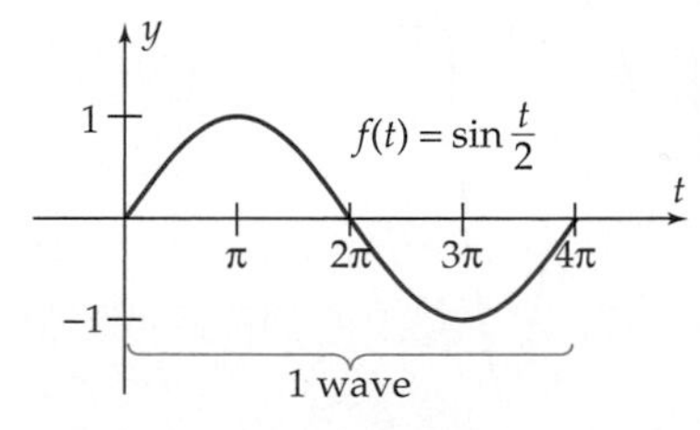

Figure 2–54

EXAMPLE 3

Computers and calculators are incapable of accurately graphing $f(t) = \sin bt$ or $g(t) = \cos bt$ when b is large. For instance, we know that the graph of

$$f(t) = \sin 500t$$

should show 500 complete waves between 0 and 2π. Depending on the model, however, a computer or calculator will produce either garbage (Figure 2–55) or a graph with far fewer than 500 waves (Figure 2–56). For the reason why, see Exercises 59 and 60. ■

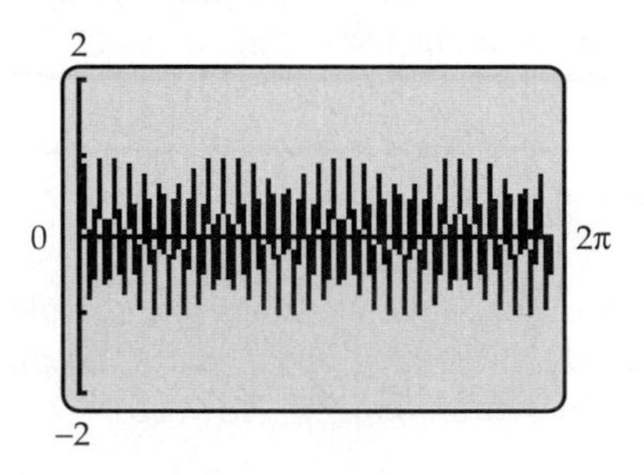

Figure 2–55

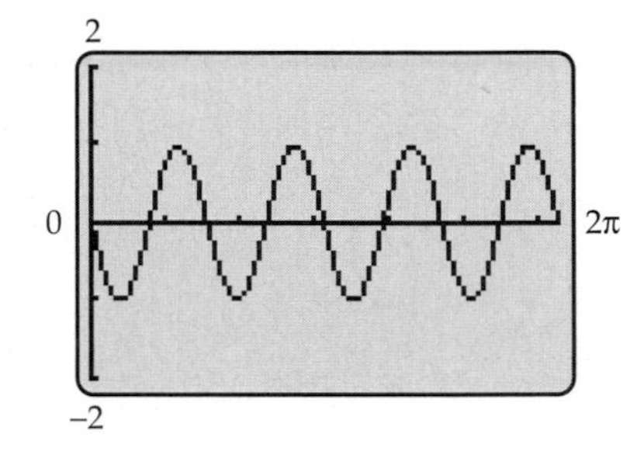

Figure 2–56

AMPLITUDE

As we saw in Section 0.6, multiplying the rule of a function by a positive constant has the effect of stretching its graph away from or shrinking it toward the horizontal axis.

EXAMPLE 4

The function $g(t) = 7 \cos 3t$ is just the function $k(t) = \cos 3t$ multiplied by 7. Consequently, the graph of g is just the graph of k (which was obtained in Example 1) stretched away from the horizontal axis by a factor of 7, as shown in Figure 2–57.

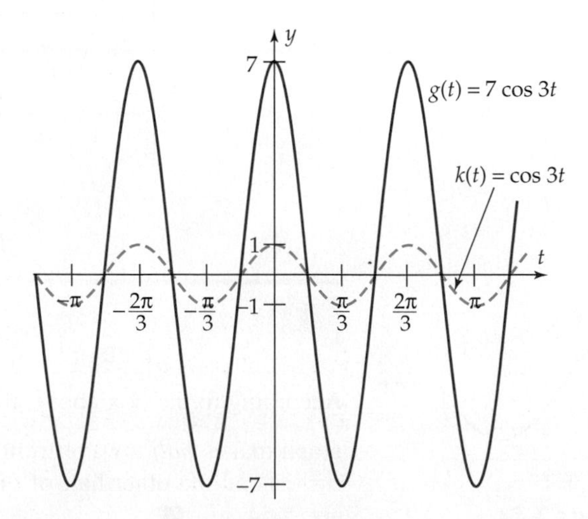

Figure 2–57

Stretching the graph affects only the height of the waves, not the period of the function: Both graphs have period $2\pi/3$, and each full wave has length $2\pi/3$. ■

The waves of the graph of $g(t) = 7 \cos 3t$ in Figure 2–57 rise 7 units above the t-axis and drop 7 units below the axis. More generally, the waves of the graph of $f(t) = A \sin bt$ or $g(t) = A \cos bt$ move a distance of $|A|$ units above and below the t-axis, and we say that these functions have **amplitude** $|A|$. In summary, we have the following.

Amplitude and Period

If $A \neq 0$ and $b > 0$, then each of the functions

$$f(t) = A \sin bt \qquad \text{or} \qquad g(t) = A \cos bt$$

has amplitude $|A|$ and period $2\pi/b$.

EXAMPLE 5

The function $f(t) = -2 \sin 4t$ has amplitude $|-2| = 2$ and period $2\pi/4 = \pi/2$. So the graph consists of waves of length $\pi/2$ that rise and fall between -2 and 2. But be careful: The waves in the graph of $2 \sin 4t$ (like the waves of $\sin t$) begin at height 0, rise, and then fall. But the graph of $f(t) = -2 \sin 4t$ is the graph of $2 \sin 4t$ reflected in the horizontal axis (see page 66). So its waves start at height 0, move *downward,* and then rise, as shown in Figure 2–58. ■

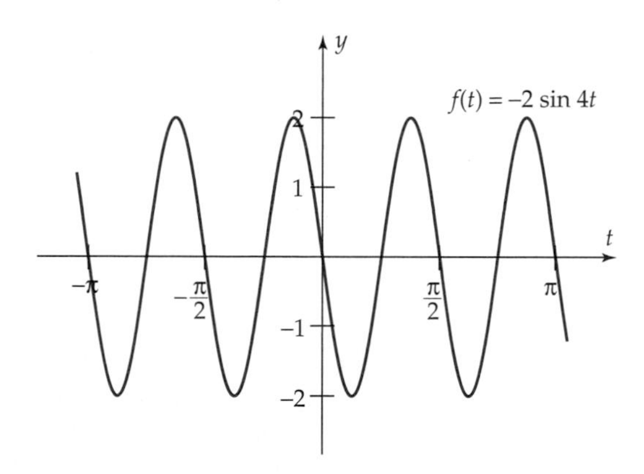

Figure 2–58

PHASE SHIFT

Next, we consider horizontal shifts. As we saw in Section 0.6, the graph of $\sin(t - 3)$ is the graph of $\sin t$ shifted 3 units to the right, and the graph of $\sin(t + 3)$ is the graph of $\sin t$ shifted 3 units to the left.

EXAMPLE 6

(a) Find a sine function whose graph looks like Figure 2–59.
(b) Find a cosine function whose graph looks like Figure 2–59.

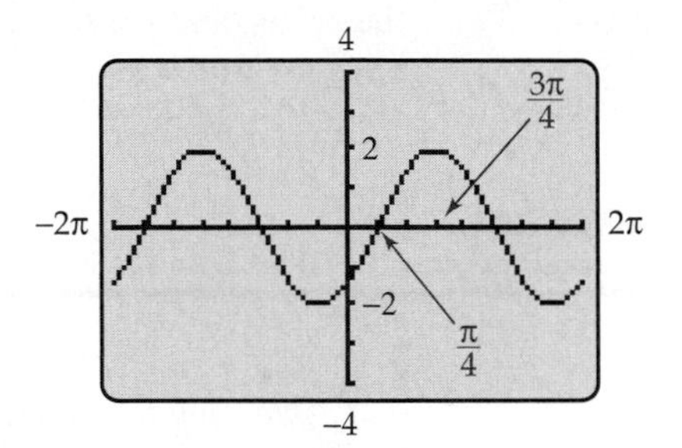

Figure 2–59

SOLUTION

(a) Since each wave has height 2, Figure 2–59 looks like the graph of 2 sin t shifted $\pi/4$ units to the right (so that a sine wave starts at $t = \pi/4$). Since the graph of $2\sin(t - \pi/4)$ is the graph of 2 sin t shifted $\pi/4$ units to the right (see page 63), we conclude that Figure 2–59 closely resembles the graph of $f(t) = 2\sin(t - \pi/4)$.

(b) Figure 2–59 also looks like the graph of 2 cos t shifted $3\pi/4$ units to the right (so that a cosine wave starts at $t = 3\pi/4$). Hence, Figure 2–59 could also be the graph of $g(t) = 2\cos(t - 3\pi/4)$. ■

EXAMPLE 7

(a) Find the amplitude and the period of

$$f(t) = 3\sin(2t + 5).$$

(b) Do the same for the function $f(t) = A\sin(bt + c)$, where A, b, c are constants.

SOLUTION The analysis of $f(t) = 3\sin(2t + 5)$ is in the left-hand column below, and the analysis of the general case $f(t) = A\sin(bt + c)$ is in the right-hand column. Observe that exactly the same procedure is used in both cases: Just change 3 to A, 2 to b, and 5 to c.

(a) Rewrite the rule of $f(t) = 3\sin(2t + 5)$ as

$$f(t) = 3\sin(2t + 5) = 3\sin\left(2\left(t + \frac{5}{2}\right)\right).$$

Thus, the rule of f can be obtained from the rule of the function $k(t) = 3\sin 2t$ by replacing t with $t + \frac{5}{2}$. Therefore, the graph of f is just the graph of k shifted horizontally $5/2$ units to the left, as shown in Figure 2–60 on the next page.

(b) Rewrite the rule of $f(t) = A\sin(bt + c)$ as

$$f(t) = A\sin(bt + c) = A\sin\left(b\left(t + \frac{c}{b}\right)\right).$$

Thus, the rule of f can be obtained from the rule of the function $k(t) = A\sin bt$ by replacing t with $t + \frac{c}{b}$. Therefore, the graph of f is just the graph of k shifted horizontally by c/b units.

Hence, $f(t) = 3\sin(2t + 5)$ has the same amplitude as $k(t) = 3\sin 2t$, namely, 3, and the same period, namely, $2\pi/2 = \pi$.

On the graph of $k(t) = 3\sin 2t$, a wave begins when $t = 0$. On the graph of

$$f(t) = 3\sin 2\left(t + \frac{5}{2}\right),$$

the shifted wave begins when $t + 5/2 = 0$, that is, when $t = -5/2$.

Hence, $f(t) = A\sin(bt + c)$ has the same amplitude as $k(t) = A\sin bt$, namely, $|A|$, and the same period, namely, $2\pi/b$.

On the graph of $k(t) = A\sin bt$, a wave begins when $t = 0$. On the graph of

$$f(t) = A\sin b\left(t + \frac{c}{b}\right),$$

the shifted wave begins when $t + c/b = 0$, that is, when $t = -c/b$. ■

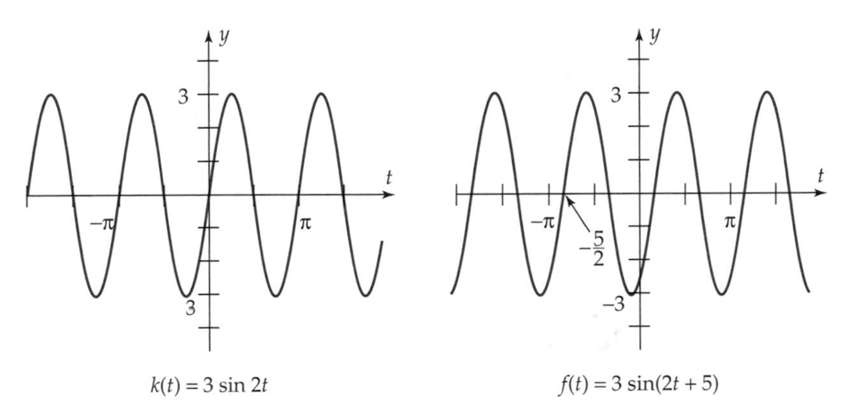

Figure 2–60

We say that the function $f(t) = A\sin(bt + c)$ has **phase shift** $-c/b$. A similar analysis applies to the function $g(t) = \cos(bt + c)$ and leads to this conclusion.

Amplitude, Period, and Phase Shift

> If $A \neq 0$ and $b > 0$, then each of the functions
>
> $$f(t) = A\sin(bt + c) \qquad \text{and} \qquad g(t) = A\cos(bt + c)$$
>
> has
>
> amplitude $|A|$, period $2\pi/b$, phase shift $-c/b$.
>
> A wave of the graph begins at $t = -c/b$.

EXAMPLE 8

Describe the graph of $g(t) = 2\cos(3t - 4)$.

SOLUTION The rule of g can be rewritten as

$$g(t) = 2\cos(3t + (-4)).$$

This is the case described in the preceding box with $A = 2$, $b = 3$, and $c = -4$. Therefore, the function g has

$$\text{amplitude } |A| = |2| = 2, \qquad \text{period } \frac{2\pi}{b} = \frac{2\pi}{3},$$

$$\text{phase shift } -\frac{c}{b} = -\frac{-4}{3} = \frac{4}{3}.$$

Hence, the graph of g consists of waves of length of $2\pi/3$ that run vertically between 2 and -2. A wave begins at $t = 4/3$.

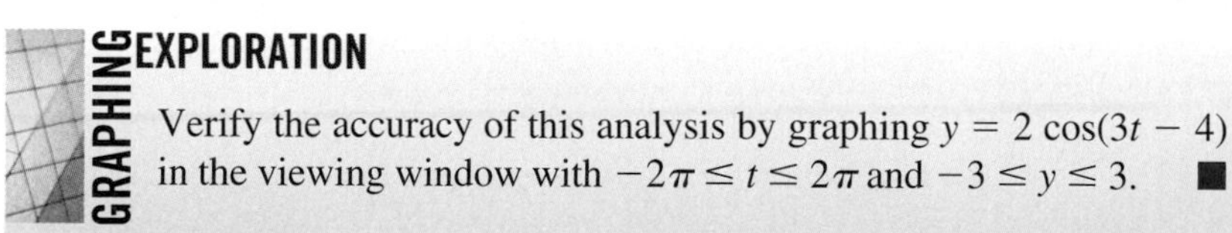

GRAPHING EXPLORATION

Verify the accuracy of this analysis by graphing $y = 2\cos(3t - 4)$ in the viewing window with $-2\pi \le t \le 2\pi$ and $-3 \le y \le 3$. ■

Many other types of trigonometric graphs, including those consisting of waves of varying height and length, are considered in Special Topics 2.5.A.

APPLICATIONS

The sine and cosine functions, or variations of them, can be used to describe many different phenomena.

EXAMPLE 9

A typical person's blood pressure can be modeled by the function

$$f(t) = 22\cos(2.5\pi t) + 95,$$

where t is time (in seconds) and $f(t)$ is in millimeters of mercury. The highest pressure (systolic) occurs when the heart beats, and the lowest pressure (diastolic) occurs when the heart is at rest between beats. The blood pressure is the ratio systolic/diastolic.

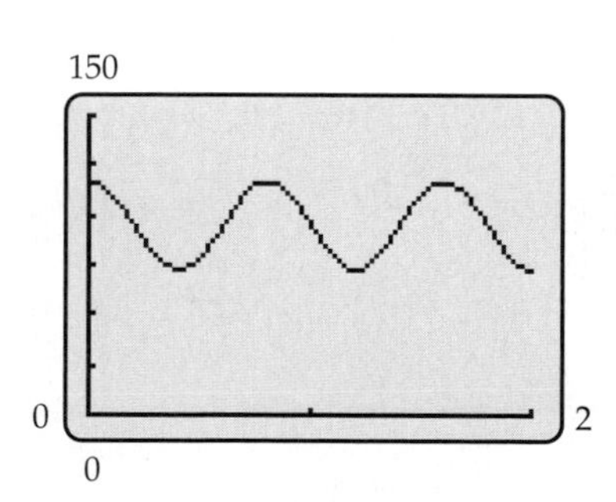

Figure 2–61

(a) Graph the blood pressure function over a period of two seconds and determine the person's blood pressure.

(b) Find the person's pulse rate (number of heartbeats per minute).

SOLUTION

(a) The graph of f is shown in Figure 2–61. The systolic pressure occurs at each local maximum of the graph and the diastolic pressure at each local minimum. Their heights can be determined by using our knowledge of periodic functions. The graph of f is the graph of $22\cos(2.5\pi t)$ shifted upward by 95 units (as explained in Section 0.6). Since the amplitude of $22\cos(2.5\pi t)$ is 22, its graph rises 22 units above and falls 22 units below the x-axis. When this graph is shifted 95 units upward, it rises and falls 22 units above and below the horizontal line $y = 95$ (see Figure 2–62), that is

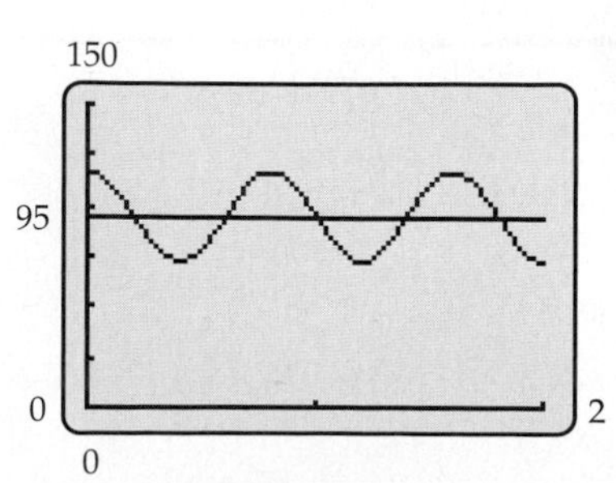

Figure 2–62

$$\text{from a high of } 95 + 22 = 117 \text{ to a low of } 95 - 22 = 73.$$

In other words, the systolic pressure is 117 and the diastolic pressure is 73. So the person's blood pressure is 117/73.

GRAPHING EXPLORATION

Use a maximum/minimum finder to confirm that the local maxima of the graph in Figure 2–61 occur when $y = 117$ and the local minima when $y = 73$.

(b) The time between heartbeats is the horizontal distance between peaks of the graph, that is, the period of the function. The period of $\cos(2.5\pi t)$ is

$$\frac{2\pi}{b} = \frac{2\pi}{2.5\pi} = .8 \text{ second.}$$

Since one minute is 60 seconds, the number of beats per minute (pulse rate) is

$$\frac{60}{.8} = 75. \quad ■$$

EXAMPLE 10

A wheel of radius 2 centimeters is rotating counterclockwise at 3 radians per second. A free-hanging rod 10 centimeters long is connected to the edge of the wheel at point P and remains vertical as the wheel rotates (Figure 2–63). Assuming that the center of the wheel is at the origin and that P is at (2, 0) at time $t = 0$, find a function that describes the y-coordinate of the tip E of the rod at time t.

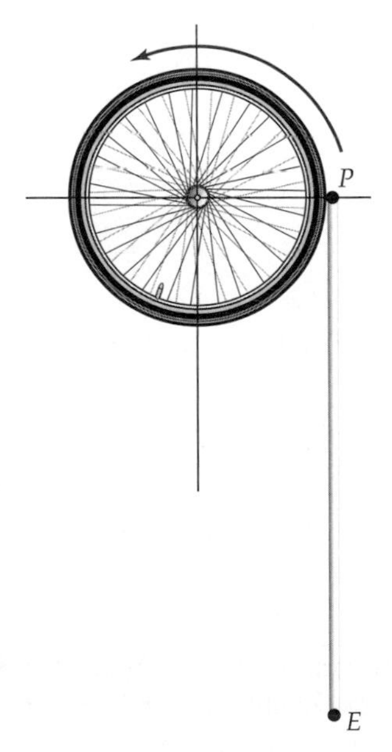

Figure 2–63

SOLUTION The wheel is rotating at 3 radians per second, so after t seconds, the point P has moved through an angle of $3t$ radians and is 2 units from the origin, as shown in Figure 2–64. By the point-in-the-plane description, the coordinates (x, y) of P satisfy

$$\frac{x}{2} = \cos 3t \qquad \frac{y}{2} = \sin 3t$$

$$x = 2\cos 3t \qquad y = 2\sin 3t.$$

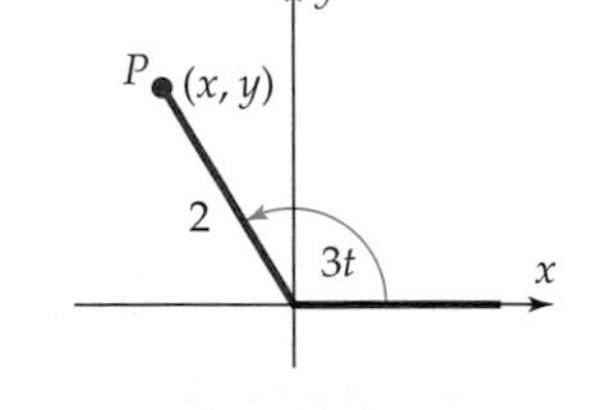

Figure 2–64

Since E lies 10 centimeters directly below P, its y-coordinate is 10 less than the y-coordinate of P. Hence, the function giving the y-coordinate of E at time t is

$$f(t) = y - 10 = 2\sin 3t - 10. \quad ■$$

EXAMPLE 11

Suppose that a weight hanging from a spring is set in motion by an upward push (Figure 2–65) and that it takes 5 seconds for it to move from its equilibrium position to 8 centimeters above, then drop to 8 centimeters below, and finally return to its equilibrium position. [We consider an idealized situation in which the spring has perfect elasticity and friction, air resistance, etc., are negligible.]

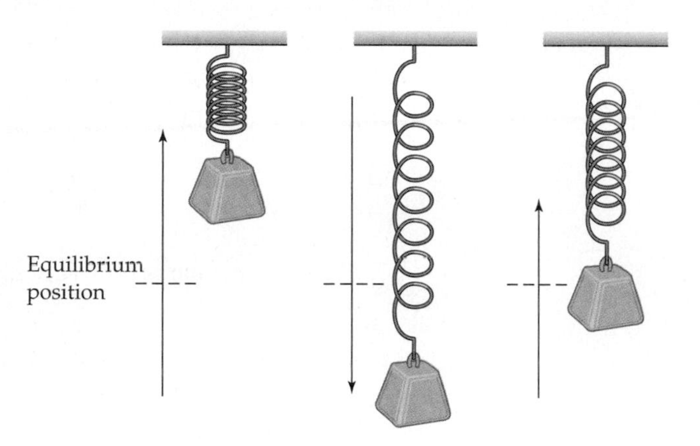

Figure 2–65

Let $h(t)$ denote the distance of the weight above (+) or below (−) its equilibrium position at time t. Then $h(t)$ is 0 when $t = 0$. As t runs from 0 to 5, $h(t)$ increases from 0 to 8, decreases to -8, and increases again to 0. In the next 5 seconds, it repeats the same pattern, and so on. Thus, the graph of h has some kind of wave shape. Two possibilities are shown in Figure 2–66.

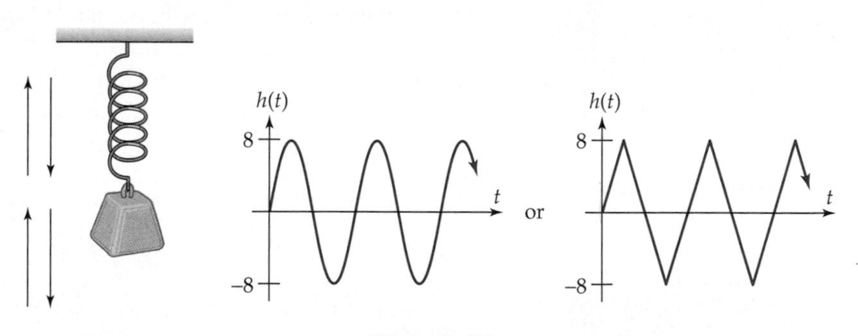

Figure 2–66

Careful physical experiment suggests that the left-hand curve in Figure 2–66, which resembles the sine graphs studied above, is a reasonably accurate model of this process. Facts from physics, calculus, and differential equations show that the rule of the function h has the form $h(t) = A\sin(bt + c)$ for some constants A, b, c. Since the amplitude of h is 8, its period is 5, and its phase shift is 0, the constants A, b, and c must satisfy

$$A = 8, \qquad \frac{2\pi}{b} = 5, \qquad -\frac{c}{b} = 0$$

or, equivalently,

$$A = 8 \qquad b = \frac{2\pi}{5}, \qquad c = 0.$$

Therefore, the motion of the moving spring can be described by the function

$$h(t) = A\sin(bt + c) = 8\sin\left(\frac{2\pi}{5}t + 0\right) = 8\sin\frac{2\pi t}{5}. \quad ■$$

Motion that can be described by a function of the form $f(t) = A\sin(bt + c)$ or $f(t) = A\cos(bt + c)$ is called **simple harmonic motion.** Many kinds of physical motion are simple harmonic motions. Other periodic phenomena, such as sound waves, are more complicated to describe. Their graphs consist of waves of varying amplitude. Such graphs are discussed in Special Topics 2.5.A.

EXAMPLE 12*

The table shows the average monthly temperature in Cleveland, Ohio, based on data from 1971 to 2000.† Since average temperatures are not likely to vary much from year to year, the data essentially repeats the same pattern in subsequent years. So a periodic model is appropriate.

Month	Temperature (°F)
Jan.	25.7
Feb.	28.4
March	37.5
April	47.6
May	58.5
June	67.5

Month	Temperature (°F)
July	71.9
Aug.	70.2
Sept.	63.3
Oct.	52.2
Nov.	41.8
Dec.	31.1

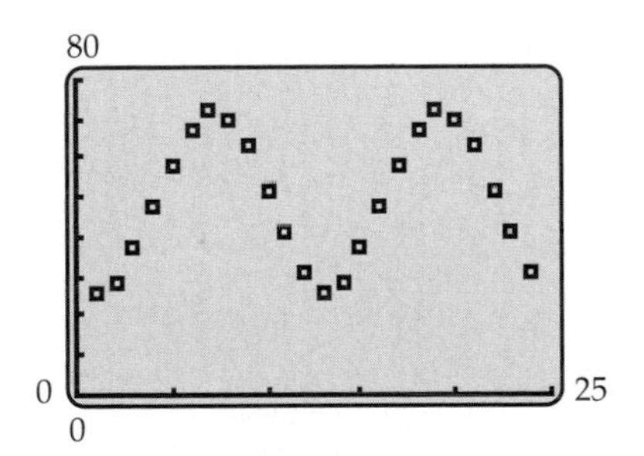

Figure 2–67

The data for a two-year period is plotted in Figure 2–67 (with $x = 1$ corresponding to January, $x = 2$ to February, and so on). The sine regression feature on a calculator produces this model from the 24 data points.

$$y = 22.7\sin(.5219x - 2.1842) + 49.5731.$$

The period of this function is $2\pi/.5219 \approx 12.04$ slightly off from the 12-month period we would expect. However, its graph in Figure 2–68 appears to fit the data well. ■

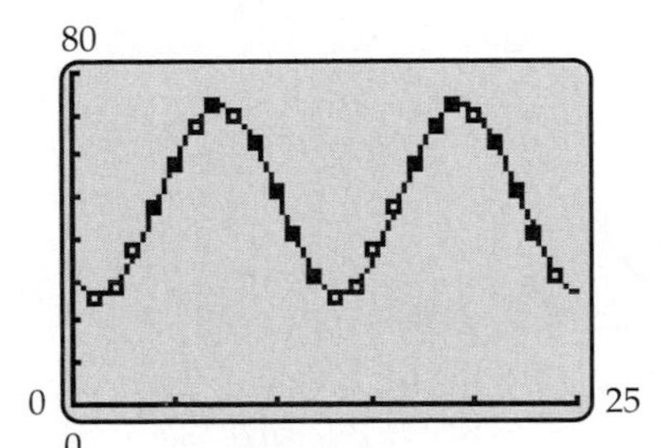

Figure 2–68

*This optional example assumes that you know how to find regression models on a calculator. Check your manual for details on how to do this.

†National Climatic Data Center.

EXERCISES 2.5

In Exercises 1–7, state the amplitude, period, and phase shift of the function.

1. $g(t) = 3\sin(2t - \pi)$
2. $h(t) = -4\cos(3t - \pi/6)$
3. $q(t) = -7\sin(7t + 1/7)$
4. $g(t) = 97\cos(14t + 5)$
5. $f(t) = \cos 2\pi t$
6. $k(t) = \cos(2\pi t/3)$
7. $p(t) = 6\cos(3\pi t + 1)$
8. (a) What is the period of $f(t) = \sin 2\pi t$?
   (b) For what values of t (with $0 \le t \le 2\pi$) is $f(t) = 0$?
   (c) For what values of t (with $0 \le t \le 2\pi$) is $f(t) = 1$? or $f(t) = -1$?

In Exercises 9–14, give the rule of a periodic function with the given numbers as amplitude, period, and phase shift (in this order).

9. $3, \pi/4, \pi/5$
10. $2, 3, 0$
11. $2/3, 1, 0$
12. $4/5, 2, 3$
13. $7, 5/3, -\pi/2$
14. $19, 4, -5$

In Exercises 15–18, state the rule of a function of the form $f(t) = A\sin bt$ or $g(t) = A\cos bt$ whose graph appears to be identical to the given graph.

15.

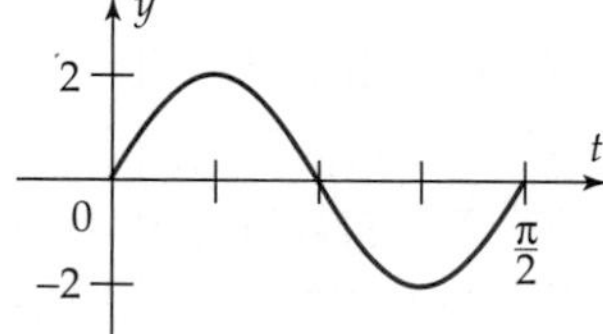

16.

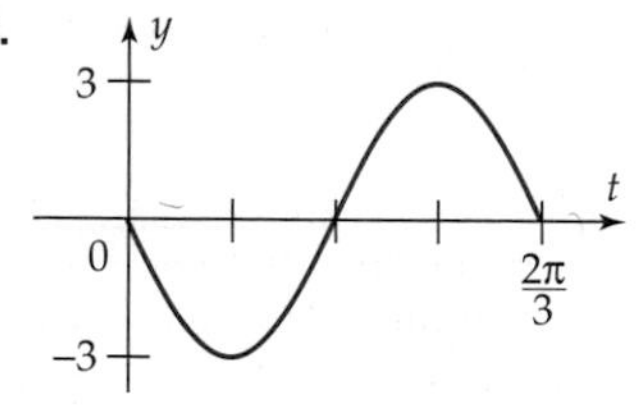

17.

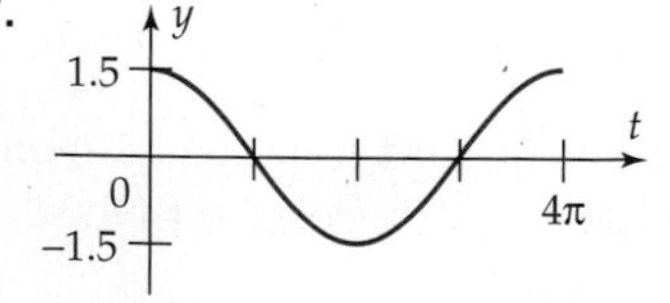

18.

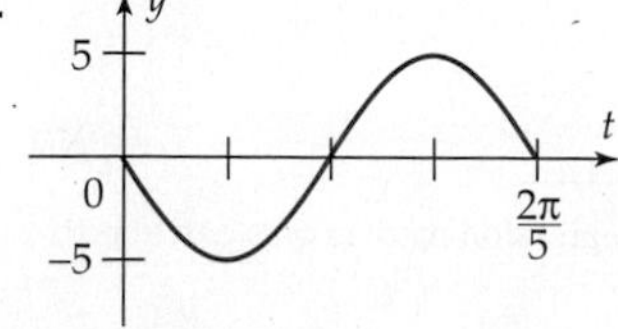

In Exercises 19–22,
(a) State the period of the function.
(b) Describe the graph of the function between 0 and 2π.
(c) Find a viewing window that accurately shows exactly four complete waves of the graph.

19. $f(t) = \sin 300t$
20. $f(t) = \sin 1200t$
21. $g(t) = \cos 900t$
22. $g(t) = \cos 575t$

In Exercises 23–26,
(a) State the rule of a function of the form

$$f(t) = A\sin(bt + c)$$

whose graph appears to be identical with the given graph.
(b) State the rule of a function of the form

$$g(t) = A\cos(bt + c)$$

whose graph appears to be identical with the given graph.

23.

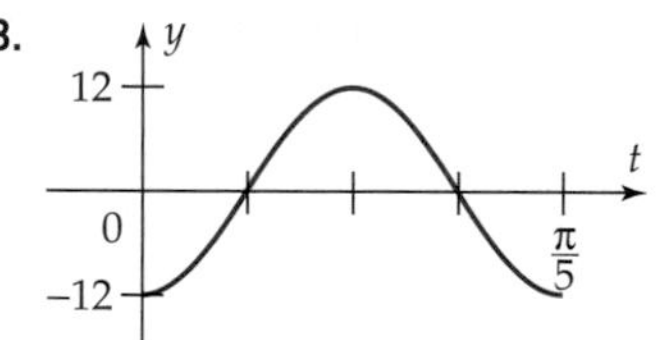

24.

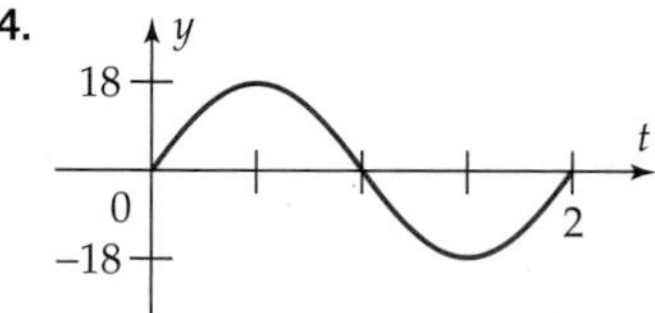

25.

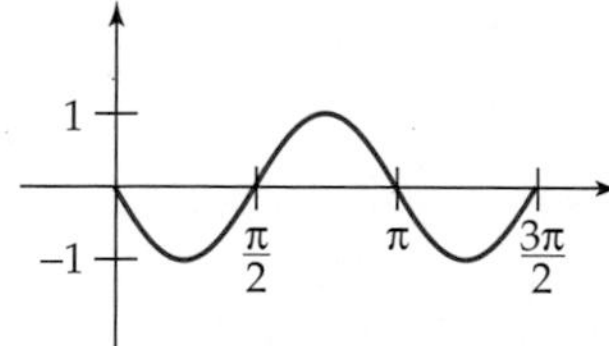

26.

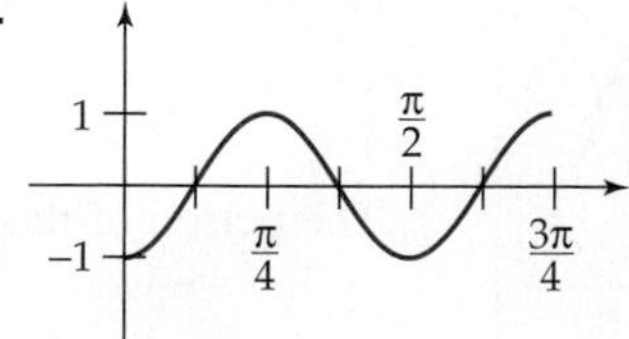

In Exercises 27–32, sketch a complete graph of the function.

27. $k(t) = -3\sin t$
28. $y(t) = -2\cos 3t$
29. $p(t) = -\dfrac{1}{2}\sin 2t$
30. $q(t) = \dfrac{2}{3}\cos\dfrac{3}{2}t$

31. $h(t) = 3\sin(2t + \pi/2)$ **32.** $p(t) = 3\cos(3t - \pi)$

In Exercises 33–36, graph the function for $0 \le t \le 2\pi$ and determine the location of all local maxima and minima. [This can be done either graphically or algebraically.]

33. $f(t) = \frac{1}{2}\sin\left(t - \frac{\pi}{3}\right)$

34. $g(t) = 2\sin(2t/3 - \pi/9)$

35. $f(t) = -2\sin(3t - \pi)$

36. $h(t) = \frac{1}{2}\cos\left(\frac{\pi}{2}t - \frac{\pi}{8}\right) + 1$

In Exercises 37–40, graph $f(t)$ in a viewing window with $-2\pi \le t \le 2\pi$. Use the trace feature to determine constants A, b, c such that the graph of $f(t)$ appears to coincide with the graph of $g(t) = A\sin(bt + c)$.

37. $f(t) = 2\sin t + 5\cos t$

38. $f(t) = -3\sin t + 2\cos t$

39. $f(t) = 3\sin(4t + 2) + 2\cos(4t - 1)$

40. $f(t) = 2\sin(3t - 5) - 3\cos(3t + 2)$

In Exercises 41 and 42, explain why there could not possibly be constants A, b, and c such that the graph of $g(t) = A\sin(bt + c)$ coincides with the graph of $f(t)$.

41. $f(t) = \sin 2t + \cos 3t$

42. $f(t) = 2\sin(3t - 1) + 3\cos(4t + 1)$

43. Do parts (a) and (b) of Example 9 for a person whose blood pressure is given by

$$g(t) = 21\cos(2.5\pi t) + 113.$$

According to current guidelines, someone with systolic pressure above 140 or diastolic pressure above 90 has high blood pressure and should see a doctor about it. What would you advise the person in this case?

44. Find the function in Example 10 if the wheel has a radius of 13 centimeters and the rod is 18 centimeters long.

45. The volume $V(t)$ of air (in cubic inches) in an adult's lungs t seconds after exhaling is approximately

$$V(t) = 55 + 24.5\sin\left(\frac{\pi x}{2} - \frac{\pi}{2}\right).$$

(a) Find the maximum and minimum amount of air in the lungs.
(b) How often does the person exhale?
(c) How many breaths per minute does the person take?

46. The brightness of the binary star Beta Lyrae (as seen from the earth) varies. Its visual magnitude $M(t)$ after t days is approximately

$$M(t) = .55\cos(.97t) + 3.85.$$

The visual magnitude scale is reversed from what you would expect: The lower the number, the brighter the star. With this in mind, answer the following questions.
(a) Graph the function M when $0 \le t \le 21$.
(b) What is the visual magnitude when the star is brightest? When it is dimmest?
(c) What is the period of the magnitude (the interval between its brightest times)?

47. The current generated by an AM radio transmitter is given by a function of the form $f(t) = A\sin 2000\pi mt$, where $550 \le m \le 1600$ is the location on the broadcast dial and t is measured in seconds. For example, a station at 980 on the AM dial has a function of the form

$$f(t) = A\sin 2000\pi(980)t = A\sin 1{,}960{,}000\pi t.$$

Sound information is added to this signal by varying (modulating) A, that is, by changing the amplitude of the waves being transmitted. (*AM* means "amplitude modulation.") For a station at 980 on the dial, what is the period of function f? What is the frequency (number of complete waves per second)?

48. The number of hours of daylight in Winnipeg, Manitoba, can be approximated by

$$d(t) = 4.15\sin(.0172t - 1.377) + 12,$$

where t is measured in days, with $t = 1$ being January 1.
(a) On what day is there the most daylight? The least? How much daylight is there on these days?
(b) On which days are there 11 hours or more of daylight?
(c) What do you think the period of this function is? Why?

In Exercises 49–52, suppose there is a weight hanging from a spring (under the same idealized conditions as described in Example 11). The weight is given a push to start it moving. At any time t, let h(t) be the height (or depth) of the weight above (or below) its equilibrium point. Assume that the maximum distance the weight moves in either direction from the equilibrium point is six centimeters and that it moves through a complete cycle every four seconds. Express h(t) in terms of the sine or cosine function under the stated conditions.

49. Initial push is *upward* from the equilibrium point.

50. Initial push is *downward* from the equilibrium point. [*Hint:* What does the graph of $A \sin bt$ look like when $A < 0$?]

51. Weight is pulled six centimeters above equilibrium, and the initial movement (at $t = 0$) is downward. [*Hint:* Think cosine.]

52. Weight is pulled six centimeters below equilibrium, and the initial movement is upward.

53. The original Ferris wheel, built by George Ferris for the Columbian Exposition of 1893, was much larger and slower than its modern counterparts: It had a diameter of 250 feet and contained 36 cars, each of which held 60 people; it made one revolution every 10 minutes. Imagine that the Ferris wheel revolves counterclockwise in the x-y plane with its center at the origin. A car had coordinates (125, 0) at time $t = 0$. Find the rule of a function that gives the y-coordinate of the car at time t.

54. Do Exercise 53 if the wheel turns at 2 radians per minute and the car is at $(0, -125)$ at time $t = 0$.

55. A circular wheel of radius one foot rotates counterclockwise. A four-foot-long rod has one end attached to the edge of this wheel and the other end to the base of a piston (see the figure). It transfers the rotary motion of the wheel into a back-and-forth linear motion of the piston. If the wheel is rotating at 10 revolutions per second, point W is at (1, 0) at time $t = 0$, and point P is always on the x-axis, find the rule of a function that gives the x-coordinate of P at time t.

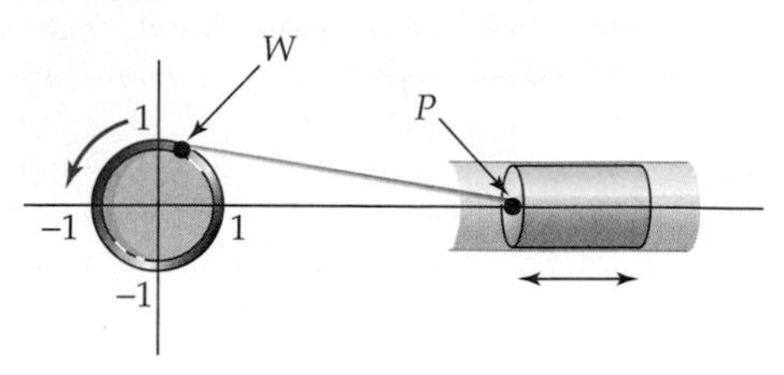

56. Do Exercise 55 if the wheel has a radius of two feet, rotates at 50 revolutions per second, and is at (2, 0) when $t = 0$.

57. A pendulum swings uniformly back and forth, taking two seconds to move from the position directly above point A to the position directly above point B.

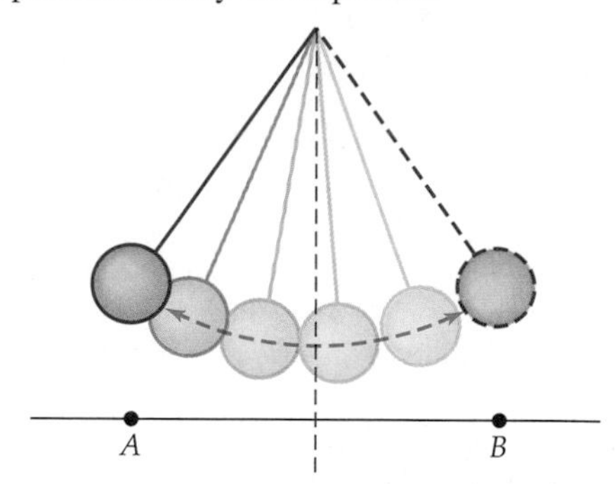

The distance from A to B is 20 centimeters. Let $d(t)$ be the horizontal distance from the pendulum to the (dashed) center line at time t seconds (with distances to the right of the line measured by positive numbers and distances to the left by negative ones). Assume that the pendulum is on the center line at time $t = 0$ and moving to the right. Assume that the motion of the pendulum is simple harmonic motion. Find the rule of the function $d(t)$.

58. The diagram shows a merry-go-round that is turning counterclockwise at a constant rate, making two revolutions in one minute. On the merry-go-round are horses A, B, C, and D at four meters from the center and horses E, F, and G at eight meters from the center. There is a function $a(t)$ that gives the distance the horse A is from the y-axis (this is the x-coordinate of the position A is in) as a function of time t (measured in minutes). Similarly, $b(t)$ gives the x-coordinate for B as a function of time, and so on. Assume that the diagram shows the situation at time $t = 0$.

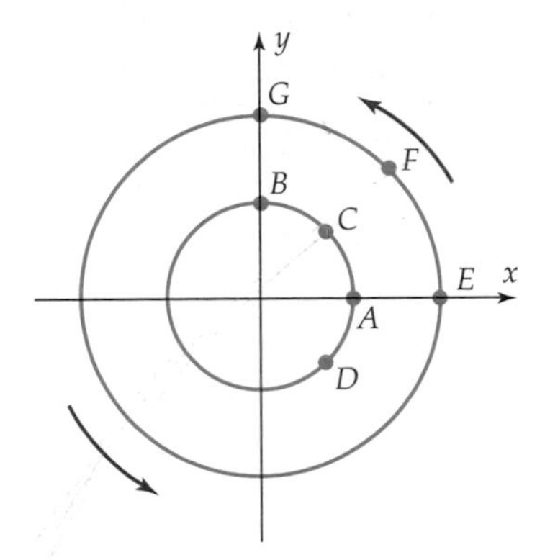

(a) Which of the following functions does $a(t)$ equal?

$$4\cos t,\quad 4\cos \pi t,\quad 4\cos 2t,\quad 4\cos 2\pi t,$$
$$4\cos\left(\tfrac{1}{2}t\right),\quad 4\cos((\pi/2)t),\quad 4\cos 4\pi t$$

Explain.

(b) Describe the functions $b(t)$, $c(t)$, $d(t)$, and so on using the cosine function:

$$b(t) = \_\_\_\_,\ c(t) = \_\_\_\_,\ d(t) = \_\_\_\_.$$
$$e(t) = \_\_\_\_,\ f(t) = \_\_\_\_,\ g(t) = \_\_\_\_.$$

(c) Suppose the x-coordinate of a horse S is given by the function $4\cos(4\pi t - (5\pi/6))$ and the x-coordinate of another horse T is given by $8\cos(4\pi t - (\pi/3))$. Where are these horses located in relation to the rest of the horses? Mark the positions of T and S at $t = 0$ into the figure.

Exercises 59 and 60 explore various ways in which a calculator can produce inaccurate or misleading graphs of trigonometric functions.

59. (a) If you were going to draw a rough picture of a full wave of the sine function by plotting some points and connecting them with straight-line segments, approximately how many points would you have to plot?

(b) If you were drawing a rough sketch of the graph of $f(t) = \sin 100t$ when $0 \le t \le 2\pi$, according to the method in part (a), approximately how many points would have to be plotted?

(c) How wide (in pixels) is your calculator screen? Your answer to this question is the maximum number of points that your calculator plots when graphing any function.

(d) Use parts (a)–(c) to explain why your calculator cannot possibly produce an accurate graph of $f(t) = \sin 100t$ in any viewing window with $0 \le t \le 2\pi$.

60. (a) Using a viewing window with $0 \le t \le 2\pi$, use the trace feature to move the cursor along the horizontal axis. [On some calculators, it may be necessary to graph $y = 0$ to do this.] What is the distance between one pixel and the next (to the nearest hundredth)?

(b) What is the period of $f(t) = \sin 300t$? Since the period is the length of one full wave of the graph, approximately how many waves should there be between two adjacent pixels? What does this say about the possibility of your calculator's producing an accurate graph of this function between 0 and 2π?

61. The table below shows the number of unemployed people in the labor force (in millions) for 1984–2002.*

(a) Sketch a scatter plot of the data, with $x = 0$ corresponding to 1980.

(b) Does the data appear to be periodic? If so, find an appropriate model.

(c) Do you think this model is likely to be accurate much beyond 2002? Why?

Year	Unemployed
1984	8.539
1985	8.312
1986	8.237
1987	7.425
1988	6.701
1989	6.528
1990	7.047
1991	8.628
1992	9.613
1993	8.940

Year	Unemployed
1994	7.996
1995	7.404
1996	7.236
1997	6.739
1998	6.210
1999	5.880
2000	5.655
2001	6.742
2002	8.234

*Bureau of Labor Statistics, U.S. Department of Labor.

In Exercises 62 and 63, do the following.

(a) Use 12 data points (with $x = 1$ corresponding to January) to find a periodic model of the data.

(b) What is the period of the function found in part (a)? Is this reasonable?

(c) Plot 24 data points (two years) and graph the function from part (a) on the same screen. Is the function a good model in the second year?

(d) Use the 24 data points in part (c) to find another periodic model for the data.

(e) What is the period of the function in part (d)? Does its graph fit the data well?

62. The table shows the average monthly temperature in Chicago, Illinois, based on data from 1971 to 2000.*

Month	Temperature (°F)
Jan.	22.0
Feb.	27.0
March	37.3
April	47.8
May	58.7
June	68.2
July	73.3
Aug.	71.7
Sept.	63.8
Oct.	52.1
Nov.	39.3
Dec.	27.4

63. The table shows the average monthly precipitation (in inches) in San Francisco, California, based on data from 1971 to 2000.†

Month	Precipitation
Jan.	4.45
Feb.	4.01
March	3.26
April	1.17
May	.38
June	.11
July	.03
Aug.	.07
Sept.	.2
Oct.	1.04
Nov.	2.49
Dec.	2.89

Thinkers

64. On the basis of the results of Exercises 37–42, under what conditions on the constants a, k, h, d, r, s does it appear that the graph of

$$f(t) = a \sin(kt + h) + d \cos(rt + s)$$

coincides with the graph of the function

$$g(t) = A \sin(bt + c)?$$

65. A grandfather clock has a pendulum length of k meters and its swing is given (as in Exercise 57) by the function $f(t) = .25 \sin(\omega t)$, where

$$\omega = \sqrt{\frac{9.8}{k}}.$$

(a) Find k such that the period of the pendulum is two seconds.

(b) The temperature in the summer months causes the pendulum to increase its length by .01%. How much time will the clock lose in June, July, and August? [*Hint:* These three months have a total of 92 days (7,948,800 seconds). If k is increased by .01%, what is $f(2)$?]

*National Climatic Data Center.

†National Climatic Data Center.

2.5.A SPECIAL TOPICS Other Trigonometric Graphs

A graphing calculator or computer enables you to explore with ease a wide variety of trigonometric functions.

GRAPHING EXPLORATION

Graph

$$g(t) = \cos t \quad \text{and} \quad f(t) = \sin(t + \pi/2)$$

on the same screen. Is there any apparent difference between the two graphs?

This exploration suggests that the equation $\cos t = \sin(t + \pi/2)$ is an identity and hence that the graph of the cosine function can be obtained by horizontally shifting the graph of the sine function. This is indeed the case, as will be proved in Section 3.2. Consequently, every graph in Section 2.5 is actually the graph of a function of the form $f(t) = A \sin(bt + c)$. In fact, considerably more is true.

EXAMPLE 1

Show that the graph of

$$g(t) = -2 \sin(t + 7) + 3 \cos(t + 2)$$

appears identical to the graph of a function of the form $f(t) = A \sin(bt + c)$ for suitable constants A, b, and c.

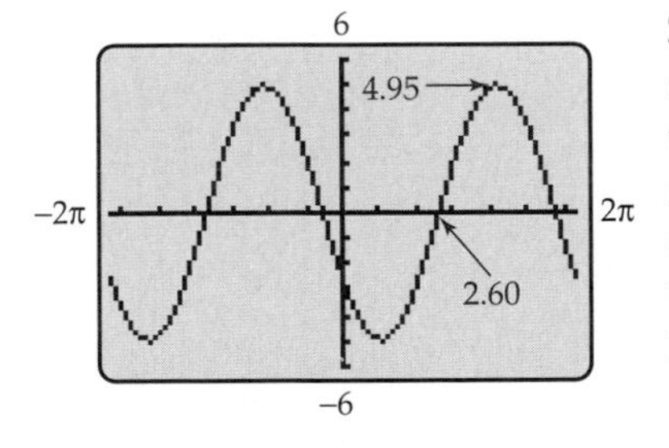

Figure 2–69

SOLUTION The function $g(t)$ has period 2π because this is the period of both $\sin(t + 7)$ and $\cos(t + 2)$. Its graph in Figure 2–69 consists of repeating waves of uniform height. By using a maximum finder and a root finder, we see that the maximum height of a wave is approximately 4.95 and that a wave similar to a sine wave begins at approximately $t = 2.60$, as indicated in Figure 2–69. Thus, the graph looks very much like a sine wave with amplitude 4.95 and phase shift 2.60. As we saw in Section 2.5, the function

$$f(t) = 4.95 \sin(t - 2.60)$$

has amplitude of 4.95, period $2\pi/1 = 2\pi$, and phase shift $-(-2.60)/1 = 2.60$.

GRAPHING EXPLORATION

Graph

$$g(t) = -2 \sin(t + 7) + 3 \cos(t + 2)$$

and

$$f(t) = 4.95 \sin(t - 2.60)$$

on the same screen. Do the graphs look identical? ■

Example 1 is an illustration (but not a proof) of the following fact.

Sinusoidal Graphs

> If b, D, E, r, s are constants, then the graph of the function
> $$g(t) = D\sin(bt + r) + E\cos(bt + s)$$
> is a sine curve: There exist constants A and c such that
> $$D\sin(bt + r) + E\cos(bt + s) = A\sin(bt + c).$$

EXAMPLE 2

Estimate the constants A, b, c such that

$$A\sin(bt + c) = 4\sin(3t + 2) + 2\cos(3t - 4).$$

SOLUTION The function $g(t) = 4\sin(3t + 2) + 2\cos(3t - 4)$ has period $2\pi/3$ because this is the period of both $\sin(3t + 2)$ and $\cos(3t - 4)$. The function $f(t) = A\sin(bt + c)$ has period $2\pi/b$. So we must have

$$\frac{2\pi}{b} = \frac{2\pi}{3}, \qquad \text{or equivalently,} \qquad b = 3.$$

Using a maximum finder and a root finder on the graph of

$$g(t) = 4\sin(3t + 2) + 2\cos(3t - 4)$$

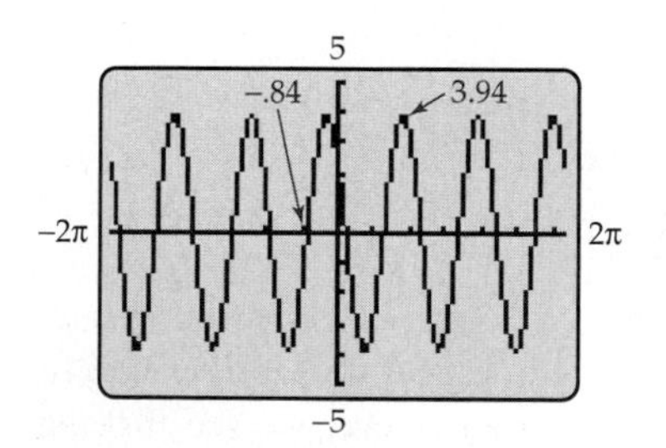

Figure 2–70

in Figure 2–70, we see that the maximum height (amplitude) of a wave is approximately 3.94 and that a sine wave begins at approximately $t = -.84$. Therefore, the graph has (approximate) amplitude 3.94 and phase shift $-.84$. Since $b = 3$ and $f(t) = A\sin(bt + c)$ has amplitude $|A|$ and phase shift $-c/b = -c/3$, we have $A \approx 3.94$ and

$$-\frac{c}{3} \approx -.84 \qquad \text{or, equivalently,} \qquad c \approx 3(.84) = 2.52.$$

Therefore,

$$3.94\sin(3t + 2.52) \approx 4\sin(3t + 2) + 2\cos(3t - 4).$$

GRAPHING EXPLORATION

Graphically confirm this fact by graphing

$$f(t) = 3.94\sin(3t + 2.52)$$

and

$$g(t) = 4\sin(3t + 2) + 2\cos(3t - 4)$$

on the same screen. Do the graphs appear identical? ■

In the preceding examples, the variable t had the same coefficient in both the sine and cosine term of the function's rule. When this is not the case, the graph will consist of waves of varying size and shape, as you can readily illustrate.

GRAPHING EXPLORATION

Graph each of the following functions separately in the viewing window with $-2\pi \leq t \leq 2\pi$ and $-6 \leq y \leq 6$.

$$f(t) = \sin 3t + \cos 2t, \qquad g(t) = -2\sin(3t + 5) + 4\cos(t + 2),$$
$$h(t) = 2\sin 2t - 3\cos 3t.$$

EXAMPLE 3

Find a complete graph of

$$f(t) = 4\sin 100\pi t + 2\cos 40\pi t.$$

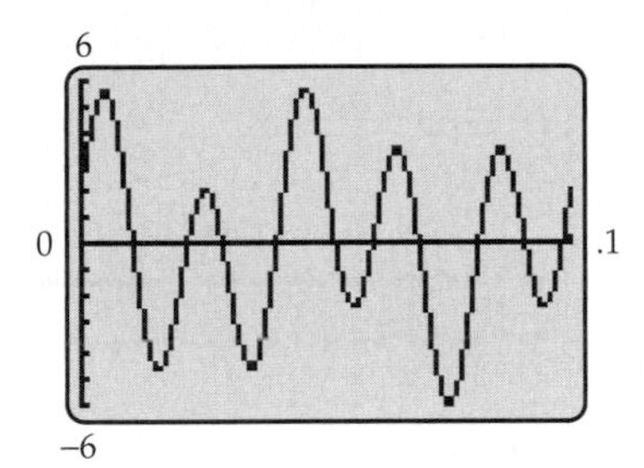

Figure 2–71

SOLUTION If you graph f in a window with $-2\pi \leq t \leq 2\pi$, you will get "garbage" on the screen (try it!). Trial and error might lead to a viewing window that shows a readable graph, but the graph might not be accurate. A better procedure is to note that this is a periodic function. Hence, we need only graph it over one period to have a complete graph. The period of $4\sin 100\pi t$ is

$$\frac{2\pi}{100\pi} = \frac{1}{50} = .02,$$

and the period of $2\cos 40\pi t$ is $2\pi/40\pi = 1/20 = .05$. So the period of their sum will be the least common multiple of .02 and .05, which is .1.* By graphing f in the viewing window with $0 \leq t \leq .1$ and $-6 \leq y \leq 6$, we obtain the complete graph in Figure 2–71. ■

DAMPED AND COMPRESSED TRIGONOMETRIC GRAPHS

Many physical situations can be described by functions whose graphs consist of waves of different heights. Other situations (for instance, sound waves in FM radio transmission) are modeled by functions whose graphs consist of waves of uniform height and varying frequency. Here are some examples of such functions.

EXAMPLE 4

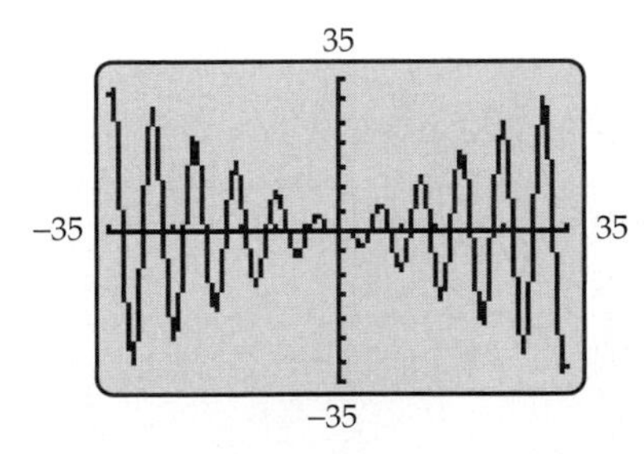

Figure 2–72

Explain why the graph of $f(t) = t\cos t$ in Figure 2–72 has the shape it does.

SOLUTION The graph appears to consist of waves that get larger and larger as you move away from the origin. To explain this situation, we analyze the situation algebraically. We know that

$$-1 \leq \cos t \leq 1 \qquad \text{for every } t.$$

*The multiples of .02 are .02, .04, .06, .08, .10, . . . , and the multiples of .05 are .05, .10, ... Hence, the smallest common multiple is .10.

If we multiply each term of this inequality by t and remember the rules for changing the direction of inequalities when multiplying by negatives, we see that

$$-t \leq t\cos t \leq t \qquad \text{when } t \geq 0$$

and

$$-t \geq t\cos t \geq t \qquad \text{when } t < 0.$$

In graphical terms, this means that the graph of $f(x) = t\cos t$ lies between the straight lines $y = t$ and $y = -t$, with the waves growing larger or smaller to fill this space. The graph touches the lines $y = \pm t$ exactly when $t\cos t = \pm t$, that is, when $\cos t = \pm 1$. This occurs when $t = 0, \pm\pi, \pm 2\pi, \pm 3\pi, \ldots$.

GRAPHING EXPLORATION

Illustrate this analysis by graphing $f(t) = t\cos t$, $y = t$, and $y = -t$ on the same screen. ■

EXAMPLE 5

No single viewing window gives a completely readable graph of $g(t) = .5^t \sin t$ (try some). To the left of the y-axis, the graph gets quite large, but to the right, it almost coincides with the horizontal axis. To get a better mental picture, note that $.5^t > 0$ for every t. Multiplying each term of the known inequality $-1 \leq \sin t \leq 1$ by $.5^t$, we see that

$$-.5^t \leq .5^t \sin t \leq .5^t \qquad \text{for every } t.$$

Hence, the graph of g lies between the graphs of the exponential functions $y = -.5^t$ and $y = .5^t$, which are shown in Figure 2–73.* The graph of g will consist of sine waves rising and falling between those exponential graphs, as indicated in the sketch in Figure 2–74 (which is not to scale).

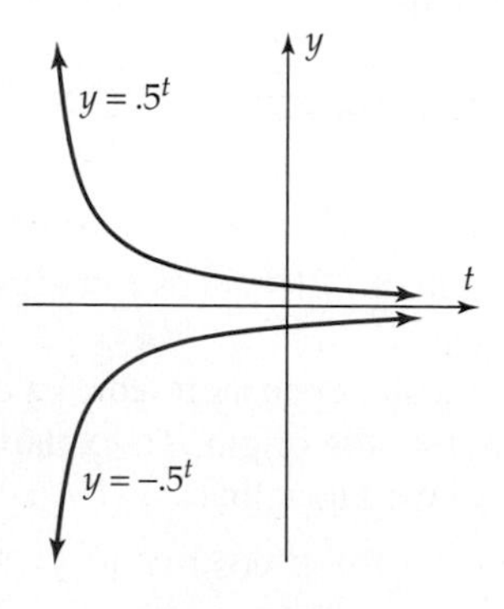

Figure 2–73

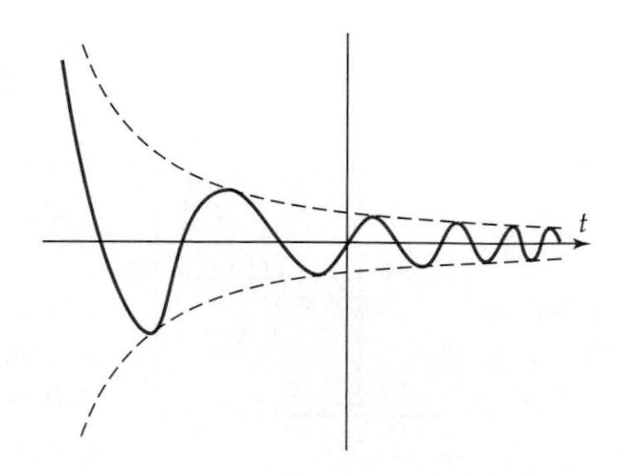

Figure 2–74

*Exponential graphs are discussed in Section 6.1.

The best you can do with a calculator is to look at various viewing windows in which a portion of the graph is readable.

GRAPHING EXPLORATION

Find viewing windows that clearly show the graph of $g(t) = .5^t \sin t$ in each of these ranges.

$$-2\pi \le t \le 0, \qquad 0 \le t \le 2\pi, \qquad 2\pi \le t \le 4\pi. \qquad ■$$

EXAMPLE 6

If you graph $f(t) = \sin(\pi/t)$ in a wide viewing window such as Figure 2–75, it is clear that the horizontal axis is an asymptote of the graph.* Near the origin, however, the graph is not very readable, even in a very narrow viewing window like Figure 2–76.

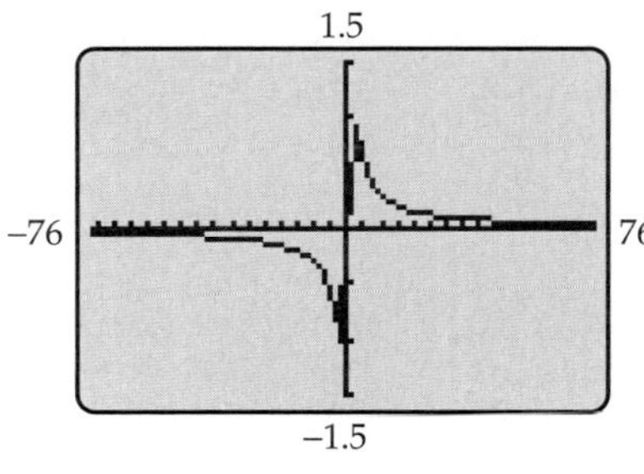

Figure 2–75

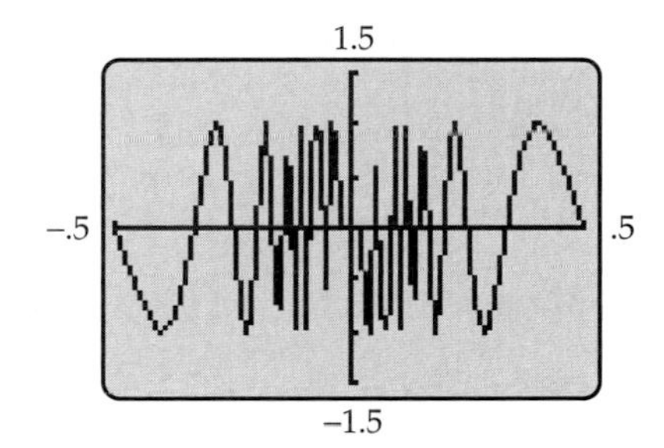

Figure 2–76

To understand the behavior of f near the origin, consider what happens as you move left from $t = 1/2$ to $t = 0$:

$$\text{As } t \text{ goes from } \frac{1}{2} \text{ to } \frac{1}{4}, \text{ then } \frac{\pi}{t} \text{ goes from } \frac{\pi}{1/2} = 2\pi \text{ to } \frac{\pi}{1/4} = 4\pi.$$

As π/t takes all values from 2π to 4π, the graph of $f(t) = \sin(\pi/t)$ makes one complete sine wave. Similarly,

$$\text{As } t \text{ goes from } \frac{1}{4} \text{ to } \frac{1}{6}, \text{ then } \frac{\pi}{t} \text{ goes from } \frac{\pi}{1/4} = 4\pi \text{ to } \frac{\pi}{1/6} = 6\pi.$$

As π/t takes all values from 4π to 6π, the graph of $f(t) = \sin(\pi/t)$ makes another complete sine wave. The same pattern continues, so the graph of f makes a complete wave from $t = 1/2$ to $t = 1/4$, another from $t = 1/4$ to $t = 1/6$, another from $t = 1/6$ to $t = 1/8$, and so on. A similar phenomenon occurs as t takes values between $-1/2$ and 0. Consequently, the graph of f near 0 oscillates

*This can also be demonstrated algebraically: When t is very large in absolute value, then π/t is very close to 0; hence, $\sin(\pi/t)$ is very close to 0 as well.

infinitely often between -1 and 1, with the waves becoming more and more compressed as t gets closer to 0, as indicated in Figure 2–77. Since the function is not defined at $t = 0$, the left and right halves of the graph are not connected. ■

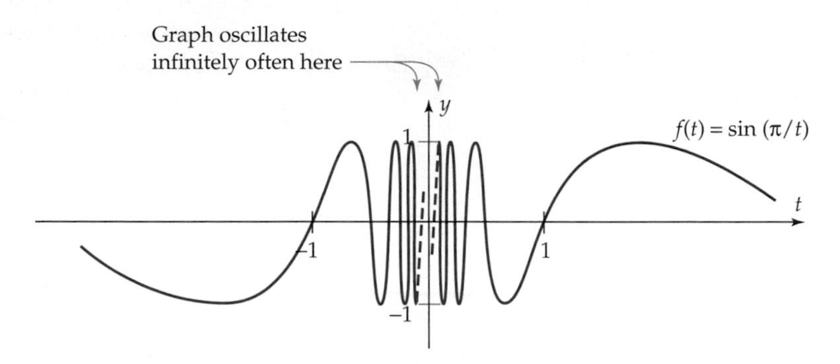

Figure 2–77

EXAMPLE 7

Describe the graph of $g(t) = \cos e^t$.

SOLUTION When t is very negative (such as $t = -100$ or $t = -1000$), then e^t is very close to 0 (why?); hence, $\cos e^t$ is very close to 1. Therefore, the horizontal line $y = 1$ is an asymptote of the half of the graph to the left of the origin. As t takes increasing positive values, the corresponding values of e^t increase at a much faster rate (as explained in Section 6.1). For instance, as t goes from 0 to 2π,

$$e^t \text{ goes from } e^0 = 1 \text{ to } e^{2\pi} \approx 535.5 \approx 170\pi = 85(2\pi).$$

Consequently, $\cos e^t$ runs through 85 periods, that is, the graph of g makes 85 full waves between 0 and 2π. As t gets larger, the graph of g oscillates more and more rapidly.

GRAPHING EXPLORATION

To see how compressed the waves become, graph $g(t)$ in three viewing windows, with

$$0 \le t \le 3.5, \qquad 4.5 \le t \le 5, \qquad 6 \le t \le 6.2,$$

and note how the number of waves increases in each succeeding window, even though the widths of the windows are getting smaller. ■

EXERCISES 2.5.A

In Exercises 1–6, estimate constants A, b, c such that

$$f(t) = A\sin(bt + c).$$

1. $f(t) = \sin t + 2\cos t$

2. $f(t) = 3\sin t + 2\cos t$

3. $f(t) = 2\sin 4t - 5\cos 4t$

4. $f(t) = 3\sin(2t - 1) + 4\cos(2t + 3)$

5. $f(t) = -5\sin(3t + 2) + 2\cos(3t - 1)$

6. $f(t) = .3\sin(2t + 4) - .4\cos(2t - 3)$

In Exercises 7–16, find a viewing window that shows a complete graph of the function.

7. $g(t) = (5\sin 2t)(\cos 5t)$

8. $h(t) = e^{\sin t}$

9. $f(t) = t/2 + \cos 2t$

10. $g(t) = \sin\left(\frac{t}{3} - 2\right) + 2\cos\left(\frac{t}{4} - 2\right)$

11. $h(t) = \sin 300t + \cos 500t$

12. $f(t) = 3\sin(200t + 1) - 2\cos(300t + 2)$

13. $g(t) = -5\sin(250\pi t + 5) + 2\cos(400\pi t - 7)$

14. $h(t) = 4\sin(600\pi t + 3) - 6\cos(500\pi t - 3)$

15. $f(t) = 4\sin .2\pi t - 5\cos .4\pi t$

16. $g(t) = 6\sin .05\pi t + 2\cos .04\pi t$

In Exercises 17–24, describe the graph of the function verbally (including such features as asymptotes, undefined points, amplitude and number of waves between 0 and 2π, etc.) as in Examples 5–7. Find viewing windows that illustrate the main features of the graph.

17. $g(t) = \sin e^t$

18. $h(t) = \dfrac{\cos 2t}{1 + t^2}$

19. $f(t) = \sqrt{|t|}\cos t$

20. $g(t) = e^{-t^2/8}\sin 2\pi t$

21. $h(t) = \dfrac{1}{t}\sin t$

22. $f(t) = t\sin\dfrac{1}{t}$

23. $g(t) = \ln|\cos t|$

24. $h(t) = \ln|\sin t + 1|$

25. At a beach in Maui, Hawaii, the level of the tides is approximated by

$$f(t) = -.7\sin(.52t - 1.3728) + .73\cos(.26t - .6864) + 1.4,$$

where t is measured in hours and $f(t)$ in feet.

(a) Graph the tide function over a three-day period.

(b) At approximately what times during the day does the highest tide occur? The lowest?

(c) What is the period of this function?

2.6 Other Trigonometric Functions

This section is divided into three parts, each of which may be covered earlier, as shown in the table.

Subsection of Section 2.6	May be covered at the end of
Definitions, Alternate Descriptions	Section 2.2
Algebra and Identities	Section 2.3
Graphs	Section 2.4

DEFINITIONS

The three remaining trigonometric functions are defined in terms of sine and cosine, as follows.

Definition of Cotangent, Secant, and Cosecant Functions

Name of Function	Value of Function at t Is Denoted	Rule of Function
cotangent	$\cot t$	$\cot t = \dfrac{\cos t}{\sin t}$
secant	$\sec t$	$\sec t = \dfrac{1}{\cos t}$
cosecant	$\csc t$	$\csc t = \dfrac{1}{\sin t}$

The domain of each function consists of all real numbers for which the denominator is not 0. The graphs of the sine and cosine function in Section 2.4 show that $\sin t = 0$ only when $t = 0, \pm\pi, \pm 2\pi, \pm 3\pi, \ldots$ and $\cos t = 0$ only when $t = \pm\pi/2, \pm 3\pi/2, \pm 5\pi/2, \ldots$. So the domains of cotangent, secant, and cosecant are as follows.

Function	Domain
$f(t) = \cot t$	All real numbers except $0, \pm\pi, \pm 2\pi, \pm 3\pi, \ldots$
$g(t) = \sec t$	All real numbers except $\pm\pi/2, \pm 3\pi/2, \pm 5\pi/2, \ldots$
$h(t) = \csc t$	All real numbers except $0, \pm\pi, \pm 2\pi, \pm 3\pi, \ldots$

CAUTION

The calculator keys labeled SIN^{-1}, COS^{-1}, and TAN^{-1} do *not* denote the functions $1/\sin t$, $1/\cos t$, and $1/\tan t$. For instance, if you key in

COS^{-1} 7 ENTER

you will get an error message, not the number sec 7, and if you key in

TAN^{-1} −5 ENTER

you will obtain −1.3734, which is *not* cot(−5).

The values of these functions may be approximated on a calculator by using the SIN and COS keys. For instance,

$$\cot(-3.1) = \frac{\cos(-3.1)}{\sin(-3.1)} \approx 24.0288, \qquad \sec 7 = \frac{1}{\cos 7} \approx 1.3264,$$

$$\csc 18.5 = \frac{1}{\sin 18.5} \approx -2.9199.$$

The cotangent function can also be evaluated with the TAN key, by using this fact:

$$\cot t = \frac{\cos t}{\sin t} = \frac{1}{\dfrac{\sin t}{\cos t}} = \frac{1}{\tan t}.^{*}$$

For example,

$$\cot(-5) = \frac{1}{\tan(-5)} \approx .2958.$$

These new trigonometric functions may be evaluated exactly at any integer multiple of $\pi/3$, $\pi/4$, or $\pi/6$.

---

*This identity is valid except for $t = \pm\pi/2, \pm 3\pi/2, \pm 5\pi/2, \ldots$. At these values, $\cos t = 0$ and $\sin t = \pm 1$, so $\cot t = 0$, but $\tan t$ is not defined.

EXAMPLE 1

Evaluate the cotangent, secant, and cosecant functions at $t = \pi/3$.

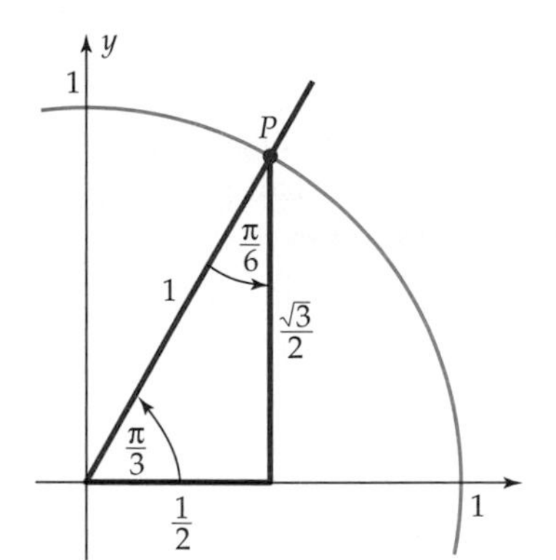

Figure 2–78

SOLUTION Let P be the point where the terminal side of an angle of $\pi/3$ radians in standard position meets the unit circle (Figure 2–78). Draw the vertical line from P to the x-axis, forming a right triangle with hypotenuse 1, angles of $\pi/3$ and $\pi/6$ radians, and sides of lengths of $1/2$ and $\sqrt{3}/2$ as explained on page 153. Then P has coordinates $(1/2, \sqrt{3}/2)$, and by definition,

$$\sin\frac{\pi}{3} = y\text{-coordinate of } P = \sqrt{3}/2$$

$$\cos\frac{\pi}{3} = x\text{-coordinate of } P = 1/2.$$

Therefore,

$$\csc\frac{\pi}{3} = \frac{1}{\sin(\pi/3)} = \frac{1}{\sqrt{3}/2} = \frac{2}{\sqrt{3}} = \frac{2\sqrt{3}}{3}$$

$$\sec\frac{\pi}{3} = \frac{1}{\cos(\pi/3)} = \frac{1}{1/2} - 2$$

$$\cot\frac{\pi}{3} = \frac{\cos(\pi/3)}{\sin(\pi/3)} = \frac{1/2}{\sqrt{3}/2} = \frac{1}{\sqrt{3}} = \frac{\sqrt{3}}{3}. \qquad \blacksquare$$

ALTERNATE DESCRIPTIONS

The point-in-the-plane description of sine, cosine, and tangent readily extends to these new functions.

Point-in-the-Plane Description

Let t be a real number and (x, y) any point (except the origin) on the terminal side of an angle of t radians in standard position. Let

$$r = \sqrt{x^2 + y^2}.$$

Then,

$$\cot t = \frac{x}{y}, \qquad \sec t = \frac{r}{x}, \qquad \csc t = \frac{r}{y}$$

for each number t in the domain of the given function.

These statements are proved by using the similar descriptions of sine and cosine. For instance,

$$\cot t = \frac{\cos t}{\sin t} = \frac{x/r}{y/r} = \frac{x}{y}.$$

The proofs of the other statements are similar.

EXAMPLE 2

Evaluate all six trigonometric functions at $t = 3\pi/4$.

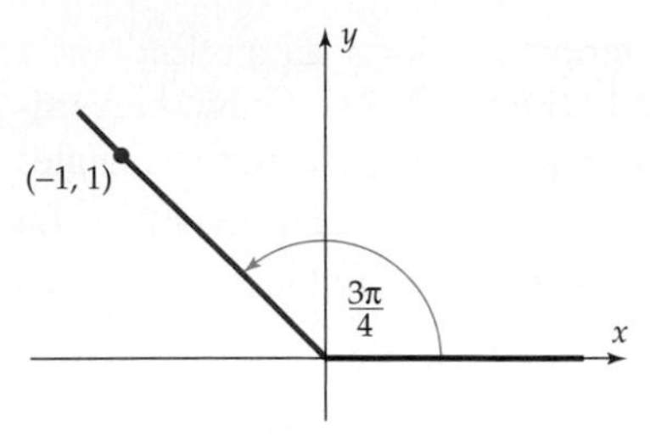

Figure 2–79

SOLUTION The terminal side of an angle of $3\pi/4$ radians in standard position lies on the line $y = -x$, as shown in Figure 2–79. We shall use the point $(-1, 1)$ on this line to compute the function values. In this case,

$$r = \sqrt{x^2 + y^2} = \sqrt{(-1)^2 + 1^2} = \sqrt{2}.$$

Therefore,

$$\sin\frac{3\pi}{4} = \frac{y}{r} = \frac{1}{\sqrt{2}} = \frac{\sqrt{2}}{2} \qquad \cos\frac{3\pi}{4} = \frac{x}{r} = \frac{-1}{\sqrt{2}} = \frac{-\sqrt{2}}{2}$$

$$\tan\frac{3\pi}{4} = \frac{y}{x} = \frac{1}{-1} = -1 \qquad \csc\frac{3\pi}{4} = \frac{r}{y} = \frac{\sqrt{2}}{1} = \sqrt{2}$$

$$\sec\frac{3\pi}{4} = \frac{r}{x} = \frac{\sqrt{2}}{-1} = -\sqrt{2} \qquad \cot\frac{3\pi}{4} = \frac{x}{y} = \frac{-1}{1} = -1.$$ ■

ALGEBRA AND IDENTITIES

We begin by noting the relationship between the cotangent and tangent functions.

Reciprocal Identities

The cotangent and tangent functions are reciprocals; that is,

$$\cot t = \frac{1}{\tan t} \qquad \text{and} \qquad \tan t = \frac{1}{\cot t}$$

for every number t in the domain of both functions.

The first of these identities was proved on page 204, and the second is proved similarly (Exercise 43).

Period of Secant, Cosecant, and Cotangent

The secant and cosecant functions are periodic with period 2π, and the cotangent function is periodic with period π. In symbols,

$$\sec(t + 2\pi) = \sec t, \qquad \csc(t + 2\pi) = \csc t,$$

$$\cot(t + \pi) = \cot t$$

for every number t in the domain of the given function.

The proof of these statements uses the fact that each of these functions is the reciprocal of a function whose period is known. For instance,

$$\csc(t + 2\pi) = \frac{1}{\sin(t + 2\pi)} = \frac{1}{\sin t} = \csc t$$

$$\cot(t + \pi) = \frac{1}{\tan(t + \pi)} = \frac{1}{\tan t} = \cot t.$$

See Exercise 44 for the remaining identity.

Pythagorean Identities

For every number t in the domain of both functions,

$$1 + \tan^2 t = \sec^2 t$$

and

$$1 + \cot^2 t = \csc^2 t.$$

Proof By the definitions of the functions and the Pythagorean identity $(\sin^2 t + \cos^2 t = 1)$,

$$1 + \tan^2 t = 1 + \frac{\sin^2 t}{\cos^2 t} = \frac{\cos^2 t + \sin^2 t}{\cos^2 t} = \frac{1}{\cos^2 t} = \left(\frac{1}{\cos t}\right)^2 = \sec^2 t.$$

The second identity is proved similarly (Exercise 45). ■

EXAMPLE 3

Simplify the expression $\frac{30 \cos^3 t \sin t}{6 \sin^2 t \cos t}$, assuming that $\sin t \neq 0$ and $\cos t \neq 0$.

SOLUTION

$$\frac{30 \cos^3 t \sin t}{6 \sin^2 t \cos t} = \frac{5 \cos^3 t \sin t}{\cos t \sin^2 t} = \frac{5 \cos^2 t}{\sin t} = 5\frac{\cos t}{\sin t}\cos t = 5 \cot t \cos t.$$ ■

EXAMPLE 4

Assume that $\cos t \neq 0$ and simplify $\cos^2 t + \cos^2 t \tan^2 t$.

SOLUTION

$$\cos^2 t + \cos^2 t \tan^2 t = \cos^2 t(1 + \tan^2 t) = \cos^2 t \sec^2 t = \cos^2 t \cdot \frac{1}{\cos^2 t} = 1.$$

■

EXAMPLE 5

If $\tan t = 3/4$ and $\sin t < 0$, find $\cot t$, $\cos t$, $\sin t$, $\sec t$, and $\csc t$.

SOLUTION First we have $\cot t = 1/\tan t = 1/(3/4) = 4/3$. Next we use the Pythagorean identity to obtain

$$\sec^2 t = 1 + \tan^2 t = 1 + \left(\frac{3}{4}\right)^2 = 1 + \frac{9}{16} = \frac{25}{16}$$

$$\sec t = \pm\sqrt{\frac{25}{16}} = \pm\frac{5}{4}$$

$$\frac{1}{\cos t} = \pm\frac{5}{4} \quad \text{or, equivalently,} \quad \cos t = \pm\frac{4}{5}.$$

Since $\sin t$ is given as negative and $\tan t = \sin t/\cos t$ is positive, $\cos t$ must be negative. Hence, $\cos t = -4/5$. Consequently,

$$\frac{3}{4} = \tan t = \frac{\sin t}{\cos t} = \frac{\sin t}{(-4/5)},$$

so

$$\sin t = \left(-\frac{4}{5}\right)\left(\frac{3}{4}\right) = -\frac{3}{5}.$$

Therefore,

$$\sec t = \frac{1}{\cos t} = \frac{1}{(-4/5)} = -\frac{5}{4} \quad \text{and} \quad \csc t = \frac{1}{\sin t} = \frac{1}{(-3/5)} = -\frac{5}{3}. \quad ■$$

GRAPHS

The graph of the secant function is shown in red in Figure 2–80.

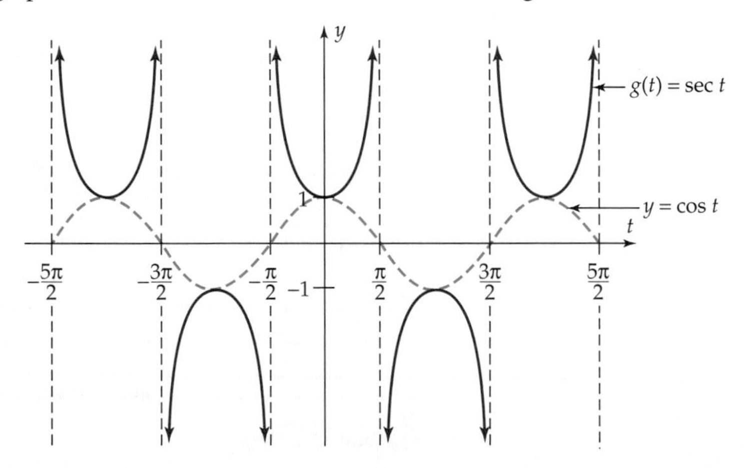

Figure 2–80

The shape of the secant graph can be understood by looking at the graph of cosine (blue in Figure 2–80) and noting these facts:

1. $\sec t = 1/\cos t$ is not defined when $\cos t = 0$, that is, when $t = \pm\pi/2$, $\pm 3\pi/2$, $\pm 5\pi/2$, and so on.
2. The graph of $\sec t$ has a vertical asymptote at $t = \pm\pi/2, \pm 3\pi/2, \pm 5\pi/2, \ldots$ The reason is that when $\cos t$ is close to 0 (graph close to t-axis), then $\sec t = 1/\cos t$ is very large in absolute value, so its graph is far from the axis.

3. When $\cos t$ is near 1 or -1 (that is, when t is near 0, $\pm\pi$, $\pm 2\pi$, $\pm 3\pi, \ldots$), then so is $\sec t = 1/\cos t$.

The graphs of $h(t) = \csc t = 1/\sin t$ and $f(t) = \cot t = 1/\tan t$ can be obtained in a similar fashion (Figure 2–81).

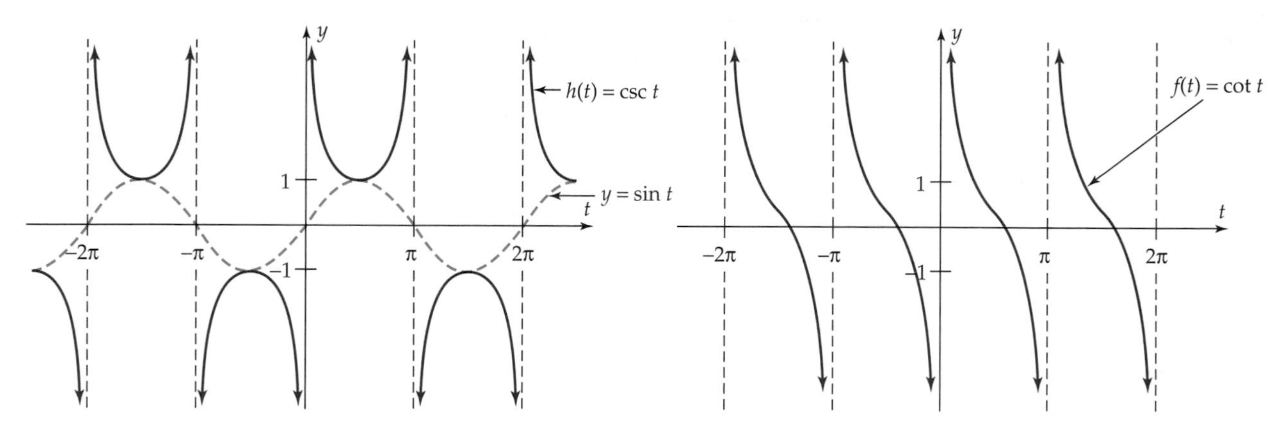

Figure 2–81

EXERCISES 2.6

Note: *The arrangement of the exercises corresponds to the subsections of this section.*

Definitions and Alternate Descriptions

In Exercises 1–6, determine the quadrant containing the terminal side of an angle of t radians in standard position under the given conditions.

1. $\cos t > 0$ and $\sin t < 0$
2. $\sin t < 0$ and $\tan t > 0$
3. $\sec t < 0$ and $\cot t < 0$
4. $\csc t < 0$ and $\sec t > 0$
5. $\sec t > 0$ and $\cot t < 0$
6. $\sin t > 0$ and $\sec t < 0$

In Exercises 7–16, evaluate all six trigonometric functions at t, where the given point lies on the terminal side of an angle of t radians in standard position.

7. $(3, 4)$
8. $(0, 6)$
9. $(-5, 12)$
10. $(-2, -3)$
11. $(-1/5, 1)$
12. $(4/5, -3/5)$
13. $(\sqrt{2}, \sqrt{3})$
14. $(-2\sqrt{3}, \sqrt{3})$
15. $(1 + \sqrt{2}, 3)$
16. $(1 + \sqrt{3}, 1 - \sqrt{3})$

In Exercises 17–19, evaluate all six trigonometric functions at the given number without using a calculator.

17. $\dfrac{4\pi}{3}$
18. $-\dfrac{7\pi}{6}$
19. $\dfrac{7\pi}{4}$

20. Fill in the missing entries in the following table. Give exact answers, not decimal approximations.

t	0	$\frac{\pi}{6}$	$\frac{\pi}{4}$	$\frac{\pi}{3}$	$\frac{\pi}{2}$	$\frac{2\pi}{3}$	$\frac{3\pi}{4}$	$\frac{5\pi}{6}$	π	$\frac{3\pi}{2}$
$\sin t$										
$\cos t$										
$\tan t$					—					—
$\cot t$	—								—	
$\sec t$					—					—
$\csc t$	—								—	

Algebra and Identities

*C Exercises 21–23 use the following concept. If f is a function, then the **average rate of change** of f from $t = a$ to $t = b$ is the number*

$$\frac{f(b) - f(a)}{b - a}.$$

21. Find the average rate of change of $f(t) = \cot t$ from $t = 1$ to $t = 3$.

22. Find the average rate of change of $g(t) = \csc t$ from $t = 2$ to $t = 3$.

23. (a) Find the average rate of change of $f(t) = \tan t$ from $t = 2$ to $t = 2 + h$, for each of these values of h: .01, .001, .0001, and .00001.
(b) Compare your answers in part (a) with the number $(\sec 2)^2$. What would you guess that the instantaneous rate of change of $f(t) = \tan t$ is at $t = 2$?

In Exercises 24–30, perform the indicated operations, then simplify your answers by using appropriate definitions and identities.

24. $\tan t(\cos t - \csc t)$

25. $\cos t \sin t(\csc t + \sec t)$

26. $(1 + \cot t)^2$

27. $(1 - \sec t)^2$

28. $(\sin t - \csc t)^2$

29. $(\cot t - \tan t)(\cot^2 t + 1 + \tan^2 t)$

30. $(\sin t + \csc t)(\sin^2 t + \csc^2 t - 1)$

In Exercises 31–36, factor and simplify the given expression.

31. $\sec t \csc t - \csc^2 t$

32. $\tan^2 t - \cot^2 t$

33. $\tan^4 t - \sec^4 t$

34. $4 \sec^2 t + 8 \sec t + 4$

35. $\cos^3 t - \sec^3 t$

36. $\csc^4 t + 4 \csc^2 t - 5$

* C indicates exercises that are relevant to calculus.

In Exercises 37–42, simplify the given expression. Assume that all denominators are nonzero and all quantities under radicals are nonnegative.

37. $\dfrac{\cos^2 t \sin t}{\sin^2 t \cos t}$

38. $\dfrac{\sec^2 t + 2 \sec t + 1}{\sec t}$

39. $\dfrac{4 \tan t \sec t + 2 \sec t}{6 \sin t \sec t + 2 \sec t}$

40. $\dfrac{\sec^2 t \csc t}{\csc^2 t \sec t}$

41. $(2 + \sqrt{\tan t})(2 - \sqrt{\tan t})$

42. $\dfrac{6 \tan t \sin t - 3 \sin t}{9 \sin^2 t + 3 \sin t}$

In Exercises 43–46, prove the given identity.

43. $\tan t = \dfrac{1}{\cot t}$ [*Hint:* See page 204.]

44. $\sec(t + 2\pi) = \sec t$ [*Hint:* See page 207.]

45. $1 + \cot^2 t = \csc^2 t$ [*Hint:* See page 207.]

46. $\cot(-t) = -\cot t$ [*Hint:* Express the left side in terms of sine and cosine; then use the negative angle identities and express the result in terms of cotangent.]

47. $\sec(-t) = \sec t$ [Adapt the hint for Exercise 46.]

48. $\csc(-t) = -\csc t$

In Exercises 49–54, find the values of all six trigonometric functions at t if the given conditions are true.

49. $\cos t = -1/2$ and $\sin t > 0$
[*Hint:* $\sin^2 t + \cos^2 t = 1$.]

50. $\cos t = \dfrac{1}{2}$ and $\sin t < 0$

51. $\cos t = 0$ and $\sin t = 1$

52. $\sin t = -2/3$ and $\sec t > 0$

53. $\sec t = -13/5$ and $\tan t < 0$

54. $\csc t = 8$ and $\cos t < 0$

Graphs

In Exercises 55–58, use graphs to determine whether the equation could possibly be an identity or is definitely not an identity.

55. $\tan t = \cot\left(\frac{\pi}{2} - t\right)$

56. $\frac{\cos t}{\cos(t - \pi/2)} = \cot t$

57. $\frac{\sin t}{1 - \cos t} = \cot t$

58. $\frac{\sec t + \csc t}{1 + \tan t} = \csc t$

59. Show graphically that the equation $\sec t = t$ has infinitely many solutions, but none between $-\pi/2$ and $\pi/2$.

Thinkers

60. In the diagram of the unit circle in the figure, find six line segments whose respective lengths are $\sin t$, $\cos t$, $\tan t$, $\cot t$, $\sec t$, $\csc t$. [*Hint:* $\sin t$ = length CA. Why? Note that OC has length 1 and various right triangles in the figure are similar.]

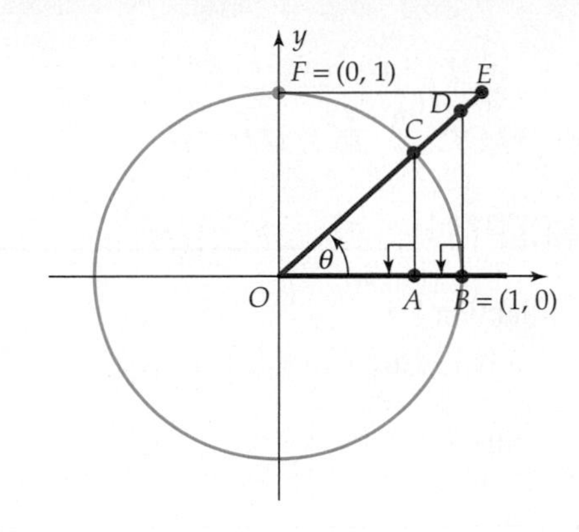

61. In the figure for Exercise 60, find the following areas in terms of θ.

(a) triangle OCA

(b) triangle ODB

(c) circular segment OCB

Chapter 2 Review

IMPORTANT CONCEPTS

Section 2.1

Angle 136
Vertex 136
Initial side 136
Terminal side 136
Coterminal angles 137, 141
Standard position 137
Degree measure 137
Radian measure 138

Special Topics 2.1.A

Arc length 144
Angular speed 144

Section 2.2 [Alternate Section 2.2]

Point-in-the-plane description of sine, cosine, and tangent 148 [162]
Unit circle description of sine, cosine, and tangent 151 [158]
Calculating functional values for multiples of $\pi/3$, $\pi/4$, and $\pi/6$ 153 [160]

Section 2.3

Algebra with trigonometric functions 164
Pythagorean identity 166
Periodicity 168
Negative angle identities 169

Section 2.4

Graphs of sine, cosine, and tangent functions 172, 175, 177
Graphs and identities 179

Section 2.5

Period 183
Amplitude 185
Phase shift 187
Simple harmonic motion 191

Special Topics 2.5.A

Sinusoidal graphs 198
Damped and compressed graphs 199

Section 2.6

Cotangent function 204
Secant function 204
Cosecant function 204
Alternative descriptions 205
Identities 206–207
Graphs of cotangent, secant, and cosecant functions 208–209

IMPORTANT FACTS & FORMULAS

- *Conversion Rules:* To convert radians to degrees, multiply by $180/\pi$. To convert degrees to radians, multiply by $\pi/180$.
- *Unit Circle Description of Trigonometric Functions:* If P is the point where the terminal side of an angle of t radians in standard position meets the unit circle, then

$$\sin t = y\text{-coordinate of } P$$

$$\cos t = x\text{-coordinate of } P$$

$$\tan t = \frac{\sin t}{\cos t}, \qquad \cot t = \frac{\cos t}{\sin t}, \qquad \sec t = \frac{1}{\cos t}, \qquad \csc t = \frac{1}{\sin t}.$$

- *Point-in-the-Plane Description:* If (x, y) is any point other than the origin on the terminal side of an angle of t radians in standard position and $r = \sqrt{x^2 + y^2}$, then

$$\sin t = \frac{y}{r} \qquad \cos t = \frac{x}{r}$$

$$\tan t = \frac{y}{x} \qquad \cot t = \frac{x}{y}$$

$$\sec t = \frac{r}{x} \qquad \csc t = \frac{r}{y}.$$

- *Basic Identities:*

$$\sin^2 t + \cos^2 t = 1 \qquad \sin(-t) = -\sin t$$
$$1 + \tan^2 t = \sec^2 t \qquad \cos(-t) = \cos t$$
$$1 + \cot^2 t = \csc^2 t \qquad \tan(-t) = -\tan t$$
$$\tan t = \frac{1}{\cot t} \qquad \cot t = \frac{1}{\tan t}$$
$$\sin(t \pm 2\pi) = \sin t \qquad \csc(t \pm 2\pi) = \csc t$$
$$\cos(t \pm 2\pi) = \cos t \qquad \sec(t \pm 2\pi) = \sec t$$
$$\tan(t \pm \pi) = \tan t \qquad \cot(t \pm \pi) = \cot t$$

- If $A \neq 0$ and $b > 0$, then each of the functions $f(t) = A\sin(bt + c)$ and $g(t) = A\cos(bt + c)$ has

$$\text{amplitude } |A|, \qquad \text{period } 2\pi/b, \qquad \text{phase shift } -c/b.$$

REVIEW QUESTIONS 

1. Find a number t between 0 and 2π such that an angle of t radians in standard position is coterminal with an angle of $-23\pi/3$ radians in standard position.
2. Through how many radians does the second hand of a clock move in 2 minutes and 40 seconds?
3. $\frac{9\pi}{5}$ radians = ____ degrees.
4. 36 degrees = ____ radians.
5. $220° =$ ____ radians.
6. $\frac{17\pi}{12}$ radians = ____ degrees.
7. $-\frac{11\pi}{4}$ radians = ____ degrees.
8. $-135° =$ ____ radians.
9. If an angle of v radians has its terminal side in the second quadrant and $\sin v = \sqrt{8/9}$, then find $\cos v$.
10. $\cos \frac{47\pi}{2} = ?$
11. $\sin(-13\pi) = ?$
12. Simplify: $\frac{\tan(t + \pi)}{\sin(t + 2\pi)}$

Use the figure on the next page in Questions 13–17.

13. $\cos \frac{\pi}{5} = ?$
14. $\sin\left(\frac{7\pi}{6}\right) = ?$
15. $\cos\left(\frac{-5\pi}{6}\right) = ?$
16. $\sin\left(\frac{16\pi}{6}\right) = ?$

REVIEW QUESTIONS

17. $\sin\left(\frac{-4\pi}{3}\right) = ?$

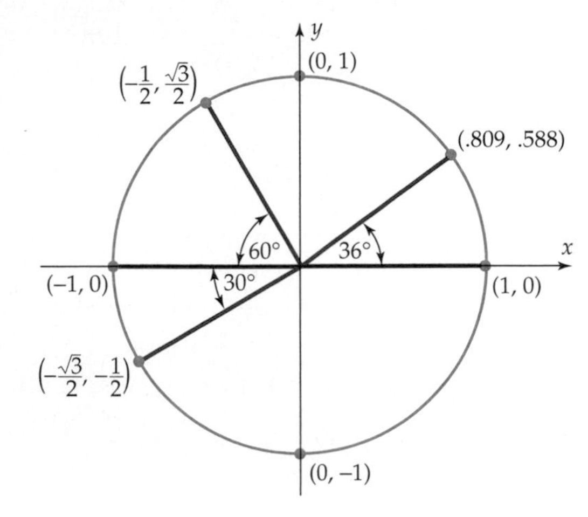

18. $\left[3 \sin\left(\frac{\pi}{5^{500}}\right)\right]^2 + \left[3 \cos\left(\frac{\pi}{5^{500}}\right)\right]^2 = ?$

19. Fill in the blanks (approximations not allowed):

t	0	$\frac{\pi}{6}$	$\frac{\pi}{4}$	$\frac{\pi}{3}$	$\frac{\pi}{2}$
sin *t*					
cos *t*					

20. Express as a single real number:

$$\cos\frac{3\pi}{4}\sin\frac{5\pi}{6} - \sin\frac{3\pi}{4}\cos\frac{5\pi}{6}$$

21. $\left(\sin\frac{\pi}{6} + 1\right)^2 = ?$

22. $\sin(\pi/2) + \sin 0 + \cos 0 = ?$

23. If $f(x) = \log_{10} x$ and $g(t) = -\cos t$, then $f(g(\pi)) = ?$

24. Cos t is negative when the terminal side of an angle of t radians in standard position lies in which quadrants?

25. If $\sin t = 1/\sqrt{3}$ and the terminal side of an angle of t radians in standard position lies in the second quadrant, then $\cos t = ?$

26. Which of the following could possibly be a true statement about a real number t?
    (a) $\sin t = -2$ and $\cos t = 1$
    (b) $\sin t = 1/2$ and $\cos t = \sqrt{2}/2$
    (c) $\sin t = -1$ and $\cos t = 1$
    (d) $\sin t = \pi/2$ and $\cos t = 1 - (\pi/2)$
    (e) $\sin t = 3/5$ and $\cos t = 4/5$

27. If $\sin t = -4/5$ and the terminal side of an angle of t radians in standard position lies in the third quadrant, then $\cos t =$ ____.

28. If $\sin(-101\pi/2) = -1$, then $\sin(-105\pi/2) = ?$

29. If $\pi/2 < t < \pi$ and $\sin t = 5/13$, then $\cos t = ?$

30. $\cos\left(-\frac{\pi}{6}\right) = ?$

31. $\cos\left(\frac{2\pi}{3}\right) = ?$

32. $\sin\left(-\frac{11\pi}{6}\right) = ?$

33. $\sin\left(\frac{\pi}{3}\right) = ?$

34. $\tan(5\pi/3) = ?$

35. Which of the following is *not* true about the graph of $f(t) = \sin t$?
    (a) It has no sharp corners.
    (b) It crosses the horizontal axis more than once.
    (c) It rises higher and higher as t gets larger.
    (d) It is periodic.
    (e) It has no vertical asymptotes.

36. Which of the following functions has the graph in the figure between $-\pi$ and π?
    (a) $f(x) = \begin{cases} \sin x, & \text{if } x \geq 0 \\ \cos x, & \text{if } x < 0 \end{cases}$
    (b) $g(x) = \cos x - 1$
    (c) $h(x) = \begin{cases} \sin x, & \text{if } x \geq 0 \\ \sin(-x), & \text{if } x < 0 \end{cases}$
    (d) $k(x) = |\cos x|$
    (e) $p(x) = \sqrt{1 - \sin^2 x}$

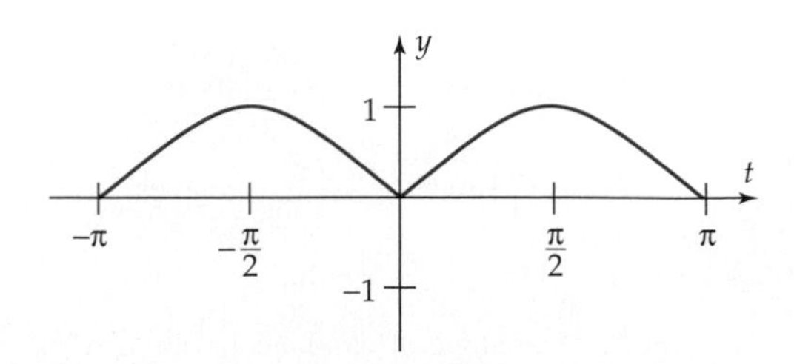

The point $(-3/\sqrt{50}, 7/\sqrt{50})$ lies on the terminal side of an angle of t radians (in standard position). Find:

37. $\sin t$

38. $\cos t$

REVIEW QUESTIONS

39. $\tan t$

40. $\sec t$

41. Find the equation of the straight line containing the terminal side of an angle of $5\pi/3$ radians (in standard position).

Suppose that an angle of w radians has its terminal side in the fourth quadrant and $\cos w = 2/\sqrt{13}$. *Find:*

42. $\sin w$

43. $\tan w$

44. $\csc w$

45. $\cos(-w)$

46. Fill in the blanks (approximations not allowed):

t	$\sin t$	$\tan t$	$\sec t$
$\pi/4$			
$2\pi/3$			
$5\pi/6$			

47. Sketch the graphs of $f(t) = \sin t$ and $h(t) = \csc t$ on the same set of coordinate axes ($-2\pi \le t \le 2\pi$).

48. *Let* θ be the angle shown in the figure. Which of the following statements is true?

(a) $\sin \theta = \dfrac{\sqrt{2}}{2}$

(b) $\cos \theta = \dfrac{\sqrt{2}}{2}$

(c) $\tan \theta = 1$

(d) $\cos \theta = \sqrt{2}$

(e) $\tan \theta = -1$

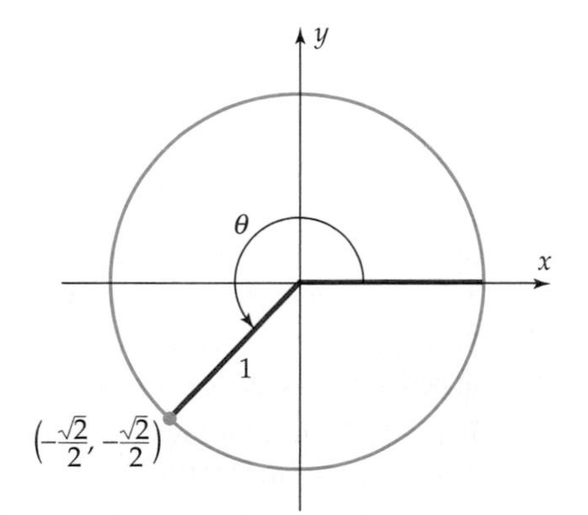

49. Let θ be as indicated in the figure in the next column. Which of the statements **(i)–(iii)** are *true*?

(i) $\cos \theta = -\dfrac{1}{3}$

(ii) $\tan \theta = \dfrac{2\sqrt{2}}{9}$

(iii) $\sin \theta = -\dfrac{2\sqrt{2}}{3}$

(a) only ii
(b) only ii and iii
(c) all of them
(d) only i and iii
(e) none of them

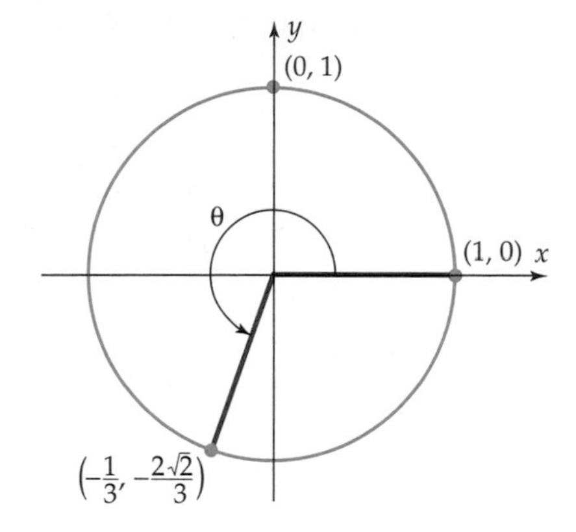

50. Between (and including) 0 and 2π, the function $h(t) = \tan t$ has

(a) 3 roots and is undefined at 2 places.
(b) 2 roots and is undefined at 3 places.
(c) 2 roots and is undefined at 2 places.
(d) 3 roots and is defined everywhere.
(e) no roots and is undefined at 3 places.

51. If the terminal side of an angle of θ radians in standard position passes through the point $(-2, 3)$, then $\tan \theta =$ ____.

52. Which of the statements **(i)–(iii)** are true?

(i) $\sin(-x) = -\sin x$
(ii) $\cos(-x) = -\cos x$
(iii) $\tan(-x) = -\tan x$

(a) (i) and (ii) only
(b) (ii) only
(c) (i) and (iii) only
(d) all of them
(e) none of them

53. If $\sec x = 1$ and $-\pi/2 < x < \pi/2$, then $x =$?

54. If $\tan t = 4/3$ and $0 < t < \pi$, what is $\cos t$?

55. Which of the following is true about $\sec t$?

(a) $\sec(0) = 0$
(b) $\sec t = 1/\sin t$
(c) Its graph has no asymptotes.
(d) It is a periodic function.
(e) It is never negative.

56. If $\cot t = 0$ and $0 < t \le \pi$, then $t =$ ____.

57. What is $\cot\left(\dfrac{2\pi}{3}\right)$?

REVIEW QUESTIONS

58. Which of the following functions has the graph in the figure?
(a) $f(t) = \tan t$
(b) $g(t) = \tan\left(t + \frac{\pi}{2}\right)$
(c) $h(t) = 1 + \tan t$
(d) $k(t) = 3 \tan t$
(e) $p(t) = -\tan t$

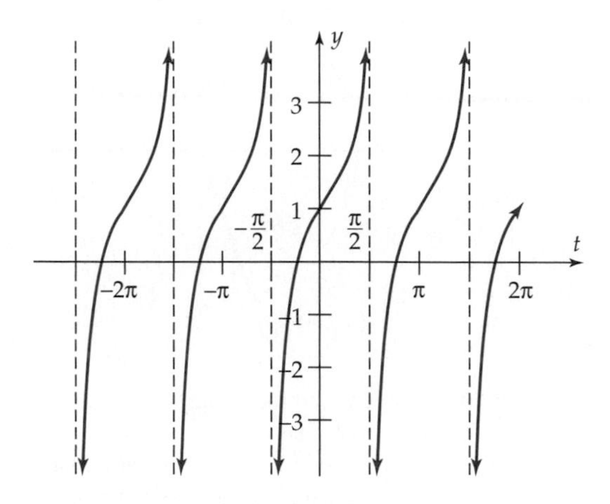

59. Let $f(t) = \frac{3}{2} \sin 5t$.
(a) What is the largest possible value of $f(x)$?
(b) Find the smallest positive number t such that $f(t) = 0$.

60. Sketch the graph of $g(t) = -2 \cos t$.

61. Sketch the graph of $f(t) = -\frac{1}{2} \sin 2t \ (-2\pi \le t \le 2\pi)$.

62. Sketch the graph of $f(t) = \sin 4t \ (0 \le t \le 2\pi)$.

In Questions 63–66, determine graphically whether the given equation could possibly be an identity.

63. $\cos t = \sin\left(t - \frac{\pi}{2}\right)$

64. $\tan \frac{t}{2} = \frac{\sin t}{1 + \cos t}$

65. $\frac{\sin t - \sin 3t}{\cos t + \cos 3t} = -\tan t$

66. $\cos 2t = \frac{1}{1 - 2 \sin^2 t}$

67. What is the period of the function $g(t) = \sin 4\pi t$?

68. What are the amplitude, period, and phase shift of the function

$$h(t) = 13 \cos(14t + 15)?$$

69. The number of hours of daylight in Boston on day t of the year is approximated by

$$d(t) = 3.1 \sin(.0172t - 1.377) + 12.$$

(a) On the day with the most daylight, how many hours of daylight are there? On the day with the least daylight, how many hours are there?
(b) On what days are there less than 11 hours of daylight?

70. A certain person's blood pressure $P(t)$ at time t seconds is approximately

$$P(t) = 22 \cos(2.5\pi t) + 98.$$

(a) The period of this function is time between heartbeats. What is this person's pulse rate (heartbeats per minute)?
(b) What is this person's blood pressure [systolic pressure (maximum) over diastolic pressure (minimum)]?

71. State the rule of a periodic function whose graph from $t = 0$ to $t = 2\pi$ closely resembles the one in the figure.

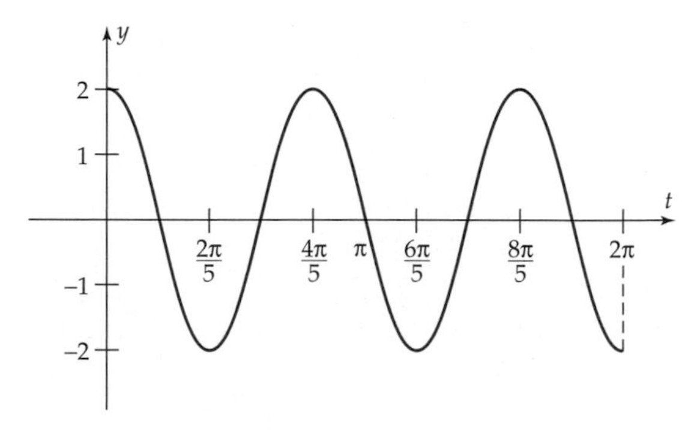

72. State the rule of a periodic function with amplitude 3, period π, and phase shift $\pi/3$.

73. State the rule of a periodic function with amplitude 8, period 5, and phase shift 14.

74. If $g(t) = 20 \sin(200t)$, for how many values of t, with $0 \le t \le 2\pi$, is it true that $g(t) = 1$?

In Exercises 75 and 76, estimate constants A, b, c such that $f(t) = A \sin(bt + c)$.

75. $f(t) = 6 \sin(4t + 7) - 5 \cos(4t + 8)$

76. $f(t) = -5 \sin(5t - 3) + 2 \cos(5t + 2)$

REVIEW QUESTIONS

In Exercises 77 and 78, find a viewing window that shows a complete graph of the function.

77. $f(t) = 3\sin(300t + 5) - 2\cos(500t + 8)$

78. $g(t) = -5\sin(400\pi t + 1) + 2\cos(150\pi t - 6)$

79. The average monthly temperatures in St. Louis, Missouri, are shown in the table.*

Month	Temperture (°F)
Jan.	30
Feb.	35
March	46
April	57
May	67
June	77
July	80
Aug.	78
Sept.	70
Oct.	58
Nov.	45
Dec.	34

(a) Let $x = 1$ correspond to January. Plot 24 data points (two years).

(b) Use regression to find a sine function that models this data.

(c) What is the period of the function in part (b)? Does it fit the data well?

*Based on 1971–2000 data from the National Climatic Data Center.

DISCOVERY PROJECT 2 Pistons and Flywheels

Andre Jenny/Focus Group/PictureQuest

A common and well-proven piece of technology is the piston and flywheel combination. It is clearly visible in photographs of steam locomotives from the middle of the nineteenth century. The structure involves a wheel (or crankshaft) connected by a rigid arm to a sliding plug in a cylinder. The axis of rotation of the wheel is perpendicular to the central axis of the cylinder. The sliding plug, the piston, moves in one dimension, in and out of the cylinder. The motion of the piston is periodic, like the basic trigonometric functions, but doesn't have the same elegant symmetry.

It is quite easy to superimpose a coordinate plane on the flywheel-piston system so that the center of the flywheel is the origin, the flywheel rotates counterclockwise, and the piston moves along the x-axis. As you can see in the diagram, the flywheel typically has a larger radius than the radial distance from the center to the point where the arm attaches. In this particular figure, the radius of the flywheel is 50 centimeters, and the radial distance to the attachment point Q is 45 centimeters. The arm is 150 centimeters long measured from the base of the piston to the attachment point.

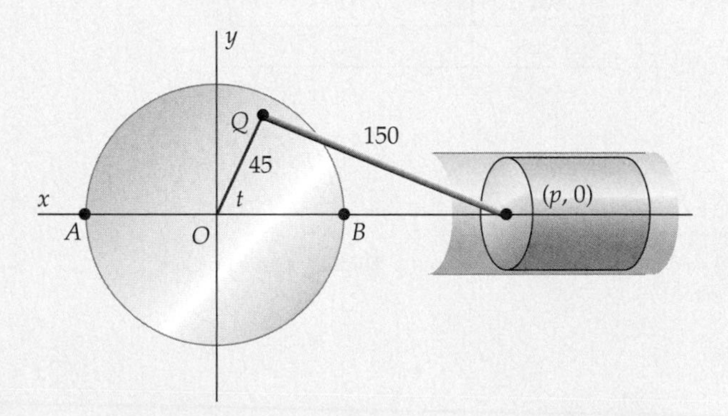

1. What are the coordinates of A and B in the figure? How close does the base of the piston come to the flywheel? What is the length of the piston stroke?

2. Show that Q has coordinates $(45 \cos t, 45 \sin t)$, where t is the radian measure of angle BOQ. [*Hint:* If Q has coordinates (x, y), use the point-in-the-plane description to compute $\cos t$ and solve for x; find y similarly.]

3. Let p be the x-coordinate of the center point of the base of the piston. Express p as a function of t. [*Hint:* Use the distance formula to express the distance from Q to $(p, 0)$ in terms of p and t. Set this expression equal to 150 (why?) and solve for p.]

4. Let $p(t)$ be the function found in Question 3 (that is, $p(t)$ = the x-coordinate of the base of the piston when angle BOQ measures t radians). What is the range of this function? How does this relate to Question 1?
5. Approximate the values of $p(0)$, $p(\pi/2)$, and $p(\pi)$. Note that $p(\pi/2)$ is not halfway between $p(0)$ and $p(\pi)$. Which side of the halfway point is it on? Does this mean that the piston moves faster on the average when it is to the right or left of the halfway point? Find the value of t that places the base of the piston at the halfway point of its motion.
6. If the flywheel spins at a constant speed, does the piston move back and forth at a constant speed? How do you know?

C H A P T E R 3

Trigonometric Identities and Equations

Photodisc Collection/Getty Images

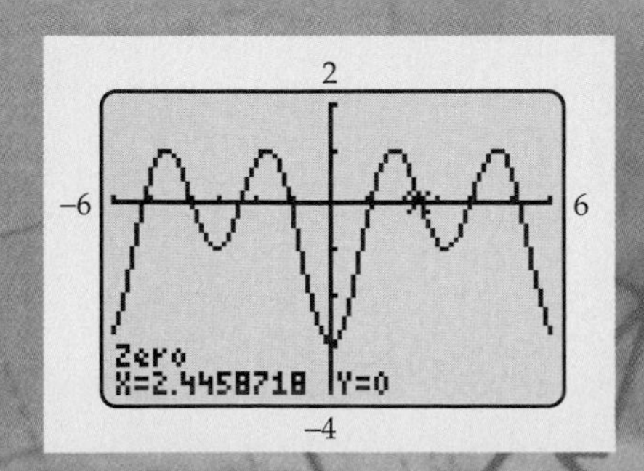

Beam me up, Scotty!

When a light beam passes from one medium to another (for instance, from air to water), it changes both its speed and its direction. If you know what some of these numbers are (say, the speed of light in air or the angle at which a light beam hits the water), then you can determine the unknown ones by solving a trigonometric equation. See Exercises 61–64 on page 268.

Chapter Outline

Interdependence of Sections

Chapters 3 and 4 are independent of each other and may be read in any order. The interdependence of the sections of this chapter is shown below. Sections 3.1, 3.4, and 3.5 are independent of one another and may be read in any order.

3.1 ⟶ 3.2 ⟶ 3.3

3.4

3.5

Until now, the variable t has been used for trigonometric functions, to avoid confusion with the x's and y's that appear in their definitions. Now that you are comfortable with these functions, we shall usually use the letter x (or occasionally y) for the variable. Unless stated otherwise, all trigonometric functions in this chapter are considered to be functions of real numbers, rather than functions of angles in degree measure.

Two kinds of trigonometric equations are considered here. *Identities* (Sections 3.1–3.3) are equations that are valid for all values of the variable for which the equation is defined, such as

$$\sin^2 x + \cos^2 x = 1 \qquad \text{and} \qquad \cot x = \frac{1}{\tan x}.$$

Identities can be used for simplifying expressions, rewriting the rule of a trigonometric function, performing numerical computations, and in other ways. *Conditional equations* (Section 3.5) are valid only for certain values of the variable, such as

$$\sin x = 0 \qquad \text{(true only when } x \text{ is an integer multiple of } \pi\text{).}$$

Inverse trigonometric functions, which have a number of uses, are considered in Section 3.4.

3.1 Basic Identities and Proofs

When you suspect that an equation might be an identity, it's a good idea to see whether there is any graphical evidence to support this conclusion.

EXAMPLE 1

Is either of the following equations an identity?

(a) $2\sin^2 x - \cos x = 2\cos^2 x + \sin x$

(b) $\dfrac{1 + \sin x - \sin^2 x}{\cos x} = \cos x + \tan x$

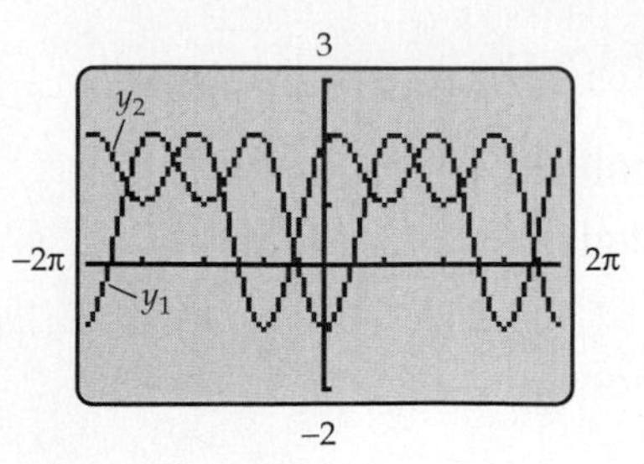

Figure 3–1

SOLUTION

(a) Test the equation by graphing these two equations on the same screen:

$$y_1 = 2\sin^2 x - \cos x \qquad \textit{[left-hand side of equation being tested]}$$

$$y_2 = 2\cos^2 x + \sin x \qquad \textit{[right-hand side of equation being tested]}$$

If the given equation is an identity (meaning that $y_1 = y_2$), then the two graphs will be identical. As Figure 3–1 shows, however, the graphs are quite different. Hence, this equation is not an identity.

(b) Test the second equation graphically.

GRAPHING EXPLORATION

Graph the functions

$$f(x) = \frac{1 + \sin x - \sin^2 x}{\cos x} \qquad \text{and} \qquad g(x) = \cos x + \tan x$$

on the same screen, using a viewing window with $-2\pi \leq x \leq 2\pi$. Do the graphs appear to be identical?

The exploration suggests that the equation *may* be an identity because the two graphs *appear* to be identical. However, this graphical evidence is *not* a proof—that must be done algebraically. ■

Any equation can be graphically tested, as in Example 1, to see whether it might be an identity. If the left-side and right-side graphs are different, then it definitely is *not* an identity. If the graphs appear to be the same, then it is *possible,* but not certain, that the equation is an identity.

The fact that two graphs appear identical on a calculator screen does *not* prove that they are actually the same, as the following Graphing Exploration shows.

GRAPHING EXPLORATION

In the viewing window with $-\pi \leq x \leq \pi$ and $-2 \leq y \leq 2$, graph both sides of the equation

$$\cos x = 1 - \frac{x^2}{2} + \frac{x^4}{24} - \frac{x^6}{720} + \frac{x^8}{40{,}320}.$$

Do the graphs appear to be identical? Now change the viewing window so that $-2\pi \leq x \leq 2\pi$. Is the equation an identity?

PROVING IDENTITIES

The phrases "prove the identity" and "verify the identity" mean "prove that the given equation is an identity." In the proofs, we shall assume the elementary identities that were proved in Chapter 2 and are summarized here.

Basic Trigonometric Identities

Reciprocal Identities

$$\sec x = \frac{1}{\cos x} \qquad \csc x = \frac{1}{\sin x}$$

$$\tan x = \frac{\sin x}{\cos x} \qquad \cot x = \frac{\cos x}{\sin x}$$

$$\tan x = \frac{1}{\cot x} \qquad \cot x = \frac{1}{\tan x}$$

Periodicity Identities

$$\sin(x + 2\pi) = \sin x \qquad \cos(x + 2\pi) = \cos x$$

$$\sec(x + 2\pi) = \sec x \qquad \csc(x + 2\pi) = \csc x$$

$$\tan(x + \pi) = \tan x \qquad \cot(x + \pi) = \cot x$$

Pythagorean Identities

$$\sin^2 x + \cos^2 x = 1 \qquad 1 + \tan^2 x = \sec^2 x \qquad 1 + \cot^2 x = \csc^2 x$$

Negative Angle Identities

$$\sin(-x) = -\sin x \qquad \cos(-x) = \cos x \qquad \tan(-x) = -\tan x$$

There are no cut-and-dried rules for simplifying trigonometric expressions or proving identities, but there are some common strategies that are often helpful. Four of these strategies are illustrated in the following examples. There are often a varicty of ways to proceed, and it will take some practice before you can easily decide which strategies are likely to be the most efficient in a particular case.

Strategy 1

Express everything in terms of sine and cosine.

EXAMPLE 2

Simplify $(\csc x + \cot x)(1 - \cos x)$.

SOLUTION Using Strategy 1, we have

$$(\csc x + \cot x)(1 - \cos x) = \left(\frac{1}{\sin x} + \frac{\cos x}{\sin x}\right)(1 - \cos x) \qquad \text{[Reciprocal identities]}$$

$$= \frac{(1 + \cos x)}{\sin x}(1 - \cos x)$$

$$= \frac{(1 + \cos x)(1 - \cos x)}{\sin x}$$

$$= \frac{1 - \cos^2 x}{\sin x} = \frac{\sin^2 x}{\sin x} \qquad \text{[Pythagorean identity]}$$

$$= \sin x. \quad \blacksquare$$

Strategy 2

> Use algebra and identities to transform the expression on one side of the equal sign into the expression on the other side.*

EXAMPLE 3

In Example 1, we verified graphically that the equation

$$\frac{1 + \sin x - \sin^2 x}{\cos x} = \cos x + \tan x$$

might be an identity. Prove that it is.

SOLUTION We use Strategy 2, beginning with the left-hand side of the equation:

$$\begin{aligned}\frac{1 + \sin x - \sin^2 x}{\cos x} &= \frac{(1 - \sin^2 x) + \sin x}{\cos x} \\ &= \frac{\cos^2 x + \sin x}{\cos x} \qquad \text{[Pythagorean identity]} \\ &= \frac{\cos^2 x}{\cos x} + \frac{\sin x}{\cos x} \\ &= \cos x + \frac{\sin x}{\cos x} \\ &= \cos x + \tan x. \qquad \blacksquare\end{aligned}$$

The strategies presented above and those to be considered below are plans of attack. By themselves they are not much help unless you also have some techniques for carrying out these plans. In the previous examples we used the techniques of basic algebra and the use of known identities to change trigonometric expressions to equivalent ones. Here is another technique that is often useful when dealing with fractions.

Rewrite a fraction in equivalent form by multiplying its numerator and denominator by the same quantity.

EXAMPLE 4

Prove that

$$\frac{\sin x}{1 + \cos x} = \frac{1 - \cos x}{\sin x}.$$

---

*That is, start with expression A on one side and use identities and algebra to produce a sequence of equalities $A = B$, $B = C$, $C = D$, $D = E$, where E is the other side of the identity to be proved; conclude that $A = E$.

SOLUTION We shall use Strategy 2, beginning with the left-hand side, whose denominator is $1 + \cos x$. Multiply its numerator and denominator by $1 - \cos x$:*

$$\frac{\sin x}{1+\cos x} = \frac{\sin x}{1+\cos x} \cdot \frac{1-\cos x}{1-\cos x} = \frac{\sin x(1-\cos x)}{(1+\cos x)(1-\cos x)}$$

$$= \frac{\sin x(1-\cos x)}{1-\cos^2 x}$$

$$= \frac{\sin x(1-\cos x)}{\sin^2 x} \qquad \text{[Pythagorean identity]}$$

$$= \frac{1-\cos x}{\sin x}.$$

ALTERNATE SOLUTION The numerators of the given equation look similar to the Pythagorean identity—with the squares missing. So we begin with the left-hand side and introduce some squares by multiplying it by $\dfrac{\sin x}{\sin x} = 1$.

$$\frac{\sin x}{1+\cos x} = 1 \cdot \frac{\sin x}{1+\cos x} = \frac{\sin x}{\sin x} \cdot \frac{\sin x}{1+\cos x} = \frac{\sin^2 x}{\sin x(1+\cos x)}$$

$$= \frac{1-\cos^2 x}{\sin x(1+\cos x)} \qquad \text{[Pythagorean identity]}$$

$$= \frac{(1-\cos x)(1+\cos x)}{\sin x(1+\cos x)} \qquad \text{[Factor numerator]}$$

$$= \frac{1-\cos x}{\sin x}. \qquad \blacksquare$$

Strategy 3

> Deal separately with each side of the equation $A = B$. First use identities and algebra to transform A into some expression C (so that $A = C$). Then use (possibly different) identities and algebra to transform B into the *same* expression C (so that $B = C$). Conclude that $A = B$.

EXAMPLE 5

Prove that $\csc x - \cot x = \dfrac{\sin x}{1+\cos x}$.

SOLUTION We use Strategy 3, together with Strategy 1, beginning with the left-hand side.

$$(*) \qquad \csc x - \cot x = \frac{1}{\sin x} - \frac{\cos x}{\sin x} = \frac{1-\cos x}{\sin x}.$$

---

*This is analogous to the process used to rationalize the denominator of a fraction by multiplying its numerator and denominator by the conjugate of the denominator, as in the following example.

$$\frac{1}{3+\sqrt{2}} = \frac{1}{3+\sqrt{2}} \cdot \frac{3-\sqrt{2}}{3-\sqrt{2}} = \frac{3-\sqrt{2}}{3^2-(\sqrt{2})^2} = \frac{3-\sqrt{2}}{7}.$$

Example 4 shows that the right-hand side of the identity to be proved can also be transformed into this same expression.

$$(**)\qquad \frac{\sin x}{1+\cos x} = \frac{1-\cos x}{\sin x}.$$

Combining the equalities (*) and (**) proves the identity.

$$\csc x - \cot x = \frac{1-\cos x}{\sin x} = \frac{\sin x}{1+\cos x}. \qquad \blacksquare$$

Proving identities involving fractions can sometimes be quite complicated. It often helps to approach a fractional identity indirectly, as in the following example.

EXAMPLE 6

Prove these identities.

(a) $\sec x(\sec x - \cos x) = \tan^2 x$ (b) $\dfrac{\sec x}{\tan x} = \dfrac{\tan x}{\sec x - \cos x}$.

SOLUTION

(a) Beginning with the left-hand side and using Strategy 1, we have

$$\begin{aligned}
\sec x(\sec x - \cos x) &= \sec^2 x - \sec x \cos x \\
&= \sec^2 x - \frac{1}{\cos x}\cos x \\
&= \sec^2 x - 1 \\
&= \tan^2 x. \qquad \text{[Pythagorean identity]}
\end{aligned}$$

Therefore, $\sec x(\sec x - \cos x) = \tan^2 x$.

(b) By part (a), we know that

$$\sec x(\sec x - \cos x) = \tan^2 x$$

Dividing both sides of this equation by $\tan x(\sec x - \cos x)$ shows that

$$\frac{\sec x(\sec x - \cos x)}{\tan x(\sec x - \cos x)} = \frac{\tan^2 x}{\tan x(\sec x - \cos x)}$$

$$\frac{\sec x(\sec x - \cos x)}{\tan x(\sec x - \cos x)} = \frac{\tan x \tan x}{\tan x(\sec x - \cos x)}$$

$$\frac{\sec x}{\tan x} = \frac{\tan x}{\sec x - \cos x}. \qquad \blacksquare$$

Look carefully at how identity (b) was proved in Example 6. We first proved identity (a), which is of the form $AD = BC$ (with $A = \sec x$, $B = \tan x$, $C = \tan x$, and $D = \sec x - \cos x$). Then we divided both sides by BD (that is, by $\tan x(\sec x - \cos x)$ to conclude that $A/B = C/D$. The same argument works in general and provides the following useful strategy for dealing with identities involving fractions.

Strategy 4

> If you can prove that $AD = BC$, with $B \neq 0$ and $D \neq 0$, then you can conclude that
> $$\frac{A}{B} = \frac{C}{D}.$$

Many students misunderstand Strategy 4: It does *not* say that you begin with a fractional equation $A/B = C/D$ and cross-multiply to eliminate the fractions. If you did that, you would be assuming what has to be proved. What the strategy says is that to prove an identity involving fractions, you need only prove a different identity that does not involve fractions. In other words, if you prove that $AD = BD$ whenever $B \neq 0$ and $D \neq 0$, then you can conclude that $A/B = C/D$. Note that you do not *assume* that $AD = BC$; you use Strategy 2 or 3 or some other means to *prove* this statement.

EXAMPLE 7

Prove that $\dfrac{\cot x - 1}{\cot x + 1} = \dfrac{1 - \tan x}{1 + \tan x}$.

SOLUTION We use Strategy 4, with

$$A = \cot x - 1,\ B = \cot x + 1,\ C = 1 - \tan x, \text{ and } D = 1 + \tan x.$$

We must prove that this equation is an identity:

$$AD = BC$$

$$(***) \qquad (\cot x - 1)(1 + \tan x) = (\cot x + 1)(1 - \tan x).$$

Strategy 3 will be used. Multiplying out the left-hand side shows that

$$\begin{aligned}(\cot x - 1)(1 + \tan x) &= \cot x - 1 + \cot x \tan x - \tan x \\ &= \cot x - 1 + \frac{1}{\tan x}\tan x - \tan x \\ &= \cot x - 1 + 1 - \tan x \\ &= \cot x - \tan x.\end{aligned}$$

Similarly, on the right-hand side of (***),

$$\begin{aligned}(\cot x + 1)(1 - \tan x) &= \cot x + 1 - \cot x \tan x - \tan x \\ &= \cot x + 1 - 1 - \tan x \\ &= \cot x - \tan x.\end{aligned}$$

Since the left-hand and right-hand sides are equal to the same expression, we have proved that (***) is an identity. Therefore, by Strategy 4, we conclude that

$$\frac{\cot x - 1}{\cot x + 1} = \frac{1 - \tan x}{1 + \tan x}$$

is also an identity. ■

TECHNOLOGY TIP Using SOLVE in the TI-89 ALGEBRA menu to solve an equation that might be an identity produces one of three responses. "True" means the equation *probably* is an identity [algebraic proof is required for certainty]. "False" means the equation is not an identity. A numerical answer is inconclusive [the equation might or might not be an identity].

It takes a good deal of practice, as well as *much* trial and error, to become proficient in proving identities. The more practice you have, the easier it will get. Since there are many correct methods, your proofs may be quite different from those of your instructor or the text answers.

If you don't see what to do immediately, try something and see where it leads: Multiply out, or factor, or multiply numerator and denominator by the same nonzero quantity. Even if this doesn't lead anywhere, it might give you some ideas on other things to try. When you do obtain a proof, check to see whether it can be done more efficiently. In your final proof, don't include the side trips that might have given you some ideas but aren't themselves part of the proof.

EXERCISES 3.1

In Exercises 1–4, test the equation graphically to determine whether it might be an identity. You need not prove those equations that seem to be identities.

1. $\dfrac{\sec x - \cos x}{\sec x} = \sin^2 x$

2. $\tan x + \cot x = (\sin x)(\cos x)$

3. $\dfrac{1 - \cos(2x)}{2} = \sin^2 x$

4. $\dfrac{\tan x + \cot x}{\csc x} = \sec x$

In Exercises 5–8, insert one of A–F on the right of the equal sign so that the resulting equation appears to be an identity when you test it graphically. You need not prove the identity.

A. $\cos x$ **B.** $\sec x$ **C.** $\sin^2 x$

D. $\sec^2 x$ **E.** $\sin x - \cos x$ **F.** $\dfrac{1}{\sin x \cos x}$

5. $\csc x \tan x =$ ____

6. $\dfrac{\sin x}{\tan x} =$ ____

7. $\dfrac{\sin^4 x - \cos^4 x}{\sin x + \cos x} =$ ____

8. $\tan^2(-x) - \dfrac{\sin(-x)}{\sin x} =$ ____

In Exercises 9–18, prove the identity.

9. $\tan x \cos x = \sin x$

10. $\cot x \sin x = \cos x$

11. $\cos x \sec x = 1$

12. $\sin x \csc x = 1$

13. $\tan x \csc x = \sec x$

14. $\sec x \cot x = \csc x$

15. $\dfrac{\tan x}{\sec x} = \sin x$

16. $\dfrac{\cot x}{\csc x} = \cos x$

17. $(1 + \cos x)(1 - \cos x) = \sin^2 x$

18. $(\csc x - 1)(\csc x + 1) = \cot^2 x$

In Exercises 19–48, state whether or not the equation is an identity. If it is an identity, prove it.

19. $\sin x = \sqrt{1 - \cos^2 x}$

20. $\cot x = \dfrac{\csc x}{\sec x}$

21. $\dfrac{\sin(-x)}{\cos(-x)} = -\tan x$

22. $\tan x = \sqrt{\sec^2 x - 1}$

23. $\cot(-x) = -\cot x$

24. $\sec(-x) = \sec x$

25. $1 + \sec^2 x = \tan^2 x$

26. $\sec^4 x - \tan^4 x = 1 + 2\tan^2 x$

27. $\sec^2 x - \csc^2 x = \tan^2 x - \cot^2 x$

28. $\sec^2 x + \csc^2 x = \sec^2 x \csc^2 x$

29. $\sin^2 x(\cot x + 1)^2 = \cos^2 x(\tan x + 1)^2$

30. $\cos^2 x(\sec x + 1)^2 = (1 + \cos x)^2$

31. $\sin^2 x - \tan^2 x = -\sin^2 x \tan^2 x$

32. $\cot^2 x - 1 = \csc^2 x$

33. $(\cos^2 x - 1)(\tan^2 x + 1) = -\tan^2 x$

34. $(1 - \cos^2 x)\csc x = \sin x$

35. $\tan x = \dfrac{\sec x}{\csc x}$

36. $\dfrac{\cos(-x)}{\sin(-x)} = -\cot x$

37. $\cos^4 x - \sin^4 x = \cos^2 x - \sin^2 x$

38. $\cot^2 x - \cos^2 x = \cos^2 x \cot^2 x$

39. $(\sin x + \cos x)^2 = \sin^2 x + \cos^2 x$

40. $(1 + \tan x)^2 = \sec^2 x$

41. $\dfrac{\sec x}{\csc x} + \dfrac{\sin x}{\cos x} = 2 \tan x$

42. $\dfrac{1 + \cos x}{\sin x} + \dfrac{\sin x}{1 + \cos x} = 2 \csc x$

43. $\dfrac{\sec x + \csc x}{1 + \tan x} = \csc x$

44. $\dfrac{\cot x - 1}{1 - \tan x} = \dfrac{\csc x}{\sec x}$

45. $\dfrac{1}{\csc x - \sin x} = \sec x \tan x$

46. $\dfrac{1 + \csc x}{\csc x} = \dfrac{\cos^2 x}{1 - \sin x}$

47. $\dfrac{\sin x \quad \cos x}{\tan x} = \dfrac{\tan x}{\sin x + \cos x}$

48. $\dfrac{\cot x}{\csc x - 1} = \dfrac{\csc x + 1}{\cot x}$

In Exercises 49–52, half of an identity is given. Graph this half in a viewing window with $-2\pi \le x \le 2\pi$ and make a conjecture as to what the right-hand side of the identity is. Then prove your conjecture.

49. $1 - \dfrac{\sin^2 x}{1 + \cos x} = ?$ [*Hint:* What familiar function has a graph that looks like this?]

50. $\dfrac{1 + \cos x - \cos^2 x}{\sin x} - \cot x = ?$

51. $(\sin x + \cos x)(\sec x + \csc x) - \cot x - 2 = ?$

52. $\cos^3 x(1 - \tan^4 x + \sec^4 x) = ?$

In Exercises 53–58, prove the identity.

53. $\dfrac{1 - \sin x}{\sec x} = \dfrac{\cos^3 x}{1 + \sin x}$

54. $\dfrac{\sin x}{1 - \cot x} + \dfrac{\cos x}{1 - \tan x} = \cos x + \sin x$

55. $\dfrac{\cos x}{1 - \sin x} = \sec x + \tan x$

56. $\dfrac{1 + \sec x}{\tan x + \sin x} = \csc x$

57. $\dfrac{\cos x \cot x}{\cot x - \cos x} = \dfrac{\cot x + \cos x}{\cos x \cot x}$

58. $\dfrac{\cos^3 x - \sin^3 x}{\cos x - \sin x} = 1 + \sin x \cos x$

Exercises 59–62 deal with parametric graphing (which was explained at the end of Section 0.5).

59. (a) Graph the curve given by

$$x = \cos t + 2 \qquad \text{and} \qquad y = \sin t + 4 \qquad (0 \le t \le 2\pi).$$

(b) Use an identity to show that every point on this curve satisfies the equation

$$(x - 2)^2 + (y - 4)^2 = 1.$$

(c) Describe the graph in words. [*Hint:* Circular thinking will help.]

60. Without graphing, show that the graph of the parametric curve

$$x = 3 \cos t + 1 \qquad \text{and} \qquad y = 3 \sin t - 2 \qquad (0 \le t \le 2\pi)$$

is a circle of radius 3 with center at $(1, -2)$.

61. Use identities to show that every point on the parametric curve

$$x = \frac{2}{\cos t} \qquad \text{and} \qquad y = 4 \tan t \qquad (0 \le t \le 2\pi)$$

satisfies the equation

$$\frac{x^2}{4} - \frac{y^2}{16} = 1.$$

[*Note:* On TI calculators, the graph of the parametric curve may include some erroneous straight lines that are *not* part of the graph.]

62. Find an equation in x and y that is satisfied by every point on the parametric curve

$$x = 5 \tan t \qquad \text{and} \qquad y = \frac{3}{\cos t} \qquad (0 \le t \le 2\pi).$$

3.2 Addition and Subtraction Identities

A common student ERROR is to write

$$\sin\left(x + \frac{\pi}{6}\right) = \sin x + \sin\frac{\pi}{6} = \sin x + \frac{1}{2}.$$

GRAPHING EXPLORATION

Verify graphically that the equation above is NOT an identity by graphing $y = \sin(x + \pi/6)$ and $y = \sin x + 1/2$ on the same screen.

The exploration shows that "$\sin(x + y) = \sin x + \sin y$" is NOT an identity (because it's false when $y = \pi/6$). There is an identity that relates these quantities, but it is a bit more complicated, as we now see.

Addition and Subtraction Identities

$$\sin(x + y) = \sin x \cos y + \cos x \sin y$$

$$\sin(x - y) = \sin x \cos y - \cos x \sin y$$

$$\cos(x + y) = \cos x \cos y - \sin x \sin y$$

$$\cos(x - y) = \cos x \cos y + \sin x \sin y$$

The addition and subtraction identities are probably the most important of all the trigonometric identities. Before reading their proofs at the end of this section, you should become familiar with the examples and special cases below.

EXAMPLE 1 

Use the addition identities to find the *exact* values of $\sin(5\pi/12)$ and $\cos(5\pi/12)$.

SOLUTION Since

$$\frac{5\pi}{12} = \frac{2\pi}{12} + \frac{3\pi}{12} = \frac{\pi}{6} + \frac{\pi}{4},$$

we apply the addition identities with $x = \pi/6$ and $y = \pi/4$.

$$\sin\frac{5\pi}{12} = \sin\left(\frac{\pi}{6} + \frac{\pi}{4}\right) = \sin\frac{\pi}{6}\cos\frac{\pi}{4} + \cos\frac{\pi}{6}\sin\frac{\pi}{4}$$

$$= \frac{1}{2}\cdot\frac{\sqrt{2}}{2} + \frac{\sqrt{3}}{2}\cdot\frac{\sqrt{2}}{2} = \frac{\sqrt{2}(\sqrt{3} + 1)}{4},$$

$$\cos\frac{5\pi}{12} = \cos\left(\frac{\pi}{6} + \frac{\pi}{4}\right) = \cos\frac{\pi}{6}\cos\frac{\pi}{4} - \sin\frac{\pi}{6}\sin\frac{\pi}{4}$$

$$= \frac{\sqrt{3}}{2}\cdot\frac{\sqrt{2}}{2} - \frac{1}{2}\cdot\frac{\sqrt{2}}{2} = \frac{\sqrt{2}(\sqrt{3} - 1)}{4}.$$ ■

EXAMPLE 2

Find $\sin(\pi - y)$.

SOLUTION Apply the subtraction identity with $x = \pi$.

$$\begin{aligned}\sin(\pi - y) &= \sin \pi \cos y - \cos \pi \sin y \\ &= (0)(\cos y) - (-1)(\sin y) \\ &= \sin y. \quad \blacksquare\end{aligned}$$

C

EXAMPLE 3*

Show that for the function $f(x) = \sin x$ and any number $h \neq 0$,

$$\frac{f(x+h) - f(x)}{h} = \sin x\left(\frac{\cos h - 1}{h}\right) + \cos x\left(\frac{\sin h}{h}\right).$$

SOLUTION Use the addition identity for $\sin(x + y)$ with $y = h$.

$$\begin{aligned}\frac{f(x+h) - f(x)}{h} &= \frac{\sin(x+h) - \sin x}{h} \\ &= \frac{\sin x \cos h + \cos x \sin h - \sin x}{h} \\ &= \frac{\sin x(\cos h - 1) + \cos x \sin h}{h} \\ &= \sin x\left(\frac{\cos h - 1}{h}\right) + \cos x\left(\frac{\sin h}{h}\right). \quad \blacksquare\end{aligned}$$

EXAMPLE 4

Prove that

$$\cos x \cos y = \frac{1}{2}[\cos(x+y) + \cos(x-y)].$$

SOLUTION We begin with the more complicated right-hand side and use the addition and subtraction identities for cosine to transform it into the left-hand side.

$$\begin{aligned}\frac{1}{2}[\cos(x+y) + \cos(x-y)] &= \frac{1}{2}[(\cos x \cos y - \sin x \sin y) \\ &\qquad + (\cos x \cos y + \sin x \sin y)] \\ &= \frac{1}{2}(\cos x \cos y + \cos x \cos y) \\ &= \frac{1}{2}(2 \cos x \cos y) = \cos x \cos y. \quad \blacksquare\end{aligned}$$

---

* C and ◯ indicate examples, exercises, and sections that are relevant to calculus.

The addition and subtraction identities for sine and cosine can be used to obtain the following identities, as outlined in Exercise 36.

Addition and Subtraction Identities for Tangent

$$\tan(x+y) = \frac{\tan x + \tan y}{1 - \tan x \tan y}$$

$$\tan(x-y) = \frac{\tan x - \tan y}{1 + \tan x \tan y}$$

It is sometimes convenient to say that x is a number *in the first quadrant* if $0 < x < \pi/2$, that x is a number *in the second quadrant* if $\pi/2 < x < \pi$, and so on.

EXAMPLE 5

Suppose x is a number in the first quadrant and y is a number in the third quadrant. If $\sin x = 3/4$ and $\cos y = -1/3$, find the exact values of $\sin(x+y)$ and $\tan(x+y)$, and determine in which quadrant $x+y$ lies.

SOLUTION Using the Pythagorean identity and the fact that $\cos x$ and $\tan x$ are positive when $0 < x < \pi/2$, we have

$$\cos x = \sqrt{1-\sin^2 x} = \sqrt{1-\left(\frac{3}{4}\right)^2} = \sqrt{1-\frac{9}{16}} = \sqrt{\frac{7}{16}} = \frac{\sqrt{7}}{4}$$

$$\tan x = \frac{\sin x}{\cos x} = \frac{3/4}{\sqrt{7}/4} = \frac{3}{4}\cdot\frac{4}{\sqrt{7}} = \frac{3}{\sqrt{7}} = \frac{3\sqrt{7}}{7}.$$

Since y lies between π and $3\pi/2$, its sine is negative; hence,

$$\sin y = -\sqrt{1-\cos^2 y} = \sqrt{1-\left(-\frac{1}{3}\right)^2} = -\sqrt{\frac{8}{9}} = -\frac{\sqrt{8}}{3} = -\frac{2\sqrt{2}}{3}$$

$$\tan y = \frac{\sin y}{\cos y} = \frac{-2\sqrt{2}/3}{-1/3} = \frac{-2\sqrt{2}}{3}\cdot\frac{3}{-1} = 2\sqrt{2}.$$

The addition identities for sine and tangent now show that

$$\begin{aligned}\sin(x+y) &= \sin x \cos y + \cos x \sin y \\ &= \frac{3}{4}\cdot\frac{-1}{3} + \frac{\sqrt{7}}{4}\cdot\frac{-2\sqrt{2}}{3} = \frac{-3}{12} - \frac{2\sqrt{14}}{12} = \frac{-3-2\sqrt{14}}{12}\end{aligned}$$

$$\begin{aligned}\tan(x+y) &= \frac{\tan x + \tan y}{1 - \tan x \tan y} \\ &= \frac{\dfrac{3\sqrt{7}}{7} + 2\sqrt{2}}{1 - \left(\dfrac{3\sqrt{7}}{7}\right)(2\sqrt{2})} = \frac{\dfrac{3\sqrt{7}+14\sqrt{2}}{7}}{\dfrac{7-6\sqrt{14}}{7}} = \frac{3\sqrt{7}+14\sqrt{2}}{7-6\sqrt{14}}.\end{aligned}$$

So both the sine and tangent of $x+y$ are negative numbers. Therefore, $x+y$ must be in the interval $(3\pi/2, 2\pi)$, since the sign chart in Exercise 53 on page 156 shows that this is the only one of the four quadrants in which both sine and tangent are negative. ■

COFUNCTION IDENTITIES

Other special cases of the addition and subtraction identities are the cofunction identities.

Cofunction Identities

$$\sin x = \cos\left(\frac{\pi}{2} - x\right) \qquad \cos x = \sin\left(\frac{\pi}{2} - x\right)$$

$$\tan x = \cot\left(\frac{\pi}{2} - x\right) \qquad \cot x = \tan\left(\frac{\pi}{2} - x\right)$$

$$\sec x = \csc\left(\frac{\pi}{2} - x\right) \qquad \csc x = \sec\left(\frac{\pi}{2} - x\right)$$

The first confunction identity is proved by using the identity for $\cos(x - y)$ with $\pi/2$ in place of x and x in place of y.

$$\cos\left(\frac{\pi}{2} - x\right) = \cos\frac{\pi}{2}\cos x + \sin\frac{\pi}{2}\sin x = (0)(\cos x) + (1)(\sin x) = \sin x.$$

Since the first cofunction identity is valid for *every* number x, it is also valid with the number $\pi/2 - x$ in place of x.

$$\sin\left(\frac{\pi}{2} - x\right) = \cos\left[\frac{\pi}{2} - \left(\frac{\pi}{2} - x\right)\right] = \cos x.$$

Thus, we have proved the second cofunction identity. The others now follow from these two. For instance,

$$\tan\left(\frac{\pi}{2} - x\right) = \frac{\sin[(\pi/2) - x]}{\cos[(\pi/2) - x]} = \frac{\cos x}{\sin x} = \cot x.$$

EXAMPLE 6 

Verify that $\dfrac{\cos(x - \pi/2)}{\cos x} = \tan x$.

SOLUTION Beginning on the left-hand side, we see that the term $\cos(x - \pi/2)$ looks almost, but not quite, like the term $\cos(\pi/2 - x)$ in the cofunction identity. But note that $-(x - \pi/2) = \pi/2 - x$. Therefore,

$$\frac{\cos\left(x - \frac{\pi}{2}\right)}{\cos x} = \frac{\cos\left[-\left(x - \frac{\pi}{2}\right)\right]}{\cos x} \qquad \text{[Negative angle identity with } x - \tfrac{\pi}{2} \text{ in place of } x\text{]}$$

$$= \frac{\cos\left(\frac{\pi}{2} - x\right)}{\cos x}$$

$$= \frac{\sin x}{\cos x} \qquad \text{[Cofunction identity]}$$

$$= \tan x. \qquad \text{[Reciprocal identity]}$$

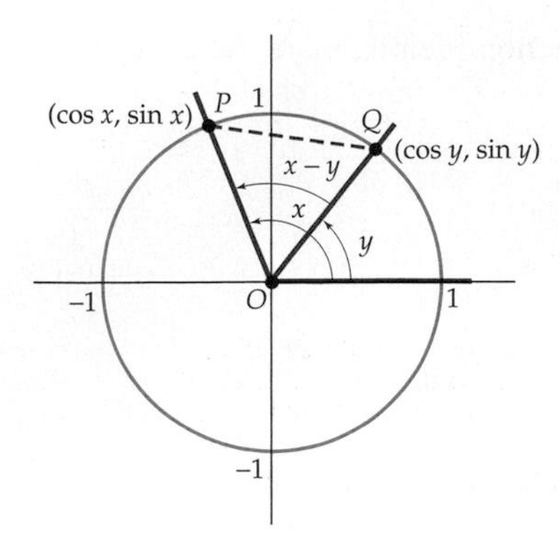

Figure 3–2

PROOF OF THE ADDITION AND SUBTRACTION IDENTITIES

We first prove that

$$\cos(x - y) = \cos x \cos y + \sin x \sin y.$$

If $x = y$, then it is true by the Pythagorean identity:

$$\cos(x - x) = \cos 0 = 1 = \cos^2 x + \sin^2 x = \cos x \cos x + \sin x \sin x.$$

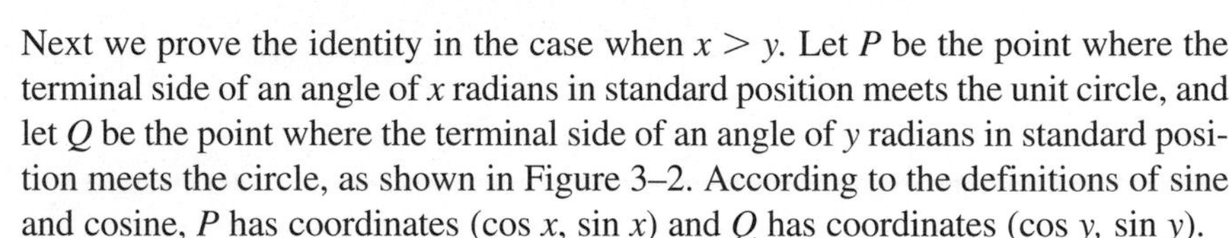

Next we prove the identity in the case when $x > y$. Let P be the point where the terminal side of an angle of x radians in standard position meets the unit circle, and let Q be the point where the terminal side of an angle of y radians in standard position meets the circle, as shown in Figure 3–2. According to the definitions of sine and cosine, P has coordinates $(\cos x, \sin x)$ and Q has coordinates $(\cos y, \sin y)$.

The angle QOP formed by the two terminal sides has radian measure $x - y$. Rotate this angle clockwise until side OQ lies on the horizontal axis, as shown in Figure 3–3. Angle QOP is now in standard position, and its terminal side meets the unit circle at P. Since angle QOP has radian measure $x - y$, the definitions of sine and cosine show that the point P, in this new location, has coordinates $(\cos(x - y), \sin(x - y))$. Q now has coordinates $(1, 0)$.

Using the coordinates of P and Q *before* the angle was rotated and the distance formula, we have

Distance from P to Q

$$= \sqrt{(\cos x - \cos y)^2 + (\sin x - \sin y)^2}$$

$$= \sqrt{\cos^2 x - 2\cos x \cos y + \cos^2 y + \sin^2 x - 2\sin x \sin y + \sin^2 y}$$

$$= \sqrt{(\cos^2 x + \sin^2 x) + (\cos^2 y + \sin^2 y) - 2\cos x \cos y - 2\sin x \sin y}$$

$$= \sqrt{1 + 1 - 2\cos x \cos y - 2\sin x \sin y} \qquad \text{[Pythagorean identity]}$$

$$= \sqrt{2 - 2\cos x \cos y - 2\sin x \sin y}.$$

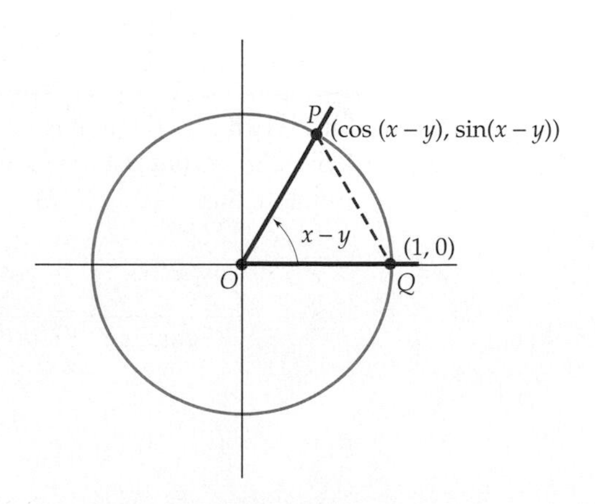

Figure 3–3

But using the coordinates of P and Q *after* the angle is rotated shows that

$$\begin{aligned}
\text{Distance from } P \text{ to } Q &= \sqrt{[\cos(x-y)-1]^2+[\sin(x-y)-0]^2} \\
&= \sqrt{\cos^2(x-y)-2\cos(x-y)+1+\sin^2(x-y)} \\
&= \sqrt{\cos^2(x-y)+\sin^2(x-y)-2\cos(x-y)+1} \\
&= \sqrt{1-2\cos(x-y)+1} \qquad \text{[Pythagorean identity]} \\
&= \sqrt{2-2\cos(x-y)}.
\end{aligned}$$

The two expressions for the distance from P to Q must be equal. Hence,

$$\sqrt{2-2\cos(x-y)} = \sqrt{2-2\cos x\cos y-2\sin x\sin y}.$$

Squaring both sides of this equation and simplifying the result yield

$$\begin{aligned}
2-2\cos(x-y) &= 2-2\cos x\cos y-2\sin x\sin y \\
-2\cos(x-y) &= -2(\cos x\cos y+\sin x\sin y) \\
\cos(x-y) &= \cos x\cos y+\sin x\sin y.
\end{aligned}$$

This completes the proof of the last addition identity when $x > y$. If $y > x$, then the proof just given is valid with the roles of x and y interchanged; it shows that

$$\begin{aligned}
\cos(y-x) &= \cos y\cos x+\sin y\sin x \\
&= \cos x\cos y+\sin x\sin y.
\end{aligned}$$

The negative angle identity with $x - y$ in place of x shows that

$$\cos(x-y) = \cos[-(x-y)] = \cos(y-x).$$

Combining this fact with the previous one shows that

$$\cos(x-y) = \cos x\cos y+\sin x\sin y$$

in this case also. Therefore, the last addition identity is proved.

The identity for $\cos(x + y)$ now follows from the one for $\cos(x - y)$ by using the negative angle identities for sine and cosine:

$$\begin{aligned}
\cos(x+y) = \cos[x-(-y)] &= \cos x\cos(-y)+\sin x\sin(-y) \\
&= \cos x\cos y+(\sin x)(-\sin y) = \cos x\cos y-\sin x\sin y.
\end{aligned}$$

The proof of the first two cofunction identities on page 233 depended only on the addition identity for $\cos(x - y)$. Since that has been proved, we can validly use the first two cofunction identities in the remainder of the proof. In particular,

$$\sin(x-y) = \cos\left[\frac{\pi}{2}-(x-y)\right] = \cos\left[\left(\frac{\pi}{2}-x\right)+y\right].$$

Applying the proven identity for $\cos(x + y)$ with $(\pi/2) - x$ in place of x and the two cofunction identities now yields

$$\begin{aligned}
\sin(x-y) &= \cos\left[\left(\frac{\pi}{2}-x\right)+y\right] \\
&= \cos\left(\frac{\pi}{2}-x\right)\cos y-\sin\left(\frac{\pi}{2}-x\right)\sin y \\
&= \sin x\cos y-\cos x\sin y.
\end{aligned}$$

This proves the second of the addition and subtraction identities. The first is obtained from the second in the same way that the third was obtained from the last.

EXERCISES 3.2

In Exercises 1–12, find the exact value.

1. $\sin \frac{\pi}{12}$ **2.** $\cos \frac{\pi}{12}$ **3.** $\tan \frac{\pi}{12}$

4. $\sin \frac{5\pi}{12}$ **5.** $\cot \frac{5\pi}{12}$ **6.** $\cos \frac{7\pi}{12}$

7. $\tan \frac{7\pi}{12}$ **8.** $\cos \frac{11\pi}{12}$ **9.** $\cot \frac{11\pi}{12}$

10. $\sin 75°$ [*Hint:* $75° = 45° + 30°$.]

11. $\sin 105°$ **12.** $\cos 165°$

In Exercises 13–18, rewrite the given expression in terms of $\sin x$ *and* $\cos x$.

13. $\sin\left(\frac{\pi}{2} + x\right)$ **14.** $\cos\left(x + \frac{\pi}{2}\right)$

15. $\cos\left(x - \frac{3\pi}{2}\right)$ **16.** $\csc\left(x + \frac{\pi}{2}\right)$

17. $\sec(x - \pi)$ **18.** $\cot(x + \pi)$

In Exercises 19–24, simplify the given expression.

19. $\sin 3 \cos 5 - \cos 3 \sin 5$

20. $\sin 37° \sin 53° - \cos 37° \cos 53°$

21. $\cos(x + y) \cos y + \sin(x + y) \sin y$

22. $\sin(x - y) \cos y + \cos(x - y) \sin y$

23. $\cos(x + y) - \cos(x - y)$

24. $\sin(x + y) - \sin(x - y)$

25. If $\sin x = \frac{1}{3}$ and $0 < x < \frac{\pi}{2}$, then $\sin\left(\frac{\pi}{4} + x\right) = ?$

26. If $\cos x = -\frac{1}{4}$ and $\frac{\pi}{2} < x < \pi$, then $\cos\left(\frac{\pi}{6} - x\right) = ?$

27. If $\cos x = -\frac{1}{5}$ and $\pi < x < \frac{3\pi}{2}$, then $\sin\left(\frac{\pi}{3} - x\right) = ?$

28. If $\sin x = -\frac{3}{4}$ and $\frac{3\pi}{2} < x < 2\pi$, then $\cos\left(\frac{\pi}{4} + x\right) = ?$

In Exercises 29–32, assume that $\sin x = .8$ *and* $\sin y = \sqrt{.75}$ *and that* x *and* y *lie between* 0 *and* $\pi/2$. *Evaluate the given expressions.*

29. $\sin(x + y)$ **30.** $\cos(x - y)$

31. $\sin(x - y)$ **32.** $\tan(x + y)$

33. The figure shows an angle of t radians. Prove that for any number x,

$$5 \sin(x + t) = 3 \sin x + 4 \cos x.$$

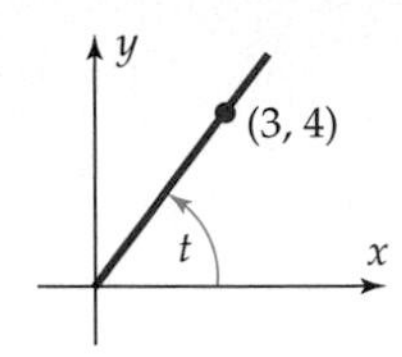

34. The figure shows an angle of t radians. Prove that for any number y,

$$13 \cos(t - y) = 12 \cos y + 5 \sin y.$$

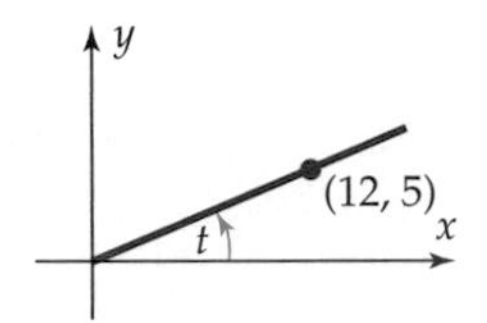

35. If $f(x) = \cos x$ and h is a fixed nonzero number, prove that

$$\frac{f(x + h) - f(x)}{h} = \cos x\left(\frac{\cos h - 1}{h}\right) - \sin x\left(\frac{\sin h}{h}\right).$$

36. Prove the addition and subtraction identities for the tangent function (page 232). [*Hint:*

$$\tan(x + y) = \frac{\sin(x + y)}{\cos(x + y)}.$$

Use the addition identities on the numerator and denominator; then divide both numerator and denominator by $\cos x \cos y$ and simplify.]

37. If x is in the first quadrant, y is in the second quadrant, $\sin x = 24/25$, and $\sin y = 4/5$, find the exact value of $\sin(x + y)$ and $\tan(x + y)$ and the quadrant in which $x + y$ lies.

38. If x and y are in the second quadrant, $\sin x = 1/3$, and $\cos y = -3/4$, find the exact value of $\sin(x + y)$, $\cos(x + y)$, $\tan(x + y)$, and find the quadrant in which $x + y$ lies.

39. If x is in the first quadrant, y is in the second quadrant, $\sin x = 4/5$, and $\cos y = -12/13$, find the exact value of $\cos(x + y)$ and $\tan(x + y)$ and the quadrant in which $x + y$ lies.

40. If x is in the fourth quadrant, y is in the first quadrant, $\cos x = 1/3$, and $\cos y = 2/3$, find the exact value of $\sin(x - y)$ and $\tan(x - y)$ and the quadrant in which $x - y$ lies.

41. Express $\sin(u + v + w)$ in terms of sines and cosines of u, v, and w. [*Hint:* First apply the addition identity with $x = u + v$ and $y = w$.]

42. Express $\cos(x + y + z)$ in terms of sines and cosines of x, y, and z.

43. If $x + y = \pi/2$, show that $\sin^2 x + \sin^2 y = 1$.

44. Prove that $\cot(x + y) = \dfrac{\cot x \cot y - 1}{\cot x + \cot y}$.

In Exercises 45–53, prove the identity.

45. $\sin(x - \pi) = -\sin x$

46. $\cos(x - \pi) = -\cos x$

47. $\cos(\pi - x) = -\cos x$

48. $\tan(\pi - x) = -\tan x$

49. $\sin(x + \pi) = -\sin x$

50. $\cos(x + \pi) = -\cos x$

51. $\tan(x + \pi) = \tan x$

52. $\cos(x + y)\cos(x - y) = \cos^2 x \cos^2 y - \sin^2 x \sin^2 y$

53. $\sin(x + y)\sin(x - y) = \sin^2 x \cos^2 y - \cos^2 x \sin^2 y$

In Exercises 54–56, use the method of Example 4 to prove the identity.

54. $\sin x \cos y = \frac{1}{2}[\sin(x + y) + \sin(x - y)]$

55. $\sin x \sin y = \frac{1}{2}[\cos(x - y) - \cos(x + y)]$

56. $\cos x \sin y = \frac{1}{2}[\sin(x + y) - \sin(x - y)]$

In Exercises 57–60, determine graphically whether the equation could possibly be an identity (by choosing a numerical value for y and graphing both sides). If it could, prove that it is.

57. $\dfrac{\cos(x - y)}{\sin x \cos y} = \cot x + \tan y$

58. $\dfrac{\cos(x + y)}{\sin x \cos y} = \cot x - \tan y$

59. $\sin(x - y) = \sin x - \sin y$

60. $\cos(x + y) = \cos x + \cos y$

3.2.A SPECIAL TOPICS Lines and Angles

If L is a nonhorizontal straight line, the **angle of inclination** of L is the angle θ formed by the part of L above the x-axis and the x-axis in the positive direction, as shown in Figure 3–4.

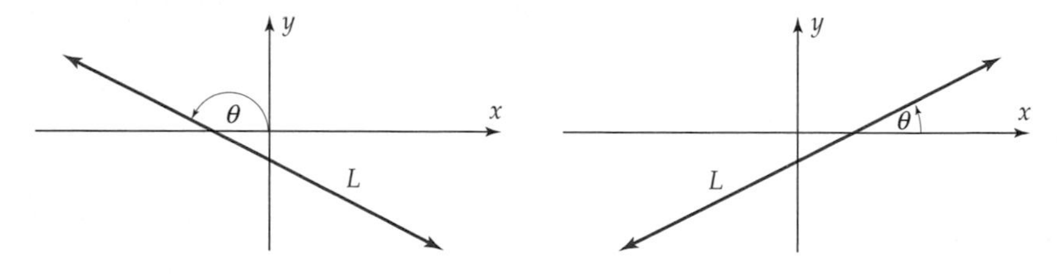

Figure 3–4

The angle of inclination of a horizontal line is defined to be $\theta = 0$. Thus, the radian measure of the angle of inclination of any line satisfies $0 \le \theta < \pi$. Furthermore,

Angle of Inclination

> If L is a nonvertical line with an angle of inclination of θ radians, then
>
> $$\tan \theta = \text{slope of } L.$$

Proof First, suppose L is horizontal. Then L has slope 0 and angle of inclination $\theta = 0$. Hence,

$$\tan \theta = \tan 0 = 0 = \text{slope } L.$$

Next, suppose L is not horizontal. L is parallel to a line M through the origin, as shown in Figure 3–5.*

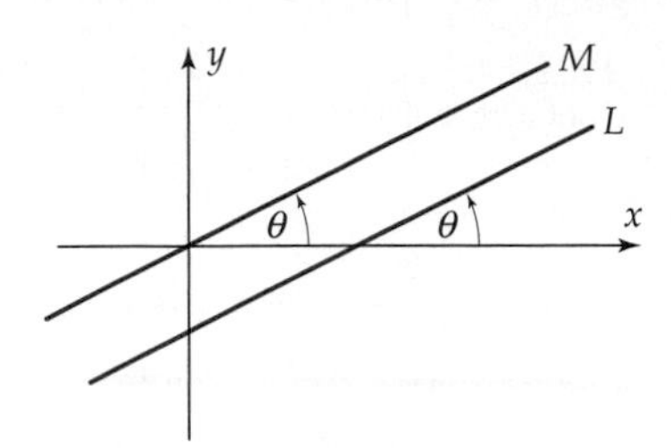

Figure 3–5

Basic facts about parallel lines show that M has the same angle of inclination θ as L. Furthermore, M lies on the terminal side of an angle of θ radians in standard position. Therefore, as we proved in Section 2.4,

$$\text{slope } M = \tan \theta.$$

Since parallel lines have the same slope, we have

$$\text{slope } L = \text{slope } M = \tan \theta. \qquad \blacksquare$$

ANGLES BETWEEN TWO LINES

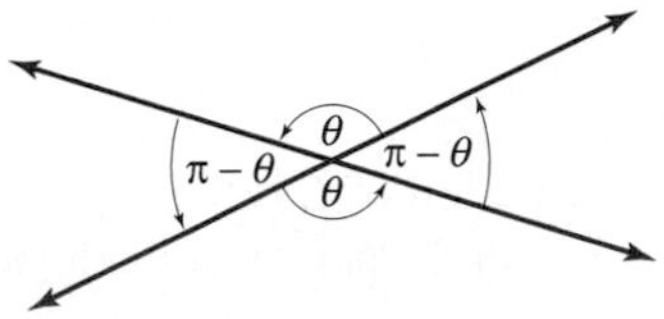

Figure 3–6

If two lines intersect, then they determine four angles with vertex at the point of intersection, as shown in Figure 3–6. If one of these angles measures θ radians, then each of the two angles adjacent to it measures $\pi - \theta$ radians. (Why?) The fourth angle also measures θ radians by the vertical angle theorem from plane geometry.

The angles between intersecting lines can be determined from the angles of inclination of the lines. Suppose L and M have angles of inclination α and β, respectively, such that $\beta \geq \alpha$. Basic facts about parallel lines, as illustrated in Figure 3–7, show that $\beta - \alpha$ is one angle between L and M and $\pi - (\beta - \alpha)$ is the other one.

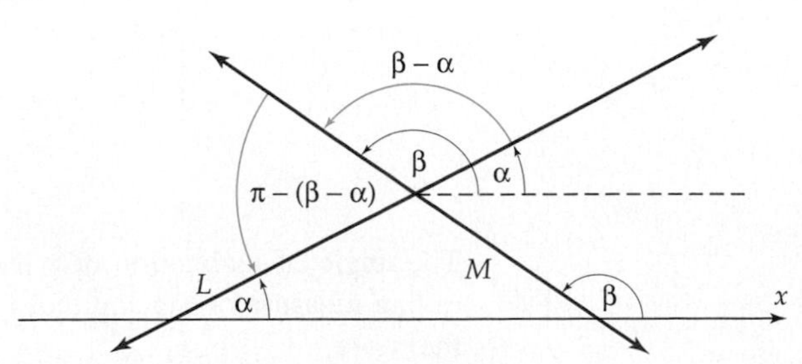

Figure 3–7

*Figure 3–5 illustrates the case when θ is acute and L lies to the right of M. The pictures are different in the other possible cases, but the argument is the same.

The angle between two lines can also be found from their slopes by using this fact.

Angle Between Two Lines

> If two nonvertical, nonperpendicular lines have slopes m and k, then one angle θ between them satisfies
>
> $$\tan\theta = \left|\frac{m-k}{1+mk}\right|.$$

Proof Suppose the line with slope k has angle of inclination α and the line with slope m has angle of inclination β. Then

$$\tan\alpha = k \qquad \text{and} \qquad \tan\beta = m.$$

If $\beta \geq \alpha$, then $\theta = \beta - \alpha$ is one angle between the lines. By the subtraction identity for tangent,

$$\tan\theta = \tan(\beta - \alpha) = \frac{\tan\beta - \tan\alpha}{1 + \tan\beta\tan\alpha} = \frac{m-k}{1+mk}.$$

The other angle between the lines is $\pi - \theta$, and

$$\begin{aligned} \tan(\pi - \theta) &= \tan(-\theta) && \text{[tangent has period } \pi\text{]} \\ &= -\tan\theta && \text{[negative angle identity]} \\ &= -\frac{m-k}{1+mk}. \end{aligned}$$

By the definition of absolute value,

$$\left|\frac{m-k}{1+mk}\right| = \pm\frac{m-k}{1+mk},$$

whichever is positive. Thus, the tangent of one of the angles θ or $\pi - \theta$ is

$$\left|\frac{m-k}{1+mk}\right|.$$

This completes the proof when $\beta \geq \alpha$. The proof when $\alpha \geq \beta$ is similar. ■

EXAMPLE 1 

Find the angle between a line L with slope 8 and a line M of slope -3.

SOLUTION One angle θ between the lines satisfies

$$\tan\theta = \left|\frac{8-(-3)}{1+8(-3)}\right| = \left|\frac{11}{-23}\right| = \frac{11}{23}.$$

Although you could solve the equation $\tan\theta = 11/23$ graphically, it is easier to use the TAN^{-1} key on your calculator. When you key in $\text{TAN}^{-1}(11/23)$, the calculator displays an angle between 0 and $\pi/2$ radians such that $\tan\theta = 11/23$, as in Figure 3–8.* Therefore, $\theta \approx .4461$ radians. ■

```
tan-1(11/23)
     .4461055489
```

Figure 3–8

---

*$\text{Tan}^{-1}(x)$ denotes the inverse tangent function, which is explained in Section 3.4. In this context, using the TAN^{-1} key is the electronic equivalent of searching through a table of tangent values until you find a number whose tangent is 11/23.

We can now prove the following fact, which is mentioned in the Geometry Review Appendix.

Slope Theorem for Perpendicular Lines

> Let L be a line with slope k and M a line with slope m. Then
>
> L and M are perpendicular exactly when $km = -1$.

Proof First, suppose L and M are perpendicular. We must show that $km = -1$. If α and β (with $\beta \geq \alpha$) are the angles of inclination of L and M, then $\beta - \alpha$ is the angle between L and M, so $\beta - \alpha = \pi/2$, or, equivalently, $\beta = \alpha + \pi/2$. Therefore, by the addition identities for sine and cosine,

$$m = \tan \beta = \tan\left(\alpha + \frac{\pi}{2}\right) = \frac{\sin[\alpha + (\pi/2)]}{\cos[\alpha + (\pi/2)]}$$

$$= \frac{\sin \alpha \cos(\pi/2) + \cos \alpha \sin(\pi/2)}{\cos \alpha \cos(\pi/2) - \sin \alpha \sin(\pi/2)}$$

$$= \frac{\sin \alpha(0) + \cos \alpha(1)}{\cos \alpha(0) - \sin \alpha(1)}$$

$$= -\frac{\cos \alpha}{\sin \alpha} = -\cot \alpha = \frac{-1}{\tan \alpha} = \frac{-1}{k}.$$

Thus, $m = -1/k$, and hence, $mk = -1$.

Now suppose that $mk = -1$. We must show that L and M are perpendicular. If L and M are *not* perpendicular, then neither of the angles between them is $\pi/2$. In this case, if θ is either of the angles between L and M, then $\tan \theta$ is a well-defined real number. But we know that one of these angles must satisfy

$$\tan \theta = \left|\frac{m - k}{1 + mk}\right| = \left|\frac{m - k}{1 + (-1)}\right| = \frac{|m - k|}{0},$$

which is *not* defined. This contradiction shows that L and M must be perpendicular. ■

✓ EXERCISES 3.2.A

In Exercises 1–6, find $\tan \theta$, *where* θ *is the angle of inclination of the line through the given points.*

1. $(-1, 2), (3, 5)$

2. $(0, 4), (5, -1)$

3. $(1, 4), (6, 0)$

4. $(4, 2), (-3, -2)$

5. $(3, -7), (3, 5)$

6. $(0, 0), (-4, -5)$

In Exercises 7–13, find one of the angles between the straight lines L and M.

7. L has slope $3/2$ and M has slope -1.

8. L has slope 1 and M has slope 3.

9. L has slope -1 and M has slope 0.

10. L has slope -2 and M has slope -3.

11. $(3, 2)$ and $(5, 6)$ are on L; $(0, 3)$ and $(4, 0)$ are on M.

12. $(-1, 2)$ and $(3, -3)$ are on L; $(3, -3)$ and $(6, 1)$ are on M.

13. L is parallel to the line with equation $y = 3x - 2$ and M is perpendicular to the line with equation $y = -.5x + 1$.

14. If θ is an angle between two nonperpendicular lines with slopes m and k, respectively, and $\tan \theta = \left|\frac{m - k}{1 + mk}\right|$, explain why θ is an acute angle. [*Hint*: For what values of θ is $\tan \theta$ positive?]

3.3 Other Identities

We now present a variety of identities that are special cases of the addition and subtraction identities of Section 3.2, beginning with

Double-Angle Identities

$$\sin 2x = 2\sin x \cos x$$

$$\cos 2x = \cos^2 x - \sin^2 x$$

$$\tan 2x = \frac{2\tan x}{1 - \tan^2 x}$$

Proof Let $x = y$ in the addition identities.

$$\sin 2x = \sin(x + x) = \sin x \cos x + \cos x \sin x = 2\sin x \cos x$$

$$\cos 2x = \cos(x + x) = \cos x \cos x - \sin x \sin x = \cos^2 x - \sin^2 x$$

$$\tan 2x = \tan(x + x) = \frac{\tan x + \tan x}{1 - \tan x \tan x} = \frac{2\tan x}{1 - \tan^2 x}.$$ ■

EXAMPLE 1

If $\pi < x < 3\pi/2$ and $\cos x = -8/17$, find $\sin 2x$ and $\cos 2x$, and show that $5\pi/2 < 2x < 3\pi$.

SOLUTION To use the double-angle identities, we first must determine $\sin x$. It can be found by using the Pythagorean identity.

$$\sin^2 x = 1 - \cos^2 x = 1 - \left(-\frac{8}{17}\right)^2 = 1 - \frac{64}{289} = \frac{225}{289}.$$

Since $\pi < x < 3\pi/2$, we know $\sin x$ is negative. Therefore,

$$\sin x = -\sqrt{\frac{225}{289}} = -\frac{15}{17}.$$

We now substitute these values in the double-angle identities.

$$\sin 2x = 2\sin x \cos x = 2\left(-\frac{15}{17}\right)\left(-\frac{8}{17}\right) = \frac{240}{289} \approx .83$$

$$\cos 2x = \cos^2 x - \sin^2 x = \left(-\frac{8}{17}\right)^2 - \left(-\frac{15}{17}\right)^2$$

$$= \frac{64}{289} - \frac{225}{289} = -\frac{161}{289} \approx -.56.$$

Since $\pi < x < 3\pi/2$, we know that $2\pi < 2x < 3\pi$. The calculations above show that at $2x$, sine is positive and cosine is negative. This can occur only if $2x$ lies between $5\pi/2$ and 3π. ■

EXAMPLE 2

Express the rule of the function $f(x) = \sin 3x$ in terms of $\sin x$ and constants.

SOLUTION We first use the addition identity for $\sin(x + y)$ with $y = 2x$.

$$f(x) = \sin 3x = \sin(x + 2x) = \sin x \cos 2x + \cos x \sin 2x.$$

Next apply the double-angle identities for $\cos 2x$ and $\sin 2x$.

$$\begin{aligned} f(x) = \sin 3x &= \sin x \cos 2x + \cos x \sin 2x \\ &= \sin x(\cos^2 x - \sin^2 x) + \cos x(2 \sin x \cos x) \\ &= \sin x \cos^2 x - \sin^3 x + 2 \sin x \cos^2 x \\ &= 3 \sin x \cos^2 x - \sin^3 x. \end{aligned}$$

Finally, use the Pythagorean identity.

$$\begin{aligned} f(x) = \sin 3x = 3 \sin x \cos^2 x - \sin^3 x &= 3 \sin x(1 - \sin^2 x) - \sin^3 x \\ = 3 \sin x - 3 \sin^3 x - \sin^3 x &= 3 \sin x - 4 \sin^3 x. \end{aligned}$$ ■

The double-angle identity for $\cos 2x$ can be rewritten in several useful ways. For instance, we can use the Pythagorean identity in the form of $\cos^2 x = 1 - \sin^2 x$ to obtain

$$\cos 2x = \cos^2 x - \sin^2 x = (1 - \sin^2 x) - \sin^2 x = 1 - 2 \sin^2 x.$$

Similarly, using the Pythagorean identity in the form $\sin^2 x = 1 - \cos^2 x$, we have

$$\cos 2x = \cos^2 x - \sin^2 x = \cos^2 x - (1 - \cos^2 x) = 2 \cos^2 x - 1.$$

In summary, we have the following identities.

More Double-Angle Identities

$$\cos 2x = 1 - 2 \sin^2 x$$

$$\cos 2x = 2 \cos^2 x - 1$$

EXAMPLE 3

Prove that

$$\frac{1 - \cos 2x}{\sin 2x} = \tan x.$$

SOLUTION The first identity in the preceding box and the double-angle identity for sine show that

$$\frac{1 - \cos 2x}{\sin 2x} = \frac{1 - (1 - 2 \sin^2 x)}{2 \sin x \cos x} = \frac{2 \sin^2 x}{2 \sin x \cos x} = \frac{\sin x}{\cos x} = \tan x.$$ ■

If we solve the first equation in the preceding box for $\sin^2 x$ and the second one for $\cos^2 x$, we obtain a useful alternate form for these identities.

Power-Reducing Identities

$$\sin^2 x = \frac{1 - \cos 2x}{2}$$

$$\cos^2 x = \frac{1 + \cos 2x}{2}$$

C

EXAMPLE 4

Express the rule of the function $f(x) = \sin^4 x$ in terms of constants and first powers of the cosine function.

SOLUTION We begin by applying the power-reducing identity.

$$f(x) = \sin^4 x = \sin^2 x \sin^2 x = \frac{1 - \cos 2x}{2} \cdot \frac{1 - \cos 2x}{2}$$

$$= \frac{1 - 2\cos 2x + \cos^2 2x}{4}.$$

Next we apply the power reducing identity for cosine to $\cos^2 2x$. Note that this means using $2x$ in place of x in the identity.

$$\cos^2 2x = \frac{1 + \cos 2(2x)}{2} = \frac{1 + \cos 4x}{2}.$$

Finally, we substitute this last result in the expression for $\sin^4 x$ above.

$$f(x) = \sin^4 x = \frac{1 - 2\cos 2x + \cos^2 2x}{4} = \frac{1 - 2\cos 2x + \dfrac{1 + \cos 4x}{2}}{4}$$

$$= \frac{1}{4} - \frac{1}{2}\cos 2x + \frac{1}{8}(1 + \cos 4x)$$

$$= \frac{3}{8} - \frac{1}{2}\cos 2x + \frac{1}{8}\cos 4x. \quad \blacksquare$$

HALF-ANGLE IDENTITIES

If we use the power-reducing identity with $x/2$ in place of x, we obtain

$$\sin^2\left(\frac{x}{2}\right) = \frac{1 - \cos 2\left(\dfrac{x}{2}\right)}{2} = \frac{1 - \cos x}{2}.$$

Consequently, we must have

$$\sin\left(\frac{x}{2}\right) = \pm\sqrt{\frac{1 - \cos x}{2}}.$$

This proves the first of the half-angle identities.

Half-Angle Identities

$$\sin\frac{x}{2} = \pm\sqrt{\frac{1-\cos x}{2}} \qquad \cos\frac{x}{2} = \pm\sqrt{\frac{1+\cos x}{2}}$$

$$\tan\frac{x}{2} = \pm\sqrt{\frac{1-\cos x}{1+\cos x}}$$

The half-angle identity for cosine is derived from a power-reducing identity, as was the half-angle identity for sine. The half-angle identity for tangent then follows immediately, since $\tan(x/2) = \sin(x/2)/\cos(x/2)$. In all cases *the sign in front of the radical depends on the quadrant in which $x/2$ lies.*

EXAMPLE 5

Find the exact value of

(a) $\cos 5\pi/8$ (b) $\sin \pi/12$.

SOLUTION

(a) Since $5\pi/8 = \frac{1}{2}(5\pi/4)$, we use the half-angle identity with $x = 5\pi/4$ and the fact that $\cos(5\pi/4) = -\sqrt{2}/2$. The sign chart in Exercise 53 on page 156 shows that $\cos(5\pi/8)$ is negative because $5\pi/8$ is in the second quadrant. So we use the negative sign in front of the radical.

$$\begin{aligned}\cos\frac{5\pi}{8} = \cos\frac{5\pi/4}{2} &= -\sqrt{\frac{1+\cos(5\pi/4)}{2}} \\ &= -\sqrt{\frac{1+(-\sqrt{2}/2)}{2}} = -\sqrt{\frac{(2-\sqrt{2})/2}{2}} \\ &= -\sqrt{\frac{2-\sqrt{2}}{4}} \\ &= \frac{-\sqrt{2-\sqrt{2}}}{2}\end{aligned}$$

(b) Since $\pi/12 = \frac{1}{2}(\pi/6)$ and $\pi/12$ is in the first quadrant, where sine is positive, we have

$$\begin{aligned}\sin\frac{\pi}{12} = \sin\frac{\pi/6}{2} &= \sqrt{\frac{1-\cos(\pi/6)}{2}} \\ &= \sqrt{\frac{1-\sqrt{3}/2}{2}} = \sqrt{\frac{(2-\sqrt{3})/2}{2}} = \sqrt{\frac{2-\sqrt{3}}{4}} \\ &= \frac{\sqrt{2-\sqrt{3}}}{2}. \quad \blacksquare\end{aligned}$$

The problem of determining signs in the half-angle formulas can be eliminated with tangent by using the following identities.

Half-Angle Identities for Tangent

$$\tan\frac{x}{2} = \frac{1 - \cos x}{\sin x}$$

$$\tan\frac{x}{2} = \frac{\sin x}{1 + \cos x}$$

Proof In the identity

$$\tan x = \frac{1 - \cos 2x}{\sin 2x},$$

which was proved in Example 3, replace x by $x/2$.

$$\tan\left(\frac{x}{2}\right) = \frac{1 - \cos 2(x/2)}{\sin 2(x/2)} = \frac{1 - \cos x}{\sin x}.$$

The second identity in the box is proved in Exercise 71. ■

EXAMPLE 6

If $\tan x = \frac{3}{2}$ and $\pi < x < \frac{3\pi}{2}$, find $\tan\frac{x}{2}$.

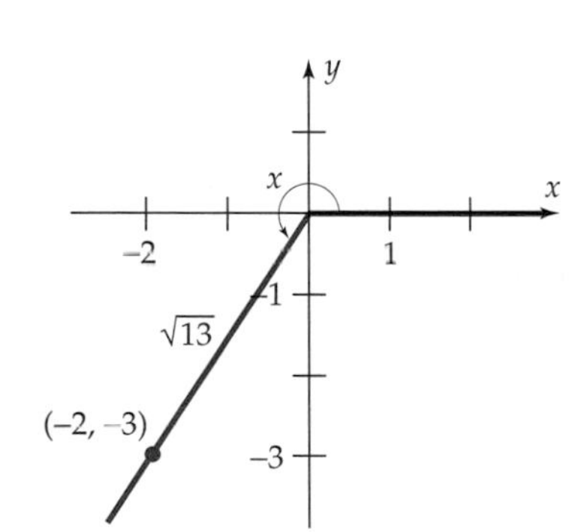

Figure 3–9

SOLUTION The terminal side of an angle of x radians in standard position lies in the third quadrant, as shown in Figure 3–9. The tangent of the angle in standard position whose terminal side passes through the point $(-2, -3)$ is $\frac{-3}{-2} = \frac{3}{2}$. Since there is only one angle in the third quadrant with tangent $3/2$, the point $(-2, -3)$ must lie on the terminal side of the angle of x radians.

Since the distance from $(-2, -3)$ to the origin is

$$\sqrt{(-2 - 0)^2 + (-3 - 0)^2} = \sqrt{13},$$

we have

$$\sin x = \frac{-3}{\sqrt{13}} \qquad \text{and} \qquad \cos x = \frac{-2}{\sqrt{13}}.$$

Therefore, by the first of the half-angle identities for tangent,

$$\tan\frac{x}{2} = \frac{1 - \cos x}{\sin x} = \frac{1 - \left(\dfrac{-2}{\sqrt{13}}\right)}{\dfrac{-3}{\sqrt{13}}} = \frac{\dfrac{\sqrt{13} + 2}{\sqrt{13}}}{\dfrac{-3}{\sqrt{13}}} = -\frac{\sqrt{13} + 2}{3}.$$ ■

SUM/PRODUCT IDENTITIES

The following identities were proved in Example 4 and Exercises 54–56 of Section 3.2.

Product to Sum Identities

$$\sin x \cos y = \frac{1}{2}[\sin(x + y) + \sin(x - y)]$$

$$\sin x \sin y = \frac{1}{2}[\cos(x - y) - \cos(x + y)]$$

$$\cos x \cos y = \frac{1}{2}[\cos(x + y) + \cos(x - y)]$$

$$\cos x \sin y = \frac{1}{2}[\sin(x + y) - \sin(x - y)]$$

If we use the first product to sum identity with $\frac{1}{2}(x + y)$ in place of x and $\frac{1}{2}(x - y)$ in place of y, we obtain

$$\begin{aligned}\sin\left[\frac{1}{2}(x + y)\right]\cos\left[\frac{1}{2}(x - y)\right] &= \frac{1}{2}\left[\sin\left(\frac{1}{2}(x + y) + \frac{1}{2}(x - y)\right)\right.\\ &\qquad \left. + \sin\left(\frac{1}{2}(x + y) - \frac{1}{2}(x - y)\right)\right]\\ &= \frac{1}{2}(\sin x + \sin y).\end{aligned}$$

Multiplying both sides of the last equation by 2 produces the first of the following identities.

Sum to Product Identities

$$\sin x + \sin y = 2\sin\left(\frac{x + y}{2}\right)\cos\left(\frac{x - y}{2}\right)$$

$$\sin x - \sin y = 2\cos\left(\frac{x + y}{2}\right)\sin\left(\frac{x - y}{2}\right)$$

$$\cos x + \cos y = 2\cos\left(\frac{x + y}{2}\right)\cos\left(\frac{x - y}{2}\right)$$

$$\cos x - \cos y = -2\sin\left(\frac{x + y}{2}\right)\sin\left(\frac{x - y}{2}\right)$$

The last three sum to product identities are proved in the same way as the first (see Exercises 52–54).

EXAMPLE 7

Prove the identity

$$\frac{\sin t + \sin 3t}{\cos t + \cos 3t} = \tan 2t.$$

SOLUTION Using the first sum to product identity with $x = t$ and $y = 3t$ yields

$$\sin t + \sin 3t = 2\sin\left(\frac{t+3t}{2}\right)\cos\left(\frac{t-3t}{2}\right) = 2\sin 2t\cos(-t).$$

Similarly,

$$\cos t + \cos 3t = 2\cos\left(\frac{t+3t}{2}\right)\cos\left(\frac{t-3t}{2}\right) = 2\cos 2t\cos(-t),$$

so

$$\frac{\sin t + \sin 3t}{\cos t + \cos 3t} = \frac{2\sin 2t\cos(-t)}{2\cos 2t\cos(-t)} = \frac{\sin 2t}{\cos 2t} = \tan 2t. \quad ■$$

EXERCISES 3.3

In Exercises 1–7, find $\sin 2x$, $\cos 2x$, *and* $\tan 2x$ *under the given conditions.*

1. $\sin x = \frac{5}{13}$ $\left(0 < x < \frac{\pi}{2}\right)$

2. $\sin x = -\frac{4}{5}$ $\left(\pi < x < \frac{3\pi}{2}\right)$

3. $\cos x = -\frac{3}{5}$ $\left(\pi < x < \frac{3\pi}{2}\right)$

4. $\cos x = -\frac{1}{3}$ $\left(\frac{\pi}{2} < x < \pi\right)$

5. $\tan x = \frac{3}{4}$ $\left(\pi < x < \frac{3\pi}{2}\right)$

6. $\tan x = -\frac{3}{2}$ $\left(\frac{\pi}{2} < x < \pi\right)$

7. $\csc x = 4$ $\left(0 < x < \frac{\pi}{2}\right)$

8. A batter hits a baseball that is caught by a fielder. If the ball leaves the bat at an angle of θ radians to the horizontal, with an initial velocity of v feet per second, then the approximate horizontal distance d traveled by the ball is given by

$$d = \frac{v^2 \sin\theta\cos\theta}{16}.$$

(a) Use an identity to show that

$$d = \frac{v^2 \sin 2\theta}{32}.$$

(b) If the initial velocity is 115 ft/second, what angle θ will produce the maximum distance? [*Hint:* Use part (a). For what value of θ is $\sin 2\theta$ as large as possible?]

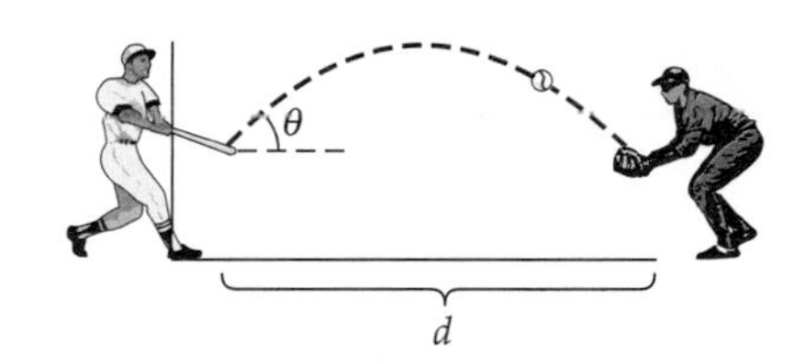

9. A rectangle is inscribed in a semicircle of radius 3 inches and the radius to the corner makes an angle of t radians with the horizontal, as shown in the figure.

(a) Express the horizontal length, vertical height, and area of the rectangle in terms of x and y.

(b) Express x and y in terms of sine and cosine.

(c) Use parts (a) and (b) and suitable identities to show that the area A of the rectangle is given by

$$A = 9\sin 2t.$$

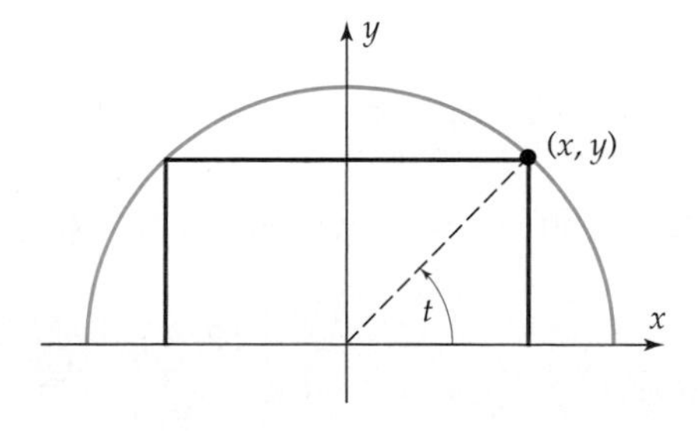

10. In Exercise 9, what angle will produce a rectangle with largest possible area? What is this maximum area?

In Exercises 11–22, use the half-angle identities to evaluate the given expression exactly.

11. $\cos\frac{\pi}{8}$ **12.** $\tan\frac{\pi}{8}$ **13.** $\sin\frac{3\pi}{8}$ **14.** $\cos\frac{3\pi}{8}$

15. $\tan\frac{\pi}{12}$ **16.** $\sin\frac{5\pi}{8}$ **17.** $\cos\frac{\pi}{12}$ **18.** $\tan\frac{5\pi}{8}$

19. $\sin\frac{7\pi}{8}$ **20.** $\cos\frac{7\pi}{8}$ **21.** $\tan\frac{7\pi}{8}$ **22.** $\cot\frac{\pi}{8}$

In Exercises 23–28, find $\sin\frac{x}{2}$, $\cos\frac{x}{2}$, *and* $\tan\frac{x}{2}$ *under the given conditions.*

23. $\cos x = .4 \quad \left(0 < x < \frac{\pi}{2}\right)$

24. $\sin x = .6 \quad \left(\frac{\pi}{2} < x < \pi\right)$

25. $\sin x = -\frac{3}{5} \quad \left(\frac{3\pi}{2} < x < 2\pi\right)$

26. $\cos x = .8 \quad \left(\frac{3\pi}{2} < x < 2\pi\right)$

27. $\tan x = \frac{1}{2} \quad \left(\pi < x < \frac{3\pi}{2}\right)$

28. $\cot x = 1 \quad \left(-\pi < x < -\frac{\pi}{2}\right)$

In Exercises 29–34, write each expression as a sum or difference.

29. $\sin 4x \cos 6x$ **30.** $\sin 5x \sin 7x$

31. $\cos 2x \cos 4x$ **32.** $\sin 3x \cos 5x$

33. $\sin 17x \sin(-3x)$ **34.** $\cos 13x \cos(-5x)$

In Exercises 35–38, write each expression as a product.

35. $\sin 3x + \sin 5x$ **36.** $\cos 2x + \cos 6x$

37. $\sin 9x - \sin 5x$ **38.** $\cos 5x - \cos 7x$

In Exercises 39–44, assume that $\sin x = .6$ *and* $0 < x < \pi/2$ *and evaluate the given expression.*

39. $\sin 2x$ **40.** $\cos 4x$ **41.** $\cos 2x$ **42.** $\sin 4x$

43. $\sin\frac{x}{2}$ **44.** $\cos\frac{x}{2}$

45. Express $\cos 3x$ in terms of $\cos x$.

46. (a) Express the rule of the function $f(x) = \cos^3 x$ in terms of constants and first powers of the cosine function as in Example 4.

(b) Do the same for $f(x) = \cos^4 x$.

In Exercises 47–51, simplify the given expression.

47. $\frac{\sin 2x}{2\sin x}$ **48.** $1 - 2\sin^2\left(\frac{x}{2}\right)$

49. $2\cos 2y \sin 2y$ (Think!)

50. $\cos^2\left(\frac{x}{2}\right) - \sin^2\left(\frac{x}{2}\right)$ **51.** $(\sin x + \cos x)^2 - \sin 2x$

In Exercises 52–54, prove the given sum to product identity. [Hint: See the proof of the first one on page 246.]

52. $\sin x - \sin y = 2\cos\left(\frac{x+y}{2}\right)\sin\left(\frac{x-y}{2}\right)$

53. $\cos x + \cos y = 2\cos\left(\frac{x+y}{2}\right)\cos\left(\frac{x-y}{2}\right)$

54. $\cos x - \cos y = -2\sin\left(\frac{x+y}{2}\right)\sin\left(\frac{x-y}{2}\right)$

In Exercises 55–68, determine graphically whether the equation could possibly be an identity. If it could, prove that it is.

55. $\sin 16x = 2\sin 8x \cos 8x$

56. $\cos 8x = \cos^2 4x - \sin^2 4x$

57. $\cos^4 x - \sin^4 x = \cos 2x$

58. $\sec 2x = \frac{1}{1 - 2\sin^2 x}$

59. $\cos 4x = 2\cos 2x - 1$

60. $\sin^2 x = \cos^2 x - 2\sin x$

61. $\frac{1 + \cos 2x}{\sin 2x} = \cot x$ **62.** $\sin 2x = \frac{2\cot x}{\csc^2 x}$

63. $\sin 3x = (\sin x)(3 - 4\sin^2 x)$

64. $\sin 4x = (4\cos x \sin x)(1 - 2\sin^2 x)$

65. $\cos 2x = \frac{2\tan x}{\sec^2 x}$

66. $\cos 3x = (\cos x)(3 - 4\cos^2 x)$

67. $\csc^2\left(\frac{x}{2}\right) = \frac{2}{1 - \cos x}$ **68.** $\sec^2\left(\frac{x}{2}\right) = \frac{2}{1 + \cos x}$

In Exercises 69 and 70, prove the identity.

69. $\frac{\sin x - \sin 3x}{\cos x + \cos 3x} = -\tan x$

70. $\frac{\sin x - \sin 3x}{\cos x - \cos 3x} = -\cot 2x$

71. (a) Prove that $\dfrac{1 - \cos x}{\sin x} = \dfrac{\sin x}{1 + \cos x}$.
(b) Use part (a) and the half-angle identity proved in the text on page 245 to prove that

$$\tan \frac{x}{2} = \frac{\sin x}{1 + \cos x}.$$

72. To avoid a steep hill, a road is being built in straight segments from P to Q and from Q to R; it makes a turn of t radians at Q, as shown in the figure. The distance from P to S is 40 miles, and the distance from R to S is 10 miles. Use suitable trigonometric functions to express:
(a) c in terms of b and t
(b) b in terms of t
(c) a in terms of t [*Hint:* $a = 40 - c$; use parts (a) and (b).]
(d) Use parts (b) and (c) and a suitable identity to show that the length $a + b$ of the road is

$$40 + 10 \tan \frac{t}{2}.$$

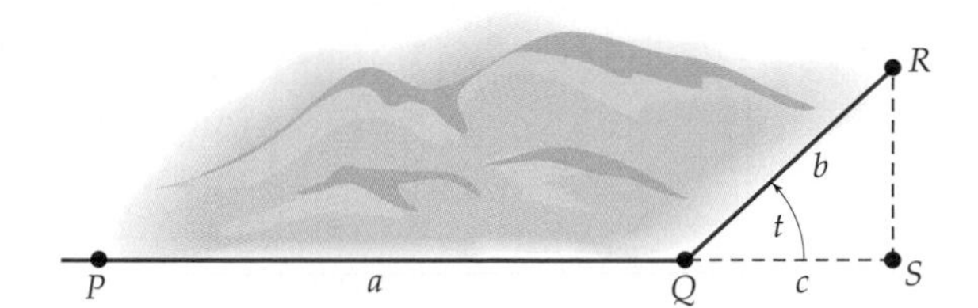

3.4 Inverse Trigonometric Functions

We begin by summarizing the basic facts about inverse functions. A function f is said to be **one-to-one** if distinct inputs always produce distinct outputs, that is,

$$\text{If } a \neq b, \text{ then } f(a) \neq f(b),$$

EXAMPLE 1

(a) $f(x) = x^3$ is one-to-one because when $a \neq b$, then $a^3 \neq b^3$.

(b) The function $f(x) = x^2$ is *not* one-to-one because, for example, both 2 and -2 produce the same output:

$$f(2) = 2^2 = 4 \qquad \text{and} \qquad f(-2) = (-2)^2 = 4. \quad ■$$

To understand the idea of an inverse function, consider the one-to-one function f given by this table:

***f*-input**	−2	−1	0	1	2
***f*-output**	−3	−2	1	4	5

Define a new function g by the following table, which simply *reverses* the rows in the f table.

***g*-input**	−3	−2	1	4	5
***g*-output**	−2	−1	0	1	2

Note that the inputs of f are the outputs of g and the outputs of f are the inputs of g. In other words,

$$\text{Domain of } f = \text{Range of } g \qquad \text{and} \qquad \text{Range of } f = \text{Domain of } g.$$

The rule of g *reverses* the action of f by taking each output of f back to the input it came from. For instance,

$$g(4) = 1 \qquad \text{and} \qquad f(1) = 4$$

$$g(-3) = -2 \qquad \text{and} \qquad f(-2) = -3$$

and in general,

$$g(y) = x \qquad \text{exactly when} \qquad f(x) = y.$$

We say that g is the *inverse function* of f.

The preceding construction works for any one-to-one function f. Each output of f comes from exactly one input (because different inputs produce different outputs). Consequently, we can define a new function g that reverses the action of f by sending each output back to the unique input it came from. For instance, if $f(7) = 11$, then $g(11) = 7$. Thus, the outputs of f become the inputs of g, and we have this definition.

Inverse Functions

Let f be a one-to-one function. Then the **inverse function** of f is the function g whose rule is

$$g(y) = x \qquad \text{exactly when} \qquad f(x) = y.$$

The domain of g is the range of f and the range of g is the domain of f.

The following theorem is often useful.

Round-Trip Theorem

A one-to-one function f and its inverse function g have these properties:

$$g(f(x)) = x \quad \text{for every } x \text{ in the domain of } f;$$

$$f(g(x)) = x \quad \text{for every } x \text{ in the domain of } g.$$

Conversely, if f and g are functions having these properties, then f is one-to-one and its inverse is g.

Proof By the definition of inverse function,

$$g(d) = c \qquad \text{exactly when} \qquad f(c) = d.$$

Consequently, for any c in the domain of f.

$$\begin{aligned} g(f(c)) &= g(d) \quad (\text{because } f(c) = d) \\ &= c \qquad (\text{because } g(d) = c). \end{aligned}$$

A similar argument shows that $f(g(d)) = d$ for any d in the domain of g. The last statement in the Theorem is proved in Exercise 77. ■

THE INVERSE SINE FUNCTION

The sine function is not one-to-one because many different inputs produce the same output. For instance,

$$0 = \sin 0 = \sin \pi = \sin 2\pi = \sin 3\pi, \text{ and so on.}$$

So it does not have an inverse function. However, a function closely related to the sine function (same rule, but smaller domain) *is* one-to-one, and hence, has an inverse function, as we now see.

The **restricted sine function** is defined as follows:

Domain: all x with $-\pi/2 \leq x \leq \pi/2$ *Rule:* $f(x) = \sin x$.

Its graph in Figure 3–10 shows that for each number v between -1 and 1, there is exactly one number u between $-\pi/2$ and $\pi/2$ such that $\sin u = v$. In other words, two different values of u cannot produce the same number v. So f is one-to-one.

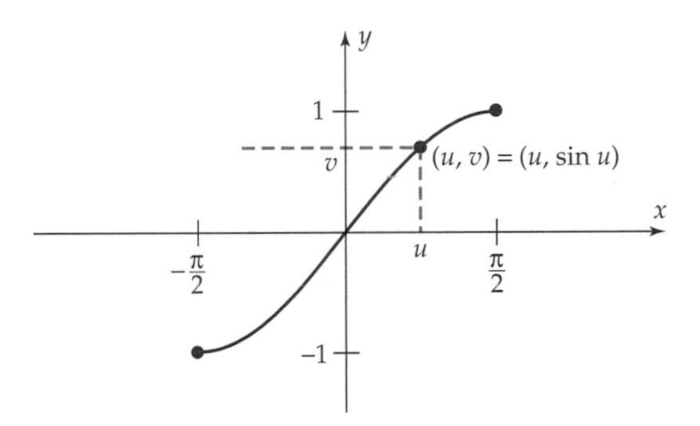

Figure 3–10

Since the restricted sine function is one-to-one, we know that it has an inverse function. This inverse function is called the **inverse sine** (or **arcsine**) **function** and is denoted by $g(x) = \sin^{-1} x$ **or** $g(x) = \arcsin x$. The domain of the inverse sine function is the interval from -1 to 1 (inclusive), and its rule is as follows.

Inverse Sine Function

For each v with $-1 \leq v \leq 1$,

$$\sin^{-1} v = \text{the unique number } u \text{ between } -\pi/2 \text{ and } \pi/2 \text{ whose sine is } v;$$

that is,

$$\sin^{-1} v = u \quad \text{exactly when} \quad \sin u = v.$$

EXAMPLE 2

Find

(a) $\sin^{-1}(1/2)$ (b) $\sin^{-1}(-\sqrt{2}/2)$.

SOLUTION

(a) $\text{Sin}^{-1}(1/2)$ is the one number between $-\pi/2$ and $\pi/2$ whose sine is $1/2$. From our study of special values, we know that $\sin \pi/6 = 1/2$, *and* $\pi/6$ is between $-\pi/2$ and $\pi/2$. Hence, $\sin^{-1}(1/2) = \pi/6$.

(b) $\text{Sin}^{-1}(-\sqrt{2}/2) = -\pi/4$ because $\sin(-\pi/4) = -\sqrt{2}/2$ and $-\pi/4$ is between $-\pi/2$ and $\pi/2$. ■

EXAMPLE 3

Except for special values, you should use the SIN^{-1} key (labeled ASIN on some calculators) in *radian mode* to evaluate the inverse sine function. For instance,

$$\sin^{-1}(-.67) = -.7342 \qquad \text{and} \qquad \sin^{-1}(.42) = .4334. \quad ■$$

EXAMPLE 4

If you key in SIN^{-1} 2 ENTER, you will get an error message, because 2 is not in the domain of the inverse sine function.* ■

CAUTION

The notation $\sin^{-1}x$ is *not* exponential notation. It does *not* mean either $(\sin x)^{-1}$ or $\frac{1}{\sin x}$. For instance, Example 1 shows that

$$\sin^{-1}(1/2) = \pi/6 \approx .5236,$$

but

$$\left(\sin\frac{1}{2}\right)^{-1} = \frac{1}{\sin\frac{1}{2}} \approx \frac{1}{.4794} \approx 2.0858.$$

Suppose $-1 \le v \le 1$ and $\sin^{-1} v = u$. Then by the definition of the inverse sine function, we know that $-\pi/2 \le u \le \pi/2$ and $\sin u = v$. Therefore,

$$\sin^{-1}(\sin u) = \sin^{-1}(v) = u \qquad \text{and} \qquad \sin(\sin^{-1} v) = \sin u = v.$$

This shows that the restricted sine function and the inverse sine function have the usual "round-trip properties" of inverse functions. In summary,

Properties of Inverse Sine

$$\sin^{-1}(\sin u) = u \quad \text{if} \quad -\frac{\pi}{2} \le u \le \frac{\pi}{2}$$

$$\sin(\sin^{-1} v) = v \quad \text{if} \quad -1 \le v \le 1$$

*TI-85/86 and HP-39+ display the complex number $(1.5707\cdots, -1.3169\cdots)$ for $\sin^{-1}(2)$. Casio FX 2.0 does the same when it is in complex mode. For our purposes, this is equivalent to an error message, since we deal only with functions whose values are real numbers.

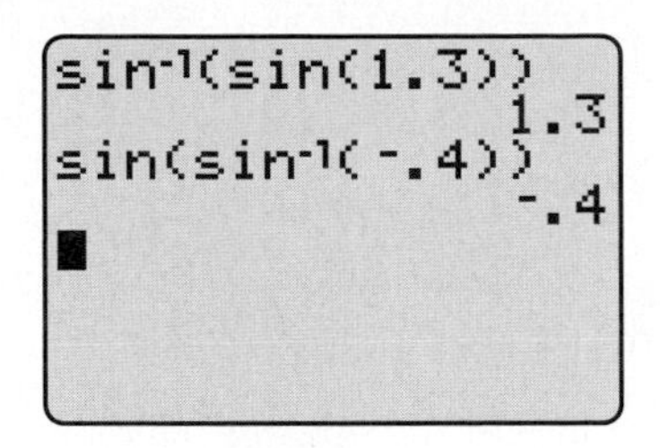

Figure 3–11

A calculator can illustrate the identities in the preceding box, as shown in Figure 3–11. Nevertheless, when special values are involved, you should be able to deal with them by hand.

EXAMPLE 5

Find (a) $\sin^{-1}(\sin \pi/6)$ (b) $\sin^{-1}(\sin 5\pi/6)$.

SOLUTION

(a) We know that $\sin \pi/6 = 1/2$. Hence,

$$\sin^{-1}\left(\sin\frac{\pi}{6}\right) = \sin^{-1}\left(\frac{1}{2}\right) = \frac{\pi}{6}$$

because $\pi/6$ is the number between $-\pi/2$ and $\pi/2$ whose sine is $1/2$.

(b) We also have $\sin 5\pi/6 = 1/2$, so the expression $\sin^{-1}(\sin 5\pi/6)$ is defined. However,

$$\sin^{-1}\left(\sin\frac{5\pi}{6}\right) \quad \text{is NOT equal to} \quad \frac{5\pi}{6}$$

because the identity in the box on page 252 is valid only when u is between $-\pi/2$ and $\pi/2$. Using the result of part (a), we see that

$$\sin^{-1}\left(\sin\frac{5\pi}{6}\right) = \sin^{-1}\left(\frac{1}{2}\right) = \frac{\pi}{6}.$$ ■

THE INVERSE COSINE FUNCTION

The **restricted cosine function** is defined as follows:

Domain: $0 \le x \le \pi$ Rule: $f(x) = \cos x$.

Its graph in Figure 3–12 shows that for each number v between -1 and 1, there is exactly one number u between 0 and π such that $\cos u = v$. In other words, two different values of u cannot produce the same number v. So f is one-to-one.

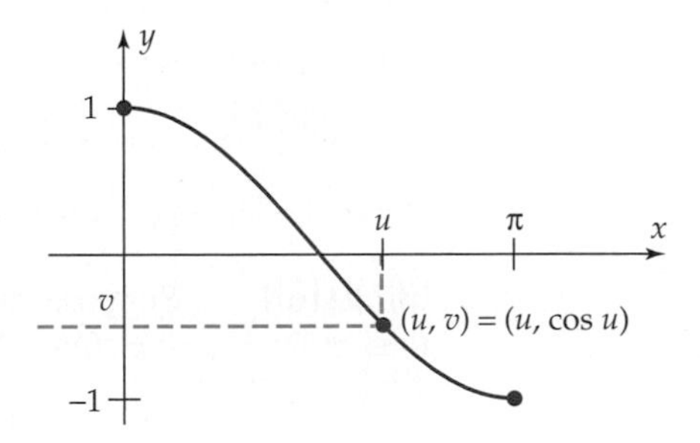

Figure 3–12

Since the restricted cosine function is one-to-one, we know that it has an inverse function. This inverse function is called the **inverse cosine** (or **arccosine**) **function** and is denoted by $\boldsymbol{h(x) = \cos^{-1} x}$ or $\boldsymbol{h(x) = \arccos x}$. The domain of the inverse cosine function is the interval from -1 to 1 (inclusive), and its rule is as follows.

Inverse Cosine Function

For each v with $-1 \leq v \leq 1$,

$$\cos^{-1} v = \text{the unique number } u \text{ between 0 and } \pi \text{ whose cosine is } v;$$

that is,

$$\cos^{-1} v = u \qquad \text{exactly when} \qquad \cos u = v.$$

The inverse cosine function has these properties.

$$\cos^{-1}(\cos u) = u \qquad \text{if} \qquad 0 \leq u \leq \pi;$$

$$\cos(\cos^{-1} v) = v \qquad \text{if} \qquad -1 \leq v \leq 1.$$

EXAMPLE 6

Find (a) $\cos^{-1}(1/2)$ (b) $\cos^{-1}(0)$ (c) $\cos^{-1}(-.63)$.

SOLUTION

(a) $\text{Cos}^{-1}(1/2) = \pi/3$ since $\pi/3$ is the unique number between 0 and π whose cosine is $1/2$.

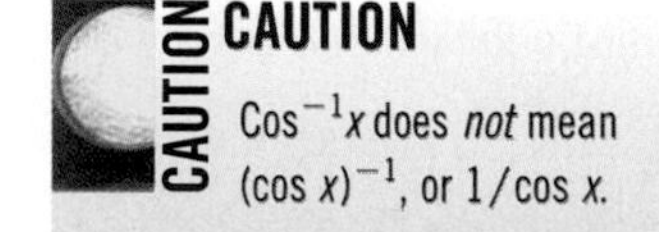

CAUTION

$\text{Cos}^{-1}x$ does *not* mean $(\cos x)^{-1}$, or $1/\cos x$.

(b) $\text{Cos}^{-1}(0) = \pi/2$ because $\cos \pi/2 = 0$ and $0 \leq \pi/2 \leq \pi$.

(c) The COS^{-1} key on a calculator in *radian mode* shows that $\cos^{-1}(-.63) = 2.2523$. ■

EXAMPLE 7

Write $\sin(\cos^{-1} v)$ as an algebraic expression in v.

SOLUTION $\text{Cos}^{-1} v = u$, where $\cos u = v$ and $0 \leq u \leq \pi$. Hence, $\sin u$ is nonnegative, and by the Pythagorean identity, $\sin u = \sqrt{\sin^2 u} = \sqrt{1 - \cos^2 u}$. Also, $\cos^2 u = v^2$. Therefore,

$$\sin(\cos^{-1} v) = \sin u = \sqrt{1 - \cos^2 u} = \sqrt{1 - v^2}. \qquad ■$$

EXAMPLE 8 

Prove the identity $\sin^{-1} x + \cos^{-1} x = \pi/2$.

SOLUTION Suppose $\sin^{-1} x = u$, with $-\pi/2 \leq u \leq \pi/2$. Verify that $0 \leq \pi/2 - u \leq \pi$ (Exercise 28). Then we have

$$\sin u = x \qquad \text{[Definition of inverse sine]}$$

$$\cos\left(\frac{\pi}{2} - u\right) = x \qquad \text{[Cofunction identity]}$$

$$\cos^{-1} x = \frac{\pi}{2} - u. \qquad \text{[Definition of inverse cosine]}$$

Therefore,

$$\sin^{-1} x + \cos^{-1} x = u + \left(\frac{\pi}{2} - u\right) = \frac{\pi}{2}. \qquad \blacksquare$$

THE INVERSE TANGENT FUNCTION

The **restricted tangent function** is defined as follows:

$$\text{Domain: } -\pi/2 < x < \pi/2 \qquad \text{Rule: } f(x) = \tan x.$$

Its graph in Figure 3–13 shows that for every real number v, there is exactly one number u between $-\pi/2$ and $\pi/2$ such that $\tan u = v$. In other words, two different values of u cannot produce the same number v. So f is one-to-one.

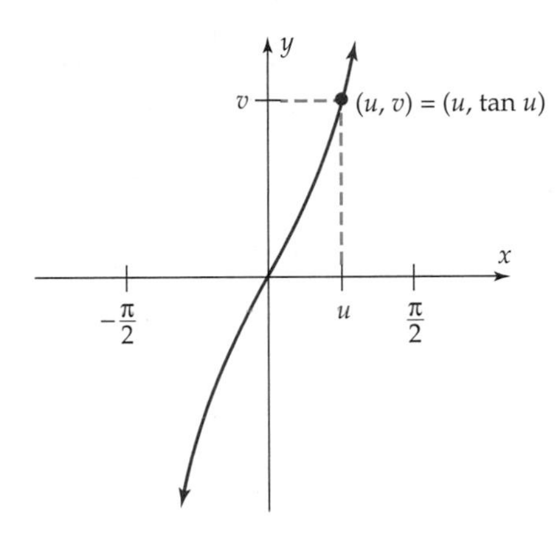

Figure 3–13

Since the restricted tangent function is one-to-one, we know that it has an inverse function. This inverse function is called the **inverse tangent** (or **arctangent) function** and is denoted by $\mathbf{g(x) = \tan^{-1} x}$ or $\mathbf{g(x) = \arctan x}$. The domain of the inverse tangent function is the set of all real numbers, and its rule is as follows.

Inverse Tangent Function

> For each real number v,
>
> $\tan^{-1} v =$ the unique number u between $-\pi/2$ and $\pi/2$ whose tangent is v;
>
> that is,
>
> $$\tan^{-1} v = u \qquad \text{exactly when} \qquad \tan u = v.$$
>
> The inverse tangent function has these properties:
>
> $$\tan^{-1}(\tan u) = u \qquad \text{if} \qquad -\frac{\pi}{2} < u < \frac{\pi}{2};$$
>
> $$\tan(\tan^{-1} v) = v \qquad \text{for every number } v.$$

CAUTION

$\text{Tan}^{-1} x$ does *not* mean $(\tan x)^{-1}$, or $1/\tan x$.

EXAMPLE 9

$\text{Tan}^{-1} 1 = \pi/4$ because $\pi/4$ is the unique number between $-\pi/2$ and $\pi/2$ such that $\tan \pi/4 = 1$. A calculator in *radian mode* shows that $\tan^{-1}(136) = 1.5634$. ■

EXAMPLE 10 

Find the exact value of $\cos[\tan^{-1}(\sqrt{5}/2)]$.

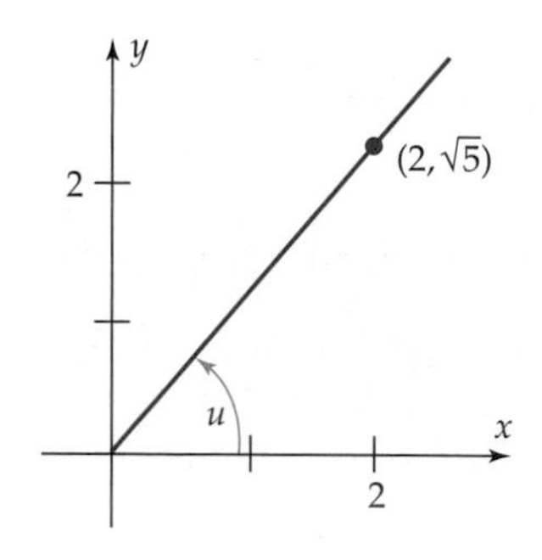

Figure 3–14

SOLUTION Consider an angle of u radians in standard position whose terminal side passes through $(2, \sqrt{5})$, as in Figure 3–14. By the point-in-the-plane description,

$$\tan u = \sqrt{5}/2.$$

Since u is between $-\pi/2$ and $\pi/2$ and $\tan u = \sqrt{5}/2$, we must have

$$u = \tan^{-1}(\sqrt{5}/2).$$

Furthermore, the distance from $(2, \sqrt{5})$ to the origin is

$$\sqrt{(2-0)^2 + (\sqrt{5}-0)^2} = \sqrt{4+5} = 3,$$

so

$$\cos u = 2/3.$$

Therefore,

$$\cos[\tan^{-1}(\sqrt{5}/2)] = \cos u = 2/3. \quad ■$$

GRAPHS OF THE INVERSE TRIG FUNCTIONS

Suppose f is a one-to-one function and g is its inverse function. Then by the definition of inverse function:

$$f(a) = b \qquad \text{exactly when} \qquad g(b) = a.$$

But $f(a) = b$ means that (a, b) is on the graph of f and $g(b) = a$ means that (b, a) is on the graph of g. Therefore,

Inverse Function Graphs

> If f is a one-to-one function and g is its inverse function, then
>
> (a, b) is on the graph of f exactly when (b, a) is on the graph of g.

Therefore, the graph of the inverse function g can be obtained by reversing the coordinates of each point on the graph of f. There are two practical ways of doing this, each of which is illustrated below.

EXAMPLE 11

Use the fact in the preceding box to graph the inverse sine function.

SOLUTION First, we use parametric graphing to graph the restricted sine function $f(x) = \sin x$. We let

$$x = t \quad \text{and} \quad y = f(t) = \sin t \quad (-\pi/2 \le t \le \pi/2),$$

as shown in Figure 3–15. According to the preceding box, the graph of the inverse function $g(x) = \sin^{-1} x$ can be obtained by taking each point on the graph of f and reversing its coordinates. Thus, g can be graphed parametrically by letting

$$x = f(t) = \sin t \quad \text{and} \quad y = t,$$

as shown in Figure 3–16. ■

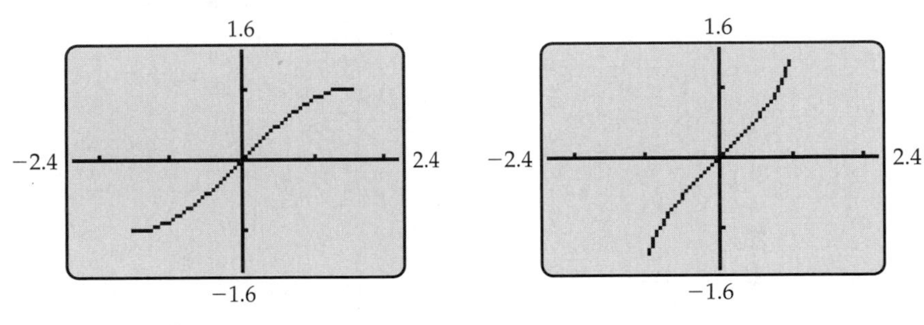

Figure 3–15 **Figure 3–16**

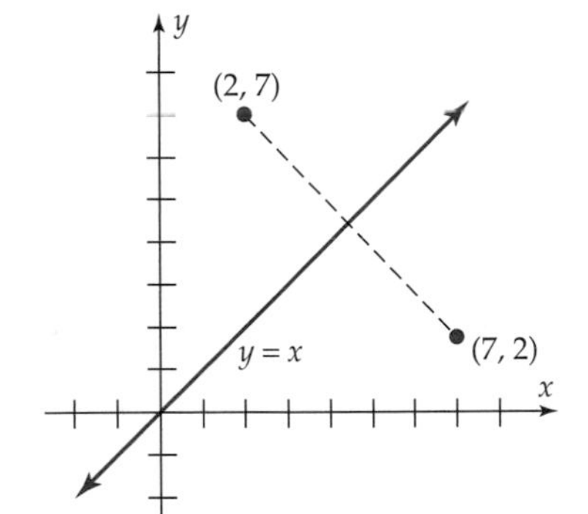

Figure 3–17

The second method of graphing inverse functions by reversing coordinates depends on this geometric fact, which is proved in Exercise 75:

> The line $y = x$ is the perpendicular bisector of the line segment from (a, b) to (b, a),

as shown in Figure 3–17 when $a = 7$, $b = 2$. Thus (a, b) and (b, a) lie on opposite sides of $y = x$, the same distance from it: They are mirror images of each other, with the line $y = x$ being the mirror.* Consequently, the graph of the inverse function g is the mirror image of the graph of f. In formal terms,

Inverse Function Graphs

> If g is the inverse function of f, then the graph of g is the reflection of the graph of f in the line $y = x$.

EXAMPLE 12

Use the fact in the preceding box to graph $h(x) = \cos^{-1} x$ and $k(x) = \tan^{-1} x$.

SOLUTION First graph the restricted cosine and tangent functions, then reflect the graphs in the line $y = x$, as shown in Figure 3–18 on the next page. ■

---

*In technical terms (a, b) and (b, a) are **symmetric with respect to the line** $y = x$.

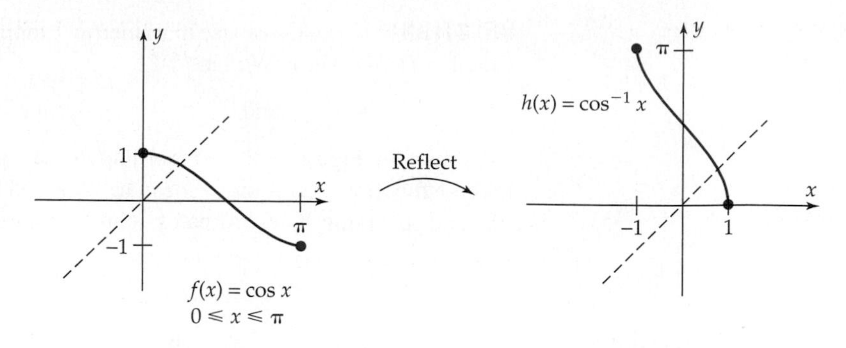

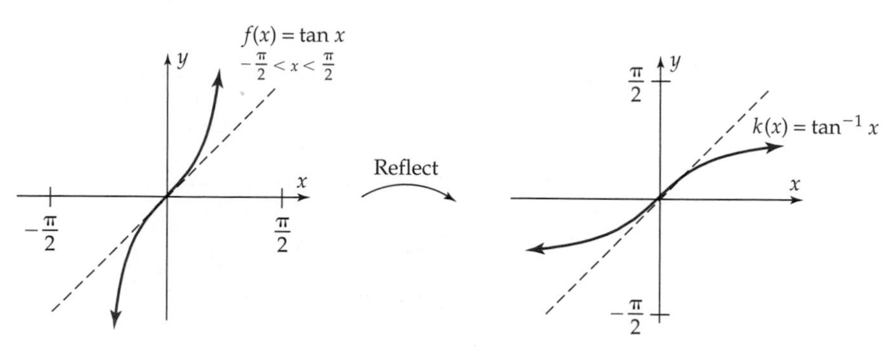

Figure 3–18

EXERCISES 3.4

In Exercises 1–4, use the Round-Trip Theorem on page 250 to show that g is the inverse of f.

1. $f(x) = x + 1, \quad g(x) = x - 1$

2. $f(x) = 2x - 6, \quad g(x) = \dfrac{x}{2} + 3$

3. $f(x) = \dfrac{1}{x+1}, \quad g(x) = \dfrac{1-x}{x}$

4. $f(x) = \dfrac{-3}{2x+5}, \quad g(x) = \dfrac{-3-5x}{2x}$

In Exercises 5–18, find the exact functional value without using a calculator.

5. $\sin^{-1} 1$ **6.** $\cos^{-1} 0$ **7.** $\tan^{-1}(-1)$

8. $\sin^{-1}(-1)$ **9.** $\cos^{-1} 1$ **10.** $\tan^{-1} 1$

11. $\tan^{-1}(\sqrt{3}/3)$ **12.** $\cos^{-1}(\sqrt{3}/2)$

13. $\sin^{-1}(-\sqrt{2}/2)$ **14.** $\sin^{-1}(\sqrt{3}/2)$

15. $\tan^{-1}(-\sqrt{3})$ **16.** $\cos^{-1}(-\sqrt{2}/2)$

17. $\cos^{-1}\left(-\dfrac{1}{2}\right)$ **18.** $\sin^{-1}\left(-\dfrac{1}{2}\right)$

In Exercises 19–27, use a calculator in radian mode to approximate the functional value.

19. $\sin^{-1} .35$ **20.** $\cos^{-1} .76$

21. $\tan^{-1}(-3.256)$ **22.** $\sin^{-1}(-.795)$

23. $\sin^{-1}(\sin 7)$ [The answer is *not* 7.]

24. $\cos^{-1}(\cos 3.5)$ **25.** $\tan^{-1}[\tan(-4)]$

26. $\sin^{-1}[\sin(-2)]$ **27.** $\cos^{-1}[\cos(-8.5)]$

28. Let u be a number such that

$$-\frac{\pi}{2} \leq u \leq \frac{\pi}{2}.$$

Prove that

$$0 \leq \frac{\pi}{2} - u \leq \pi.$$

29. Given that $u = \sin^{-1}(-\sqrt{3}/2)$, find the exact value of $\cos u$ and $\tan u$.

30. Given that $u = \tan^{-1}(4/3)$, find the exact value of $\sin u$ and $\sec u$.

In Exercises 31–46, find the exact functional vlaue without using a calculator.

31. $\sin^{-1}(\cos 0)$ **32.** $\cos^{-1}(\sin \pi/6)$

33. $\cos^{-1}(\sin 4\pi/3)$ **34.** $\tan^{-1}(\cos \pi)$

35. $\sin^{-1}(\cos 7\pi/6)$ **36.** $\cos^{-1}(\tan 7\pi/4)$

37. $\sin^{-1}(\sin 2\pi/3)$ (See Exercise 23.)

38. $\cos^{-1}(\cos 5\pi/4)$ **39.** $\cos^{-1}[\cos(-\pi/6)]$

40. $\tan^{-1}[\tan(-4\pi/3)]$

41. $\sin[\cos^{-1}(3/5)]$ (See Example 10.)

42. $\tan[\sin^{-1}(3/5)]$ **43.** $\cos[\tan^{-1}(-3/4)]$

44. $\cos[\sin^{-1}(\sqrt{3}/5)]$ **45.** $\tan[\sin^{-1}(5/13)]$

46. $\sin[\cos^{-1}(3/\sqrt{13})]$

In Exercises 47–50, write the expression as an algebraic expression in v, as in Example 7.

47. $\cos(\sin^{-1} v)$ **48.** $\cot(\cos^{-1} v)$

49. $\tan(\sin^{-1} v)$ **50.** $\sin(2 \sin^{-1} v)$

In Exercises 51 and 52, the graph of a function f is given. Sketch the graph of the inverse function of f. [Reflect carefully.]

51.

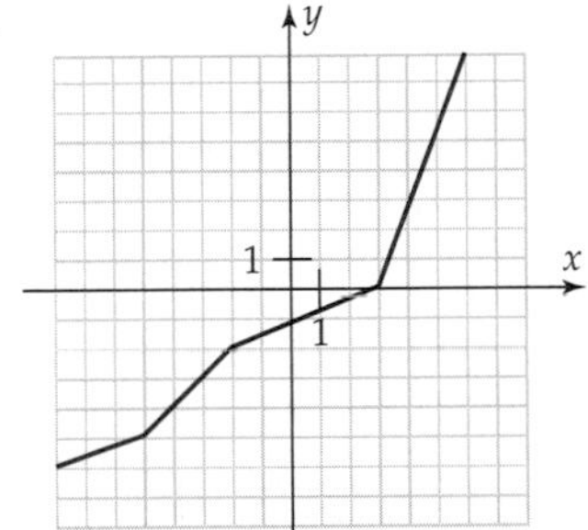

52.

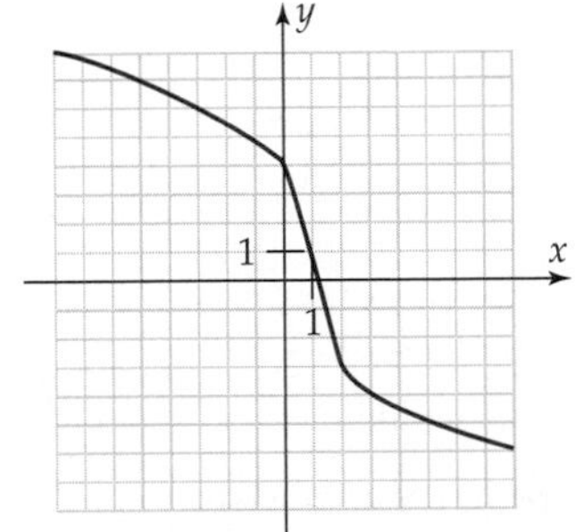

In Exercises 53–56, graph the function.

53. $f(x) = \cos^{-1}(x + 1)$ **54.** $g(x) = \tan^{-1} x + \pi$

55. $h(x) = \sin^{-1}(\sin x)$ **56.** $k(x) = \sin(\sin^{-1} x)$

57. In an alternating current circuit, the voltage is given by the formula

$$V = V_{max} \cdot \sin(2\pi ft + \phi),$$

where V_{max} is the maximum voltage, f is the frequency (in cycles per second), t is the time in seconds, and ϕ is the phase angle.

(a) If the phase angle is 0, solve the voltage equation for t.

(b) If $\phi = 0$, $V_{max} = 20$, $V = 8.5$, and $f = 120$, find the smallest positive value of t.

58. A model plane 40 feet above the ground is flying away from an observer.

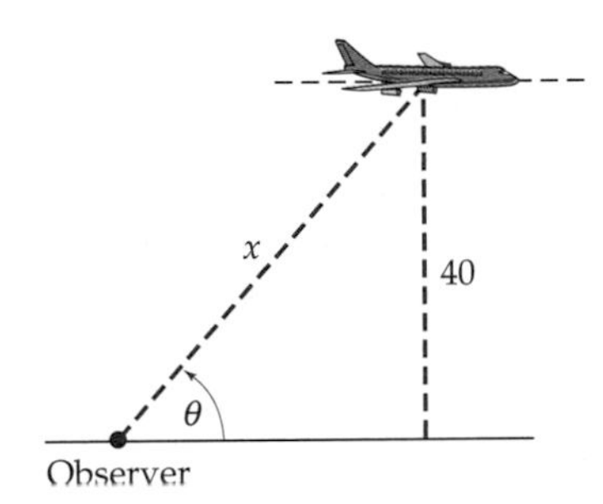

(a) Express the angle of elevation θ of the plane as a function of the distance x from the observer to the plane.

(b) What is θ when the plane is 250 feet away from the observer?

59. A 15-foot-wide highway sign is placed 10 feet from a road, perpendicular to the road (see figure). A spotlight at the edge of the road is aimed at the sign.

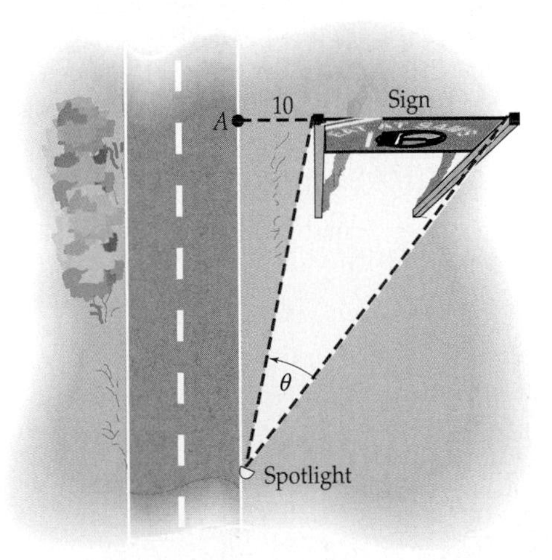

(a) Express θ as a function of the distance x from point A to the spotlight.

(b) How far from point A should the spotlight be placed so that the angle θ is as large as possible?

60. A camera on a 5-foot-high tripod is placed in front of a 6-foot-high picture that is mounted 3 feet above the floor.

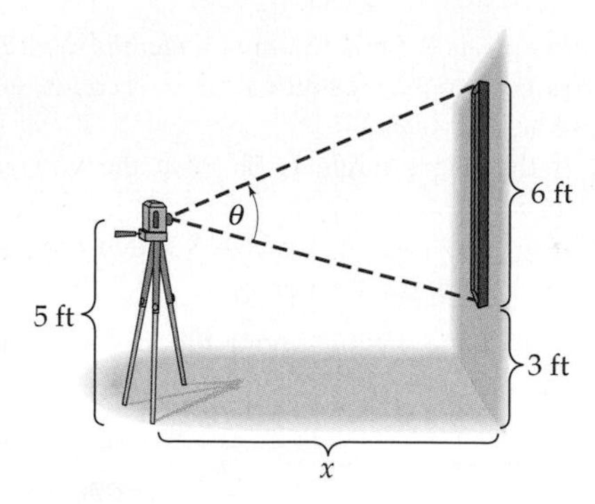

(a) Express angle θ as a function of the distance x from the camera to the wall.

(b) The photographer wants to use a particular lens, for which $\theta = 36°$ ($\pi/5$ radians). How far should she place the camera from the wall to be sure that the entire picture will show in the photograph?

61. Show that the restricted secant function, whose domain consists of all numbers x such that $0 \le x \le \pi$ and $x \ne \pi/2$, has an inverse function. Sketch its graph.

62. Show that the restricted cosecant function, whose domain consists of all numbers x such that $-x/2 \le x \le \pi/2$, and $x \ne 0$, has an inverse function. Sketch its graph.

63. Show that the restricted cotangent function, whose domain consists of all numbers x such that $0 < x < \pi$, has an inverse function. Sketch its graph.

64. (a) Show that the inverse cosine function actually has the two properties listed in the box on page 254.

(b) Show that the inverse tangent function actually has the two properties listed in the box on page 255.

In Exercises 65–72, prove the identity.

65. $\sin^{-1}(-x) = -\sin^{-1} x$ [*Hint:* Let $u = \sin^{-1}(-x)$ and show that $\sin^{-1} x = -u$.]

66. $\tan^{-1}(-x) = -\tan^{-1} x$

67. $\cos^{-1}(-x) = \pi - \cos^{-1} x$ [*Hint:* Let $u = \cos^{-1}(-x)$ and show that $0 \le \pi - u \le \pi$; use the identity

$$\cos(\pi - u) = -\cos u.]$$

68. $\sin^{-1}(\cos x) = \pi/2 - x \quad (0 \le x \le \pi)$

69. $\tan^{-1}(\cot x) = \pi/2 - x \quad (0 < x < \pi)$

70. $\tan^{-1} x + \tan^{-1}\left(\dfrac{1}{x}\right) = \dfrac{\pi}{2}$

71. $\sin^{-1} x = \tan^{-1}\left(\dfrac{x}{\sqrt{1 - x^2}}\right) \quad (-1 < x < 1)$

[*Hint:* Let $u = \sin^{-1} x$ and show that $\tan u = x/\sqrt{1 - x^2}$. Since $\sin u = x$, $\cos u = \pm\sqrt{1 - x^2}$. Show that in this case, $\cos u = \sqrt{1 - x^2}$.]

72. $\cos^{-1} x = \dfrac{\pi}{2} - \tan^{-1}\left(\dfrac{x}{\sqrt{1 - x^2}}\right) \quad (-1 < x < 1)$

[*Hint:* See Example 8 and Exercise 71.]

73. Is it true that $\tan^{-1} x = \dfrac{\sin^{-1} x}{\cos^{-1} x}$? Justify your answer.

74. Using the viewing window with $-2\pi \le x \le 2\pi$ and $-4 \le y \le 4$ graph the functions $f(x) = \cos(\cos^{-1} x)$ and $g(x) = \cos^{-1}(\cos x)$. How do you explain the shapes of the two graphs?

Thinkers

75. Show that the points $P = (a, b)$ and $Q = (b, a)$ are symmetric with respect to the line $y = x$ as follows.

(a) Find the slope of the line through P and Q.

(b) Use slopes to show that the line through P and Q is perpendicular to $y = x$.

(c) Let R be the point where the line $y = x$ intersects line segment PQ. Since R is on $y = x$, it has coordinates (c, c) for some number c, as shown in the figure. Use the distance formula to show that segment PR has the same length as segment RQ. Conclude that the line $y = x$ is the perpendicular bisector of segment PQ. Therefore, P and Q are symmetric with respect to the line $y = x$.

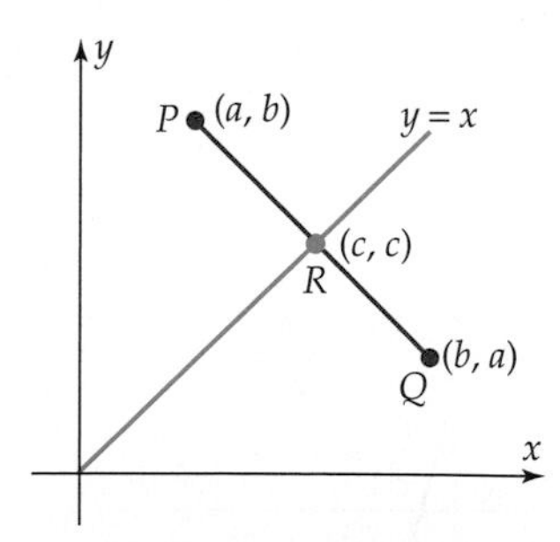

76. Prove that the function $h(x) = 1 - .2x^3$ is one-to-one by showing that it satisfies the definition:

$$\text{If } a \ne b, \text{ then } h(a) \ne h(b).$$

[*Hint:* Use the rule of h to show that when $h(a) = h(b)$, then $a = b$. If this is the case, then it is impossible to have $h(a) = h(b)$ when $a \ne b$.]

77. Suppose that functions f and g have these round-trip properties:
(1) $g(f(x)) = x$ for every x in the domain of f.
(2) $f(g(y)) = y$ for every y in the domain of g.
To complete the proof of the Round-Trip Theorem, we must show that g is the inverse function of f. Do this as follows.
(a) Prove that f is one-to-one by showing that

$$\text{if} \quad a \neq b, \qquad \text{then} \qquad f(a) \neq f(b).$$

[*Hint:* If $f(a) = f(b)$, apply g to both sides and use (1) to show that $a = b$. Consequently, if $a \neq b$, it is impossible to have $f(a) = f(b)$.]
(b) If $g(y) = x$, show that $f(x) = y$. [*Hint:* Use (2).]
(c) If $f(x) = y$, show that $g(y) = x$. [*Hint:* Use (1).]
Parts (b) and (c) prove that

$$g(y) = x \qquad \text{exactly when} \qquad f(x) = y.$$

Hence, g is the inverse function of f (see page 250).

3.5 Trigonometric Equations

Any equation that involves trigonometric functions can be solved graphically, and many can be solved algebraically. Unlike the equations solved previously, trigonometric equations typically have an infinite number of solutions. In most cases these solutions can be systematically determined by using periodicity, as we now see.

BASIC EQUATIONS

We begin with **basic equations,** such as

$$\sin x = .39, \qquad \cos x = .2, \qquad \tan x = -3.$$

Basic equations can be solved by the methods illustrated in Examples 1–3.

EXAMPLE 1

Solve $\tan x = 2$.

SOLUTION The equation can be solved graphically by graphing $y = \tan x$ and $y = 2$ on the same coordinate axes and finding the intersection points. The x-coordinate of every such point is a number whose tangent is 2, that is, a solution of the equation.

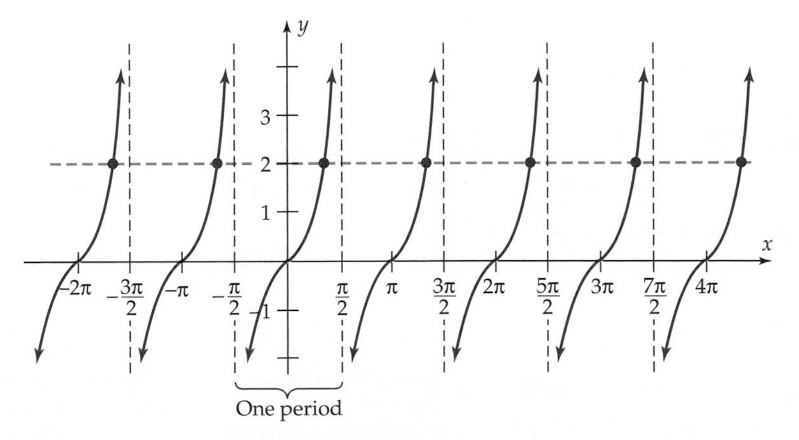

Figure 3–19

Figure 3–19 shows that there is exactly one solution in each period of tan x (for instance, between $-\pi/2$ and $\pi/2$). This solution could be found graphically, but it's faster to key in

$$\text{TAN}^{-1}\,2 \quad \text{ENTER}^{*}$$

tan⁻¹(2)
1.107148718

Figure 3–20

The calculator then displays the number between $-\pi/2$ and $\pi/2$ whose tangent is 2, namely, $x = 1.1071$, as shown in Figure 3–20.[†] Since the tangent graph repeats its pattern with period π, all the other solutions differ from this one by a multiple of π. Thus, all the solutions are

$$1.1071, \qquad 1.1071 \pm \pi, \qquad 1.1071 \pm 2\pi, \qquad 1.1071 \pm 3\pi, \qquad \text{etc.}$$

These solutions are customarily written like this:

$$x = 1.1071 + k\pi \qquad (k = 0, \pm 1, \pm 2, \pm 3, \ldots). \quad \blacksquare$$

EXAMPLE 2

Solve $\sin x = -.75$

SOLUTION The solutions are the x-coordinates of the points where the graphs of $y = \sin x$ and $y = -.75$ intersect (why?). Note that there are exactly two solutions in every period of sin x (for instance, between $-\pi/2$ and $3\pi/2$).

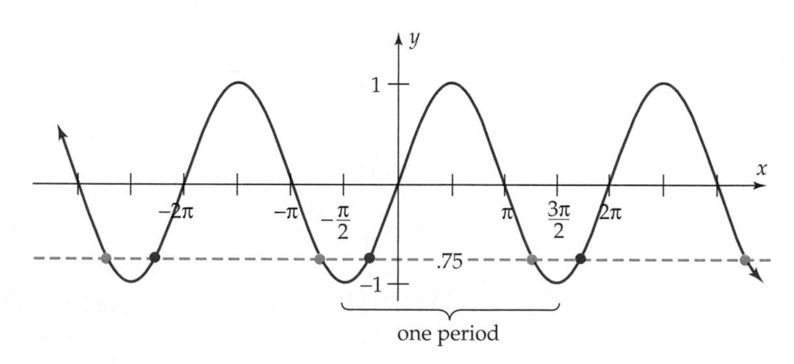

Figure 3–21

sin⁻¹(-.75)
-.848062079

Figure 3–22

Figure 3–21 shows that there is one solution between $-\pi/2$ and $\pi/2$. By keying in

$$\text{SIN}^{-1}(-.75) \quad \text{ENTER},$$

we find that this solution is $x = -.8481$, as shown in Figure 3–22. Since the sine graph repeats its pattern with period 2π, all of the following numbers are also solutions.

$$-.8481, \qquad -.8481 \pm 2\pi, \qquad -.8481 \pm 4\pi, \qquad -.8481 \pm 6\pi, \qquad \text{etc.}$$

---

*Unless stated otherwise, radian mode is used throughout this section.

[†]For convenient reading, all solutions in the text are rounded to four or fewer decimal places, but the full decimal expansion given by the calculator is used in all computations. We write = rather than ≈ even though these calculator solutions are approximations of the actual solutions.

These solutions correspond to the red intersection points in Figure 3–21, each of which is 2π units from the next red point. As you can see, there are still more solutions (corresponding to the blue points). One of them can be found by using the identity that was proved in Example 2 of Section 3.2:

$$\sin(\pi - x) = \sin x.$$

Applying this identity with the solution $x = -.8481$ shows that

$$\sin[\pi - (-.8481] = \sin(-.8481) = -.75.$$

In other words, $\pi - (-.8481) = 3.9897$ is also a solution of $\sin x = -.75$. The other solutions of the equation are

$$3.9897, \qquad 3.9897 \pm 2\pi, \qquad 3.9897 \pm 4\pi, \qquad 3.9897 \pm 6\pi, \qquad \text{etc.},$$

corresponding to the blue intersection points in Figure 3–21, each of which is 2π units from the next blue point. Therefore, all the solutions of $\sin x = -0.75$ are

$$x = -.8481 + 2k\pi \qquad \text{and} \qquad x = 3.9897 + 2k\pi$$
$$(k = 0, \pm 1, \pm 2, \pm 3, \ldots). \quad ■$$

EXAMPLE 3

Solve $\cos x = \sqrt{3}/2$.

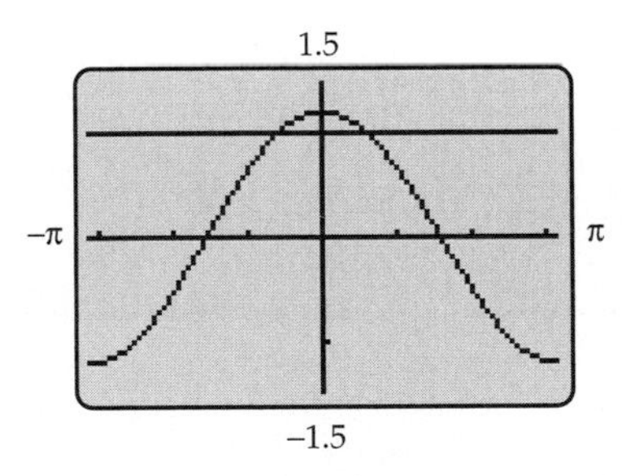

Figure 3–23

SOLUTION By graphing $y = \cos x$ and $y = \sqrt{3}/2$ on the same screen (Figure 3–23), we see that there are two solutions of the equation between $-\pi$ and π (one full period of cosine). The positive solution could be approximated by the COS^{-1} key on a calculator. In this case, however, our knowledge of special values provides an exact solution. As we saw in Section 2.2, $\cos \pi/6 = \sqrt{3}/2$. Therefore, one solution of the equation is $x = \pi/6$. The negative angle identity $\cos(-x) = \cos x$ shows that the second solution is $x = -\pi/6$, because

$$\cos\left(-\frac{\pi}{6}\right) = \cos\left(\frac{\pi}{6}\right) = \frac{\sqrt{3}}{2}.$$

Since the interval from $-\pi$ to π is one full period of cosine, all the solutions of the equation are

$$x = \frac{\pi}{6} + 2k\pi \qquad \text{and} \qquad x = -\frac{\pi}{6} + 2k\pi \qquad (k = 0, \pm 1, \pm 2, \pm 3, \ldots). \quad ■$$

The procedures in Examples 1–3 work in the general case and lead to the following conclusion.

Solution Algorithm for Basic Trigonometric Equations

If an equation of the form

$$\sin x = c, \qquad \cos x = c, \qquad \text{or} \qquad \tan x = c$$

has solutions, then one solution may be found by using the appropriate inverse function [$\sin^{-1}(x)$, $\cos^{-1}(x)$, or $\tan^{-1}(x)$] or by using your knowledge of special values.

When one solution u has been found, then the remaining solutions are as follows.

Equation	All Solutions
$\sin x = c$	$u + 2k\pi$ and $(\pi - u) + 2k\pi$ $(k = 0, \pm 1, \pm 2, \pm 3, \ldots)$
$\cos x = c$	$u + 2k\pi$ and $-u + 2k\pi$ $(k = 0, \pm 1, \pm 2, \pm 3, \ldots)$
$\tan x = c$	$u + k\pi$ $(k = 0, \pm 1, \pm 2, \pm 3, \ldots)$

EXAMPLE 4

Find all solutions of $\sec x = 8$ from 0 to 2π.

SOLUTION Note that $\sec x = 8$ exactly when

$$\frac{1}{\cos x} = 8 \qquad \text{or, equivalently,} \qquad \cos x = \frac{1}{8} = .125.$$

Since $\cos^{-1}(.125) = 1.4455$, the solutions of $\cos x = .125$, and hence of $\sec x = 8$, are

$$x = 1.4455 + 2\,k\pi \qquad \text{and} \qquad x = -1.4455 + 2k\pi$$
$$(k = 0, \pm 1, \pm 2, \pm 3, \ldots).$$

Of these solutions, the two between 0 and 2π are

$$x = 1.4455 \qquad \text{and} \qquad x = -1.4455 + 2\pi = 4.8377. \qquad \blacksquare$$

EXAMPLE 5

Solve $\sin u = \sqrt{2}/2$ without using a calculator.

SOLUTION Our knowledge of special values shows that $u = \pi/4$ is one solution. Hence, a second solution is

$$u = \pi - \frac{\pi}{4} = \frac{3\pi}{4},$$

and all solutions are

$$u = \frac{\pi}{4} + 2k\pi \qquad \text{and} \qquad u = \frac{3\pi}{4} + 2k\pi \qquad (k = 0, \pm 1, \pm 2, \pm 3, \ldots). \qquad \blacksquare$$

GRAPHICAL SOLUTION METHOD

When the techniques of the preceding examples are inadequate, trigonometric equations may be solved by the following graphical procedure.

Graphical Method for Solving Trigonometric Equations

1. Write the equation in the form $f(x) = 0$.
2. Determine the period of p of $f(x)$.
3. Graph $f(x)$ over an interval of length p.
4. Use a graphical root finder to determine the x-intercepts of the graph in this interval.
5. For each x-intercept u, all of the numbers

$$u + kp \qquad (k = 0, \pm 1, \pm 2, \pm 3, \ldots)$$

are solutions of the equation.

EXAMPLE 6

Solve $3\sin^2 x - \cos x - 2 = 0$,

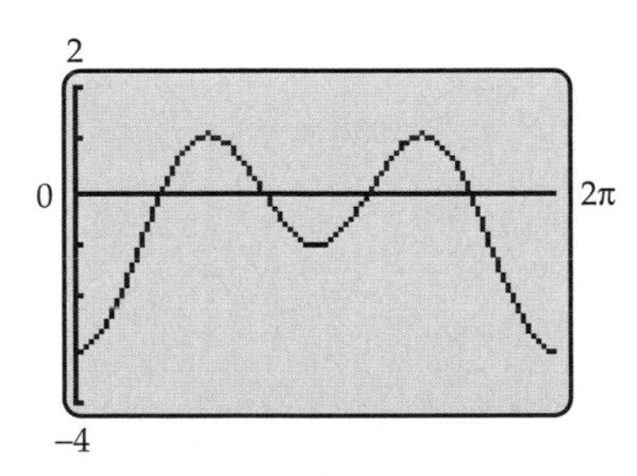

Figure 3–24

SOLUTION Both sine and cosine have period 2π, so $f(x) = 3\sin^2 x - \cos x - 2$ also has period 2π. Figure 3–24 shows one full period of the graph of f. A graphical root finder shows that the four x-intercepts (solutions of the equation) in this window are

$$x = 1.1216, \qquad x = 2.4459, \qquad x = 3.8373, \qquad x = 5.1616.$$

Since the graph repeats its pattern to the left and right, the other x-intercepts (solutions) will differ from these four by multiples of 2π. For instance, in addition to the solution $x = 1.1216$, each of the following is a solution.

$$x = 1.1216 \pm 2\pi, \qquad x = 1.1216 \pm 4\pi, \qquad x = 1.1216 \pm 6\pi, \qquad \text{etc.}$$

A similar analysis applies to the other solutions between 0 and 2π. Hence, all solutions of the equation are given by

$$x = 1.1216 + 2k\pi, \qquad x = 2.4459 + 2k\pi, \qquad x = 3.8373 + 2k\pi,$$
$$x = 5.1616 + 2k\pi, \qquad \text{where } k = 0, \pm 1, \pm 2, \pm 3, \ldots. \qquad \blacksquare$$

EXAMPLE 7

Solve $\tan x = 3\sin 2x$.

SOLUTION We first rewrite the equation as

$$\tan x - 3\sin 2x = 0.$$

Both $\tan x$ and $\sin 2x$ have period π (see pages 178 and 183). Hence, the function given by the left side of the equation, $f(x) = \tan x - 3 \sin 2x$, also has period π. The graph of f from $x = 0$ to $x = \pi$ (Figure 3–25) shows an erroneous vertical line segment at $x = \pi/2$, where tangent is not defined, as well as x-intercepts at the endpoints of the interval. Consequently, we use the more easily read graph f in Figure 3–26, which has $-\pi/2 < x < \pi/2$.

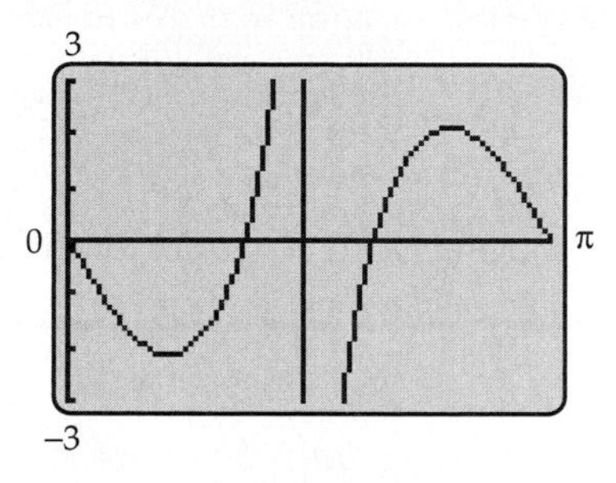

Figure 3–25

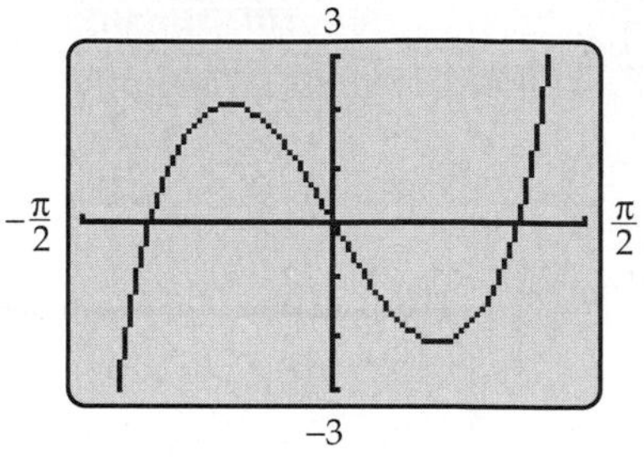

3

$-\frac{\pi}{2}$ $\frac{\pi}{2}$

–3

Figure 3–26

Even without the graph, we can verify that there is an x-intercept at the origin because

$$f(0) = \tan 0 - 3 \sin(2 \cdot 0) = 0.$$

A root finder shows that the other two x-intercepts in Figure 3–26 are

$$x = -1.1503 \qquad \text{and} \qquad x = 1.1503.$$

Since $f(x)$ has period π, all solutions of the equation are given by

$$x = -1.1503 + k\pi, \qquad x = 0 + k\pi, \qquad x = 1.1503 + k\pi$$

$$(k = 0, \pm 1, \pm 2, \pm 3, \ldots). \qquad \blacksquare$$

EXERCISES 3.5

In all exercises, find exact solutions if possible (as in Examples 3 and 5) and approximate ones otherwise. When a calculator is used, round your answers (but not any intermediate results) to four decimal places.

In Exercises 1–10, find all the solutions of the equation.

1. $\sin x = -.465$ **2.** $\sin x = -.682$

3. $\cos x = -.564$ **4.** $\cos x = -.371$

5. $\tan x = -.237$ **6.** $\tan x = -12.45$

7. $\cot x = 2.3$ [*Remember:* $\cot x = 1/\tan x$.]

8. $\cot x = -3.5$ **9.** $\sec x = -2.65$

10. $\csc x = 5.27$

In Exercises 11–14, approximate all solutions of the given equation between 0 and 2π.

11. $\sin x = .119$ **12.** $\cos x = .958$

13. $\tan x = 5$ **14.** $\tan x = 17.65$

In Exercises 15–24, use your knowledge of special values to find the exact solutions of the equation.

15. $\sin x = \sqrt{3}/2$ **16.** $2 \cos x = \sqrt{2}$

17. $\tan x = -\sqrt{3}$ **18.** $\tan x = 1$

19. $2 \cos x = -\sqrt{3}$ **20.** $\sin x = 0$

21. $2 \sin x + 1 = 0$ **22.** $\csc x = \sqrt{2}$

23. $\csc x = 2$ **24.** $-2 \sec x = 4$

In Exercises 25–36, solve the equation graphically.

25. $4 \sin 2x - 3 \cos 2x = 2$ **26.** $5 \sin 3x + 6 \cos 3x = 1$

27. $3 \sin^3 2x = 2 \cos x$

28. $\sin^2 2x - 3\cos 2x + 2 = 0$

29. $\tan x + 5\sin x = 1$

30. $2\cos^2 x + \sin x + 1 = 0$

31. $\cos^3 x - 3\cos x + 1 = 0$

32. $\tan x = 3\cos x$

33. $\cos^4 x - 3\cos^3 x + \cos x = 1$

34. $\sec x + \tan x = 3$

35. $\sin^3 x + 2\sin^2 x - 3\cos x + 2 = 0$

36. $\csc^2 x + \sec x = 1$

*In Exercises 37–46, find all angles θ with $0° \le \theta \le 360°$ that are solutions of the given equation. Find exact solutions if possible and approximations otherwise. [*Hint:* Put your calculator in degree mode, and replace π by $180°$ in the solution algorithms for basic equations.]*

37. $\sin\theta = 1/2$

38. $\sin\theta = .45$

39. $\cos\theta = \sqrt{3}/2$

40. $\cos\theta = -.45$

41. $\tan\theta = 7.95$

42. $\tan\theta = 69.4$

43. $\cos\theta = -.42$

44. $\cot\theta = -2.4$

45. $\sec\theta = \sqrt{2}$

46. $\csc\theta = 5$

Exercises 47–54, deal with a circle of radius r and a central angle of t radians ($0 < t < \pi$), as shown in the figure. The length L of the chord determined by the angle and the area A of the shaded segment are given by

$$L = 2r\sin\frac{t}{2} \quad \text{and} \quad A = \frac{r^2}{2}(t - \sin t).$$

(See Exercise 72 for a proof of the first of these formulas.)

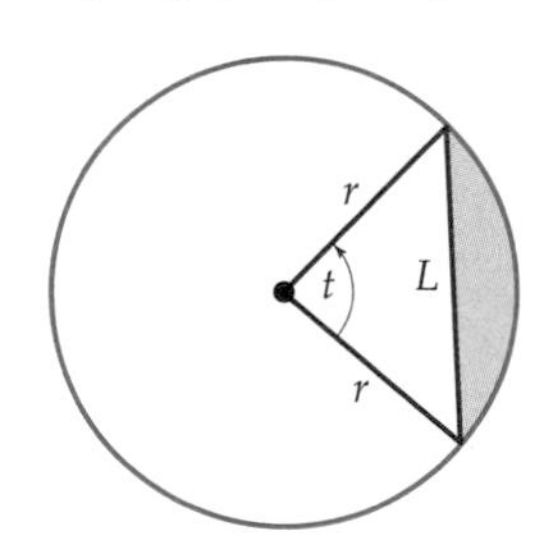

In Exercises 47–50, find the radian measure of the angle and the area of the segment under the given conditions.

47. $r = 5$ and $L = 8$

48. $r = 8$ and $L = 5$

49. $r = 1$ and $L = 1.5$

50. $r = 10$ and $L = 12$.

In Exercises 51–54, find the radian measure of the angle and the length L of the chord under the given conditions.

51. $r = 10$ and $A = 50$

52. $r = 1$ and $A = .5$

53. $r = 8$ and $A = 20$

54. $r = 5$ and $A = 2$

Exercises 55 and 56 deal with a rectangle inscribed in the segment of the graph of $f(x) = 2\cos 2x$ shown in the figure.

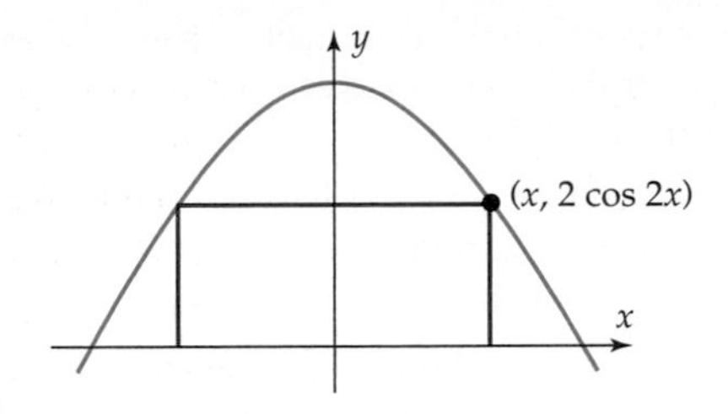

55. (a) Find a formula for the area of the rectangle in terms of x. [*Hint:* The length is $2x$.]
(b) For what values of x does the rectangle have an area of 1 square unit?

56. Use the formula in Exercise 55(a) to determine the value of x that determines the rectangle with the largest possible area. What is this maximum area?

At the instant you hear a sonic boom from an airplane overhead, your angle of elevation α to the plane is given by the equation $\sin\alpha = 1/m$, where m is the Mach number for the speed of the plane (Mach 1 is the speed of sound, Mach 2.5 is 2.5 times the speed of sound, etc.). In Exercises 57–60, find the angle of elevation (in degrees) for the given Mach number. Remember that an angle of elevation must be between $0°$ and $90°$.

57. $m = 1.1$

58. $m = 1.6$

59. $m = 2$

60. $m = 2.4$

When a light beam passes from one medium to another (for instance, from air to water), both its speed and direction change. According to Snell's Law of Refraction,

$$\frac{\sin\theta_1}{\sin\theta_2} = \frac{v_1}{v_2},$$

where v_1 is the speed of light in the first medium, v_2 is its speed in the second medium, θ_1 is the angle of incidence, and θ_2 is the angle of refraction, as shown in the figure. The number v_1/v_2 is called the index of refraction. *Use this information to do Exercises 61–64 on the next page.*

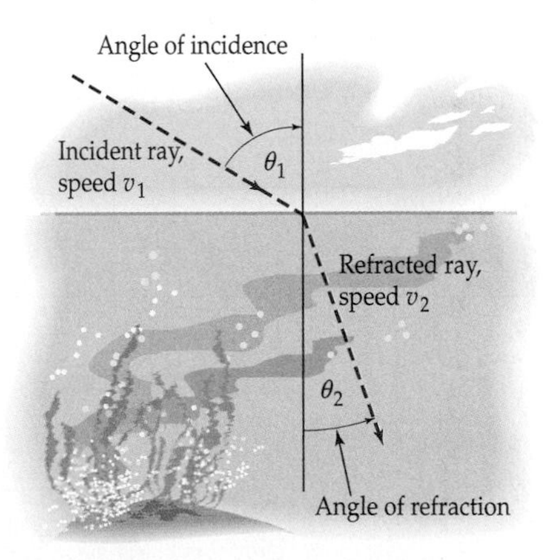

61. The index of refraction of light passing from air to water is 1.33. If the angle of incidence is 38°, find the angle of refraction.

62. The index of refraction of light passing from air to ordinary glass is 1.52. If the angle of incidence is 17°, find the angle of refraction.

63. The index of refraction of light passing from air to dense glass is 1.66. If the angle of incidence is 24°, find the angle of refraction.

64. The index of refraction of light passing from air to quartz is 1.46. If the angle of incidence is 50°, find the angle of refraction.

65. The number of hours of daylight in Detroit on day t of a non–leap year (with $t = 0$ being January 1) is given by the function

$$d(t) = 3\sin\left[\frac{2\pi}{365}(t - 80)\right] + 12.$$

(a) On what days of the year are there exactly 11 hours of daylight?
(b) What day has the maximum amount of daylight?

66. A weight hanging from a spring is set into motion (see Figure 2–65 on page 190), moving up and down. Its distance (in centimeters) above or below the equilibrium point at time t seconds is given by

$$d = 5(\sin 6t - 4\cos 6t).$$

At what times during the first 2 seconds is the weight at the equilibrium position ($d = 0$)?

In Exercises 67–70, use the following fact: When a projectile (such as a ball or a bullet) leaves its starting point at angle of elevation θ with velocity v, the horizontal distance d it travels is given by the equation

$$d = \frac{v^2}{32}\sin 2\theta,$$

where d is measured in feet and v in feet per second. Note that the horizontal distance traveled may be the same for two different angles of elevation, so some of these exercises may have more than one correct answer.

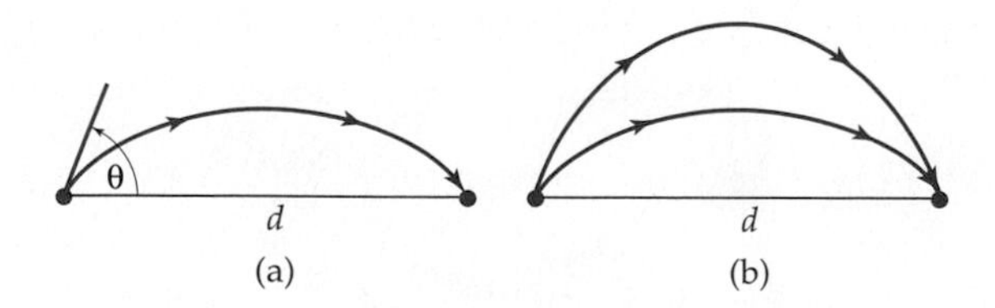

67. If muzzle velocity of a rifle is 300 feet per second, at what angle of elevation (in radians) should it be aimed for the bullet to hit a target 2500 feet away?

68. Is it possible for the rifle in Exercise 67 to hit a target that is 3000 feet away? [At what angle of elevation would it have to be aimed?]

69. A fly ball leaves the bat at a velocity of 98 mph and is caught by an outfielder 288 feet away. At what angle of elevation (in degrees) did the ball leave that bat?

70. An outfielder throws the ball at a speed of 75 mph to the catcher who is 200 feet away. At what angle of elevation was the ball thrown?

Thinkers

71. Under what conditions (on the constant) does a basic equation involving the sine or cosine function have *no* solutions?

72. Prove the formula $L = 2r\sin\frac{t}{2}$ used in Exercises 47–54 as follows.

(a) Construct the perpendicular line from the center of the circle to the chord PQ, as shown in the figure. Verify that triangles OCP and OCQ are congruent. [*Hint:* Angles P and Q are equal by the Isosceles Triangle Theorem,* and in each triangle, angle C is a right angle (why?). Use the Congruent Triangles Theorem.*]
(b) Use part (a) to explain why angle POC measures $t/2$ radians.
(c) Show that the length of PC is $r\sin\frac{t}{2}$.
(d) Use the fact that PC and QC have the same length to conclude that the length L of PC is

$$L = 2r\sin\frac{t}{2}.$$

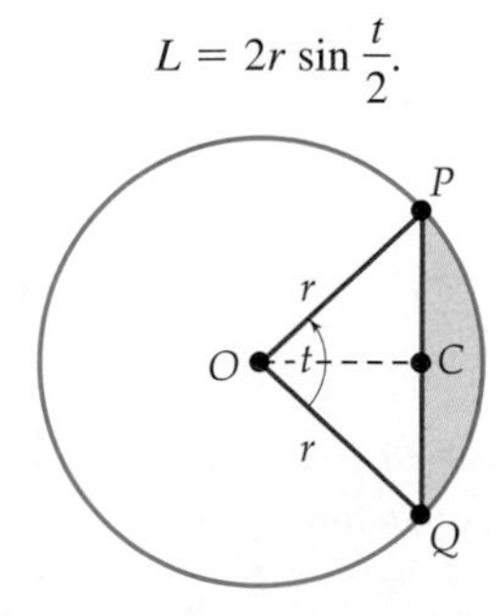

*See the Geometry Review Appendix.

3.5.A *SPECIAL TOPICS* Other Solution Methods for Trigonometric Equations

Many trigonometric equations can be solved algebraically by using factoring, the quadratic formula, and identities to reduce the problem to an equivalent one that involves only basic equations.

EXAMPLE 1

Solve exactly: $\sin 2x = \sqrt{2}/2$.

SOLUTION First, let $u = 2x$ and solve the basic equation $\sin u = \sqrt{2}/2$. As we saw in Example 5 on page 264, the solutions are

$$u = \frac{\pi}{4} + 2k\pi \quad \text{and} \quad u = \frac{3\pi}{4} + 2k\pi \qquad (k = 0, \pm 1, \pm 2, \pm 3, \ldots).$$

Since $u = 2x$, each of these solutions leads to a solution of the original equation.

$$2x = u = \frac{\pi}{4} + 2k\pi \quad \text{or, equivalently,} \quad x = \frac{1}{2}\left(\frac{\pi}{4} + 2k\pi\right) = \frac{\pi}{8} + k\pi.$$

Similarly,

$$2x = u = \frac{3\pi}{4} + 2k\pi \quad \text{or, equivalently,} \quad x = \frac{1}{2}\left(\frac{3\pi}{4} + 2k\pi\right) = \frac{3\pi}{8} + k\pi.$$

Therefore, all solutions of $\sin 2x = \sqrt{2}/2$ are given by

$$x = \frac{\pi}{8} + k\pi \quad \text{and} \quad x = \frac{3\pi}{8} + k\pi \qquad (k = 0, \pm 1, \pm 2, \pm 3, \ldots).$$

The fact that the solutions are obtained by adding multiples of π rather than 2π is a reflection of the fact that the period of $\sin 2x$ is π. ■

The technique used in Example 1 can be used to solve similar equations, such as

$$\cos 4x = .8, \qquad \tan\frac{x}{2} = 5, \qquad \sin 3x = .45.$$

For example, we would use the substitution $u = 4x$ in the first equation above and the substitution $u = x/2$ in the second equation.

Trigonometric identities can sometimes be used to replace an equation with an equivalent one that is easier to solve. This is often true when the original equation involves several trigonometric functions.

EXAMPLE 2

Solve $-10\cos^2 x - 3\sin x + 9 = 0$.

SOLUTION We first use the Pythagorean identity to rewrite the equation in terms of the sine function.

$$-10\cos^2 x - 3\sin x + 9 = 0$$
$$-10(1 - \sin^2 x) - 3\sin x + 9 = 0$$
$$-10 + 10\sin^2 x - 3\sin x + 9 = 0$$
$$10\sin^2 x - 3\sin x - 1 = 0$$

Now factor the left-hand side:*

$$(2\sin x - 1)(5\sin x + 1) = 0$$

$$2\sin x - 1 = 0 \qquad \text{or} \qquad 5\sin x + 1 = 0$$
$$2\sin x = 1 \qquad\qquad 5\sin x = -1$$
$$\sin x = 1/2 \qquad\qquad \sin x = -1/5 = -.2.$$

Each of these basic equations is readily solved. We note that $\sin(\pi/6) = 1/2$, so $x = \pi/6$ and $x = \pi - \pi/6 = 5\pi/6$ are solutions of the first one. Since $\sin^{-1}(-.2) = -.2014$, both $x = -.2014$ and $x = \pi - (-.2014) = 3.3430$ are solutions of the second equation. Therefore, all solutions of the original equation are given by

$$x = \frac{\pi}{6} + 2k\pi, \qquad x = \frac{5\pi}{6} + 2k\pi, \qquad x = -.2014 + 2k\pi,$$
$$x = 3.3430 + 2k\pi,$$

where $k = 0, \pm 1, \pm 2, \pm 3, \ldots$. ■

EXAMPLE 3

Solve $\sec^2 x + 5\tan x = -2$.

SOLUTION We use the Pythagorean identity $\sec^2 x = 1 + \tan^2 x$ to obtain an equivalent equation.

$$\sec^2 x + 5\tan x = -2$$
$$(1 + \tan^2 x) + 5\tan x + 2 = 0$$
$$\tan^2 x + 5\tan x + 3 = 0.$$

If we let $u = \tan x$, this last equation becomes $u^2 + 5u + 3 = 0$. Since the left-hand side does not readily factor, we use the quadratic formula to solve the equation.

$$u = \frac{-5 \pm \sqrt{5^2 - 4 \cdot 1 \cdot 3}}{2} = \frac{-5 \pm \sqrt{13}}{2}.$$

*The factorization may be easier to see if you first substitute v for $\sin x$, so that $10\sin^2 x - 3\sin x - 1$ becomes $10v^2 - 3v - 1 = (2v - 1)(5v + 1)$.

Since $u = \tan x$, the original equation is equivalent to

$$\tan x = \frac{-5 + \sqrt{13}}{2} \approx -.6972 \qquad \text{or} \qquad \tan x = \frac{-5 - \sqrt{13}}{2} \approx -4.3028.$$

Solving these basic equations as above, we find that $x = -.6089$ is a solution of the first and $x = -1.3424$ is a solution of the second. Hence, the solutions of the original equation are

$$x = -.6089 + k\pi \qquad \text{and} \qquad x = -1.3424 + k\pi$$
$$(k = 0, \pm 1, \pm 2, \pm 3, \ldots). \quad ■$$

EXAMPLE 4

Solve $5 \cos x + 3 \cos 2x = 3$.

SOLUTION We use the double-angle identity: $\cos 2x = 2\cos^2 x - 1$ as follows.

$$5 \cos x + 3 \cos 2x = 3$$

Use double-angle identity: $5 \cos x + 3(2\cos^2 x - 1) = 3$

Multiply out left-hand side: $5 \cos x + 6\cos^2 x - 3 = 3$

Rearrange terms: $6\cos^2 x + 5\cos x - 6 = 0$

Factor left-hand side: $(2\cos x + 3)(3\cos x - 2) = 0$

$$2\cos x + 3 = 0 \qquad \text{or} \qquad 3\cos x - 2 = 0$$
$$2\cos x = -3 \qquad\qquad 3\cos x = 2$$
$$\cos x = -\frac{3}{2} \qquad\qquad \cos x = \frac{2}{3}.$$

The equation $\cos x = -3/2$ has no solutions because $\cos x$ always lies between -1 and 1. A calculator shows that the solutions of $\cos x = 2/3$ are

$$x = .8411 + 2k\pi \qquad \text{and} \qquad x = -.8411 + 2k\pi$$
$$(k = 0, \pm 1, \pm 2, \pm 3, \ldots). \quad ■$$

✓ EXERCISES 3.5.A

Directions: *Find exact solutions if possible and approximate ones otherwise. Round approximate answers (but not any intermediate computations) to four decimal places.*

In Exercises 1–10, use substitution to solve the equation.

1. $\sin 2x = -\sqrt{3}/2$
2. $\cos 2x = \sqrt{2}/2$
3. $2 \cos \frac{x}{2} = \sqrt{2}$
4. $2 \sin \frac{x}{3} = 1$
5. $\tan 3x = -\sqrt{3}$
6. $5 \sin 2x = 2$
7. $5 \cos 3x = -3$
8. $2 \tan 4x = 16$
9. $4 \tan \frac{x}{2} = 8$
10. $5 \sin \frac{x}{4} = 4$

In Exercises 11–36, use factoring, the quadratic formula, or identities to solve the equation. Find all solutions between 0 and 2π.

11. $3\sin^2 x - 8 \sin x - 3 = 0$

12. $5\cos^2 x + 6\cos x = 8$

13. $2\tan^2 x + 5\tan x + 3 = 0$

14. $3\sin^2 x + 2\sin x = 5$

15. $\cot x \cos x = \cos x$ [Be careful; see Exercise 37.]

16. $\tan x \cos x = \cos x$

17. $\cos x \csc x = 2\cos x$

18. $\tan x \sec x + 3\tan x = 0$

19. $4\sin x \tan x - 3\tan x + 20\sin x - 15 = 0$
[*Hint:* One factor is $\tan x + 5$.]

20. $25\sin x \cos x - 5\sin x + 20\cos x = 4$

21. $\sin^2 x + 2\sin x - 2 = 0$

22. $\cos^2 x + 5\cos x = 1$

23. $\tan^2 x + 1 = 3\tan x$

24. $4\cos^2 x - 2\cos x = 1$

25. $2\tan^2 x - 1 = 3\tan x$

26. $6\sin^2 x + 4\sin x = 1$

27. $\sin^2 x + 3\cos^2 x = 0$

28. $\sec^2 x - 2\tan^2 x = 0$

29. $\sin 2x + \cos x = 0$

30. $\cos 2x - \sin x = 1$

31. $9 - 12\sin x = 4\cos^2 x$

32. $\sec^2 x + \tan x = 3$

33. $\cos^2 x - \sin^2 x + \sin x = 0$

34. $2\tan^2 x + \tan x = 5 - \sec^2 x$

35. $\sin\frac{x}{2} = 1 - \cos x$

36. $4\sin^2\left(\frac{x}{2}\right) + \cos^2 x = 2$

37. What is wrong with this so-called solution?

$$\sin x \tan x = \sin x$$

$$\tan x = 1$$

$$x = \frac{\pi}{4} \quad \text{or} \quad \frac{5\pi}{4}.$$

[*Hint:* Solve the original equation by moving all terms to one side and factoring. Compare your answers with the ones shown here.]

38. Let n be a fixed positive integer. Describe *all* solutions of the equation $\sin nx = 1/2$. [*Hint:* See Exercises 1–10.]

Chapter 3 Review

IMPORTANT

CONCEPTS

IMPORTANT

FACTS & FORMULAS

- All identities in the Chapter 2 Review
- *Addition and Subtraction Identities:*

$$\sin(x + y) = \sin x \cos y + \cos x \sin y$$

$$\sin(x - y) = \sin x \cos y - \cos x \sin y$$

$$\cos(x + y) = \cos x \cos y - \sin x \sin y$$

$$\cos(x - y) = \cos x \cos y + \sin x \sin y$$

$$\tan(x + y) = \frac{\tan x + \tan y}{1 - \tan x \tan y}$$

$$\tan(x - y) = \frac{\tan x - \tan y}{1 + \tan x \tan y}$$

- *Cofunction Identities:*

$$\sin x = \cos\left(\frac{\pi}{2} - x\right) \qquad \cos x = \sin\left(\frac{\pi}{2} - x\right)$$

$$\tan x = \cot\left(\frac{\pi}{2} - x\right) \qquad \cot x = \tan\left(\frac{\pi}{2} - x\right)$$

$$\sec x = \csc\left(\frac{\pi}{2} - x\right) \qquad \csc x = \sec\left(\frac{\pi}{2} - x\right)$$

- *Double-Angle Identities:*

$$\sin 2x = 2\sin x\cos x$$

$$\cos 2x = \cos^2 x - \sin^2 x$$

$$\cos 2x = 2\cos^2 x - 1 \qquad \cos 2x = 1 - 2\sin^2 x$$

$$\tan 2x = \frac{2\tan x}{1-\tan^2 x}$$

- *Half-Angle Identities:*

$$\sin\frac{x}{2} = \pm\sqrt{\frac{1-\cos x}{2}}$$

$$\cos\frac{x}{2} = \pm\sqrt{\frac{1+\cos x}{2}}$$

$$\tan\frac{x}{2} = \frac{1-\cos x}{\sin x} \qquad \tan\frac{x}{2} = \frac{\sin x}{1+\cos x} \qquad \tan\frac{x}{2} = \pm\sqrt{\frac{1-\cos x}{1+\cos x}}$$

- $\sin^{-1} v = u$ exactly when $\sin u = v$ $\left(-\frac{\pi}{2} \le u \le \frac{\pi}{2}, -1 \le v \le 1\right)$
- $\cos^{-1} v = u$ exactly when $\cos u = v$ $(0 \le u \le \pi, -1 \le v \le 1)$
- $\tan^{-1} v = u$ exactly when $\tan u = v$ $\left(-\frac{\pi}{2} < u < \frac{\pi}{2}, \text{any } v\right)$

REVIEW QUESTIONS 

In Questions 1–4, simplify the given expression.

1. $\dfrac{\sin^2 t + (\tan^2 t + 2\tan t - 4) + \cos^2 t}{3\tan^2 t - 3\tan t}$

2. $\dfrac{\sec^2 t \csc t}{\csc^2 t \sec t}$

3. $\dfrac{\tan^2 x - \sin^2 x}{\sec^2 x}$

4. $\dfrac{(\sin x + \cos x)(\sin x - \cos x) + 1}{\sin^2 x}$

In Questions 5–12, determine graphically whether the equation could possibly be an identity. If it could, prove that it is.

5. $\sin^4 t - \cos^4 t = 2\sin^2 t - 1$

6. $1 + 2\cos^2 t + \cos^4 t = \sin^4 t$

7. $\dfrac{\sin t}{1-\cos t} = \dfrac{1+\cos t}{\sin t}$

8. $\dfrac{\sin^2 t}{\cos^2 t} + 1 = \dfrac{1}{\cos^2 t}$

9. $\dfrac{\cos^2(\pi + t)}{\sin^2(\pi + t)} - 1 = \dfrac{1}{\sin^2 t}$

10. $\tan x + \cot x = \sec x \csc x$

11. $(\sin x + \cos x)^2 - \sin 2x = 1$

12. $\dfrac{1-\cos 2x}{\tan x} = \sin 2x$

In Questions 13–22, prove the given identity.

13. $\dfrac{\tan x - \sin x}{2\tan x} = \sin^2\left(\dfrac{x}{2}\right)$

14. $2\cos x - 2\cos^3 x = \sin x \sin 2x$

15. $\cos(x+y)\cos(x-y) = \cos^2 x - \sin^2 y$

16. $\dfrac{\cos(x-y)}{\cos x \cos y} = 1 + \tan x \tan y$

17. $\dfrac{\sec x + 1}{\tan x} = \dfrac{\tan x}{\sec x - 1}$

18. $\dfrac{\cos^4 x - \sin^4 x}{1 - \tan^4 x} = \cos^4 x$

19. $\dfrac{1 + \tan^2 x}{\tan^2 x} = \csc^2 x$

20. $\sec x - \cos x = \sin x \tan x$

21. $\tan^2 x - \sec^2 x = \cot^2 x - \csc^2 x$

22. $\sin 2x = \dfrac{1}{\tan x + \cot 2x}$

REVIEW QUESTIONS

23. If $\tan x = 5/12$ and $\sin x > 0$, find $\sin 2x$.
24. If $\cos x = 15/17$ and $0 < x < \pi/2$, find $\sin(x/2)$.
25. If $\tan x = 4/3$ with $\pi < x < 3\pi/2$, and $\cot y = -5/12$ with $3\pi/2 < y < 2\pi$, find $\sin(x - y)$.
26. If $\sin x = -12/13$ with $\pi < x < 3\pi/2$, and $\sec y = 13/12$ with $3\pi/2 < y < 2\pi$, find $\cos(x + y)$.
27. If $\sin x = 1/4$ and $0 < x < \pi/2$, then $\sin(\pi/3 + x) = ?$
28. If $\sin x = -2/5$ and $3\pi/2 < x < 2\pi$, then $\cos(\pi/4 + x) = ?$
29. If $\sin x = 0$, is it true that $\sin 2x = 0$? Justify your answer.
30. If $\cos x = 0$, is it true that $\cos 2x = 0$? Justify your answer.
31. Show that
$$\sqrt{2 + \sqrt{3}} = \frac{\sqrt{2} + \sqrt{6}}{2}$$
by computing $\cos(\pi/12)$ in two ways, using the half-angle identity and the subtraction identity for cosine.
32. True or false: $2 \sin x = \sin 2x$. Justify your answer.
33. $\sin(5\pi/12) = ?$
34. Express $\sec(x - \pi)$ in terms of $\sin x$ and $\cos x$.
35. $\sqrt{\dfrac{1 - \cos^2 x}{1 - \sin^2 x}} = $ ____.
    (a) $|\tan x|$ (b) $|\cot x|$
    (c) $\sqrt{\dfrac{1 - \sin^2 x}{1 - \cos^2 x}}$ (d) $\sec x$ (e) undefined
36. $\dfrac{1}{(\csc x)(\sec^2 x)} = $ ____.
    (a) $\dfrac{1}{(\sin x)(\cos^2 x)}$ (b) $\sin x - \sin^3 x$
    (c) $\dfrac{1}{(\sin x)(1 + \tan^2 x)}$ (d) $\sin x - \dfrac{1}{1 + \tan^2 x}$
    (e) $1 + \tan^3 x$
37. If $\sin x = .6$ and $0 < x < \pi/2$, find $\sin 2x$.
38. If $\sin x = .6$ and $0 < x < \pi/2$, find $\sin(x/2)$.
39. Find the angle of inclination of the straight line through the points (2, 6) and (−2, 2).
40. Find one of the angles between the line L through the points (−3, 2) and (5, 1) and the line M, which has slope 2.
41. $\cos^{-1}(\sqrt{2}/2) = ?$
42. $\sin^{-1}(\sqrt{3}/2) = ?$
43. $\tan^{-1} \sqrt{3} = ?$
44. $\sin^{-1}(\cos 11\pi/6) = ?$
45. $\cos^{-1}(\sin 5\pi/3) = ?$
46. $\tan^{-1}(\cos 7\pi/2) = ?$
47. $\sin^{-1}(\sin .75) = ?$
48. $\cos^{-1}(\cos 2) = ?$
49. $\sin^{-1}(\sin 8\pi/3) = ?$
50. $\cos^{-1}(\cos 13\pi/4) = ?$
51. Sketch the graph of $f(x) = \tan^{-1} x - \pi$.
52. Sketch the graph of $g(x) = \sin^{-1}(x - 2)$.
53. Find the exact value of $\sin[\cos^{-1}(1/4)]$.
54. Find the exact value of $\sin[\tan^{-1}(1/2) - \cos^{-1}(4/5)]$.

In Questions 55–70, solve the equation by any means. Find exact solutions when possible and approximate ones otherwise.

55. $2 \sin x = 1$
56. $\cos x = \sqrt{3}/2$
57. $\tan x = -1$
58. $\sin 3x = -\sqrt{3}/2$
59. $\sin x = .7$
60. $\cos x = -.8$
61. $\tan x = 13$
62. $\cot x = .4$
63. $2 \sin^2 x + 5 \sin x = 3$
64. $4 \cos^2 x - 2 = 0$
65. $2 \sin^2 x - 3 \sin x = 2$
66. $\cos 2x = \cos x$ [*Hint:* First use an identity.]
67. $\sec^2 x + 3 \tan^2 x = 13$
68. $\sec^2 x = 4 \tan x - 2$
69. $2 \sin^2 x + \sin x - 2 = 0$
70. $\cos^2 x - 3 \cos x - 2 = 0$

In Questions 71–74, solve the equation graphically.

71. $5 \tan x = 2 \sin 2x$
72. $\sin^3 x + \cos^2 x - \tan x = 2$
73. $\sin x + \sec^2 x = 3$
74. $\cos^2 x - \csc^2 x + \tan(x - \pi/2) + 5 = 0$
75. Find all angles θ with $0° \le \theta \le 360°$ such that $\sin \theta = -.7133$.
76. Find all angles θ with $0° \le \theta \le 360°$ such that $\tan \theta = 3.7321$.
77. A cannon has a muzzle velocity of 600 feet per second. At what angle of elevation should it be fired in order to hit a target 3500 feet away? [*Hint:* Use the projectile equation preceding Exercise 67 of Section 3.5.]
78. A weight hanging from a spring is set into motion (see Figure 2–65 on page 190), moving up and down. Its distance (in centimeters) above or below the equilibrium point at time t seconds is given by $d = 5 \sin 3t - 3 \cos 3t$. At what times during the first 2 seconds is the weight at the equilibrium position ($d = 0$)?

DISCOVERY PROJECT 3 The Sun and the Moon

Jack Hollingsworth/Getty Images

It has long been known that the cycles of the sun and the moon are periodic; that is, the moon is full at regular intervals and new at regular intervals. The same is true of the sun; the interval between the summer and winter solstices is also regular. It is therefore quite natural to use the sun and the moon to keep track of time, and to study the interaction between the two to predict events such as full moons, new moons, solstices, equinoxes, and eclipses. Indeed, these solar and lunar events have consequences on the earth, including the succession of the seasons and the severity of tides.

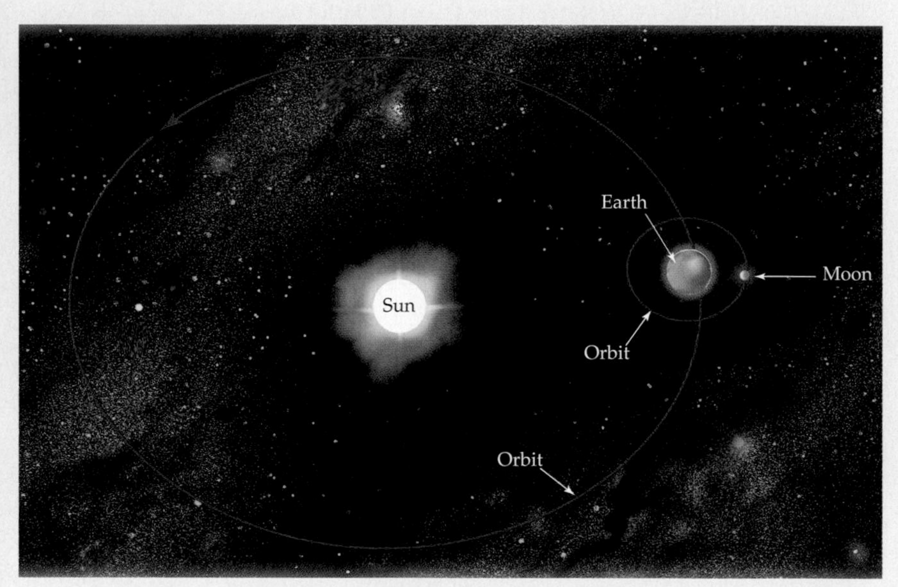

1. The following is a list of the days in 1999 when the moon was full. The length of the lunar month is the length of time between full moons. Use the data to approximate the length of the lunar month.

January 2	May 30	October 24
January 31	June 28	November 23
March 2	July 28	December 22
March 31	August 26	
April 30	September 25	

2. Write a function that has value 1 when the moon is full and 0 when the moon is new. Measure the independent variable in days with January 2, 1999, set as time 0. Use a function of the form

$$m(x) = \frac{\cos kt + 1}{2}$$

with period equal to the length of the lunar month.

3. Use your function to predict the date of the first full moon of the 21st century (in January of the year 2001). Does your function agree with the actual date of January 9? If not, what could have caused the discrepancy?

4. The solar year is approximately 365.24 days long. Write a function $s(x)$ with period equal to the length of the solar year so that on the date of the summer solstice, $s(x) = 1$ and on the day of the winter solstice, $s(x) = 0$. The summer solstice in 1999 was June 21, and the winter solstice falls midway between summer solstices.

5. If your functions $s(x)$ and $m(x)$ were accurate, when would you expect to see the next full moon on the summer solstice?

6. How would you go about making your models $s(x)$ and $m(x)$ more accurate?

C H A P T E R 4

Applications of Trigonometry

Is this bridge safe?

When planning a bridge or a building, architects and engineers must determine the stress on cables and other parts of the structure to be sure that all parts are adequately supported. Problems like these can be modeled and solved by using vectors. See Exercise 73 on page 314.

Chapter Outline

Interdependence of Sections

This chapter has two independent parts, which may be read in either order.

4.1 ⟶ 4.2 ⟶ 4.3

4.4 ⟶ 4.5

In this chapter we investigate the applications of trigonometry to complex numbers (Sections 4.1–4.3) and to vectors, which are used in physics and engineering (Sections 4.4 and 4.5).

4.1 Complex Numbers

The equation $x^2 = 2$ has no solutions in the rational number system but has $\sqrt{2}$ and $-\sqrt{2}$ as solutions in the real number system. So the idea of enlarging a number system to solve an equation that can't be solved in the present system is a natural one.

Equations such as $x^2 = -1$ and $x^2 = -4$ have no solutions in the real number system because $\sqrt{-1}$ and $\sqrt{-4}$ are not real numbers. To solve such equations (or, equivalently, to find square roots of negative numbers), we must enlarge the number system again. We claim that there is a number system, called the **complex number system,** with the following properties.

Properties of the Complex Number System

1. The complex number system contains all real numbers.
2. Addition, subtraction, multiplication, and division of complex numbers obey the same rules of arithmetic that hold in the real number system, with one exception: The exponent laws hold for *integer* exponents, but not necessarily for fractional ones.
3. The complex number system contains a number (usually denoted by i) such that $i^2 = -1$.
4. Every complex number can be written in the **standard form** $a + bi$, where a and b are real numbers.*
5. Two complex numbers $a + bi$ and $c + di$ are equal exactly when $a = c$ and $b = d$.

In view of our past experience with enlarging the number system, this claim *ought* to appear plausible. But the mathematicians who invented the complex numbers in the seventeenth century were very uneasy about a number i such that $i^2 = -1$ (that is, $i = \sqrt{-1}$). Consequently, they called numbers of the form bi (b any real number), such as $5i$ and $-\frac{1}{4}i$, **imaginary numbers.** The old familiar numbers (integers, rationals, irrationals) were called **real numbers.** Sums of real and imaginary numbers, numbers of the form $a + bi$, such as

$$5 + 2i, \qquad 7 - 4i, \qquad 18 + \frac{3}{2}i, \qquad \sqrt{3} - 12i$$

were called **complex numbers.**†

Every real number is a complex number; for instance, $7 = 7 + 0i$. Similarly, every imaginary number bi is a complex number, since $bi = 0 + bi$. Since the usual laws of arithmetic still hold, it's easy to add, subtract, and multiply complex numbers. As the following examples demonstrate, *all symbols can be treated as if they were real numbers, provided that* i^2 *is replaced by* -1. Unless directed otherwise, express your answers in the standard form $a + bi$.

EXAMPLE 1

(a) $(1 + i) + (3 - 7i) = 1 + i + 3 - 7i$
$= (1 + 3) + (i - 7i) = 4 - 6i.$

(b) $(4 + 3i) - (8 - 6i) = 4 + 3i - 8 - (-6i)$
$= (4 - 8) + (3i + 6i) = -4 + 9i.$

---

*Hereafter, whenever we write $a + bi$ or $c + di$, it is assumed that a, b, c, d are real numbers and $i^2 = -1$.

†This terminology is still used, even though there is nothing complicated, unreal, or imaginary about complex numbers—they are just as valid mathematically as are real numbers.

(c) $4i\left(2+\frac{1}{2}i\right) = 4i \cdot 2 + 4i\left(\frac{1}{2}i\right) = 8i + 4 \cdot \frac{1}{2} \cdot i^2$
$= 8i + 2i^2 = 8i + 2(-1) = -2 + 8i.$

(d) $(2+i)(3-4i) = 2 \cdot 3 + 2(-4i) + i \cdot 3 + i(-4i)$
$= 6 - 8i + 3i - 4i^2 = 6 - 8i + 3i - 4(-1)$
$= (6+4) + (-8i + 3i) = 10 - 5i.$ ■

The familiar multiplication patterns and exponent laws for integer exponents hold in the complex number system.

EXAMPLE 2

(a) $(3+2i)(3-2i) = 3^2 - (2i)^2$
$= 9 - 4i^2 = 9 - 4(-1) = 9 + 4 = 13.$

(b) $(4+i)^2 = 4^2 + 2 \cdot 4 \cdot i + i^2 = 16 + 8i + (-1) = 15 + 8i.$

(c) To find i^{54}, we first note that $i^4 = i^2 i^2 = (-1)(-1) = 1$ and that

$$54 = 52 + 2 = 4 \cdot 13 + 2.$$

Consequently,

$$i^{54} = i^{52+2} = i^{52} i^2 = i^{4 \cdot 13} i^2 = (i^4)^{13} i^2 = 1^{13}(-1) = -1.$$ ■

TECHNOLOGY TIP

To do complex arithmetic on TI-86 and HP-39+, enter $a + bi$ as (a, b). On other calculators, use the special i key whose location is

TI-83+/89: keyboard

Casio FX 2.0: keyboard

Casio: 9850: OPTN/CPLX

The **conjugate** of the complex number $a + bi$ is the number $a - bi$, and the conjugate of $a - bi$ is $a + bi$. For example, the conjugate of $3 + 4i$ is $3 - 4i$ and the conjugate of $-3i = 0 - 3i$ is $0 + 3i = 3i$. *Every real number is its own conjugate;* for instance, the conjugate of $17 = 17 + 0i$ is $17 - 0i = 17$.

For any complex number $a + bi$, we have

$$(a+bi)(a-bi) = a^2 - (bi)^2 = a^2 - b^2 i^2 = a^2 - b^2(-1) = a^2 + b^2.$$

Since a^2 and b^2 are nonnegative real numbers, so is $a^2 + b^2$. Therefore, *the product of a complex number and its conjugate is a nonnegative real number.* This fact enables us to express quotients of complex numbers in standard form.

EXAMPLE 3

To express $\frac{3+4i}{1+2i}$ in the form $a + bi$, *multiply both numerator and denominator by the conjugate of the denominator,* namely, $1 - 2i$:

$$\frac{3+4i}{1+2i} = \frac{3+4i}{1+2i} \cdot \frac{1-2i}{1-2i} = \frac{(3+4i)(1-2i)}{(1+2i)(1-2i)}$$
$$= \frac{3+4i-6i-8i^2}{1^2-(2i)^2} = \frac{3+4i-6i-8(-1)}{1-4i^2} = \frac{11-2i}{1-4(-1)}$$
$$= \frac{11-2i}{5} = \frac{11}{5} - \frac{2}{5}i.$$

This is the form $a + bi$ with $a = 11/5$ and $b = -2/5$. ■

EXAMPLE 4

To express $\dfrac{1}{1-i}$ in standard form, note that the conjugate of the denominator is $1+i$, and therefore

$$\frac{1}{1-i} = \frac{1\cdot(1+i)}{(1-i)(1+i)} = \frac{1+i}{1^2-i^2} = \frac{1+i}{1-(-1)} = \frac{1+i}{2} = \frac{1}{2}+\frac{1}{2}i.$$

We can check this result by multiplying $\frac{1}{2}+\frac{1}{2}i$ by $1-i$ to see whether the product is 1 $\left(\text{which it should be if } \frac{1}{2}+\frac{1}{2}i = \frac{1}{1-i}\right)$:

$$\left(\frac{1}{2}+\frac{1}{2}i\right)(1-i) = \frac{1}{2}\cdot 1 - \frac{1}{2}i + \frac{1}{2}i\cdot 1 - \frac{1}{2}i^2 = \frac{1}{2} - \frac{1}{2}(-1) = 1. \quad \blacksquare$$

Since $i^2 = -1$, we define $\sqrt{-1}$ to be the complex number i. Similarly, since $(5i)^2 = 5^2i^2 = 25(-1) = -25$, we define $\sqrt{-25}$ to be $5i$. In general,

Square Roots of Negative Numbers

Let b be a positive real number.

$$\sqrt{-b} \text{ is defined to be } \sqrt{b}\,i$$

because $(\sqrt{b}\,i)^2 = (\sqrt{b})^2 i^2 = b(-1) = -b$.

CAUTION

$\sqrt{b}\,i$ is *not* the same as $\sqrt{bi}$. To avoid confusion, it may help to write $\sqrt{b}\,i$ as $i\sqrt{b}$.

EXAMPLE 5

(a) $\sqrt{-3} = \sqrt{3}i = i\sqrt{3}$.

(b) $\dfrac{1-\sqrt{-7}}{3} = \dfrac{1-\sqrt{7}i}{3} = \dfrac{1}{3} - \dfrac{\sqrt{7}}{3}i.$ ■

CAUTION

The property $\sqrt{cd} = \sqrt{c}\sqrt{d}$ (or equivalently in exponential notation, $(cd)^{1/2} = c^{1/2}d^{1/2}$), which is valid for positive real numbers, does *not hold* when both c and d are negative. For example,

$$\sqrt{-20}\sqrt{-5} = \sqrt{20}i\cdot\sqrt{5}i = \sqrt{20}\sqrt{5}\cdot i^2 = \sqrt{20\cdot 5}(-1)$$
$$= \sqrt{100}(-1) = -10,$$

but $\sqrt{(-20)(-5)} = \sqrt{100} = 10$. So

$$\sqrt{(-20)(-5)} \neq \sqrt{-20}\sqrt{-5}.$$

To avoid difficulty, *always write square roots of negative numbers in terms of i before doing any simplification.*

TECHNOLOGY TIP

Most calculators that do complex number arithmetic automatically return a complex number when asked for the square root of a negative number. On TI-83+/89, however, the MODE must be set to "rectangular" or "$a + bi$." On Casio FX 2.0, the COMPLEX MODE (in the SET UP menu) must be set to "$a + bi$."

EXAMPLE 6

$$\begin{aligned}(7 - \sqrt{-4})(5 + \sqrt{-9}) &= (7 - \sqrt{4}i)(5 + \sqrt{9}i) \\ &= (7 - 2i)(5 + 3i) \\ &= 35 + 21i - 10i - 6i^2 \\ &= 35 + 11i - 6(-1) = 41 + 11i. \quad \blacksquare\end{aligned}$$

Since every negative real number has a square root in the complex number system, we can now find complex solutions for equations that have no real solutions. For example, the solutions of $x^2 = -25$ are $x = \pm\sqrt{-25} = \pm 5i$. In fact,

Every quadratic equation with real coefficients has solutions in the complex number system.

EXAMPLE 7

To solve the equation $2x^2 + x + 3 = 0$, we apply the quadratic formula.

$$x = \frac{-1 \pm \sqrt{1^2 - 4 \cdot 2 \cdot 3}}{2 \cdot 2} = \frac{-1 \pm \sqrt{-23}}{4}.$$

Since $\sqrt{-23}$ is not a real number, this equation has no real number solutions. But $\sqrt{-23}$ *is* a complex number, namely, $\sqrt{-23} = \sqrt{23}i$. Thus, the equation does have solutions in the complex number system.

$$x = \frac{-1 \pm \sqrt{-23}}{4} = \frac{-1 \pm \sqrt{23}i}{4} = -\frac{1}{4} \pm \frac{\sqrt{23}}{4}i.$$

Note that the two solutions, $-\frac{1}{4} + \frac{\sqrt{23}}{4}i$ and $-\frac{1}{4} - \frac{\sqrt{23}}{4}i$, are conjugates of each other. ■

TECHNOLOGY TIP

The polynomial solvers on TI-86, HP-39+ and Casio produce all real and complex solutions of any polynomial equation that they can solve. See Calculator Investigation 4 on page 10 for details.

On TI-89, use cSOLVE in the COMPLEX submenu of the ALGEBRA menu to find all solutions.

EXAMPLE 8

To find *all* solutions of $x^3 = 1$, we rewrite the equation and factor:

$$\begin{aligned} x^3 &= 1 \\ x^3 - 1 &= 0 \\ (x - 1)(x^2 + x + 1) &= 0 \\ x - 1 = 0 \quad &\text{or} \quad x^2 + x + 1 = 0. \end{aligned}$$

The solution of the first equation is $x = 1$. The solutions of the second can be obtained from the quadratic formula.

$$x = \frac{-1 \pm \sqrt{1^2 - 4 \cdot 1 \cdot 1}}{2 \cdot 1} = \frac{-1 \pm \sqrt{-3}}{2} = \frac{-1 \pm \sqrt{3}i}{2} = -\frac{1}{2} \pm \frac{\sqrt{3}}{2}i.$$

Therefore, the equation $x^3 = 1$ has one real solution ($x = 1$) and two nonreal complex solutions [$x = -1/2 + (\sqrt{3}/2)i$ and $x = -1/2 - (\sqrt{3}/2)i$]. Each of these solutions is said to be a **cube root of 1** or a **cube root of unity.** Observe that the two nonreal complex cube roots of unity are conjugates of each other. ■

The preceding examples illustrate this useful fact.

Conjugate Solutions

If $a + bi$ is a solution of a polynomial equation with *real* coefficients, then its conjugate $a - bi$ is also a solution of this equation.

✓ EXERCISES 4.1

In Exercises 1–54, perform the indicated operation and write the result in the form $a + bi$.

1. $(2 + 3i) + (6 - i)$
2. $(-5 + 7i) + (14 + 3i)$
3. $(2 - 8i) - (4 + 2i)$
4. $(3 + 5i) - (3 - 7i)$
5. $\frac{5}{4} - \left(\frac{7}{4} + 2i\right)$
6. $(\sqrt{3} + i) + (\sqrt{5} - 2i)$
7. $\left(\frac{\sqrt{2}}{2} + i\right) - \left(\frac{\sqrt{3}}{2} - i\right)$
8. $\left(\frac{1}{2} + \frac{\sqrt{3}i}{2}\right) + \left(\frac{3}{4} - \frac{5\sqrt{3}i}{2}\right)$
9. $(2 + i)(3 + 5i)$
10. $(2 - i)(5 + 2i)$
11. $(-3 + 2i)(4 - i)$
12. $(4 + 3i)(4 - 3i)$
13. $(2 - 5i)^2$
14. $(1 + i)(2 - i)i$
15. $(\sqrt{3} + i)(\sqrt{3} - i)$
16. $\left(\frac{1}{2} - i\right)\left(\frac{1}{4} + 2i\right)$
17. i^{15}
18. i^{26}
19. i^{33}
20. $(-i)^{53}$
21. $(-i)^{107}$
22. $(-i)^{213}$
23. $\frac{1}{5 - 2i}$
24. $\frac{1}{i}$
25. $\frac{1}{3i}$
26. $\frac{i}{2 + i}$
27. $\frac{3}{4 + 5i}$
28. $\frac{2 + 3i}{i}$
29. $\frac{1}{i(4 + 5i)}$
30. $\frac{1}{(2 - i)(2 + i)}$
31. $\frac{2 + 3i}{i(4 + i)}$
32. $\frac{2}{(2 + 3i)(4 + i)}$
33. $\frac{2 + i}{1 - i} + \frac{1}{1 + 2i}$
34. $\frac{1}{2 - i} + \frac{3 + i}{2 + 3i}$
35. $\frac{i}{3 + i} - \frac{3 + i}{4 + i}$
36. $6 + \frac{2i}{3 + i}$
37. $\sqrt{-36}$
38. $\sqrt{-81}$
39. $\sqrt{-14}$
40. $\sqrt{-50}$
41. $-\sqrt{-16}$
42. $-\sqrt{-12}$
43. $\sqrt{-16} + \sqrt{-49}$
44. $\sqrt{-25} - \sqrt{-9}$
45. $\sqrt{-15} - \sqrt{-18}$
46. $\sqrt{-12}\sqrt{-3}$
47. $\sqrt{-16}/\sqrt{-36}$
48. $-\sqrt{-64}/\sqrt{-4}$
49. $(\sqrt{-25} + 2)(\sqrt{-49} - 3)$
50. $(5 - \sqrt{-3})(-1 + \sqrt{-9})$
51. $(2 + \sqrt{-5})(1 - \sqrt{-10})$
52. $\sqrt{-3}(3 - \sqrt{-27})$
53. $1/(1 + \sqrt{-2})$
54. $(1 + \sqrt{-4})(3 - \sqrt{-9})$

In Exercises 55–58, find x and y. Remember that

$$a + bi = c + di$$

exactly when $a = c$ and $b = d$.

55. $3x - 4i = 6 + 2yi$
56. $8 - 2yi = 4x + 12i$
57. $3 + 4xi = 2y - 3i$
58. $8 - xi = \frac{1}{2}y + 2i$

In Exercises 59–70, solve the equation and express each solution in the form $a + bi$.

59. $3x^2 - 2x + 5 = 0$
60. $5x^2 + 2x + 1 = 0$
61. $x^2 + x + 2 = 0$
62. $5x^2 - 6x + 2 = 0$
63. $2x^2 - x = -4$
64. $x^2 + 1 = 4x$
65. $2x^2 + 3 = 6x$
66. $3x^2 + 4 = -5x$
67. $x^3 - 8 = 0$
68. $x^3 + 125 = 0$
69. $x^4 - 1 = 0$
70. $x^4 - 81 = 0$
71. Simplify: $i + i^2 + i^3 + \cdots + i^{15}$
72. Simplify: $i - i^2 + i^3 - i^4 + i^5 - \cdots + i^{15}$

Thinkers

73. If $z = a + bi$ is a complex number, then its conjugate is usually denoted $\bar{z}$, that is $\bar{z} = a - bi$. Verify that

$$\bar{\bar{z}} = z.$$

74. The **real part** of the complex number $a + bi$ is defined to be the real number a. The **imaginary part** of $a + bi$ is defined to the real number b (*not* bi).

(a) Show that the real part of $z = a + bi$ is $\dfrac{z + \bar{z}}{2}$.

(b) Show that the imaginary part of $z = a + bi$ is $\dfrac{z - \bar{z}}{2i}$.

75. If $z = a + bi$ (with a, b real numbers, not both 0), express $1/z$ in standard form.

4.2 The Complex Plane and Polar Form for Complex Numbers

The complex number system can be represented geometrically by the coordinate plane:

The complex number $a + bi$ corresponds to the point (a, b) in the plane.

For example, the point (2, 3) in Figure 4–1 is labeled by $2 + 3i$, and similarly for the other points shown.

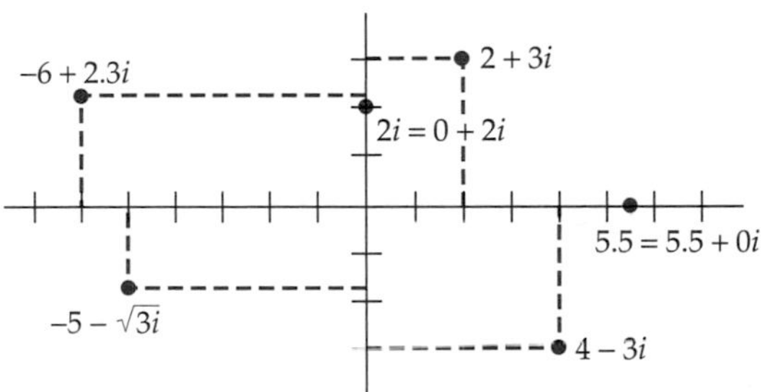

Figure 4–1

TECHNOLOGY TIP

To do complex arithmetic on TI-86 and HP-39+, enter $a + bi$ as (a, b). On other calculators, use the special i key whose location is

TI-83+/89: keyboard

Casio FX 2.0: keyboard

Casio 9850: OPTN/CPLX

When the coordinate plane is labeled by complex numbers in this way, it is called the **complex plane.** Each real number $a = a + 0i$ corresponds to the point $(a, 0)$ on the horizontal axis; so this axis is called the **real axis.** The vertical axis is called the **imaginary axis** because every imaginary number $bi = 0 + bi$ corresponds to the point $(0, b)$ on the vertical axis.

The absolute value of a real number c is the distance from c to 0 on the number line (see page 6). So we define the **absolute value** (or **modulus**) of the complex number $a + bi$ to be the distance from $a + bi$ to the origin in the complex plane:

$$|a + bi| = \text{distance from } (a, b) \text{ to } (0, 0) = \sqrt{(a - 0)^2 + (b - 0)^2}.$$

Therefore, we have the following.

Absolute Value

The **absolute value** (or **modulus**) of the complex number $a + bi$ is

$$|a + bi| = \sqrt{a^2 + b^2}.$$

TIP

To find the absolute value of a complex number, use the ABS key, which is in this menu/submenu:

TI-83+: MATH/CPX

TI-86: CPLX

TI-89: MATH/COMPLEX

Casio: OPTN/CPLX

HP-39+: Keyboard

EXAMPLE 1

(a) $|3 + 2i| = \sqrt{3^2 + 2^2} = \sqrt{13}.$

(b) $|4 - 5i| = \sqrt{4^2 + (-5)^2} = \sqrt{41}.$ ■

Let $a + bi$ be a nonzero complex number, and denote $|a + bi|$ by r. Then r is the length of the line segment joining (a, b) and $(0, 0)$ in the plane. Let θ be the angle in standard position with this line segment as its terminal side (Figure 4–2).

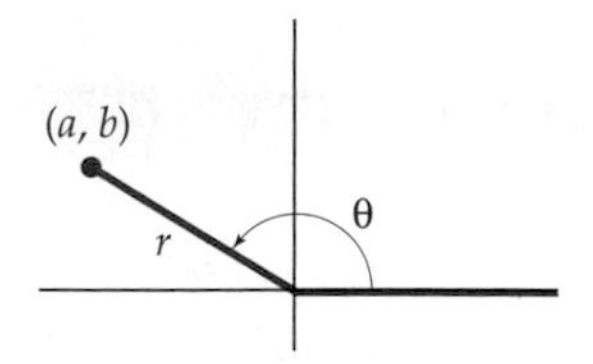

Figure 4–2

According to the point-in-the-plane description of sine and cosine,

$$\cos\theta = \frac{a}{r} \qquad \text{and} \qquad \sin\theta = \frac{b}{r},$$

so

$$a = r\cos\theta \qquad \text{and} \qquad b = r\sin\theta.$$

Consequently,

$$a + bi = r\cos\theta + (r\sin\theta)i = r(\cos\theta + i\sin\theta).^*$$

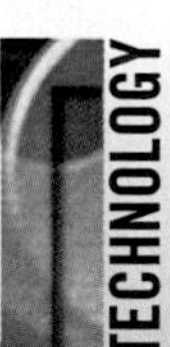

TIP

To find the argument θ, use the ANGLE or ARG key in the same menu/submenu as the ABS key [except on HP-39+, where it is in MATH/COMPLEX].

When a complex number $a + bi$ is written in this way, it is said to be in **polar form** or **trigonometric form.** The angle θ is called the **argument** and is usually expressed in radian measure. The number $r = |a + bi|$ is sometimes called the **modulus** (plural: moduli). The number 0 can also be written in polar notation by letting $r = 0$ and letting θ be any angle. Thus we have the following.

Polar Form

Every complex number $a + bi$ can be written in polar form:

$$r(\cos\theta + i\sin\theta),$$

where $r = |a + bi| = \sqrt{a^2 + b^2}$, $a = r\cos\theta$, and $b = r\sin\theta$.

When a complex number is written in polar form, the argument θ is not uniquely determined, since θ, $\theta \pm 2\pi$, $\theta \pm 4\pi$, etc., all satisfy the conditions in the box.

*It is customary to place i in front of $\sin\theta$ rather than after it. Some books abbreviate $r(\cos\theta + i\sin\theta)$ as $r \text{ cis } \theta$.

TECHNOLOGY TIP

The complex number

$r(\cos\theta + i\sin\theta)$

is entered in a calculator as follows.

TI-83+/Casio FX 2.0: $re^{i\theta}$
[Use the special i key.]

TI-86/89: $(r\angle\,\theta)$
[$\angle$ is on the keyboard]

Converting a number in polar form to rectangular form $a + bi$ is easy: Just evaluate and multiply out. For instance,

$$8\left(\cos\frac{\pi}{3} + i\sin\frac{\pi}{3}\right) = 8\left(\frac{1}{2} + i\frac{\sqrt{3}}{2}\right) = 4 + 4\sqrt{3}i.$$

To convert $a + bi$ into polar form $r(\cos\theta + i\sin\theta)$, you must find a suitable angle θ. Note that

$$\frac{b}{a} = \frac{r\sin\theta}{r\cos\theta} = \frac{\sin\theta}{\cos\theta} = \tan\theta,$$

so θ is a solution of the equation

$$\tan\theta = \frac{b}{a}.$$

EXAMPLE 2

Express $-\sqrt{3} + i$ in polar form.

SOLUTION Here $a = -\sqrt{3}$ and $b = 1$, so

$$r = \sqrt{a^2 + b^2} = \sqrt{(-\sqrt{3})^2 + 1^2} = \sqrt{3 + 1} = 2.$$

The angle θ must satisfy

$$\tan\theta = \frac{b}{a} = \frac{1}{-\sqrt{3}} = -\frac{1}{\sqrt{3}}.$$

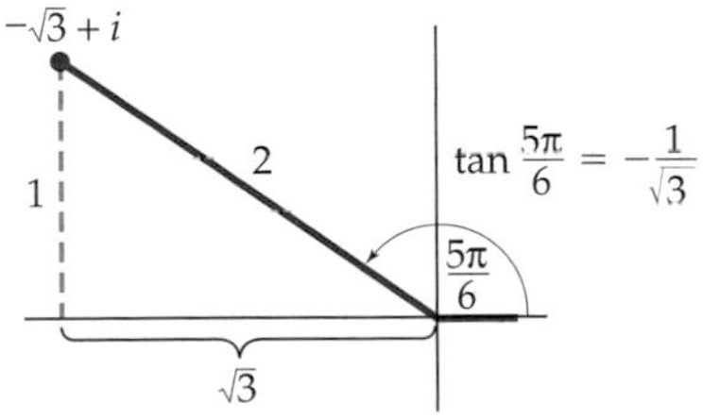

Since $-\sqrt{3} + i$ lies in the second quadrant (Figure 4–3), θ must be a second-quadrant angle. Our knowledge of special angles and Figure 4–3 show that $\theta = 5\pi/6$ satisfies these conditions. Hence,

$$-\sqrt{3} + i = 2\left(\cos\frac{5\pi}{6} + i\sin\frac{5\pi}{6}\right). \quad \blacksquare$$

Figure 4–3

In Example 2 we solved the equation $\tan\theta = \dfrac{b}{a}$ by using our knowledge of special angles. When this isn't possible, we must use the solution methods discussed in Section 3.5. On the basis of the facts developed there, we have this useful result.

Conversion to Polar Form

If $a + bi$ lies in the first or fourth quadrant, then it has polar form

$$r(\cos\theta + i\sin\theta),$$

where $r = \sqrt{a^2 + b^2}$ and $\theta = \tan^{-1}\dfrac{b}{a}$.

If $a + bi$ lies in the second or third quadrant, then it has polar form

$$r(\cos\theta + i\sin\theta),$$

where $r = \sqrt{a^2 + b^2}$ and $\theta = \tan^{-1}\dfrac{b}{a} + \pi$.

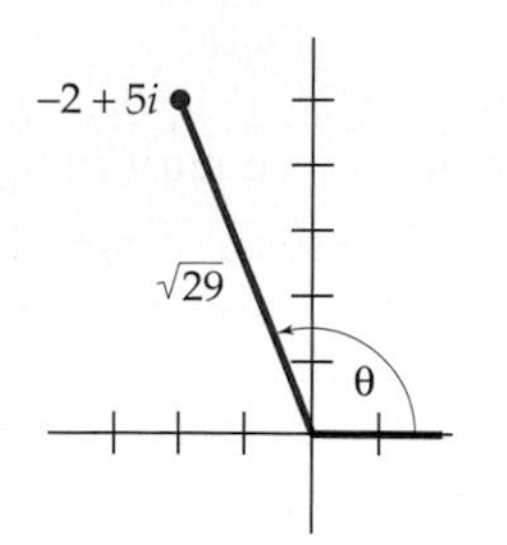

Figure 4–4

EXAMPLE 3

Express $-2 + 5i$ in polar form.

SOLUTION Since $a = -2$ and $b = 5$,

$$r = \sqrt{a^2 + b^2} = \sqrt{(-2)^2 + 5^2} = \sqrt{29}.$$

Now $-2 + 5i$ lies in the second quadrant (Figure 4–4). So

$$\theta = \tan^{-1}\left(\frac{b}{a}\right) + \pi = \tan^{-1}\left(\frac{5}{-2}\right) + \pi \approx 1.9513.$$

Therefore,

$$-2 + 5i \approx \sqrt{29}\,(\cos 1.9513 + i \sin 1.9513). \quad ■$$

TECHNOLOGY TIP

To convert from rectangular to polar form, or vice versa, use ►POL or ►RECT in this menu/submenu:

TI-83+: MATH/CPX

TI-86: CPLX

TI-89: MATH/MATRIX/VECTOR OPS

On Casio FX 2.0, use ►*re*^*θ i* or ►*a* + *bi* in the OPTN/CPLX menu.

POLAR MULTIPLICATION AND DIVISION

Multiplication and division of complex numbers in polar form are done by the following rules, which are proved at the end of the section.

Polar Multiplication and Division Rules

If $z_1 = r_1(\cos\theta_1 + i\sin\theta_1)$ and $z_2 = r_2(\cos\theta_2 + i\sin\theta_2)$ are any two complex numbers, then

$$z_1 z_2 = r_1 r_2[\cos(\theta_1 + \theta_2) + i\sin(\theta_1 + \theta_2)]$$

and

$$\frac{z_1}{z_2} = \frac{r_1}{r_2}[\cos(\theta_1 - \theta_2) + i\sin(\theta_1 - \theta_2)] \qquad (z_2 \neq 0).$$

In other words, to multiply two numbers in polar form, just *multiply the moduli and add the arguments.* To divide, just *divide the moduli and subtract the arguments.* Before proving the statements in the box, we will illustrate them with some examples.

EXAMPLE 4 

Find z_1z_2, when

$$z_1 = 2[\cos(5\pi/6) + i\sin(5\pi/6)] \quad \text{and} \quad z_2 = 3[\cos(7\pi/4) + i\sin(7\pi/4)].$$

TECHNOLOGY TIP Complex arithmetic can be done with numbers in polar form on TI calculators and Casio FX 2.0. Some answers may be expressed in rectangular form.

SOLUTION Here r_1 is the number 2, and $\theta_1 = 5\pi/6$; similarly, $r_2 = 3$, and $\theta_2 = 7\pi/4$, and we have

$$\begin{aligned} z_1z_2 &= r_1r_2[\cos(\theta_1 + \theta_2) + i\sin(\theta_1 + \theta_2)] \\ &= 2 \cdot 3\left[\cos\left(\frac{5\pi}{6} + \frac{7\pi}{4}\right) + i\sin\left(\frac{5\pi}{6} + \frac{7\pi}{4}\right)\right] \\ &= 6\left[\cos\left(\frac{10\pi}{12} + \frac{21\pi}{12}\right) + i\sin\left(\frac{10\pi}{12} + \frac{21\pi}{12}\right)\right] \\ &= 6\left(\cos\frac{31\pi}{12} + i\sin\frac{31\pi}{12}\right). \quad ■ \end{aligned}$$

EXAMPLE 5 

Find z_1/z_2, where

$$z_1 = 10[\cos(\pi/3) + i\sin(\pi/3)] \quad \text{and} \quad z_2 = 2[\cos(\pi/4) + i\sin(\pi/4)].$$

SOLUTION

$$\begin{aligned} \frac{z_1}{z_2} = \frac{10\left(\cos\frac{\pi}{3} + i\sin\frac{\pi}{3}\right)}{2\left(\cos\frac{\pi}{4} + i\sin\frac{\pi}{4}\right)} &= \frac{10}{2}\left[\cos\left(\frac{\pi}{3} - \frac{\pi}{4}\right) + i\sin\left(\frac{\pi}{3} - \frac{\pi}{4}\right)\right] \\ &= 5\left(\cos\frac{\pi}{12} + i\sin\frac{\pi}{12}\right). \quad ■ \end{aligned}$$

PROOF OF THE POLAR MULTIPLICATION RULE

If $z_1 = r_1(\cos\theta_1 + i\sin\theta_1)$ and $z_2 = r_2(\cos\theta_2 + i\sin\theta_2)$, then

$$\begin{aligned} z_1z_2 &= r_1(\cos\theta_1 + i\sin\theta_1)r_2(\cos\theta_2 + i\sin\theta_2) \\ &= r_1r_2(\cos\theta_1 + i\sin\theta_1)(\cos\theta_2 + i\sin\theta_2) \\ &= r_1r_2(\cos\theta_1\cos\theta_2 + i\sin\theta_1\cos\theta_2 + i\cos\theta_1\sin\theta_2 + i^2\sin\theta_1\sin\theta_2) \\ &= r_1r_2[(\cos\theta_1\cos\theta_2 - \sin\theta_1\sin\theta_2) + i(\sin\theta_1\cos\theta_2 + \cos\theta_1\sin\theta_2)]. \end{aligned}$$

But the addition identities for sine and cosine (page 230) show that

$$\begin{aligned} \cos\theta_1\cos\theta_2 - \sin\theta_1\sin\theta_2 &= \cos(\theta_1 + \theta_2) \\ \sin\theta_1\cos\theta_2 + \cos\theta_1\sin\theta_2 &= \sin(\theta_1 + \theta_2). \end{aligned}$$

Therefore,

$$z_1z_2 = r_1r_2[(\cos\theta_1 \cos\theta_2 - \sin\theta_1 \sin\theta_2) + i(\sin\theta_1 \cos\theta_2 + \cos\theta_1 \sin\theta_2)]$$
$$= r_1r_2[\cos(\theta_1 + \theta_2) + i\sin(\theta_1 + \theta_2)].$$

This completes the proof of the multiplication rule. The division rule is proved similarly (Exercise 77).

EXERCISES 4.2

In Exercises 1–8, plot the point in the complex plane corresponding to the number.

1. $3 + 2i$
2. $-7 + 6i$
3. $-\frac{8}{3} - \frac{5}{3}i$
4. $\sqrt{2} - 7i$
5. $(1 + i)(1 - i)$
6. $(2 + i)(1 - 2i)$
7. $2i\left(3 - \frac{5}{2}i\right)$
8. $\frac{4i}{3}(-6 - 3i)$

In Exercises 9–14, find the absolute value.

9. $|5 - 12i|$
10. $|2i|$
11. $|1 + \sqrt{2}i|$
12. $|2 - 3i|$
13. $|-12i|$
14. $|i^7|$
15. Give an example of complex numbers z and w such that $|z + w| \neq |z| + |w|$.
16. If $z = 3 - 4i$, find $|z|^2$ and $z\bar{z}$, where $\bar{z}$ is the conjugate of z (see page 281).

In Exercises 17–24, sketch the graph of the equation in the complex plane (z denotes a complex number of the form $a + bi$).

17. $|z| = 4$ [*Hint:* The graph consists of all points that lie 4 units from the origin.]
18. $|z| = 1$
19. $|z - 1| = 10$ [*Hint:* 1 corresponds to (1, 0) in the complex plane. What does the equation say about the distance from z to 1?]
20. $|z + 3| = 1$
21. $|z - 2i| = 4$
22. $|z - 3i + 2| = 9$ [*Hint:* Rewrite it as $|z - (-2 + 3i)| = 9$.]
23. $\text{Re}(z) = 2$ [The **real part** of the complex number $z = a + bi$ is defined to be the number a and is denoted $\text{Re}(z)$.]
24. $\text{Im}(z) = -5/2$ [The **imaginary part** of $z = a + bi$ is defined to be the number b *(not bi)* and is denoted $\text{Im}(z)$.]

In Exercises 25–36, express the number in the form $a + bi$.

25. $2\left(\cos\frac{\pi}{4} + i\sin\frac{\pi}{4}\right)$
26. $3\left(\cos\frac{\pi}{3} + i\sin\frac{\pi}{3}\right)$
27. $\cos\frac{\pi}{2} + i\sin\frac{\pi}{2}$
28. $4\left(\cos\frac{3\pi}{4} + i\sin\frac{3\pi}{4}\right)$
29. $5\left(\cos\frac{2\pi}{3} + i\sin\frac{2\pi}{3}\right)$
30. $2\left(\cos\frac{7\pi}{6} + i\sin\frac{7\pi}{6}\right)$
31. $1.5\left(\cos\frac{\pi}{6} + i\sin\frac{\pi}{6}\right)$
32. $5(\cos 3 + i\sin 3)$
33. $2(\cos 4 + i\sin 4)$
34. $3(\cos 5 + i\sin 5)$
35. $4(\cos 2 + i\sin 2)$
36. $2(\cos 1.5 + i\sin 1.5)$

In Exercises 37–52, express the number in polar form.

37. $3 + 3i$
38. $5 - 5i$
39. $2 + 2\sqrt{3}i$
40. $5\sqrt{3} + 5i$
41. $3\sqrt{3} - 3i$
42. $-4 - 4\sqrt{3}i$
43. $-\sqrt{3} - \sqrt{3}i$
44. $2\sqrt{5} - 2\sqrt{5}i$
45. $3 + 4i$
46. $-4 + 3i$
47. $5 - 12i$
48. $-\sqrt{7} - 3i$
49. $1 + 2i$
50. $3 - 5i$
51. $-\frac{5}{2} + \frac{7}{2}i$
52. $\sqrt{5} + \sqrt{11}i$

In Exercises 53–64, perform the indicated multiplication or division. Express your answer in both polar form $r(\cos\theta + i\sin\theta)$ and rectangular form $a + bi$.

53. $\left(\cos\frac{\pi}{2} + i\sin\frac{\pi}{2}\right) \cdot (\cos\pi + i\sin\pi)$
54. $2\left(\cos\frac{\pi}{6} + i\sin\frac{\pi}{6}\right) \cdot 5\left(\cos\frac{\pi}{3} + i\sin\frac{\pi}{3}\right)$
55. $4\left(\cos\frac{\pi}{4} + i\sin\frac{\pi}{4}\right) \cdot 3\left(\cos\frac{\pi}{12} + i\sin\frac{\pi}{12}\right)$
56. $\left(\cos\frac{\pi}{12} + i\sin\frac{\pi}{12}\right) \cdot 2\left(\cos\frac{7\pi}{12} + i\sin\frac{7\pi}{12}\right)$
57. $3\left(\cos\frac{\pi}{8} + i\sin\frac{\pi}{8}\right) \cdot 12\left(\cos\frac{3\pi}{8} + i\sin\frac{3\pi}{8}\right)$

58. $12\left(\cos\frac{11\pi}{12} + i\sin\frac{11\pi}{12}\right)\cdot\frac{7}{2}\left(\cos\frac{\pi}{4} + i\sin\frac{\pi}{4}\right)$

59. $\dfrac{\cos\pi + i\sin\pi}{\cos\frac{2\pi}{3} + i\sin\frac{2\pi}{3}}$

60. $\dfrac{\cos\frac{3\pi}{4} + i\sin\frac{3\pi}{4}}{\cos\frac{\pi}{4} + i\sin\frac{\pi}{4}}$

61. $\dfrac{8\left(\cos\frac{4\pi}{3} + i\sin\frac{4\pi}{3}\right)}{4\left(\cos\frac{7\pi}{6} + i\sin\frac{7\pi}{6}\right)}$

62. $\dfrac{8\left(\cos\frac{5\pi}{18} + i\sin\frac{5\pi}{18}\right)}{4\left(\cos\frac{\pi}{9} + i\sin\frac{\pi}{9}\right)}$

63. $\dfrac{6\left(\cos\frac{7\pi}{20} + i\sin\frac{7\pi}{20}\right)}{4\left(\cos\frac{\pi}{10} + i\sin\frac{\pi}{10}\right)}$

64. $\dfrac{\sqrt{54}\left(\cos\frac{9\pi}{4} + i\sin\frac{9\pi}{4}\right)}{\sqrt{6}\left(\cos\frac{7\pi}{12} + i\sin\frac{7\pi}{12}\right)}$

In Exercises 65–72, convert to polar form and then multiply or divide. Express your answer in polar form.

65. $(1 + i)(1 + \sqrt{3}i)$

66. $(1 - i)(3 - 3i)$

67. $\dfrac{1 + i}{1 - i}$

68. $\dfrac{2 - 2i}{-1 - i}$

69. $3i(2\sqrt{3} + 2i)$

70. $\dfrac{-4i}{\sqrt{3} + i}$

71. $i(i + 1)(-\sqrt{3} + i)$

72. $(1 - i)(2\sqrt{3} - 2i)(-4 - 4\sqrt{3}i)$

73. Explain what is meant by saying that multiplying a complex number $z = r(\cos\theta + i\sin\theta)$ by i amounts to rotating z 90° counterclockwise around the origin. [*Hint:* Express i and iz in polar form. What are their relative positions in the complex plane?]

74. Describe what happens geometrically when you multiply a complex number by 2.

Thinkers

75. The sum of two distinct complex numbers, $a + bi$ and $c + di$, can be found geometrically by means of the so-called **parallelogram rule:** Plot the points $a + bi$ and $c + di$ in the complex plane, and form the parallelogram, three of whose vertices are 0, $a + bi$, and $c + di$, as in the figure. Then the fourth vertex of the parallelogram is the point whose coordinate is the sum

$$(a + bi) + (c + di) = (a + c) + (b + d)i.$$

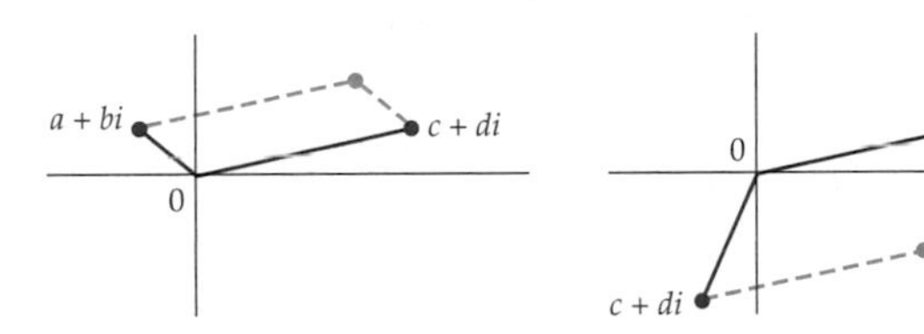

Complete the following *proof* of the parallelogram rule when $a \neq 0$ and $c \neq 0$.

(a) Find the *slope* of the line K from 0 to $a + bi$. [*Hint: K* contains the points (0, 0) and (a, b).]

(b) Find the *slope* of the line N from 0 to $c + di$.

(c) Find the *equation* of the line L through $a + bi$ and parallel to line N of part (b). [*Hint:* The point (a, b) is on L; find the slope of L by using part (b) and facts about the slope of parallel lines.]

(d) Find the *equation* of the line M through $c + di$ and parallel to line K of part (a).

(e) Label the lines K, L, M, and N in the figure.

(f) Show by using substitution that the point $(a + c, b + d)$ satisfies both the equation of line L and the equation of line M. Therefore, $(a + c, b + d)$ lies on both L and M. Since the only point on both L and M is the fourth vertex of the parallelogram (see the figure), this vertex must be $(a + c, b + d)$. Hence, this vertex has coordinate

$$(a + c) + (b + d)i = (a + bi) + (c + di).$$

76. Let $z = a + bi$ be a complex number and denote its conjugate $a - bi$ by $\bar{z}$. Prove that $|z|^2 = z\bar{z}$.

(Exercises continue on next page.)

77. *Proof of the polar division rule.* Let $z_1 = r_1(\cos \theta_1 + i \sin \theta_1)$ and $z_2 = r_2(\cos \theta_2 + i \sin \theta_2)$. Then

$$\frac{z_1}{z_2} = \frac{r_1(\cos \theta_1 + i \sin \theta_1)}{r_2(\cos \theta_2 + i \sin \theta_2)}$$

$$= \frac{r_1(\cos \theta_1 + i \sin \theta_1)}{r_2(\cos \theta_2 + i \sin \theta_2)} \cdot \frac{\cos \theta_2 - i \sin \theta_2}{\cos \theta_2 - i \sin \theta_2}.$$

(a) Multiply out the denominator on the right side and use the Pythagorean identity to show that it is just the number r_2.

(b) Multiply out the numerator on the right side; use the subtraction identities for sine and cosine (page 230) to show that it is

$$r_1[\cos(\theta_1 - \theta_2) + i \sin(\theta_1 - \theta_2)].$$

Therefore,

$$\frac{z_1}{z_2} = \left(\frac{r_1}{r_2}\right)[\cos(\theta_1 - \theta_2) + i \sin(\theta_1 - \theta_2)].$$

78. (a) If $s(\cos \beta + i \sin \beta) = r(\cos \theta + i \sin \theta)$, (with $r > 0, s > 0$), explain why we must have $s = r$. [*Hint:* Think distance.]

(b) If $r(\cos \beta + i \sin \beta) = r(\cos \theta + i \sin \theta)$, explain why $\cos \beta = \cos \theta$ and $\sin \beta = \sin \theta$. [*Hint:* See property 5 of the complex numbers on page 280.]

(c) If $\cos \beta = \cos \theta$ and $\sin \beta = \sin \theta$, show that angles β and θ in standard position have the same terminal side. [*Hint:* $(\cos \beta, \sin \beta)$ and $(\cos \theta, \sin \theta)$ are points on the unit circle.]

(d) Use parts (a)–(c) to prove this **equality rule for polar form:**

$$s(\cos \beta + i \sin \beta) = r(\cos \theta + i \sin \theta)$$

exactly when $s = r$ and $\beta = \theta + 2k\pi$ for some integer k. [*Hint:* Angles with the same terminal side must differ by an integer multiple of 2π.]

4.3 DeMoivre's Theorem and *n*th Roots of Complex Numbers

Polar form provides a convenient way to calculate both powers and roots of complex numbers. If $z = r(\cos \theta + i \sin \theta)$, then the multiplication formula on page 288 shows that

$$z^2 = z \cdot z = r \cdot r[\cos(\theta + \theta) + i \sin(\theta + \theta)]$$
$$= r^2(\cos 2\theta + i \sin 2\theta).$$

Similarly,

$$z^3 = z^2 \cdot z = r^2 \cdot r[\cos(2\theta + \theta) + i \sin(2\theta + \theta)]$$
$$= r^3(\cos 3\theta + i \sin 3\theta).$$

Repeated application of the multiplication formula proves the following theorem.

DeMoivre's Theorem

For any complex number $z = r(\cos \theta + i \sin \theta)$ and any positive integer n,

$$z^n = r^n(\cos n\theta + i \sin n\theta).$$

EXAMPLE 1

Compute $(-\sqrt{3} + i)^5$.

SOLUTION We first express $-\sqrt{3} + i$ in polar form (see Example 2 on page 287):

$$-\sqrt{3} + i = 2\left(\cos \frac{5\pi}{6} + i \sin \frac{5\pi}{6}\right).$$

By DeMoivre's Theorem,

$$(-\sqrt{3} + i)^5 = 2^5\left[\cos\left(5 \cdot \frac{5\pi}{6}\right) + i \sin\left(5 \cdot \frac{5\pi}{6}\right)\right] = 32\left(\cos \frac{25\pi}{6} + i \sin \frac{25\pi}{6}\right).$$

Since $25\pi/6 = (\pi/6) + (24\pi/6) = (\pi/6) + 4\pi$, we have

$$(-\sqrt{3} + i)^5 = 32\left(\cos \frac{25\pi}{6} + i \sin \frac{25\pi}{6}\right) = 32\left(\cos \frac{\pi}{6} + i \sin \frac{\pi}{6}\right)$$

$$= 32\left(\frac{\sqrt{3}}{2} + \frac{1}{2}i\right) = 16\sqrt{3} + 16i. \quad ■$$

EXAMPLE 2

Find $(1 + i)^{10}$.

SOLUTION First verify that the polar form of $1 + i$ is

$$1 + i = \sqrt{2}\left(\cos \frac{\pi}{4} + i \sin \frac{\pi}{4}\right).$$

Therefore, by DeMoivre's Theorem,

$$(1 + i)^{10} = (\sqrt{2})^{10}\left(\cos \frac{10\pi}{4} + i \sin \frac{10\pi}{4}\right)$$

$$= (2^{1/2})^{10}\left(\cos \frac{5\pi}{2} + i \sin \frac{5\pi}{2}\right) = 2^5(0 + i \cdot 1) = 32i.$$

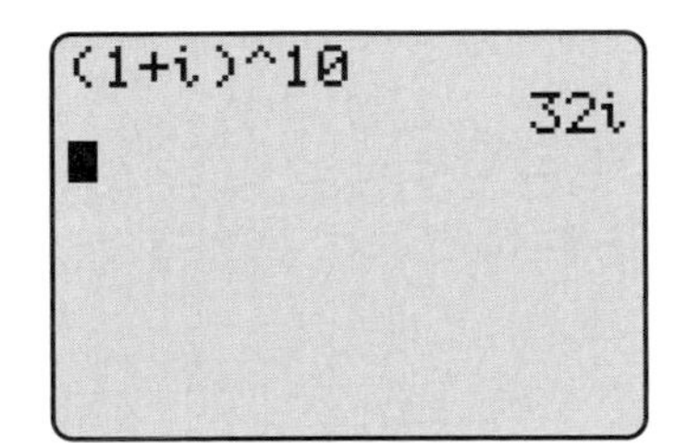

Figure 4–5

A calculator that can do complex arithmetic confirms this result (Figure 4–5). ■

*n*TH ROOTS

Recall that for a positive real number c, we called the positive solution of the equation $x^n = c$ the nth root of c. In the complex number system, however, "positive" and "negative" are not meaningful terms (for instance, should $3 - 2i$ be called positive or negative?). Consequently, if $a + bi$ is a complex number, then *any* solution of the equation

$$z^n = a + bi$$

is called an ***n*th root** of $a + bi$.

Every real number is, of course, a complex number. When the definition of an nth root of a complex number is applied to a real number, we must change our previous terminology. For instance, 16 now has *four* fourth roots, since each of 2, -2, $2i$, and $-2i$ is a solution of $z^4 = 16$, whereas we previously said that 2 was *the* fourth root of 16. In the context of complex numbers, this change will not cause any confusion.

Although nth roots are no longer unique, the radical symbol will be used only for nonnegative real numbers and will have the same meaning as before: If r is a nonnegative real number, then $\sqrt[n]{r}$ denotes the unique nonnegative real number whose nth power is r.

All nth roots of a complex number $a + bi$ can easily be found if $a + bi$ is written in polar form, as illustrated in the next example.

EXAMPLE 3

Find the fourth roots of $-8 + 8\sqrt{3}i$.

SOLUTION To solve $z^4 = -8 + 8\sqrt{3}i$, first verify that the polar form of $-8 + 8\sqrt{3}i$ is $16\left(\cos\frac{2\pi}{3} + i\sin\frac{2\pi}{3}\right)$. We must find numbers s and β such that

$$[s(\cos\beta + i\sin\beta)]^4 = 16\left(\cos\frac{2\pi}{3} + i\sin\frac{2\pi}{3}\right).$$

By DeMoivre's Theorem, we have

$$s^4(\cos 4\beta + i\sin 4\beta) = 16\left(\cos\frac{2\pi}{3} + i\sin\frac{2\pi}{3}\right).$$

The equality rules for complex numbers in polar form (Exercise 78 in Section 4.2) show that this can happen only when

$$s^4 = 16 \qquad \text{and} \qquad 4\beta = \frac{2\pi}{3} + 2k\pi \qquad (k \text{ an integer})$$

$$s = \sqrt[4]{16} = 2 \qquad\qquad \beta = \frac{2\pi/3 + 2k\pi}{4}.$$

Substituting these values in $s(\cos\beta + i\sin\beta)$ shows that the solutions of $z^4 = 16\left(\cos\frac{2\pi}{3} + i\sin\frac{2\pi}{3}\right)$ are

$$z = 2\left(\cos\frac{2\pi/3 + 2k\pi}{4} + i\sin\frac{2\pi/3 + 2k\pi}{4}\right) \qquad (k = 0, \pm1, \pm2, \pm3, \ldots).$$

which can be simplified as

$$z = 2\left[\cos\left(\frac{\pi}{6} + \frac{k\pi}{2}\right) + i\sin\left(\frac{\pi}{6} + \frac{k\pi}{2}\right)\right] \qquad (k = 0, \pm1, \pm2, \pm3, \ldots).$$

Letting $k = 0, 1, 2, 3$, produces four distinct solutions.

$$k = 0: \quad z = 2\left(\cos\frac{\pi}{6} + i\sin\frac{\pi}{6}\right) = \sqrt{3} + i$$

$$k = 1: \quad z = 2\left[\cos\left(\frac{\pi}{6} + \frac{\pi}{2}\right) + i\sin\left(\frac{\pi}{6} + \frac{\pi}{2}\right)\right] = 2\left(\cos\frac{2\pi}{3} + i\sin\frac{2\pi}{3}\right)$$
$$= -1 + \sqrt{3}i.$$

$$k = 2: \quad z = 2\left[\cos\left(\frac{\pi}{6} + \pi\right) + i\sin\left(\frac{\pi}{6} + \pi\right)\right] = 2\left(\cos\frac{7\pi}{6} + i\sin\frac{7\pi}{6}\right)$$
$$= -\sqrt{3} - i.$$

$$k = 3: \quad z = 2\left[\cos\left(\frac{\pi}{6} + \frac{3\pi}{2}\right) + i\sin\left(\frac{\pi}{6} + \frac{3\pi}{2}\right)\right] = 2\left(\cos\frac{5\pi}{3} + i\sin\frac{5\pi}{3}\right).$$
$$= 1 - \sqrt{3}i.$$

Any other value of k produces an angle β with the same terminal side as one of the four angles used above, and hence leads to the same solution. For instance, when $k = 4$, then $\beta = \frac{\pi}{6} + \frac{4\pi}{2} = \frac{\pi}{6} + 2\pi$ and β has the same terminal side as $\pi/6$. Therefore, we have found *all* the solutions—the four fourth roots of $-8 + 8\sqrt{3}i$. ■

The general equation $z^n = r(\cos\theta + i\sin\theta)$ can be solved by exactly the same method used in the preceding example: Just substitute n for 4, r for 16, and θ for $2\pi/3$, as follows. A solution is a number $s(\cos\beta + i\sin\beta)$ such that

$$[s(\cos\beta + i\sin\beta)]^n = r(\cos\theta + i\sin\theta)$$
$$s^n(\cos n\beta + i\sin n\beta) = r(\cos\theta + i\sin\theta).$$

Therefore,

$$s^n = r \qquad \text{and} \qquad n\beta = \theta + 2k\pi \qquad (k \text{ any integer})$$
$$s = \sqrt[n]{r} \qquad\qquad \beta = \frac{\theta + 2k\pi}{n}$$

Taking $k = 0, 1, 2, \ldots, n - 1$ produces n distinct angles β. Any other value of k leads to an angle β with the same terminal side as one of these. Hence, we have the following.

Formula for nth Roots

For each positive integer n, the nonzero complex number

$$r(\cos\theta + i\sin\theta)$$

has exactly n distinct nth roots. They are given by

$$\sqrt[n]{r}\left[\cos\left(\frac{\theta + 2k\pi}{n}\right) + i\sin\left(\frac{\theta + 2k\pi}{n}\right)\right],$$

where $k = 0, 1, 2, 3, \ldots n - 1$.

TECHNOLOGY TIP

The polynomial solvers on TI-86, HP-39+, and Casio FX 2.0 can solve

$$z^5 = 4 + 4i$$

and similar equations. See Calculator Investigation 4 on page 10 for details.

On TI-89, use cSOLVE in the COMPLEX submenu of the ALGEBRA menu.

EXAMPLE 4

Find the fifth roots of $4 + 4i$.

SOLUTION First write it in polar form as $4\sqrt{2}\left(\cos\frac{\pi}{4} + i\sin\frac{\pi}{4}\right)$. Now apply the root formula with $n = 5$, $r = 4\sqrt{2}$, $\theta = \pi/4$, and $k = 0, 1, 2, 3, 4$. Note that

$$\sqrt[5]{r} = \sqrt[5]{4\sqrt{2}} = (4\sqrt{2})^{1/5} = (2^2 2^{1/2})^{1/5} = (2^{5/2})^{1/5} = 2^{5/10} = 2^{1/2} = \sqrt{2}.$$

Therefore, the fifth roots are

$$\sqrt{2}\left[\cos\left(\frac{\pi/4 + 2k\pi}{5}\right) + i\sin\left(\frac{\pi/4 + 2k\pi}{5}\right)\right] \qquad k = 0, 1, 2, 3, 4,$$

that is,

$$k = 0: \sqrt{2}\left[\cos\left(\frac{\pi/4 + 0}{5}\right) + i\sin\left(\frac{\pi/4 + 0}{5}\right)\right] = \sqrt{2}\left(\cos\frac{\pi}{20} + i\sin\frac{\pi}{20}\right)$$

$$k = 1: \sqrt{2}\left[\cos\left(\frac{\pi/4 + 2\pi}{5}\right) + i\sin\left(\frac{\pi/4 + 2\pi}{5}\right)\right] = \sqrt{2}\left(\cos\frac{9\pi}{20} + i\sin\frac{9\pi}{20}\right)$$

$$k = 2: \sqrt{2}\left[\cos\left(\frac{\pi/4 + 4\pi}{5}\right) + i\sin\left(\frac{\pi/4 + 4\pi}{5}\right)\right] = \sqrt{2}\left(\cos\frac{17\pi}{20} + i\sin\frac{17\pi}{20}\right)$$

$$k = 3: \sqrt{2}\left[\cos\left(\frac{\pi/4 + 6\pi}{5}\right) + i\sin\left(\frac{\pi/4 + 6\pi}{5}\right)\right] = \sqrt{2}\left(\cos\frac{25\pi}{20} + i\sin\frac{25\pi}{20}\right)$$

$$k = 4: \sqrt{2}\left[\cos\left(\frac{\pi/4 + 8\pi}{5}\right) + i\sin\left(\frac{\pi/4 + 8\pi}{5}\right)\right] = \sqrt{2}\left(\cos\frac{33\pi}{20} + i\sin\frac{33\pi}{20}\right).$$

■

ROOTS OF UNITY

The n distinct nth roots of 1 (the solutions of $z^n = 1$) are called the ***n*th roots of unity.** Since $\cos 0 = 1$ and $\sin 0 = 0$, the polar form of the number 1 is $\cos 0 + i\sin 0$. Applying the root formula with $r = 1$ and $\theta = 0$ shows that

Roots of Unity

For each positive integer n, there are n distinct nth roots of unity.

$$\cos\frac{2k\pi}{n} + i\sin\frac{2k\pi}{n} \qquad (k = 0, 1, 2, \ldots, n - 1).$$

EXAMPLE 5

Find the cube roots of unity.

SOLUTION Apply the formula with $n = 3$ and $k = 0, 1, 2$.

$$k = 0: \qquad \cos 0 + i \sin 0 = 1$$

$$k = 1: \qquad \cos \frac{2\pi}{3} + i \sin \frac{2\pi}{3} = -\frac{1}{2} + \frac{\sqrt{3}}{2}i$$

$$k = 2: \qquad \cos \frac{4\pi}{3} + i \sin \frac{4\pi}{3} = -\frac{1}{2} - \frac{\sqrt{3}}{2}i$$ ■

Denote by ω the first complex cube root of unity obtained in Example 5:

$$\omega = \cos \frac{2\pi}{3} + i \sin \frac{2\pi}{3}.$$

If we use DeMoivre's Theorem to find ω^2 and ω^3, we see that these numbers are the other two cube roots of unity found in Example 5:

$$\omega^2 = \left(\cos \frac{2\pi}{3} + i \sin \frac{2\pi}{3}\right)^2 = \cos \frac{4\pi}{3} + i \sin \frac{4\pi}{3}$$

$$\omega^3 = \left(\cos \frac{2\pi}{3} + i \sin \frac{2\pi}{3}\right)^3 = \cos \frac{6\pi}{3} + i \sin \frac{6\pi}{3} = \cos 2\pi + i \sin 2\pi$$

$$= 1 + 0 \cdot i = 1.$$

In other words, all the cube roots of unity are powers of ω. The same thing is true in the general case.

Roots of Unity

> Let n be a positive integer with $n > 1$. Then the number
>
> $$z = \cos \frac{2\pi}{n} + i \sin \frac{2\pi}{n}$$
>
> is an nth root of unity and all the nth roots of unity are
>
> $$z, z^2, z^3, z^4, \ldots z^{n-1}, z^n = 1.$$

The nth roots of unity have an interesting geometric interpretation. Every nth root of unity has absolute value 1 by the Pythagorean identity.

$$\left|\cos \frac{2k\pi}{n} + i \sin \frac{2k\pi}{n}\right| = \left(\cos \frac{2k\pi}{n}\right)^2 + \left(\sin \frac{2k\pi}{n}\right)^2$$

$$= \cos^2\left(\frac{2k\pi}{n}\right) + \sin^2\left(\frac{2k\pi}{n}\right) = 1$$

Therefore, in the complex plane, every nth root of unity is exactly 1 unit from the origin. In other words, the nth roots of unity all lie on the unit circle.

EXAMPLE 6

Find the fifth roots of unity.

SOLUTION They are

$$\cos\frac{2k\pi}{5} + i\sin\frac{2k\pi}{5} \qquad (k = 0, 1, 2, 3, 4),$$

that is,

$$\cos 0 + i\sin 0 = 1, \qquad \cos\frac{2\pi}{5} + i\sin\frac{2\pi}{5}, \qquad \cos\frac{4\pi}{5} + i\sin\frac{4\pi}{5},$$

$$\cos\frac{6\pi}{5} + i\sin\frac{6\pi}{5}, \qquad \cos\frac{8\pi}{5} + i\sin\frac{8\pi}{5}.$$

These five roots can be plotted in the complex plane by starting at $1 = 1 + 0i$ and moving counterclockwise around the unit circle, moving through an angle of $2\pi/5$ at each step, as shown in Figure 4–6. If you connect these five roots, they form the vertices of a regular pentagon (Figure 4–7). ■

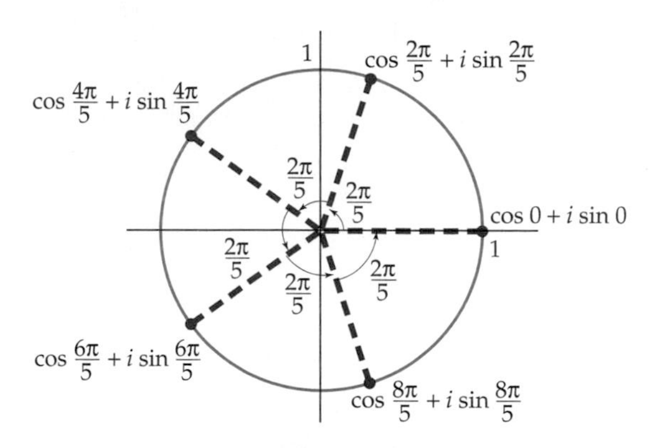

Figure 4–6

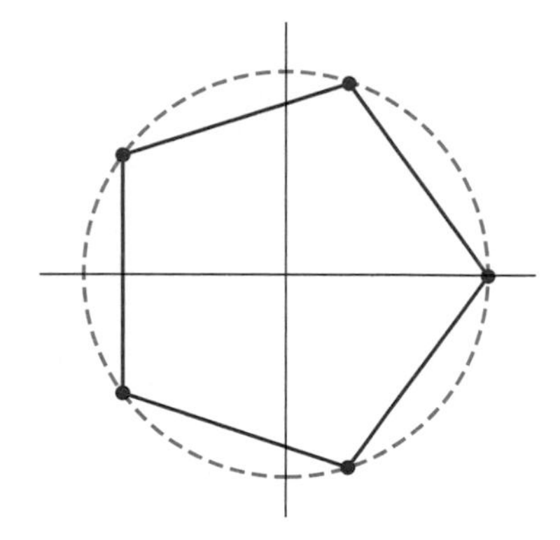

Figure 4–7

GRAPHING EXPLORATION

With your calculator in parametric graphing mode, set these range values:

$$0 \le t \le 2\pi, \qquad t\text{-step} \approx .067,$$

$$-1.5 \le x \le 1.5, \qquad -1 \le y \le 1$$

and graph the unit circle, whose parametric equations are

$$x = \cos t \qquad \text{and} \qquad y = \sin t.^{*}$$

Reset the t-step to be $2\pi/5$ and graph again. Your screen now looks exactly like the red lines in Figure 4–7 because the calculator plotted only the five points corresponding to $t = 0, 2\pi/5, 4\pi/5, 6\pi/5, 8\pi/5^{\dagger}$ and connected them with the shortest possible segments. Use the trace feature to move along the graph. The cursor will jump from vertex to vertex, that is, it will move from one fifth root of unity to the next.

---

*On wide-screen calculators, use $-2 \le x \le 2$ or $-1.7 \le x \le 1.7$ so that the unit circle looks like a circle.

†The point corresponding to $t = 10\pi/5 = 2\pi$ is the same as the one corresponding to $t = 0$.

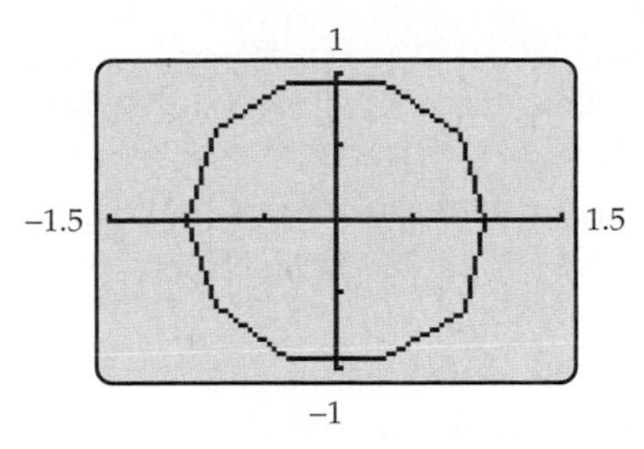

Figure 4–8

EXAMPLE 7

Find the tenth roots of unity graphically.

SOLUTION Graph the unit circle as in the preceding exploration, but use $2\pi/10$ as the t-step. The result (Figure 4–8) is a regular decagon whose vertices are the tenth roots of unity. By using the trace feature, you can approximate each of them.

GRAPHING EXPLORATION

Verify that the two tenth roots of unity in the first quadrant are (approximately) $.8090 + .5878i$ and $.3090 + .9511i$. ■

EXERCISES 4.3

In Exercises 1–6, calculate the given product and express your answer in the form $a + bi$.

1. $\left(\cos\frac{\pi}{12} + i\sin\frac{\pi}{12}\right)^6$ **2.** $\left(\cos\frac{\pi}{5} + i\sin\frac{\pi}{5}\right)^{20}$

3. $\left[2\left(\cos\frac{\pi}{24} + i\sin\frac{\pi}{24}\right)\right]^8$ **4.** $\left[\sqrt{2}\left(\cos\frac{\pi}{60} + i\sin\frac{\pi}{60}\right)\right]^{10}$

5. $\left[3\left(\cos\frac{7\pi}{30} + i\sin\frac{7\pi}{30}\right)\right]^5$ **6.** $\left[\sqrt[3]{4}\left(\cos\frac{7\pi}{36} + i\sin\frac{7\pi}{36}\right)\right]^{12}$

In Exercises 7–14, calculate the product by expressing the number in polar form and using DeMoivre's Theorem. Express your answer in the form $a + bi$.

7. $\left(\frac{1}{2} + \frac{\sqrt{3}}{2}i\right)^3$ **8.** $\left(-\frac{\sqrt{2}}{2} + \frac{\sqrt{2}}{2}i\right)^4$

9. $(1 - i)^{12}$ **10.** $(2 + 2i)^8$

11. $\left(\frac{\sqrt{3}}{2} + \frac{1}{2}i\right)^{10}$ **12.** $\left(-\frac{1}{2} + \frac{\sqrt{3}}{2}i\right)^{20}$

13. $\left(\frac{-1}{\sqrt{2}} + \frac{i}{\sqrt{2}}\right)^{14}$ **14.** $(-1 + \sqrt{3}i)^8$

In Exercises 15 and 16, find the indicated roots of unity and express your answers in the form $a + bi$.

15. Fourth roots of unity **16.** Sixth roots of unity

In Exercises 17–30, find the nth roots in polar form.

17. $36\left(\cos\frac{\pi}{3} + i\sin\frac{\pi}{3}\right)$; $n = 2$

18. $64\left(\cos\frac{\pi}{4} + i\sin\frac{\pi}{4}\right)$; $n = 2$

19. $64\left(\cos\frac{\pi}{5} + i\sin\frac{\pi}{5}\right)$; $n = 3$

20. $8\left(\cos\frac{\pi}{10} + i\sin\frac{\pi}{10}\right)$; $n = 3$

21. $81\left(\cos\frac{\pi}{12} + i\sin\frac{\pi}{12}\right)$; $n = 4$

22. $16\left(\cos\frac{\pi}{7} + i\sin\frac{\pi}{7}\right)$; $n = 5$

23. -1; $n = 5$ **24.** 1; $n = 7$

25. i; $n = 5$ **26.** $-i$; $n = 6$

27. $1 + i$; $n = 2$ **28.** $1 - \sqrt{3}i$; $n = 3$

29. $8\sqrt{3} + 8i$; $n = 4$ **30.** $-16\sqrt{2} - 16\sqrt{2}i$; $n = 5$

In Exercises 31–40, solve the given equation in the complex number system.

31. $x^6 = -1$ **32.** $x^6 + 64 = 0$ **33.** $x^3 = i$

34. $x^4 = i$ **35.** $x^3 + 27i = 0$ **36.** $x^6 + 729 = 0$

37. $x^5 - 243i = 0$ **38.** $x^7 = 1 - i$

39. $x^4 = -1 + \sqrt{3}i$ **40.** $x^4 = -8 - 8\sqrt{3}i$

In Exercises 41–46, represent the roots of unity graphically. Then use the trace feature to obtain approximations of the form $a + bi$ for each root (round to four places).

41. Seventh roots of unity **42.** Fifth roots of unity

43. Eighth roots of unity

44. Twelfth roots of unity

45. Ninth roots of unity

46. Tenth roots of unity

47. Solve the equation $x^3 + x^2 + x + 1 = 0$. [*Hint:* First find the quotient when $x^4 - 1$ is divided by $x - 1$ and then consider solutions of $x^4 - 1 = 0$.]

48. Solve the equation $x^4 + x^3 + x^2 + x + 1 = 0$. [*Hint:* Consider $x^5 - 1$ and $x - 1$ and see Exercise 47.]

49. Solve $x^5 + x^4 + x^3 + x^2 + x + 1 = 0$. [*Hint:* Consider $x^6 - 1$ and $x - 1$ and see Exercise 47.]

50. What do you think are the solutions of $x^{n-1} + x^{n-2} + \cdots + x^3 + x^2 + x + 1 = 0$? (See Exercises 47–49.)

Thinkers

51. In the complex plane, identify each point with its complex number label. The unit circle consists of all numbers (points) z such that $|z| = 1$. Suppose v and w are two points (numbers) that move around the unit circle in such a way that $v = w^{12}$ at all times. When w has made one complete trip around the circle, how many trips has v made? [*Hint:* Think polar and DeMoivre.]

52. Suppose u is an nth root of unity. Show that $1/u$ is also an nth root of unity. [*Hint:* Use the definition, *not* polar form.]

53. Let $u_1, u_2, \ldots, u_n$ be the distinct nth roots of unity and suppose v is a nonzero solution of the equation

$$z^n = r(\cos\theta + i\sin\theta).$$

Show that $vu_1, vu_2, \ldots, vu_n$ are n distinct solutions of the equation. [*Remember:* Each u_i is a solution of $x^n = 1$.]

54. Use the formula for nth roots and the identities

$$\cos(x + \pi) = -\cos x \qquad \sin(x + \pi) = -\sin x$$

to show that the nonzero complex number $r(\cos\theta + i\sin\theta)$ has two square roots and that these square roots are negatives of each other.

4.4 Vectors in the Plane

Once a unit of measure has been agreed upon, quantities such as area, length, time, and temperature can be described by a single number. Other quantities, such as an east wind of 10 mph, require two numbers to describe them because they involve both *magnitude* and *direction.* Such quantities are called **vectors** and are represented geometrically by a directed line segment or arrow, as in Figure 4–9.

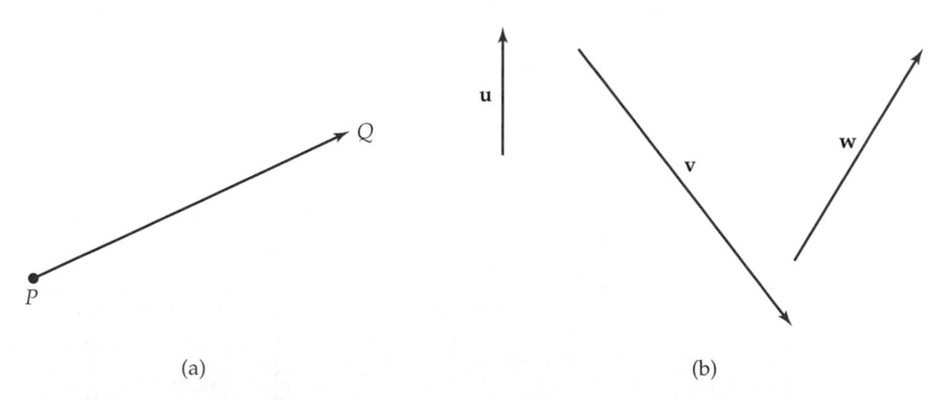

Figure 4–9

When a vector extends from a point P to a point Q, as in Figure 4–9(a), P is called the **initial point** of the vector and Q is called the **terminal point,** and the vector is written $\overrightarrow{PQ}$. Its **length** is denoted by $\|\overrightarrow{PQ}\|$. When the endpoints are not specified, as in Figure 4–9(b), vectors are denoted by boldface letters such as **u, v,** and **w.** The length of a vector **u** is denoted by $\|\mathbf{u}\|$ and is called the **magnitude** of **u.**

If **u** and **v** are vectors with the same magnitude and direction, we say that **u** and **v** are **equivalent** and write **u** = **v**. Some examples are shown in Figure 4–10.

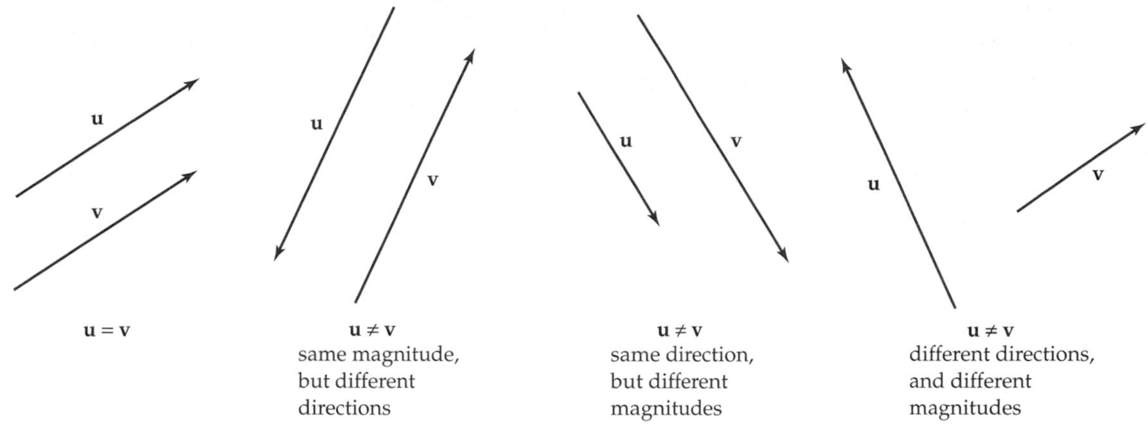

Figure 4–10

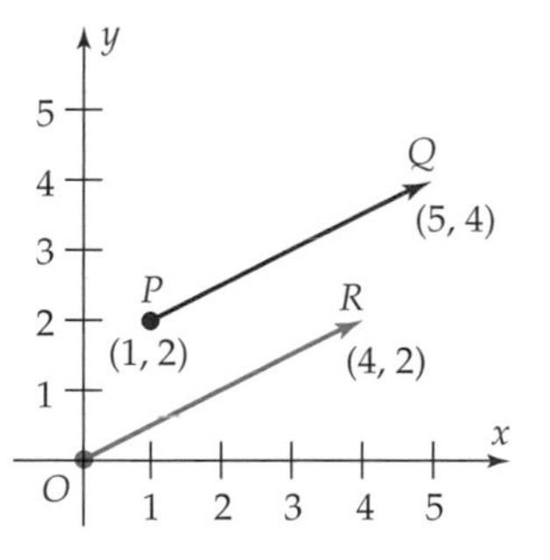

Figure 4–11

EXAMPLE 1

Let $P = (1, 2)$, $Q = (5, 4)$, $O = (0, 0)$, and $R = (4, 2)$, as in Figure 4–11. Show that $\overrightarrow{PQ} = \overrightarrow{OR}$.

SOLUTION The distance formula shows that $\overrightarrow{PQ}$ and $\overrightarrow{OR}$ have the *same length.*

$$\|\overrightarrow{PQ}\| = \sqrt{(5-1)^2 + (4-2)^2} = \sqrt{4^2 + 2^2} = \sqrt{20}$$
$$\|\overrightarrow{OR}\| = \sqrt{(4-0)^2 + (2-0)^2} = \sqrt{4^2 + 2^2} = \sqrt{20}.$$

Furthermore, the lines through PQ and OR have the same slope.

$$\text{slope } PQ = \frac{4-2}{5-1} = \frac{2}{4} = \frac{1}{2} \qquad \text{slope } OR = \frac{2-0}{4-0} = \frac{2}{4} = \frac{1}{2}$$

Since $\overrightarrow{PQ}$ and $\overrightarrow{OR}$ both point to the upper right on lines of the same slope, $\overrightarrow{PQ}$ and $\overrightarrow{OR}$ have the *same direction*. Therefore, $\overrightarrow{PQ} = \overrightarrow{OR}$. ■

According to the definition of equivalence, a vector may be moved from one location to another, provided that its magnitude and direction are not changed. Consequently, we have the following.

Equivalent Vectors

Every vector $\overrightarrow{PQ}$ is equivalent to a vector $\overrightarrow{OR}$ with initial point at the origin: If $P = (x_1, y_1)$ and $Q = (x_2, y_2)$, then

$$\overrightarrow{PQ} = \overrightarrow{OR}, \qquad \text{where} \qquad R = (x_2 - x_1, y_2 - y_1).$$

Proof The proof is similar to the one used in Example 1. It follows from the fact that $\overrightarrow{PQ}$ and $\overrightarrow{OR}$ have the same length.

$$\begin{aligned}\|\overrightarrow{OR}\| &= \sqrt{[(x_2 - x_1) - 0]^2 + [(y_2 - y_1) - 0]^2} \\ &= \sqrt{(x_2 - x_1)^2 + (y_2 - y_1)^2} = \|\overrightarrow{PQ}\|,\end{aligned}$$

and that either the line segments PQ and OR are both vertical or they have the same slope,

$$\text{slope } OR = \frac{(y_2 - y_1) - 0}{(x_2 - x_1) - 0} = \frac{y_2 - y_1}{x_2 - x_1} = \text{slope } PQ,$$

as shown in Figure 4–12. ■

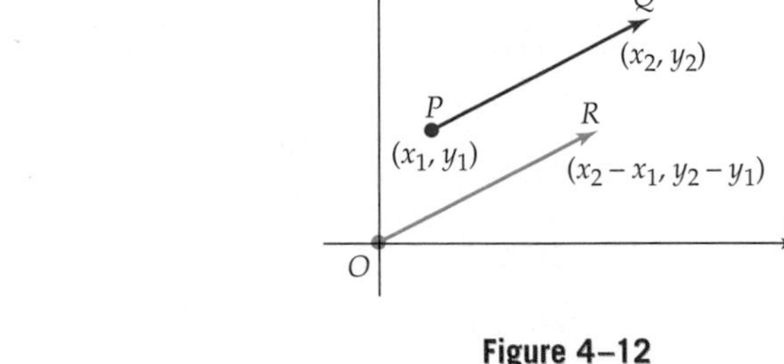

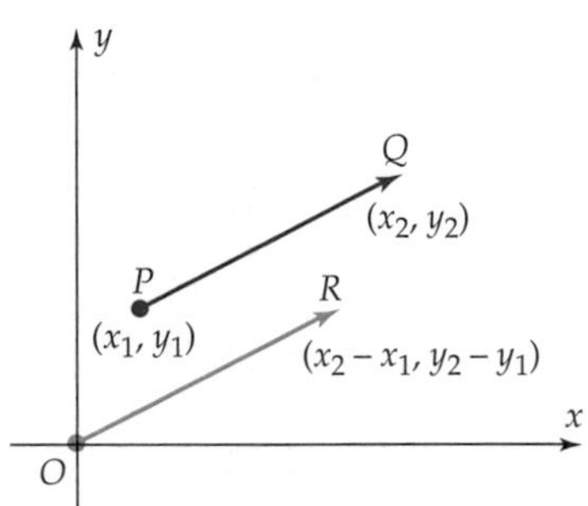

Figure 4–12

TECHNOLOGY TIP

Vectors in component form can be entered on TI-86/89 and HP-39+ by using $[a, b]$ in place of $\langle a, b\rangle$.

The magnitude and direction of a vector with the origin as initial point are completely determined by the coordinates of its terminal point. Consequently, we denote the vector with initial point (0, 0) and terminal point (a, b) by $\langle a, b\rangle$. The numbers a and b are called the **components** of the vector $\langle a, b\rangle$.

Since the length of the vector $\langle a, b\rangle$ is the distance from (0, 0) to (a, b), the distance formula shows the following.

Magnitude

> The **magnitude** (or **norm**) of the vector $\mathbf{v} = \langle a, b\rangle$ is
>
> $$\|\mathbf{v}\| = \sqrt{a^2 + b^2}.$$

EXAMPLE 2

Find the components and the magnitude of the vector with initial point $P = (-2, 6)$ and terminal point $Q = (4, -3)$.

SOLUTION According to the fact in the box on page 301 (with $x_1 = -2$, $y_1 = 6$, $x_2 = 4$, $y_2 = -3$),

$$\overrightarrow{PQ} = \overrightarrow{OR}, \qquad \text{where} \qquad R = (4 - (-2), -3 - 6) = (6, -9),$$

that is,

$$\overrightarrow{PQ} = \overrightarrow{OR} = \langle 6, -9\rangle.$$

Therefore,

$$\|\overrightarrow{PQ}\| = \|\overrightarrow{OR}\| = \sqrt{6^2 + (-9)^2} = \sqrt{36 + 81} = \sqrt{117}. \quad ■$$

TECHNOLOGY TIP

To find the magnitude of a vector, use NORM in this menu/submenu:

TI-86: VECTOR/MATH

TI-89: MATH/MATRIX/NORMS

VECTOR ARITHMETIC

When dealing with vectors, it is customary to refer to ordinary real numbers as **scalars. Scalar multiplication** is an operation in which a scalar k is "multiplied" by a vector $\mathbf{v}$ to produce another *vector* denoted by $k\mathbf{v}$. Here is the formal definition.

Scalar Multiplication

> If k is a real number and $\mathbf{v} = \langle a, b\rangle$ is a vector, then
>
> $$k\mathbf{v} \text{ is the vector } \langle ka, kb\rangle.$$
>
> The vector $k\mathbf{v}$ is called a **scalar multiple** of $\mathbf{v}$.

EXAMPLE 3

If $\mathbf{v} = \langle 3, 1\rangle$, then

$$3\mathbf{v} = 3\langle 3, 1\rangle = \langle 3 \cdot 3, 3 \cdot 1\rangle = \langle 9, 3\rangle$$
$$-2\mathbf{v} = -2\langle 3, 1\rangle = \langle -2 \cdot 3, -2 \cdot 1\rangle = \langle -6, -2\rangle,$$

as shown in Figure 4–13.

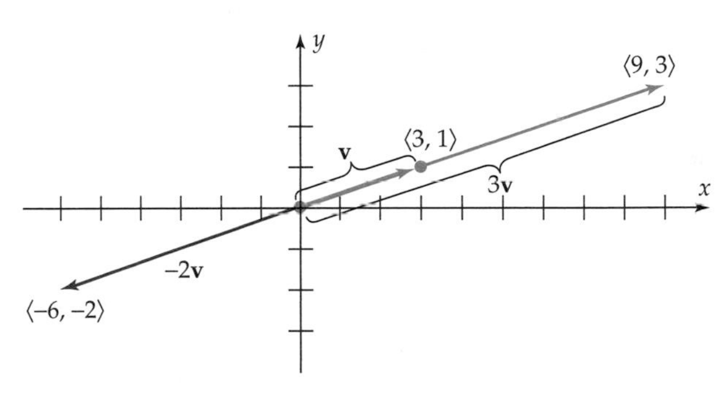

Figure 4–13

Figure 4–13 shows that $3\mathbf{v}$ has the same direction as $\mathbf{v}$, while $-2\mathbf{v}$ has the opposite direction. Also note that

$$\|\mathbf{v}\| = \|\langle 3, 1\rangle\| = \sqrt{3^2 + 1^2} = \sqrt{10}$$
$$\|-2\mathbf{v}\| = \|\langle -6, -2\rangle\| = \sqrt{(-6)^2 + (-2)^2} = \sqrt{40} = 2\sqrt{10}.$$

Therefore,

$$\|-2\mathbf{v}\| = 2\sqrt{10} = 2\|\mathbf{v}\| = |-2| \cdot \|\mathbf{v}\|.$$

Similarly, you can verify that $\|3\mathbf{v}\| = |3| \cdot \|\mathbf{v}\| = 3\|\mathbf{v}\|$. ■

Example 3 is an illustration of the following facts.

Geometric Interpretation of Scalar Multiplication

The *magnitude* of the vector $k\mathbf{v}$ is $|k|$ times the length of $\mathbf{v}$, that is,

$$\|k\mathbf{v}\| = |k| \cdot \|\mathbf{v}\|.$$

The *direction of* $k\mathbf{v}$ is the same as that of $\mathbf{v}$ when k is positive and opposite that of $\mathbf{v}$ when k is negative.

See Exercise 77 for a proof of this statement.

Vector addition is an operation in which two vectors $\mathbf{u}$ and $\mathbf{v}$ are added to produce a new vector denoted $\mathbf{u} + \mathbf{v}$. Formally, we have the following.

Vector Addition

If $\mathbf{u} = \langle a, b \rangle$ and $\mathbf{v} = \langle c, d \rangle$, then

$$\mathbf{u} + \mathbf{v} = \langle a + c, b + d \rangle.$$

EXAMPLE 4

If $\mathbf{u} = \langle -5, 2 \rangle$ and $\mathbf{v} = \langle 3, 1 \rangle$, find $\mathbf{u} + \mathbf{v}$.

SOLUTION

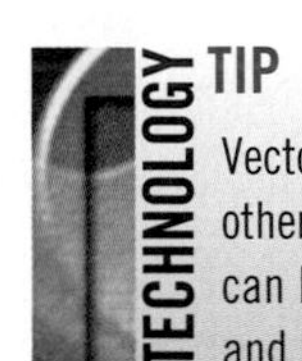

TECHNOLOGY TIP

Vector arithmetic and other vector operations can be done on TI-86/89 and HP-39+.

$$\mathbf{u} + \mathbf{v} = \langle -5, 2 \rangle + \langle 3, 1 \rangle = \langle -5 + 3, 2 + 1 \rangle = \langle -2, 3 \rangle$$

as shown in Figure 4–14. ■

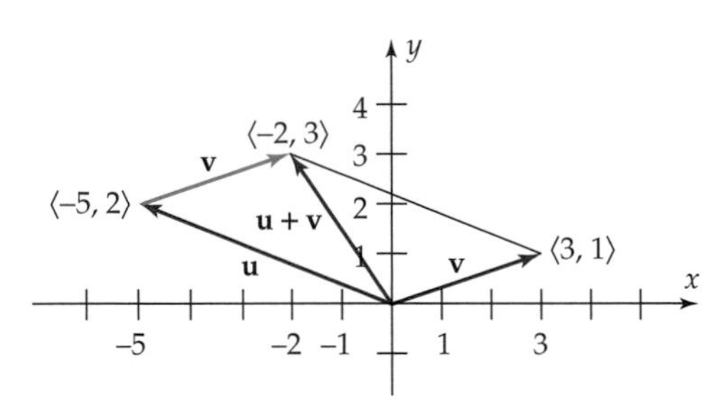

Figure 4–14

Example 4 is an illustration of these facts.

Geometric Interpretations of Vector Addition

1. If $\mathbf{u}$ and $\mathbf{v}$ are vectors with the same initial point P, then $\mathbf{u} + \mathbf{v}$ is the vector $\overrightarrow{PQ}$, where $\overrightarrow{PQ}$ is the diagonal of the parallelogram with adjacent sides $\mathbf{u}$ and $\mathbf{v}$.
2. If the vector $\mathbf{v}$ is moved (without changing its magnitude or direction) so that its initial point lies on the endpoint of the vector $\mathbf{u}$, then $\mathbf{u} + \mathbf{v}$ is the vector with the same initial point P as $\mathbf{u}$ and the same terminal point Q as $\mathbf{v}$.

See Exercise 78 for a proof of these statements.

The **negative** of a vector $\mathbf{v} = \langle c, d\rangle$ is defined to be the vector $(-1)\mathbf{v} = (-1)\langle c, d\rangle = \langle -c, -d\rangle$ and is denoted $-\mathbf{v}$. **Vector subtraction** is then defined as follows.

Vector Subtraction

If $\mathbf{u} = \langle a, b\rangle$ and $\mathbf{v} = \langle c, d\rangle$, then $\mathbf{u} - \mathbf{v}$ is the vector

$$\begin{aligned}\mathbf{u} + (-\mathbf{v}) &= \langle a, b\rangle + \langle -c, -d\rangle \\ &= \langle a - c, b - d\rangle.\end{aligned}$$

A geometric interpretation of vector subtraction is given in Exercise 79.

EXAMPLE 5

If $\mathbf{u} = \langle 2, 5\rangle$ and $\mathbf{v} = \langle 6, 1\rangle$, find $\mathbf{u} - \mathbf{v}$.

SOLUTION

$$\mathbf{u} - \mathbf{v} = \langle 2, 5\rangle - \langle 6, 1\rangle = \langle 2 - 6, 5 - 1\rangle = \langle -4, 4\rangle,$$

as shown in Figure 4–15. ■

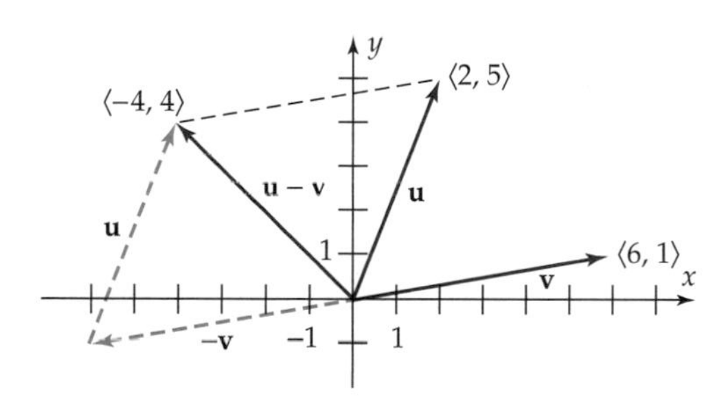

Figure 4–15

The vector $\langle 0, 0\rangle$ is called the **zero vector** and is denoted **0**.

EXAMPLE 6

If $\mathbf{u} = \langle -1, 6\rangle$, $\mathbf{v} = \langle 2/3, -4\rangle$, and $\mathbf{w} = \langle 2, 5/2\rangle$, find $2\mathbf{u} + 3\mathbf{v}$ and $4\mathbf{w} - 2\mathbf{u}$.

SOLUTION

$$\begin{aligned}2\mathbf{u} + 3\mathbf{v} &= 2\langle -1, 6\rangle + 3\left\langle \frac{2}{3}, -4\right\rangle = \langle -2, 12\rangle + \langle 2, -12\rangle \\ &= \langle 0, 0\rangle = \mathbf{0},\end{aligned}$$

and

$$\begin{aligned} 4\mathbf{w} - 2\mathbf{u} &= 4\left\langle 2, \frac{5}{2} \right\rangle - 2\langle -1, 6 \rangle \\ &= \langle 8, 10 \rangle - \langle -2, 12 \rangle \\ &= \langle 8 - (-2), 10 - 12 \rangle = \langle 10, -2 \rangle. \end{aligned}$$ ■

Operations on vectors share many of the same properties as arithmetical operations on numbers.

Properties of Vector Addition and Scalar Multiplication

For any vectors **u, v,** and **w** and any scalars r and s,

1. $\mathbf{u} + (\mathbf{v} + \mathbf{w}) = (\mathbf{u} + \mathbf{v}) + \mathbf{w}$
2. $\mathbf{u} + \mathbf{v} = \mathbf{v} + \mathbf{u}$
3. $\mathbf{v} + \mathbf{0} = \mathbf{v} = \mathbf{0} + \mathbf{v}$
4. $\mathbf{v} + (-\mathbf{v}) = \mathbf{0}$
5. $r(\mathbf{u} + \mathbf{v}) = r\mathbf{u} + r\mathbf{v}$
6. $(r + s)\mathbf{v} = r\mathbf{v} + s\mathbf{v}$
7. $(rs)\mathbf{v} = r(s\mathbf{v}) = s(r\mathbf{v})$
8. $1\mathbf{v} = \mathbf{v}$
9. $0\mathbf{v} = \mathbf{0}$ and $r\mathbf{0} = \mathbf{0}$

Proof If $\mathbf{u} = \langle a, b \rangle$ and $\mathbf{v} = \langle c, d \rangle$, then because addition of real numbers is commutative, we have

$$\begin{aligned} \mathbf{u} + \mathbf{v} = \langle a, b \rangle + \langle c, d \rangle &= \langle a + c, b + d \rangle \\ &= \langle c + a, d + b \rangle = \langle c, d \rangle + \langle a, b \rangle = \mathbf{v} + \mathbf{u}. \end{aligned}$$

The other properties are proved similarly; see Exercises 53–58. ■

UNIT VECTORS

A vector with length 1 is called a **unit vector.** For instance, $\langle 3/5, 4/5 \rangle$ is a unit vector, since

$$\left\| \left\langle \frac{3}{5}, \frac{4}{5} \right\rangle \right\| = \sqrt{\left(\frac{3}{5}\right)^2 + \left(\frac{4}{5}\right)^2} = \sqrt{\frac{9}{25} + \frac{16}{25}} = \sqrt{\frac{25}{25}} = 1.$$

EXAMPLE 7

Find a unit vector that has the same direction as the vector $\mathbf{v} = \langle 5, 12 \rangle$.

SOLUTION The length of **v** is

$$\|\mathbf{v}\| = \|\langle 5, 12 \rangle\| = \sqrt{5^2 + 12^2} = \sqrt{169} = 13.$$

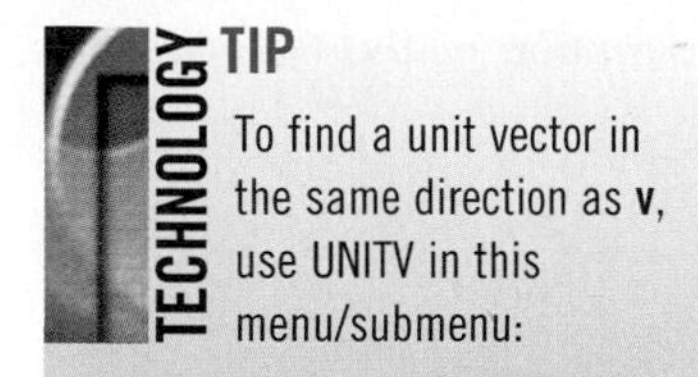

TI-86: VECTOR/MATH

TI-89: MATH/MATRIX/VECTOR OPS

The vector

$$\mathbf{u} = \frac{1}{13}\mathbf{v} = \left\langle \frac{5}{13}, \frac{12}{13} \right\rangle$$

has the same direction as **v** (since it is a scalar multiple by a positive number), and **u** is a unit vector because

$$\|\mathbf{u}\| = \left\|\frac{1}{13}\mathbf{v}\right\| = \left|\frac{1}{13}\right| \cdot \|\mathbf{v}\| = \frac{1}{13} \cdot 13 = 1. \quad ■$$

The procedure used in Example 7 (multiplying a vector by the reciprocal of its length) works in the general case.

Unit Vectors

> If **v** is a nonzero vector, then $\dfrac{1}{\|\mathbf{v}\|}\mathbf{v}$ is a unit vector with the same direction as **v**.

You can easily verify that the vectors $\mathbf{i} = \langle 1, 0 \rangle$ and $\mathbf{j} = \langle 0, 1 \rangle$ are unit vectors. The vectors **i** and **j** play a special role because they lead to a useful alternate notation for vectors. For example, if $\mathbf{u} = \langle 5, -7 \rangle$, then

$$\mathbf{u} = \langle 5, 0 \rangle + \langle 0, -7 \rangle = 5\langle 1, 0 \rangle - 7\langle 0, 1 \rangle = 5\mathbf{i} - 7\mathbf{j}.$$

Similarly, if $\mathbf{v} = \langle a, b \rangle$ is any vector, then

$$\mathbf{v} = \langle a, b \rangle = \langle a, 0 \rangle + \langle 0, b \rangle = a\langle 1, 0 \rangle + b\langle 0, 1 \rangle = a\mathbf{i} + b\mathbf{j}.$$

The vector **v** is said to be a **linear combination** of **i** and **j**. When vectors are written as linear combinations of **i** and **j**, then the properties in the box on page 306 can be used to write the rules for vector addition and scalar multiplication in this form.

$$(a\mathbf{i} + b\mathbf{j}) + (c\mathbf{i} + d\mathbf{j}) = (a + c)\mathbf{i} + (b + d)\mathbf{j}$$

and

$$c(a\mathbf{i} + b\mathbf{j}) = ca\mathbf{i} + cb\mathbf{j}.$$

EXAMPLE 8

If $\mathbf{u} = 2\mathbf{i} - 6\mathbf{j}$ and $\mathbf{v} = -5\mathbf{i} + 2\mathbf{j}$, find $3\mathbf{u} - 2\mathbf{v}$.

SOLUTION

$$3\mathbf{u} - 2\mathbf{v} = 3(2\mathbf{i} - 6\mathbf{j}) - 2(-5\mathbf{i} + 2\mathbf{j}) = 6\mathbf{i} - 18\mathbf{j} + 10\mathbf{i} - 4\mathbf{j}$$
$$= 16\mathbf{i} - 22\mathbf{j} \quad ■$$

DIRECTION ANGLES

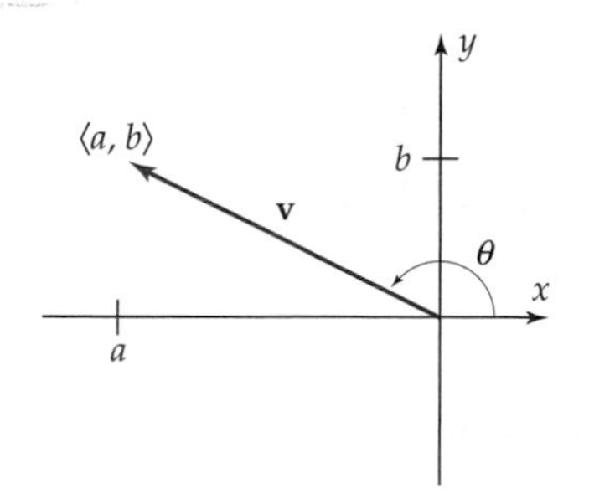

Figure 4–16

If $\mathbf{v} = \langle a, b \rangle = a\mathbf{i} + b\mathbf{j}$ is a vector, then the direction of **v** is completely determined by the standard position angle θ between 0° and 360° whose terminal side is **v**, as shown in Figure 4–16. The angle θ is called the **direction angle** of the vector **v**. According to the point-in-the-plane description of the trigonometric functions,

$$\cos\theta = \frac{a}{\|\mathbf{v}\|} \qquad \text{and} \qquad \sin\theta = \frac{b}{\|\mathbf{v}\|}.$$

Rewriting each of these equations produces the following result.

Components and the Direction Angle

If $\mathbf{v} = \langle a, b\rangle = a\mathbf{i} + b\mathbf{j}$, then

$$a = \|\mathbf{v}\|\cos\theta \qquad \text{and} \qquad b = \|\mathbf{v}\|\sin\theta$$

where θ is the direction angle of **v**.

EXAMPLE 9

Find the component form of the vector that represents the velocity of an airplane at the instant its wheels leave the ground if the plane is going 60 mph and the body of the plane makes a 7° angle with the horizontal.

SOLUTION The velocity vector $\mathbf{v} = a\mathbf{i} + b\mathbf{j}$ has magnitude 60 and direction angle $\theta = 7°$, as shown in Figure 4–17. Hence,

$$\begin{aligned}\mathbf{v} &= (\|\mathbf{v}\|\cos\theta)\mathbf{i} + (\|\mathbf{v}\|\sin\theta)\mathbf{j}\\ &= (60\cos 7°)\mathbf{i} + (60\sin 7°)\mathbf{j}\\ &\approx (60 \cdot .9925)\mathbf{i} + (60 \cdot .1219)\mathbf{j}\\ &\approx 59.55\mathbf{i} + 7.31\mathbf{j} = \langle 59.55, 7.31\rangle.\end{aligned}$$ ■

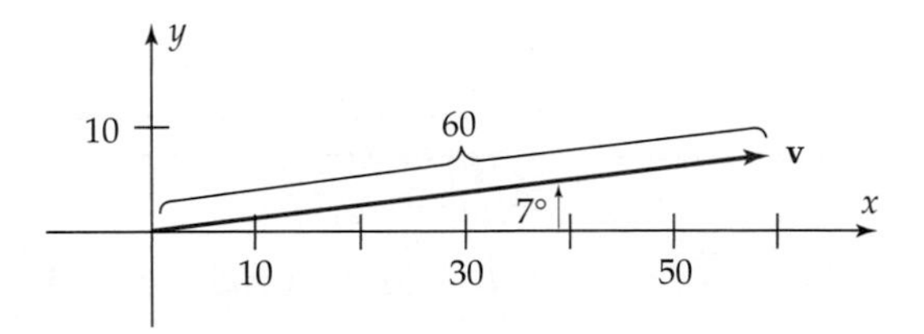

Figure 4–17

If $\mathbf{v} = a\mathbf{i} + b\mathbf{j}$ is a nonzero vector with direction angle θ, then

$$\tan\theta = \frac{\sin\theta}{\cos\theta} = \frac{b/\|\mathbf{v}\|}{a/\|\mathbf{v}\|} = \frac{b}{a}.$$

This fact provides a convenient way to find the direction angle of a vector.

EXAMPLE 10

Find the direction angle of

(a) $\mathbf{u} = 5\mathbf{i} + 13\mathbf{j}$ (b) $\mathbf{v} = -10\mathbf{i} + 7\mathbf{j}$.

SOLUTION

(a) The direction angle θ of $\mathbf{u}$ satisfies $\tan\theta = b/a = 13/5 = 2.6$. Using the TAN^{-1} key on a calculator, we find that $\theta \approx 68.96°$, as shown in Figure 4–18(a).

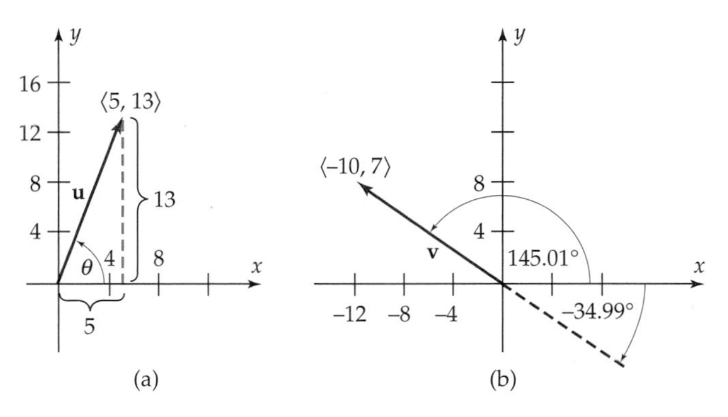

Figure 4–18

(b) The direction angle of $\mathbf{v}$ satisfies $\tan\theta = -7/10 = -.7$. Since $\mathbf{v}$ lies in the second quadrant, θ must be between 90° and 180°. A calculator shows that $-34.99°$ is an angle with tangent (approximately) $-.7$. Since tangent has period $\pi(= 180°)$, we know that $\tan t = \tan(t + 180°)$ for every t. Therefore, $\theta = -34.99° + 180° = 145.01°$ is the angle between 90° and 180° such that $\tan\theta \approx -.7$. See Figure 4–18(b). ■

EXAMPLE 11

An object at the origin is acted upon by two forces. A 150-pound force makes an angle of 20° with the positive x-axis, and the other force of 100 pounds makes an angle of 70°, as shown in Figure 4–19. Find the direction and magnitude of the resultant force.

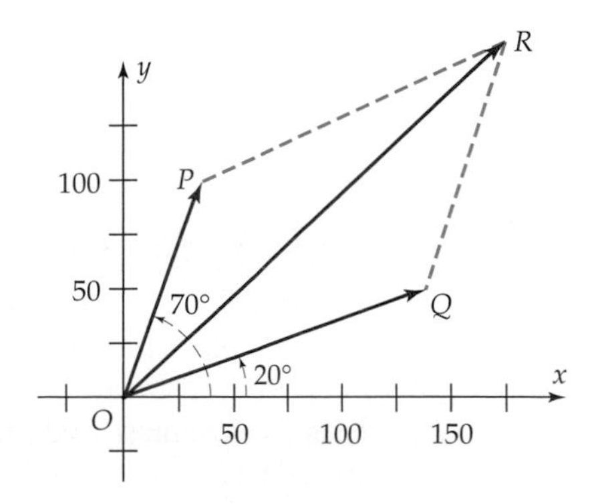

Figure 4–19

SOLUTION The forces acting on the object are

$$\overrightarrow{OP} = (100 \cos 70°)\mathbf{i} + (100 \sin 70°)\mathbf{j}$$
$$\overrightarrow{OQ} = (150 \cos 20°)\mathbf{i} + (150 \sin 20°)\mathbf{j}.$$

The resultant force $\overrightarrow{OR}$ is the sum of $\overrightarrow{OP}$ and $\overrightarrow{OQ}$. Hence,

$$\overrightarrow{OR} = (100 \cos 70° + 150° \cos 20°)\mathbf{i} + (100 \sin 70° + 150 \sin 20°)\mathbf{j}$$
$$\approx 175.16\mathbf{i} + 145.27\mathbf{j}.$$

Therefore, the magnitude of the resultant force is

$$\|\overrightarrow{OR}\| \approx \sqrt{(175.16)^2 + (145.27)^2} \approx 227.56.$$

The direction angle θ of the resultant force satisfies

$$\tan \theta \approx 145.27/175.16 \approx .8294.$$

A calculator shows that $\theta \approx 39.67°$. ■

APPLICATIONS

EXAMPLE 12

A 200-pound box lies on a ramp that makes an angle of 24° with the horizontal. A rope is tied to the box from a post at the top of the ramp to keep it in position. Ignoring friction, how much force is being exerted on the rope by the box?

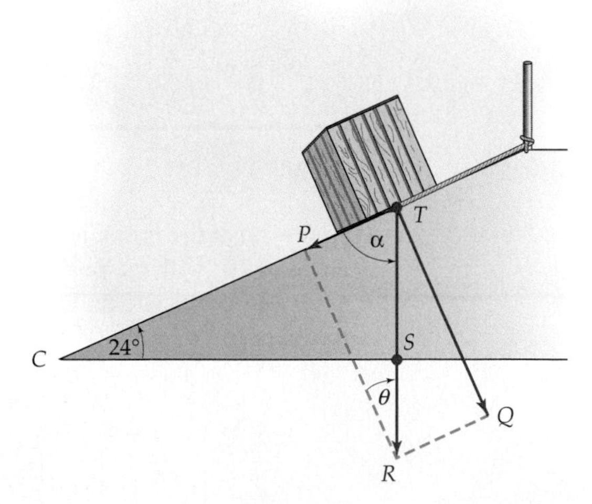

Figure 4–20

SOLUTION Because of gravity, the box exerts a 200-pound weight straight down (vector $\overrightarrow{TR}$). As Figure 4–20 shows, $\overrightarrow{TR}$ is the sum of $\overrightarrow{TP}$ and $\overrightarrow{TQ}$. The force on the rope is exerted by $\overrightarrow{TP}$, the vector of the force pulling the box down the ramp, so we must find $\|\overrightarrow{TP}\|$. In right triangle TSC, $\alpha + 24° = 90°$, and in right triangle TPR, $\alpha + \theta = 90°$. Hence,

$$\alpha + \theta = \alpha + 24°; \qquad \text{hence,} \qquad \theta = 24°.$$

Therefore,

$$\frac{\|\overrightarrow{TP}\|}{\|\overrightarrow{TR}\|} = \sin\theta$$

$$\frac{\|\overrightarrow{TP}\|}{200} = \sin 24°$$

$$\|\overrightarrow{TP}\| = 200 \sin 24° \approx 81.35.$$

So the force on the rope is 81.35 pounds. ■

In aerial navigation, directions are given in terms of the angle measured in degrees clockwise from true north. Thus, north is 0°, east is 90°, and so on.

EXAMPLE 13

An airplane is traveling in the direction 50° with an air speed of 300 mph, and there is a 35-mph wind from the direction 120°, as represented by the vectors **p** and **w** in Figure 4–21(a). Find the *course* and *ground speed* of the plane (that is, its direction and speed relative to the ground).

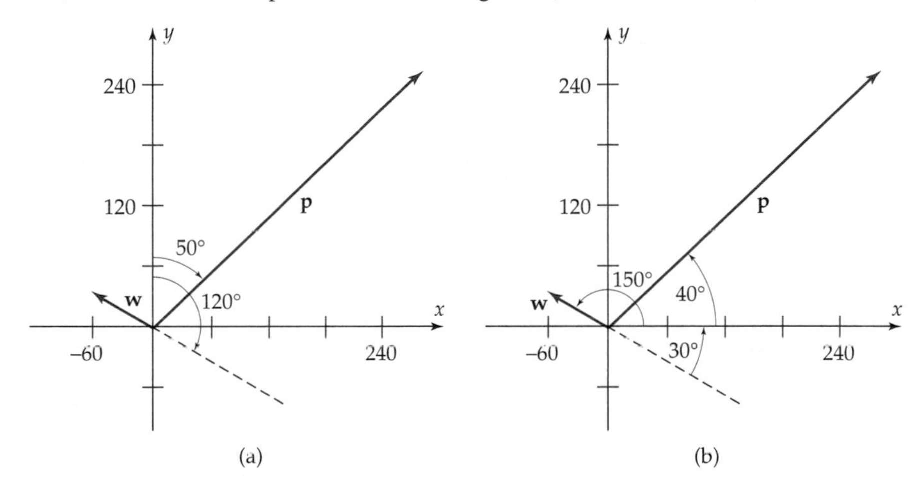

Figure 4–21

SOLUTION The course of the plane is the direction of the vector **p** + **w**, and its ground speed is the magnitude of **p** + **w**. Figure 4–21(b) shows that the direction angle of **p** (the angle it makes with the positive x-axis) is 40° and that the direction angle of **w** is 150°. Therefore,

$$\begin{aligned}\mathbf{p} + \mathbf{w} &= [(300 \cos 40°)\mathbf{i} + (300 \sin 40°)\mathbf{j}] + [(35 \cos 150°)\mathbf{i} + (35 \sin 150°)\mathbf{j}] \\ &= (300 \cos 40° + 35 \cos 150°)\mathbf{i} + (300 \sin 40° + 35 \sin 150°)\mathbf{j} \\ &\approx 199.50\mathbf{i} + 210.34\mathbf{j}.\end{aligned}$$

The direction angle of **p** + **w** satisfies $\tan\theta = 210.34/199.50 \approx 1.0543$, and a calculator shows that $\theta \approx 46.5°$. This is the angle **p** + **w** makes with the positive x-axis; hence, the course of the plane (the angle between true north and **p** + **w**) is $90° - 46.5° = 43.5°$. The ground speed of the plane is

$$\|\mathbf{p} + \mathbf{w}\| \approx \sqrt{(199.5)^2 + (210.34)^2} \approx 289.9 \text{ mph}.$$ ■

EXERCISES 4.4

In Exercises 1–4, find the magnitude of the vector $\overrightarrow{PQ}$.

1. $P = (2, 3), Q = (5, 9)$
2. $P = (-3, 5), Q = (7, -11)$
3. $P = (-7, 0), Q = (-4, -5)$
4. $P = (30, 12), Q = (25, 5)$

In Exercises 5–10, find a vector with the origin as initial point that is equivalent to the vector $\overrightarrow{PQ}$.

5. $P = (1, 5), Q = (7, 11)$
6. $P = (2, 7), Q = (-2, 9)$
7. $P = (-4, -8), Q = (-10, 2)$
8. $P = (-5, 6), Q = (-7, -9)$
9. $P = \left(\frac{4}{5}, -2\right), Q = \left(\frac{17}{5}, -\frac{12}{5}\right)$
10. $P = (\sqrt{2}, 4), Q = (\sqrt{3}, -1)$

In Exercises 11–20, find $\mathbf{u} + \mathbf{v}$, $\mathbf{u} - \mathbf{v}$, and $3\mathbf{u} - 2\mathbf{v}$.

11. $\mathbf{u} = \langle -2, 4 \rangle, \mathbf{v} = \langle 6, 1 \rangle$
12. $\mathbf{u} = \langle 4, 0 \rangle, \mathbf{v} = \langle 1, -3 \rangle$
13. $\mathbf{u} = \langle 3, 3\sqrt{2} \rangle, \mathbf{v} = \langle 4\sqrt{2}, 1 \rangle$
14. $\mathbf{u} = \left\langle \frac{2}{3}, 4 \right\rangle, \mathbf{v} = \left\langle -7, \frac{19}{3} \right\rangle$
15. $\mathbf{u} = 2\langle -2, 5 \rangle, \mathbf{v} = \frac{1}{4}\langle -7, 12 \rangle$
16. $\mathbf{u} = \mathbf{i} - \mathbf{j}, \mathbf{v} = 2\mathbf{i} + \mathbf{j}$
17. $\mathbf{u} = 8\mathbf{i}, \mathbf{v} = 2(3\mathbf{i} - 2\mathbf{j})$
18. $\mathbf{u} = -4(-\mathbf{i} + \mathbf{j}), \mathbf{v} = -3\mathbf{i}$
19. $\mathbf{u} = -\left(2\mathbf{i} + \frac{3}{2}\mathbf{j}\right), \mathbf{v} = \frac{3}{4}\mathbf{i}$
20. $\mathbf{u} = \sqrt{2}\mathbf{j}, \mathbf{v} = \sqrt{3}\mathbf{i}$

In Exercises 21–26, find the components of the given vector, where $\mathbf{u} = \mathbf{i} - 2\mathbf{j}$, $\mathbf{v} = 3\mathbf{i} + \mathbf{j}$, $\mathbf{w} = -4\mathbf{i} + \mathbf{j}$.

21. $\mathbf{u} + 2\mathbf{w}$
22. $\frac{1}{2}(3\mathbf{v} + \mathbf{w})$
23. $\frac{1}{2}\mathbf{w}$
24. $-2\mathbf{u} + 3\mathbf{v}$
25. $\frac{1}{4}(8\mathbf{u} + 4\mathbf{v} - \mathbf{w})$
26. $3(\mathbf{u} - 2\mathbf{v}) - 6\mathbf{w}$

In Exercises 27–34, find the component form of the vector $\mathbf{v}$ whose magnitude and direction angle θ are given.

27. $\|\mathbf{v}\| = 4, \theta = 0°$
28. $\|\mathbf{v}\| = 5, \theta = 30°$
29. $\|\mathbf{v}\| = 10, \theta = 225°$
30. $\|\mathbf{v}\| = 20, \theta = 120°$
31. $\|\mathbf{v}\| = 6, \theta = 40°$
32. $\|\mathbf{v}\| = 8, \theta = 160°$
33. $\|\mathbf{v}\| = 1/2, \theta = 250°$
34. $\|\mathbf{v}\| = 3, \theta = 310°$

In Exercises 35–42, find the magnitude and direction angle of the vector $\mathbf{v}$.

35. $\mathbf{v} = \langle 4, 4 \rangle$
36. $\mathbf{v} = \langle 5, 5\sqrt{3} \rangle$
37. $\mathbf{v} = \langle -8, 0 \rangle$
38. $\mathbf{v} = \langle 4, 5 \rangle$
39. $\mathbf{v} = 6\mathbf{j}$
40. $\mathbf{v} = 4\mathbf{i} - 8\mathbf{j}$
41. $\mathbf{v} = -2\mathbf{i} + 8\mathbf{j}$
42. $\mathbf{v} = -15\mathbf{i} - 10\mathbf{j}$

In Exercises 43–46, find a unit vector that has the same direction as $\mathbf{v}$.

43. $\langle 4, -5 \rangle$
44. $-7\mathbf{i} + 8\mathbf{j}$
45. $5\mathbf{i} + 10\mathbf{j}$
46. $-3\mathbf{i} - 9\mathbf{j}$

In Exercises 47–50, an object at the origin is acted upon by two forces, $\mathbf{u}$ and $\mathbf{v}$, with direction angle θ_u and θ_v, respectively. Find the direction and magnitude of the resultant force.

47. $\|\mathbf{u}\| = 30$ pounds, $\theta_{\mathbf{u}} = 0°$; $\|\mathbf{v}\| = 90$ pounds, $\theta_{\mathbf{v}} = 60°$
48. $\|\mathbf{u}\| = 6$ pounds, $\theta_{\mathbf{u}} = 45°$; $\|\mathbf{v}\| = 6$ pounds, $\theta_{\mathbf{v}} = 120°$
49. $\|\mathbf{u}\| = 12$ newtons, $\theta_{\mathbf{u}} = 130°$; $\|\mathbf{v}\| = 20$ newtons $\theta_{\mathbf{v}} = 250°$
50. $\|\mathbf{u}\| = 30$ newtons, $\theta_{\mathbf{u}} = 300°$; $\|\mathbf{v}\| = 80$ newtons, $\theta_{\mathbf{v}} = 40°$

If forces $\mathbf{u}_1, \mathbf{u}_2, \ldots, \mathbf{u}_k$ act on an object at the origin, the resultant force *is the sum $\mathbf{u}_1 + \mathbf{u}_2 + \cdots + \mathbf{u}_k$. The forces are said to be in* equilibrium *if their resultant force is $\mathbf{0}$. In Exercises 51 and 52, find the resultant force and find an additional force $\mathbf{v}$ that, if added to the system, produces equilibrium.*

51. $\mathbf{u}_1 = \langle 2, 5 \rangle, \mathbf{u}_2 = \langle -6, 1 \rangle, \mathbf{u}_3 = \langle -4, -8 \rangle$
52. $\mathbf{u}_1 = \langle 3, 7 \rangle, \mathbf{u}_2 = \langle 8, -2 \rangle, \mathbf{u}_3 = \langle -9, 0 \rangle, \mathbf{u}_4 = \langle -5, 4 \rangle$

In Exercises 53–58, let $\mathbf{u} = \langle a, b \rangle$ and $\mathbf{v} = \langle c, d \rangle$, and let r and s be scalars. Prove that the stated property holds by calculating the vector on each side of the equal sign.

53. $\mathbf{v} + \mathbf{0} = \mathbf{v} = \mathbf{0} + \mathbf{v}$
54. $\mathbf{v} + (-\mathbf{v}) = \mathbf{0}$
55. $r(\mathbf{u} + \mathbf{v}) = r\mathbf{u} + r\mathbf{v}$
56. $(r + s)\mathbf{v} = r\mathbf{v} + s\mathbf{v}$
57. $(rs)\mathbf{v} = r(s\mathbf{v}) = s(r\mathbf{v})$
58. $1\mathbf{v} = \mathbf{v}$ and $0\mathbf{v} = \mathbf{0}$

59. Two ropes are tied to a wagon. A child pulls one with a force of 20 pounds while another child pulls the other with a force of 30 pounds (see the figure). If the angle between the two ropes is 28°, how much force must be exerted by a third child, standing behind the wagon, to keep the wagon from moving? [*Hint:* Assume that the wagon is at the origin and one rope runs along the positive x-axis. Proceed as in Example 11 to find the resultant force on the wagon from the ropes. The third child must use the same amount in the opposite direction.]

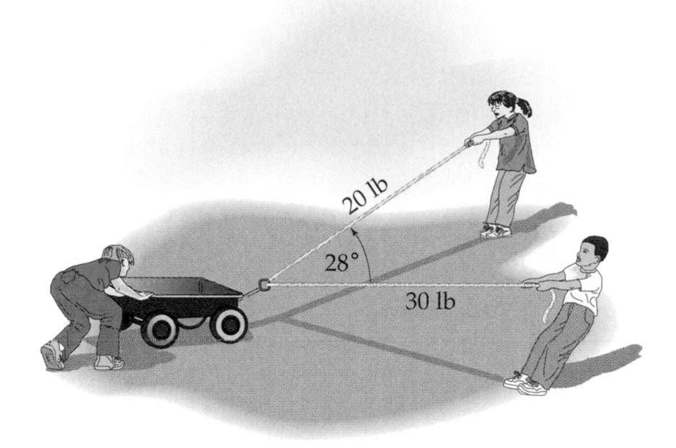

60. Two circus elephants, Bessie and Maybelle, are dragging a large wagon, as shown in the figure. If Bessie pulls with a force of 2200 pounds, Maybelle pulls with a force of 1500 pounds and the wagon moves along the dashed line, what is angle θ?

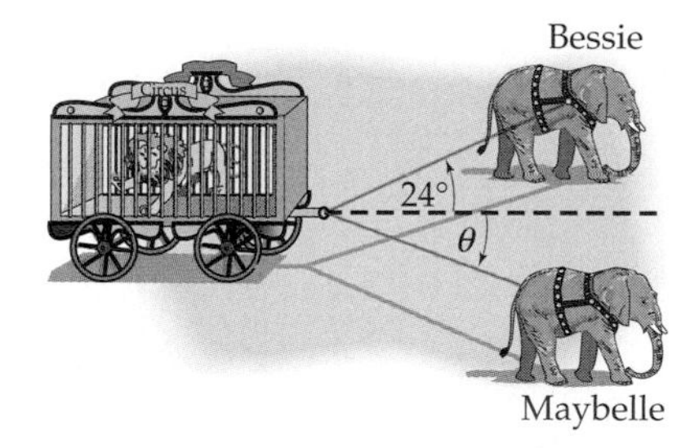

Exercises 61–64 deal with an object on an inclined plane. The situation is similar to that in Figure 4–20 of Example 12, where $\|\overrightarrow{TP}\|$ *is the component of the weight of the object parallel to the plane and* $\|\overrightarrow{TQ}\|$ *is the component of the weight perpendicular to the plane.*

61. An object weighing 50 pounds lies on an inclined plane that makes a 40° angle with the horizontal. Find the components of the weight parallel and perpendicular to the plane. [*Hint:* Solve an appropriate triangle.]

62. Do Exercise 61 when the object weighs 200 pounds and the inclined plane makes a 20° angle with the horizontal.

63. If an object on an inclined plane weighs 150 pounds and the component of the weight perpendicular to the plane is 60 pounds, what angle does the plane make with the horizontal?

64. A force of 500 pounds is needed to pull a cart up a ramp that makes a 15° angle with the ground. Assume that no friction is involved, and find the weight of the cart. [*Hint:* Draw a picture similar to Figure 4–20; the 500-pound force is parallel to the ramp.]

In Exercises 65–70, find the course and ground speed of the plane under the given conditions. (See Example 13.)

65. Air speed 250 mph in the direction 60°; wind speed 40 mph from the direction 330°.

66. Air speed 400 mph in the direction 150°; wind speed 30 mph from the direction 60°.

67. Air speed 300 mph in the direction 300°; wind speed 50 mph in (*not* from) the direction 30°.

68. Air speed 500 mph in the direction 180°; wind speed 70 mph in the direction 40°.

69. The course and ground speed of a plane are 70° and 400 mph, respectively. There is a 60-mph wind blowing south. Find the (approximate) direction and air speed of the plane.

70. A plane is flying in the direction 200° with an air speed of 500 mph. Its course and ground speed are 210° and 450 mph, respectively. What are the direction and speed of the wind?

71. A river flows from east to west. A swimmer on the south bank wants to swim to a point on the opposite shore directly north of her starting point. She can swim at 2.8 mph, and there is a 1-mph current in the river. In what direction should she head so as to travel directly north (that is, what angle should her path make with the south bank of the river)?

72. A river flows from west to east. A swimmer on the north bank swims at 3.1 mph along a straight course that makes a 75° angle with the north bank of the river and reaches the south bank at a point directly south of his starting point. How fast is the current in the river?

(Exercises continue on next page.)

73. A 400-pound weight is suspended by two cables (see the figure). What is the force (tension) on each cable? [*Hint:* Imagine that the weight is at the origin and that the dashed line is the x-axis. Then cable $\mathbf{v}$ is represented by the vector $(c \cos 65°)\mathbf{i} + (c \sin 65°)\mathbf{j}$, which has magnitude c (why?). Represent cable $\mathbf{u}$ similarly, denoting its magnitude by d. Use the fact that $\mathbf{u} + \mathbf{v} = 0\mathbf{i} + 400\mathbf{j}$ (why?) to set up a system of two equations in the unknowns c and d.]

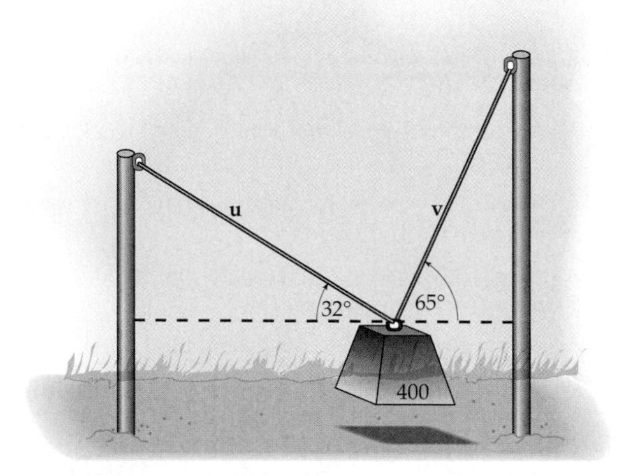

74. An 800-pound weight is suspended from two cables, as shown in the figure. What is the tension (force) on each cable? [See the Hint for Exercise 73.]

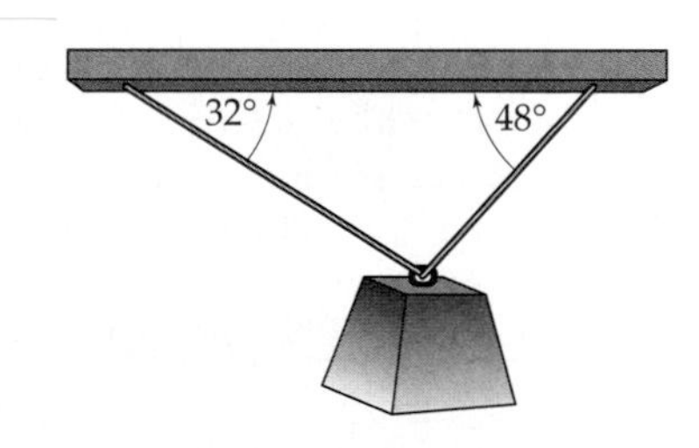

75. Do Exercise 74 when the weight is 600 pounds and the angles are 28° and 38°.

76. A 175-pound high-wire artist stands balanced on a tightrope, which sags slightly at the point where he is standing. The rope in front of him makes a 6° angle with the horizontal, and the rope behind him makes a 4° angle with the horizontal. Find the force on each end of the rope. [*Hint:* Use a picture and procedure similar to that in Exercise 73.]

77. Let $\mathbf{v}$ be the vector with initial point (x_1, y_1) and terminal point (x_2, y_2), and let k be any real number.
(a) Find the component form of $\mathbf{v}$ and $k\mathbf{v}$.
(b) Calculate $\|\mathbf{v}\|$ and $\|k\mathbf{v}\|$.
(c) Use the fact that $\sqrt{k^2} = |k|$ to verify that $\|k\mathbf{v}\| = |k| \cdot \|\mathbf{v}\|$.
(d) Show that $\tan \theta = \tan \beta$, where θ is the direction angle of $\mathbf{v}$ and β is the direction angle of $k\mathbf{v}$. Use the fact that $\tan t = \tan(t + 180°)$ to conclude that $\mathbf{v}$ and $k\mathbf{v}$ have either the same or opposite directions.
(e) Use the fact that (c, d) and $(-c, -d)$ lie on the same straight line on opposite sides of the origin (Exercise 63 in Section 0.2) to verify that $\mathbf{v}$ and $k\mathbf{v}$ have the same direction if $k > 0$ and opposite directions if $k < 0$.

78. Let $\mathbf{u} = \langle a, b\rangle$, $\mathbf{v} = \langle c, d\rangle$. Verify the accuracy of the two geometric interpretations of vector addition given on page 304 as follows.
(a) Show that the distance from (a, b) to $(a + c, b + d)$ is the same as $\|\mathbf{v}\|$.
(b) Show that the distance from (c, d) to $(a + c, b + d)$ is the same as $\|\mathbf{u}\|$.
(c) Show that the line through (a, b) and $(a + c, b + d)$ is parallel to $\mathbf{v}$ by showing they have the same slope.
(d) Show that the line through (c, d) and $(a + c, b + d)$ is parallel to $\mathbf{u}$.

79. Let $\mathbf{u} = \langle a, b\rangle$, and let $\mathbf{v} = \langle c, d\rangle$. Show that $\mathbf{u} - \mathbf{v}$ is equivalent to the vector $\mathbf{w}$ with initial point (c, d) and terminal point (a, b) as follows. (See the figure.)
(a) Show that $\|\mathbf{u} - \mathbf{v}\| = \|\mathbf{w}\|$.
(b) Show that $\mathbf{u} - \mathbf{v}$ and $\mathbf{w}$ have the same direction.

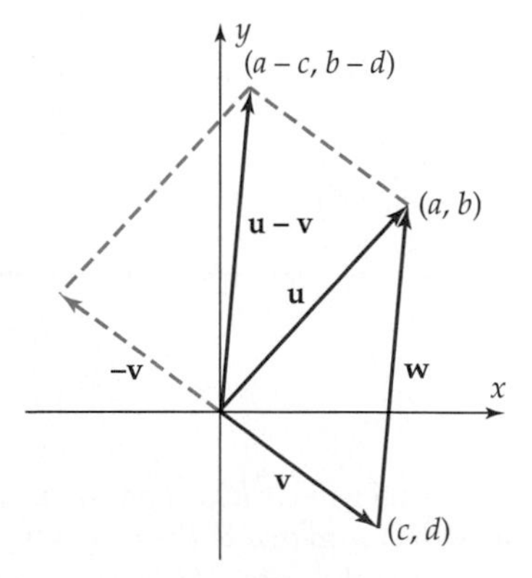

4.5 The Dot Product

When two vectors are added, their sum is another vector, but the situation is different with products. The dot product of two *vectors* is the *real number* defined as follows.

Dot Product

> The **dot product** of vectors $\mathbf{u} = \langle a, b\rangle = a\mathbf{i} + b\mathbf{j}$ and $\mathbf{v} = \langle c, d\rangle = c\mathbf{i} + d\mathbf{j}$ is denoted $\mathbf{u} \cdot \mathbf{v}$ and is defined to be the real number $ac + bd$. Thus,
>
> $$\mathbf{u} \cdot \mathbf{v} = ac + bd.$$

TECHNOLOGY TIP

To compute dot products use DOT or DOTP in this menu/submenu:

TI-86: VECTOR/MATH

TI-89: MATH/MATRIX/VECTOR OPS

HP-39+: MATH/MATRIX

EXAMPLE 1

(a) If $\mathbf{u} = \langle 5, 3\rangle$ and $\mathbf{v} = \langle -2, 6\rangle$, then

$$\mathbf{u} \cdot \mathbf{v} = 5(-2) + 3 \cdot 6 = 8.$$

(b) If $\mathbf{u} = 4\mathbf{i} - 2\mathbf{j}$ and $\mathbf{v} = 3\mathbf{i} - \mathbf{j}$, then

$$\mathbf{u} \cdot \mathbf{v} = 4 \cdot 3 + (-2)(-1) = 14.$$

(c) $\langle 2, -4\rangle \cdot \langle 6, 3\rangle = 2 \cdot 6 + (-4)3 = 0.$ ■

The dot product has a number of useful properties.

Properties of the Dot Product

> If **u**, **v**, **w** are vectors and k is a real number, then
>
> 1. $\mathbf{u} \cdot \mathbf{u} = \|\mathbf{u}\|^2$.
> 2. $\mathbf{u} \cdot \mathbf{v} = \mathbf{v} \cdot \mathbf{u}$.
> 3. $\mathbf{u} \cdot (\mathbf{v} + \mathbf{w}) = \mathbf{u} \cdot \mathbf{v} + \mathbf{u} \cdot \mathbf{w}$.
> 4. $k\mathbf{u} \cdot \mathbf{v} = k(\mathbf{u} \cdot \mathbf{v}) = \mathbf{u} \cdot k\mathbf{v}$.
> 5. $\mathbf{0} \cdot \mathbf{u} = 0$.

Proof

1. If $\mathbf{u} = \langle a, b\rangle$, then

$$\|\mathbf{u}\| = \sqrt{a^2 + b^2}.$$

Hence,

$$\mathbf{u} \cdot \mathbf{u} = \langle a, b\rangle \cdot \langle a, b\rangle = a \cdot a + b \cdot b = a^2 + b^2 = (\sqrt{a^2 + b^2})^2 = \|\mathbf{u}\|^2.$$

2. If $\mathbf{u} = \langle a, b\rangle$ and $\mathbf{v} = \langle c, d\rangle$, then

$$\mathbf{u} \cdot \mathbf{v} = \langle a, b\rangle \cdot \langle c, d\rangle = ac + bd = ca + db = \langle c, d\rangle \cdot \langle a, b\rangle = \mathbf{v} \cdot \mathbf{u}.$$

The last three statements are proved similarly (Exercises 37–39.) ■

Figure 4–22

ANGLES

If $\mathbf{u} = \langle a, b\rangle$ and $\mathbf{v} = \langle c, d\rangle$ are nonzero vectors, then the **angle between u and v** is the smallest angle θ formed by these two line segments, as shown in Figure 4–22. We ignore clockwise or counterclockwise rotation and consider the angle between **v** and **u** to be the same as the angle between **u** and **v**. Thus, the radian measure of θ ranges from 0 to π.

Nonzero vectors **u** and **v** are said to be **parallel** if the angle between them is either 0 or π radians (that is, **u** and **v** lie on the same straight line through the origin and have either the same or opposite directions). The zero vector **0** is considered to be parallel to every vector.

Any scalar multiple of **u** is parallel to **u,** since it lies on the same straight line as **u** (see Example 3 in Section 4.4). Conversely, if **v** is parallel to **u**, it is easy to show that **v** must be a scalar multiple of **u** (Exercise 40). Hence, we have the following.

Parallel Vectors

> Vectors **u** and **v** are parallel exactly when
>
> $\mathbf{v} = k\mathbf{u}$ for some real number k.

EXAMPLE 2

The vectors $\langle 2, 3\rangle$ and $\langle 8, 12\rangle$ are parallel because $\langle 8, 12\rangle = 4\langle 2, 3\rangle$. ■

The angle between nonzero vectors **u** and **v** is closely related to their dot product.

Angle Theorem

> If θ is the angle between the nonzero vectors **u** and **v,** then
>
> $$\mathbf{u} \cdot \mathbf{v} = \|\mathbf{u}\|\,\|\mathbf{v}\| \cos \theta$$
>
> or, equivalently,
>
> $$\cos \theta = \frac{\mathbf{u} \cdot \mathbf{v}}{\|\mathbf{u}\|\,\|\mathbf{v}\|}.$$

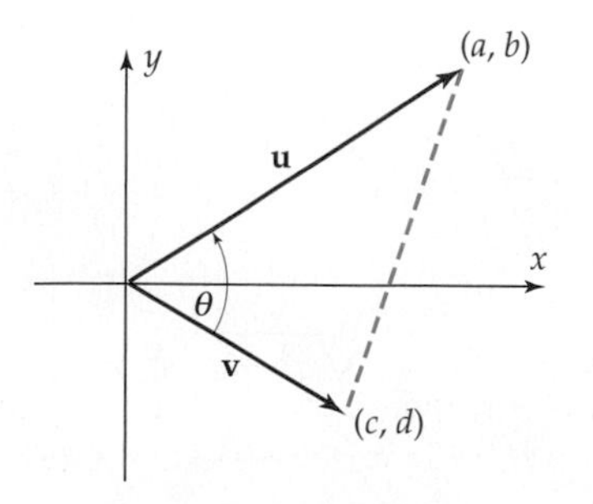

Figure 4–23

Proof If $\mathbf{u} = \langle a, b\rangle$, $\mathbf{v} = \langle c, d\rangle$, and the angle θ is not 0 or π, then **u** and **v** form two sides of a triangle, as shown in Figure 4–23.

The lengths of two sides of the triangle are $\|\mathbf{u}\| = \sqrt{a^2 + b^2}$ and $\|\mathbf{v}\| = \sqrt{c^2 + d^2}$. The distance formula shows that the length of the third side (opposite angle θ) is $\sqrt{(a - c)^2 + (b - d)^2}$. Therefore, by the Law of Cosines,

$$\left[\sqrt{(a - c)^2 + (b - d)^2}\right]^2 = \|\mathbf{u}\|^2 + \|\mathbf{v}\|^2 - 2\|\mathbf{u}\|\,\|\mathbf{v}\| \cos \theta$$

$$(a - c)^2 + (b - d)^2 = (a^2 + b^2) + (c^2 + d^2) - 2\|\mathbf{u}\|\,\|\mathbf{v}\| \cos \theta$$

$$a^2 - 2ac + c^2 + b^2 - 2bd + d^2 = (a^2 + c^2) + (b^2 + d^2) - 2\|\mathbf{u}\|\,\|\mathbf{v}\| \cos \theta$$

$$-2ac - 2bd = -2\|\mathbf{u}\|\,\|\mathbf{v}\| \cos \theta.$$

Dividing both sides by -2 shows that

$$ac + bd = \|\mathbf{u}\| \|\mathbf{v}\| \cos \theta.$$

Since the left side of this equation is precisely $\mathbf{u} \cdot \mathbf{v}$, the proof is complete in this case. The proof when θ is 0 or π is left to the reader (Exercise 41). ■

EXAMPLE 3

Find the angle θ between the vectors $\langle -3, 1\rangle$ and $\langle 5, 2\rangle$ shown in Figure 4–24.

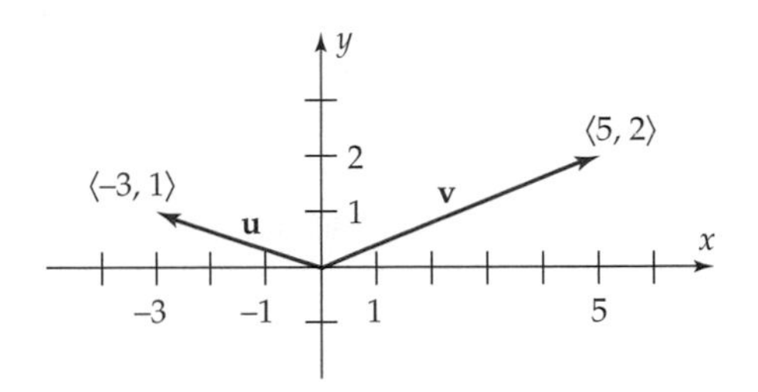

Figure 4–24

SOLUTION Apply the formula in the box above with $\mathbf{u} = \langle -3, 1\rangle$ and $\mathbf{v} = \langle 5, 2\rangle$.

$$\cos \theta = \frac{\mathbf{u} \cdot \mathbf{v}}{\|\mathbf{u}\| \|\mathbf{v}\|} = \frac{(-3)5 + 1 \cdot 2}{\sqrt{(-3)^2 + 1^2}\sqrt{5^2 + 2^2}} = \frac{-13}{\sqrt{10}\sqrt{29}} = \frac{-13}{\sqrt{290}}.$$

Using the COS^{-1} key, we see that

$$\theta \approx 2.4393 \text{ radians } (\approx 139.76°). \quad ■$$

The Angle Theorem has several useful consequences. For instance, by taking absolute values on both sides of $\mathbf{u} \cdot \mathbf{v} = \|\mathbf{u}\| \|\mathbf{v}\| \cos \theta$ and using the fact that $\big|\|\mathbf{u}\| \|\mathbf{v}\|\big| = \|\mathbf{u}\| \|\mathbf{v}\|$ (because $\|\mathbf{u}\| \|\mathbf{v}\| \geq 0$), we see that

$$|\mathbf{u} \cdot \mathbf{v}| = \big|\|\mathbf{u}\| \|\mathbf{v}\| \cos \theta\big| = \big|\|\mathbf{u}\| \|\mathbf{v}\|\big| \, |\cos \theta| = \|\mathbf{u}\| \|\mathbf{v}\| \, |\cos \theta|.$$

But for any angle θ, $|\cos \theta| \leq 1$, so

$$|\mathbf{u} \cdot \mathbf{v}| = \|\mathbf{u}\| \|\mathbf{v}\| \, |\cos \theta| \leq \|\mathbf{u}\| \|\mathbf{v}\|.$$

This proves the Schwarz inequality.

Schwarz Inequality

For any vectors **u** and **v**,

$$|\mathbf{u} \cdot \mathbf{v}| \leq \|\mathbf{u}\| \|\mathbf{v}\|.$$

Vectors **u** and **v** are said to be **orthogonal** (or **perpendicular**) if the angle between them is $\pi/2$ radians (90°), or if at least one of them is **0**. Here is the key fact about orthogonal vectors.

Orthogonal Vectors

Let **u** and **v** be vectors. Then

$$\mathbf{u} \text{ and } \mathbf{v} \text{ are orthogonal exactly when } \mathbf{u} \cdot \mathbf{v} = 0.$$

Proof If **u** or **v** is **0**, then $\mathbf{u} \cdot \mathbf{v} = 0$, and if **u** and **v** are nonzero orthogonal vectors, then by the Angle Theorem,

$$\mathbf{u} \cdot \mathbf{v} = \|\mathbf{u}\|\,\|\mathbf{v}\| \cos\theta = \|\mathbf{u}\|\,\|\mathbf{v}\| \cos(\pi/2) = \|\mathbf{u}\|\,\|\mathbf{v}\|\,(0) = 0.$$

Conversely, if **u** and **v** are vectors such that $\mathbf{u} \cdot \mathbf{v} = 0$, then Exercise 42 shows that **u** and **v** are orthogonal. ■

EXAMPLE 4

(a) The vectors $\mathbf{u} = \langle 2, -6\rangle$ and $\mathbf{v} = \langle 9, 3\rangle$ are orthogonal because

$$\mathbf{u} \cdot \mathbf{v} = \langle 2, -6\rangle \cdot \langle 9, 3\rangle = 2 \cdot 9 + (-6)3 = 18 - 18 = 0.$$

(b) The vectors $\frac{1}{2}\mathbf{i} + 5\mathbf{j}$ and $10\mathbf{i} - \mathbf{j}$ are orthogonal, since

$$\left(\frac{1}{2}\mathbf{i} + 5\mathbf{j}\right) \cdot (10\mathbf{i} - \mathbf{j}) = \frac{1}{2}(10) + 5(-1) = 5 - 5 = 0. \quad ■$$

PROJECTIONS AND COMPONENTS

If **u** and **v** are nonzero vectors and θ is the angle between them, construct the perpendicular line segment from the terminal point P of **u** to the straight line on which **v** lies. This perpendicular segment intersects the line at a point Q, as shown in Figure 4–25.

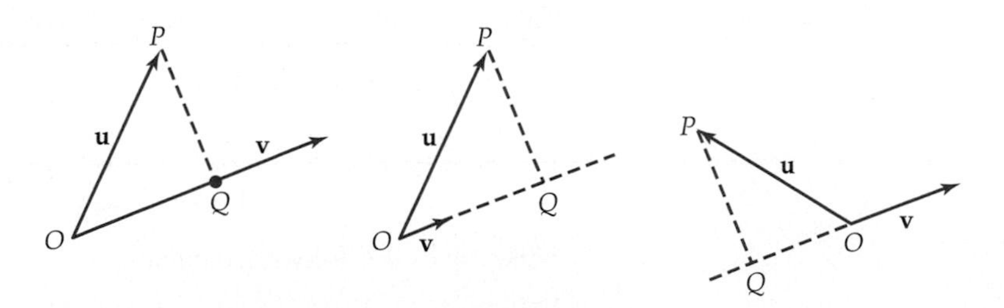

Figure 4–25

The vector $\overrightarrow{OQ}$ is called the **projection of u on v** and is denoted $\text{proj}_{\mathbf{v}}\mathbf{u}$. Here is a useful description of $\text{proj}_{\mathbf{v}}\mathbf{u}$.

Projection of u on v

If **u** and **v** are nonzero vectors, then the projection of **u** on **v** is the vector

$$\text{proj}_{\mathbf{v}}\mathbf{u} = \left(\frac{\mathbf{u} \cdot \mathbf{v}}{\|\mathbf{v}\|^2}\right)\mathbf{v}.$$

Proof Since $\text{proj}_{\mathbf{v}}\mathbf{u}$ and $\mathbf{v}$ lie on the same straight line, they are parallel, and hence, $\text{proj}_{\mathbf{v}}\mathbf{u} = k\mathbf{v}$ for some real number k. Let $\mathbf{w}$ be the vector with initial point at the origin and the same length and direction as $\overrightarrow{QP}$, as in the two cases shown in Figure 4–26.

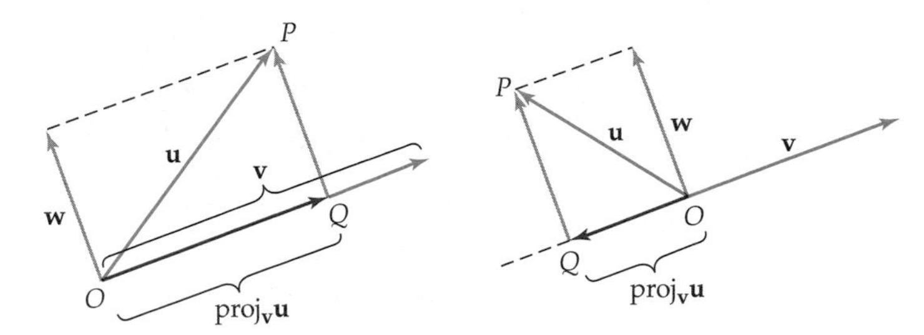

Figure 4–26

Note that $\mathbf{w}$ is parallel to $\overrightarrow{QP}$ and hence is orthogonal to $\mathbf{v}$. As is shown in Figure 4–26, $\mathbf{u} = \text{proj}_{\mathbf{v}}\mathbf{u} + \mathbf{w} = k\mathbf{v} + \mathbf{w}$. Consequently, by the properties of the dot product,

$$\mathbf{u} \cdot \mathbf{v} = (k\mathbf{v} + \mathbf{w}) \cdot \mathbf{v} = (k\mathbf{v}) \cdot \mathbf{v} + \mathbf{w} \cdot \mathbf{v}$$

$$= k(\mathbf{v} \cdot \mathbf{v}) + \mathbf{w} \cdot \mathbf{v} = k\|\mathbf{v}\|^2 + \mathbf{w} \cdot \mathbf{v}.$$

But $\mathbf{w} \cdot \mathbf{v} = 0$ because $\mathbf{w}$ and $\mathbf{v}$ are orthogonal. Hence,

$$\mathbf{u} \cdot \mathbf{v} = k\|\mathbf{v}\|^2 \qquad \text{or, equivalently,} \qquad k = \frac{\mathbf{u} \cdot \mathbf{v}}{\|\mathbf{v}\|^2}.$$

Therefore,

$$\text{proj}_{\mathbf{v}}\mathbf{u} = k\mathbf{v} = \left(\frac{\mathbf{u} \cdot \mathbf{v}}{\|\mathbf{v}\|^2}\right)\mathbf{v},$$

and the proof is complete. ■

EXAMPLE 5

If $\mathbf{u} = 8\mathbf{i} + 3\mathbf{j}$ and $\mathbf{v} = 4\mathbf{i} - 2\mathbf{j}$, find $\text{proj}_{\mathbf{v}}\mathbf{u}$ and $\text{proj}_{\mathbf{u}}\mathbf{v}$.

SOLUTION

$$\mathbf{u} \cdot \mathbf{v} = 8 \cdot 4 + 3(-2) = 26, \qquad \text{and} \qquad \|\mathbf{v}\|^2 = \mathbf{v} \cdot \mathbf{v} = 4^2 + (-2)^2 = 20.$$

Therefore,

$$\text{proj}_{\mathbf{v}}\mathbf{u} = \left(\frac{\mathbf{u} \cdot \mathbf{v}}{\|\mathbf{v}\|^2}\right)\mathbf{v} = \frac{26}{20}(4\mathbf{i} - 2\mathbf{j}) = \frac{26}{5}\mathbf{i} - \frac{13}{5}\mathbf{j},$$

as is shown in Figure 4–27. We can find the projection of **v** on **u** by noting that $\|\mathbf{u}\|^2 = \mathbf{u} \cdot \mathbf{u} = 8^2 + 3^2 = 73$, and hence,

$$\text{proj}_{\mathbf{u}}\mathbf{v} = \left(\frac{\mathbf{v} \cdot \mathbf{u}}{\|\mathbf{u}\|^2}\right)\mathbf{u} = \frac{26}{73}(8\mathbf{i} + 3\mathbf{j}) = \frac{208}{73}\mathbf{i} + \frac{78}{73}\mathbf{j}. \quad \blacksquare$$

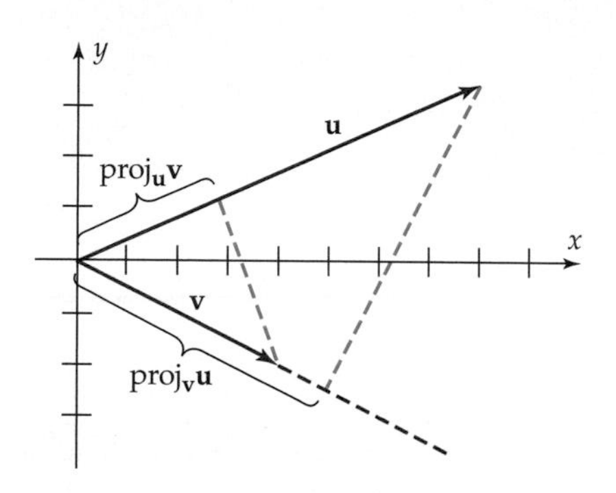

Figure 4–27

Recall that $\frac{1}{\|\mathbf{v}\|}\mathbf{v}$ is a unit vector in the direction of **v** (see page 307). We can express $\text{proj}_{\mathbf{v}}\mathbf{u}$ as a scalar multiple of this unit vector as follows.

$$\text{proj}_{\mathbf{v}}\mathbf{u} = \left(\frac{\mathbf{u} \cdot \mathbf{v}}{\|\mathbf{v}\|^2}\right)\mathbf{v} = \left(\frac{\mathbf{u} \cdot \mathbf{v}}{\|\mathbf{v}\|}\right)\left(\frac{1}{\|\mathbf{v}\|}\mathbf{v}\right).$$

The scalar $\frac{\mathbf{u} \cdot \mathbf{v}}{\|\mathbf{v}\|}$ is called the **component of u along v** and is denoted $\text{comp}_{\mathbf{v}}\mathbf{u}$. Thus,

$$\text{proj}_{\mathbf{v}}\mathbf{u} = \left(\frac{\mathbf{u} \cdot \mathbf{v}}{\|\mathbf{v}\|}\right)\left(\frac{1}{\|\mathbf{v}\|}\mathbf{v}\right) = \text{comp}_{\mathbf{v}}\mathbf{u}\left(\frac{1}{\|\mathbf{v}\|}\mathbf{v}\right).$$

Since $\frac{1}{\|\mathbf{v}\|}\mathbf{v}$ is a unit vector, the length of $\text{proj}_{\mathbf{v}}\mathbf{u}$ is

$$\|\text{proj}_{\mathbf{v}}\mathbf{u}\| = \left\|\text{comp}_{\mathbf{v}}\mathbf{u}\left(\frac{1}{\|\mathbf{v}\|}\mathbf{v}\right)\right\| = |\text{comp}_{\mathbf{v}}\mathbf{u}| \left\|\frac{1}{\|\mathbf{v}\|}\mathbf{v}\right\| = |\text{comp}_{\mathbf{v}}\mathbf{u}|.$$

Furthermore, since $\mathbf{u} \cdot \mathbf{v} = \|\mathbf{u}\|\,\|\mathbf{v}\| \cos\theta$, where θ is the angle between **u** and **v**, we have

$$\text{comp}_{\mathbf{v}}\mathbf{u} = \frac{\mathbf{u} \cdot \mathbf{v}}{\|\mathbf{v}\|} = \frac{\|\mathbf{u}\|\,\|\mathbf{v}\| \cos\theta}{\|\mathbf{v}\|}.$$

Cancelling $\|\mathbf{v}\|$ on the right side produces this result.

Projections and Components

If **u** and **v** are nonzero vectors and θ is the angle between them, then

$$\text{comp}_{\mathbf{v}}\mathbf{u} = \frac{\mathbf{u} \cdot \mathbf{v}}{\|\mathbf{v}\|} = \|\mathbf{u}\| \cos\theta$$

and

$$\|\text{proj}_{\mathbf{v}}\mathbf{u}\| = |\text{comp}_{\mathbf{v}}\mathbf{u}|.$$

EXAMPLE 6

If $\mathbf{u} = 2\mathbf{i} + 3\mathbf{j}$ and $\mathbf{v} = -5\mathbf{i} + 2\mathbf{j}$, find $\text{comp}_{\mathbf{v}}\mathbf{u}$ and $\text{comp}_{\mathbf{u}}\mathbf{v}$.

SOLUTION

$$\text{comp}_{\mathbf{v}}\mathbf{u} = \frac{\mathbf{u} \cdot \mathbf{v}}{\|\mathbf{v}\|} = \frac{2(-5) + 3 \cdot 2}{\sqrt{(-5)^2 + 2^2}} = \frac{-4}{\sqrt{29}}.$$

$$\text{comp}_{\mathbf{u}}\mathbf{v} = \frac{\mathbf{v} \cdot \mathbf{u}}{\|\mathbf{u}\|} = \frac{-4}{\sqrt{2^2 + 3^2}} = \frac{-4}{\sqrt{13}}.$$ ■

APPLICATIONS

Vectors and the dot product can be used to solve a variety of physical problems.

EXAMPLE 7

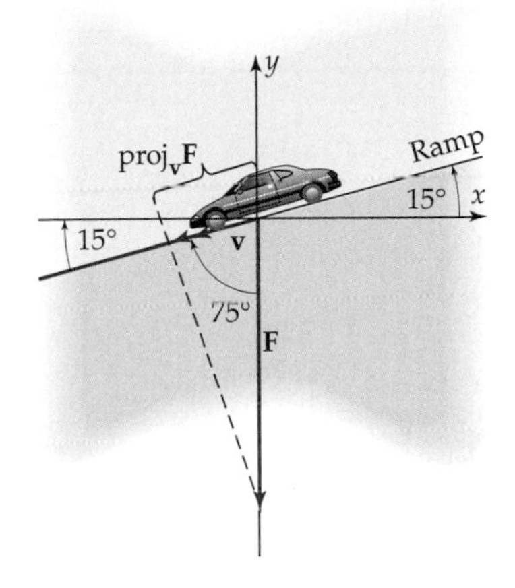

Figure 4–28

A 4000-pound automobile is on an inclined ramp that makes a 15° angle with the horizontal. Find the force required to keep it from rolling down the ramp, assuming that the only force that must be overcome is that due to gravity.

SOLUTION The situation is shown in Figure 4–28, where the coordinate system is chosen so that the car is at the origin, the vector **F** representing the downward force of gravity is on the y-axis, and **v** is a unit vector from the origin down the ramp. Since the car weighs 4000 pounds, $\mathbf{F} = -4000\mathbf{j}$. Figure 4–28 shows that the angle between **v** and **F** is 75°. The vector $\text{proj}_{\mathbf{v}}\mathbf{F}$ is the force pulling the car down the ramp, so a force of the same magnitude in the opposite direction is needed to keep the car motionless. As we saw in the preceding box,

$$\|\text{proj}_{\mathbf{v}}\mathbf{F}\| = |\text{comp}_{\mathbf{v}}\mathbf{F}| = \big|\|\mathbf{F}\| \cos 75°\big|$$
$$\approx 4000(.25882) \approx 1035.3.$$

Therefore, a force of 1035.3 pounds is required to hold the car in place. ■

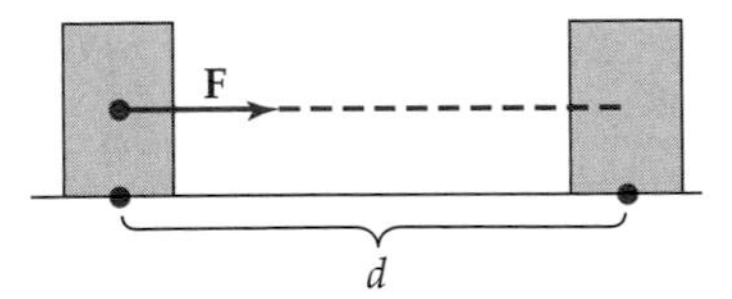

Figure 4–29

If a constant force **F** is applied to an object, pushing or pulling it a distance d in the direction of the force as shown in Figure 4–29, the amount of **work** done by the force is defined to be the product

$$W = (\text{magnitude of force})(\text{distance}) = \|\mathbf{F}\| \cdot d.$$

If the magnitude of **F** is measured in pounds and d in feet, then the units for W are foot-pounds. For example, if you push a car for 35 feet along a level driveway by exerting a constant force of 110 pounds, the amount of work done is $110 \cdot 35 = 3850$ foot-pounds.

When a force **F** moves an object in the direction of a vector **d** rather than in the direction of **F**, as shown in Figure 4–30, then the motion of the object can be considered as the result of the vector $\text{proj}_{\mathbf{d}}\mathbf{F}$, which is a force in the same direction as **d**.

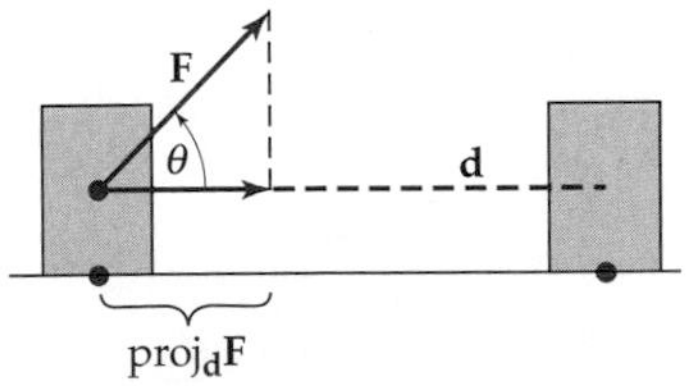

Figure 4–30

Therefore, the amount of work done by **F** is the same as the amount of work done by $\text{proj}_{\mathbf{d}}\mathbf{F}$, namely,

$$W = (\text{magnitude of } \text{proj}_{\mathbf{d}}\mathbf{F})(\text{length of } \mathbf{d}) = \|\text{proj}_{\mathbf{d}}\mathbf{F}\| \cdot \|\mathbf{d}\|.$$

The box on page 320 and the Angle Theorem (page 316) show that

$$\begin{aligned} W = \|\text{proj}_{\mathbf{d}}\mathbf{F}\| \cdot \|\mathbf{d}\| &= |\text{comp}_{\mathbf{d}}\mathbf{F}| \cdot \|\mathbf{d}\| \\ &= \|\mathbf{F}\|(\cos\theta)\|\mathbf{d}\| \\ &= \mathbf{F} \cdot \mathbf{d}.^* \end{aligned}$$

Consequently, we have these descriptions of work.

Work

> The work W done by a constant force F as its point of application moves along the vector **d** is
>
> $$W = |\text{comp}_{\mathbf{u}}\mathbf{F}| \cdot \|\mathbf{d}\| \qquad \text{or, equivalently,} \qquad W = \mathbf{F} \cdot \mathbf{d}.$$

EXAMPLE 8

How much work is done by a child who pulls a sled 100 feet over level ground by exerting a constant 20-pound force on a rope that makes a 45° angle with the ground?

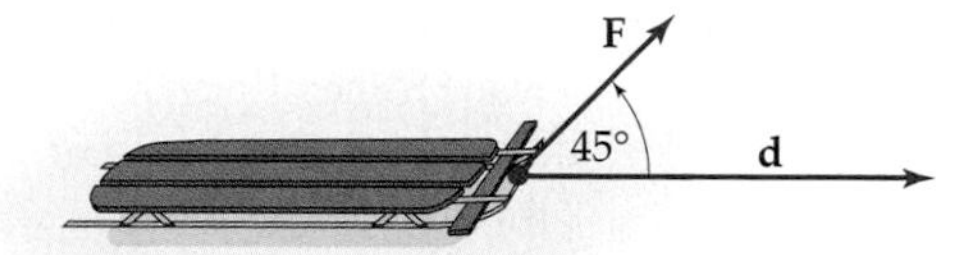

Figure 4–31

SOLUTION The situation is shown in Figure 4–31, where the force **F** on the rope has magnitude 20 and the sled moves along vector **d** of length 100. The work done is

$$\begin{aligned} W = \mathbf{F} \cdot \mathbf{d} = \|\mathbf{F}\|\,\|\mathbf{d}\| \cdot \cos\theta &= 20 \cdot 100 \cdot \frac{\sqrt{2}}{2} \\ &= 1000\sqrt{2} \approx 1414.2 \text{ foot-pounds.} \quad \blacksquare \end{aligned}$$

---

*This formula reduces to the previous one when **F** and **d** have the same direction because in that case, $\cos\theta = \cos 0 = 1$, so $W = \|\mathbf{F}\| \cdot \|\mathbf{d}\|$ = (magnitude of force)(distance moved).

EXERCISES 4.5

In Exercises 1–6, find $\mathbf{u} \cdot \mathbf{v}$, $\mathbf{u} \cdot \mathbf{u}$, *and* $\mathbf{v} \cdot \mathbf{v}$.

1. $\mathbf{u} = \langle 3, 4\rangle$, $\mathbf{v} = \langle -5, 2\rangle$
2. $\mathbf{u} = \langle -1, 6\rangle$, $\mathbf{v} = \langle -4, 1/3\rangle$
3. $\mathbf{u} = 2\mathbf{i} + \mathbf{j}$, $\mathbf{v} = 3\mathbf{i}$
4. $\mathbf{u} = \mathbf{i} - \mathbf{j}$, $\mathbf{v} = 5\mathbf{j}$
5. $\mathbf{u} = 3\mathbf{i} + 2\mathbf{j}$, $\mathbf{v} = 2\mathbf{i} + 3\mathbf{j}$
6. $\mathbf{u} = 4\mathbf{i} - \mathbf{j}$, $\mathbf{v} = -\mathbf{i} + 2\mathbf{j}$

In Exercises 7–12, find the dot product when $\mathbf{u} = \langle 2, 5\rangle$, $\mathbf{v} = \langle -4, 3\rangle$, *and* $\mathbf{w} = \langle 2, -1\rangle$.

7. $\mathbf{u} \cdot (\mathbf{v} + \mathbf{w})$
8. $\mathbf{u} \cdot (\mathbf{v} - \mathbf{w})$
9. $(\mathbf{u} + \mathbf{v}) \cdot (\mathbf{v} + \mathbf{w})$
10. $(\mathbf{u} + \mathbf{v}) \cdot (\mathbf{u} - \mathbf{v})$
11. $(3\mathbf{u} + \mathbf{v}) \cdot (2\mathbf{w})$
12. $(\mathbf{u} + 4\mathbf{v}) \cdot (2\mathbf{u} + \mathbf{w})$

In Exercises 13–18, find the angle between the two vectors.

13. $\langle 4, -3\rangle$, $\langle 1, 2\rangle$
14. $\langle 2, 4\rangle$, $\langle 0, -5\rangle$
15. $2\mathbf{i} - 3\mathbf{j}$, $-\mathbf{i}$
16. $2\mathbf{j}$, $4\mathbf{i} + \mathbf{j}$
17. $\sqrt{2}\mathbf{i} + \sqrt{2}\mathbf{j}$, $\mathbf{i} - \mathbf{j}$
18. $3\mathbf{i} - 5\mathbf{j}$, $-2\mathbf{i} + 3\mathbf{j}$

In Exercises 19–24, determine whether the given vectors are parallel, orthogonal, or neither.

19. $\langle 2, 6\rangle$, $\langle 3, -1\rangle$
20. $\langle -5, 3\rangle$, $\langle 2, 6\rangle$
21. $\langle 9, -6\rangle$, $\langle -6, 4\rangle$
22. $-\mathbf{i} + 2\mathbf{j}$, $2\mathbf{i} - 4\mathbf{j}$
23. $2\mathbf{i} - 2\mathbf{j}$, $5\mathbf{i} + 8\mathbf{j}$
24. $6\mathbf{i} - 4\mathbf{j}$, $2\mathbf{i} + 3\mathbf{j}$

In Exercises 25–28, find a real number k such that the two vectors are orthogonal.

25. $2\mathbf{i} + 3\mathbf{j}$, $3\mathbf{i} - k\mathbf{j}$
26. $-3\mathbf{i} + \mathbf{j}$, $2k\mathbf{i} - 4\mathbf{j}$
27. $\mathbf{i} - \mathbf{j}$, $k\mathbf{i} + \sqrt{2}\mathbf{j}$
28. $-4\mathbf{i} + 5\mathbf{j}$, $2\mathbf{i} + 2k\mathbf{j}$

In Exercises 29–32, find $\text{proj}_{\mathbf{u}}\mathbf{v}$ *and* $\text{proj}_{\mathbf{v}}\mathbf{u}$.

29. $\mathbf{u} = 3\mathbf{i} - 5\mathbf{j}$, $\mathbf{v} = 6\mathbf{i} + 2\mathbf{j}$
30. $\mathbf{u} = 2\mathbf{i} - 3\mathbf{j}$, $\mathbf{v} = \mathbf{i} + 2\mathbf{j}$
31. $\mathbf{u} = \mathbf{i} + \mathbf{j}$, $\mathbf{v} = \mathbf{i} - \mathbf{j}$
32. $\mathbf{u} = 5\mathbf{i} + \mathbf{j}$, $\mathbf{v} = -2\mathbf{i} + 3\mathbf{j}$

In Exercises 33–36, find $\text{comp}_{\mathbf{v}}\mathbf{u}$.

33. $\mathbf{u} = 10\mathbf{i} + 4\mathbf{j}$, $\mathbf{v} = 3\mathbf{i} - 2\mathbf{j}$
34. $\mathbf{u} = \mathbf{i} - 2\mathbf{j}$, $\mathbf{v} = 3\mathbf{i} + \mathbf{j}$
35. $\mathbf{u} = 3\mathbf{i} + 2\mathbf{j}$, $\mathbf{v} = -\mathbf{i} + 3\mathbf{j}$
36. $\mathbf{u} = \mathbf{i} + \mathbf{j}$, $\mathbf{v} = -3\mathbf{i} - 2\mathbf{j}$

In Exercises 37–39, let $\mathbf{u} = \langle a, b\rangle$, $\mathbf{v} = \langle c, d\rangle$, *and* $\mathbf{w} = \langle r, s\rangle$. *Verify that the given property of dot products is valid by calculating the quantities on each side of the equal sign.*

37. $\mathbf{u} \cdot (\mathbf{v} + \mathbf{w}) = \mathbf{u} \cdot \mathbf{v} + \mathbf{u} \cdot \mathbf{w}$
38. $k\mathbf{u} \cdot \mathbf{v} = k(\mathbf{u} \cdot \mathbf{v}) = \mathbf{u} \cdot k\mathbf{v}$
39. $\mathbf{0} \cdot \mathbf{u} = 0$
40. Suppose $\mathbf{u} = \langle a, b\rangle$ and $\mathbf{v} = \langle c, d\rangle$ are nonzero parallel vectors.
    (a) If $c \neq 0$, show that $\mathbf{u}$ and $\mathbf{v}$ lie on the same nonvertical straight line through the origin.
    (b) If $c \neq 0$, show that $\mathbf{v} = \frac{a}{c}\mathbf{u}$ (that is, $\mathbf{v}$ is a scalar multiple of $\mathbf{u}$). [*Hint:* The equation of the line on which $\mathbf{u}$ and $\mathbf{v}$ lie is $y = mx$ for some constant m (why?), which implies that $b = ma$ and $d = mc$.]
    (c) If $c = 0$, show that $\mathbf{v}$ is a scalar multiple of $\mathbf{u}$. [*Hint:* If $c = 0$, then $a = 0$ (why?), and hence, $b \neq 0$ (otherwise, $\mathbf{u} = \mathbf{0}$).]
41. Prove the Angle Theorem in the case when θ is 0 or π.
42. If $\mathbf{u}$ and $\mathbf{v}$ are nonzero vectors such that $\mathbf{u} \cdot \mathbf{v} = 0$, show that $\mathbf{u}$ and $\mathbf{v}$ are orthogonal. [*Hint:* If θ is the angle between $\mathbf{u}$ and $\mathbf{v}$, what is $\cos\theta$ and what does this say about θ?]
43. Show that (1, 2), (3, 4), (5, 2) are the vertices of a right triangle by considering the sides of the triangle as vectors.
44. Find a number x such that the angle between the vectors $\langle 1, 1\rangle$ and $\langle x, 1\rangle$ is $\pi/4$ radians.
45. Find nonzero vectors $\mathbf{u}$, $\mathbf{v}$, and $\mathbf{w}$ such that $\mathbf{u} \cdot \mathbf{v} = \mathbf{u} \cdot \mathbf{w}$ and $\mathbf{v} \neq \mathbf{w}$ and neither $\mathbf{v}$ nor $\mathbf{w}$ is orthogonal to $\mathbf{u}$.
46. If $\mathbf{u}$ and $\mathbf{v}$ are nonzero vectors, show that the vectors $\|\mathbf{u}\|\mathbf{v} + \|\mathbf{v}\|\mathbf{u}$ and $\|\mathbf{u}\|\mathbf{v} - \|\mathbf{v}\|\mathbf{u}$ are orthogonal.

47. A 600-pound trailer is on an inclined ramp that makes a 30° angle with the horizontal. Find the force required to keep it from rolling down the ramp, assuming that the only force that must be overcome is that due to gravity.

48. In Example 7, find the vector that represents the force necessary to keep the car motionless.

In Exercises 49–52, find the work done by a constant force $\mathbf{F}$ *as the point of application of* $\mathbf{F}$ *moves along the vector* $\overrightarrow{PQ}$.

49. $\mathbf{F} = 2\mathbf{i} + 5\mathbf{j}$, $P = (0, 0)$, $Q = (4, 1)$

50. $\mathbf{F} = \mathbf{i} - 2\mathbf{j}$, $P = (0, 0)$, $Q = (-5, 2)$

51. $\mathbf{F} = 2\mathbf{i} + 3\mathbf{j}$, $P = (2, 3)$, $Q = (5, 9)$ [*Hint:* Find the component form of $\overrightarrow{PQ}$.]

52. $\mathbf{F} = 5\mathbf{i} + \mathbf{j}$, $P = (-1, 2)$, $Q = (4, -3)$

53. A lawn mower handle makes an angle of 60° with the ground. A woman pushes on the handle with a force of 30 pounds. How much work is done in moving the lawn mower a distance of 75 feet on level ground?

54. A child pulls a wagon along a level sidewalk by exerting a force of 18 pounds on the wagon handle, which makes an angle of 25° with the horizontal. How much work is done in pulling the wagon 200 feet?

55. A 40-pound cart is pushed 100 feet up a ramp that makes a 20° angle with the horizontal (see the figure). How much work is done against gravity? [*Hint:* The amount of work done against gravity is the negative of the amount of work done *by* gravity. Coordinatize the situation so that the cart is at the origin. Then the cart moves along vector $\mathbf{d} = (100 \cos 20°)\mathbf{i} + (100 \sin 20°)\mathbf{j}$, and the downward force of gravity is $\mathbf{F} = 0\mathbf{i} - 40\mathbf{j}$.]

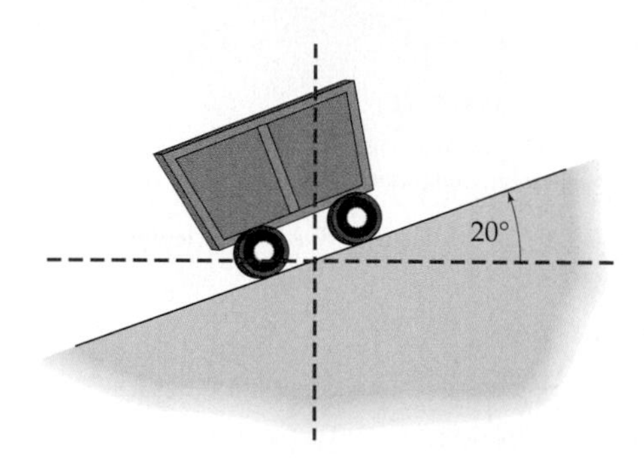

56. Suppose the child in Exercise 54 is pulling the wagon up a hill that makes an angle of 20° with the horizontal and all other facts remain the same. How much work is done in pulling the wagon 150 feet?

Chapter 4 Review

IMPORTANT CONCEPTS

IMPORTANT FACTS & FORMULAS

- $|a + bi| = \sqrt{a^2 + b^2}$
- $a + bi = r(\cos\theta + i\sin\theta)$, where

$$r = \sqrt{a^2 + b^2}, \qquad a = r\cos\theta, \qquad b = r\sin\theta.$$

- $r_1(\cos\theta_1 + i\sin\theta_1) \cdot r_2(\cos\theta_2 + i\sin\theta_2) = r_1r_2[\cos(\theta_1 + \theta_2) + i\sin(\theta_1 + \theta_2)]$
- $\dfrac{r_1(\cos\theta_1 + i\sin\theta_1)}{r_2(\cos\theta_2 + i\sin\theta_2)} = \dfrac{r_1}{r_2}[\cos(\theta_1 - \theta_2) + i\sin(\theta_1 - \theta_2)]$
- *DeMoivre's Theorem:*

$$[r(\cos\theta + i\sin\theta)]^n = r^n[\cos(n\theta) + i\sin(n\theta)]$$

- The distinct nth roots of $r(\cos\theta + i\sin\theta)$ are

$$\sqrt[n]{r}\left[\cos\left(\frac{\theta + 2k\pi}{n}\right) + i\sin\left(\frac{\theta + 2k\pi}{n}\right)\right] \quad (k = 0, 1, 2, \ldots, n - 1).$$

- The distinct nth roots of unity are

$$\cos\frac{2k\pi}{n} + i\sin\frac{2k\pi}{n} \quad (k = 0, 1, 2, \ldots, n - 1).$$

- If $P = (x_1, y_1)$ and $Q = (x_2, y_2)$, then $\overrightarrow{PQ} = \langle x_2 - x_1, y_2 - y_1\rangle$.
- $\|\langle a, b\rangle\| = \sqrt{a^2 + b^2}$
- If $\mathbf{u} = \langle a, b\rangle$ and k is a scalar, then $k\mathbf{u} = \langle ka, kb\rangle$.
- If $\mathbf{u} = \langle a, b\rangle$ and $\mathbf{v} = \langle c, d\rangle$, then

$$\mathbf{u} + \mathbf{v} = \langle a + c, b + d\rangle \qquad \text{and} \qquad \mathbf{u} - \mathbf{v} = \langle a - c, b - d\rangle.$$

- *Properties of Vector Addition and Scalar Multiplication:* For any vectors **u**, **v**, and **w** and any scalars r and s,
  1. $\mathbf{u} + (\mathbf{v} + \mathbf{w}) = (\mathbf{u} + \mathbf{v}) + \mathbf{w}$
  2. $\mathbf{u} + \mathbf{v} = \mathbf{v} + \mathbf{u}$
  3. $\mathbf{v} + \mathbf{0} = \mathbf{v} = \mathbf{0} + \mathbf{v}$
  4. $\mathbf{v} + (-\mathbf{v}) = \mathbf{0}$
  5. $r(\mathbf{u} + \mathbf{v}) = r\mathbf{u} + r\mathbf{v}$
  6. $(r + s)\mathbf{v} = r\mathbf{v} + s\mathbf{v}$
  7. $(rs)\mathbf{v} = r(s\mathbf{v}) = s(r\mathbf{v})$
  8. $1\mathbf{v} = \mathbf{v}$
  9. $0\mathbf{v} = \mathbf{0} = r\mathbf{0}$
- If $\mathbf{v} = \langle a, b \rangle = a\mathbf{i} + b\mathbf{j}$, then

$$a = \|\mathbf{v}\| \cos \theta \qquad \text{and} \qquad b = \|\mathbf{v}\| \sin \theta,$$

  where θ is the direction angle of **v**.
- If $\mathbf{u} = \langle a, b \rangle = a\mathbf{i} + b\mathbf{j}$ and $\mathbf{v} = \langle c, d \rangle = c\mathbf{i} + d\mathbf{j}$, then

$$\mathbf{u} \cdot \mathbf{v} = ac + bd.$$

- If θ is the angle between nonzero vectors **u** and **v**, then

$$\mathbf{u} \cdot \mathbf{v} = \|\mathbf{u}\| \, \|\mathbf{v}\| \cos \theta.$$

- *Schwarz Inequality:* $|\mathbf{u} \cdot \mathbf{v}| \leq \|\mathbf{u}\| \, \|\mathbf{v}\|$.
- Vectors **u** and **v** are orthogonal exactly when $\mathbf{u} \cdot \mathbf{v} = 0$.
- $\text{proj}_{\mathbf{v}}\mathbf{u} = \left(\dfrac{\mathbf{u} \cdot \mathbf{v}}{\|\mathbf{v}\|^2}\right)\mathbf{v}$
- $\text{comp}_{\mathbf{v}}\mathbf{u} = \dfrac{\mathbf{u} \cdot \mathbf{v}}{\|\mathbf{v}\|} = \|\mathbf{u}\| \cos \theta$, where θ is the angle between **u** and **v**.

REVIEW QUESTIONS

In Questions 1–4, perform the indicated operations and write the result in the form $a + bi$.

1. $(\sqrt{2} + i) + (\sqrt{3} - 2i)$

2. $(3 - i)(4 + 2i)$

3. i^{27}

4. $\dfrac{1}{3 - 4i}$

In Exercises 5 and 6, find all solutions of the equation. Express complex solutions in the form $a + bi$.

5. $x^2 + 5 = 4x$

6. $3x^2 + 2x + 1 = 0$

7. Simplify: $|i(4 + 2i)| + |3 - i|$.

8. Simplify: $|3 + 2i| - |1 - 2i|$.

9. Graph the equation $|z| = 2$ in the complex plane.

10. Graph the equation $|z - 3| = 1$ in the complex plane.

11. Express in polar form: $1 + \sqrt{3}i$.

12. Express in polar form: $4 - 5i$.

In Questions 13–17, multiply or divide, and express the answer in the form $a + bi$.

13. $2\left(\cos \dfrac{\pi}{12} + i \sin \dfrac{\pi}{12}\right) \cdot 4\left(\cos \dfrac{\pi}{6} + i \sin \dfrac{\pi}{6}\right)$

REVIEW QUESTIONS 

14. $3\left(\cos\frac{\pi}{8} + i\sin\frac{\pi}{8}\right) \cdot 2\left(\cos\frac{3\pi}{8} + i\sin\frac{3\pi}{8}\right)$

15. $\dfrac{12\left(\cos\frac{7\pi}{12} + i\sin\frac{7\pi}{12}\right)}{3\left(\cos\frac{5\pi}{12} + i\sin\frac{5\pi}{12}\right)}$

16. $\left(\cos\frac{\pi}{12} + i\sin\frac{\pi}{12}\right)^{18}$

17. $\left[\sqrt[3]{3}\left(\cos\frac{5\pi}{36} + i\sin\frac{5\pi}{36}\right)\right]^{12}$

In Questions 18–22, solve the given equation in the complex number system, and express your answers in polar form.

18. $x^3 = i$
19. $x^6 = 1$
20. $x^8 = -\sqrt{3} - 3i$
21. $x^4 = i$
22. $x^3 = 1 - i$

In Questions 23–26, let $\mathbf{u} = \langle 3, -2\rangle$ *and* $\mathbf{v} = \langle 8, 1\rangle$. *Find*

23. $\mathbf{u} + \mathbf{v}$
24. $\|-3\mathbf{v}\|$
25. $\|2\mathbf{v} - 4\mathbf{u}\|$
26. $3\mathbf{u} \quad \frac{1}{2}\mathbf{v}$

In Questions 27–30, let $\mathbf{u} = -2\mathbf{i} + \mathbf{j}$ *and* $\mathbf{v} = 3\mathbf{i} - 4\mathbf{j}$. *Find*

27. $4\mathbf{u} - \mathbf{v}$
28. $\mathbf{u} + 2\mathbf{v}$
29. $\|\mathbf{u} + \mathbf{v}\|$
30. $\|\mathbf{u}\| + \|\mathbf{v}\|$
31. Find the components of the vector $\mathbf{v}$ such that $\|\mathbf{v}\| = 5$ and the direction angle of $\mathbf{v}$ is 45°.
32. Find the magnitude and direction angle of $3\mathbf{i} + 4\mathbf{j}$.
33. Find a unit vector whose direction is *opposite* the direction of $3\mathbf{i} - 6\mathbf{j}$.
34. An object at the origin is acted upon by a 10-pound force with direction angle 90° and a 20-pound force with direction angle 30°. Find the magnitude and direction of the resultant force.
35. A plane flies in the direction 120° with an air speed of 300 mph. The wind is blowing from north to south at 40 mph. Find the course and ground speed of the plane.
36. An object weighing 40 pounds lies on an inclined plane that makes a 30° angle with the horizontal. Find the components of the weight parallel and perpendicular to the plane.

In Questions 37–40, $\mathbf{u} = \langle 3, -4\rangle$, $\mathbf{v} = \langle -2, 5\rangle$, *and* $\mathbf{w} = \langle 0, 3\rangle$. *Find*

37. $\mathbf{u} \cdot \mathbf{v}$
38. $\mathbf{u} \cdot \mathbf{u} - \mathbf{v} \cdot \mathbf{v}$
39. $(\mathbf{u} + \mathbf{v}) \cdot \mathbf{w}$
40. $(\mathbf{u} + \mathbf{w}) \cdot (\mathbf{w} - 3\mathbf{v})$
41. What is the angle between the vectors $5\mathbf{i} - 2\mathbf{j}$ and $3\mathbf{i} + \mathbf{j}$?
42. Is $3\mathbf{i} - 2\mathbf{j}$ orthogonal to $4\mathbf{i} + 6\mathbf{j}$?

In Questions 43 and 44, $\mathbf{u} = 4\mathbf{i} - 3\mathbf{j}$ *and* $\mathbf{v} = 2\mathbf{i} + \mathbf{j}$. *Find*

43. $\text{proj}_{\mathbf{v}}\mathbf{u}$
44. $\text{comp}_{\mathbf{u}}\mathbf{v}$
45. If $\mathbf{u}$ and $\mathbf{v}$ have the same magnitude, show that $\mathbf{u} + \mathbf{v}$ and $\mathbf{u} - \mathbf{v}$ are orthogonal.
46. If $\mathbf{u}$ and $\mathbf{v}$ are nonzero vectors, show that the vector $\mathbf{u} - k\mathbf{v}$ is orthogonal to $\mathbf{v}$, where $k = \dfrac{\mathbf{u} \cdot \mathbf{v}}{\|\mathbf{v}\|^2}$.
47. A 3500-pound automobile is on an inclined ramp that makes a 30° angle with the horizontal. Find the force required to keep it from rolling down the ramp, assuming that the only force that must be overcome is that due to gravity.
48. A sled is pulled along level ground by a rope that makes a 50° angle with the horizontal. If a force of 40 pounds is used to pull the sled, how much work is done in pulling it 100 feet?

DISCOVERY PROJECT 4 Surveying

Arthur S. Aubry/Getty Images

You might ask, how far apart are two points A and B? Under ordinary circumstances, the easiest thing to do would be to measure the distance. However, sometimes it is impractical to make such a direct measurement because of intervening obstacles. Two historical methods for measuring the distance between the two mutually invisible points A and B are given below.

1. Find the distance from A to B, using the classic surveyor's method shown in the figure below. A transit is used to measure the angle between two distant objects, and a reasonably short baseline is measured. Angles are reported in degrees. Triangles are drawn, and the law of sines (page 116) is used to calculate the length of unknown edges. To find the distance from A to B, draw an additional triangle that has AB as one of the sides. Caution: The picture is not drawn to scale.

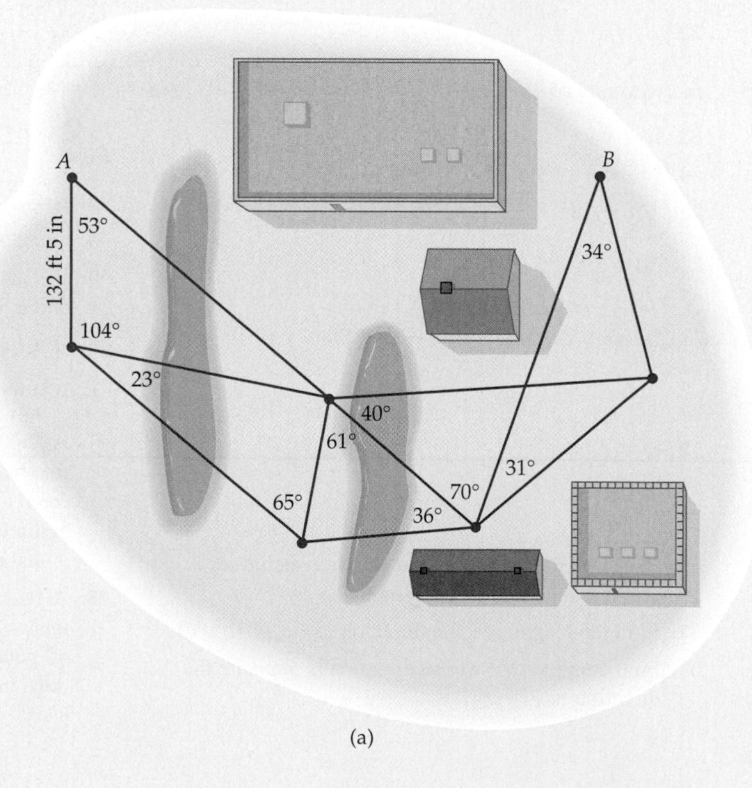

(a)

2. Find the distance from A to B, using a modern laser range finder. Laser technology removes the necessity of drawing triangles; the tool provides a vector (angle and distance) from point to point, so the distance from A to B can be found by using vector arithmetic in place of repeated application of the law of sines. Angles are taken from the internal compass and are reported in degrees measured clockwise from North (0°), so you have to convert compass direction angles to angles in standard position. For example, the angle from point A (straight South) is reported as 180°, but in standard position, you would list the angle as 270°.

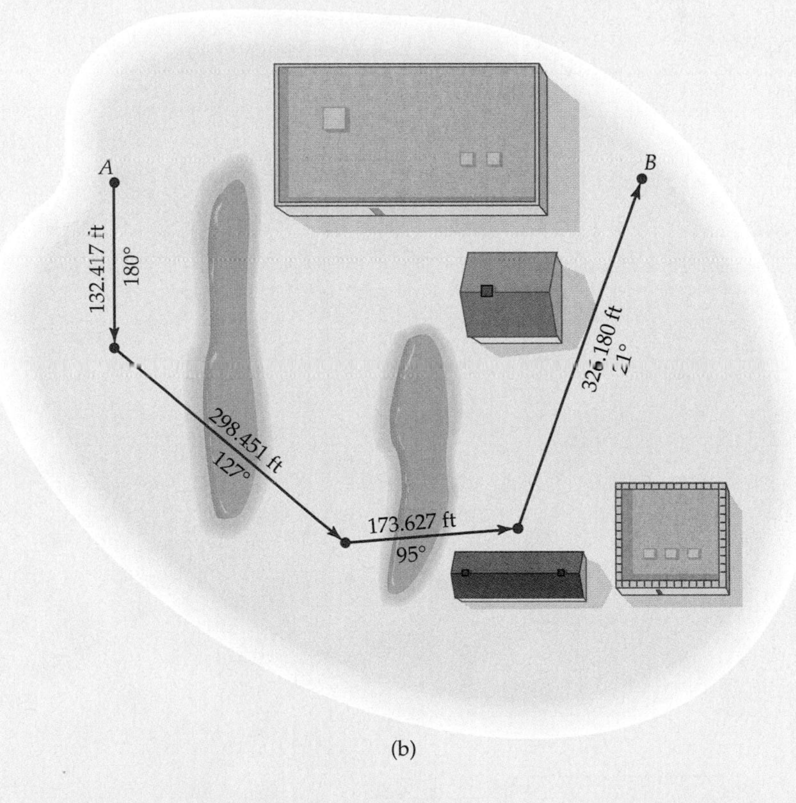

(b)

CHAPTER 5

Analytic Geometry and Trigonometry

Peter Menzel/Stock, Boston Inc./PictureQuest

Calling all ships!

The planets travel in elliptical orbits around the sun, and satellites travel in elliptical orbits around the earth. Parabolic reflectors are used in spotlights, radar antennas, and satellite dishes. The long-range navigation system (loran) uses hyperbolas to enable a ship to determine its exact location. See Exercise 56 on page 359.

Chapter Outline

Interdependence of Sections

Sections 5.1–5.3, 5.5, and 5.6 may be read in any order, but a few exercises in Sections 5.2 and 5.3 assume knowledge of preceding sections. Section 5.4 depends on Sections 5.1–5.3, and Section 5.7 depends on Section 5.6.

The discussion of analytic geometry that was begun in Section 0.2 is continued in Sections 5.1–5.4 with an examination of conic sections (which have played a significant role in mathematics since ancient times).

When a right circular cone is cut by a plane, the intersection is a curve called a **conic section,** as shown in Figure 5–1.* Conic sections were studied by the ancient Greeks and are still of interest. For instance, planets travel in elliptical orbits, parabolic mirrors are used in telescopes, and certain atomic particles follow hyperbolic paths.

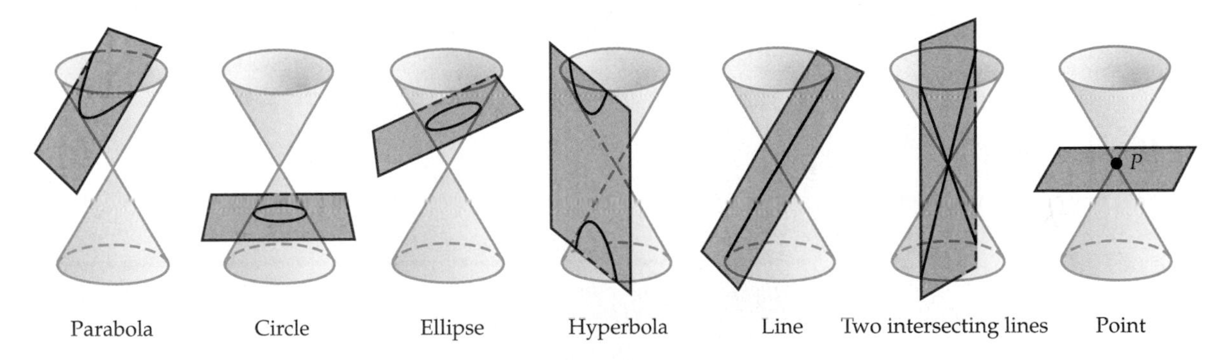

Figure 5–1

Although the Greeks studied conic sections from a purely geometric point of view, the modern approach is to describe them in terms of the coordinate plane and distance or as the graphs of certain types of equations. This was done for circles in Section 0.2 and will be done here for ellipses, hyperbolas, and parabolas.

Sections 5.5–5.7 present a more thorough treatment of parametric graphing and an alternate method of coordinatizing the plane and graphing.

---

*A point, a line, or two intersecting lines are sometimes called **degenerate** conic sections.

5.1 Circles and Ellipses

Let P be a point in the plane, and let r be a positive number. Then the **circle** with center P and radius r consists of all points X in the plane such that

$$\text{Distance from } X \text{ to } P = r,$$

as shown in Figure 5–2.

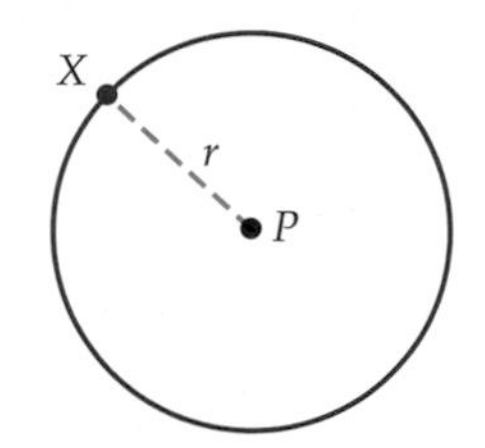

Figure 5–2

In Section 0.2, we proved the following result.

Equation of a Circle

> The circle with center (0, 0) and radius r is the graph of the equation
> $$x^2 + y^2 = r^2.$$
> The circle with center (h, k) and radius r is the graph of
> $$(x - h)^2 + (y - k)^2 = r^2.$$

EXAMPLE 1

Find the center and radius of the circle whose equation is given.

(a) $(x - 2)^2 + (y - 7)^2 = 25$

(b) $(x - 4)^2 + (y + 3)^2 = 36$

SOLUTION

(a) The right side of the equation is 5^2. So the center is at (2, 7), and the radius is 5.

(b) First, we rewrite the equation to match the form in the box above:

$$(x - 4)^2 + (y + 3)^2 = 36$$
$$(x - 4)^2 + [y - (-3)]^2 = 6^2.$$

Hence, the circle has center (4, −3) and radius 6. ■

The left-hand side of the equation of the circle can be thought of as the sum of two perfect squares [in Example 1(a), for instance, $(x - 2)^2 + (y - 7)^2$]. To find the center and radius of a circle whose equation is multiplied out, such as

$$x^2 + y^2 - 8x - 6y = -9,$$

we must first express the left-hand side as the sum of two perfect squares. The technique for doing this is based on a simple algebraic fact:

$$\left(x \pm \frac{b}{2}\right)^2 = x^2 \pm bx + \left(\frac{b}{2}\right)^2,$$

which you can easily verify by multiplying out the left-hand side. This equation says that if you add $\left(\frac{b}{2}\right)^2$ to $x^2 \pm bx$, the result is a perfect square. Note how $\left(\frac{b}{2}\right)^2$ is related to the expression $x^2 \pm bx$.

$$\left(\frac{b}{2}\right)^2 \text{ is the square of one-half the coefficient of } x.$$

EXAMPLE 2

Show that the graph of

$$3x^2 + 3y^2 - 12x - 30y + 45 = 0$$

is a circle and find its center and radius.

SOLUTION The technique to be used below requires that x^2 and y^2 each have coefficient 1. So we begin by dividing both sides of the equation by 3 and regrouping the terms.

$$x^2 + y^2 - 4x - 10y + 15 = 0$$
$$(x^2 - 4x) + (y^2 - 10y) = -15.$$

We want to add a constant to the expression $x^2 - 4x$ so that the result will be a perfect square. Using the fact discussed before the example, we take half the coefficient of x, namely, -2, and square it, obtaining 4. Adding 4 to both sides of the equation, we obtain

$$(x^2 - 4x + 4) + (y^2 - 10y) = -15 + 4,$$

which factors as

$$(x - 2)^2 + (y^2 - 10y) = -11.$$

Similarly, in the expression $y^2 - 10y$, we take half the coefficient of y, namely, -5, and square it, obtaining 25. Now add 25 to both sides of the equation and factor.

$$(x - 2)^2 + (y^2 - 10y + 25) = -11 + 25$$
$$(x - 2)^2 + (y - 5)^2 = 14$$
$$(x - 2)^2 + (y - 5)^2 = (\sqrt{14})^2$$

Now we see that the graph is a circle with center (2, 5) and radius $\sqrt{14}$. ■

The technique of adding the square of half the coefficient of x to $x^2 - 4x$ to obtain the perfect square $(x - 2)^2$ in Example 2 is called **completing the square.** It will be used frequently in this chapter.

PARAMETRIC EQUATIONS

In addition to the Cartesian equations, circles can also be described by parametric equations.

EXAMPLE 3

Show that the curve given by the parametric equations

$$x = 2 \cos t + 3 \quad \text{and} \quad y = 2 \sin t + 1 \quad (0 \leq t \leq 2\pi)$$

is the circle with center (3, 1) and radius 2.

SOLUTION The circle with center (3, 1) and radius 2 is the graph of the equation

$$(x - 3)^2 + (y - 1)^2 = 2^2.$$

We must show that the coordinates given by the parametric equations satisfy this equation.

$$\begin{aligned}(x - 3)^2 + (y - 1)^2 &= [(2 \cos t + 3) - 3]^2 + [(2 \sin t + 1) - 1]^2 \\ &= [2 \cos t]^2 + [2 \sin t]^2 \\ &= 4 \cos^2 t + 4 \sin^2 t \\ &= 4(\cos^2 t + \sin^2 t) \\ &= 4(1) = 4. \qquad \text{[Pythagorean Identity]}\end{aligned}$$

Therefore, the curve given by the parametric equations is the circle with center (3, 1) and radius 2. ■

If you replace 3, 1, and 2 in Example 3 by h, k, and r, respectively, then the same computation leads to the following result.

Parametric Equations of a Circle

The circle with center (h, k) and radius r is given by the parametric equations

$$x = r \cos t + h \quad \text{and} \quad y = r \sin t + k \quad (0 \leq t \leq 2\pi).$$

The preceding discussion of circles provides the model for our discussion of the other conic sections. In each case, the conic is defined in terms of points and distances, and its Cartesian equation is determined. The standard form of the equation of a conic includes the key information necessary for a rough sketch of its graph, just as the standard form of the equation of a circle tells you its center and radius. Finally, appropriate Pythagorean identities are used to develop parametric equations for each conic section.

ELLIPSES

Definition. Let P and Q be points in the plane, and let r be a number greater than the distance from P to Q. The **ellipse** with **foci*** P and Q is the set of all points X such that

$$(\text{Distance from } X \text{ to } P) + (\text{Distance from } X \text{ to } Q) = r.$$

To draw this ellipse, take a piece of string of length r and pin its ends on P and Q. Put your pencil point against the string and move it, keeping the string taut. You will trace out the ellipse, as shown in Figure 5–3.

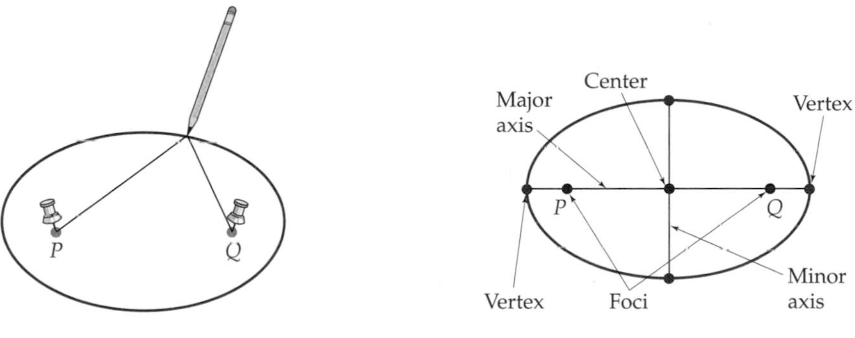

Figure 5–3

Figure 5–4

The midpoint of the line segment from P to Q is the **center** of the ellipse. The points where the straight line through the foci intersects the ellipse are its **vertices.** The **major axis** of the ellipse is the line segment joining the vertices; its **minor axis** is the line segment through the center, perpendicular to the major axis, as shown in Figure 5–4.

Equation. Suppose that the foci P and Q are on the x-axis, with coordinates

$$P = (-c, 0) \qquad \text{and} \qquad Q = (c, 0) \qquad \text{for some } c > 0.$$

Let $a = r/2$, so that $2a = r$. Then the point (x, y) is on the ellipse exactly when

$$[\text{Distance from } (x, y) \text{ to } P] + [\text{Distance from } (x, y) \text{ to } Q] = r$$

$$\sqrt{(x + c)^2 + (y - 0)^2} + \sqrt{(x - c)^2 + (y - 0)^2} = 2a$$

$$\sqrt{(x + c)^2 + y^2} = 2a - \sqrt{(x - c)^2 + y^2}.$$

*"Foci" is the plural of "focus."

Squaring both sides and simplifying (Exercise 82), we obtain

$$a\sqrt{(x - c)^2 + y^2} = a^2 - cx.$$

Again squaring both sides and simplifying, we have

$$(a^2 - c^2)x^2 + a^2y^2 = a^2(a^2 - c^2).$$

To simplify the form of this equation, let $b = \sqrt{a^2 - c^2}$* so that $b^2 = a^2 - c^2$ and the equation becomes

$$b^2x^2 + a^2y^2 = a^2b^2.$$

Dividing both sides by a^2b^2 shows that the coordinates of every point on the ellipse satisfy the equation

$$\frac{x^2}{a^2} + \frac{y^2}{b^2} = 1.$$

Conversely, it can be shown that every point whose coordinates satisfy this equation is on the ellipse. When the equation is in this form the x- and y-intercepts of the graph are easily found. For instance, to find the x-intercepts, we set $y = 0$ and solve

$$\frac{x^2}{a^2} + \frac{0^2}{b^2} = 1$$

$$x^2 = a^2$$

$$x = \pm a.$$

Similarly, the y-intercepts are $\pm b$.

A similar argument applies when the foci are on the y-axis and leads to this conclusion.

*The distance between the foci is $2c$. Since $r = 2a$ and $r > 2c$ by definition, we have $2a > 2c$, and hence, $a > c$. Therefore, $a^2 - c^2$ is a positive number and has a real square root.

Standard Equations of Ellipses Centered at the Origin

Let a and b be real numbers with $a > b > 0$. Then the graph of each of the following equations is an ellipse centered at the origin:

$$\frac{x^2}{a^2} + \frac{y^2}{b^2} = 1 \begin{cases} x\text{-intercepts: } \pm a \qquad y\text{-intercepts: } \pm b \\ \text{major axis on the } x\text{-axis, with vertices } (a, 0) \text{ and } (-a, 0) \\ \text{foci: } (c, 0) \text{ and } (-c, 0), \text{ where } c = \sqrt{a^2 - b^2} \end{cases}$$

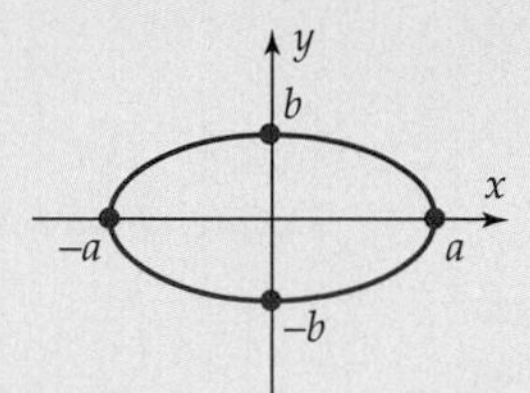

$$\frac{x^2}{b^2} + \frac{y^2}{a^2} = 1 \begin{cases} x\text{-intercepts: } \pm b \qquad y\text{-intercepts: } \pm a \\ \text{major axis on the } y\text{-axis, with vertices } (0, a) \text{ and } (0, -a) \\ \text{foci: } (0, c) \text{ and } (0, -c), \text{ where } c = \sqrt{a^2 - b^2} \end{cases}$$

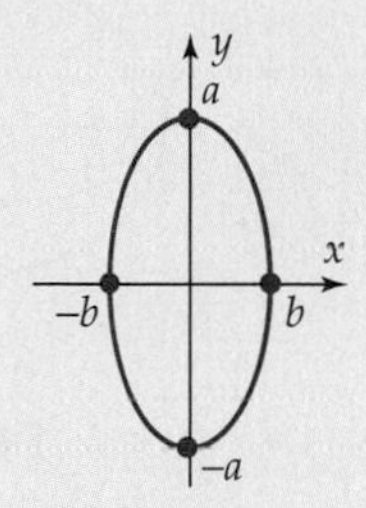

In the preceding box $a > b$, but don't let all the letters confuse you: When the equation is in standard form, the denominator of the x term tells you the x-intercepts, the denominator of the y term tells you the y-intercepts, and the major axis is the longer one, as illustrated in the following examples.

EXAMPLE 4

Identify and sketch the graph of the equation $4x^2 + 9y^2 = 36$.

SOLUTION To identify the graph, we put the equation in standard form.

$$4x^2 + 9y^2 = 36$$

Divide both sides by 36: $$\frac{4x^2}{36} + \frac{9y^2}{36} = \frac{36}{36}$$

Simplify: $$\frac{x^2}{9} + \frac{y^2}{4} = 1$$

$$\frac{x^2}{3^2} + \frac{y^2}{2^2} = 1$$

The graph is now in the form of the first equation in the preceding box, with $a = 3$ and $b = 2$. So its graph is an ellipse with x-intercepts ± 3 and y-intercepts ± 2. Its major axis and foci lie on the x-axis, as do its vertices $(3, 0)$ and $(-3, 0)$. A hand-sketched graph is shown in Figure 5–5. To graph this ellipse on a calculator, we first solve its equation for y.

$$4x^2 + 9y^2 = 36$$

Subtract $4x^2$ from both sides: $$9y^2 = 36 - 4x^2$$

Divide both sides by 9: $$y^2 = \frac{36 - 4x^2}{9}.$$

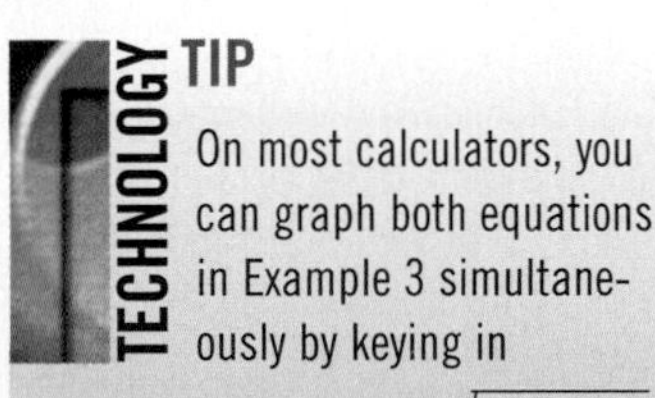

TIP

On most calculators, you can graph both equations in Example 3 simultaneously by keying in

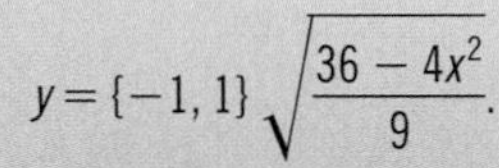

$$y = \{-1, 1\}\sqrt{\frac{36 - 4x^2}{9}}.$$

Taking square roots on both sides, we see that

$$y = \sqrt{\frac{36 - 4x^2}{9}} \quad \text{or} \quad y = -\sqrt{\frac{36 - 4x^2}{9}}.$$

Graphing both of these equations on the same screen, we obtain Figure 5–6. ■

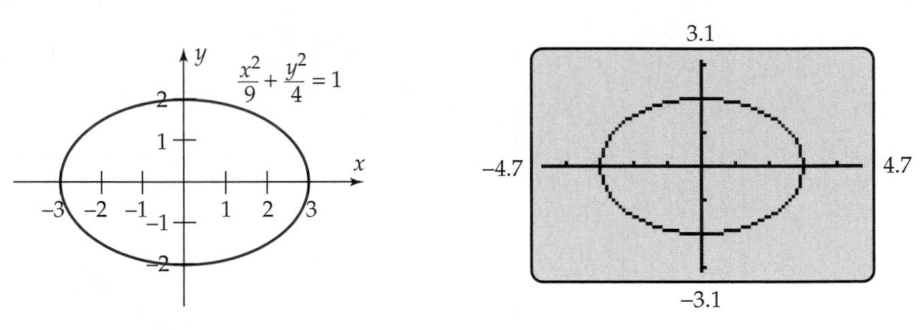

Figure 5–5 **Figure 5–6**

EXAMPLE 5

Find the equation of the ellipse with vertices $(0, \pm 6)$ and foci $(0, \pm 2\sqrt{6})$, and sketch its graph.

SOLUTION Since the foci are $(0, 2\sqrt{6})$ and $(0, -2\sqrt{6})$, the center of the ellipse is $(0, 0)$, and its major axis lies on the y-axis. Hence, its equation is of the form

$$\frac{x^2}{b^2} + \frac{y^2}{a^2} = 1.$$

From the box on page 337, we see that $a = 6$ and $c = 2\sqrt{6}$. Since $c = \sqrt{a^2 - b^2}$, we have $c^2 = a^2 - b^2$, so

$$b^2 = a^2 - c^2 = 6^2 - (2\sqrt{6})^2 = 36 - 4 \cdot 6 = 12.$$

Hence, $b = \sqrt{12}$, and the equation of the ellipse is

$$\frac{x^2}{(\sqrt{12})^2} + \frac{y^2}{6^2} = 1 \quad \text{or, equivalently,} \quad \frac{x^2}{12} + \frac{y^2}{36} = 1.$$

The graph has x-intercepts $\pm\sqrt{12} \approx \pm 3.46$ and y-intercepts ± 6, as sketched in Figure 5–7. ■

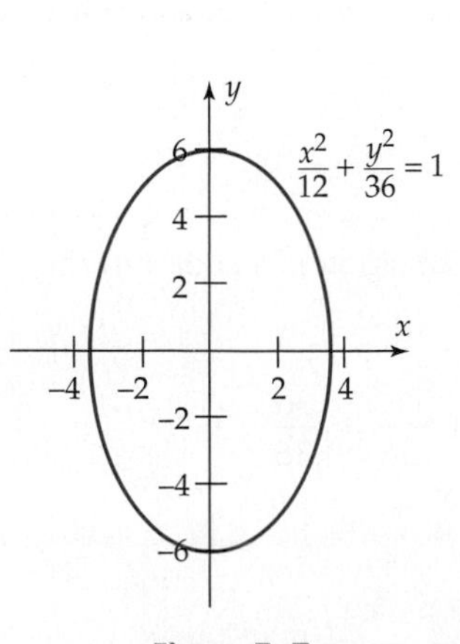

Figure 5–7

VERTICAL AND HORIZONTAL SHIFTS

We now examine ellipses that have foci on a line parallel to one of the coordinate axes. Recall that in Section 0.6, we saw that replacing the variable x by $x - 5$ in the rule of the function $y = f(x)$ shifts the graph horizontally 5 units to the right, whereas replacing x by $x + 5$ [that is, $x - (-5)$] shifts the graph horizontally 5 units to the left (see the box on page 63).

Similarly, if the rule of a function is given by $y = f(x)$, then replacing y by $y - 4$ shifts the graph 4 units vertically upward because

$$y - 4 = f(x) \qquad \text{is equivalent to} \qquad y = f(x) + 4$$

(see the box on page 62). For arbitrary equations, we have similar results.

Vertical and Horizontal Shifts

> Let h and k be constants. Replacing x by $x - h$ and y by $y - k$ in an equation shifts the graph of the equation
>
> $|h|$ units horizontally (right for positive h, left for negative h) and
>
> $|k|$ units vertically (upward for positive k, downward for negative k).

EXAMPLE 6

Identify and sketch the graph of

$$\frac{(x-5)^2}{9} + \frac{(y+4)^2}{36} = 1.$$

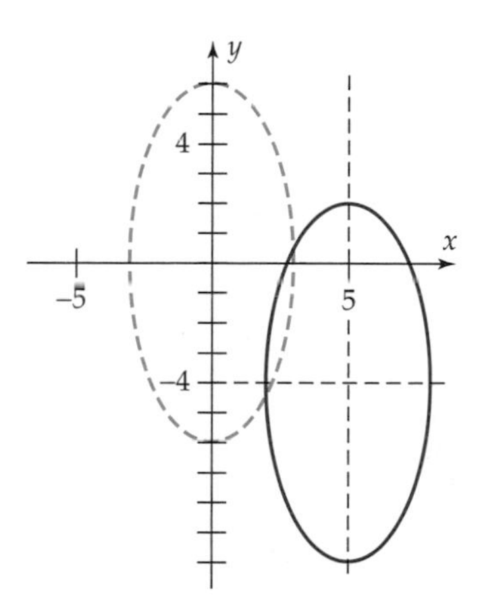

Figure 5–8

SOLUTION This equation can be obtained from the equation $\frac{x^2}{9} + \frac{y^2}{36} = 1$ (whose graph is known to be an ellipse) as follows.

$$\text{Replace } x \text{ by } x - 5 \qquad \text{and} \qquad \text{replace } y \text{ by } y - (-4) = y + 4.$$

This is the situation described in the previous box, with $h = 5$ and $k = -4$. Therefore, the graph is the ellipse $\frac{x^2}{9} + \frac{y^2}{36} = 1$ shifted horizontally 5 units to the right and vertically 4 units downward, as shown in Figure 5–8. The center of the ellipse is at $(5, -4)$. Its major axis (the longer one) lies on the vertical line $x = 5$, as do its foci. The minor axis is on the horizontal line $y = -4$. ■

EXAMPLE 7

Identify and sketch the graph of

$$4x^2 + 9y^2 - 32x - 90y + 253 = 0.$$

SOLUTION We first rewrite the equation.

$$(4x^2 - 32x) + (9y^2 - 90y) = -253$$
$$4(x^2 - 8x) + 9(y^2 - 10y) = -253$$

CAUTION

Completing the square works only when the coefficient of x^2 is 1. In an expression such as $4x^2 - 32x$, you must first factor out the 4,

$$4(x^2 - 8x),$$

and then complete the square on the expression in parentheses.

The factoring in the last equation was done in preparation for completing the square (see the Caution in the margin). To complete the square on $x^2 - 8x$, we add 16 (the square of half the coefficient of x), and to complete the square on $y^2 - 10y$, we add 25 (the square of half the coefficient of y).

$$4(x^2 - 8x + 16) + 9(y^2 - 10y + 25) = -253 + ? + ?.$$

Be careful here: On the left-hand side, we haven't just added 16 and 25. When the left-hand side is multiplied out, we have actually added $4 \cdot 16 = 64$ and $9 \cdot 25 = 225$. To leave the equation unchanged, we must add these numbers on the right.

$$4(x^2 - 8x + 16) + 9(y^2 - 10y + 25) = -253 + 64 + 225$$

Factor and simplify:
$$4(x - 4)^2 + 9(y - 5)^2 = 36$$

Divide both sides by 36:
$$\frac{4(x - 4)^2}{36} + \frac{9(y - 5)^2}{36} = \frac{36}{36}$$

Simplify:
$$\frac{(x - 4)^2}{9} + \frac{(y - 5)^2}{4} = 1.$$

The graph of this equation is the ellipse $\frac{x^2}{9} + \frac{y^2}{4} = 1$ shifted 4 units to the right and 5 units upward. Its center is at (4, 5). Its major axis lies on the horizontal line $y = 5$, and its minor axis lies on the vertical line $x = 4$, as shown in Figure 5–9. ■

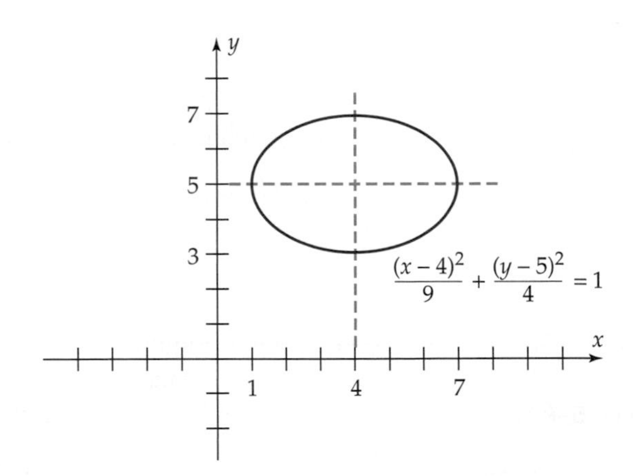

Figure 5–9

By translating the information about ellipses centered at the origin in the box on page 337, we obtain the following.

Standard Equations of Ellipses with Center at (h, k)

Let (h, k) be any point in the plane. If a and b are real numbers with $a > b > 0$, then the graph of each of the following equations is an ellipse with center (h, k).

$$\frac{(x-h)^2}{a^2} + \frac{(y-k)^2}{b^2} = 1 \quad \begin{cases} \text{major axis on the horizontal line } y = k \\ \text{minor axis on the vertical line } x = h \\ \text{foci: } (h - c, k) \text{ and } (h + c, k), \text{ where} \\ \qquad c = \sqrt{a^2 - b^2} \end{cases}$$

$$\frac{(x-h)^2}{b^2} + \frac{(y-k)^2}{a^2} = 1 \quad \begin{cases} \text{major axis on the vertical line } x = h \\ \text{minor axis on the horizontal line } y = k \\ \text{foci: } (h, k - c) \text{ and } (h, k + c), \text{ where} \\ \qquad c = \sqrt{a^2 - b^2} \end{cases}$$

TECHNOLOGY TIP

Casio 9850 and FX 2.0 have conic section graphers (on the main menu) that produce the graphs of equations in standard form when the various coefficients are entered. TI-83+ users can download a conic grapher from TI.

EXAMPLE 8

Find an appropriate viewing window and graph the ellipse

$$\frac{(x-3)^2}{40} + \frac{(y+2)^2}{120} = 1.$$

SOLUTION To graph the ellipse, we first solve its equation for y.

Subtract $\frac{(x-3)^2}{40}$ from both sides: $\quad \frac{(y+2)^2}{120} = 1 - \frac{(x-3)^2}{40}$

Multiply both sides by 120. $\quad (y+2)^2 = 120\left[1 - \frac{(x-3)^2}{40}\right]$

Multiply out right side: $\quad (y+2)^2 = 120 - 3(x-3)^2$

Take square roots on both sides: $\quad y + 2 = \pm\sqrt{120 - 3(x-3)^2}$

$$y = \sqrt{120 - 3(x-3)^2} - 2 \quad \text{or} \quad y = -\sqrt{120 - 3(x-3)^2} - 2$$

So we should graph both of these last two equations on the same screen.

To determine an appropriate window, look at the original form of the equation:

$$\frac{(x-3)^2}{40} + \frac{(y+2)^2}{120} = 1.$$

The center of the ellipse is at $(3, -2)$. Its graph extends a distance of $\sqrt{40}$ to the left and right of the center and a distance of $\sqrt{120}$ above and below the center. Since $\sqrt{40}$ is a bit less than 7 and $\sqrt{120}$ is a bit less than 11 (why?), our window should include x-values from $3 - 7$ to $3 + 7$ (that is, $-4 \leq x \leq 10$) and y-values from $-2 - 11$ to $-2 + 11$ (that is, $-13 \leq y \leq 9$). So we first try the window in Figure 5–10 on the next page. In this window, the ellipse looks longer horizontally than vertically. We know from its equation, however, that the major (longer) axis is vertical because the larger constant 120 is the denominator of the y-term. So we change to a square window and obtain the more accurate graph in

Figure 5–11.* Even this graph has gaps that shouldn't be there (an ellipse is a connected figure), but this is unavoidable because of the limited resolution of the calculator screen. ■

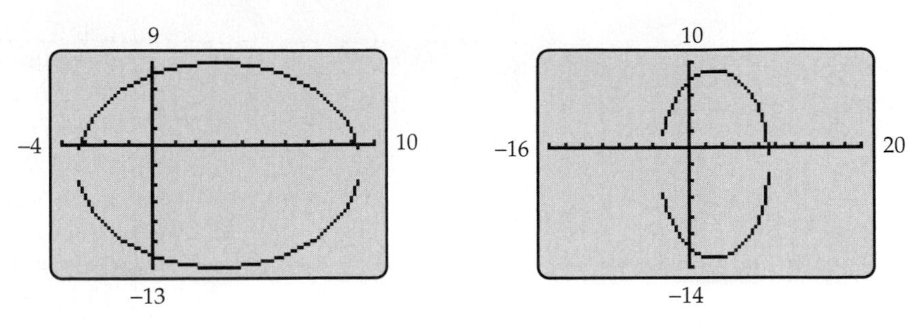

Figure 5–10

Figure 5–11

EXAMPLE 9

Graph the equation $x^2 + 8y^2 + 6x + 9y + 4 = 0$ without first putting it in standard form.

SOLUTION Rewrite it like this:

$$8y^2 + 9y + (x^2 + 6x + 4) = 0.$$

This is a quadratic equation of the form $ay^2 + by + c = 0$, with

$$a = 8, \qquad b = 9, \qquad c = x^2 + 6x + 4$$

and hence can be solved by using the quadratic formula.

$$y = \frac{-b \pm \sqrt{b^2 - 4ac}}{2a}$$

$$y = \frac{-9 \pm \sqrt{9^2 - 4 \cdot 8 \cdot (x^2 + 6x + 4)}}{2 \cdot 8}$$

$$y = \frac{-9 \pm \sqrt{81 - 32(x^2 + 6x + 4)}}{16}$$

TECHNOLOGY TIP

TI users can save keystrokes by entering the first equation in Example 9 as y_1 and then using the RCL key to copy the text of y_1 to y_2. [On TI-89, use COPY and PASTE in place of RCL.] Then only one sign needs to be changed to make y_2 into the second equation of Example 9.

GRAPHING EXPLORATION

Find a complete graph of the original equation by graphing both of the following functions on the same screen. The Technology Tip in the margin may be helpful.

$$y = \frac{-9 + \sqrt{81 - 32(x^2 + 6x + 4)}}{16}$$

$$y = \frac{-9 - \sqrt{81 - 32(x^2 + 6x + 4)}}{16}.$$ ■

*Figure 5–11 shows a square window for a TI-83+. On wide-screen calculators, a longer x-axis is needed for a square window.

PARAMETRIC EQUATIONS

Parametric equations for an ellipse can be obtained in the same way that the ones for the circle were.

EXAMPLE 10

Show that the curve given by the parametric equations

$$x = 2\cos t + 3 \qquad \text{and} \qquad y = \sqrt{5}\sin t + 6 \quad (0 \le t \le 2\pi)$$

is the ellipse with equation

$$\frac{(x-3)^2}{4} + \frac{(y-6)^2}{5} = 1.$$

SOLUTION We use the Pythagorean identity to show that the coordinates given by the parametric equations satisfy the equation

$$\begin{aligned}\frac{(x-3)^2}{4} + \frac{(y-6)^2}{5} &= \frac{[(2\cos t + 3) - 3]^2}{4} + \frac{[(\sqrt{5}\sin t + 6) - 6]^2}{5}\\ &= \frac{[2\cos t]^2}{4} + \frac{[\sqrt{5}\sin t]^2}{5}\\ &= \frac{4\cos^2 t}{4} + \frac{5\sin^2 t}{5}\\ &= \cos^2 t + \sin^2 t = 1. \quad \blacksquare\end{aligned}$$

If you use, h, k, a, and b in place of 3, 6, 2, and $\sqrt{5}$, respectively, in Example 10, then the same computation leads to this result.

Parametric Equations for Ellipses

The ellipse with equation

$$\frac{(x-h)^2}{a^2} + \frac{(y-k)^2}{b^2} = 1$$

is given by the parametric equations

$$x = a\cos t + h \qquad \text{and} \qquad y = b\sin t + k \quad (0 \le t \le 2\pi).$$

APPLICATIONS

Elliptical surfaces have interesting reflective properties. If a sound or light ray passes through one focus and reflects off an ellipse, the ray will pass through the other focus, as shown in Figure 5–12. Exactly this situation occurs under the elliptical dome of the U.S. Capitol. A person who stands at one focus and whispers can be clearly heard by anyone at the other focus. Before this fact was widely known, when Congress used to sit under the dome, several political secrets were inadvertently revealed by congressmen to members of the other party.

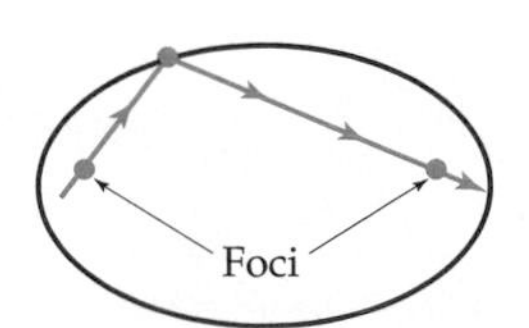

Figure 5–12

The planets and many comets have elliptical orbits, with the sun as one focus. The moon travels in an elliptical orbit with the earth as one focus. Satellites are usually put into elliptical orbits around the earth.

EXAMPLE 11

The earth's orbit around the sun is an ellipse that is almost a circle. The sun is one focus, and the major and minor axes have lengths 186,000,000 miles and 185,974,062 miles, respectively. What are the minimum and maximum distances from the earth to the sun?

SOLUTION The orbit is shown in Figure 5–13. If we use a coordinate system with the major axis on the x-axis and the sun having coordinates $(c, 0)$, then we obtain Figure 5–14.

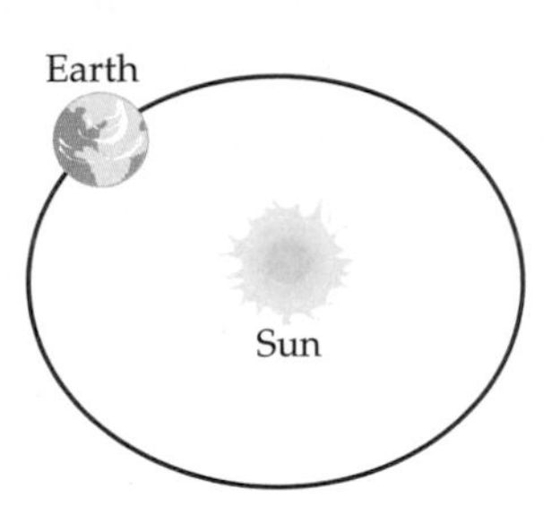

Figure 5–13

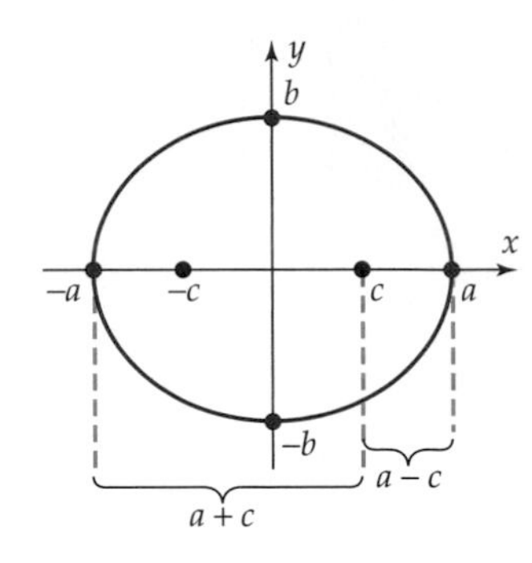

Figure 5–14

The length of the major axis is $2a = 186{,}000{,}000$, so that $a = 93{,}000{,}000$. Similarly, $2b = 185{,}974{,}062$, so $b = 92{,}987{,}031$. As was shown earlier, the equation of the orbit is

$$\frac{x^2}{a^2} + \frac{y^2}{b^2} = 1, \qquad \text{where}$$

$$c = \sqrt{a^2 - b^2} = \sqrt{(93{,}000{,}000)^2 - (92{,}987{,}031)^2} \approx 1{,}553{,}083.$$

Figure 5–14 suggests a fact that can also be proven algebraically: The minimum and maximum distances from a point on the ellipse to the focus $(c, 0)$ occur at the endpoints of the major axis:

$$\text{Minimum distance} = a - c \approx 93{,}000{,}000 - 1{,}553{,}083 = 91{,}446{,}917 \text{ miles}$$

$$\text{Maximum distance} = a + c \approx 93{,}000{,}000 + 1{,}553{,}083 = 94{,}553{,}083 \text{ miles.}$$

■

EXERCISES 5.1

In Exercises 1–6, determine which of the following equations could possibly have the given graph.

$2x^2 + y^2 = 12,$ $\quad (x - 4)^2 + (y - 3)^2 = 4,$

$x^2 + 6y^2 = 18,$ $\quad (x + 3)^2 + (y + 4)^2 = 6,$

$(x + 3)^2 + y^2 = 9,$ $\quad x^2 + (y - 3)^2 = 4,$

$(x - 3)^2 + (y + 4)^2 = 5,$ $\quad (x + 2)^2 + (y - 3)^2 = 2$

1.

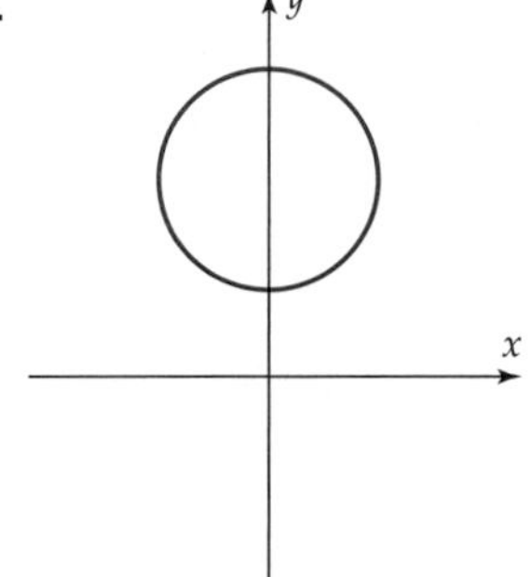

2.

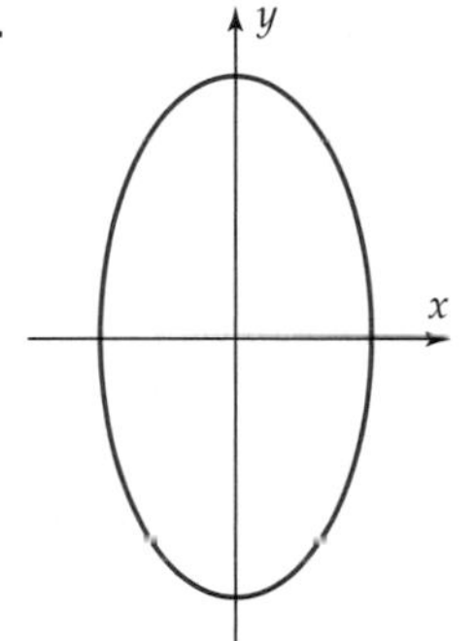

3.

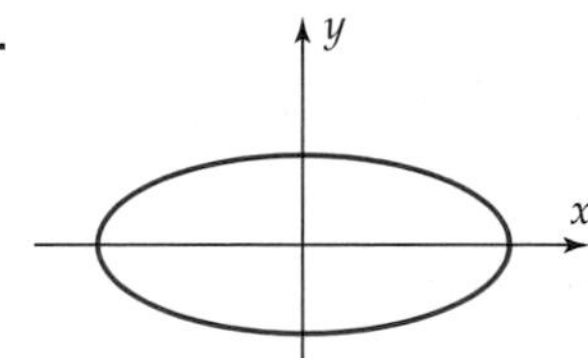

4.

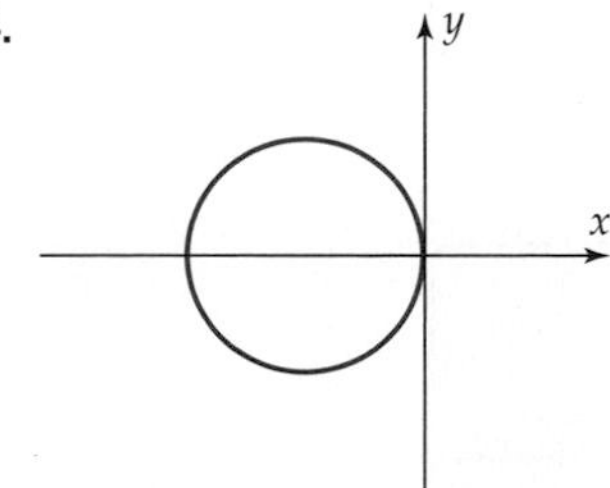

5.

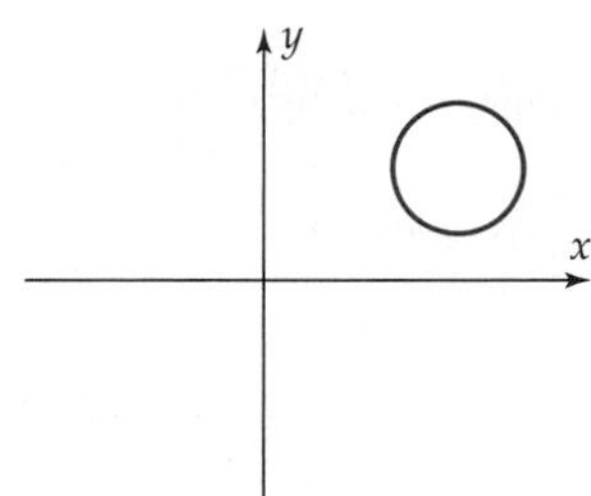

6.

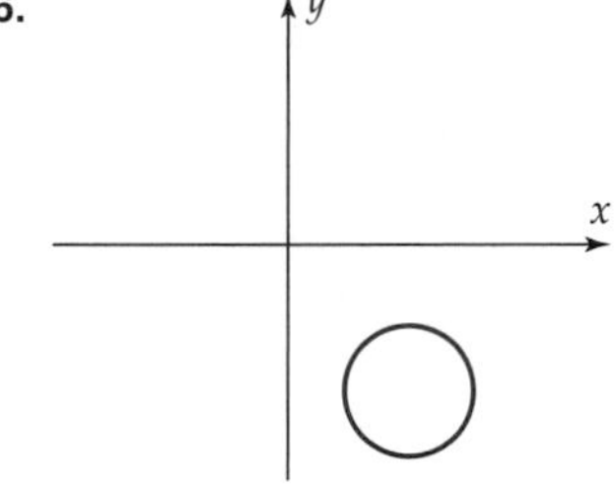

In Exercises 7–12, find the center and radius of the circle whose equation is given, as in Example 2.

7. $x^2 + y^2 + 8x - 6y - 15 = 0$

8. $15x^2 + 15y^2 = 10$

9. $x^2 + y^2 + 6x - 4y - 15 = 0$

10. $x^2 + y^2 + 10x - 75 = 0$

11. $x^2 + y^2 + 25x + 10y = -12$

12. $3x^2 + 3y^2 + 12x + 12 = 18y$

In Exercises 13–20, identify the conic section whose equation is given and find a complete graph of the equation as in Example 4.

13. $\dfrac{x^2}{25} + \dfrac{y^2}{4} = 1$

14. $\dfrac{x^2}{6} + \dfrac{y^2}{16} = 1$

15. $4x^2 + 3y^2 = 12$

16. $9x^2 + 4y^2 = 72$

17. $\dfrac{x^2}{10} - 1 = \dfrac{-y^2}{36}$

18. $\dfrac{y^2}{49} + \dfrac{x^2}{81} = 1$

19. $4x^2 + 4y^2 = 1$

20. $x^2 + 4y^2 = 1$

In Exercises 21–26, find the equation of the ellipse that satisfies the given conditions.

21. Center (0, 0); foci on x-axis; x-intercepts ± 7; y-intercepts ± 2.

22. Center (0, 0); foci on y-axis; x-intercepts ± 1; y-intercepts ± 8.

23. Center (0, 0); foci on x-axis; major axis of length 12; minor axis of length 8.

24. Center (0, 0); foci on y-axis; major axis of length 20; minor axis of length 18.

25. Center (0, 0); endpoints of major and minor axes: $(0, -7)$, $(0, 7)$, $(-3, 0)$, $(3, 0)$.

26. Center (0, 0); vertices (8, 0) and $(-8, 0)$; minor axis of length 8.

Calculus can be used to show that the area of the ellipse with equation $\frac{x^2}{a^2} + \frac{y^2}{b^2} = 1$ *is* πab. *Use this fact to find the area of each ellipse in Exercises 27–32.*

27. $\frac{x^2}{16} + \frac{y^2}{4} = 1$

28. $\frac{x^2}{9} + \frac{y^2}{5} = 1$

29. $3x^2 + 4y^2 = 12$

30. $7x^2 + 5y^2 = 35$

31. $6x^2 + 2y^2 = 14$

32. $5x^2 + y^2 = 5$

In Exercises 33–42, identify the conic section whose equation is given, list its center, and find its graph.

33. $\frac{(x-1)^2}{4} + \frac{(y-5)^2}{9} = 1$

34. $\frac{(x-2)^2}{16} + \frac{(y+3)^2}{12} = 1$

35. $\frac{(x+1)^2}{16} + \frac{(y-4)^2}{8} = 1$

36. $\frac{(x+5)^2}{4} + \frac{(y+2)^2}{12} = 1$

37. $9x^2 + 4y^2 + 54x - 8y + 49 = 0$

38. $4x^2 + 5y^2 - 8x + 30y + 29 = 0$

39. $x^2 + y^2 + 6x - 8y + 5 = 0$

40. $x^2 + y^2 - 4x + 2y - 7 = 0$

41. $4x^2 + y^2 + 24x - 4y + 36 = 0$

42. $9x^2 + y^2 - 36x + 10y + 52 = 0$

In Exercises 43–50, find parametric equations for the curve whose equation is given, and use them to find a complete graph of the curve.

43. $\frac{x^2}{10} - 1 = \frac{-y^2}{36}$

44. $\frac{y^2}{49} + \frac{x^2}{81} = 1$

45. $4x^2 + 4y^2 = 1$

46. $x^2 + 4y^2 = 1$

47. $\frac{(x-1)^2}{4} + \frac{(y-5)^2}{9} = 1$

48. $\frac{(x-2)^2}{16} + \frac{(y+3)^2}{12} = 1$

49. $\frac{(x+1)^2}{16} + \frac{(y-4)^2}{8} = 1$

50. $\frac{(x+5)^2}{4} + \frac{(y+2)^2}{12} = 1$

In Exercises 51–56, find the equation of the ellipse that satisfies the given conditions.

51. Center (2, 3); endpoints of major and minor axes: $(2, -1)$, $(0, 3)$, $(2, 7)$, $(4, 3)$.

52. Center $(-5, 2)$; endpoints of major and minor axes: $(0, 2)$, $(-5, 17)$, $(-10, 2)$, $(-5, -13)$.

53. Center $(7, -4)$; foci on the line $x = 7$; major axis of length 12; minor axis of length 5.

54. Center $(-3, -9)$; foci on the line $y = -9$; major axis of length 15; minor axis of length 7.

55. Center $(3, -2)$; passing through $(3, -6)$ and $(9, -2)$.

56. Center (2, 5); passing through (2, 4) and $(-3, 5)$.

In Exercises 57 and 58, find the equations of two distinct ellipses satisfying the given conditions.

57. Center at $(-5, 3)$; major axis of length 14; minor axis of length 8.

58. Center at $(2, -6)$; major axis of length 15; minor axis of length 6.

In Exercises 59–64, determine which of the following equations could possibly have the given graph. Assume that all viewing windows are square.

$\frac{(x+3)^2}{4} + \frac{(y+3)^2}{8} = 1$, $\quad \frac{(x-3)^2}{9} + \frac{(y+4)^2}{4} = 1$,

$2x^2 + 2y^2 - 8 = 0$, $\quad 4x^2 + 2y^2 - 8 = 0$,

$2x^2 + y^2 - 8x - 6y + 9 = 0$,

$x^2 + 3y^2 + 6x - 12y + 17 = 0$.

59.

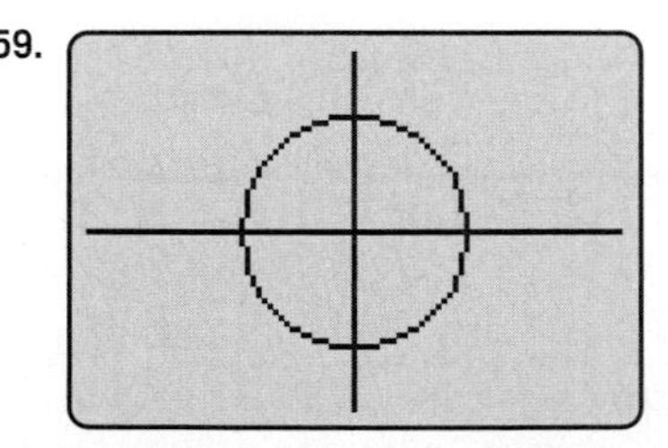

60.

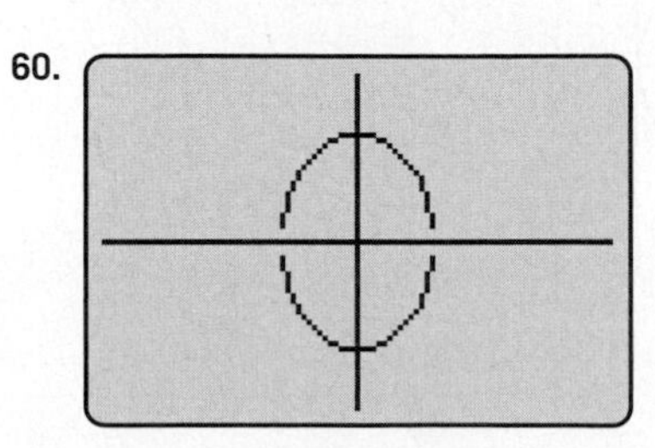

61.

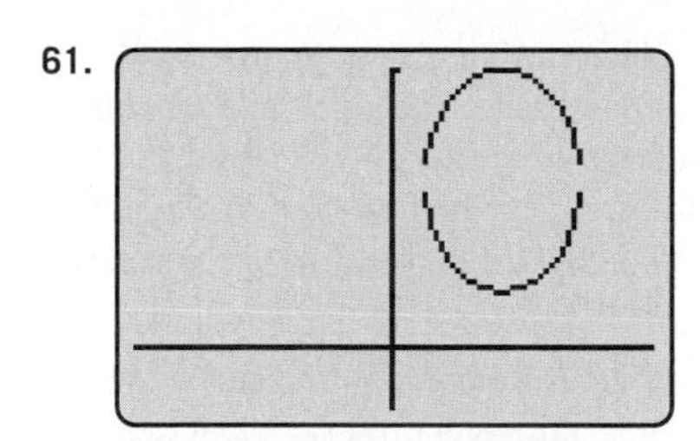

62.

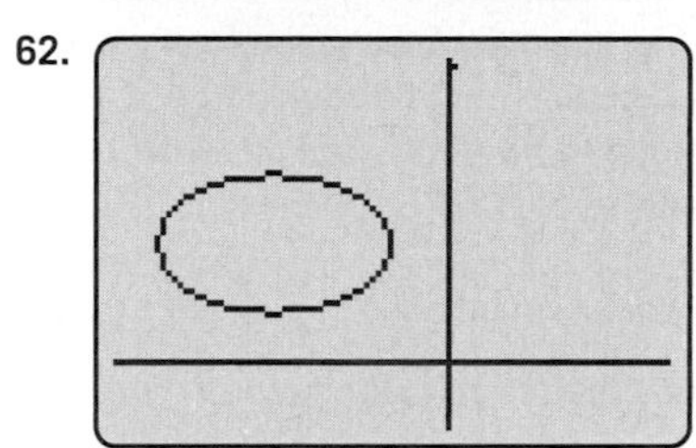

63.

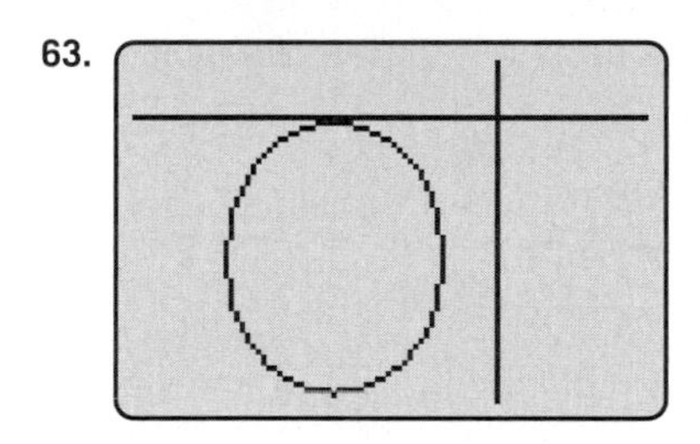

64.

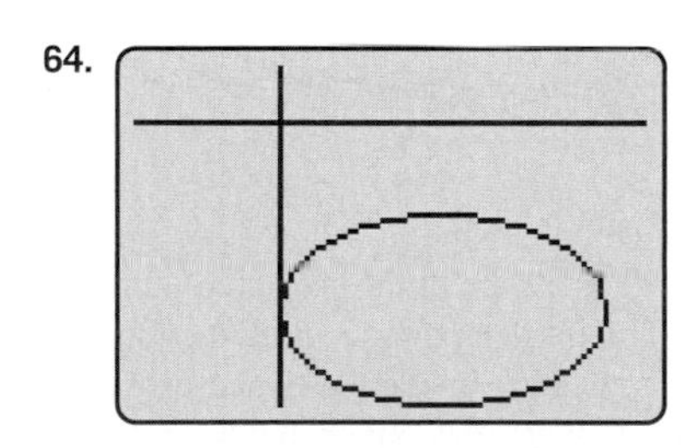

65. A circle is inscribed in the ellipse with equation

$$x^2 + 4y^2 = 64.$$

A diameter of the circle lies on the minor axis of the ellipse. Find the equation of the circle.

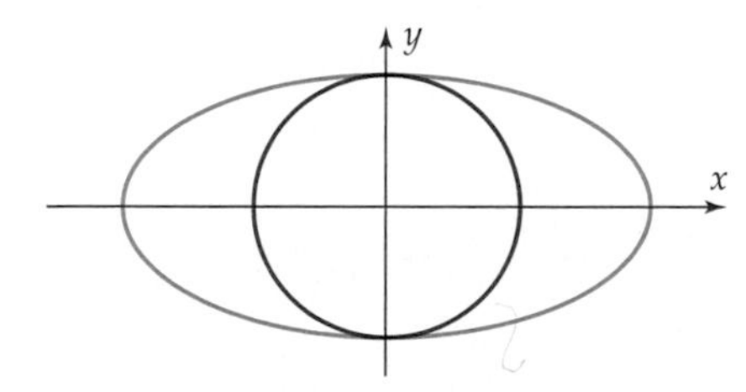

66. If (c, d) is a point on the ellipse in Exercise 65, prove that $(c/2, d)$ is a point on the circle.

67. The orbit of the moon around the earth is an ellipse with the earth as one focus. If the length of the major axis of the orbit is 477,736 miles and the length of the minor axis is 477,078 miles, find the minimum and maximum distances from the earth to the moon.

68. Halley's Comet has an elliptical orbit with the sun as one focus and a major axis that is 1,636,484,848 miles long. The closest the comet comes to the sun is 54,004,000 miles. What is the maximum distance from the comet to the sun?

69. A cross section of the ceiling of a "whispering room" at a museum is half of an ellipse. Two people stand so that their heads are approximately at the foci of the ellipse. If one whispers upward, the sound waves are reflected off the elliptical ceiling to the other person, as indicated in the figure. (For the reason why, see Figure 5–12 and the accompanying text.) How far from the center are the two people standing? [*Hint:* Use the rectangular coordinate system suggested in the figure to find the equation of the ellipse; then find the foci.]

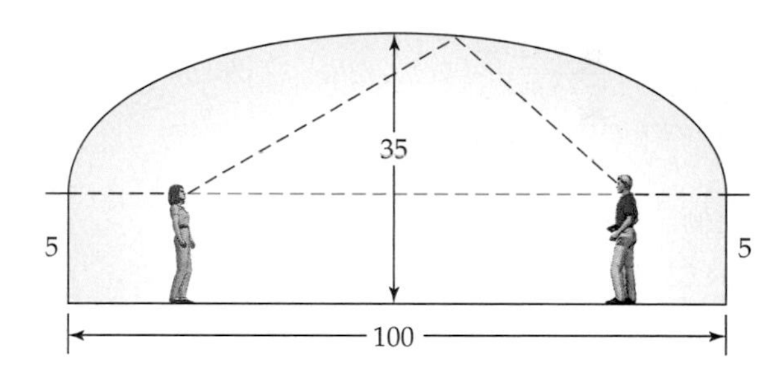

70. Suppose that the whispering room in Exercise 69 is 100 feet long and that the two people at the foci are 80 feet apart. How high is the center of the roof?

71. The bottom of a bridge is shaped like half an ellipse and is 20 feet above the river at the center, as shown in the figure. Find the height of the bridge bottom over a point on the river 25 feet from the center of the river. [*Hint:* Use a coordinate system, with the x-axis on the surface of the water and the y-axis running through the center of the bridge, to find an equation for the ellipse.]

72. The stained glass window in the figure is shaped like the top half of an ellipse. The window is 10 feet wide, and the two figures in the window are located 3 feet from the center. If the figures are 1.6 feet high, find the height of the window at the center. [*Hint:* Use a coordinate system, with the bottom of the window as the x-axis and the vertical line through the center of the window as the y-axis.]

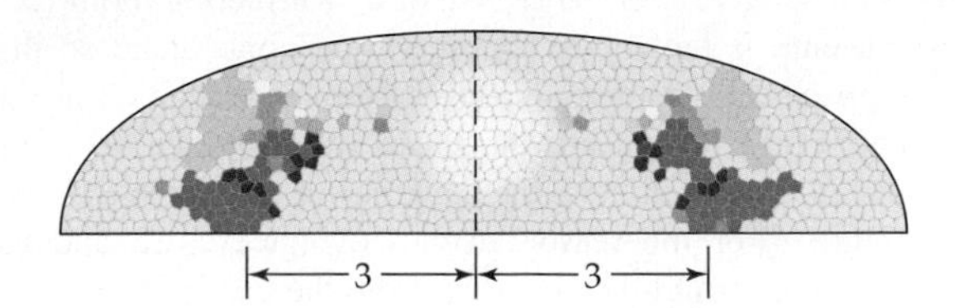

If $a > b > 0$, then the ***eccentricity*** *of the ellipse*

$$\frac{(x-h)^2}{a^2} + \frac{(y-k)^2}{b^2} = 1 \quad \text{or} \quad \frac{(x-h)^2}{b^2} + \frac{(y-k)^2}{a^2} = 1$$

is the number $\frac{\sqrt{a^2 - b^2}}{a}$. *In Exercises 73–76, find the eccentricity of the ellipse whose equation is given.*

73. $\frac{x^2}{100} + \frac{y^2}{99} = 1$

74. $\frac{x^2}{18} + \frac{y^2}{25} = 1$

75. $\frac{(x-3)^2}{10} + \frac{(y-9)^2}{40} = 1$

76. $\frac{(x+5)^2}{12} + \frac{(y-4)^2}{8} = 1$

77. On the basis of your answers to Exercises 73–76, how is the eccentricity of an ellipse related to its graph? [*Hint:* What is the shape of the graph when the eccentricity is close to 0? When it is close to 1?]

78. Assuming that the viewing windows are square, which of these ellipses has the larger eccentricity?

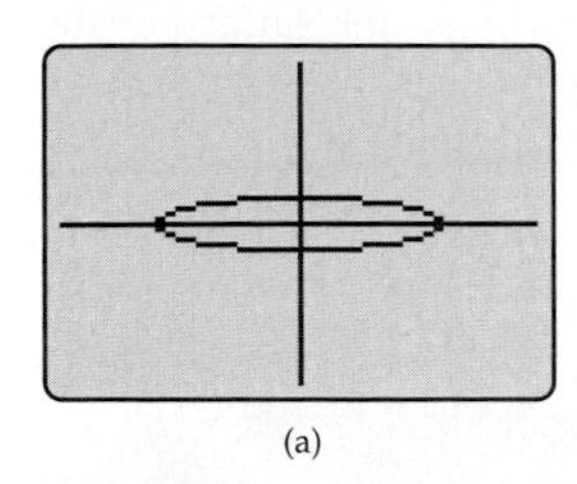

(a)

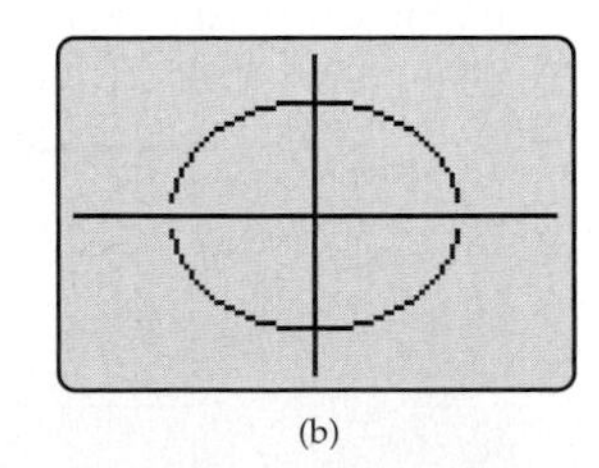

(b)

79. A satellite is to be placed in an elliptical orbit, with the center of the earth as one focus. The satellite's maximum distance from the surface of the earth is to be 22,380 km, and its minimum distance is to be 6540 km. Assume that the radius of the earth is 6400 km, and find the eccentricity of the satellite's orbit.

80. The first step in landing Apollo 11 on the moon was to place the spacecraft in an elliptical orbit such that the minimum distance from the *surface* of the moon to the spacecraft was 110 km and the maximum distance was 314 km. If the radius of the moon is 1740 km, find the eccentricity of the Apollo 11 orbit.

81. Consider the ellipse whose equation is $\frac{x^2}{a^2} + \frac{y^2}{b^2} = 1$. Show that if $a = b$, then the graph is actually a circle.

82. Complete the derivation of the equation of the ellipse on pages 335–336 as follows.

(a) By squaring both sides, show that the equation

$$\sqrt{(x+c)^2 + y^2} = 2a - \sqrt{(x-c)^2 + y^2}$$

may be simplified as

$$a\sqrt{(x-c)^2 + y^2} = a^2 - cx.$$

(b) Show that the last equation in part (a) may be further simplified as

$$(a^2 - c^2)x^2 + a^2y^2 = a^2(a^2 - c^2).$$

Thinker

83. The punch bowl and a table holding the punch cups are placed 50 feet apart at a garden party. A portable fence is then set up so that any guest inside the fence can walk straight to the table, then to the punch bowl, and then return to his or her starting point without traveling more than 150 feet. Describe the longest possible such fence that encloses the largest possible area.

5.2 Hyperbolas

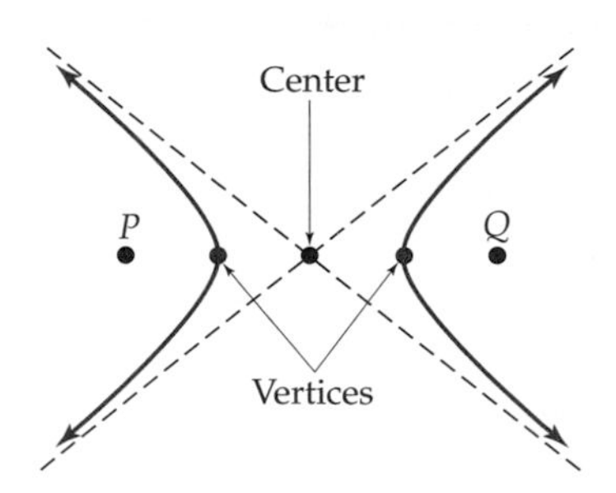

Figure 5–15

Definition. Let P and Q be points in the plane, and let r be a positive number. The set of all points X such that

$$|(\text{Distance from } P \text{ to } X) - (\text{Distance from } Q \text{ to } X)| = r$$

is the **hyperbola** with **foci** P and Q; r will be called the **distance difference.** Every hyperbola has the general shape shown by the red curve in Figure 5–15. The dashed straight lines are the **asymptotes** of the hyperbola; it gets closer and closer to the asymptotes, but never touches them. The asymptotes intersect at the midpoint of the line segment from P to Q; this point is called the **center** of the hyperbola. The **vertices** of the hyperbola are the points where it intersects the line segment from P to Q. The line through P and Q is called the **focal axis.**

Equation. Another complicated exercise in the use of the distance formula, which will be omitted here, leads to the following algebraic description.

Standard Equations of Hyperbolas Centered at the Origin

Let a and b be positive real numbers. Then the graph of each of the following equations is a hyperbola centered at the origin.

$$\frac{x^2}{a^2} - \frac{y^2}{b^2} = 1 \begin{cases} x\text{-intercepts: } \pm a \qquad y\text{-intercepts: none} \\ \text{focal axis on the } x\text{-axis, with vertices } (a, 0) \text{ and } (-a, 0) \\ \text{foci: } (c, 0) \text{ and } (-c, 0), \text{ where } c = \sqrt{a^2 + b^2}. \\ \text{asymptotes: } y = \dfrac{b}{a}x \text{ and } y = -\dfrac{b}{a}x \end{cases}$$

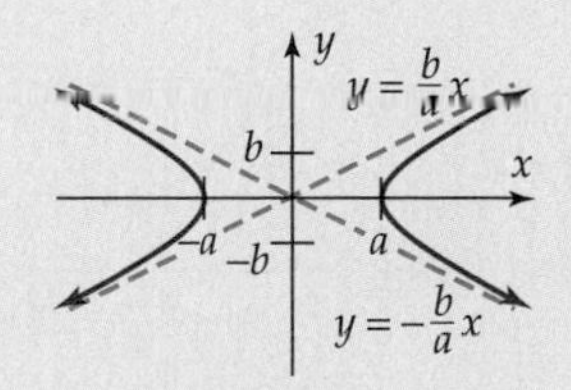

$$\frac{y^2}{a^2} - \frac{x^2}{b^2} = 1 \begin{cases} x\text{-intercepts: none} \qquad y\text{-intercepts: } \pm a \\ \text{focal axis on the } y\text{-axis, with vertices } (0, a) \text{ and } (0, -a) \\ \text{foci: } (0, c) \text{ and } (0, -c), \text{ where } c = \sqrt{a^2 + b^2}. \\ \text{asymptotes: } y = \dfrac{a}{b}x \text{ and } y = -\dfrac{a}{b}x \end{cases}$$

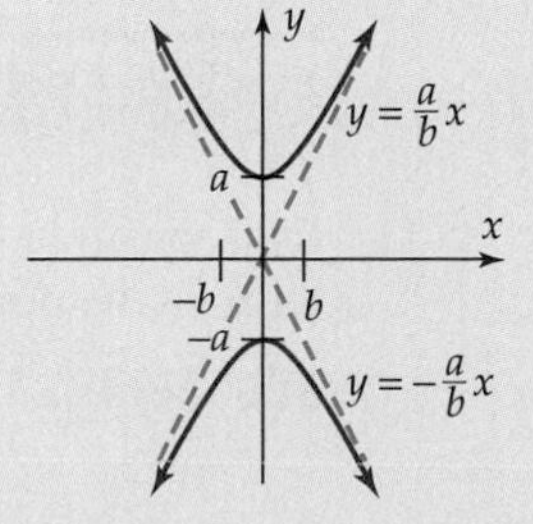

Once again, don't worry about all the letters in the box. When the equation is in standard form with the x term positive and y term negative, the hyperbola intersects the x-axis and opens from side to side. When the x term is negative and the y term is positive, the hyperbola intersects the y-axis and opens up and down.

EXAMPLE 1

Identify and sketch the graph of the equation $9x^2 - 4y^2 = 36$.

SOLUTION We first put the equation in standard form.

$$9x^2 - 4y^2 = 36$$

Divide both sides by 36:
$$\frac{9x^2}{36} - \frac{4y^2}{36} = \frac{36}{36}$$

Simplify:
$$\frac{x^2}{4} - \frac{y^2}{9} = 1$$

$$\frac{x^2}{2^2} - \frac{y^2}{3^2} = 1$$

Applying the fact in the box with $a = 2$ and $b = 3$ shows that the graph is a hyperbola with vertices (2, 0) and (−2, 0) and asymptotes $y = \frac{3}{2}x$ and $y = -\frac{3}{2}x$. We first plot the vertices and sketch the rectangle determined by the vertical lines $x = \pm 2$ and the horizontal lines $y = \pm 3$. The asymptotes go through the origin and the corners of this rectangle, as shown on the left in Figure 5–16. It is then easy to sketch the hyperbola. ■

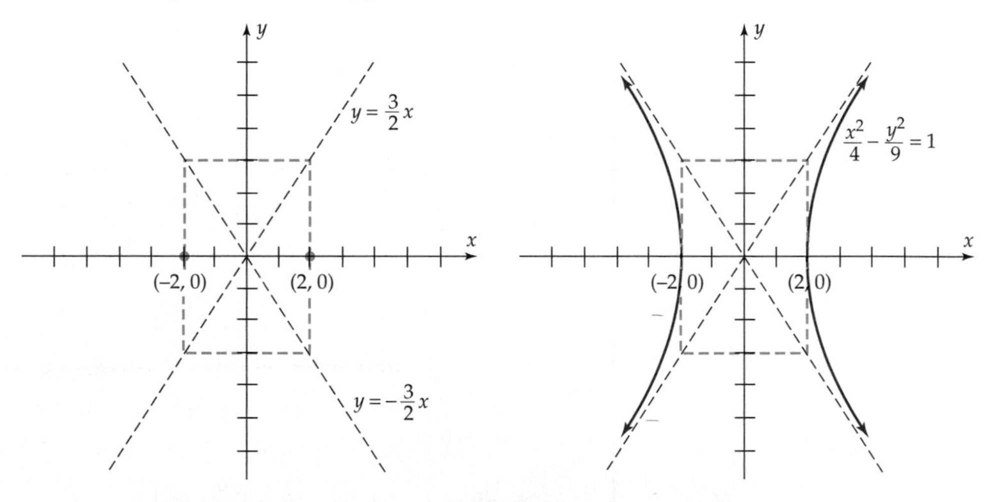

Figure 5–16

EXAMPLE 2

Find the equation of the hyperbola with vertices (0, 1) and (0, −1) that passes through the point $(3, \sqrt{2})$. Then sketch its graph.

SOLUTION The vertices are on the y-axis, and the equation is of the form

$$\frac{y^2}{a^2} - \frac{x^2}{b^2} = 1$$

with $a = 1$. Since $(3, \sqrt{2})$ is on the graph, we have

$$\frac{(\sqrt{2})^2}{1^2} - \frac{3^2}{b^2} = 1$$

Simplify:
$$2 - \frac{9}{b^2} = 1$$

Subtract 2 from both sides:
$$-\frac{9}{b^2} = -1$$

Multiply both sides by $-b^2$:
$$9 = b^2.$$

Therefore, $b = 3$, and the equation is

$$\frac{y^2}{1^2} - \frac{x^2}{3^2} = 1 \qquad \text{or, equivalently,} \qquad y^2 - \frac{x^2}{9} = 1.$$

The asymptotes of the hyperbola are the lines $y = \pm\frac{1}{3}x$. To sketch the graph, we first solve the equation for y.

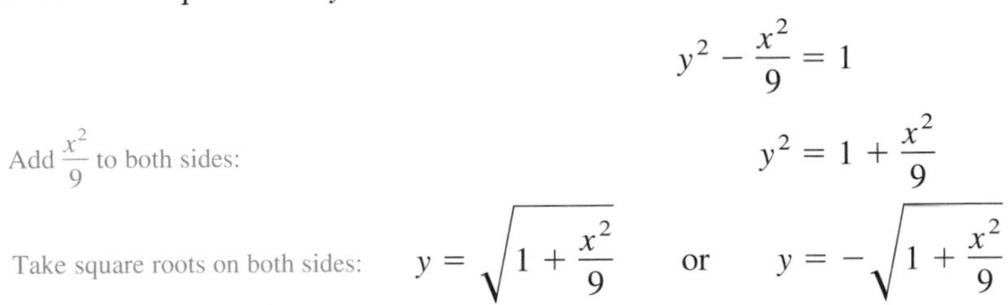

$$y^2 - \frac{x^2}{9} = 1$$

Add $\frac{x^2}{9}$ to both sides:
$$y^2 = 1 + \frac{x^2}{9}$$

Take square roots on both sides:
$$y = \sqrt{1 + \frac{x^2}{9}} \qquad \text{or} \qquad y = -\sqrt{1 + \frac{x^2}{9}}$$

Figure 5–17

Graphing these last two equations on the same screen, we obtain Figure 5–17. ■

VERTICAL AND HORIZONTAL SHIFTS

Let h and k be constants. As we saw on page 339, replacing x by $x - h$ and y by $y - k$ in an equation shifts the graph of the equation $|h|$ units horizontally and $|k|$ units vertically. If the graph of the equation is a hyperbola with center at the origin, then the shifted graph has its center at (h, k). Using this fact and translating the information in the box on page 349, we obtain the following.

Standard Equations of Hyperbolas with Center at (h, k)

If a and b are positive real numbers, then the graph of each of the following equations is a hyperbola with center (h, k).

$$\frac{(x-h)^2}{a^2} - \frac{(y-k)^2}{b^2} = 1$$

- focal axis on the horizontal line $y = k$
- foci: $(h - c, k)$ and $(h + c, k)$, where $c = \sqrt{a^2 + b^2}$
- vertices: $(h - a, k)$ and $(h + a, k)$
- asymptotes: $y = \pm\frac{b}{a}(x - h) + k$

$$\frac{(y-k)^2}{a^2} - \frac{(x-h)^2}{b^2} = 1$$

- focal axis on the vertical line $x = h$
- foci: $(h, k - c)$ and $(h, k + c)$, where $c = \sqrt{a^2 + b^2}$
- vertices: $(h, k - a)$ and $(h, k + a)$
- asymptotes: $y = \pm\frac{a}{b}(x - h) + k$

EXAMPLE 3

Identify and sketch the graph of

$$\frac{(x-3)^2}{4} - \frac{(y+2)^2}{9} = 1.$$

SOLUTION If we rewrite the equation as

$$\frac{(x-3)^2}{2^2} - \frac{(y-(-2))^2}{3^2} = 1,$$

then it has the first form in the preceding box with $a = 2$, $b = 3$, $h = 3$, and $k = -2$. Its graph is a hyperbola with center $(3, -2)$. There are several ways to obtain the graph.

Method 1. The equation of this hyperbola can be obtained from

$$(*) \qquad \frac{x^2}{4} - \frac{y^2}{9} = 1$$

by replacing x by $x - 3$ and y by $y + 2 = y - (-2)$. So its graph is just the graph of $(*)$ (see Figure 5–16) shifted 3 units to the right and 2 units downward, as shown in Figure 5–18.

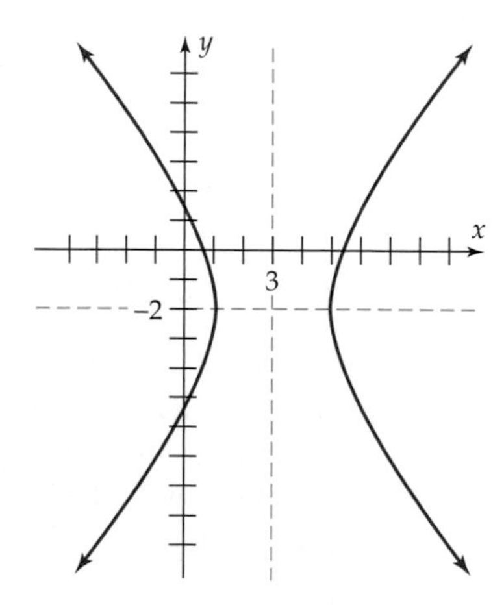

Figure 5–18

Method 2. Solve the original equation for y.

$$\frac{(x-3)^2}{4} - \frac{(y+2)^2}{9} = 1$$

Multiply both sides by 36: $9(x-3)^2 - 4(y+2)^2 = 36$

Rearrange terms: $4(y+2)^2 = 9(x-3)^2 - 36$

Divide both sides by 4: $(y+2)^2 = \dfrac{9(x-3)^2 - 36}{4}$

Take square roots of both sides: $y + 2 = \pm\sqrt{\dfrac{9(x-3)^2 - 36}{4}}$

$$y = -2 + \sqrt{\frac{9(x-3)^2 - 36}{4}} \quad \text{or} \quad y = -2 - \sqrt{\frac{9(x-3)^2 - 36}{4}}$$

GRAPHING EXPLORATION

Graph these last two equations on the same screen in a square window with $-4 \le x \le 11$. How does your graph compare with Figure 5–18? What is the explanation for the gaps? ■

EXAMPLE 4

Sketch the graph of $6y^2 - 8x^2 - 24y - 48x - 96 = 0$.

SOLUTION We first solve the equation for y. Begin by rewriting it as

$$6y^2 - 24y + (-8x^2 - 48x - 96) = 0.$$

This is a quadratic equation of the form $ay^2 + by + c = 0$, with

$$a = 6, \qquad b = -24, \qquad c = -8x^2 - 48x - 96.$$

We use the quadratic formula to solve it.

$$y = \frac{-b \pm \sqrt{b^2 - 4ac}}{2a}$$

$$y = \frac{-(-24) \pm \sqrt{(-24)^2 - 4 \cdot 6(-8x^2 - 48x - 96)}}{2 \cdot 6}$$

$$y = \frac{24 \pm \sqrt{576 - 24(-8x^2 - 48x - 96)}}{12}$$

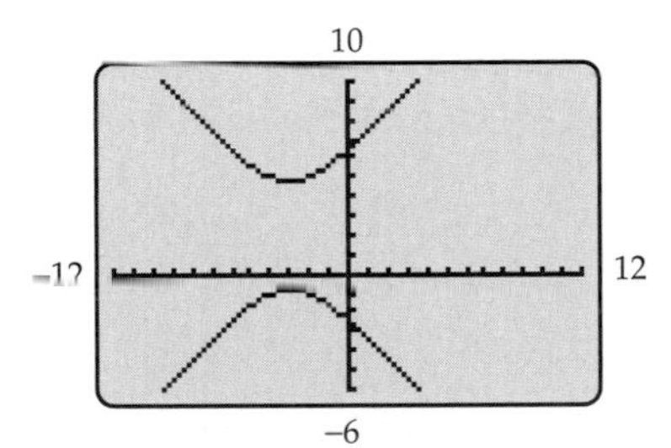

Figure 5–19

Now we graph both

$$y = \frac{24 + \sqrt{576 - 24(-8x^2 - 48x - 96)}}{12} \qquad \text{and}$$

$$y = \frac{24 - \sqrt{576 - 24(-8x^2 - 48x - 96)}}{12}$$

on the screen to obtain the hyperbola in Figure 5–19. ■

EXAMPLE 5

Find the center of the hyperbola in Example 4.

SOLUTION Begin by rearranging the equation.

	$6y^2 - 8x^2 - 24y - 48x - 96 = 0$
Add 96 to both sides:	$6y^2 - 24y - 8x^2 - 48x = 96$
Group x- and y-terms:	$(6y^2 - 24y) - (8x^2 + 48x) = 96$
Factor out coefficients of y^2 and x^2:	$6(y^2 - 4y) - 8(x^2 + 6x) = 96.$

Complete the square in the expression $y^2 - 4y$ by adding 4 (the square of half the coefficient of y), and complete the square in $x^2 + 6x$ by adding 9 (the square of half the coefficient of x).

$$6(y^2 - 4y + 4) - 8(x^2 + 6x + 9) = 96 + ? + ?$$

CAUTION

Completing the square only works when the coefficient of y^2 is 1. In an expression such as $6y^2 - 24y$, you must first factor out the 6,

$$6(y^2 - 4y),$$

and then complete the square on the expression in parentheses.

On the left side, we have actually added $6 \cdot 4 = 24$ and $-8 \cdot 9 = -72$, so we must add these numbers on the right to keep the equation unchanged.

$$6(y^2 - 4y + 4) - 8(x^2 + 6x + 9) = 96 + 24 - 72$$

Factor and simplify:
$$6(y - 2)^2 - 8(x + 3)^2 = 48$$

Divide both sides by 48:
$$\frac{(y-2)^2}{8} - \frac{(x+3)^2}{6} = 1$$

$$\frac{(y-2)^2}{8} - \frac{(x-(-3))^2}{6} = 1$$

In this form, we can see that the graph is a hyperbola with center at $(-3, 2)$. ■

PARAMETRIC EQUATIONS

Consider the hyperbola with equation

$$\frac{(x-h)^2}{a^2} - \frac{(y-k)^2}{b^2} = 1.$$

Its graph can be obtained from these parametric equations:

$$x = a \sec t + h \quad \text{and} \quad y = b \tan t + k \quad (0 \le t \le 2\pi)$$

because, by the Pythagorean identity for secant and tangent,

$$\begin{aligned}\frac{(x-h)^2}{a^2} - \frac{(y-k)^2}{b^2} &= \frac{[(a \sec t + h) - h]^2}{a^2} - \frac{[(b \tan t + k) - k]^2}{b^2} \\ &= \frac{[a \sec t]^2}{a} - \frac{[b \tan t]^2}{b^2} \\ &= \frac{a^2 \sec^2 t}{a^2} - \frac{b^2 \tan^2 t}{b^2} = \sec^2 t - \tan^2 t = 1.\end{aligned}$$

This proves the first of the following statements; the second is proved similarly.

Parametric Equations for Hyperbolas

Hyperbola	Parametric Equations
$\frac{(x-h)^2}{a^2} - \frac{(y-k)^2}{b^2} = 1$	$x = a \sec t + h = \frac{a}{\cos t} + h$ and $y = b \tan t + k$ $(0 \le t \le 2\pi)$
$\frac{(y-k)^2}{a^2} - \frac{(x-h)^2}{b^2} = 1$	$x = b \tan t + h$ and $y = a \sec t + k = \frac{a}{\cos t} + k$ $(0 \le t \le 2\pi)$

EXAMPLE 6

Find parametric equations for the hyperbola with equation

$$\frac{y^2}{9} - \frac{x^2}{16} = 1.$$

SOLUTION This equation has the form of the second equation in the preceding box, with $h = 0$, $k = 0$, $a = 3$, and $b = 4$. So the parametric equations are

$$x = 4\tan t \qquad \text{and} \qquad y = \frac{3}{\cos t} \quad (0 \leq t \leq 2\pi).$$

EXPLORATION

Use these parametric equations to graph the hyperbola. How is it traced out? Now change the t range so that $-\pi/2 \leq t \leq 3\pi/2$. Now how is the graph traced out? ■

APPLICATIONS

The reflective properties of hyperbolas are used in the design of camera and telescope lenses. If a light ray passes through one focus of a hyperbola and reflects off the hyperbola at a point P, then the reflected ray moves along the straight line determined by P and the other focus, as shown in Figure 5–20.

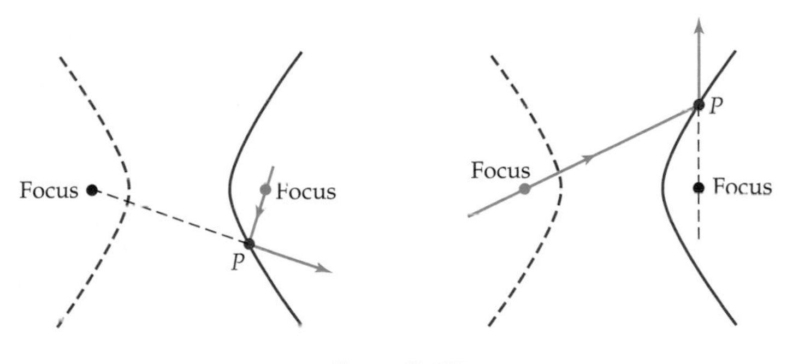

Figure 5–20

Hyperbolas are also the basis of the long-range navigation system (loran), which enables a ship to determine its exact location by radio, as illustrated in the next example.

EXAMPLE 7

Three loran transmitters Q, P, and R are located 200 miles apart along a straight shoreline and simultaneously transmit signals at regular intervals. These signals travel at a speed of 980 feet per microsecond. A ship S receives a signal from P and 305 microseconds later a signal from R. It also receives a signal from Q 528 microseconds after the one from P. Determine the ship's location.

SOLUTION Take the line through the loran stations as the x-axis, with the origin located midway between Q and P, so that the situation looks like Figure 5–21.

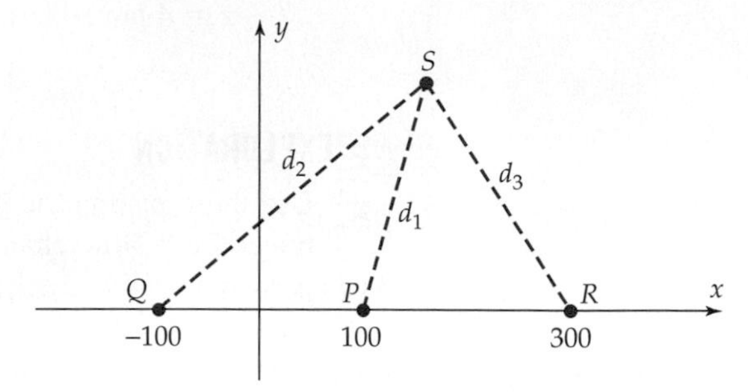

Figure 5–21

If the signal takes t microseconds to go from P to S and $t + 528$ microseconds to go from Q to S, then

$$d_1 = 980t \qquad \text{and} \qquad d_2 = 980(t + 528),$$

so

$$|d_1 - d_2| = |980t - 980(t + 528)| = 980 \cdot 528 = 517{,}440 \text{ feet}.$$

Since 1 mile is 5280 feet, this means that

$$|d_1 - d_2| = 517{,}440/5{,}280 \text{ miles} = 98 \text{ miles}.$$

In other words,

$$|(\text{Distance from } P \text{ to } S) - (\text{Distance from } Q \text{ to } S)| = |d_1 - d_2| = 98.$$

This is precisely the situation described in the definition of "hyperbola" on page 349: S is on the hyperbola with foci $P = (100, 0)$, $Q = (-100, 0)$, and distance difference $r = 98$. This hyperbola has an equation of the form

$$\frac{x^2}{a^2} - \frac{y^2}{b^2} = 1,$$

where $(\pm a, 0)$ are the vertices, $(\pm c, 0) = (\pm 100, 0)$ are the foci, and $c^2 = a^2 + b^2$. Figure 5–22 and the fact that the vertex $(a, 0)$ is on the hyperbola show that

$$\begin{aligned}
|[\text{Distance from } P \text{ to } (a, 0)] - [\text{Distance from } Q \text{ to } (a, 0)]| &= r = 98 \\
|(100 - a) - (100 + a)| &= 98 \\
|-2a| &= 98 \\
|a| &= 49.
\end{aligned}$$

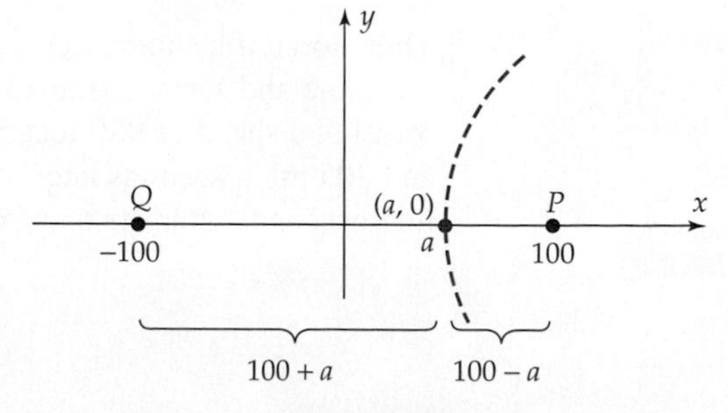

Figure 5–22

Consequently, $a^2 = 49^2 = 2401$, and hence, $b^2 = c^2 - a^2 = 100^2 - 49^2 = 7599$. Thus, the ship lies on the hyperbola

$$(*) \qquad \frac{x^2}{2401} - \frac{y^2}{7599} = 1.$$

A similar argument using P and R as foci shows that the ship also lies on the hyperbola with foci $P = (100, 0)$ and $R = (300, 0)$ and center $(200, 0)$, whose distance difference r is

$$|d_1 - d_3| = 980 \cdot 305 = 298{,}900 \text{ feet} \approx 56.61 \text{ miles}.$$

As before, you can verify that $a = 56.61/2 = 28.305$, and hence, $a^2 = 28.305^2 = 801.17$. This hyperbola has center $(200, 0)$, and its foci are $(200 - c, k) = (100, 0)$ and $(200 + c, k) = (300, 0)$, which implies that $c = 100$. Hence, $b^2 = c^2 - a^2 = 100^2 - 801.17 = 9198.83$, and the ship also lies on the hyperbola

$$(**) \qquad \frac{(x - 200)^2}{801.17} - \frac{y^2}{9198.83} = 1.$$

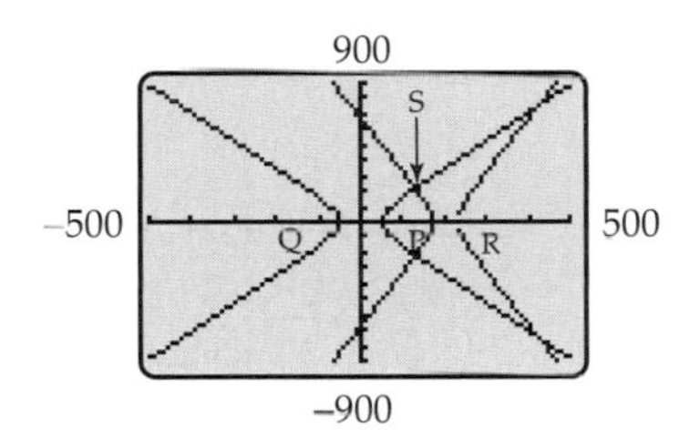

Figure 5–23

Since the ship lies on both hyperbolas, its coordinates are solutions of both the equations (*) and (**). They can be found algebraically by solving each of the equations for y^2, setting the results equal, and solving for x. They can be found geometrically by graphing both hyperbolas and finding the intersection point. As shown in Figure 5–23, there are actually four points of intersection. However, the two below the x-axis represent points on land in our situation. Furthermore, since the signal from P was received first, the ship is closest to P. So it is located at the point S in Figure 5–23. A graphical intersection finder shows that this point is approximately $(130.48, 215.14)$, where the coordinates are in miles from the origin. ■

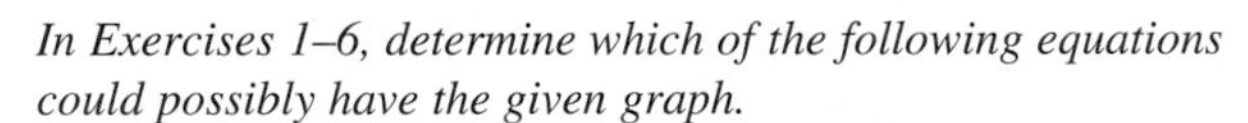

EXERCISES 5.2

In Exercises 1–6, determine which of the following equations could possibly have the given graph.

$3x^2 + 3y^2 = 12$, $\qquad 6y^2 - x^2 = 6$,

$x^2 + 4y^2 = 1$, $\qquad 4x^2 + 4(y + 2)^2 = 12$,

$4(x + 4)^2 + 4y^2 = 12$, $\qquad 6x^2 + 2y^2 = 18$,

$2x^2 - y^2 = 8$, $\qquad 3x^2 - y = 6$

1.

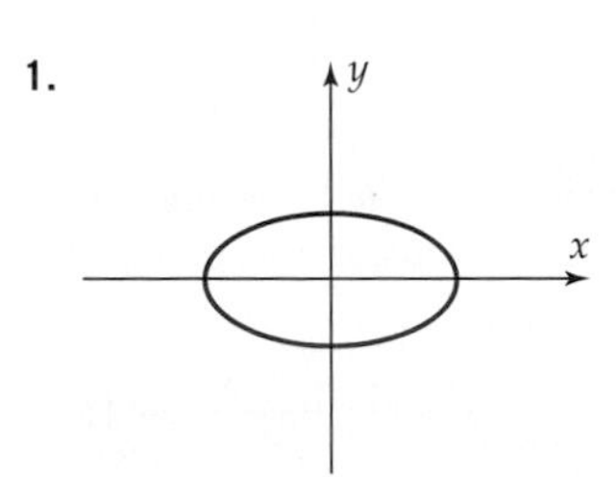

2.

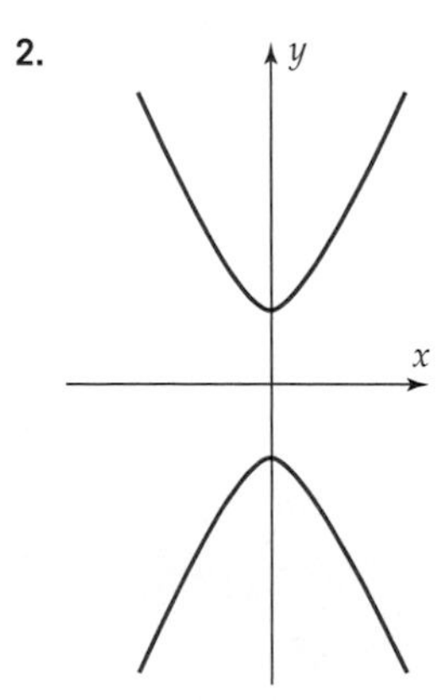

3.

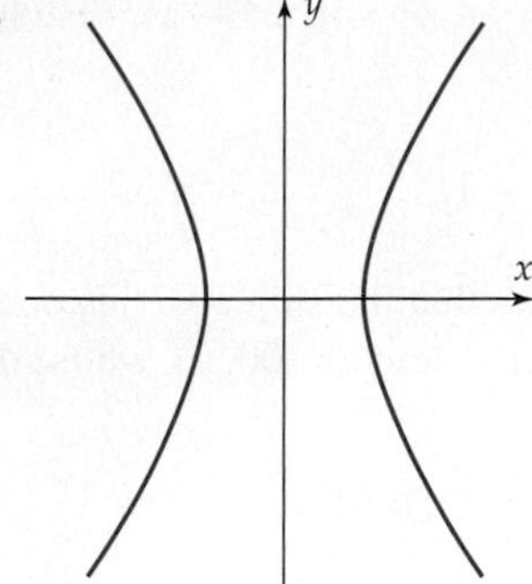

4.

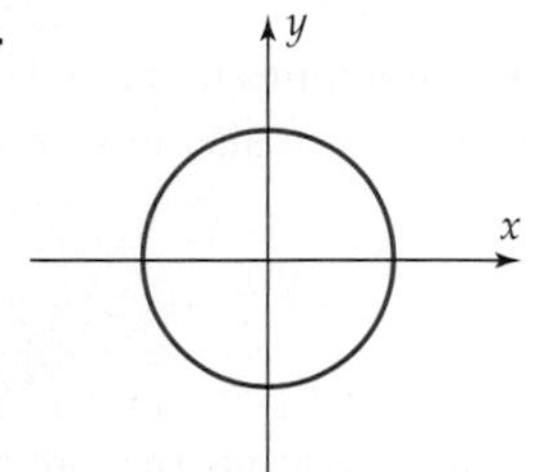

5.

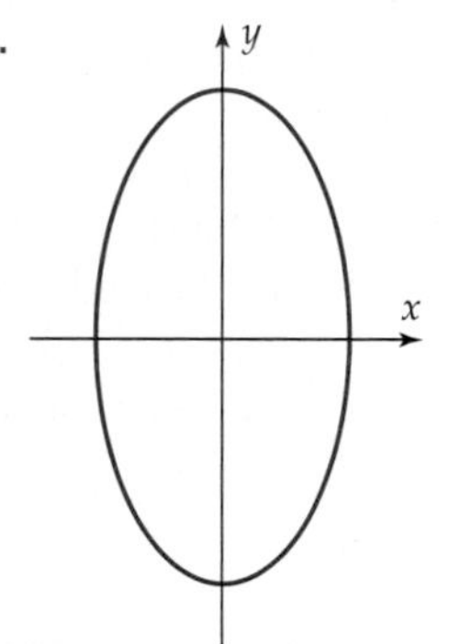

6.

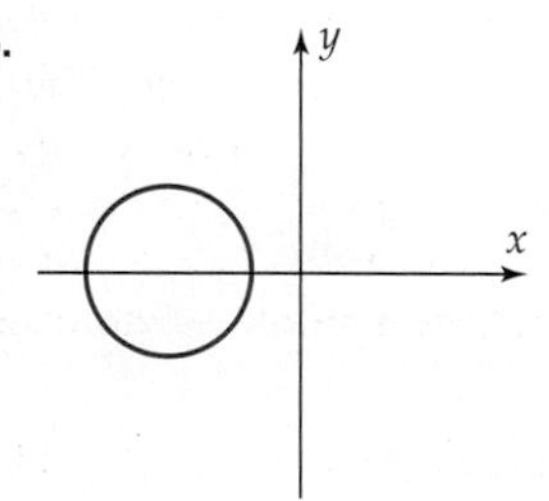

In Exercises 7–14, identify the conic section whose equation is given and find a complete graph of the equation, as in Example 1.

7. $\frac{x^2}{6} - \frac{y^2}{16} = 1$

8. $\frac{x^2}{4} - y^2 = 1$

9. $4x^2 - y^2 = 16$

10. $3y^2 - 5x^2 = 15$

11. $\frac{x^2}{10} - \frac{y^2}{36} = 1$

12. $\frac{y^2}{9} - \frac{x^2}{16} = 1$

13. $x^2 - 4y^2 = 1$

14. $2x^2 - y^2 = 4$

In Exercises 15–18, find the equation of the hyperbola that satisfies the given conditions.

15. Center (0, 0); x-intercepts ± 3; asymptote $y = 2x$.

16. Center (0, 0); y-intercepts ± 12; asymptote $y = 3x/2$.

17. Center (0, 0); vertex (2, 0); passing through $(4, \sqrt{3})$.

18. Center (0, 0); vertex $(0, \sqrt{12})$; passing through $(2\sqrt{3}, 6)$.

In Exercises 19–32, identify the conic section whose equation is given, list its center, and find its graph.

19. $\frac{(y+3)^2}{25} - \frac{(x+1)^2}{16} = 1$

20. $\frac{(y+1)^2}{9} - \frac{(x-1)^2}{25} = 1$

21. $\frac{(x+3)^2}{1} - \frac{(y-2)^2}{4} = 1$

22. $\frac{(y+5)^2}{9} - \frac{(x-2)^2}{1} = 1$

23. $(y+4)^2 - 8(x-1)^2 = 8$

24. $(x-3)^2 + 12(y-2)^2 = 24$

25. $4y^2 - x^2 + 6x - 24y + 11 = 0$

26. $x^2 - 16y^2 = 0$

27. $2x^2 + 2y^2 - 12x - 16y + 26 = 0$

28. $3x^2 + 3y^2 + 12x + 6y = 0$

29. $2x^2 + 3y^2 - 12x - 24y + 54 = 0$

30. $x^2 + 2y^2 + 4x - 4y = 8$

31. $x^2 - 3y^2 + 4x + 12y = 20$

32. $2x^2 + 16x = y^2 - 6y - 55$

In Exercises 33–36, find the equation of the hyperbola that satisfies the given conditions.

33. Center $(-2, 3)$; vertex $(-2, 1)$; passing through $(-2 + 3\sqrt{10}, 11)$.

34. Center $(-5, 1)$; vertex $(-3, 1)$; passing through $(-1, 1 - 4\sqrt{3})$.

35. Center (4, 2); vertex (7, 2); asymptote $3y = 4x - 10$.

36. Center $(-3, -5)$; vertex $(-3, 0)$; asymptote $6y = 5x - 15$.

In Exercises 37–42, determine which of the following equations could possibly have the given graph. Assume that all viewing windows are square.

$\frac{(y-2)^2}{4} - \frac{(x-3)^2}{9} = 1$, $\frac{(x+3)^2}{3} - \frac{(y+3)^2}{4} = 1$,

$4x^2 - 2y^2 = 8$, $9(y-2)^2 = 36 + 4(x+3)^2$,

$3(y+3)^2 = 4(x-3)^2 - 12$, $y^2 - 2x^2 = 6$.

37.

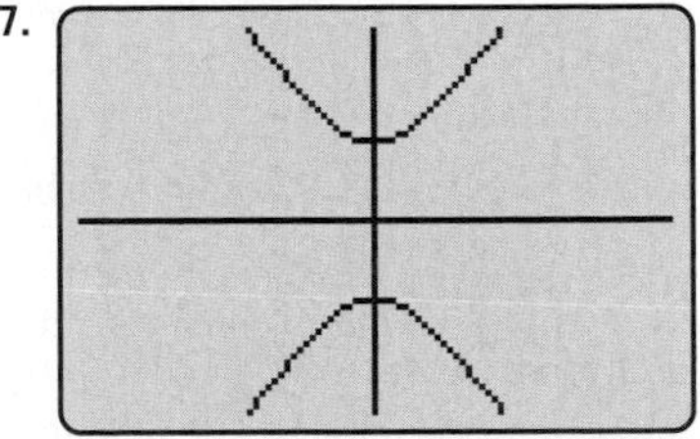

38.

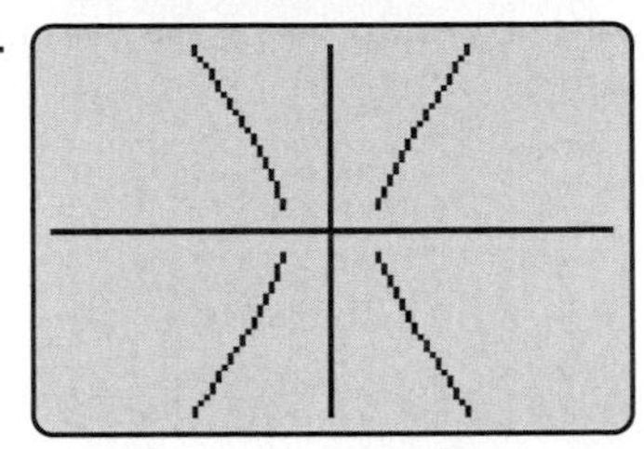

39.

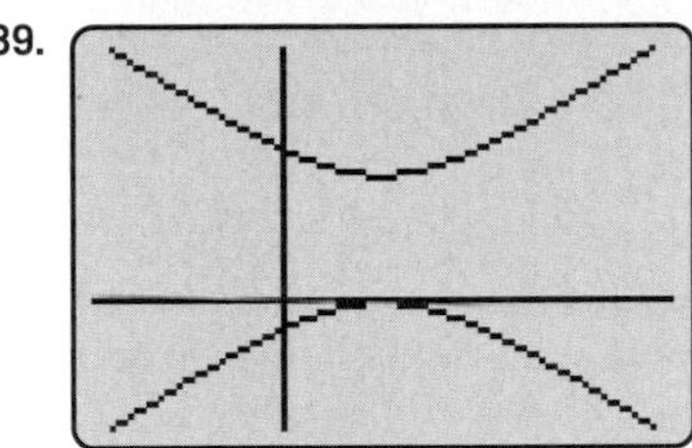

40.

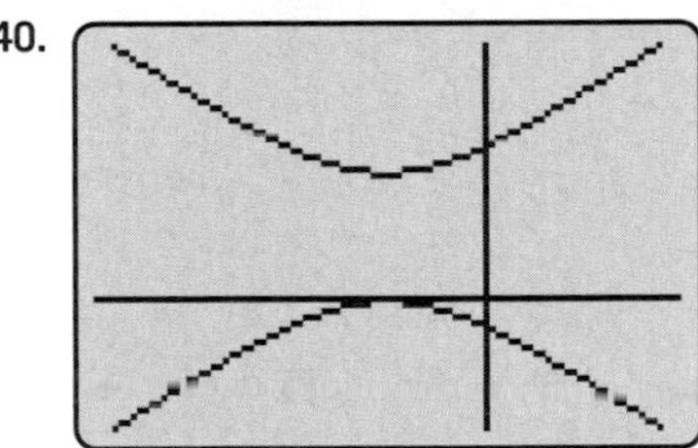

41.

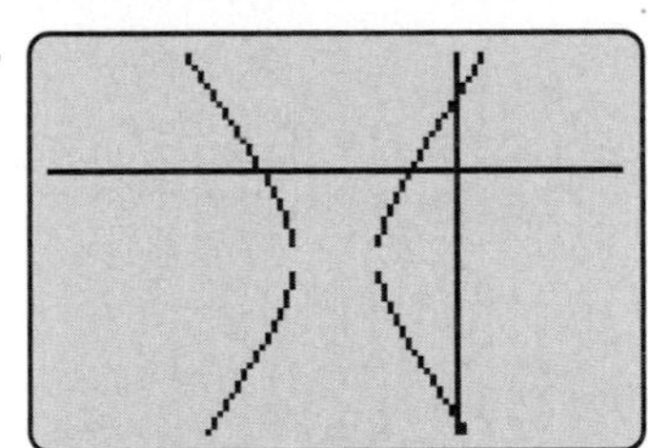

42.

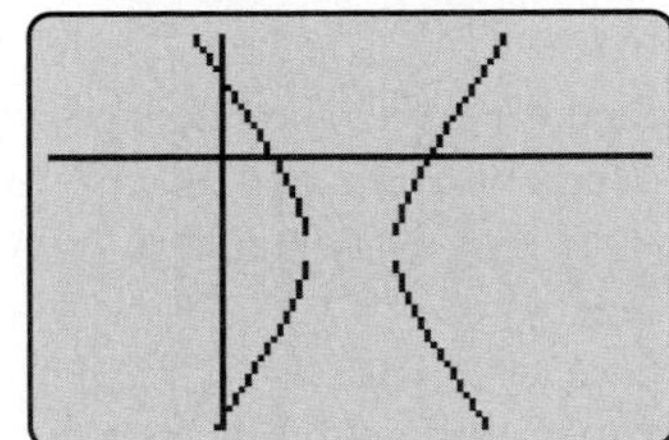

In Exercises 43–50, find parametric equations for the curve whose equation is given, and use them to find a complete graph of the curve.

43. $\dfrac{x^2}{10} - \dfrac{y^2}{36} = 1$

44. $\dfrac{y^2}{9} - \dfrac{x^2}{16} = 1$

45. $x^2 - 4y^2 = 1$

46. $2x^2 - y^2 = 4$

47. $\dfrac{(y+3)^2}{25} - \dfrac{(x+1)^2}{16} = 1$

48. $\dfrac{(y+1)^2}{9} - \dfrac{(x-1)^2}{25} = 1$

49. $\dfrac{(x+3)^2}{1} - \dfrac{(y-2)^2}{4} = 1$

50. $\dfrac{(y+5)^2}{9} - \dfrac{(x-2)^2}{1} = 1$

51. Sketch the graph of $\dfrac{y^2}{4} - \dfrac{x^2}{b^2} = 1$ for $b = 2$, $b = 4$, $b = 8$, $b = 12$, and $b = 20$. What happens to the hyperbola as b takes larger and larger values? Could the graph ever degenerate into a pair of horizontal lines?

52. Find a number k such that $(-2, 1)$, is on the graph of $3x^2 + ky^2 = 4$. Then graph the equation.

53. Show that the asymptotes of the hyperbola $\dfrac{x^2}{a^2} - \dfrac{y^2}{a^2} = 1$ are perpendicular to each other.

54. Find the approximate coordinates of the points where these hyperbolas intersect:

$$\frac{(x-1)^2}{4} - \frac{(y+1)^2}{8} = 1 \quad \text{and} \quad 4y^2 - x^2 = 1.$$

55. Two listening stations that are 1 mile apart record an explosion. One microphone receives the sound 2 seconds after the other does. Use the line through the microphones as the x-axis, with the origin midway between the microphones, and the fact that sound travels at 1100 feet per second to find the equation of a hyperbola on which the explosion is located. Can you determine the exact location of the explosion?

56. Two transmission stations P and Q are located 200 miles apart on a straight shoreline. A ship 50 miles from shore is moving parallel to the shoreline. A signal from Q reaches the ship 400 microseconds after a signal from P. If the signals travel at 980 feet per microsecond, find the location of the ship (in terms of miles) in the coordinate system with x-axis through P and Q and origin midway between them.

If $a > 0$ and $b > 0$, then the ***eccentricity*** *of the hyperbola*

$$\frac{(x-h)^2}{a^2} - \frac{(y-k)^2}{b^2} = 1 \quad \text{or} \quad \frac{(y-k)^2}{a^2} - \frac{(x-h)^2}{b^2} = 1$$

is the number $\dfrac{\sqrt{a^2+b^2}}{a}$. *In Exercises 57–61, find the eccentricity of the hyperbola whose equation is given.*

57. $\dfrac{(x-6)^2}{10} - \dfrac{y^2}{40} = 1$ **58.** $\dfrac{y^2}{18} - \dfrac{x^2}{25} = 1$

59. $6(y-2)^2 = 18 + 3(x+2)^2$

60. $16x^2 - 9y^2 - 32x + 36y + 124 = 0$

61. $4x^2 - 5y^2 - 16x - 50y + 71 = 0$

62. (a) Graph these hyperbolas (on the same screen if possible).

$$\frac{y^2}{4} - \frac{x^2}{1} = 1, \quad \frac{y^2}{4} - \frac{x^2}{12} = 1, \quad \frac{y^2}{4} - \frac{x^2}{96} = 1.$$

(b) Compute the eccentricity of each hyperbola in part (a).

(c) On the basis of parts (a) and (b), how is the shape of a hyperbola related to its eccentricity?

5.3 Parabolas

Definition. Let L be a line in the plane, and let P be a point not on L. If X is any point not on L, the distance from X to L is defined to be the length of the perpendicular line segment from X to L. The **parabola** with **focus** P and **directrix** L is the set of all points X such that

$$\text{Distance from } X \text{ to } P = \text{Distance from } X \text{ to } L$$

as shown in Figure 5–24.

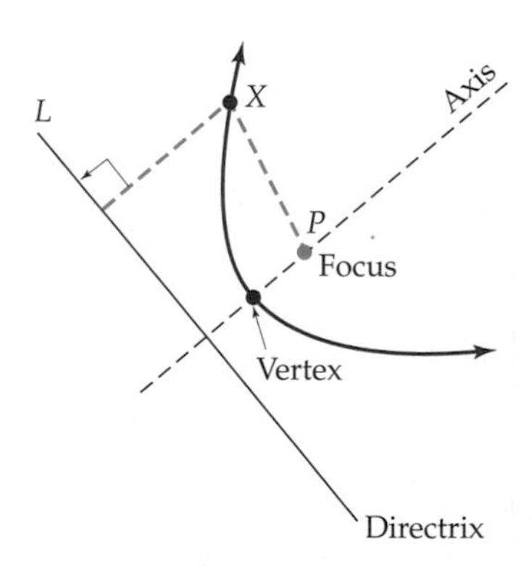

Figure 5–24

The line through P perpendicular to L is called the **axis.** The intersection of the axis with the parabola (the midpoint of the segment of the axis from P to L) is the **vertex** of the parabola, as illustrated in Figure 5–24. The parabola is symmetric with respect to its axis.

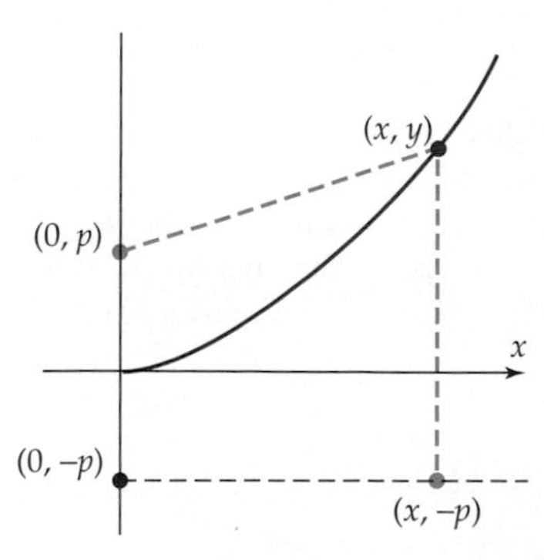

Figure 5–25

Equation. Suppose that the focus is on the y-axis at the point $(0, p)$, where p is a nonzero constant, and that the directrix is the horizontal line $y = -p$. If (x, y) is any point on the parabola, then the distance from (x, y) to the horizontal line $y = -p$ is the length of the vertical segment from (x, y) to $(x, -p)$ as shown in Figure 5–25.

By the definition of the parabola,

$$\text{Distance from } (x, y) \text{ to } (0, p) = \text{Distance from } (x, y) \text{ to line } y = -p$$

$$\text{Distance from } (x, y) \text{ to } (0, p) = \text{Distance from } (x, y) \text{ to } (x, -p)$$

$$\sqrt{(x-0)^2 + (y-p)^2} = \sqrt{(x-x)^2 + [y-(-p)]^2}.$$

Squaring both sides and simplifying, we have

$$(x-0)^2 + (y-p)^2 = (x-x)^2 + (y+p)^2$$

$$x^2 + y^2 - 2py + p^2 = 0^2 + y^2 + 2py + p^2$$

$$x^2 = 4py.$$

Conversely, it can be shown that every point whose coordinates satisfy this equation is on the parabola.

A similar argument works for the parabola with focus $(p, 0)$ on the x-axis and directrix the vertical line $x = -p$, and leads to this conclusion.

Standard Equations of Parabolas with Vertex at the Origin

Let p be a nonzero real number. Then the graph of each of the following equations is a parabola with vertex at the origin.

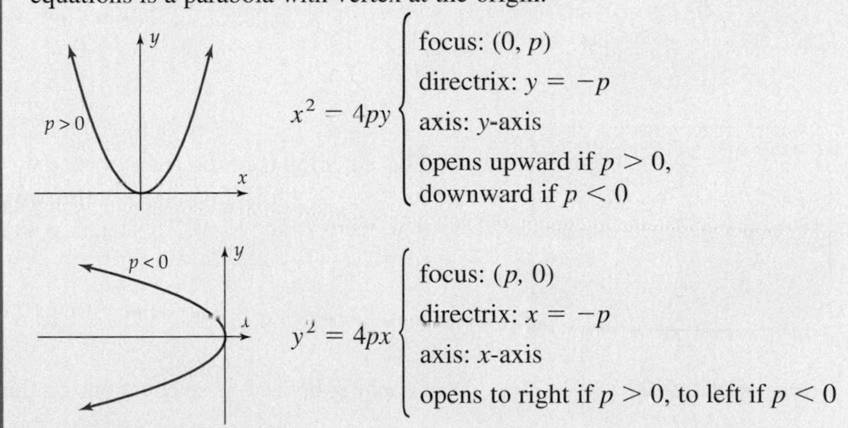

EXAMPLE 1

Show that the graph of the equation is a parabola, and find its focus and directrix; then sketch the graph.

(a) $y = -x^2/8$ (b) $x = 3y^2$

SOLUTION

(a) We rewrite the equation so that it matches one of the forms in the preceding box.

$$y = -\frac{x^2}{8}$$

Multiply both sides by -8: $\quad -8y = x^2$

The equation $x^2 = -8y$ is of the form $x^2 = 4py$, with $4p = -8$, so $p = -2$. Hence, the graph is a downward-opening parabola with focus $(0, p) = (0, -2)$ and directrix $y = -p = -(-2) = 2$, as shown in Figure 5–26.

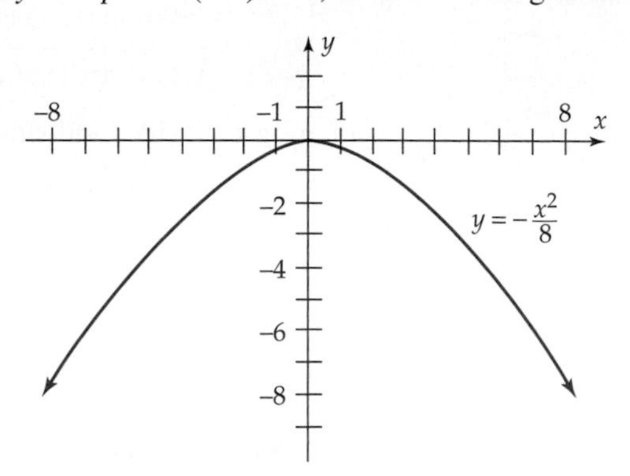

Figure 5–26

(b) Divide both sides of $x = 3y^2$ by 3 so that the equation becomes

$$y^2 = \frac{x}{3}$$

$$y^2 = \frac{1}{3}x.$$

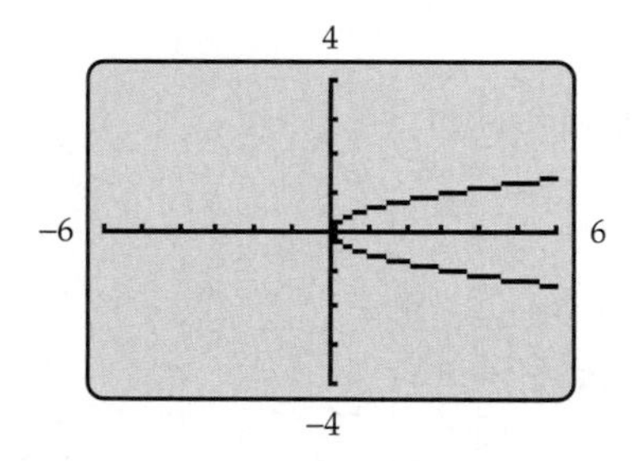

Figure 5–27

This equation is of the form $y^2 = 4px$, with $4p = 1/3$, so $p = 1/12$. Therefore, the graph is a parabola with focus $(1/12, 0)$ and directrix $x = -1/12$ that opens to the right. To sketch its graph, we solve the equation $y^2 = x/3$ for y.

$$y = \sqrt{\frac{x}{3}} \quad \text{or} \quad y = -\sqrt{\frac{x}{3}}$$

Graphing both of these equations on the same screen produces Figure 5–27. ■

EXAMPLE 2 

Find the focus, directrix, and equation of the parabola that passes through the point (8, 2), has vertex (0, 0), and focus on the x-axis.

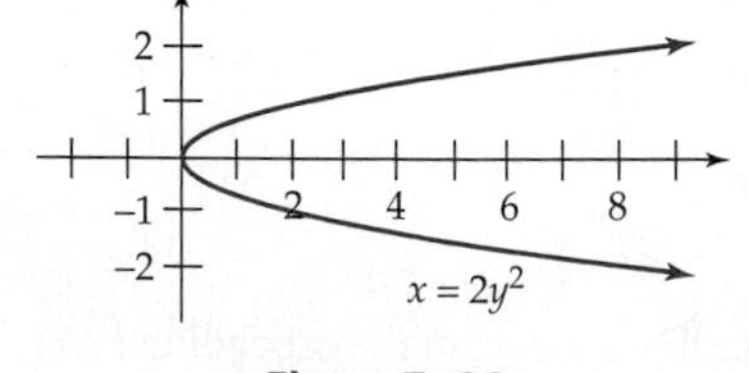

Figure 5–28

SOLUTION The equation is of the form $y^2 = 4px$. Since (8, 2) is on the graph, we have $2^2 = 4p \cdot 8$, so $p = \frac{1}{8}$. Therefore, the focus is $\left(\frac{1}{8}, 0\right)$, and the directrix is the vertical line $x = -\frac{1}{8}$. The equation is $y^2 = 4\left(\frac{1}{8}\right)x = \frac{1}{2}x$ or, equivalently, $x = 2y^2$. Its graph is sketched in Figure 5–28. ■

VERTICAL AND HORIZONTAL SHIFTS

Let h and k be constants. If we replace x by $x - h$ and y by $y - k$ in the equation of a parabola with vertex at the origin, then the graph of the new equation can be obtained by shifting the parabola vertically and horizontally so that its vertex is at (h, k), as explained on page 339. Using this fact and translating the information in the box on page 361, we obtain the following.

Standard Equations of Parabolas with Vertex at (h, k)

If p is a nonzero real number, then the graph of each of the following equations is a parabola with vertex (h, k).

$$(x-h)^2 = 4p(y-k) \begin{cases} \text{focus: } (h, k+p) \\ \text{directrix: the horizontal line } y = k - p \\ \text{axis: the vertical line } x = h \\ \text{opens upward if } p > 0, \text{ downward if } p < 0 \end{cases}$$

$$(y-k)^2 = 4p(x-h) \begin{cases} \text{focus: } (h+p, k) \\ \text{directrix: the vertical line } x = h - p \\ \text{axis: the horizontal line } y = k \\ \text{opens to right if } p > 0, \text{ to left if } p < 0 \end{cases}$$

EXAMPLE 3

Identify and sketch the graph of $y = 2(x-3)^2 + 1$.

SOLUTION We first rewrite the equation.

$$y = 2(x-3)^2 + 1$$

Subtract 1 from both sides: $$y - 1 = 2(x-3)^2$$

Divide both sides by 2: $$\frac{1}{2}(y-1) = (x-3)^2$$

$$(x-3)^2 = \frac{1}{2}(y-1)$$

This is the first form in the preceding box, with $h = 3$, $k = 1$, and $4p = 1/2$. Hence, $p = 1/8$, and the graph is an upward-opening parabola with vertex $(3, 1)$, focus $(3, 1 + 1/8) = (3, 9/8)$, and directrix the horizontal line $y = 1 - 1/8 = 7/8$. The graph of

$$(x-3)^2 = \frac{1}{2}(y-1) \qquad \text{or, equivalently,} \qquad y = 2(x-3)^2 + 1$$

is the graph of $y = 2x^2$ shifted 3 units to the right and 1 unit upward, as shown in Figure 5–29. ■

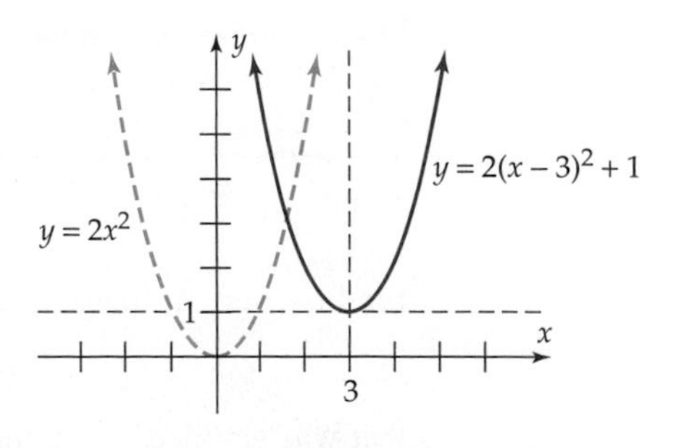

Figure 5–29

EXAMPLE 4

Graph the equation $x = 5y^2 + 30y + 41$ without putting it in standard form.

SOLUTION There are two methods of graphing this equation on a calculator.

Method 1. Rewrite the equation as

$$5y^2 + 30y + 41 - x = 0.$$

This is a quadratic equation of the form $ay^2 + by + c = 0$, with $a = 5$, $b = 30$, and $c = 41 - x$. It can be solved by using the quadratic formula.

$$y = \frac{-30 \pm \sqrt{30^2 - 4 \cdot 5(41 - x)}}{2 \cdot 5} = \frac{-30 \pm \sqrt{900 - 20(41 - x)}}{10}$$

Now graph both

$$y = \frac{-30 + \sqrt{900 - 20(41 - x)}}{10} \quad \text{and} \quad y = \frac{-30 - \sqrt{900 - 20(41 - x)}}{10}$$

on the same screen to obtain the parabola in Figure 5–30.

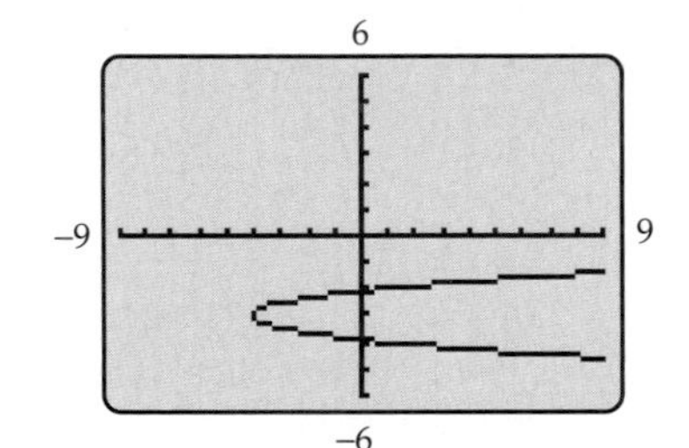

Figure 5–30

Method 2. Since the equation $x = 5y^2 + 30y + 41$ defines x as a function of y, we can use parametric graphing, as explained at the end of Section 0.5. The parametric equations are

$$x = 5t^2 + 30t + 41$$
$$y = t.$$

Graphing these equations in parametric mode also produces Figure 5–30. ■

The parametric graphing technique used in Example 4 can be applied to the parabola with equation $(y - k)^2 = 4p(x - h)$ by solving the equation for x and letting $y = t$.

Parametric Equations for Parabolas

The parabola with vertex (h, k) and equation

$$(y - k)^2 = 4p(x - h)$$

is given by the parametric equations

$$x = \frac{(t - k)^2}{4p} + h \quad \text{and} \quad y = t \quad (t \text{ any real number}).$$

EXAMPLE 5

Find the vertex, focus, and directrix of the parabola

$$x = 5y^2 + 30y + 41$$

that was graphed in Example 4.

SOLUTION We first rewrite the equation.

$$5y^2 + 30y + 41 = x$$

Subtract 41 from both sides: $$5y^2 + 30y = x - 41$$

Factor out 5 on left-hand side: $$5(y^2 + 6y) = x - 41$$

Complete the square on the expression $y^2 + 6y$ by adding 9 (the square of half the coefficient of y).

$$5(y^2 + 6y + 9) = x - 41 + ?$$

On the left-hand side, we have actually added $5 \cdot 9 = 45$, so we must add the same amount to the right-hand side.

$$5(y^2 + 6y + 9) = x - 41 + 45$$

Factor left-hand side: $$5(y + 3)^2 = x + 4$$

Divide both sides by 5: $$(y + 3)^2 = \frac{1}{5}(x + 4)$$

$$[y - (-3)]^2 = \frac{1}{5}[x - (-4)]$$

Thus, the graph is a parabola with vertex $(-4, -3)$. In this case, $4p = 1/5$, so $p = 1/20 = .05$. Hence, the focus is $(-4 + .05, -3) = (-3.95, -3)$, and the directrix is $x = -4 - .05 = -4.05$.

GRAPHING EXPLORATION

Use the parametric equations in the preceding box, with $h = -4$, $k = -3$, and $p = 1/20$, to graph this parabola in the window with

$$-9 \le x \le 9 \quad \text{and} \quad -6 \le y \le 6 \quad (-6 \le t \le 6).$$

Your graph should be identical to Figure 5–30. ■

APPLICATIONS

Certain laws of physics show that sound waves or light rays from a source at the focus of a parabola will reflect off the parabola in rays parallel to the axis of the parabola, as shown in Figure 5–31. This is the reason that parabolic reflectors are used in automobile headlights and searchlights.

Conversely, a light ray coming toward a parabola will be reflected into the focus, as shown in Figure 5–32. This fact is used in the design of radar antennas, satellite dishes, and field microphones used at outdoor sporting events to pick up conversation on the field.

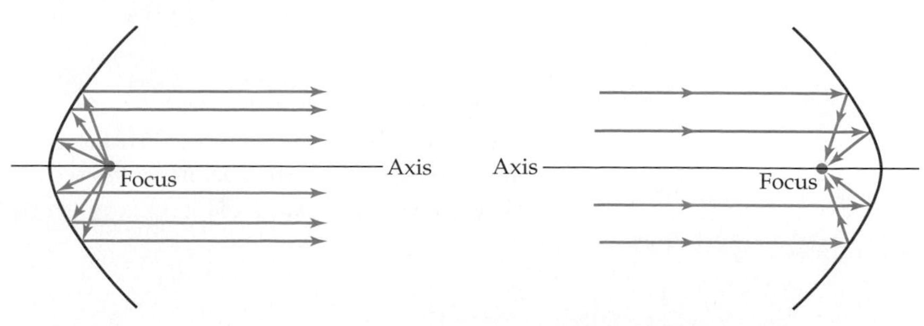

Figure 5–31 **Figure 5–32**

Projectiles follow a parabolic curve, a fact that is used in the design of water slides in which the rider slides down a sharp incline, then up and over a hill, before plunging downward into a pool. At the peak of the hill, the rider shoots up along a parabolic arc several inches above the slide, experiencing a sensation of weightlessness.

EXAMPLE 6

The radio telescope in Figure 5–33 has the shape of a parabolic dish (a cross section through the center of the dish is a parabola). It is 30 feet deep at the center and has a diameter of 200 feet. How far from the vertex of the parabolic dish should the receiver be placed to catch all the rays that hit the dish?

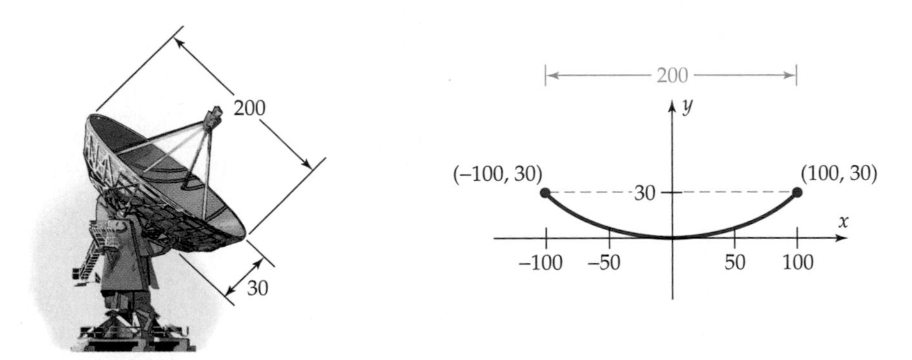

Figure 5–33

Figure 5–34

SOLUTION Rays hitting the dish are reflected into the focus, as explained above. So the radio receiver must be located at the focus. To find the focus, draw a cross section of the dish, with vertex at the origin, as in Figure 5–34. The equation of this parabola is of the form $x^2 = 4py$. Since the point (100, 30) is on the parabola, we have

$$x^2 = 4py$$

Substitute: $$100^2 = 4p(30)$$

Simplify: $$120p = 100^2$$

Divide both sides by 120: $$p = \frac{100^2}{120}$$

Simplify: $$p = \frac{250}{3}.$$

As we saw in the box on page 361, the focus is the point $(0, p)$, which is p units from the vertex (0, 0). Therefore, the receiver should be placed $250/3 \approx 83.33$ feet from the vertex. ■

EXERCISES 5.3

In Exercises 1–6, determine which of the following equations could possibly have the given graph.

$y = x^2/4,\quad x^2 = -8y,\quad 6x = y^2,\quad y^2 = -4x,$

$2x^2 + y^2 = 12,\quad x^2 + 6y^2 = 18,$

$6y^2 - x^2 = 6,\quad 2x^2 - y^2 = 8$

1. 

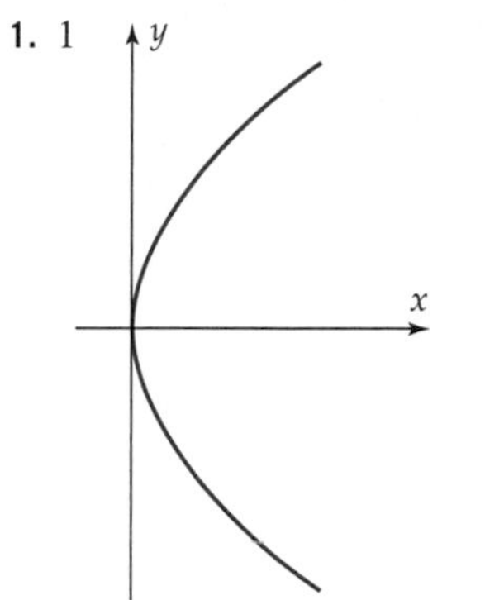

2.

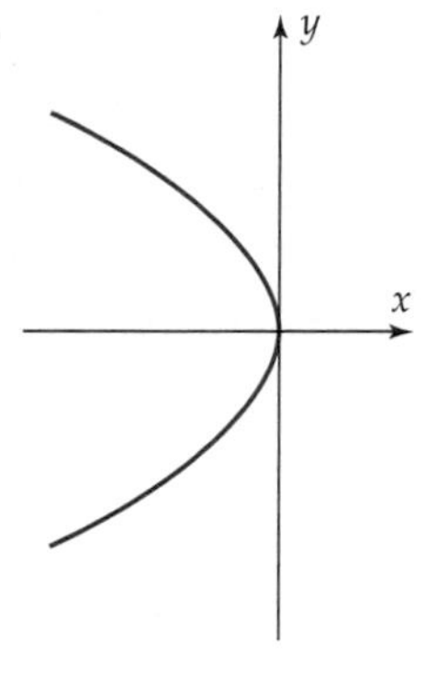

3.

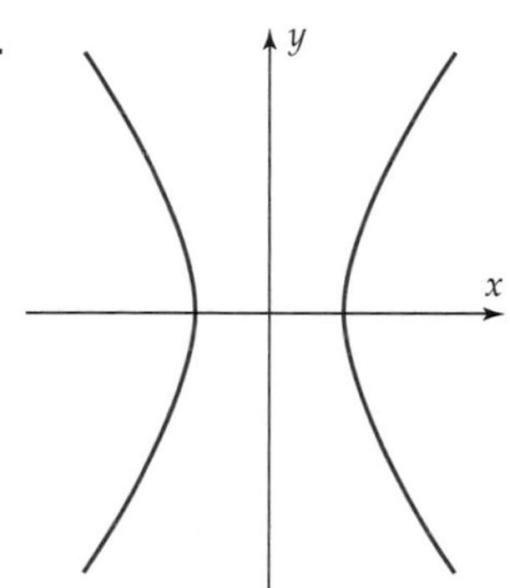

4.

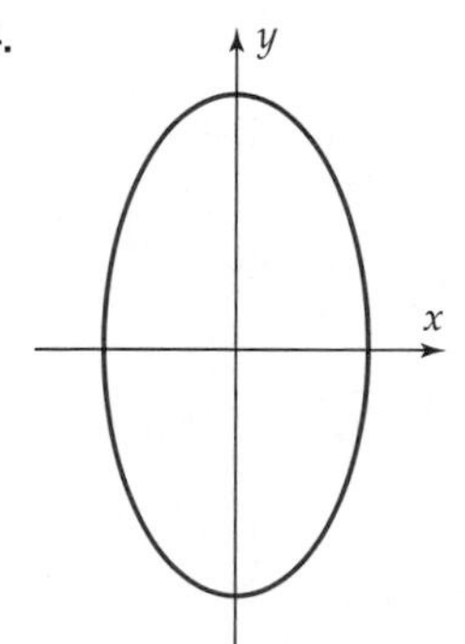

5.

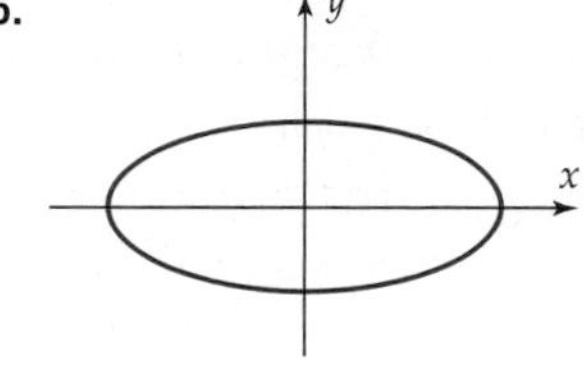

6.

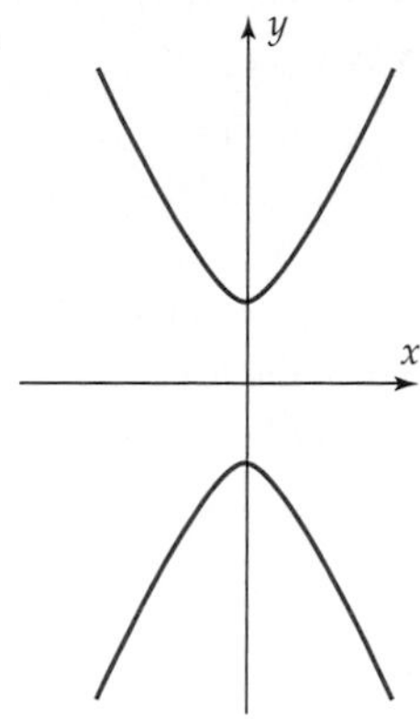

In Exercises 7–10, find the focus and directrix of the parabola.

7. $y = 3x^2$

8. $x = .5y^2$

9. $y = .25x^2$

10. $x = -6y^2$

In Exercises 11–22, determine the vertex, focus, and directrix of the parabola without graphing *and state whether it opens upward, downward, left, or right.*

11. $x - y^2 = 2$

12. $y - 3 = x^2$

13. $x + (y + 1)^2 = 2$

14. $y + (x + 2)^2 = 3$

15. $3x - 2 = (y + 3)^2$

16. $2x - 1 = -6(y + 1)^2$

17. $x = y^2 - 9y$

10. $x = y^2 + y + 1$

19. $y = 3x^2 + x - 4$

20. $y = -3x^2 + 4x - 1$

21. $y = -3x^2 + 4x + 5$

22. $y = 2x^2 - x - 1$

In Exercises 23–30, sketch the graph of the equation and label the vertex.

23. $y = 4(x - 1)^2 + 2$

24. $y = 3(x - 2)^2 - 3$

25. $x = 2(y - 2)^2$

26. $x = -3(y - 1)^2 - 2$

27. $y = x^2 - 4x - 1$

28. $y = x^2 + 8x + 6$

29. $y = x^2 + 2x$

30. $x = y^2 - 3y$

In Exercises 31–42, find the equation of the parabola satisfying the given conditions.

31. Vertex $(0, 0)$; axis $x = 0$; $(2, 12)$ on graph

32. Vertex $(0, 1)$; axis $x = 0$; $(2, -7)$ on graph

33. Vertex $(1, 0)$; axis $x = 1$; $(2, 13)$ on graph

34. Vertex $(-3, 0)$; axis $y = 0$; $(-1, 1)$ on graph

35. Vertex $(2, 1)$; axis $y = 1$; $(5, 0)$ on graph

36. Vertex $(1, -3)$; axis $y = -3$; $(-1, -4)$ on graph

37. Vertex $(-3, -2)$; focus $(-47/16, -2)$

38. Vertex $(-5, -5)$; focus $(-5, -99/20)$

39. Vertex $(1, 1)$; focus $(1, 9/8)$

40. Vertex $(-4, -3)$; $(-6, -2)$ and $(-6, -4)$ on graph

41. Vertex $(-1, 3)$; $(8, 0)$ and $(0, 4)$ on graph

42. Vertex $(1, -3)$; $(0, -1)$ and $(-1, 5)$ on graph

In Exercises 43–48, find parametric equations for the curve whose equation is given, and use them to find a complete graph of the curve.

43. $8x = 2y^2$

44. $4y = x^2$

45. $y = 4(x - 1)^2 + 2$

46. $y = 3(x - 2)^2 - 3$

47. $x = 2(y - 2)^2$

48. $x = -3(y - 1)^2 - 2$

In Exercises 49–56, identify the conic section whose equation is given, list its vertex, or vertices, if any, and find its graph.

49. $x^2 = 6x - y - 5$

50. $y^2 = x - 2y - 2$

51. $3y^2 = x - 1 + 2y$

52. $2y^2 = x - 4y - 5$

53. $3x^2 + 3y^2 - 6x - 12y - 6 = 0$

54. $2x^2 + 3y^2 + 12x - 6y + 9 = 0$

55. $2x^2 - y^2 + 16x + 4y + 24 = 0$

56. $4x^2 - 40x - 2y + 105 = 0$

In Exercises 57–62, determine which of the following equations could possibly have the given graph. Assume that all viewing windows are square.

$$y = (x + 5)^2 - 3, \qquad x = (y + 3)^2 + 2,$$

$$y = (x - 4)^2 + 2, \qquad x = -(y - 3)^2 - 2.$$

$$y^2 = 4y + x - 1, \qquad y = -x^2 - 8x - 18$$

57.

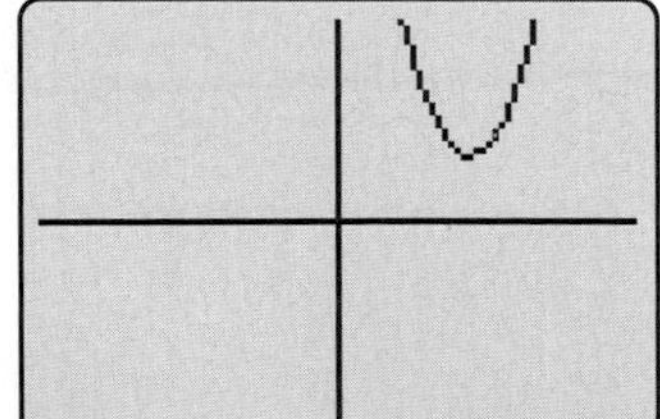

58.

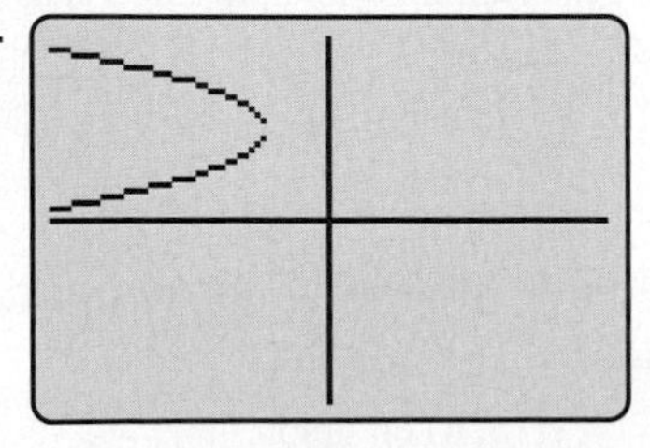

59.

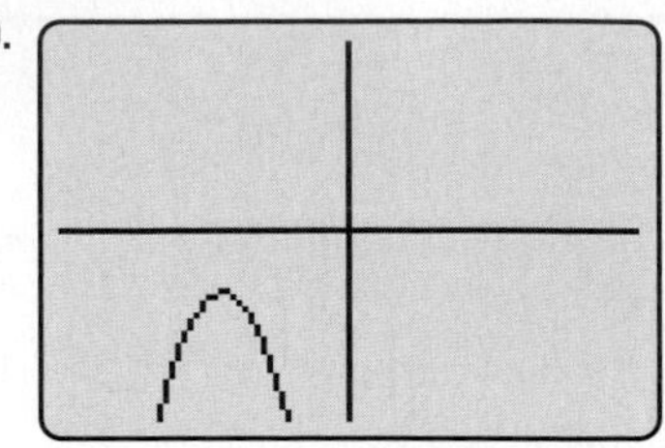

60.

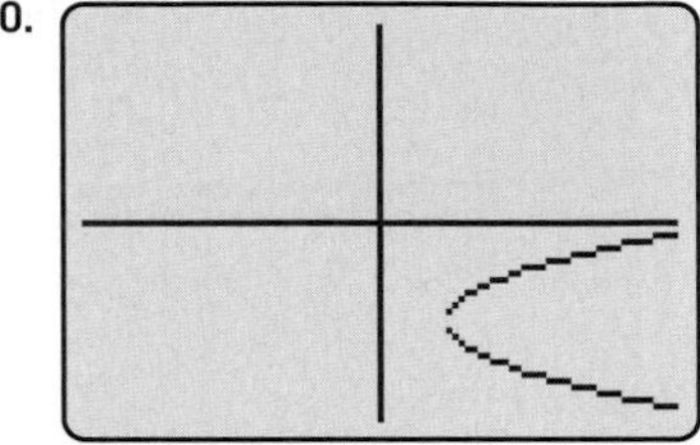

61.

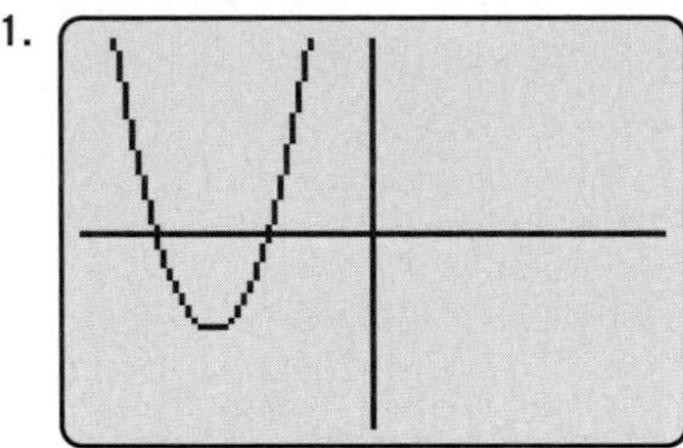

62.

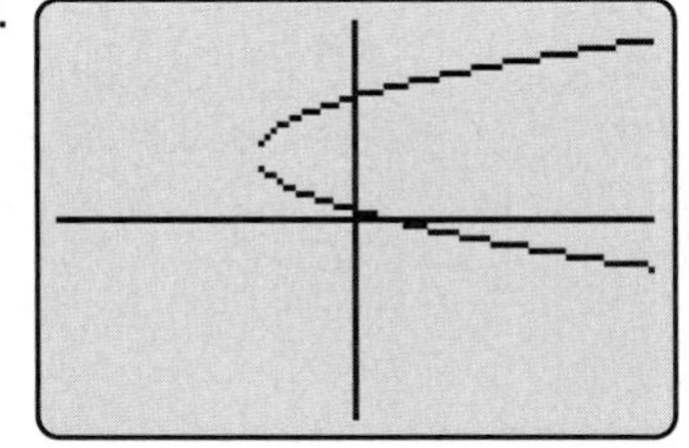

63. Find the number b such that the vertex of the parabola $y = x^2 + bx + c$ lies on the y-axis.

64. Find the number d such that the parabola $(y + 1)^2 = dx + 4$ passes through $(-6, 3)$.

65. Find the points of intersection of the parabola $4y^2 + 4y = 5x - 12$ and the line $x = 9$.

66. Find the points of intersection of the parabola $4x^2 - 8x = 2y + 5$ and the line $y = 15$.

67. A parabolic satellite dish is 4 feet in diameter and 1.5 feet deep. How far from the vertex should the receiver be placed to catch all the signals that hit the dish?

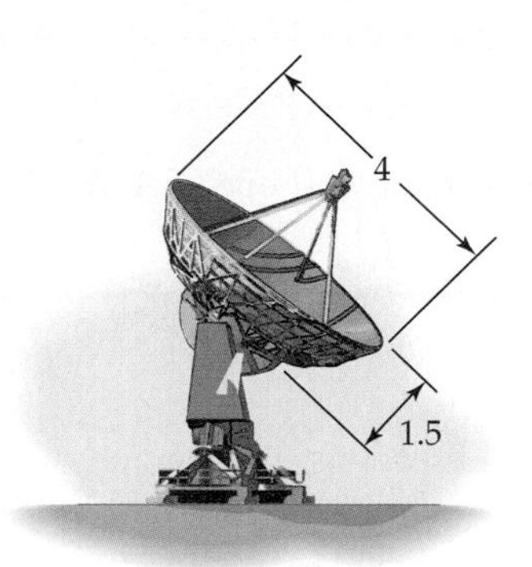

68. A flashlight has a parabolic reflector that is 3 inches in diameter and 1.5 inches deep. For the light from the bulb to reflect in beams that are parallel to the center axis of the flashlight, how far from the vertex of the reflector should the bulb be located? [*Hint:* See Figure 5–31 and the preceding discussion.]

69. A radio telescope has a parabolic dish with a diameter of 300 feet. Its receiver (focus) is located 130 feet from the vertex. How deep is the dish at its center? [*Hint:* Position the dish as in Figure 5–34, and find the equation of the parabola.]

70. The 6.5-meter MMT telescope on top of Mount Hopkins in Arizona has a parabolic mirror. The focus of the parabola is 8.125 meters from the vertex of the parabola. Find the depth of the mirror.

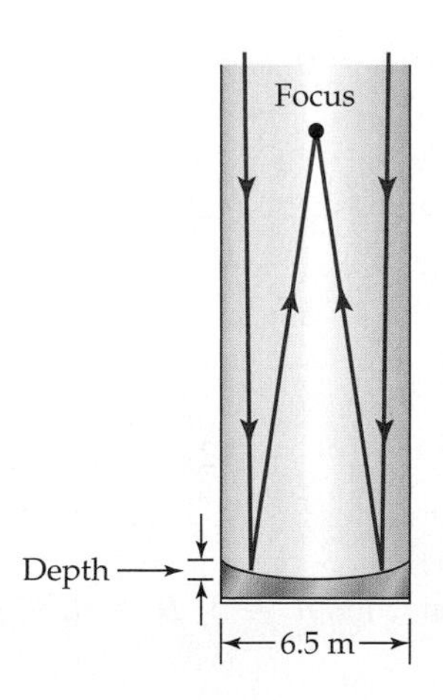

71. The Hale telescope at Mount Palomar in California also has a parabolic mirror, whose depth is .096 meter (see the figure for Exercise 70). The focus of the parabola is 16.75 meters from the vertex. Find the diameter of the mirror.

72. A large spotlight has a parabolic reflector that is 3 feet deep at its center. The light source is located $1\frac{1}{3}$ feet from the vertex. What is the diameter of the reflector?

73. The cables of a suspension bridge are shaped like parabolas. The cables are attached to the towers 100 feet from the bridge surface, and the towers are 420 feet apart. The cables touch the bridge surface at the center (midway between the towers). At a point on the bridge 100 feet from one of the towers, how far is the cable from the bridge surface?

74. At a point 120 feet from the center of a suspension bridge, the cables are 24 feet above the bridge surface. Assume that the cables are shaped like parabolas and touch the bridge surface at the center (which is midway between the towers). If the towers are 600 feet apart, how far above the surface of the bridge are the cables attached to the towers?

5.4 Rotations and Second-Degree Equations

A **second-degree equation** in x and y is one that can be written in the form

$$Ax^2 + Bxy + Cy^2 + Dx + Ey + F = 0$$

for some constants A, B, C, D, E, F, with at least one of A, B, C nonzero.

EXAMPLE 1

Show that each of the following conic sections is the graph of a second-degree equation.

(a) Ellipse: $\dfrac{x^2}{6} + \dfrac{y^2}{5} = 1$ (b) Hyperbola: $\dfrac{(x+1)^2}{4} - \dfrac{(y-3)^2}{6} = 1$

SOLUTION We need only show that each of these equations is in fact a second-degree equation. In each case, eliminate denominators, multiply out all terms, and gather them on one side of the equal sign.

(a)

$$\frac{x^2}{6} + \frac{y^2}{5} = 1$$

Multiply both sides by 30: $$5x^2 + 6y^2 = 30$$

Rearrange terms: $$5x^2 + 6y^2 - 30 = 0$$

This equation is a second-degree equation because it has the form

$$Ax^2 + Bxy + Cy^2 + Dx + Ey + F = 0$$

with $A = 5$, $B = 0$, $C = 6$, $D = 0$, $E = 0$, and $F = -30$.

(b)

$$\frac{(x+1)^2}{4} - \frac{(y-3)^2}{6} = 1$$

Multiply both sides by 12: $$3(x+1)^2 - 2(y-3)^2 = 12$$

Multiply out left side: $$3(x^2 + 2x + 1) - 2(y^2 - 6y + 9) = 12$$

$$3x^2 + 6x + 3 - 2y^2 + 12y - 18 = 12$$

Rearrange terms: $$3x^2 - 2y^2 + 6x + 12y - 27 = 0$$

This is a second-degree equation with $A = 3$, $B = 0$, $C = -2$, $D = 6$, $E = 12$, and $F = -27$. ■

Calculations like those in Example 1 can be used on the equation of any conic section to show that it is the graph of a second-degree equation. Conversely, it can be shown that

The graph of every second-degree equation is a conic section

(possibly degenerate). When the second-degree equation has no xy-term (that is, $B = 0$), as was the case in Example 1, the graph is a conic section in standard position (axis or axes parallel to the coordinate axes). When $B \neq 0$, however, the conic is rotated from standard position, so its axis or axes are not parallel to the coordinate axes.

EXAMPLE 2

Graph the equation

$$3x^2 + 6xy + y^2 + x - 2y + 7 = 0.$$

SOLUTION We first rewrite it as

$$y^2 + 6xy - 2y + 3x^2 + x + 7 = 0$$
$$y^2 + (6x - 2)y + (3x^2 + x + 7) = 0.$$

This equation has the form $ay^2 + by + c = 0$, with $a = 1$, $b = 6x - 2$, and $c = 3x^2 + x + 7$. It can be solved by using the quadratic formula.

$$y = \frac{-b \pm \sqrt{b^2 - 4ac}}{2a} = \frac{-(6x - 2) \pm \sqrt{(6x - 2)^2 - 4 \cdot 1 \cdot (3x^2 + x + 7)}}{2 \cdot 1}$$

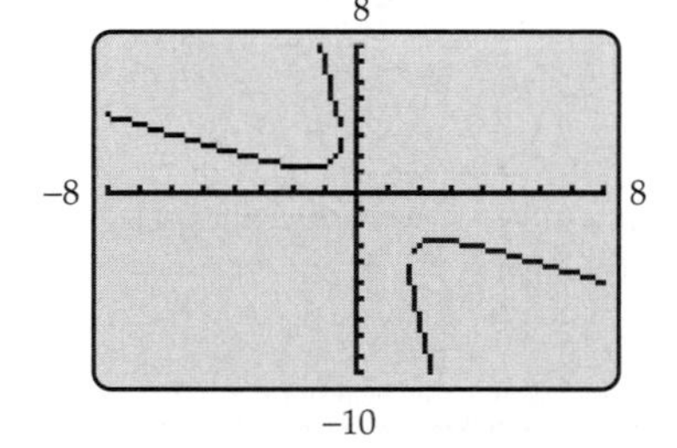

Figure 5–35

The top half of the graph is obtained by graphing

$$y = \frac{-6x + 2 + \sqrt{(6x - 2)^2 - 4(3x^2 + x + 7)}}{2},$$

and the bottom half is obtained by graphing

$$y = \frac{-6x + 2 - \sqrt{(6x - 2)^2 - 4(3x^2 + x + 7)}}{2}.$$

The graph is a hyperbola whose focal axis tilts upward to the left, as shown in Figure 5–35. ■

THE DISCRIMINANT

The following fact, whose proof is outlined in Exercise 15 of Special Topics 5.4.A, makes it easy to identify the graphs of second-degree equations without graphing them.

Graphs of Second-Degree Equations

If the equation

$$Ax^2 + Bxy + Cy^2 + Dx + Ey + F = 0 \qquad (A, B, C \text{ not all } 0)$$

has a graph, then that graph is

A circle or an ellipse (or a point), if $B^2 - 4AC < 0$;

A parabola (or a line or two parallel lines), if $B^2 - 4AC = 0$;

A hyperbola (or two intersecting lines), if $B^2 - 4AC > 0$.

The expression $B^2 - 4AC$ is called the **discriminant** of the equation.

EXAMPLE 3

Identify the graph of

$$2x^2 - 4xy + 3y^2 + 5x + 6y - 8 = 0$$

and sketch the graph.

SOLUTION We compute the discriminant with $A = 2$, $B = -4$, and $C = 3$.

$$B^2 - 4AC = (-4)^2 - 4 \cdot 2 \cdot 3 = 16 - 24 = -8.$$

Hence, the graph is an ellipse (possibly a circle or a single point). To find the graph, we rewrite the equation as

$$3y^2 - 4xy + 6y + 2x^2 + 5x - 8 = 0$$

$$3y^2 + (-4x + 6)y + (2x^2 + 5x - 8) = 0.$$

The equation has the form $ay^2 + by + c = 0$ and can be solved by the quadratic formula.

$$y = \frac{-b \pm \sqrt{b^2 - 4ac}}{2a}$$

$$= \frac{-(-4x + 6) \pm \sqrt{(-4x + 6)^2 - 4 \cdot 3 \cdot (2x^2 + 5x - 8)}}{2 \cdot 3}.$$

The graph can now be found by graphing the last two equations on the same screen.

GRAPHING EXPLORATION

Find a viewing window that shows a complete graph of the equation. In what direction does the major axis run? ■

EXAMPLE 4

Does Figure 5–36 show a complete graph of

$$3x^2 + 5xy + 2y^2 - 8y - 1 = 0?$$

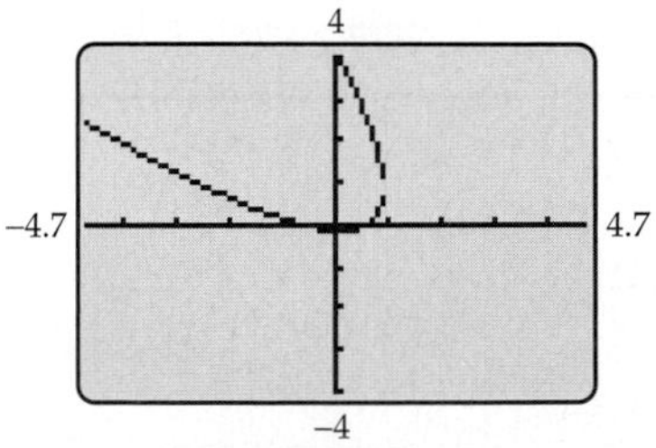

SOLUTION Although the graph in the figure looks like a parabola, appearances are deceiving. The discriminant of the equation is

$$B^2 - 4AC = 5^2 - 4 \cdot 3 \cdot 2 = 1,$$

which means that the graph is a hyperbola. So Figure 5–36 cannot possibly be a complete graph of the equation.*

Figure 5–36

GRAPHING EXPLORATION

Solve the equation for y, as in Example 3. Then find a viewing window that clearly shows both branches of the hyperbola. ■

*When I asked students to graph this equation on an examination, many of them did not bother to compute the discriminant and produced something similar to Figure 5–36. They didn't get much credit for this answer. *Moral:* Use the discriminant to identify the conic so that you know the shape of the graph you are looking for.

EXAMPLE 5

Sketch the graph of

$$3x^2 + 6xy + 3y^2 + 13x + 9y + 53 = 0.$$

SOLUTION The discriminant is $B^2 - 4AC = 6^2 - 4 \cdot 3 \cdot 3 = 0$. Hence, the graph is a parabola (or a line or parallel lines in the degenerate case).

GRAPHING EXPLORATION

Find a viewing window that shows a complete graph of the equation. ■

EXERCISES 5.4

In Exercises 1–6, assume that the graph of the equation is a nondegenerate conic section. Without graphing, determine whether the graph is an ellipse, hyperbola, or parabola.

1. $x^2 - 2xy + 3y^2 - 1 = 0$
2. $xy - 1 = 0$
3. $x^2 + 2xy + y^2 + 2\sqrt{2}x - 2\sqrt{2}y = 0$
4. $2x^2 - 4xy + 5y^2 - 6 = 0$
5. $17x^2 - 48xy + 31y^2 + 50 = 0$
6. $2x^2 - 4xy - 2y^2 + 3x + 5y - 10 = 0$

In Exercises 7–24, use the discriminant to identify the conic section whose equation is given, and find a viewing window that shows a complete graph.

7. $9x^2 + 4y^2 + 54x - 8y + 49 = 0$
8. $4x^2 + 5y^2 - 8x + 30y + 29 = 0$
9. $4y^2 - x^2 + 6x - 24y + 11 = 0$
10. $x^2 - 16y^2 = 0$
11. $3y^2 - x - 2y + 1 = 0$
12. $x^2 - 6x + y + 5 = 0$
13. $41x^2 - 24xy + 34y^2 - 25 = 0$
14. $x^2 + 2\sqrt{3}xy + 3y^2 + 8\sqrt{3}x - 8y + 32 = 0$
15. $17x^2 - 48xy + 31y^2 + 49 = 0$
16. $52x^2 - 72xy + 73y^2 = 200$
17. $9x^2 + 24xy + 16y^2 + 90x - 130y = 0$
18. $x^2 + 10xy + y^2 + 1 = 0$
19. $23x^2 + 26\sqrt{3}xy - 3y^2 - 16x + 16\sqrt{3}y + 128 = 0$
20. $x^2 + 2xy + y^2 + 12\sqrt{2}x - 12\sqrt{2}y = 0$
21. $17x^2 - 12xy + 8y^2 - 80 = 0$
22. $11x^2 - 24xy + 4y^2 + 30x + 40y - 45 = 0$
23. $3x^2 + 2\sqrt{3}xy + y^2 + 4x - 4\sqrt{3}y - 16 = 0$
24. $3x^2 + 2\sqrt{2}xy + 2y^2 - 12 = 0$

5.4.A *SPECIAL TOPICS* Rotation of Axes

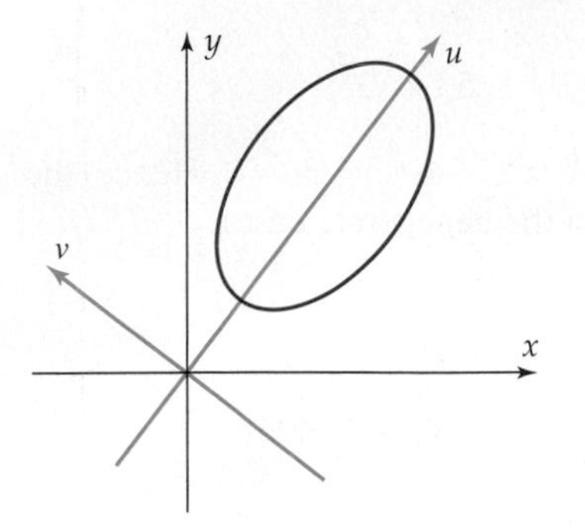

Figure 5–37

We have seen that the graph of the second-degree equation

$$Ax^2 + Bxy + Cy^2 + Dx + Ey + F = 0 \quad (B \neq 0)$$

is a conic section whose axes are not parallel to the coordinate axes, as in Figure 5–37. Although we can graph the equation on a calculator (as in Section 5.4), we cannot read off useful information about the center, vertices, etc. from the equation, as we can from an equation in standard form.

The key is to replace the *xy* coordinate system by a new coordinate system, as indicated by the blue *u*- and *v*-axes in Figure 5–37. The conic is not rotated in the new coordinate system, so it has an equation (in *u* and *v*) in standard form that will provide the desired information.

To use this approach, we must first determine the relationship between the *xy* coordinates of a point and its coordinates in the *uv* system. Suppose the *uv* coordinate system is obtained by rotating the *xy* axes about the origin, counterclockwise through an angle θ.* If a point *P* has coordinates (x, y) in the *xy* system, we can find its coordinates (u, v) in the rotated coordinate system by using Figure 5–38.

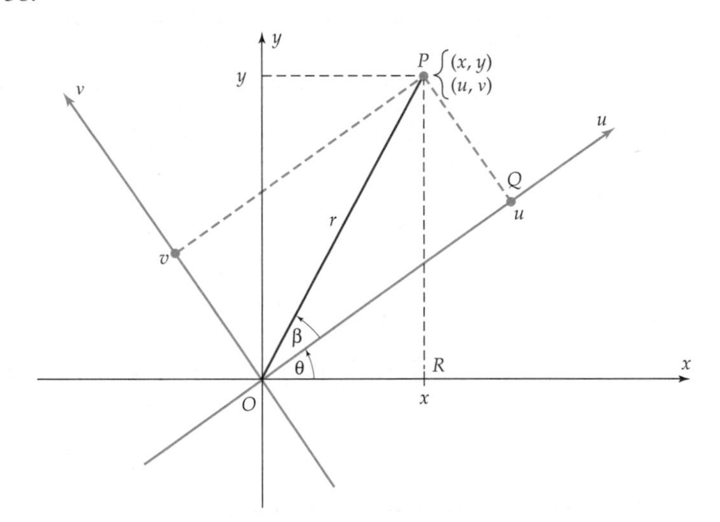

Figure 5–38

Triangle *OPQ* shows that

$$\cos\beta = \frac{OQ}{OP} = \frac{u}{r} \quad \text{and} \quad \sin\beta = \frac{PQ}{OP} = \frac{v}{r}.$$

Therefore,

$$u = r\cos\beta \quad \text{and} \quad v = r\sin\beta.$$

Similarly, triangle *OPR* shows that

$$\cos(\theta + \beta) = \frac{OR}{OP} = \frac{x}{r} \quad \text{and} \quad \sin(\theta + \beta) = \frac{PR}{OP} = \frac{y}{r},$$

---

*All rotations in this section are counterclockwise about the origin, with $0° < \theta < 90°$.

so

$$x = r\cos(\theta + \beta) \qquad \text{and} \qquad y = r\sin(\theta + \beta).$$

Applying the addition identity for cosine (page 230) shows that

$$\begin{aligned} x &= r\cos(\theta + \beta) \\ &= r(\cos\theta\cos\beta - \sin\theta\sin\beta) \\ &= (r\cos\beta)\cos\theta - (r\sin\beta)\sin\theta \\ &= u\cos\theta - v\sin\theta. \end{aligned}$$

A similar argument with $y = r\sin(\theta + \beta)$ and the addition identity for sine leads to this result.

The Rotation Equations

> If the xy coordinate axes are rotated through an angle θ to produce the uv coordinate axes, then the coordinates (x, y) and (u, v) of a point are related by these equations:
>
> $$x = u\cos\theta - v\sin\theta,$$
> $$y = u\sin\theta + v\cos\theta.$$

EXAMPLE 1

If the xy axes are rotated 30°, find the equation relative to the uv axes of the graph of

$$3x^2 + 2\sqrt{3}xy + y^2 + x - \sqrt{3}y = 0,$$

and graph the equation.

SOLUTION Since $\sin 30° = 1/2$ and $\cos 30° = \sqrt{3}/2$, the rotation equations are

$$x = u\cos 30° - v\sin 30° = \frac{\sqrt{3}}{2}u - \frac{1}{2}v,$$

$$y = u\sin 30° + v\cos 30° = \frac{1}{2}u + \frac{\sqrt{3}}{2}v.$$

Substitute these expressions in the original equation.

$$3x^2 + 2\sqrt{3}xy + y^2 + x - \sqrt{3}y = 0$$

$$3\left(\frac{\sqrt{3}}{2}u - \frac{1}{2}v\right)^2 + 2\sqrt{3}\left(\frac{\sqrt{3}}{2}u - \frac{1}{2}v\right)\left(\frac{1}{2}u + \frac{\sqrt{3}}{2}v\right)$$
$$+ \left(\frac{1}{2}u + \frac{\sqrt{3}}{2}v\right)^2 + \left(\frac{\sqrt{3}}{2}u - \frac{1}{2}v\right) - \sqrt{3}\left(\frac{1}{2}u + \frac{\sqrt{3}}{2}v\right) = 0$$

$$3\left(\frac{3}{4}u^2 - \frac{\sqrt{3}}{2}uv + \frac{1}{4}v^2\right) + 2\sqrt{3}\left(\frac{\sqrt{3}}{4}u^2 + \frac{1}{2}uv - \frac{\sqrt{3}}{4}v^2\right)$$
$$+ \left(\frac{1}{4}u^2 + \frac{\sqrt{3}}{2}uv + \frac{3}{4}v^2\right) + \left(\frac{\sqrt{3}}{2}u - \frac{1}{2}v\right) - \sqrt{3}\left(\frac{1}{2}u + \frac{\sqrt{3}}{2}v\right) = 0.$$

Verify that the last equation simplifies as

$$4u^2 - 2v = 0 \qquad \text{or, equivalently,} \qquad v = 2u^2.$$

In the uv system, $v = 2u^2$ is the equation of an upward-opening parabola with vertex at (0, 0) as shown in Figure 5–39. ■

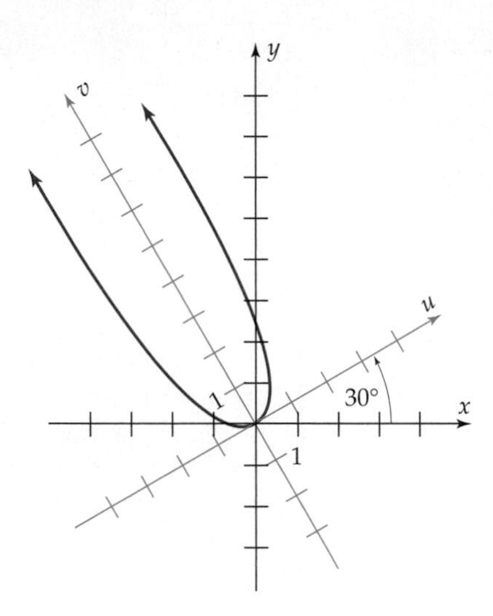

Figure 5–39

Rotating the axes in the preceding example changed the original equation, which included an xy term, to an equation that had no uv term. That enabled us to identify the graph of the new equation as a conic section. This can be done for any second-degree equation by choosing an angle of rotation that will eliminate the xy term. The choice of the angle is determined by this fact, which is proved in Exercise 13.

Rotation Angle

The equation $Ax^2 + Bxy + Cy^2 + Dx + Ey + F = 0$ $(B \neq 0)$ can be rewritten as $A'u^2 + C'v^2 + D'u + E'v + F' = 0$ by rotating the xy axes through an angle θ such that

$$\cot 2\theta = \frac{A - C}{B} \qquad (0° < \theta < 90°)$$

and using the rotation equations.

EXAMPLE 2

What angle of rotation will eliminate the xy term in the equation

$$153x^2 + 192xy + 97y^2 - 1710x - 1470y + 5625 = 0,$$

and what are the rotation equations in this case?

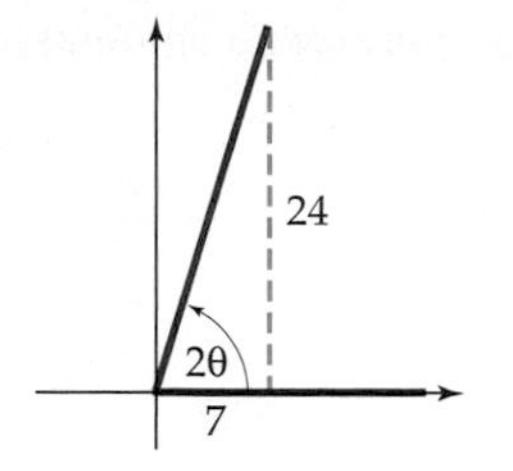

Figure 5–40

SOLUTION According to the fact in the box with $A = 153$, $B = 192$, and $C = 97$, we should rotate through an angle of θ, where

$$\cot 2\theta = \frac{153 - 97}{192} = \frac{56}{192} = \frac{7}{24}.$$

Since $0° < 2\theta < 180°$ and $\cot 2\theta$ is positive, the terminal side of the angle 2θ lies in the first quadrant, as shown in Figure 5–40. The hypotenuse of this triangle has length $\sqrt{7^2 + 24^2} = \sqrt{625} = 25$. Hence, $\cos 2\theta = 7/25$. The half-angle identities (page 244) show that

$$\sin\theta = \sqrt{\frac{1 - \cos 2\theta}{2}} = \sqrt{\frac{1 - 7/25}{2}} = \sqrt{\frac{9}{25}} = \frac{3}{5},$$

$$\cos\theta = \sqrt{\frac{1 + \cos 2\theta}{2}} = \sqrt{\frac{1 + 7/25}{2}} = \sqrt{\frac{16}{25}} = \frac{4}{5}.$$

Using $\sin\theta = 3/5$ and the SIN^{-1} key on a calculator, we find that the angle θ of rotation is approximately 36.87°. The rotation equations are

$$x = u\cos\theta - v\sin\theta = \frac{4}{5}u - \frac{3}{5}v,$$

$$y = u\sin\theta + v\cos\theta = \frac{3}{5}u + \frac{4}{5}v. \quad \blacksquare$$

EXAMPLE 3

Graph the equation without using a calculator.

$$153x^2 + 192xy + 97y^2 - 1710x - 1470y + 5625 = 0.$$

SOLUTION The angle θ and the rotation equations for eliminating the xy term were found in the preceding example. Substitute the rotation equations in the given equation.

$$153x^2 + 192xy + 97y^2 - 1710x - 1470y + 5625 = 0$$

$$153\left(\frac{4}{5}u - \frac{3}{5}v\right)^2 + 192\left(\frac{4}{5}u - \frac{3}{5}v\right)\left(\frac{3}{5}u + \frac{4}{5}v\right)$$
$$+ 97\left(\frac{3}{5}u + \frac{4}{5}v\right)^2 - 1710\left(\frac{4}{5}u - \frac{3}{5}v\right) - 1470\left(\frac{3}{5}u + \frac{4}{5}v\right) + 5625 = 0$$

$$153\left(\frac{16}{25}u^2 - \frac{24}{25}uv + \frac{9}{25}v^2\right) + 192\left(\frac{12}{25}u^2 + \frac{7}{25}uv - \frac{12}{25}v^2\right)$$
$$+ 97\left(\frac{9}{25}u^2 + \frac{24}{25}uv + \frac{16}{25}v^2\right) - 2250u - 150v + 5625 = 0$$

$$225u^2 + 25v^2 - 2250u - 150v + 5625 = 0$$

$$9u^2 + v^2 - 90u - 6v + 225 = 0$$

$$9(u^2 - 10u) + (v^2 - 6v) = -225.$$

Finally, complete the square in u and v (adding the appropriate amounts to the right side so as not to change the equation).

$$9(u^2 - 10u + 25) + (v^2 - 6v + 9) = -225 + 9 \cdot 25 + 9$$
$$9(u - 5)^2 + (v - 3)^2 = 9$$
$$\frac{(u - 5)^2}{1} + \frac{(v - 3)^2}{9} = 1.$$

Therefore the graph is an ellipse centered at (5, 3) in the uv coordinate system, as shown in Figure 5–41. ■

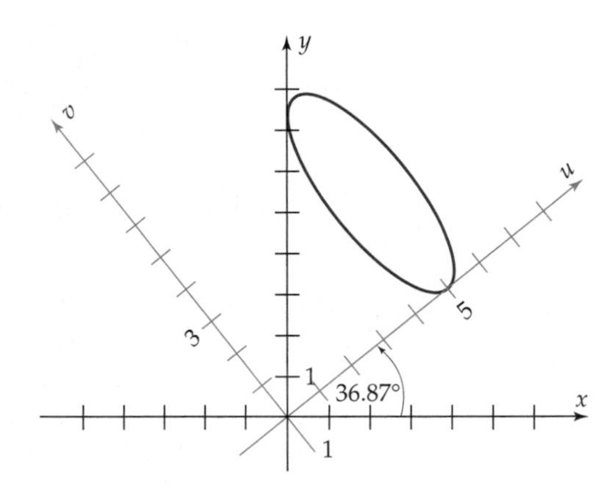

Figure 5–41

EXERCISES 5.4.A

In Exercises 1–4, find the new coordinates of the point when the coordinate axes are rotated through the given angle.

1. (3, 2); $\theta = 45°$

2. (−2, 4); $\theta = 60°$

3. (1, 0); $\theta = 30°$

4. (3, 3); $\sin \theta = 5/13$

In Exercises 5–8, rotate the axes through the given angle to form the uv coordinate system. Express the given equation in terms of the uv coordinate system.

5. $\theta = 45°$; $xy = 1$

6. $\theta = 45°$; $13x^2 + 10xy + 13y^2 = 72$

7. $\theta = 30°$; $7x^2 - 6\sqrt{3}xy + 13y^2 - 16 = 0$

8. $\sin \theta = 1/\sqrt{5}$; $x^2 - 4xy + 4y^2 + 5\sqrt{5}y + 1 = 0$

In Exercises 9–12, find the angle of rotation that will eliminate the xy term of the equation and list the rotation equations in this case.

9. $41x^2 - 24xy + 34y^2 - 25 = 0$

10. $x^2 + 2\sqrt{3}xy + 3y^2 + 8\sqrt{3}x - 8y + 32 = 0$

11. $17x^2 - 48xy + 31y^2 + 49 = 0$

12. $52x^2 - 72xy + 73y^2 = 200$

Thinkers

13. (a) Given an equation $Ax^2 + Bxy + Cy^2 + Dx + Ey + F = 0$ with $B \neq 0$ and an angle θ, use the rotation equations in the box on page 375 to rewrite the equation in the form

$$A'u^2 + B'uv + C'v^2 + D'u + E'v + F' = 0,$$

where $A', \ldots, F'$ are expressions involving $\sin \theta$, $\cos \theta$, and the constants $A, \ldots, F$.

(b) Verify that $B' = 2(C - A) \sin \theta \cos \theta + B(\cos^2\theta - \sin^2\theta)$.

(c) Use the double-angle identities to show that $B' = (C - A) \sin 2\theta + B \cos 2\theta$.

(d) If θ is chosen so that $\cot 2\theta = (A - C)/B$, show that $B' = 0$. This proves the statement in the box on page 376.

14. Assume that the graph of $A'u^2 + C'v^2 + D'u + E'v + F' = 0$ (with at least one of A', C' nonzero) in the uv coordinate system is a nondegenerate conic. Show that its graph is an ellipse if $A'C' > 0$ (A' and C' have the same sign), a hyperbola if $A'C' < 0$ (A' and C' have opposite signs), or a parabola if $A'C' = 0$.

15. Assume the graph of $Ax^2 + Bxy + Cy^2 + Dx + Ey + F = 0$ is a nondegenerate conic section. Prove the statement in the box on page 371 as follows.

(a) In Exercise 13(a), show that $(B')^2 - 4A'C' = B^2 - 4AC$.

(b) Assume that θ has been chosen so that $B' = 0$. Use Exercise 14 to show that the graph of the original equation is an ellipse if $B^2 - 4AC < 0$, a parabola if $B^2 - 4AC = 0$, and a hyperbola if $B^2 - 4AC > 0$.

5.5 Plane Curves and Parametric Equations

There are many curves in the plane that cannot be represented as the graph of a function $y = f(x)$. Parametric graphing enables us to represent such curves in terms of functions and also provides a formal definition of a curve in the plane.

Consider, for example, an object moving in the plane during a particular time interval. To describe both the path of the object and its location at any particular time, three variables are needed: the time t and the coordinates (x, y) of the object at time t. For instance, the coordinates x and y might be given by

$$x = 4\cos t + 5\cos(3t) \qquad \text{and} \qquad y = \sin(3t) + t.$$

From $t = 0$ to $t = 12.5$ the object traces out the curve shown in Figure 5–42. The points marked on the graph show the location of the object at various times. Note that the object may be at the same location at different times (the points where the graph crosses itself).

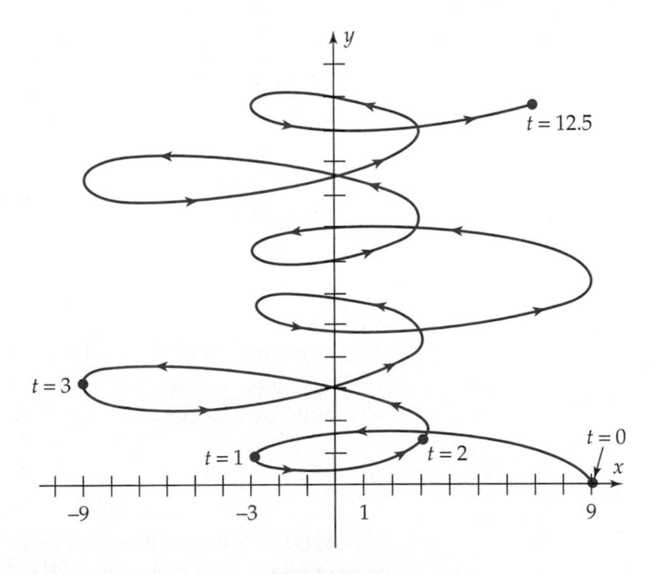

Figure 5–42

In the preceding example both x and y were determined by continuous functions of t, with t taking values in the interval from 0 to 12.5.* The example suggests the following definition.

Definition of Plane Curve

> Let f and g be continuous functions of t on an interval I. The set of all points (x, y) where
>
> $$x = f(t) \qquad \text{and} \qquad y = g(t)$$
>
> is called a **plane curve.** The variable t is called a **parameter,** and the equations defining x and y are **parametric equations.**

In this general definition of "curve," the variable t need not represent time. As the following examples illustrate, different pairs of parametric equations may produce the same curve. Each such pair of parametric equations is called a **parameterization** of the curve.

A curve given by parametric equations can be graphed by hand [choose values of t, plot the corresponding points (x, y), and make an educated guess about the shape of the curve]. A calculator or computer in parametric mode is more likely to produce an accurate graph, however.

TECHNOLOGY TIP To change to parametric graphing mode, select PAR(AMETRIC) in the following menu/submenu:

TI: MODE

Casio: GRAPH/TYPE

HP-39+: APLET

For example, to graph the curve shown in Figure 5–42, which is given by the parametric equations

$$x = 4\cos t + 5\cos(3t) \qquad \text{and} \qquad y = \sin(3t) + t,$$

put your calculator or computer in parametric graphing mode. Then enter the equations (Figure 5–43) and the viewing window, as partially shown in Figure 5–44 (scroll down to see the rest).

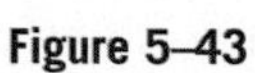
Figure 5–43

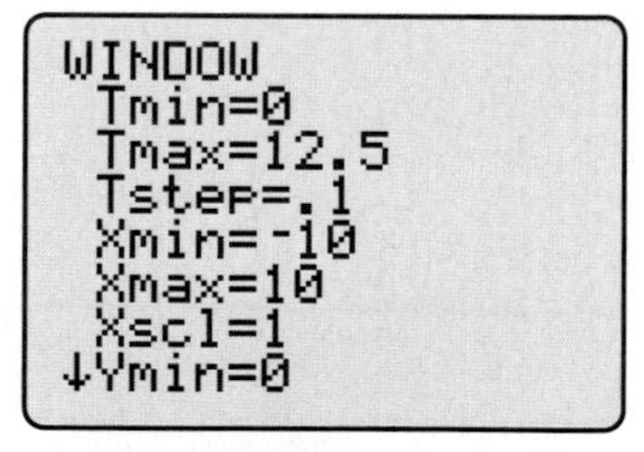

Figure 5–44

In parametric graphing, failure to make appropriate choices for the range of t-values and the t-step (or t-pitch) may result in an inaccurate graph, as the next example illustrates.

*Intuitively, "continuous" means that the graph of the function that determines x, namely, $f(t) = 4\cos t + 5\cos 3t$, is a connected curve with no gaps or holes and similarly for the function that determines y.

EXAMPLE 1

In the window with $-10 \le x \le 10$ and $0 \le y \le 15$, graph the curve given by

$$x = 4\cos t + 5\cos(3t) \qquad \text{and} \qquad y = \sin(3t) + t$$

(a) with $0 \le t \le 12.5$ and each of these t-steps: 2, 1, and .5;

(b) with t-step = .1 and each of these ranges: $2.4 \le t \le 3.9$, $4.6 \le t \le 8$, and $0 \le t \le 12.5$.

How do these graphs compare with Figure 5–42?

SOLUTION

(a) The graphs are shown in Figure 5–45. Only the last one looks even remotely like Figure 5–42.

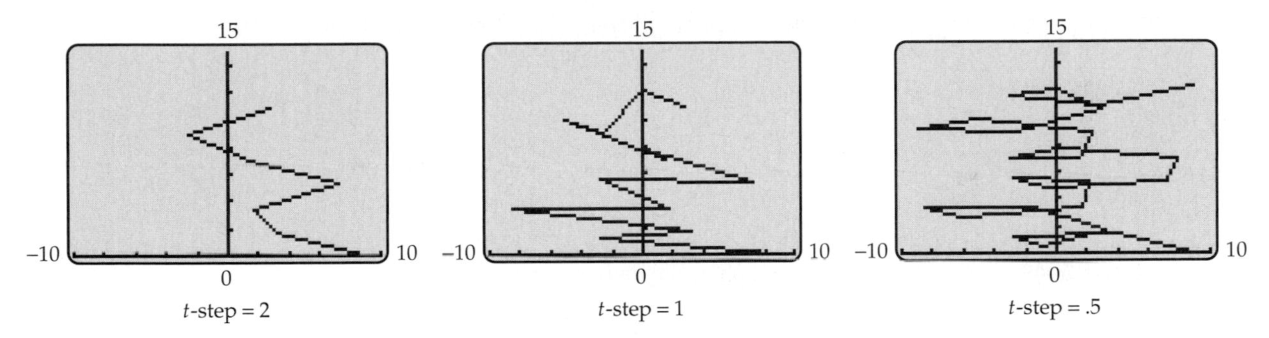

Figure 5–45

(b) The graphs are shown in Figure 5–46. The first two show only a small portion of Figure 5–42, but the last one is a complete graph that closely resembles Figure 5–42. ■

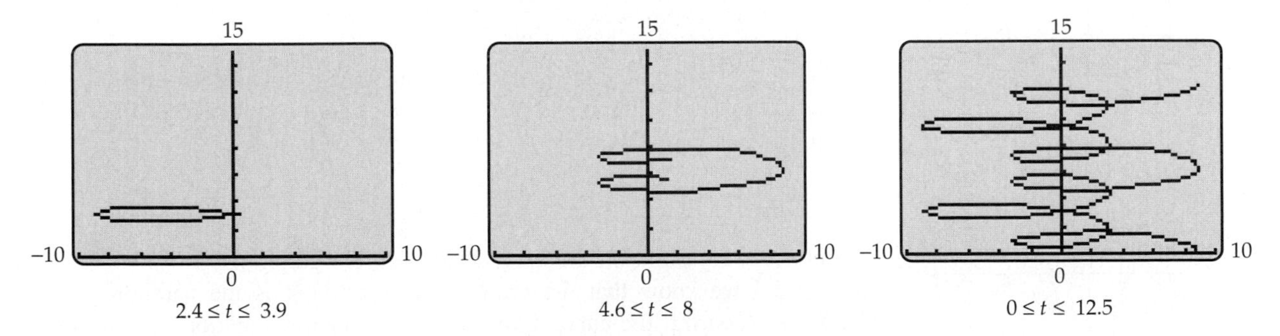

Figure 5–46

As Example 1 suggests, a t-step between .05 and .15 usually produces a reasonably smooth graph. Larger t-steps may produce jagged or totally inaccurate graphs. Smaller t-steps may involve very long graphing times and usually don't increase smoothness very much.

EXAMPLE 2

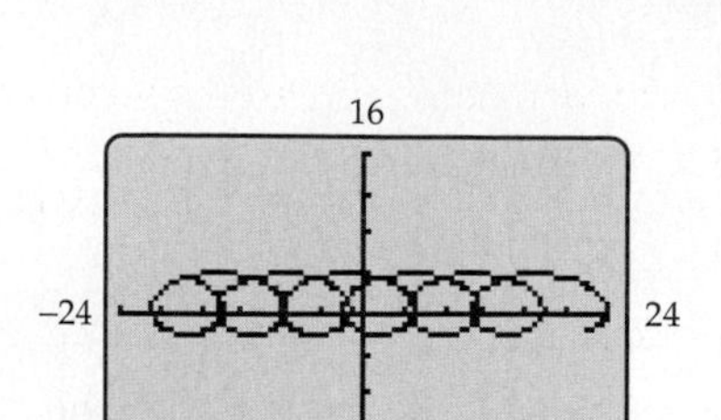

Figure 5–47

Graph the curve given by

$$x = t + 5\cos t \quad \text{and} \quad y = 1 - 3\sin t$$

in the viewing window with

$$-24 \le x \le 24, \qquad -16 \le y \le 16, \qquad -20 \le t \le 20.$$

SOLUTION See Figure 5–47. Reproduce this graph yourself to see how the curve spirals along the x-axis. ■

EXAMPLE 3 

Figure 5–48

Consider the curve given by

$$x = -2t \quad \text{and} \quad y = 4t^2 - 4 \qquad (-1 \le t \le 2).$$

(a) Graph the curve.

(b) Find an equation in x and y whose graph includes the graph in part (a).

SOLUTION

(a) The graph is shown in Figure 5–48.

GRAPHING EXPLORATION

Graph this curve on your calculator, using the same viewing window and range of t values as in Figure 5–48. In what direction is the curve traced out (that is, what is the first point graphed (corresponding to $t = -1$) and what is the last point graphed (corresponding to $t = 2$))?

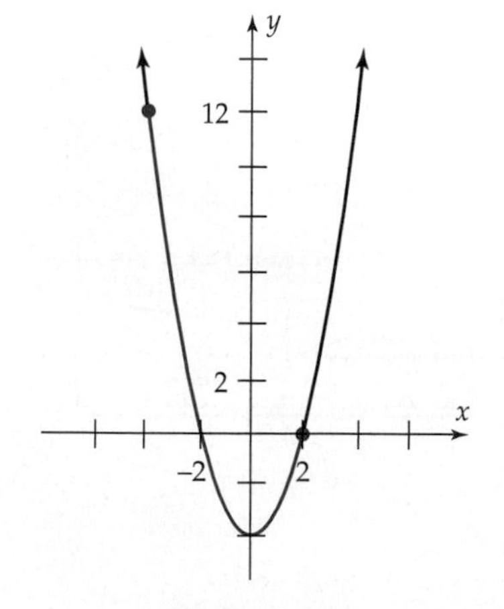

Figure 5–49

(b) We shall solve one of the parametric equations for t and substitute the result in the other one. Solving $x = -2t$ for t, we have $t = -x/2$. Substituting this in the equation for y, we have

$$y = 4t^2 - 4 = 4\left(-\frac{x}{2}\right)^2 - 4 = 4\left(\frac{x^2}{4}\right) - 4 = x^2 - 4.$$

Therefore, every point on the curve is also on the graph of $y = x^2 - 4$. From Section 5.3 we know that the graph of $y = x^2 - 4$ is the parabola in Figure 5–49. However, the curve given by the parametric equations is *not* the entire parabola, but only the part shown in red, which joins the points (2, 0) and (−4, 12) that correspond to the minimum and maximum values of t, namely, $t = -1$ and $t = 2$. ■

Expressing a parametric curve as (part of) the graph of an equation in x and y is called **eliminating the parameter.** As is illustrated in Example 3,

To eliminate the parameter, solve one of the parametric equations for t and substitute this result in the other parametric equation.

FINDING PARAMETRIC EQUATIONS FOR CURVES

Having seen how to graph a curve given by parametric equations, we now consider the reverse problem: finding a parametric respresentation for the graph of an equation in x and y. This is easy when the equation defines y as a function of x, such as

$$y = x^3 + 5x^2 - 3x + 4.$$

A parametric description of this function can be obtained by changing the variable:

$$x = t \qquad \text{and} \qquad y = t^3 + 5t^2 - 3t + 4.$$

The same thing is true for equations that define x as a function of y, such as $x = y^2 + 3$, which can be parameterized by letting

$$x = t^2 + 3 \qquad \text{and} \qquad y = t.$$

In other cases, different techniques may be needed.

EXAMPLE 4

In Section 5.1 we saw that a parameterization of the circle

$$(x - 4)^2 + (y - 1)^2 = 9$$

is given by

(*) $\qquad x = 3\cos t + 4 \qquad \text{and} \qquad y = 3\sin t + 1 \quad (0 \le t \le 2\pi).$

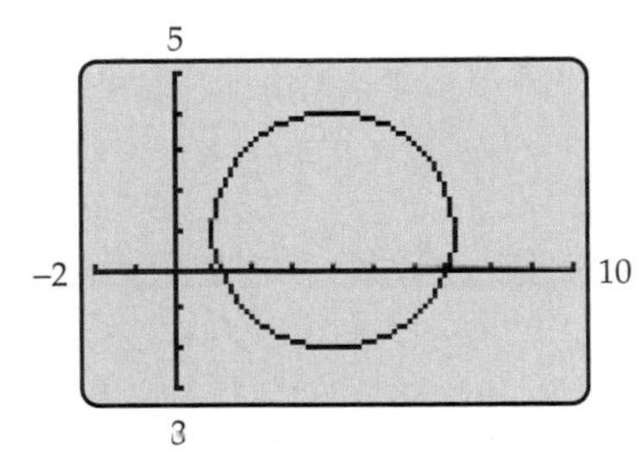

Figure 5–50

With this parameterization, the circle is traced out in a counterclockwise direction from the point (7, 1), as shown in Figure 5–50. Another parameterization is given by

$$x = 3\cos 2t + 4 \qquad \text{and} \qquad y = -3\sin 2t + 1 \quad (0 \le t \le \pi).$$

> **GRAPHING EXPLORATION**
>
> Verify that this parameterization traces out the circle in a clockwise direction, twice as fast as the parameterization given by (*), since t runs from 0 to π rather than 2π. ■

EXAMPLE 5

Find three parameterizations of the straight line through $(1, -3)$ with slope -2.

SOLUTION The point-slope form of the equation of this line is

$$y - (-3) = -2(x - 1) \qquad \text{or, equivalently,} \qquad y = -2x - 1.$$

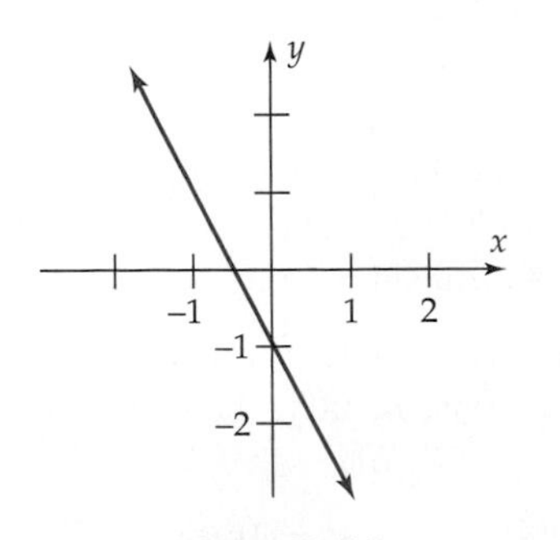

Figure 5–51

Its graph is shown in Figure 5–51. Since this equation defines y as a function of x, one parameterization is

$$x = t \qquad \text{and} \qquad y = -2t - 1 \qquad (t \text{ any real number}).$$

A second parameterization is given by letting $x = t + 1$; then

$$y = -2x - 1 = -2(t + 1) - 1 = -2t - 3 \qquad (t \text{ any real number}).$$

A third parameterization can be obtained by letting

$$x = \tan t \qquad \text{and} \qquad y = -2x - 1 = -2 \tan t - 1 \qquad (-\pi/2 < t < \pi/2).$$

When t runs from $-\pi/2$ to $\pi/2$, then $x = \tan t$ takes all possible real number values, and hence so does y. ■

CAUTION

Some substitutions in an equation $y = f(x)$ do *not* lead to a parameterization of the entire graph. For instance, in Example 5, letting $x = t^2$ and substituting in the equation $y = -2x - 1$ lead to

$$x = t^2 \qquad \text{and} \qquad y = -2t^2 - 1 \qquad (\text{any real number } t).$$

Thus, x is always nonnegative, and y is always negative. So the parameterization produces only the half of the line $y = -2x - 1$ to the right of the y-axis in Figure 5–51.

APPLICATIONS

In the following applications, we ignore air resistance and assume some facts about gravity that are proved in physics.

EXAMPLE 6

Bill Hoffman hits a golf ball with an initial velocity of 140 feet per second so that its path as it leaves the ground makes an angle of 31° with the horizontal.

(a) When does the ball hit the ground?

(b) How far from its starting point does it land?

(c) What is the maximum height of the ball during its flight?

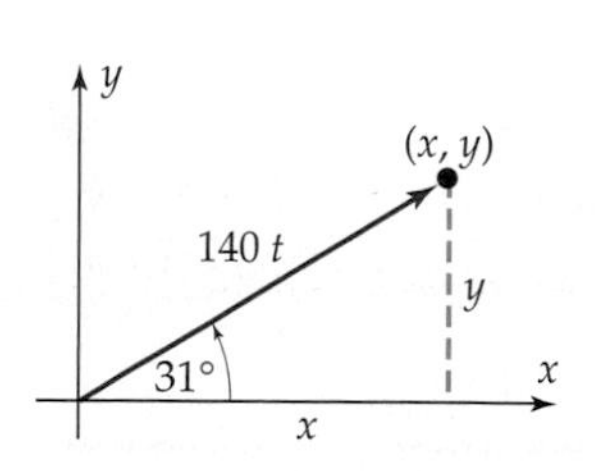

Figure 5–52

SOLUTION Imagine that the golf ball starts at the origin and travels in the direction of the positive x-axis. If there were no gravity, the distance traveled by the ball in t seconds would be $140t$ feet. As Figure 5–52 shows, the coordinates (x, y) of the ball would satisfy

$$\frac{x}{140t} = \cos 31° \qquad \frac{y}{140t} = \sin 31°$$

$$x = (140 \cos 31°)t \qquad y = (140 \sin 31°)t.$$

However, there *is* gravity, and at time t it exerts a force of $16t^2$ downward (that is, in the negative direction on the y-coordinate). Consequently, the coordinates of the golf ball at time t are

$$x = (140 \cos 31°)t \qquad \text{and} \qquad y = (140 \sin 31°)t - 16t^2.$$

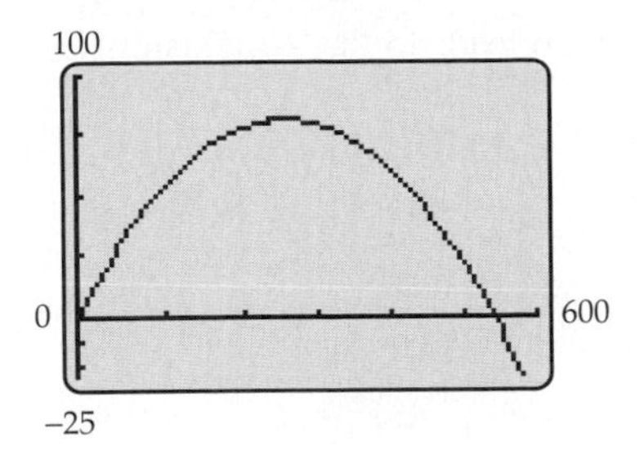

Figure 5–53

The path given by these parametric equations is graphed in Figure 5–53.*

(a) The ball is on the ground when $y = 0$, that is, at the x-intercepts of the graph. They can be found geometrically by using trace and zoom-in (the graphical root finder does not operate in parametric mode), but this is very time-consuming. To find the intercepts algebraically, we need only set $y = 0$ and solve for t.

$$(140 \sin 31°)t - 16t^2 = 0$$

$$t(140 \sin 31° - 16t) = 0$$

$$t = 0 \quad \text{or} \quad 140 \sin 31° - 16t = 0$$

$$t = \frac{140 \sin 31°}{16} \approx 4.5066.$$

Thus, the ball hits the ground after approximately 4.5066 seconds.

(b) The horizontal distance traveled by the ball is given by the x-coordinate of the intercept. The x-coordinate when $t \approx 4.5066$ is

$$x = (140 \cos 31°)(4.5066) \approx 540.81 \text{ feet.}$$

(c) The graph in Figure 5–53 looks like a parabola, and it is, as you can verify by eliminating the parameter t (Exercise 44). The y-coordinate of the vertex is the maximum height of the ball. It can be found geometrically by using trace and zoom-in (the maximum finder doesn't work in parametric mode) or algebraically as follows. The vertex occurs halfway between its two x-intercepts ($x = 0$ and $x \approx 540.81$), that is, when $x \approx 270.405$. Hence,

$$(140 \cos 31°)t = x = 270.405,$$

so

$$t = \frac{270.45}{140 \cos 31°} \approx 2.2533.$$

Therefore, the y-coordinate of the vertex (the maximum height of the ball) is

$$y = (140 \sin 31°)(2.2533) - 16(2.2533)^2 \approx 81.237 \text{ feet.} \quad ■$$

The argument used in Example 6 also applies when the initial position of the golf ball is k feet above the ground (for instance, if the golfer were on a platform at a driving range). In that case the ball begins at $(0, k)$ instead of $(0, 0)$, and its position at time t is k feet higher than before, so its coordinates are

$$x = (140 \cos 31°)t \quad \text{and} \quad y = (140 \sin 31°)t - 16t^2 + k.$$

---

*Only the part of the graph on or above the x-axis represents the ball's path, since the ball does not go underground after it lands.

Then replacing 140 by v and 31° by θ in Example 6 leads to this conclusion.

Projectile Motion

> When a projectile is fired from the position $(0, k)$ on the positive y-axis at an angle θ with the horizontal, in the direction of the positive x-axis, with initial velocity v feet per second, with negligible air resistance, then its position at time t seconds is given by the parametric equations
>
> $$x = (v \cos \theta)t \qquad \text{and} \qquad y = (v \sin \theta)t - 16t^2 + k.$$

EXAMPLE 7

A batter hits a ball that is three feet above the ground and leaves the bat with an initial velocity of 138 feet per second, making an angle of 26° with the horizontal and heading toward a 25-foot-high fence that is 400 feet away. Will the ball go over the fence?

SOLUTION According to the preceding box (with $v = 138$, $\theta = 26°$, and $k = 3$), the path of the ball is given by the parametric equations

$$x = (138 \cos 26°)t \qquad \text{and} \qquad y = (138 \sin 26°)t - 16t^2 + 3.$$

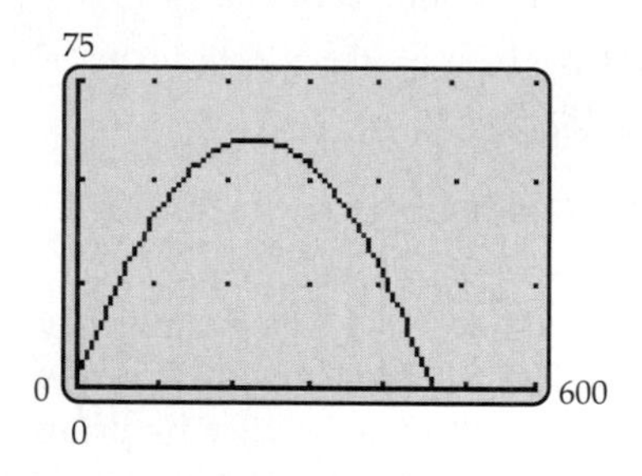

Figure 5–54

The graph of the ball's path in Figure 5–54 was made with the grid-on feature and vertical tick marks 25 units apart. It shows that the y-coordinate of the ball is greater than 25 when its x-coordinate is 400. So the ball goes over the fence. ■

TECHNOLOGY TIP In parametric graphing, zoom-in can be very time-consuming. It's often more effective to limit the t range to the values near the points you are interested in and set the t step very small. The picture may be hard to read, but trace can be used to determine coordinates.

GRAPHING EXPLORATION

Will the ball in Example 7 go over the fence if its initial velocity is 135 feet per second? Use degree mode and the viewing window of Figure 5–54 (with $0 \le t \le 4$ and t step = .1) to graph the ball's path. You might need to use trace if the graph is hard to read. If the answer still isn't clear, try changing the t step to .02.

Our final example is a curve that has several interesting applications.

EXAMPLE 8

Choose a point P on a circle of radius 3, and find a parametric description of the curve that is traced out by P as the circle rolls along the x-axis, as shown in Figure 5–55.

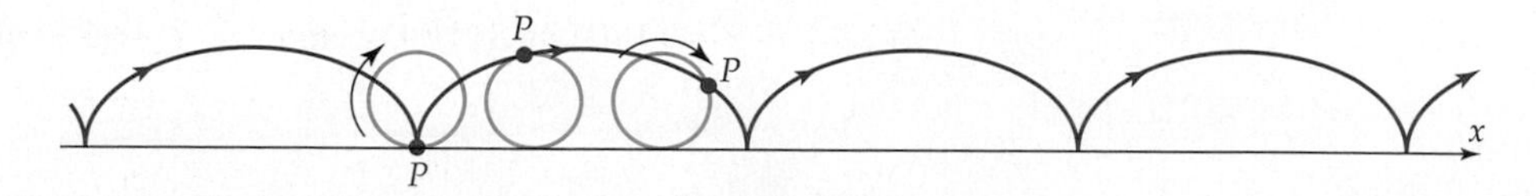

Figure 5–55

SOLUTION This curve is called a **cycloid.** Begin with P at the origin and the center C of the circle at (0, 3). As the circle rolls along the x-axis, the line segment CP moves from vertical through an angle of t radians, as shown in Figure 5–56.

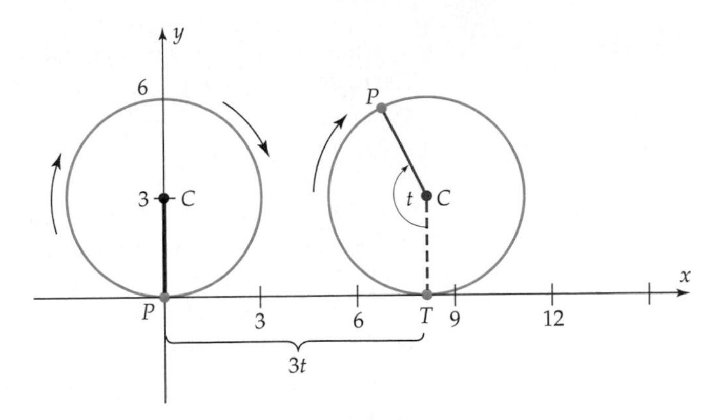

Figure 5–56

The distance from point T to the origin is the length of arc of the circle from T to P. As shown on page 144, this arc has length $3t$. Therefore the center C has coordinates $(3t, 3)$. When $0 < t < \pi/2$, the situation looks like Figure 5–57.

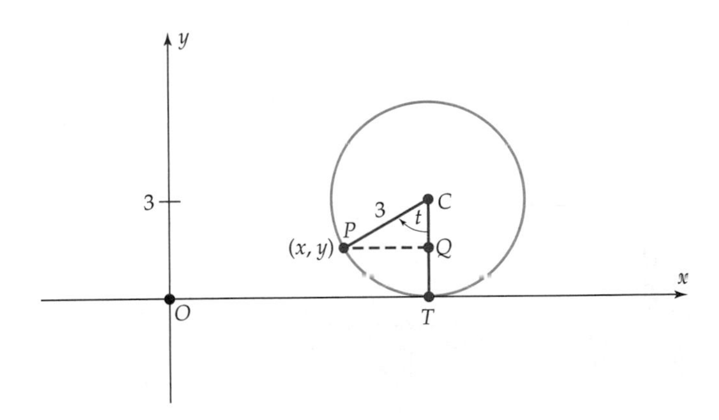

Figure 5–57

Right triangle PQC shows that

$$\sin t = \frac{PQ}{3} \quad \text{or, equivalently,} \quad PQ = 3 \sin t$$

and

$$\cos t = \frac{CQ}{3} \quad \text{or, equivalently,} \quad CQ = 3 \cos t.$$

In Figure 5–57, P has coordinates (x, y) and we have

$$x = OT - PQ = 3t - 3 \sin t = 3(t - \sin t),$$

$$y = CT - CQ = 3 - 3 \cos t = 3(1 - \cos t).$$

A similar analysis for other values of t (Exercises 45–47) shows that these equations are valid for every t. Therefore, the parametric equations of this cycloid are

$$x = 3(t - \sin t) \quad \text{and} \quad y = 3(1 - \cos t) \quad (t \text{ any real number}).$$ ■

If a cycloid is traced out by a circle of radius r, then the argument given in Example 8, with r in place of 3, shows that the parametric equations of the cycloid are

$$x = r(t - \sin t) \quad \text{and} \quad y = r(1 - \cos t) \quad (t \text{ any real number}).$$

Cycloids have a number of interesting applications. For example, among all the possible paths joining points P and Q in Figure 5–58, an arch of an inverted cycloid (shown in red) is the curve along which a particle (subject only to gravity) will slide from P to Q in the shortest possible time. This fact was first proved by J. Bernoulli in 1696.

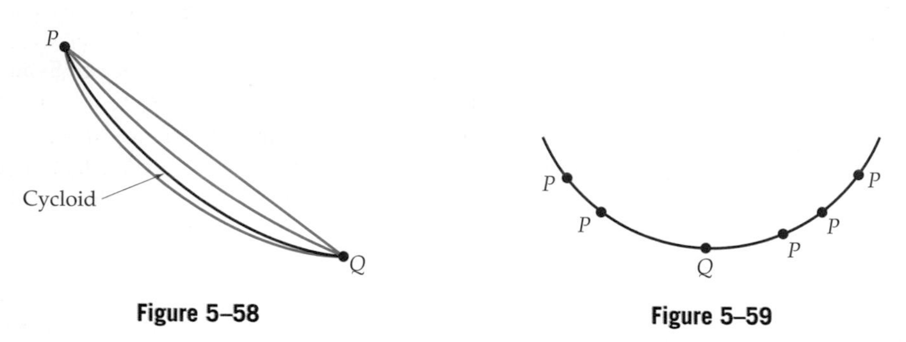

Figure 5–58 **Figure 5–59**

The Dutch physicist Christiaan Huygens (who invented the pendulum clock) proved that a particle takes the same time to slide to the bottom point Q of an inverted cycloid arch (as in Figure 5–59) from *any* point P on the curve.

EXERCISES 5.5

In Exercises 1–14, find a viewing window that shows a complete graph of the curve.

1. $x = t^2 - 4, \quad y = t/2, \quad -2 \le t \le 3$
2. $x = 3t^2, \quad y = 2 + 5t, \quad 0 \le t \le 2$
3. $x = 2t, \quad y = t^2 - 1, \quad -1 \le t \le 2$
4. $x = t - 1, \quad y = \dfrac{t+1}{t-1}, \quad t \ge 1$
5. $x = 4 \sin 2t + 9, \quad y = 6 \cos t - 8, \quad 0 \le t \le 2\pi$
6. $x = t^3 - 3t - 8, \quad y = 3t^2 - 15, \quad -4 \le t \le 4$
7. $x = 6 \cos t + 12 \cos^2 t, \quad y = 8 \sin t + 8 \sin t \cos t, \quad 0 \le t \le 2\pi$
8. $x = 12 \cos t, \quad y = 12 \sin 2t, \quad 0 \le t \le 2\pi$
9. $x = 6 \cos t + 5 \cos 3t, \quad y = 6 \sin t - 5 \sin 3t, \quad 0 \le t \le 2\pi$
10. $x = 3t^2 + 10, \quad y = 4t^3, \quad$ any real number t
11. $x = 12 \cos 3t \cos t + 6, \quad y = 12 \cos 3t \sin t - 7, \quad 0 \le t \le 2\pi$
12. $x = 2 \cos 3t - 6, \quad y = 2 \cos 3t \sin t + 7, \quad 0 \le t \le 2\pi$
13. $x = t \sin t, \quad y = t \cos t, \quad 0 \le t \le 8\pi$
14. $x = 9 \sin t, \quad y = 9t \cos t, \quad 0 \le t \le 20$

In Exercises 15–22, the given curve is part of the graph of an equation in x and y. Find the equation by eliminating the parameter.

15. $x = t - 3, \quad y = 2t + 1, \quad t \ge 0$
16. $x = t + 5, \quad y = \sqrt{t}, \quad t \ge 0$
17. $x = -2 + t^2, \quad y = 1 + 2t^2, \quad$ any real number t

18. $x = t^2 + 1$, $y = t^2 - 1$, any real number t

19. $x = e^t$, $y = t$, any real number t

20. $x = 2e^t$, $y = 1 - e^t$, $t \geq 0$

21. $x = 3 \cos t$, $y = 3 \sin t$, $0 \leq t \leq 2\pi$

22. $x = 4 \sin 2t$, $y = 2 \cos 2t$, $0 \leq t \leq 2\pi$

In Exercises 23–26, find a parameterization of the given curve. Confirm your answer by graphing.

23. circle with center $(7, -4)$ and radius 6

24. circle with center $(9, 12)$ and radius 5

25. $x^2 + y^2 - 14x + 8y + 29 = 0$ [*Hint:* Complete the square in both x and y.]

26. $x^2 + y^2 - 4x - 6y + 9 = 0$

27. (a) What is the slope of the line through (a, b) and (c, d)?
(b) Use the slope from part (a) and the point (a, b) to write the equation of the line. Do not simplify.
(c) Show that the curve with parametric equations

$$x = a + (c - a)t \quad \text{and} \quad y = b + (d - b)t$$
$$(t \text{ any real number})$$

is the line through (a, b) and (c, d). [*Hint:* Solve both equations for t, and set the results equal to each other; compare with the equation in part (b).]

28. Find parametric equations whose graph is the line segment joining the points (a, b) and (c, d). [*Hint:* Adjust the range of t values in Exercise 27(c).]

In Exercises 29–32, use Exercise 28 to find a parameterization of the line segment joining the two points. Confirm your answer by graphing.

29. $(-6, 12)$ and $(12, -10)$ **30.** $(14, -5)$ and $(5, -14)$

31. $(18, 4)$ and $(-16, 14)$

32. (a) Find a parameterization of the line segment joining $(-5, -3)$ and $(7, 4)$, as in Exercises 28–31.
(b) Explain why another parameterization of this line segment is given by

$$x = -5 + 12 \sin t \quad \text{and}$$
$$y = -3 + 7 \sin t \quad (0 \leq t \leq \pi/2).$$

(c) Use the trace feature to verify that the segment is traced out twice when the t-range in part (b) is changed to $0 \leq t \leq \pi$ (use t-step $= \pi/20$). Explain why.
(d) What happens when $0 \leq t \leq 2\pi$?

33. (a) Graph the curve given by

$$x = \sin kt \quad \text{and} \quad y = \cos t \quad (0 \leq t \leq 2\pi)$$

when $k = 1, 2, 3$, and 4. Use the window with

$$-1.5 \leq x \leq 1.5 \quad \text{and} \quad -1.5 \leq y \leq 1.5$$

and t-step $= \pi/30$.
(b) Without graphing, predict the shape of the graph when $k = 5$ and $k = 6$. Then verify your predictions graphically.

34. (a) Graph the curve given by

$$x = 3 \sin 2t \quad \text{and} \quad y = 2 \cos kt \quad (0 \leq t \leq 2\pi)$$

when $k = 1, 2, 3, 4$. Use the window with $-3.5 \leq x \leq 3.5$ and $-2.5 \leq y \leq 2.5$ and t-step $= \pi/30$.
(b) Predict the shape of the graph when $k = 5, 6, 7, 8$. Verify your predictions graphically.

In Exercises 35–42, use a calculator in degree mode and assume that air resistance is negligible.

35. A skeet target is fired from the ground with an initial velocity of 110 feet per second at an angle of 28°.
(a) Graph the target's path.
(b) How long is the target in the air?
(c) How high does it go?

36. A ball is thrown from a height of 5 feet above the ground with an initial velocity of 60 feet per second at an angle of 50° with the horizontal.
(a) Graph the ball's path.
(b) When and where does the ball hit ground?

37. A medieval bowman shoots an arrow, which leaves the bow 4 feet above the ground with an initial velocity of 88 feet per second at an angle of 48° with the horizontal.
(a) Graph the arrow's path.
(b) Will the arrow go over the 40-foot-high castle wall that is 200 feet from the archer?

38. A golfer at a driving range stands on a platform two feet above the ground and hits the ball with an initial velocity of 120 feet per second at an angle of 39° with the horizontal. There is a 32-foot-high fence 400 feet away. Will the ball fall short, hit the fence, or go over it?

39. A golf ball is hit off the tee at an angle of 30° and lands 300 feet away. What was its initial velocity? [*Hint:* The ball lands when $x = 300$ and $y = 0$. Use this fact and the parametric equations for the ball's path to find two equations in the variables t and v. Solve for v.]

40. A football kicked from the ground has an initial velocity of 75 feet per second.
(a) Set up the parametric equations that describe the ball's path. Experiment graphically with different angles to find the smallest angle (within one degree) needed so that the ball travels at least 150 feet.
(b) Use algebra and trigonometry to find the angle needed for the ball to travel exactly 150 feet. [*Hint:* The ball lands when $x = 150$ and $y = 0$. Use this fact and the parametric equations for the ball's path to find two equations in the variables t and θ. Solve the "x equation" for t, and substitute this result in the other one; then solve for θ. The double-angle identity may be

helpful for putting this equation in a form that is easy to solve.]

41. A golf ball is hit off the ground at an angle of θ degrees with an initial velocity of 100 feet per second.
(a) Graph the path of the ball when $\theta = 20°$, $\theta = 40°$, $\theta = 60°$, and $\theta = 80°$.
(b) For what angle in part (a) does the ball land farthest from where it started?
(c) Experiment with different angles, as in parts (a) and (b), and make a conjecture as to which angle results in the ball landing farthest from its starting point.

42. A golf ball is hit off the ground at an angle of θ degrees with an initial velocity of 100 feet per second.
(a) Graph the path of the ball when $\theta = 30°$ and when $\theta = 60°$. In which case does the ball land farthest away?
(b) Do part (a) when $\theta = 25°$ and $\theta = 65°$.
(c) Experiment further, and make a conjecture as to the results when the sum of the two angles is 90°.
(d) Prove your conjecture algebraically. [*Hint:* Find the value of t at which a ball hit at angle θ hits the ground (which occurs when $y = 0$); this value of t will be an expression involving θ. Find the corresponding value of x (which is the distance of the ball from the starting point). Then do the same for an angle of $90° - \theta$ and use the cofunction identities (in degrees) to show that you get the same value of x.]

Thinkers

43. Sketch the graphs of these curves and compare them. Do they differ? If so, how?
(a) $x = -4 + 6t$ and $y = 7 - 12t$ $(0 \le t \le 1)$
(b) $x = 2 - 6t$ and $y = -5 + 12t$ $(0 \le t \le 1)$

44. Show that the ball's path in Example 6 is a parabola by eliminating the parameter in the parametric equations

$$x = (140 \cos 31°)t \quad \text{and} \quad y = (140 \sin 31°)t - 16t^2.$$

[*Hint:* Solve the first equation for t, and substitute the result in the second equation.]

In Exercises 45–47, complete the derivation of the parametric equations of the cycloid in Example 8.

45. (a) If $\pi/2 < t < \pi$, verify that angle θ in the figure has measure $t - \pi/2$ and that

$$x = OT - CQ = 3t - 3\cos\left(t - \frac{\pi}{2}\right),$$

$$y = CT + PQ = 3 + 3\sin\left(t - \frac{\pi}{2}\right).$$

(b) Use the addition and subtraction identities for sine and cosine to show that in this case,

$$x = 3(t - \sin t) \quad \text{and} \quad y = 3(1 - \cos t).$$

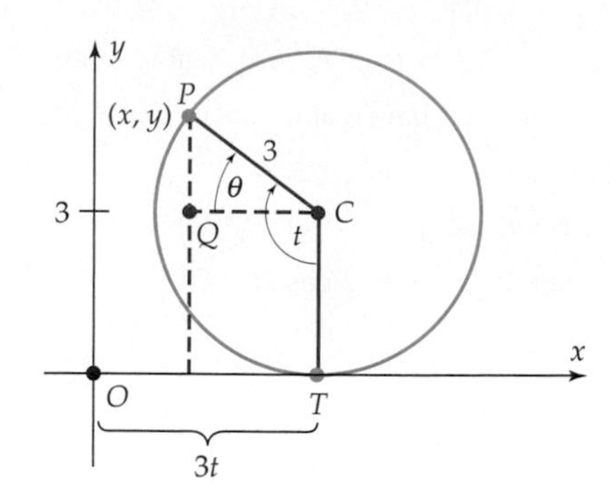

46. (a) If $\pi < t < 3\pi/2$, verify that angle θ in the figure measures $3\pi/2 - t$ and that

$$x = OT + CQ = 3t + 3\cos\left(\frac{3\pi}{2} - t\right),$$

$$y = CT + PQ = 3 + 3\sin\left(\frac{3\pi}{2} - t\right).$$

(b) Use the addition and subtraction identities for sine and cosine to show that in this case

$$x = 3(t - \sin t) \quad \text{and} \quad y = 3(1 - \cos t).$$

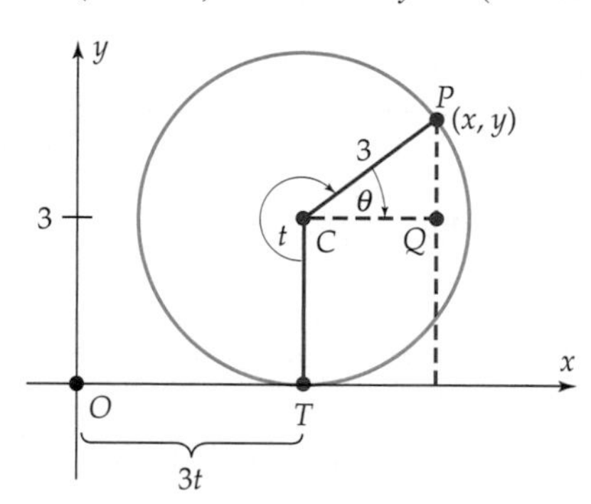

47. (a) If $3\pi/2 < t < 2\pi$, verify that angle θ in the figure has measure $t - 3\pi/2$ and that

$$x = OT + CQ = 3t + 3\cos\left(t - \frac{3\pi}{2}\right),$$

$$y = CT - PQ = 3 - 3\sin\left(y - \frac{3\pi}{2}\right).$$

(b) Use the addition and subtraction identities for sine and cosine to show that in this case

$$x = 3(t - \sin t) \quad \text{and} \quad y = 3(1 - \cos t).$$

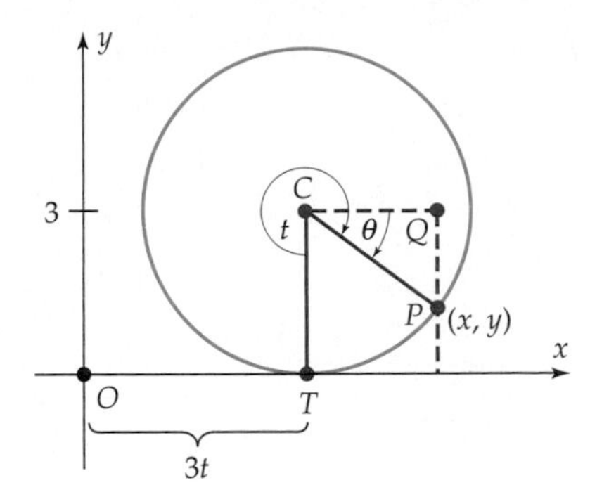

48. Let P be a point at distance k from the center of a circle of radius r. As the circle rolls along the x-axis, P traces out a curve called a **trochoid.** [When $k \leq r$, it might help to think of the circle as a bicycle wheel and P as a point on one of the spokes.]

(a) Assume that P is on the y-axis as close as possible to the x-axis when $t = 0$, and show that the parametric equations of the trochoid are

$$x = rt - k \sin t \qquad \text{and} \qquad y = r - k \cos t.$$

Note that when $k = r$, these are the equations of a cycloid.

(b) Sketch the graph of the trochoid with $r = 3$ and $k = 2$.

(c) Sketch the graph of the trochoid with $r = 3$ and $k = 4$.

49. Set your calculator for radian mode and for simultaneous graphing mode [check your instruction manual for how to do this]. Particles A, B, and C are moving in the plane, with their positions at time t seconds given by

$$A: \quad x = 8 \cos t \quad \text{and} \quad y = 5 \sin t$$

$$B: \quad x = 3t \quad \text{and} \quad y = 5t$$

$$C: \quad x = 3t \quad \text{and} \quad y = 4t.$$

(a) Graph the paths of A and B in the window with $0 \leq x \leq 12$, $0 \leq y \leq 6$, and $0 \leq t \leq 2$. The paths intersect, but do the particles actually collide? That is, are they at the same point at the same time? [For slow motion, choose a very small t step, such as .01.]

(b) Set t step $= .05$ and use trace to estimate the time at which A and B are closest to each other.

(c) Graph the paths of A and C and determine geometrically [as in part (b)] whether they collide. Approximately when are they closest?

(d) Confirm your answers in part (c) as follows. Explain why the distance between particles A and C at time t is given by

$$d = \sqrt{(8 \cos t - 3t)^2 + (5 \sin t - 4t)^2}.$$

A and C will collide if $d = 0$ at some time. Using function graphing mode, graph this distance function when $0 \leq t \leq 2$, and using zoom-in if necessary, show that d is always positive. Find the value of t for which d is smallest.

50. A particle moves on the horizontal line $y = 3$. Its x-coordinate at time t seconds is given by

$$x = 2t^3 - 13t^2 + 23t - 8.$$

This exercise explores the motion of the particle.

(a) Graph the path of the particle in the viewing window with $-10 \leq x \leq 10$, $-2 \leq y \leq 4$, $0 \leq t \leq 4.3$, and t step $= .05$. Note that the calculator seems to pause before completing the graph.

(b) Use trace (starting with $t = 0$) and watch the path of the particle as you press the right arrow key at regular intervals. How many times does it change direction? When does it appear to be moving the fastest?

(c) At what times t does the particle change direction? What are its x-coordinates at these times?

51. A circle of radius b rolls along the inside of a larger circle of radius a. The curve traced out by a fixed point P on the smaller circle is called a **hypocycloid.**

(a) Assume that the larger circle has center at the origin and that the smaller circle starts with P located at $(a, 0)$. Use the figure to show that the parametric equations of the hypocycloid are

$$x = (a - b) \cos t + b \cos\left(\frac{a - b}{b} t\right)$$

$$y = (a - b) \sin t - b \sin\left(\frac{a - b}{b} t\right).$$

(b) Sketch the graph of the hypocycloid with $a = 5$, $b = 1$, and $0 \leq t \leq 2\pi$.

(c) Sketch the graph of the hypocycloid with $a = 5$, $b = 2$, and $0 \leq t \leq 4\pi$.

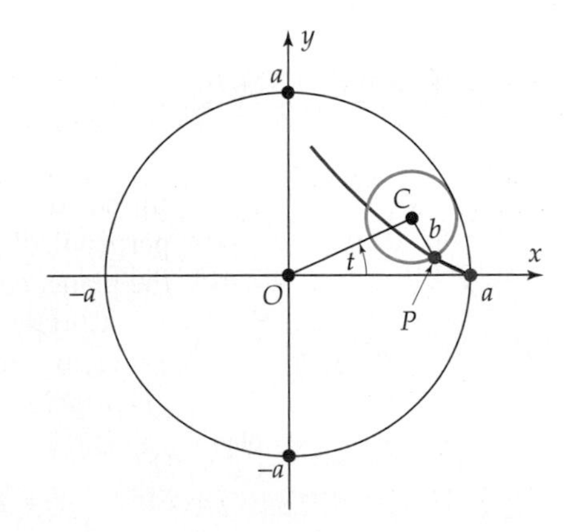

52. Jill is on a ferris wheel of radius 20 feet whose bottom just grazes the ground. The wheel is rotating counterclockwise at the rate of .7 radians per second. Jack is standing 100 feet from the bottom of the ferris wheel. When Jill is at point P, he throws a ball toward the wheel (see the figure). The ball leaves Jack's hand 5 feet above the ground with an initial velocity of 62 feet/second at an angle of 1.2 radians with the horizontal. Will Jill be able to catch the ball? Follow the steps below to answer the question.

(a) Imagine that Jack is at the origin and the bottom of the ferris wheel is at (100, 0). Then the ball leaves his hand from the point (0, 5), and the wheel is a circle with center (100, 20) and radius 20 (see the diagram in the next column). Therefore, the wheel is the graph of the equation

$$(x - 100)^2 + (y - 20)^2 = 20^2.$$

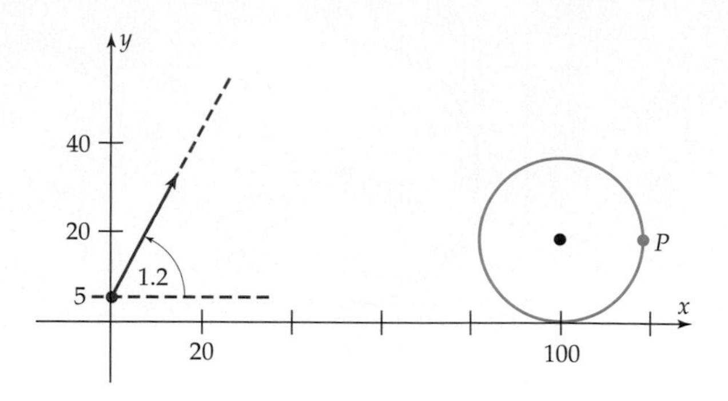

Show that Jill's movement around the wheel is given by the parametric equations

$$x = 20\cos(.7t) + 100$$

$$y = 20\sin(.7t) + 20$$

by verifying that these equations give a parameterization of the circle [as in Example 4].

(b) Find the parametric equations that describe the position of the ball at time t.

(c) Set your calculator for parametric mode, radian mode, and simultaneous graphing mode [check your instruction manual for how to do this]. Using a square viewing window with $0 \leq x \leq 130$, $0 \leq t \leq 9$, and t step $= .1$ to graph both sets of parametric equations (Jill's motion and the ball's) simultaneously. [For slow motion, make the t step smaller.] Assuming that Jill can reach 2 feet in any direction, can she catch the ball? If not, use the trace to estimate the time at which Jill is closest to the ball.

(d) Experiment by changing the angle or initial velocity (or both) of the ball to find values that will allow Jill to catch the ball.

5.6 Polar Coordinates

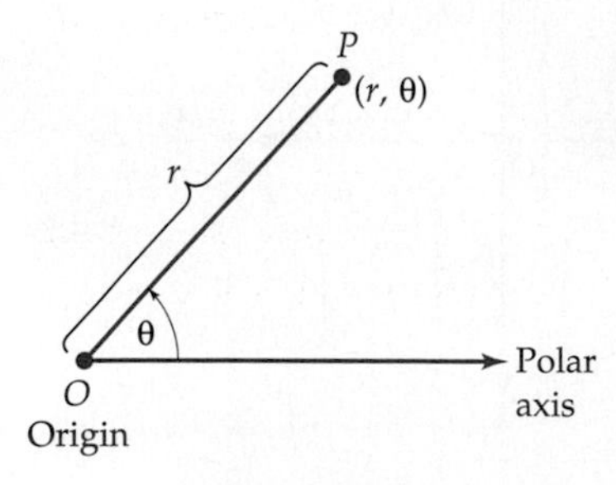

Figure 5–60

In the past we used a rectangular coordinate system in the plane, based on two perpendicular coordinate axes. Now we introduce another coordinate system for the plane, based on angles.

Choose a point O in the plane (called the **origin** or **pole**) and a half-line extending from O (called the **polar axis**). As shown in Figure 5–60, a point P is given **polar coordinates** (r, θ), where

r = Distance from P to O

θ = Angle with polar axis as initial side and OP as terminal side.

We shall usually measure the angle θ in radians; it may be either positive or negative, depending on whether it is generated by a counterclockwise or clockwise rotation. Some typical points are shown in Figure 5–61, which also illustrates the "circular grid" that a polar coordinate system imposes on the plane.

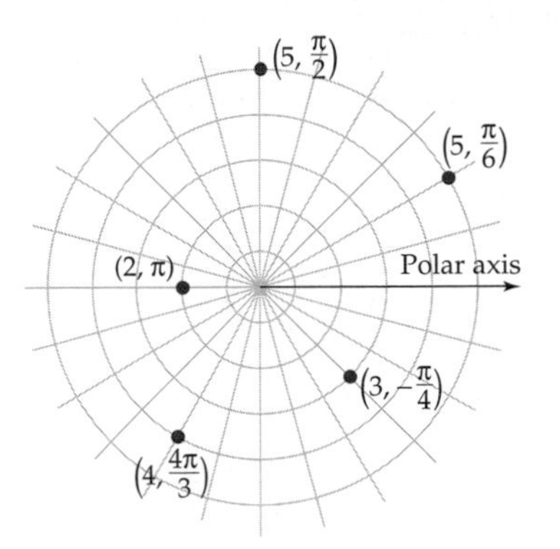

Figure 5–61

The polar coordinates of a point P are *not* unique. The angle θ may be replaced by any angle that has the same terminal side as θ, such as $\theta \pm 2\pi$. For instance, the coordinates $(2, \pi/3)$, $(2, 7\pi/3)$, and $(2, -5\pi/3)$ all represent the same point, as shown in Figure 5–62.

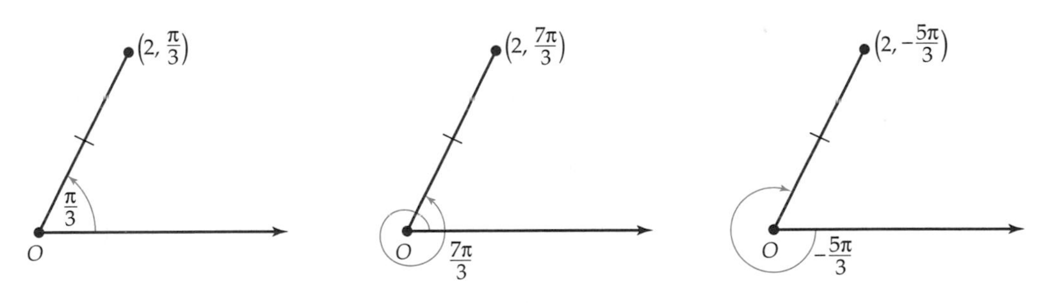

Figure 5–62

We shall consider the coordinates of the origin to be $(0, \theta)$, where θ is *any* angle.

Negative values for the first coordinate will be allowed according to this convention: For each positive r, the point $(-r, \theta)$ lies on the straight line containing the terminal side of θ, at distance r from the origin, on the *opposite* side of the origin from the point (r, θ), as shown in Figure 5–63 on the next page.

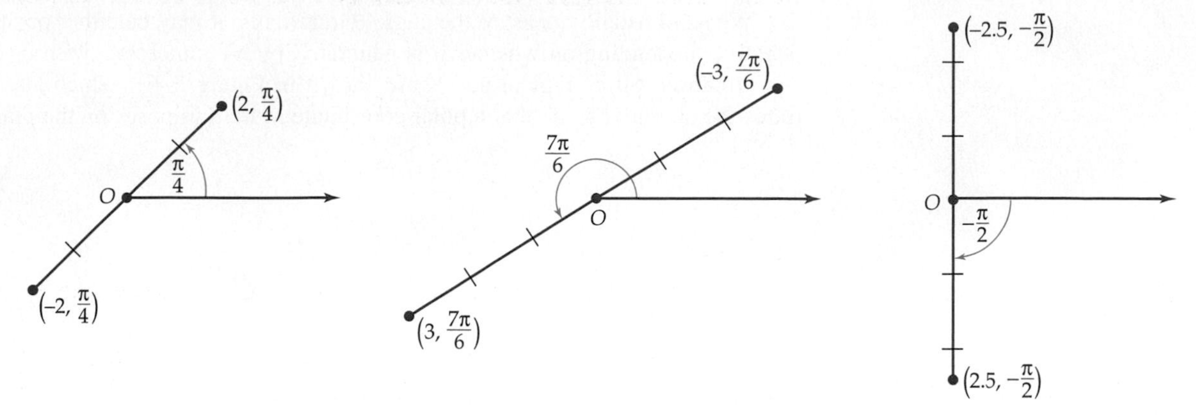

Figure 5–63

It is sometimes convenient to use both a rectangular and a polar coordinate system in the plane, with the polar axis coinciding with the positive x-axis. Then the y-axis is the polar line $\theta = \pi/2$. Suppose P has rectangular coordinates (x, y) and polar coordinates (r, θ), with $r > 0$, as in Figure 5–64.

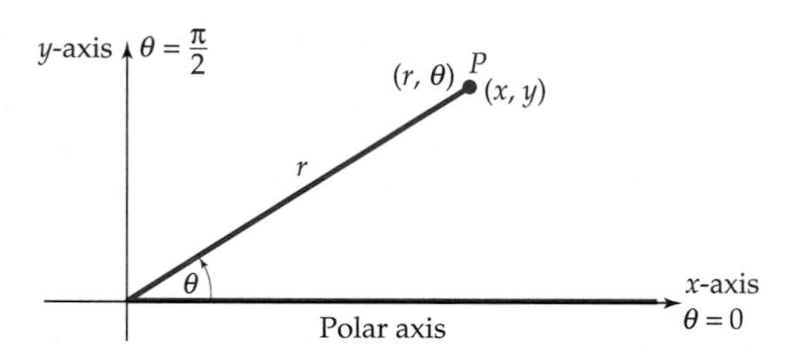

Figure 5–64

Since r is the distance from (x, y) to $(0, 0)$, the distance formula shows that $r = \sqrt{x^2 + y^2}$, and hence, $r^2 = x^2 + y^2$. The point-in-the-plane description of the trigonometric functions shows that

$$\cos\theta = \frac{x}{r} \qquad \sin\theta = \frac{y}{r} \qquad \tan\theta = \frac{y}{x}.$$

Solving the first two equations for x and y, we obtain the relationship between polar and rectangular coordinates.*

Coordinate Conversion Formulas

> If a point has polar coordinates (r, θ), then its rectangular coordinates (x, y) are
>
> $$x = r\cos\theta \quad \text{and} \quad y = r\sin\theta.$$
>
> If a point has rectangular coordinates (x, y), then its polar coordinates (r, θ) satisfy
>
> $$r^2 = x^2 + y^2 \quad \text{and} \quad \tan\theta = \frac{y}{x}.$$

*The conclusions in the box are also true when $r < 0$ (see Exercise 86).

EXAMPLE 1

Convert each of the following points in polar coordinates to rectangular coordinates.

(a) $(2, \pi/6)$ (b) $(3, 4)$

SOLUTION

TECHNOLOGY TIP

Keys to convert from rectangular to polar coordinates, or vice versa, are in this menu/submenu:

TI-83+:	ANGLE
TI-86:	VECTOR/OPS
TI-89:	MATH/ANGLE
Casio:	OPTN/ANGLE

(a) Apply the first set of equations in the box with $r = 2$ and $\theta = \pi/6$.

$$x = 2\cos\frac{\pi}{6} = 2 \cdot \frac{\sqrt{3}}{2} = \sqrt{3} \quad \text{and} \quad y = 2\sin\frac{\pi}{6} = 2 \cdot \frac{1}{2} = 1$$

So the rectangular coordinates are $(\sqrt{3}, 1)$.

(b) The point with polar coordinates $(3, 4)$ has $r = 3$ and $\theta = 4$ radians. Therefore, its rectangular coordinates are

$$(r\cos\theta, r\sin\theta) = (3\cos 4, 3\sin 4) \approx (-1.9609, -2.2704). \quad ■$$

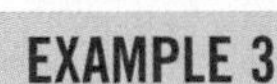

EXAMPLE 2

Find the polar coordinates of the point with rectangular coordinates $(2, -2)$.

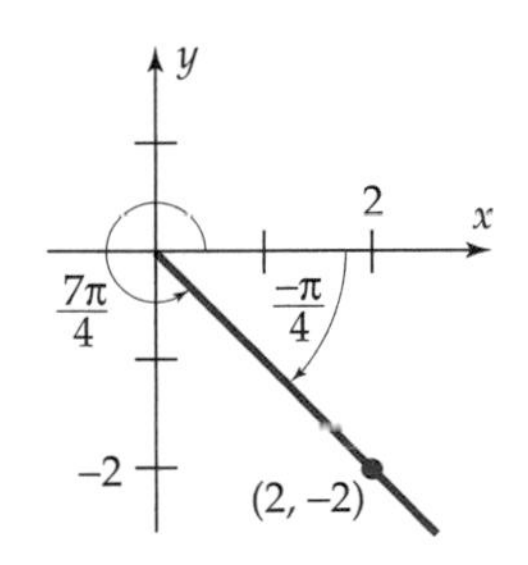

Figure 5–65

SOLUTION

The second set of equations in the box, with $x = 2$, $y = -2$, shows that

$$r = \sqrt{2^2 + (-2)^2} = \sqrt{8} = 2\sqrt{2} \quad \text{and} \quad \tan\theta = -2/2 = -1.$$

We must find an angle θ whose terminal side passes through $(2, -2)$ and whose tangent is -1. Figure 5–65 shows that two of the many possibilities are

$$\theta = -\frac{\pi}{4} \quad \text{and} \quad \theta = \frac{7\pi}{4}.$$

So one pair of polar coordinates is $\left(2\sqrt{2}, -\frac{\pi}{4}\right)$, and another is $\left(2\sqrt{2}, \frac{7\pi}{4}\right)$. ■

Rectangular to polar conversion is relatively easy when special angles are involved, as in Example 2. In other cases technology may be necessary.

EXAMPLE 3

Find the polar coordinates of the points whose rectangular coordinates are

(a) $(3, 5)$ (b) $(-2, 4)$

SOLUTION

(a) Applying the second set of equations in the box, with $x = 3$, $y = 5$, we have

$$r = \sqrt{3^2 + 5^2} = \sqrt{34} \qquad \text{and} \qquad \tan\theta = 5/3.$$

The TAN^{-1} key on a calculator shows that $\theta \approx 1.0304$ radians is an angle between 0 and $\pi/2$ with tangent $5/3$. Since $(3, 5)$ is in the first quadrant, one pair of (approximate) polar coordinates is $(\sqrt{34}, 1.0304)$.

(b) In this case,

$$r = \sqrt{(-2)^2 + 4^2} = \sqrt{20} = 2\sqrt{5} \qquad \text{and} \qquad \tan\theta = \frac{4}{-2} = -2.$$

Using the TAN^{-1} key, we find that $\theta \approx -1.1071$ is an angle between $-\pi/2$ and 0 with tangent -2. However, we want an angle between $\pi/2$ and π because $(-2, 4)$ is in the second quadrant. Since tangent has period π,

$$\tan(-1.1071 + \pi) = \tan(-1.1071) = -2.$$

Thus, $-1.1071 + \pi \approx 2.0344$ is an angle between $\pi/2$ and π whose tangent is -2. Therefore, one pair of polar coordinates is $(2\sqrt{5}, 2.0344)$. ■

The technique used in Example 3 may be summarized as follows.

Rectangular to Polar Conversion

If the rectangular coordinates of a point are (x, y), let $r = \sqrt{x^2 + y^2}$. If (x, y) lies in the first or fourth quadrant, then its polar coordinates are

$$\left(r, \tan^{-1}\frac{y}{x}\right).$$

If (x, y) lies in the second or third quadrant, then its polar coordinates are

$$\left(r, \tan^{-1}\left(\frac{y}{x}\right) + \pi\right).$$

EQUATION CONVERSION

When an equation in rectangular coordinates is given, it can be converted to polar coordinates by making the substitutions $x = r\cos\theta$ and $y = r\cos\theta$. Converting a polar equation to rectangular coordinates, however, is a bit trickier.

EXAMPLE 4

Find an equivalent rectangular equation for the given polar equation, and use it to identify the shape of the graph.

(a) $r = 4\cos\theta$ (b) $r = \dfrac{1}{1 - \sin\theta}$

SOLUTION

(a) Rewrite the equation $r = 4\cos\theta$ as follows.

Multiply both sides by r: $r^2 = 4r\cos\theta$

Substitute $r^2 = x^2 + y^2$ and $r\cos\theta = x$: $x^2 + y^2 = 4x$

$x^2 - 4x + y^2 = 0$

Add 4 to both sides: $(x^2 - 4x + 4) + y^2 = 4$

Factor: $(x - 2)^2 + y^2 = 2^2$

As we saw in Section 0.2, the graph of this equation is the circle with center (2, 0) and radius 2.

(b) Begin by eliminating fractions.

$$r = \frac{1}{1 - \sin\theta}$$

$$r(1 - \sin\theta) = 1$$

$$r - r\sin\theta = 1$$

Substitute $r = \sqrt{x^2 + y^2}$ and $y = r\sin\theta$: $\sqrt{x^2 + y^2} - y = 1$

Rearrange terms: $\sqrt{x^2 + y^2} = y + 1$

Square both sides: $x^2 + y^2 = (y + 1)^2$

Simplify: $x^2 + y^2 = y^2 + 2y + 1$

$$x^2 = 2y + 1$$

$$y = \frac{1}{2}(x^2 - 1)$$

As we saw in Section 5.3, the graph is an upward-opening parabola. ■

POLAR GRAPHS

The graphs of a few polar coordinate equations can be easily determined from the appropriate definitions.

EXAMPLE 5

Graph the equations.

(a) $r = 3$* (b) $\theta = \pi/6$*

SOLUTION

(a) The graph consists of all points (r, θ) with first coordinate 3, that is, all points whose distance from the origin is 3. So the graph is a circle with center O and radius 3, as shown in Figure 5–66.

(b) The graph consists of all points $(r, \pi/6)$. If $r \geq 0$, then $(r, \pi/6)$ lies on the terminal side of an angle of $\pi/6$ radians, whose initial side is the polar axis. If $r < 0$, then $(r, \pi/6)$ lies on the extension of this terminal side across the origin. So the graph is the straight line in Figure 5–67. ■

> **NOTE**
> The graphs in Example 5 can also be found by converting the polar equations to equivalent rectangular ones (see Exercises 41 and 43).

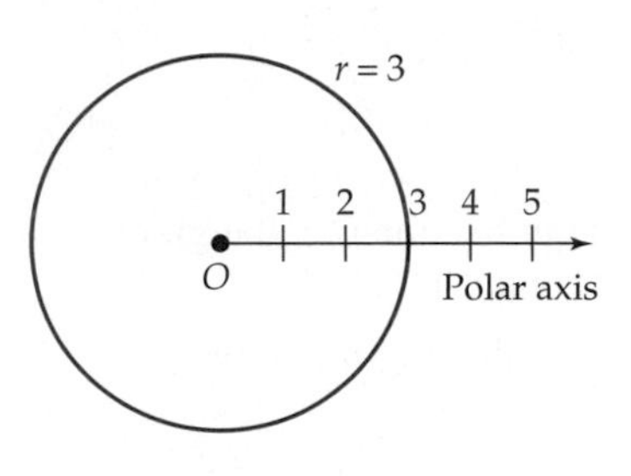

Figure 5–66

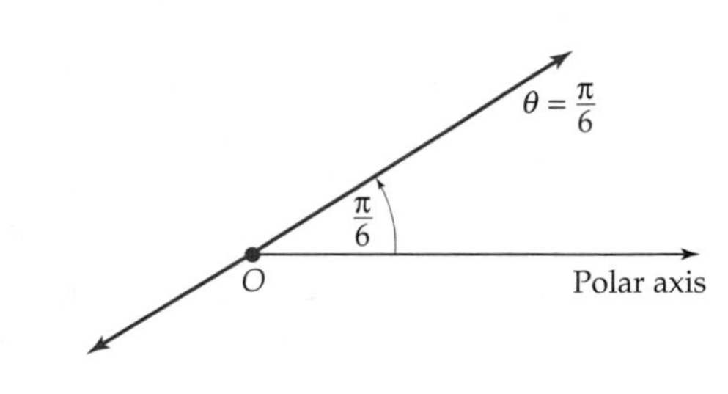

Figure 5–67

Some polar graphs can be sketched by hand by using basic facts about trigonometric functions.

EXAMPLE 6

Graph $r = 1 + \sin \theta$.

*Every equation is understood to involve two variables, but one may have coefficient 0, as is the case here: $r = 3 + 0 \cdot \theta$ and $\theta = 0 \cdot r + \pi/6$. This is analogous to equations such as $y = 5$ and $x = 2$ in rectangular coordinates.

SOLUTION Remember the behavior of $\sin\theta$ between 0 and 2π.

As θ increases from 0 to $\pi/2$, $\sin\theta$ increases from 0 to 1. So $r = 1 + \sin\theta$ increases from 1 to 2.

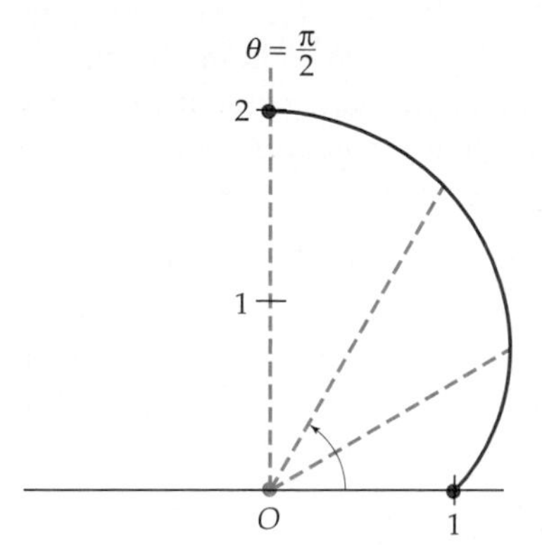

As θ increases from $\pi/2$ to π, $\sin\theta$ decreases from 1 to 0. So $r = 1 + \sin\theta$ decreases from 2 to 1.

As θ increases from π to $3\pi/2$, $\sin\theta$ decreases from 0 to -1. So $r = 1 + \sin\theta$ decreases from 1 to 0.

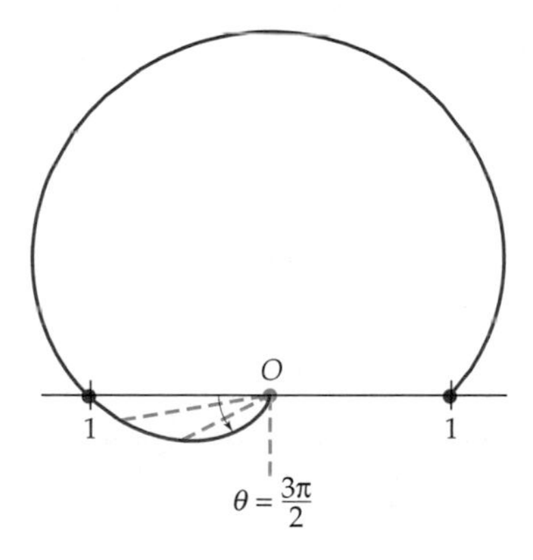

As θ increases from $3\pi/2$ to 2π, $\sin\theta$ increases from -1 to 0. So $r = 1 + \sin\theta$ increases from 0 to 1.

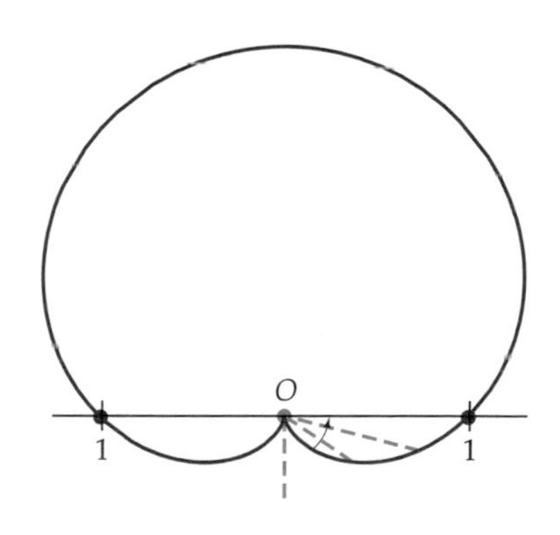

Figure 5–68

As θ takes values larger than 2π, $\sin\theta$ repeats the same pattern, and hence, so does $r = 1 + \sin\theta$. The same is true for negative values of θ. The full graph (called a **cardioid**) is at the lower right in Figure 5–68. ■

The easiest way to graph polar equation $r = f(\theta)$ is to use a calculator in polar graphing mode. A second way is to use parametric graphing mode, with the coordinate converison formulas as a parameterization.

$$x = r\cos\theta = f(\theta)\cos\theta$$

$$y = r\sin\theta = f(\theta)\sin\theta.$$

EXAMPLE 7

Graph $r = 2 + 4\cos\theta$.

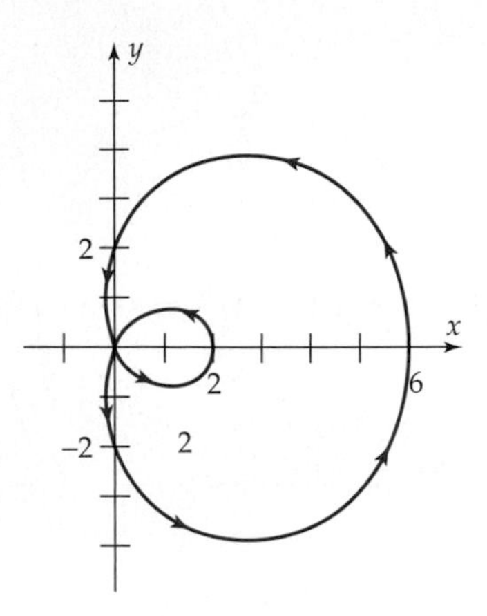

Figure 5–69

SOLUTION *Polar Method:* Put your calculator in polar graphing mode, and enter $r = 2 + 4\cos\theta$ in the function memory. Set the viewing window by entering minimum and maximum values for x, y, and θ. Since cosine has period 2π, a complete graph can be obtained by taking $0 \le \theta \le 2\pi$. You must also set the θ step (or θ pitch), which determines how many values of θ the calculator uses to plot the graph. With an appropriate θ step, the graph should look like Figure 5–69.

Parametric Method: Put your calculator in parametric graphing mode. The parametric equations for $r = 2 + 4\cos\theta$ are as follows (using t as the variable instead of θ with $0 \le t \le 2\pi$).

$$x = r\cos t = (2 + 4\cos t)\cos t = 2\cos t + 4\cos^2 t$$

$$y = r\sin t = (2 + 4\cos t)\sin t = 2\sin t + 4\sin t\cos t$$

They also produce the graph in Figure 5–69. ■

EXAMPLE 8

The graph of $r = \sin 2\theta$ in Figure 5–70 can be obtained either by graphing directly in polar mode or by using parametric mode and the equations

$$x = r\cos t = \sin 2t\cos t \quad \text{and} \quad y = r\sin t = \sin 2t\sin t \quad (0 \le t \le 2\pi).$$ ■

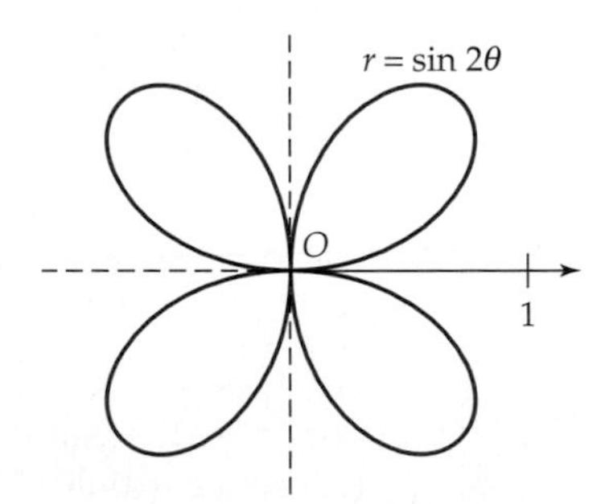

Figure 5–70

Here is a summary of commonly encountered polar graphs (in each case, a and b are constants).

Equation	Name of Graph	Shape of Graph*
$r = a\theta\ (\theta \geq 0)$ $r = a\theta\ (\theta \leq 0)$	Archimedean spiral	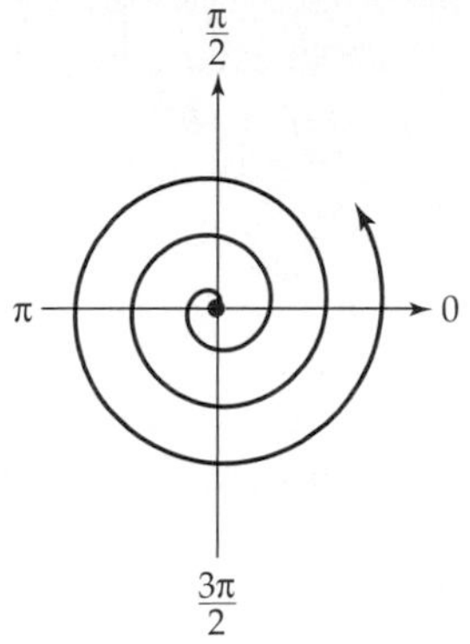 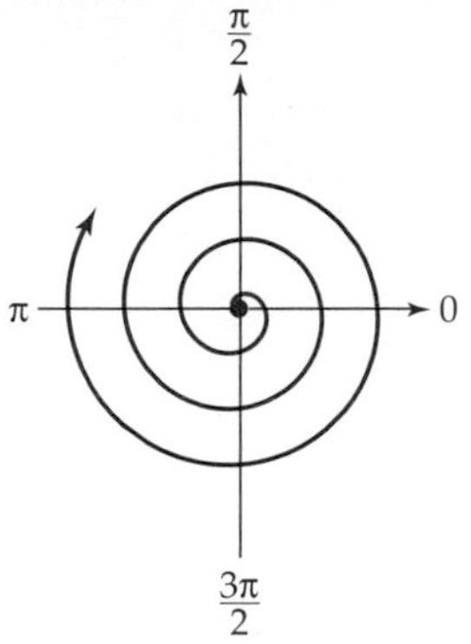 $r = a\theta\ (\theta \geq 0)$ $r = a\theta\ (\theta \leq 0)$
$r = a(1 \pm \sin\theta)$ $r = a(1 \pm \cos\theta)$	cardioid	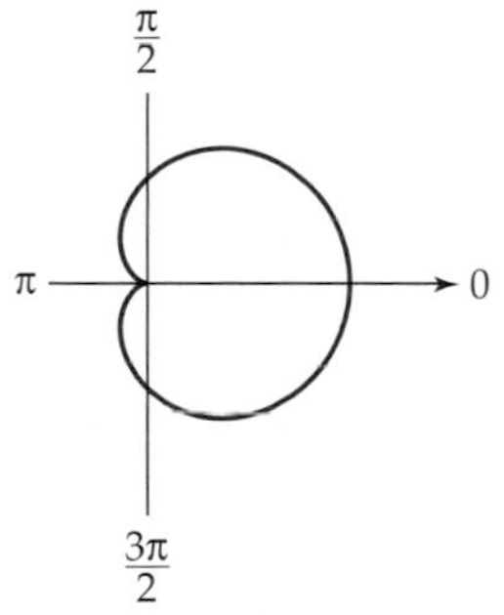 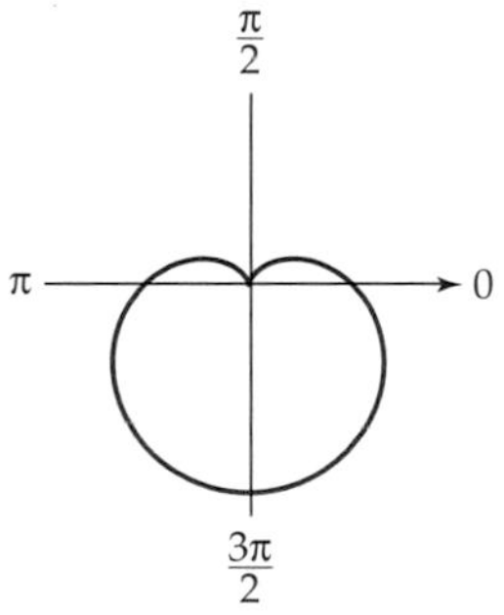 $r = a(1 + \cos\theta)$ $r = a(1 - \sin\theta)$
$r = a\sin n\theta$ $r = a\cos n\theta$ $(n \geq 2)$	rose (There are n petals when n is odd and $2n$ petals when n is even.)	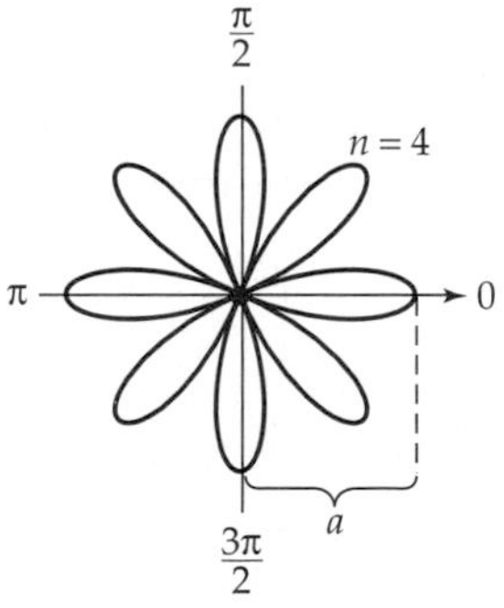 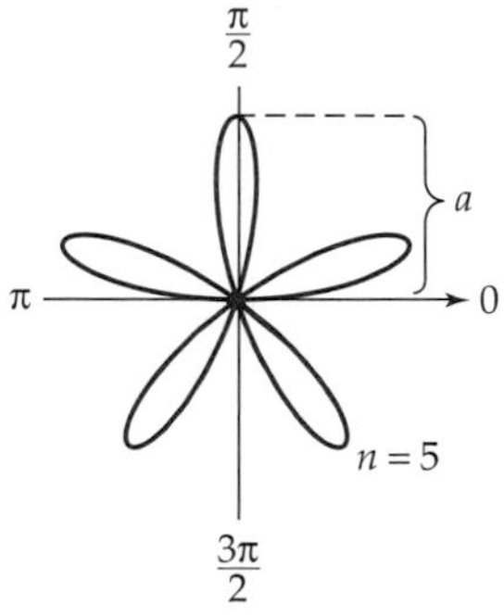 $r = a\cos n\theta$ ($n = 4$) $r = a\sin n\theta$ ($n = 5$)

Continued on next page

*Depending on the plus or minus sign and whether sine or cosine is involved, the basic shape of a specific graph may differ from those shown by a rotation, reflection, or horizontal or vertical shift.

Equation	Name of Graph	Shape of Graph
$r = a \sin \theta$ $r = a \cos \theta$	circle	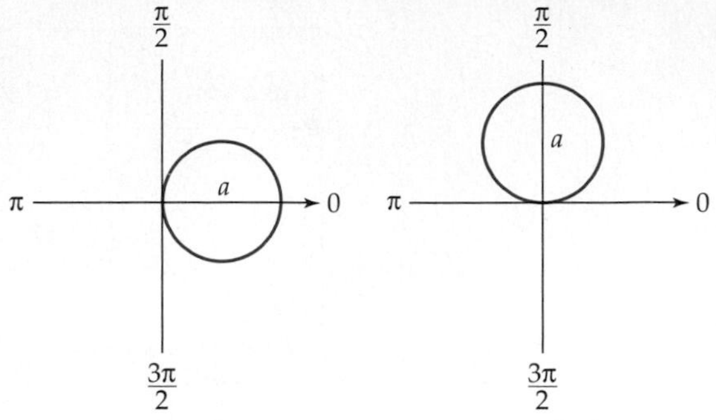 $r = a \cos \theta$ $r = a \sin \theta$
$r^2 = \pm a^2 \sin 2\theta$ $r^2 = \pm a^2 \cos 2\theta$	lemniscate	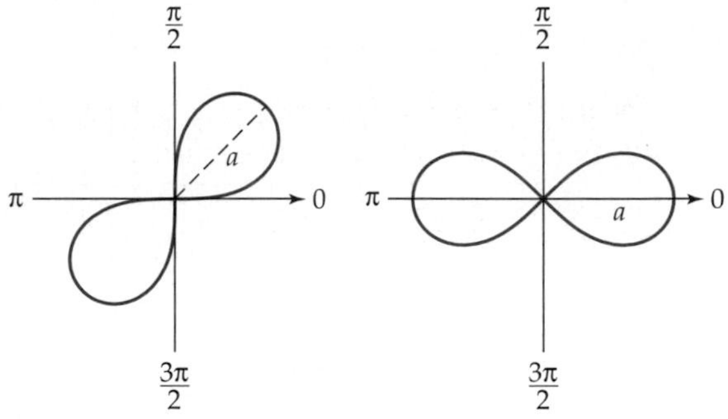 $r^2 = a^2 \sin 2\theta$ $r^2 = a^2 \cos 2\theta$
$r = a \pm b \sin \theta$ $r = a \pm b \cos \theta$ $(a, b > 0; a \neq b)$	limaçon	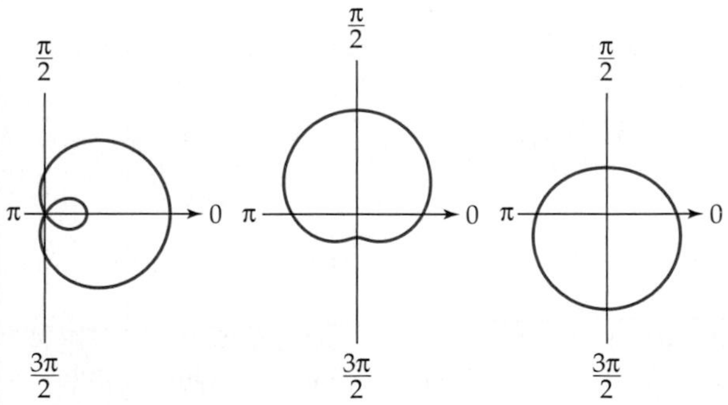 $a < b$, $r = a + b \cos \theta$ $b < a < 2b$, $r = a + b \sin \theta$ $a \geq 2b$, $r = a - b \sin \theta$

EXERCISES 5.6

1. What are the polar coordinates of the points P, Q, R, S, T, U, V in the figure?

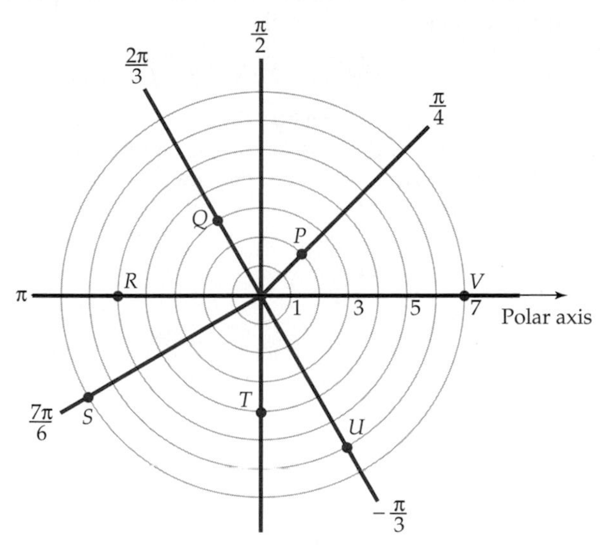

In Exercises 2–6, plot the point whose polar coordinates are given.

2. $(1, \pi/4)$
3. $(2, -3\pi/4)$
4. $(-2, 2\pi/3)$
5. $(-3, -5\pi/3)$
6. $(3, \pi/6)$

In Exercises 7–12, list four other pairs of polar coordinates for the given point, each with a different combination of signs (that is, $r > 0$, $\theta > 0$; $r > 0$, $\theta < 0$; $r < 0$, $\theta > 0$; $r < 0$, $\theta < 0$).

7. $(3, \pi/3)$
8. $(-5, \pi)$
9. $(2, -2\pi/3)$
10. $(-1, -\pi/6)$
11. $(\sqrt{3}, 3\pi/4)$
12. $(-3, 7\pi/6)$

In Exercises 13–20, convert the polar coordinates to rectangular coordinates.

13. $(3, \pi/3)$
14. $(-2, \pi/4)$
15. $(-1, 5\pi/6)$
16. $(2, 0)$
17. $(1.5, 5)$
18. $(2.2, -2.2)$
19. $(-4, -\pi/7)$
20. $(-1, 1)$

In Exercises 21–34, convert the rectangular coordinates to polar coordinates.

21. $(3\sqrt{3}, -3)$
22. $(2\sqrt{3}, 2)$
23. $(1, 1)$
24. $(\sqrt{2}, -\sqrt{2})$
25. $(3, 3\sqrt{3})$
26. $(-\sqrt{2}, \sqrt{6})$
27. $(2, 4)$
28. $(3, -2)$
29. $(-5, 2.5)$
30. $(-6.2, -3)$
31. $(0, -2)$
32. $(.5, 3.5)$
33. $(-2, 4)$
34. $(\sqrt{5}, \sqrt{10})$

In Exercises 35–40, find a polar equation that is equivalent to the given rectangular equation.

35. $x^2 + y^2 = 25$
36. $4xy = 1$
37. $x = 12$
38. $y = 4$
39. $y = 2x + 1$
40. $y = x - 2$

In Exercises 41–52, find a rectangular equation that is equivalent to the given polar equation.

41. $r = 3$ [*Hint:* Square both sides, then substitute.]
42. $r = 5$
43. $\theta = \pi/6$ {*Hint:* Take the tangent of both sides, then substitute.]
44. $\theta = -\pi/4$
45. $r = \sec\theta$ [*Hint:* Express the right side in terms of cosine.]
46. $r = \csc\theta$
47. $r^2 = \tan\theta$
48. $r^2 = \sin\theta$
49. $r = 2\sin\theta$
50. $r = 3\cos\theta$
51. $r = \dfrac{4}{1 + \sin\theta}$
52. $r = \dfrac{6}{1 - \cos\theta}$

In Exercises 53–58, sketch the graph of the equation without using a calculator.

53. $r = 4$
54. $r = -1$
55. $\theta = -\pi/3$
56. $\theta = 5\pi/6$
57. $\theta = 1$
58. $\theta = -4$

In Exercises 59–82, sketch the graph of the equation.

59. $r = \theta \quad (\theta \leq 0)$
60. $r = 3\theta \quad (\theta \geq 0)$
61. $r = 1 - \sin\theta$
62. $r = 3 - 3\cos\theta$
63. $r = -2\cos\theta$
64. $r = -6\sin\theta$
65. $r = \cos 2\theta$
66. $r = \cos 3\theta$
67. $r = \sin 3\theta$
68. $r = \sin 4\theta$
69. $r^2 = 4\cos 2\theta$
70. $r^2 = \sin 2\theta$
71. $r = 2 + 4\cos\theta$
72. $r = 1 + 2\cos\theta$
73. $r = \sin\theta + \cos\theta$
74. $r = 4\cos\theta + 4\sin\theta$
75. $r = \sin(\theta/2)$
76. $r = 4\tan\theta$
77. $r = \sin\theta\tan\theta$ (cissoid)
78. $r = 4 + 2\sec\theta$ (conchoid)
79. $r = e^\theta$ (logarithmic spiral)
80. $r^2 = 1/\theta$
81. $r = 1/\theta \quad (\theta > 0)$
82. $r^2 = \theta$

83. (a) Find a complete graph of $r = 1 - 2 \sin 3\theta$.
(b) Predict what the graph of $r = 1 - 2 \sin 4\theta$ will look like. Then check your prediction with a calculator.
(c) Predict what the graph of $r = 1 - 2 \sin 5\theta$ will look like. Then check your prediction with a calculator.

84. (a) Find a complete graph of $r = 1 - 3 \sin 2\theta$.
(b) Predict what the graph of $r = 1 - 3 \sin 3\theta$ will look like. Then check your prediction with a calculator.
(c) Predict what the graph of $r = 1 - 3 \sin 4\theta$ will look like. Then check your prediction with a calculator.

85. If a, b are constants such that $ab \neq 0$, show that the graph of $r = a \sin \theta + b \cos \theta$ is a circle. [*Hint:* Multiply both sides by r and convert to rectangular coordinates.]

86. Prove that the coordinate conversion formulas are valid when $r < 0$. [*Hint:* If P has coordinates (x, y) and (r, θ), with $r < 0$, verify that the point Q with rectangular coordinates $(-x, -y)$ has polar coordinates $(-r, \theta)$. Since $r < 0$, $-r$ is positive and the conversion formulas proved in the text apply to Q. For instance, $-x = -r \cos \theta$, which implies that $x = r \cos \theta$.]

87. ***Distance Formula for Polar Coordinates:*** Prove that the distance from (r, θ) to (s, β) is

$$\sqrt{r^2 + s^2 - 2rs \cos(\theta - \beta)}$$

[*Hint:* If $r > 0$, $s > 0$, and $\theta > \beta$, then the triangle with vertices (r, θ), (s, β), $(0, 0)$ has an angle of $\theta - \beta$, whose sides have lengths r and s. Use the Law of Cosines.]

5.7 Polar Equations of Conics

In a rectangular coordinate system, each type of conic section has a different definition. By using polar coordinates, it is possible to give a unified treatment of conics and their equations. Before doing this, we must first introduce a concept that will play a key role in the development.

Recall that both ellipses and hyperbolas are defined in terms of two foci and both have two vertices that lie on the line through the foci (see pages 335 and 349). The **eccentricity** of an ellipse or a hyperbola is denoted e and is defined to be the ratio

$$e = \frac{\text{distance between the foci}}{\text{distance between the vertices}}.$$

For conics centered at the origin, with foci on the x-axis, the situation is as follows.

Ellipse

$$\frac{x^2}{a^2} + \frac{y^2}{b^2} = 1 \quad (a > b)$$

foci: $(\pm c, 0)$ vertices: $(\pm a, 0)$

$$c = \sqrt{a^2 - b^2}$$

Hyperbola

$$\frac{x^2}{a^2} - \frac{y^2}{b^2} = 1$$

foci: $(\pm c, 0)$ vertices: $(\pm a, 0)$

$$c = \sqrt{a^2 + b^2}$$

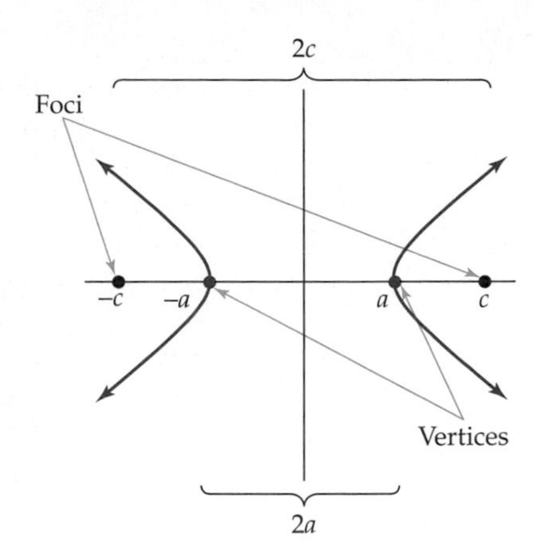

$$e = \frac{2c}{2a} = \frac{c}{a} = \frac{\sqrt{a^2 - b^2}}{a} \qquad\qquad e = \frac{2c}{2a} = \frac{c}{a} = \frac{\sqrt{a^2 + b^2}}{a}$$

A similar analysis shows that the formulas for e are also valid for conics whose equations are of the form

$$\frac{x^2}{b^2} + \frac{y^2}{a^2} = 1 \quad (a > b) \qquad \text{or} \qquad \frac{y^2}{a^2} - \frac{x^2}{b^2} = 1.$$

These formulas can be used to compute the eccentricity of any ellipse or hyperbola whose equation can be put in standard form.

EXAMPLE 1

Find the eccentricity of the conic with equation

(a) $\frac{y^2}{4} \quad \frac{x^2}{21} - 1$; (b) $4x^2 + 9y^2 - 32x - 90y + 253 = 0$.

SOLUTION

(a) In this case, $a^2 = 4$ (so $a = 2$) and $b^2 = 21$. Hence, the eccentricity is

$$e = \frac{\sqrt{a^2 + b^2}}{a} = \frac{\sqrt{4 + 21}}{2} = \frac{\sqrt{25}}{2} = \frac{5}{2} = 2.5.$$

(b) In Example 7 of Section 5.1, we saw that the equation can be put into this standard form:

$$\frac{(x - 4)^2}{9} + \frac{(y - 5)^2}{4} = 1.$$

Hence, its graph is just the ellipse

$$\frac{x^2}{9} + \frac{y^2}{4} = 1$$

shifted vertically and horizontally. Since the shifting does not change the distances between foci or vertices, both ellipses have the same eccentricity, which can be computed by using $a^2 = 9$ and $b^2 = 4$.

$$e = \frac{\sqrt{a^2 - b^2}}{a} = \frac{\sqrt{9 - 4}}{3} = \frac{\sqrt{5}}{3} \approx .745. \quad \blacksquare$$

Example 1 and the preceding pictures illustrate the following fact. For ellipses, the distance between the foci (numerator of e) is less than the distance between the vertices (denominator), so $e < 1$. For hyperbolas, however, $e > 1$ because the distance between the foci is greater than that between the vertices.

The eccentricity of an ellipse measures its "roundness." An ellipse whose eccentricity is close to 0 is almost circular (Exercise 19). The eccentricity of a hyperbola measures how "flat" its branches are. The branches of a hyperbola with large eccentricity look almost like parallel lines (Exercise 20).

CONICS AND POLAR EQUATIONS

The polar analogues of the standard equations of ellipses, parabolas, and hyperbolas are given in the following chart. The proof of these statements is given at the end of the section. In the chart, e and d are constants, with $e > 0$. Remember that in a rectangular coordinate system whose positive x-axis coincides with the polar axis, a point with polar coordinates (r, θ) is on the x-axis when $\theta = 0$ or π and on the y-axis when $\theta = \pi/2$ or $3\pi/2$.

Polar Equations for Conic Sections

Equation		Graph
$r = \dfrac{ed}{1 + e\cos\theta}$ or $r = \dfrac{ed}{1 - e\cos\theta}$	$0 < e < 1$	*Ellipse* with eccentricity e One of the foci: (0, 0) Vertices at $\theta = 0$ and $\theta = \pi$
	$e = 1$	*Parabola* with focus (0, 0) Vertex at $\theta = 0$ or $\theta = \pi$; (r is not defined for the other value of θ)
	$e > 1$	*Hyperbola* with eccentricity e One of the foci: (0, 0) Vertices at $\theta = 0$ and $\theta = \pi$
$r = \dfrac{ed}{1 + e\sin\theta}$ or $r = \dfrac{ed}{1 - e\sin\theta}$	$0 < e < 1$	*Ellipse* with eccentricity e One of the foci: (0, 0) Vertices at $\theta = \pi/2$ and $\theta = 3\pi/2$
	$e = 1$	*Parabola* with focus (0, 0) Vertex at $\theta = \pi/2$ or $\theta = 3\pi/2$; (r is not defined for the other value of θ)
	$e > 1$	*Hyperbola* with eccentricity e One of the foci: (0, 0) Vertices at $\theta = \pi/2$ and $\theta = 3\pi/2$

EXAMPLE 2

Find a complete graph of

$$r = \frac{3e}{1 + e\cos\theta}$$

when

(a) $e = .7$ (b) $e = 1$ (c) $e = 2$.

SOLUTION From the first equation in the preceding chart (with $d = 3$), we know that the graphs are an ellipse, a parabola, and a hyperbola, respectively, as shown in Figure 5–71. ■

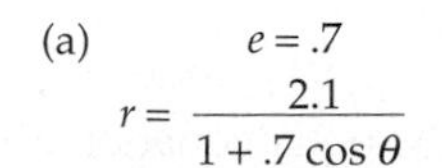

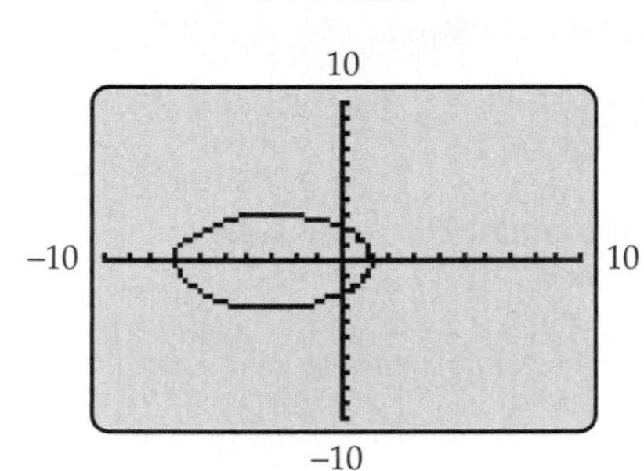

(b) $e = 1$

$r = \dfrac{3}{1 + \cos\theta}$

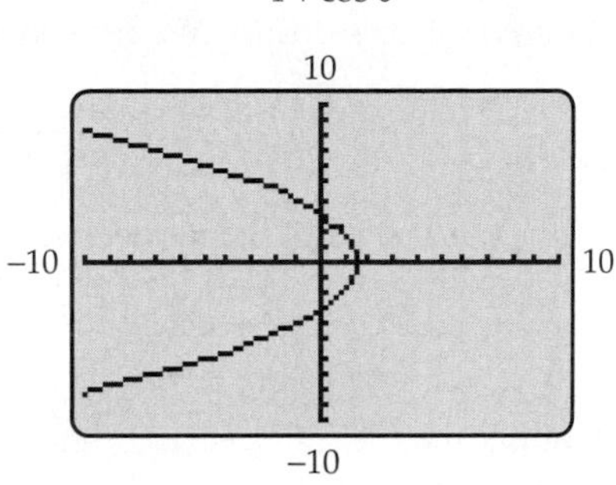

(c) $e = 2$

$r = \dfrac{6}{1 + 2\cos\theta}$

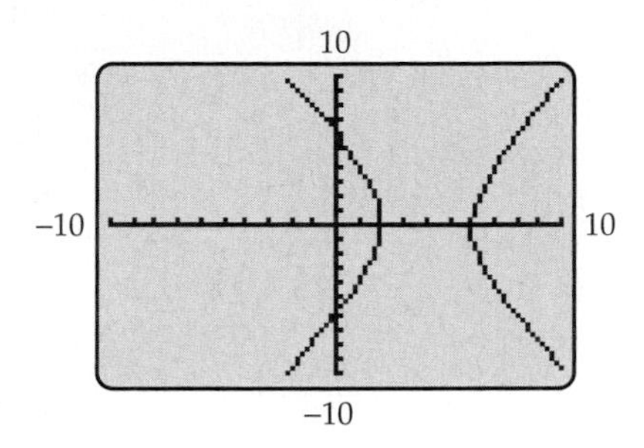

Figure 5–71

EXAMPLE 3

Identify the conic section that is the graph of

$$r = \frac{20}{4 - 10\sin\theta},$$

and find its vertices and eccentricity.

SOLUTION First, rewrite the equation in one of the forms listed in the preceding box.

$$r = \frac{20}{4 - 10\sin\theta} = \frac{20}{4\left(1 - \frac{10}{4}\sin\theta\right)} = \frac{5}{1 - 2.5\sin\theta}.$$

This is one such form, with $e = 2.5$ and $ed = 5$ (so $d = 2$). Consequently, the graph is a hyperbola with eccentricity $e = 2.5$ whose vertices are at

$$\theta = \frac{\pi}{2}, \qquad r = \frac{20}{4 - 10\sin\frac{\pi}{2}} = \frac{20}{4 - 10 \cdot 1} = -\frac{20}{6} = -\frac{10}{3}$$

and

$$\theta = \frac{3\pi}{2}, \qquad r = \frac{20}{4 - 10\sin\frac{3\pi}{2}} = \frac{20}{4 - 10(-1)} = \frac{20}{14} = \frac{10}{7}.$$

GRAPHING EXPLORATION

Find a viewing window that shows a complete graph of this hyperbola. ■

EXAMPLE 4

Find a polar equation of the ellipse with (0, 0) as a focus and vertices (3, 0) and $(6, \pi)$.

SOLUTION Because of the location of the vertices, the polar equation is of the form $r = ed/(1 \pm e\cos\theta)$. We first consider the equation

$$r = \frac{ed}{1 + e\cos\theta}.$$

Since the coordinates of the vertices satisfy the equation, we must have

$$3 = \frac{ed}{1 + e\cos 0} = \frac{ed}{1+e} \quad \text{and} \quad 6 = \frac{ed}{1 + e\cos\pi} = \frac{ed}{1-e},$$

which imply that

$$3(1 + e) = ed \quad \text{and} \quad 6(1 - e) = ed.$$

Therefore,

$$\begin{aligned} 3(1 + e) &= 6(1 - e) \\ 3 + 3e &= 6 - 6e \\ 9e &= 3 \\ e &= 1/3. \end{aligned}$$

Substituting $e = 1/3$ in either of the original equations shows that $d = 12$. So an equation of the ellipse is

$$r = \frac{ed}{1 + e\cos\theta} = \frac{\frac{1}{3}\cdot 12}{1 + \frac{1}{3}\cos\theta} = \frac{12}{3 + \cos\theta}.$$

If we had started instead with the equation $r = ed/(1 - e\cos\theta)$ and solved for e as above, we would have obtained $e = -1/3$, which is impossible, since $e > 0$.

ALTERNATE SOLUTION Verify that the vertex (3, 0) also has polar coordinates $(-3, \pi)$. Similarly, $(6, \pi)$ also has polar coordinates $(-6, 0)$. If you begin with the equation $r = ed/(1 - e\cos\theta)$ and the vertices $(-3, \pi)$ and $(-6, 0)$ and proceed as before to find e and d, you obtain the equation

$$r = \frac{-12}{3 - \cos\theta}. \quad ■$$

ALTERNATE DEFINITION OF CONICS

The theorem stated in the following box is sometimes used as a definition of the conic sections because it provides a unified approach instead of the variety of descriptions given in Sections 5.1–5.3. Its proof also provides a proof of the statements in the box on page 406.

The basic idea is to describe every conic in terms of a straight line L (the **directrix**) and a point P not on L (the **focus**), in much the same way that parabolas were defined in Section 5.3. The number e in the theorem turns out to be the eccentricity of the conic.

Conic Section Theorem

> Let L be a straight line, P a point not on L, and e a positive constant. The set of all points X in the plane such that
>
> $$\frac{\text{distance from } X \text{ to } P}{\text{distance from } X \text{ to } L} = e$$
>
> is a conic section with P as one of the foci.* The conic is an ellipse if $0 < e < 1$, a parabola if $e = 1$,† and a hyperbola if $e > 1$.

Proof Coordinatize the plane so that the pole is the point P, the polar axis is horizontal, and the directrix L is a vertical line to the left of the pole, as in Figure 5–72. Let d be the distance from P to L, and let (r, θ) be the polar coordinates of X.

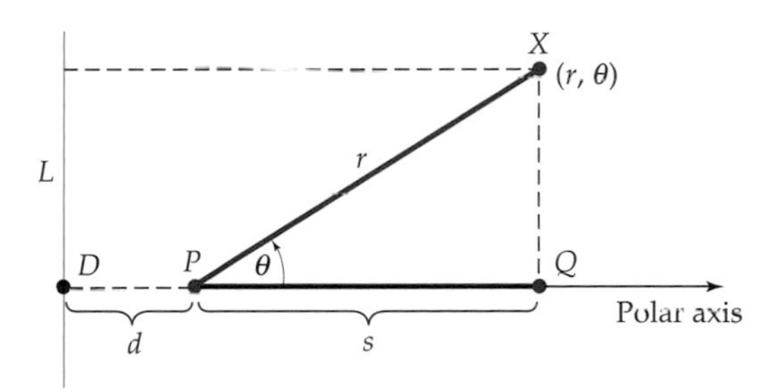

Figure 5–72

If X satisfies the condition

$$\frac{\text{distance from } X \text{ to } P}{\text{distance from } X \text{ to } L} = e,$$

then

$$(*) \qquad \text{distance from } X \text{ to } P = e(\text{distance from } X \text{ to } L).$$

Figure 5–72 shows that r is the distance from P to X and that the distance from X to L is the same as that from D to Q, namely, $d + s$. Furthermore, $\cos\theta = s/r$,

---

*The distance from X to L is measured along the line through X that is perpendicular to L.
†When $e = 1$, the given condition is equivalent to

$$\text{distance from } X \text{ to } P = \text{distance from } X \text{ to } L$$

which is the definition of a parabola given in Section 5.3.

so $s = r\cos\theta$. Consequently, equation (*) can be written in polar coordinates as follows.

$$\begin{aligned} \text{distance from } X \text{ to } P &= e(\text{distance from } X \text{ to } L) \\ r &= e(d+s) \\ r &= e(d + r\cos\theta) \\ r - er\cos\theta &= ed \\ r(1 - e\cos\theta) &= ed \\ r &= \frac{ed}{1 - e\cos\theta}. \end{aligned}$$

To show that this is actually the equation of a conic, we translate it into rectangular coordinates using the conversion formulas from Section 5.6.

$$r^2 = x^2 + y^2 \quad \text{and} \quad \cos\theta = \frac{x}{r} = \frac{x}{\pm\sqrt{x^2+y^2}}.$$

Then the polar coordinate equation becomes

$$\begin{aligned} \pm\sqrt{x^2+y^2} &= \frac{ed}{1 - e\left(\dfrac{x}{\pm\sqrt{x^2+y^2}}\right)} \\ \pm\sqrt{x^2+y^2}\left(1 - \frac{ex}{\pm\sqrt{x^2+y^2}}\right) &= ed \\ \pm\sqrt{x^2+y^2} - ex &= ed \\ \pm\sqrt{x^2+y^2} &= ed + ex. \end{aligned}$$

Squaring both sides and rearranging terms, we have

$$x^2 + y^2 = e^2d^2 + 2de^2x + e^2x^2$$

$$(**) \qquad (1 - e^2)x^2 - 2de^2x + y^2 = e^2d^2.$$

Now we consider the two possibilities $e = 1$ and $e \neq 1$.

Case 1. If $e = 1$, then equation (**) becomes

$$\begin{aligned} -2dx + y^2 &= d^2 \\ y^2 &= 2dx + d^2 \\ (y-0)^2 &= 2d\left(x + \frac{d}{2}\right) \\ (y-0)^2 &= 4\left(\frac{d}{2}\right)\left(x - \left(-\frac{d}{2}\right)\right). \end{aligned}$$

The box on page 363 (with $k = 0$, $p = d/2$, $h = -d/2$) shows that this is the standard equation of a parabola with

$$\text{vertex}\left(-\frac{d}{2}, 0\right), \qquad \text{focus } (0, 0), \qquad \text{directrix } x = -d.$$

Case 2. If $e \neq 1$, then we can divide both sides of equation (**) by the nonzero number $1 - e^2$.

$$\left(x^2 - \frac{2de^2}{1-e^2}x\right) + \frac{y^2}{1-e^2} = \frac{e^2d^2}{1-e^2}.$$

Next, we complete the square on the expression in parentheses by adding the square of half of the coefficient of x to both sides of the equation and simplify the result.

$$\left[x^2 - \frac{2de^2}{1-e^2}x + \left(\frac{de^2}{1-e^2}\right)^2\right] + \frac{y^2}{1-e^2} = \frac{e^2d^2}{1-e^2} + \left(\frac{de^2}{1-e^2}\right)^2$$

$$\left(x - \frac{de^2}{1-e^2}\right)^2 + \frac{y^2}{1-e^2} = \frac{(1-e^2)e^2d^2 + (de^2)^2}{(1-e^2)^2}$$

$$\left(x - \frac{de^2}{1-e^2}\right)^2 + \frac{y^2}{1-e^2} = \frac{e^2d^2}{(1-e^2)^2}.$$

Dividing both sides of the last equation by $e^2d^2/(1-e^2)^2$ produces the equation

$$(***) \qquad \frac{\left(x - \frac{de^2}{1-e^2}\right)^2}{\frac{e^2d^2}{(1-e^2)^2}} + \frac{y^2}{\frac{e^2d^2}{1-e^2}} = 1.$$

Now we consider the two possibilities $e < 1$ and $e > 1$.

Case 2A. If $e < 1$, then $1 - e^2 > 0$ and the constants in the denominators on the left side of equation (***) are positive. Therefore, equation (***) is of the form

$$\frac{(x-h)^2}{a^2} + \frac{(y-k)^2}{b^2} = 1$$

with $h = de^2/(1-e^2)$, $k = 0$, and a and b positive numbers such that

$$a^2 = \frac{e^2d^2}{(1-e^2)^2} \quad \text{and} \quad b^2 = \frac{e^2d^2}{1-e^2}.$$

In this case, $a > b$ by Exercise 47. According to the box on page 341, this is the standard equation of an ellipse with center $(h, 0)$ and foci $(h - c, 0)$ and $(h + c, 0)$, where $c^2 = a^2 - b^2$. Its eccentricity is the number

$$\frac{c}{a} = \sqrt{\frac{c^2}{a^2}} = \sqrt{\frac{a^2 - b^2}{a^2}} = \sqrt{1 - b^2 \cdot \frac{1}{a^2}}$$

$$= \sqrt{1 - \frac{e^2d^2}{1-e^2} \cdot \frac{(1-e^2)^2}{e^2d^2}} = \sqrt{e^2} = e.$$

Case 2B. If $e > 1$, then $1 - e^2 < 0$. Therefore,

$$e^2 - 1 = -(1 - e^2) > 0,$$

so equation (∗∗∗) may be written as

$$\frac{\left(x - \frac{de^2}{1-e^2}\right)^2}{\frac{e^2d^2}{(1-e^2)^2}} - \frac{y^2}{\frac{e^2d^2}{e^2-1}} = 1.$$

This is an equation of the form

$$\frac{(x-h)^2}{a^2} - \frac{(y-k)^2}{b^2} = 1$$

with a and b positive. The box on page 351 shows that it is the standard equation of a hyperbola with foci $(h - c, 0)$ and $(h + c, 0)$, where $c^2 = a^2 + b^2$. Exercise 48 shows that its eccentricity is e.

The preceding argument depends on coordinatizing the plane in a certain way and taking d to be the distance from the pole to L. Similar arguments, in which d is the distance from the pole to L and the plane is coordinatized so that L is to the right of the pole or parallel to the polar axis, lead to the other polar equations shown in Figure 5–73.

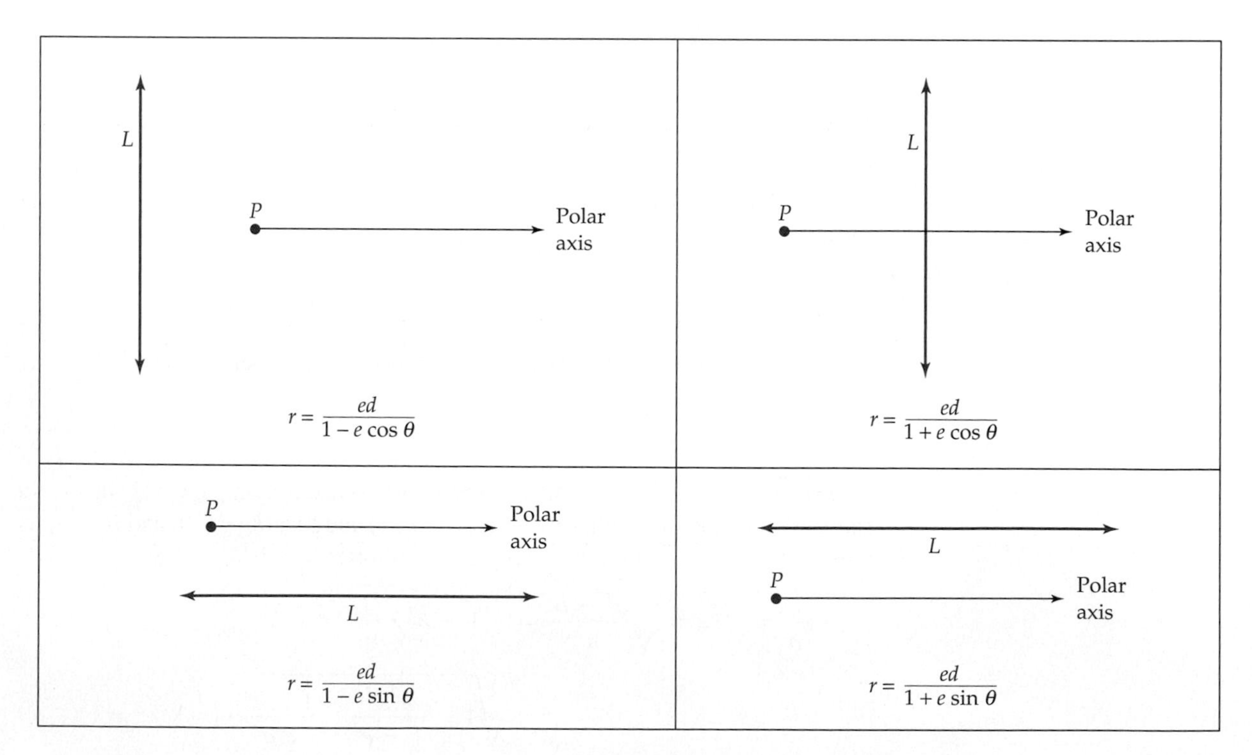

Figure 5–73

Analogous arguments when $d < 0$ (using $-d$ as the distance from the pole to L) then complete the proof. ■

EXAMPLE 5

Find the polar equation of the hyperbola with focus at the pole, directrix

$$r = -4 \csc \theta,$$

and eccentricity 3.

SOLUTION The equation of the directrix can be written as

$$r = \frac{-4}{\sin \theta} \qquad \text{or, equivalently,} \qquad r \sin \theta = -4.$$

With the conversion formulas for a rectangular coordinate system whose positive x-axis coincides with the polar axis, this equation becomes $y = -4$. So the directrix is a line parallel to the polar axis and 4 units below it. Using Figure 5–73, we see that $d = 4$, and the equation is

$$r = \frac{ed}{1 - e \sin \theta} = \frac{3 \cdot 4}{1 - 3 \sin \theta} = \frac{12}{1 - 3 \sin \theta}. \qquad \blacksquare$$

EXERCISES 5.7

In Exercises 1–6, which of the graphs (a)–(f) at the bottom of the page could possibly be the graph of the equation?

1. $r = \dfrac{3}{1 - \cos \theta}$

2. $r = \dfrac{6}{2 + \cos \theta}$

3. $r = \dfrac{6}{2 - 4 \sin \theta}$

4. $r = \dfrac{15}{1 + 4 \cos \theta}$

5. $r = \dfrac{6}{3 - 2 \sin \theta}$

6. $r = \dfrac{6}{\frac{3}{2} + \frac{3}{2} \sin \theta}$

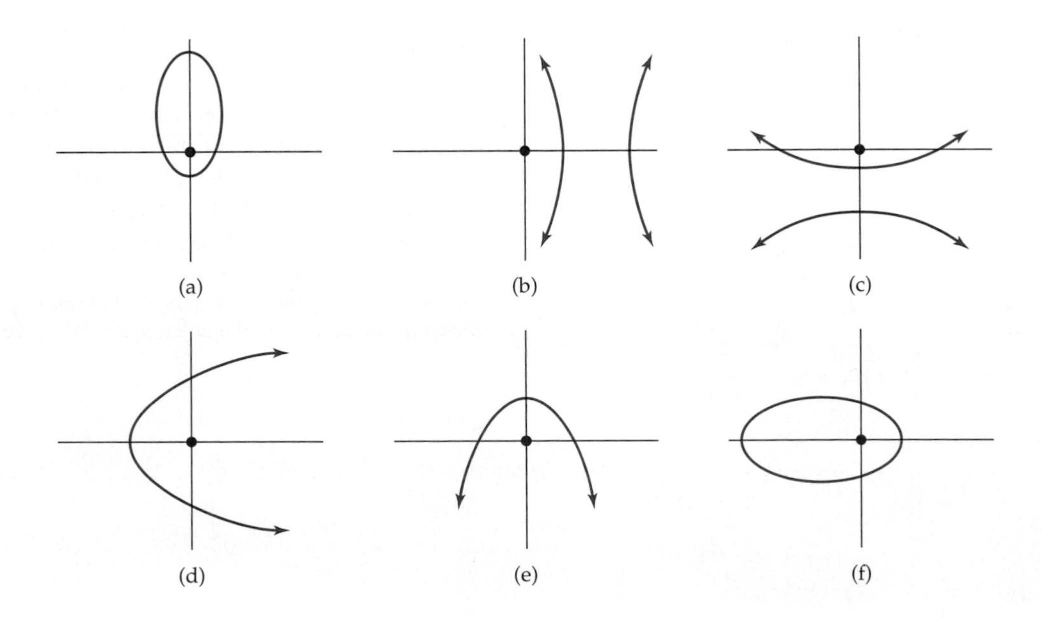

In Exercises 7–12, identify the conic section whose equation is given; if it is an ellipse or hyperbola, state its eccentricity.

7. $r = \dfrac{12}{3 + 4\sin\theta}$ **8.** $r = \dfrac{-10}{2 + 3\cos\theta}$

9. $r = \dfrac{8}{3 + 3\sin\theta}$ **10.** $r = \dfrac{20}{5 - 10\sin\theta}$

11. $r = \dfrac{2}{6 - 4\cos\theta}$ **12.** $r = \dfrac{-6}{5 + 2\cos\theta}$

In Exercises 13–18, find the eccentricity of the conic whose equation is given.

13. $\dfrac{x^2}{100} + \dfrac{y^2}{99} = 1$ **14.** $\dfrac{(x-4)^2}{18} + \dfrac{(y+5)^2}{25} = 1$

15. $\dfrac{(x-6)^2}{10} - \dfrac{y^2}{40} = 1$

16. $4x^2 + 9y^2 - 24x + 36y + 36 = 0$

17. $16x^2 - 9y^2 - 32x + 36y + 124 = 0$

18. $4x^2 - 5y^2 - 16x - 50y + 71 = 0$

19. (a) Using a square viewing window (so that circles look like circles), graph these ellipses (on the same screen if possible):

$$\frac{x^2}{16} + \frac{y^2}{1} = 1, \quad \frac{x^2}{16} + \frac{y^2}{6} = 1, \quad \frac{x^2}{16} + \frac{y^2}{14} = 1$$

(b) Compute the eccentricity of each ellipse in part (a).

(c) On the basis of parts (a) and (b), how is the shape of an ellipse related to its eccentricity?

20. (a) Graph these hyperbolas (on the same screen if possible):

$$\frac{y^2}{4} - \frac{x^2}{1} = 1, \quad \frac{y^2}{4} - \frac{x^2}{12} = 1, \quad \frac{y^2}{4} - \frac{x^2}{96} = 1$$

(b) Compute the eccentricity of each hyperbola in part (a).

(c) On the basis of parts (a) and (b), how is the shape of a hyperbola related to its eccentricity?

In Exercises 21–32, sketch the graph of the equation and label the vertices.

21. $r = \dfrac{8}{1 - \cos\theta}$ **22.** $r = \dfrac{5}{3 + 2\sin\theta}$

23. $r = \dfrac{4}{2 - 4\cos\theta}$ **24.** $r = \dfrac{5}{1 + \cos\theta}$

25. $r = \dfrac{10}{4 - 3\sin\theta}$ **26.** $r = \dfrac{12}{3 + 4\sin\theta}$

27. $r = \dfrac{15}{3 - 2\cos\theta}$ **28.** $r = \dfrac{32}{3 + 5\sin\theta}$

29. $r = \dfrac{3}{1 + \sin\theta}$ **30.** $r = \dfrac{10}{3 + 2\cos\theta}$

31. $r = \dfrac{10}{2 + 3\sin\theta}$ **32.** $r = \dfrac{15}{4 - 4\cos\theta}$

In Exercises 33–46, find the polar equation of the conic section that has focus $(0, 0)$ *and satisfies the given conditions.*

33. Parabola; vertex $(3, \pi)$

34. Parabola; vertex $(2, \pi/2)$

35. Ellipse; vertices $(2, \pi/2)$ and $(8, 3\pi/2)$

36. Ellipse; vertices $(2, 0)$ and $(4, \pi)$

37. Hyperbola; vertices $(1, 0)$ and $(-3, \pi)$

38. Hyperbola; vertices $(-2, \pi/2)$ and $(4, 3\pi/2)$

39. Eccentricity 4; directrix; $r = -2\sec\theta$

40. Eccentricity 2; directrix: $r = 4\csc\theta$

41. Eccentricity 1; directrix: $r = -3\csc\theta$

42. Eccentricity 1; directrix: $r = 5\sec\theta$

43. Eccentricity $1/2$; directrix: $r = 2\sec\theta$

44. Eccentricity $4/5$; directrix: $r = 3\csc\theta$

45. Hyperbola; vertical directrix to the left of the pole; eccentricity 2; $(1, 2\pi/3)$ is on the graph.

46. Hyperbola; horizontal directrix above the pole; eccentricity 2; $(1, 2\pi/3)$ is on the graph.

47. In Case 2A of the proof of the Conic Section Theorem, show that $a > b$.

48. In Case 2B of the proof of the Conic Section Theorem, show that the hyperbola has eccentricity e.

49. A comet travels in a parabolic orbit with the sun as focus. When the comet is 60 million miles from the sun, the line segment from the sun to the comet makes an angle of $\pi/3$ radians with the axis of the parabolic orbit. Using the sun as the pole and assuming the axis of the orbit lies along the polar axis, find a polar equation for the orbit.

50. Halley's Comet has an elliptical orbit, with eccentricity .97 and the sun as a focus. The length of the major axis of the orbit is 3364.74 million miles. Using the sun as the pole and assuming the major axis of the orbit is perpedic–ular to the polar axis, find a polar equation for the orbit.

Chapter 5 Review

IMPORTANT CONCEPTS

IMPORTANT FACTS & FORMULAS

■ Equations of conic sections

Conic Section	Cartesian Equation	Parametric Equations
Circle center (h, k) radius r	$(x-h)^2 + (y-k)^2 = r^2$	$x = r\cos t + h$ $y = r\sin t + k$ $(0 \le t \le 2\pi)$
Ellipse center (h, k) axes on the lines $x = h, y = k$	$\dfrac{(x-h)^2}{a^2} + \dfrac{(y-k)^2}{b^2} = 1$	$x = a\cos t + h$ $y = b\sin t + k$ $(0 \le t \le 2\pi)$
Hyperbola center (h, k) vertices on the line $y = k$	$\dfrac{(x-h)^2}{a^2} - \dfrac{(y-k)^2}{b^2} = 1$	$x = \dfrac{a}{\cos t} + h$ $y = b\tan t + k$ $(0 \le t \le 2\pi)$

Hyperbola center (h, k) vertices on the line $x = h$	$\frac{(y-k)^2}{a^2} - \frac{(x-h)^2}{b^2} = 1$	$x = b \tan t + h$ $y = \frac{a}{\cos t} + k$ $(0 \leq t \leq 2\pi)$
Parabola vertex (h, k) axis $x = h$	$(x-h)^2 = 4p(y-k)$	$x = t$ $y = \frac{(t-h)^2}{4p} + k$ (t any real)
Parabola vertex (h, k) axis $y = k$	$(y-k)^2 = 4p(x-h)$	$x = \frac{(t-k)^2}{4p} + h$ $y = t$ (t any real)

- The discriminant of the equation $Ax^2 + Bxy + Cy^2 + Dx + Ey + F = 0$ (where A, B, C are not all zero) is $B^2 - 4AC$.

  If $B^2 - 4AC < 0$, the graph is a circle or ellipse (or a point).
  If $B^2 - 4AC = 0$, the graph is a parabola (or a line or two parallel lines).
  If $B^2 - 4AC > 0$, the graph is a hyperbola (or two intersecting lines).

- Rotation Equations:

$$x = u \cos \theta - v \sin \theta,$$

$$y = u \sin \theta + v \cos \theta.$$

- To eliminate the xy term in $Ax^2 + Bxy + Cy^2 + Dx + Ey + F = 0$, rotate the axes through an angle θ such that $\cot 2\theta = \frac{A - C}{B}$.

- The rectangular and polar coordinates of a point are related by

$$x = r \cos \theta \quad \text{and} \quad y = r \sin \theta;$$

$$r^2 = x^2 + y^2 \quad \text{and} \quad \tan \theta = \frac{y}{x}.$$

- If e and d are constants with $e > 0$, then the graph of a polar equation of the form

$$r = \frac{ed}{1 \pm e \cos \theta} \quad \text{or} \quad r = \frac{ed}{1 \pm e \sin \theta}$$

  is an ellipse if $0 < e < 1$, a parabola if $e = 1$, and a hyperbola if $e > 1$.

REVIEW QUESTIONS

In Questions 1–4, find the foci and vertices of the conic, and state whether it is an ellipse or a hyperbola.

1. $\dfrac{x^2}{16} + \dfrac{y^2}{20} = 1$ **2.** $\dfrac{x^2}{9} - \dfrac{y^2}{16} = 1$

3. $\dfrac{(x-1)^2}{7} + \dfrac{(y-3)^2}{16} = 1$ **4.** $3x^2 = 1 + 2y^2$

5. Find the focus and directrix of the parabola $10y = 7x^2$.

6. Find the focus and directrix of the parabola

$$3y^2 - x - 4y + 4 = 0.$$

In Questions 7–20, sketch the graph of the equation. If there are asymptotes, give their equations.

7. $\dfrac{x^2}{4} + \dfrac{y^2}{25} = 1$ **8.** $25x^2 + 4y^2 = 100$

9. $\dfrac{(x-3)^2}{9} + \dfrac{(y+5)^2}{4} = 1$ **10.** $\dfrac{x^2}{9} - \dfrac{y^2}{16} = 1$

11. $\dfrac{(y+4)^2}{25} - \dfrac{(x-1)^2}{4} = 1$ **12.** $4x^2 - 9y^2 = 144$

13. $x^2 + 4y^2 - 10x + 9 = 0$

14. $9x^2 - 4y^2 - 36x + 24y - 36 = 0$

15. $2y = 4(x-3)^2 + 6$ **16.** $3y = 6(x+1)^2 - 9$

17. $x = y^2 + 2y + 2$ **18.** $y = x^2 - 2x + 3$

19. $x^2 + y^2 - 6x + 5 = 0$

20. $x^2 + y^2 - 4x + 6y + 4 = 0$

21. Find the center and radius of the circle whose equation is

$$x^2 + y^2 + 8x + 10y + 33 = 0.$$

22. Find the equation of the circle with center $(-2, 3)$ that passes through the point $(1, 7)$.

23. What is the center of the ellipse

$$4x^2 + 3y^2 - 32x + 36y + 124 = 0?$$

24. Find the equation of the ellipse with center at the origin, one vertex at $(0, 4)$, passing through $(\sqrt{3}, 2\sqrt{3})$.

25. Find the equation of the ellipse with center at $(3, 1)$, one vertex at $(1, 1)$, passing through $(2, 1 + \sqrt{3/2})$.

26. Find the equation of the hyperbola with center at the origin, one vertex at $(0, 5)$, passing through $(1, 3\sqrt{5})$.

27. Find the equation of the hyperbola with center at $(3, 0)$, one vertex at $(3, 2)$, passing through $(1, \sqrt{5})$.

28. Find the equation of the parabola with vertex $(2, 5)$, axis $x = 2$, and passing through $(3, 12)$.

29. Find the equation of the parabola with vertex $(3/2, -1/2)$, axis $y = -1/2$, and passing through $(-3, 1)$.

30. Find the equation of the parabola with vertex $(5, 2)$ that passes through the points $(7, 3)$ and $(9, 6)$.

In Questions 31–36, use parametric graphing to sketch the graph of the equation.

31. $\dfrac{x^2}{4} + \dfrac{y^2}{25} = 1$ **32.** $25x^2 + 4y^2 = 100$

33. $\dfrac{(x-3)^2}{9} + \dfrac{(y+5)^2}{4} = 1$ **34.** $\dfrac{x^2}{9} - \dfrac{y^2}{16} = 1$

35. $\dfrac{(y+4)^2}{25} - \dfrac{(x-1)^2}{4} = 1$ **36.** $4x^2 - 9y^2 = 144$

37. The arch shown in the figure has the shape of half of an ellipse. How wide is the arch at a point 8 feet above the ground? [*Hint:* Think of the arch as sitting on the x-axis, with its center at the origin, and find its equation.]

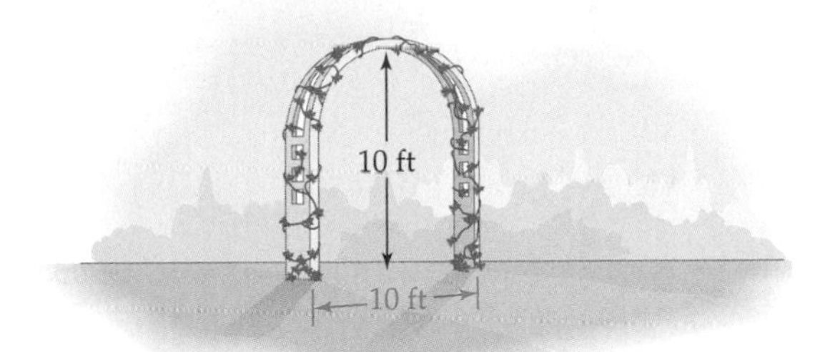

38. A bridge, whose bottom is shaped like half an ellipse, spans a 52-foot wide canal. The bottom of the bridge is 8 feet above the water at a point 20 feet from the center of the canal. Find the height of the bridge bottom above the water at the center of the canal.

39. A parabolic satellite dish is 3 feet in diameter and 1 foot deep at the vertex. How far from the vertex should the receiver be placed to catch all the signals that hit the dish?

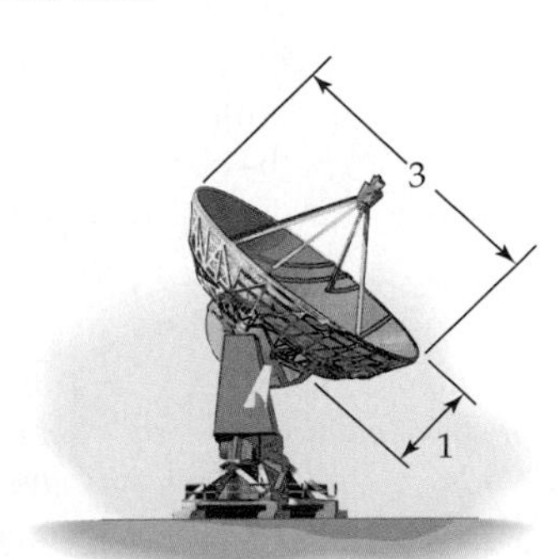

REVIEW QUESTIONS 

40. A satellite dish with a parabolic cross section is 6 feet in diameter. The receiver is located on the center axis, 1 foot from the base of the dish. How deep is the dish at its center?

In Questions 41–44, assume that the graph of the equation is a nondegenerate conic. Use the discriminant to identify the graph.

41. $3x^2 + 2\sqrt{2}xy + 2y^2 - 12 = 0$
42. $x^2 + y^2 - xy - 4y = 0$
43. $4xy - 3x^2 - 20 = 0$
44. $4x^2 - 4xy + y^2 - \sqrt{5}x - 2\sqrt{5}y = 0$

In Questions 45–50, find a viewing window that shows a complete graph of the equation.

45. $x^2 - xy + y^2 - 6 = 0$
46. $x^2 + xy + y^2 - 3y - 6 = 0$
47. $x^2 + xy - 2 = 0$
48. $x^2 - 4xy + y^2 + 5 = 0$
49. $x^2 + 3xy + y^2 - 2\sqrt{2}x + 2\sqrt{2}y = 0$
50. $x^2 + 2xy + y^2 - 4\sqrt{2}y = 0$

In Questions 51–52, find the rotation equations when the x-y axes are rotated through the given angle.

51. 60° 52. 45°

In Questions 53–54, find the angle through which the x-y axes should be rotated to eliminate the xy term in the equation.

53. $x^2 + xy + y^2 - 3y - 6 = 0$
54. $x^2 - 4xy + y^2 + 5 = 0$

In Questions 55–58, find a viewing window that shows a complete graph of the curve whose parametric equations are given.

55. $x = 8\cos t + \cos 8t$
$y = 8\sin t - \sin 8t \quad (0 \le t \le 2\pi)$
56. $x = [64\cos(\pi/6)]t$
$y = -16t^2 + [64\sin(\pi/6)]t \quad (0 \le t \le \pi)$
57. $x = t^3 + t + 1$ and $y = t^2 + 2t \quad (-3 \le t \le 3)$
58. $x = t^2 - t + 3$ and $y = t^3 - 5t \quad (-3 \le t \le 3)$

In Questions 59–62, sketch the graph of the curve whose parametric equations are given and find an equation in x and y whose graph contains the given curve.

59. $x = 2t - 1, \quad y = 2 - t, \quad -3 \le t \le 3$
60. $x = 3\cos t, \quad y = 5\sin t, \quad 0 \le t \le 2\pi$
61. $x = \cos t, \quad y = 2\sin^2 t, \quad 0 \le t \le 2\pi$
62. $x = e^t, \quad y = \sqrt{t + 1}, \quad t \ge 1$ (Knowledge of logarithms is a prerequisite for this exercise.)
63. Which of the following are *not* parameterizations of the curve $x = y^2 + 1$?
(a) $x = t^2 + 1, \quad y = t,$ any real number t
(b) $x = \sin^2 t + 1, \quad y = \sin t,$ any real number t
(c) $x = t^4 + 1, \quad y = t^2,$ any real number t
(d) $x = t^6 + 1, \quad y = t^3,$ any real number t
64. Which of the curves in Questions 55–58 appear to be the graphs of functions of the form $y = f(x)$?
65. Plot the points $(2, 3\pi/4)$ and $(-3, -2\pi/3)$ on a polar coordinate graph.
66. List four other pairs of polar coordinates for the point $(-2, \pi/4)$.

In Questions 67–76, sketch the graph of the equation in a polar coordinate system.

67. $r = 5$ 68. $r = -2$
69. $\theta = 2\pi/3$ 70. $\theta = -5\pi/6$
71. $r = 2\theta \quad (\theta \le 0)$ 72. $r = 4\cos\theta$
73. $r = 2 - 2\sin\theta$ 74. $r = \cos 3\theta$
75. $r^2 = \cos 2\theta$ 76. $r = 1 + 2\sin\theta$

77. Convert $(3, -2\pi/3)$ from polar to rectangular coordinates.
78. Convert $(3, \sqrt{3})$ from rectangular to polar coordinates.
79. What is the eccentricity of the ellipse $3x^2 + y^2 = 84$?
80. What is the eccentricity of the ellipse $24x^2 + 30y^2 = 120$?

In Questions 81–84, sketch the graph of the equation, labeling the vertices and identifying the conic.

81. $r = \dfrac{12}{2 - \sin\theta}$ 82. $r = \dfrac{14}{7 + 7\cos\theta}$
83. $r = \dfrac{-24}{3 - 9\cos\theta}$ 84. $r = \dfrac{10}{3 + 4\sin\theta}$

In Questions 85–88, find a polar equation of the conic that has focus (0, 0) and satisfies the given conditions.

85. Ellipse; vertices $(4, 0)$ and $(6, \pi)$
86. Hyperbola; vertices $(5, \pi/2)$ and $(-3, 3\pi/2)$
87. Eccentricity 1; directrix $r = 2\sec\theta$
88. Eccentricity .75; directrix $r = -3\csc\theta$

DISCOVERY PROJECT 5 Designing Machines to Make Designs

The Studio Dog/Getty Images

Parametric equations are extremely useful for modeling the behavior of moving parts, particularly when the action can be decomposed into two or more discrete movements. You saw this in Section 5.5 in the form of the cycloid curves—the rolling of a circle was decomposed into the spinning of the circle and the linear motion of the center of the circle. Similarly, the motion of machine parts can be examined from the reverse point of view. That is, you can look at how two or more discrete parts act in concert to create a desirable pattern. In the case of the cycloid, the parametric equations $x = r(t - \sin t)$ and $y = r(1 - \cos t)$ model the motion of a point on the rim of a wheel of radius r which is rolling across the x-axis. The first component of each formula, the t and the 1, show the linear movement of the axle or center of the circle, while the $\sin t$ and $\cos t$ take care of the rotation of the rim around the axle. What happens if the motion of the center and rim are no longer synchronized?

1. Graph the parametric equations $x = t + \cos(kt)$ and $y = 5 + \sin(kt)$ for various values of k greater than 1. What does it mean physically when k is larger than 1?
2. Graph the parametric equations $x = t + \cos(kt)$ and $y = 5 + 5\sin(kt)$ for various values of k less than 1. What does it mean physically when k is less than 1?
3. The formulas from the previous questions can also be used to design a machine in which objects move at a constant speed below a device that spins in a circle above the objects. This is a popular method for decorating mass-produced pastries. Find a value of k so that the graph of the parametric system looks like the picture to the right. Interpret what the value of k means in this instance.

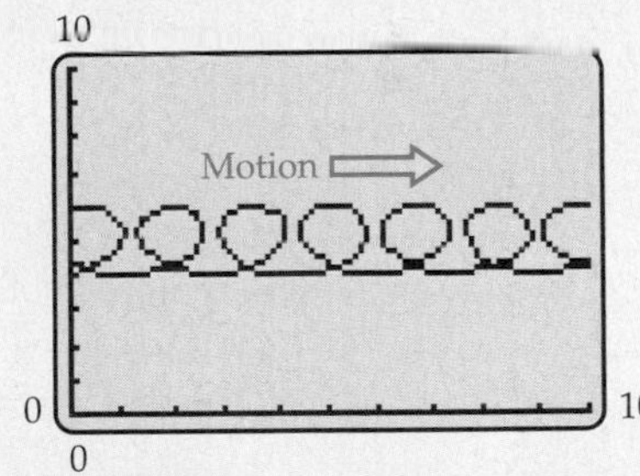

4. On the basis of your answer to the previous question, modify the parametric equations so that the pastries will have a strip of decorative icing approximately 1 cm wide, measured perpendicularly to the linear motion. Be sure to give time and length units in your answer.

DISCOVERY PROJECT 5 Continued

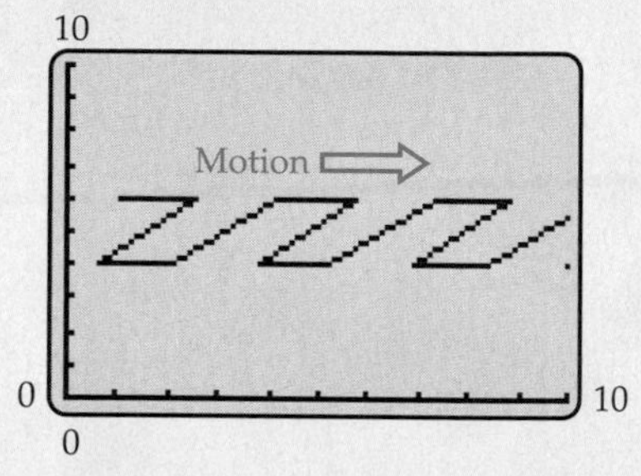

5. Sewing machines use similar devices, called feed dogs, to move the cloth beneath the presser foot. The motion of the feed dogs can be simulated using functions of the type $\frac{|\sin kt|}{\sin kt}$ in place of sines and cosines. Use this to design a set of parametric equations that would produce stitches like the ones shown to the right.

6. How would you change your answer to question 5 if the width of the stitches, measured perpendicularly to the direction of motion, needs to be 5 times as wide but the length (when measured from peak to peak in the direction of motion) remains the same?

C H A P T E R 6

Exponential and Logarithmic Functions

Walter Stuart/Index Stock Imagery/PictureQuest

15,000

−6 16

−3000

How old is that dinosaur?

Population growth (of humans, fish, bacteria, etc.), compound interest, radioactive decay, and a host of other phenomena can be mathematically described by exponential functions (see Exercises 50 on page 433, 21 on page 444, and 45 on page 445). Archeologists sometimes use carbon-14 dating to determine the approximate age of an artifact (such as a dinosaur skeleton, a mummy, or a wooden statue). This involves using logarithms to solve an appropriate exponential equation. See Exercise 53 on page 482.

Chapter Outline

Interdependence of Sections

6.1 → 6.2
6.1 → 6.3 → 6.4 → 6.5

Exponential and logarithmic functions are essential for the mathematical description of a variety of phenomena in the physical sciences, economics, and engineering. Although a calculator is necessary to evaluate these functions at most numbers, you will not be able to use your calculator efficiently or interpret its answers unless you understand the properties of these functions. When calculations can readily be done by hand, you will be expected to do them without a calculator.

6.1 Exponential Functions

We begin with a brief review of exponents. Let c be a positive real number. Then

$$\mathbf{c^0 \text{ is defined to be the number } 1.}$$

If n is a positive integer, then

$$\mathbf{c^n \text{ denotes the product } ccc\cdots c \qquad (n \text{ factors}).}$$

For example,

$$5^0 = 1, \qquad \text{and} \qquad 3^4 = 3 \cdot 3 \cdot 3 \cdot 3 = 81.$$

If $c > 0$ and $n > 0$ then

$$\mathbf{c^{-n} \text{ is defined to be the number } \frac{1}{c^n}.}$$

$$\mathbf{c^{1/n} \text{ is defined to be the number } \sqrt[n]{c},}$$

where $\sqrt[n]{c}$ is the nth root of c (the positive number whose nth power is c). For instance,

$$3^{-4} = \frac{1}{3^4} = \frac{1}{81}, \qquad \text{and} \qquad 6^{1/4} = \sqrt[4]{6} \approx 1.5651.$$

Other fractional exponents are defined as follows. If c is a positive real number and t/k is a rational number in lowest terms, with positive denominator, then

$$c^{t/k} \text{ is defined to be the number } \sqrt[k]{c^t}.$$

For example,

$$8^{2/3} = \sqrt[3]{8^2} = \sqrt[3]{64} = 4.$$

In most cases, we use a calculator to approximate rational powers of a number. For instance,

$$5^{1.4} = 5^{14/10} = \sqrt[10]{5^{14}} \approx 9.51827.$$

When exponents are defined in this way, they have all the familiar properties summarized here.

Exponent Laws

Let c and d be positive real numbers and let r and s be any rational numbers in lowest terms (with positive denominators). Then

1. $c^r c^s = c^{r+s}$
2. $\dfrac{c^r}{c^s} = c^{r-s}$
3. $(c^r)^s = c^{rs}$
4. $(cd)^r = c^r d^r$
5. $\left(\dfrac{c}{d}\right)^r = \dfrac{c^r}{d^r}$
6. $c^{-r} = \dfrac{1}{c^r}$

The next step is to define powers of a positive number when the exponent is an irrational real number. A mathematically rigorous definition is beyond the scope of this book, but we can illustrate the underlying idea.

To compute $10^{\sqrt{2}}$, for example, we use the infinite decimal expansion $\sqrt{2} \approx 1.414213562 \cdots$. Each of

$$1.4, 1.41, 1.414, 1.4142, 1.41421, \ldots$$

is a rational number approximation of $\sqrt{2}$, and each is a more accurate approximation than the preceding one. We know how to raise 10 to each of these rational numbers:

$$\begin{array}{ll} 10^{1.4} \approx 25.1189 & 10^{1.4142} \approx 25.9537 \\ 10^{1.41} \approx 25.7040 & 10^{1.41421} \approx 25.9543 \\ 10^{1.414} \approx 25.9418 & 10^{1.414213} \approx 25.9545. \end{array}$$

It appears that as the exponent r gets closer and closer to $\sqrt{2}$, 10^r gets closer and closer to a real number whose decimal expansion begins $25.954 \cdots$. We define $10^{\sqrt{2}}$ to be this number.

Similarly, for any $a > 0$,

a^t is a well-defined *positive* number for each real exponent t.

We shall also assume this fact.

The exponent laws are valid for *all* real exponents.

Consequently, for each positive real number a, the rule $f(x) = a^x$ defines a function whose domain is the set of all real numbers. The function $f(x) = a^x$ is called the **exponential function with base a.** For example, we have exponential functions:

$$f(x) = 10^x, \qquad g(x) = 2^x, \qquad h(x) = \left(\frac{1}{2}\right)^x, \qquad k(x) = \left(\frac{3}{2}\right)^x.$$

To see how the graph of $f(x) = a^x$ depends on the size of the base a, do the following Graphing Exploration.

GRAPHING EXPLORATION

Using viewing window with $-3 \le x \le 7$ and $-2 \le y \le 18$, graph

$$f(x) = 1.3^x, \qquad g(x) = 2^x, \qquad \text{and} \qquad h(x) = 10^x$$

on the same screen. How is the steepness of the graph of $f(x) = a^x$ related to the size of a?

The Graphing Exploration illustrates these facts.

The Exponential Function $f(x) = a^x$ $(a > 1)$

When $a > 1$, the graph of $f(x) = a^x$ has the shape shown here and the properties listed below.

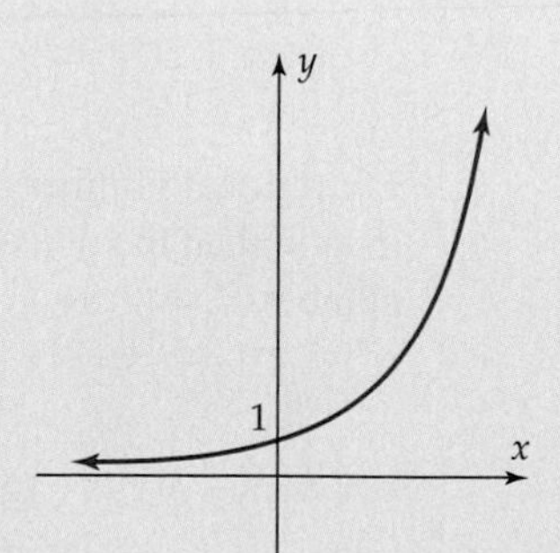

The graph is above the x-axis.

The y-intercept is 1.

$f(x)$ is an increasing function.

The negative x-axis is a horizontal asymptote.

The larger the base a, the more steeply the graph rises to the right.

The preceding figure doesn't really show how explosively an exponential graph rises.

EXAMPLE 1

Graph $f(x) = 2^x$ and estimate the height of the graph when $x = 50$.

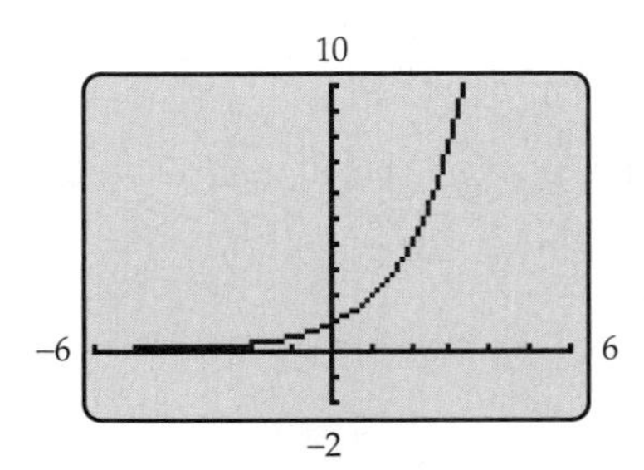

Figure 6–1

SOLUTION A small portion of the graph is shown in Figure 6–1. If the x-axis were to be extended with the same scale, $x = 50$ would be at the right edge of the page. At that point, the height of the graph is $f(50) = 2^{50}$. Now the y-axis scale in Figure 6–1 is approximately 12 units to the inch, which is equivalent to 760,320 units per mile, as you can readily verify. Therefore, the height of the graph at $x = 50$ is

$$\frac{2^{50}}{760{,}320} = 1{,}480{,}823{,}741 \text{ MILES},$$

which would put that part of the graph well beyond the planet Saturn! ■

EXAMPLE 2

Use the graph of $f(x) = 2^x$ in Figure 6–1 to describe the graphs of the following functions without actually graphing them.

(a) $g(x) = 2^{x+3}$ (b) $h(x) = 2^{x-3} - 4$

SOLUTION

(a) Since $f(x) = 2^x$, we have $f(x + 3) = 2^{x+3} = g(x)$. So the graph of $g(x)$ is just the graph of $f(x) = 2^x$ shifted horizontally 3 units to the left. [See page 63.]

GRAPHING EXPLORATION

Verify the preceding statement graphically by graphing $f(x)$ and $g(x)$ on the same screen.

(b) In this case, $f(x - 3) - 4 = 2^{x-3} - 4 = h(x)$. So the graph of $h(x)$ is the graph of $f(x) = 2^x$ shifted horizontally 3 units to the right and vertically 4 units downward. [See pages 63 and 62.]

GRAPHING EXPLORATION

Verify the preceding statement graphically by graphing $f(x)$ and $h(x)$ on the same screen. ■

When the base a is between 0 and 1, then the graph of $f(x) = a^x$ has a different shape.

GRAPHING EXPLORATION

Using viewing window with $-4 \le x \le 4$ and $-1 \le y \le 4$, graph

$f(x) = .2^x$, $g(x) = .4^x$, $h(x) = .6^x$, and $k(x) = .8^x$

on the same screen. How is the steepness of the graph of $f(x) = a^x$ related to the size of a?

The exploration supports this conclusion.

The Exponential Function $f(x) = a^x$ $(0 < a < 1)$

When $0 < a < 1$, the graph of $f(x) = a^x$ has the shape shown here and the properties listed below.

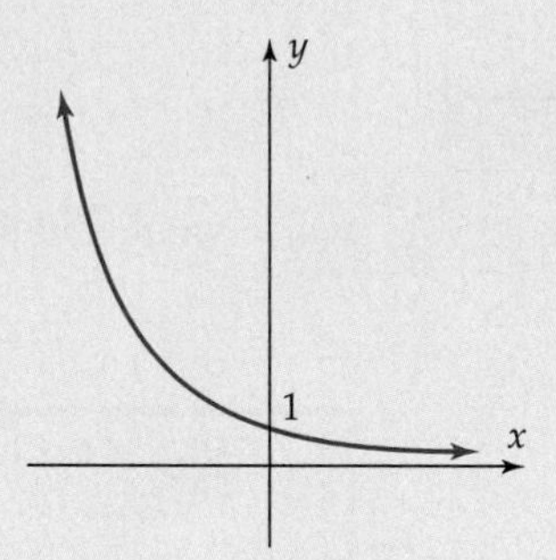

The graph is above the x-axis.

The y-intercept is 1.

$f(x)$ is a decreasing function.

The positive x-axis is a horizontal asymptote.

The closer the base a is to 0, the more steeply the graph falls to the right.

EXAMPLE 3 

Without graphing, describe the graph of $g(x) = 3^{-x}$.

SOLUTION Note that

$$g(x) = 3^{-x} = (3^{-1})^x = \left(\frac{1}{3}\right)^x.$$

So $g(x)$ is an exponential function with a positive base less than 1. Its graph has the shape shown in the preceding box: It falls quickly to the right and rises very steeply to the left of the y-axis. ■

GRAPHING EXPLORATION

Verify the analysis in Example 3 by graphing $g(x) = 3^{-x}$ in the viewing window with $-4 \le x \le 4$ and $-2 \le y \le 18$.

Since most quantities that grow exponentially don't increase or decrease as dramatically as the graphs of $f(x) = a^x$, exponential functions that model real-life situations generally have the form $f(x) = Pa^{kx}$, such as

$$f(x) = 5 \cdot 2^{.45x}, \qquad g(x) = 3.5(10^{-.03x}), \qquad h(x) = (-6)(1.076^{2x}).$$

Their graphs have the same basic shape as the graph of $f(x) = a^x$, but rise or fall at different rates, depending on the constants P, a, and k.

EXAMPLE 4

Figure 6–2 shows the graphs of

$$f(x) = 3^x, \qquad g(x) = 3^{.15x}, \qquad h(x) = 3^{.35x}, \qquad k(x) = 3^{-x}, \qquad p(x) = 3^{-.4x}.$$

Note how the coefficient of x determines the steepness of the graph. When this coefficient is positive, the graph rises; when it is negative, the graph falls from left to right.

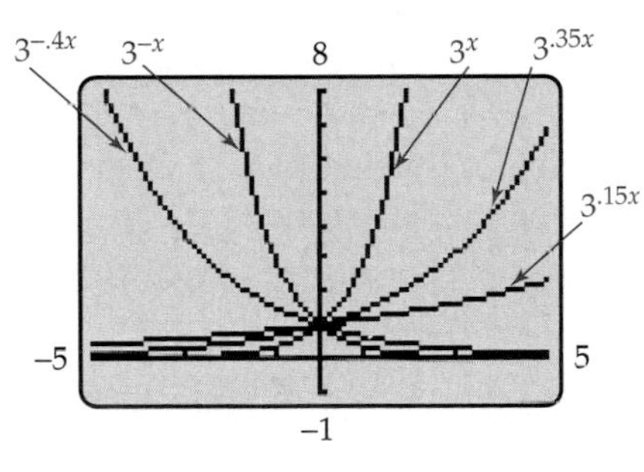

Figure 6–2

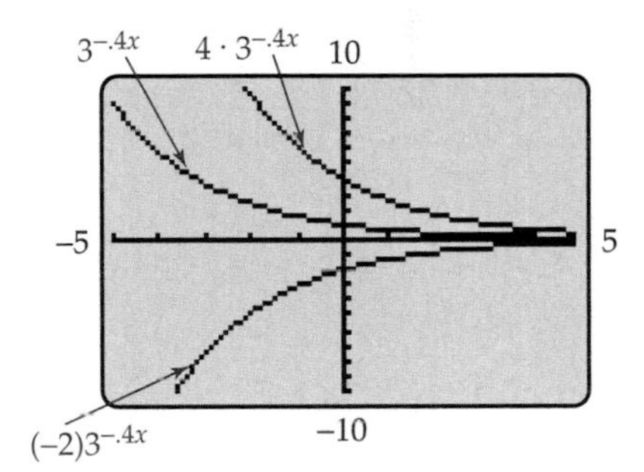

Figure 6–3

Figure 6–3 shows the graphs of

$$p(x) = 3^{-.4x}, \qquad q(x) = 4 \cdot 3^{-.4x}, \qquad r(x) = (-2)3^{-.4x}.$$

As we saw in Section 0.6, the graph of $q(x) = 4 \cdot 3^{-.4x}$ is the graph of $p(x) = 3^{-.4x}$ stretched away from the x-axis by a factor of 4. The graph of $r(x) = (-2)3^{-.4x}$ is the graph of $p(x) = 3^{-.4x}$ stretched away from the x-axis by a factor of 2 *and* reflected in the x-axis. ■

EXPONENTIAL GROWTH AND DECAY

Exponential functions have a number of practical applications, as illustrated in the following examples. We shall learn how to construct functions such as the ones in these examples in Section 6.2, but for now we concentrate on using them.

EXAMPLE 5 **Finance**

Bill and Mary Ann Burger invest $5000 in a high-flying stock that is increasing in value at the rate of 12% per year. The value of their stock is given by the function $f(x) = 5000(1.12^x)$, where x is measured in years.

(a) Assuming that the value of the stock continues growing at this rate, how much will their investment be worth in four years?

(b) When will their investment be worth $14,000?

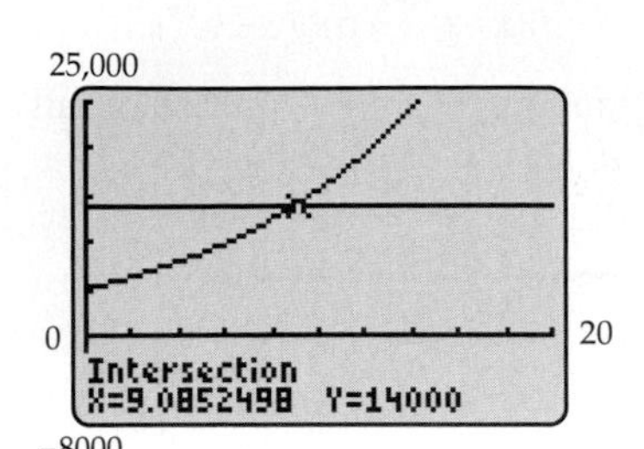

Figure 6–4

SOLUTION

(a) In four years ($x = 4$), their stock will be worth

$$f(4) = 5000(1.12^4) \approx \$7867.60.$$

(b) To determine when the stock will be worth $14,000, we must find the value of x for which $f(x) = 14{,}000$. In other words, we must solve the equation

$$5000(1.12^x) = 14{,}000.$$

The solution is the x-coordinate of the intersection point of the graphs of $f(x) = 5000(1.12^x)$ and $y = 14{,}000$. Figure 6–4 shows that this point is approximately (9.085, 14,000). Therefore, the stock will be worth $14,000 in about 9.085 years. ■

EXAMPLE 6 Population Growth

Based on figures from the past 25 years, the world population (in billions) can be approximated by the function $g(x) = 4.5(1.015^x)$, where $x = 0$ corresponds to 1980.

(a) Estimate the world population in 2009.

(b) In what year will the world population be double what it is in 2009?

SOLUTION

(a) Since 2009 corresponds to $x = 29$, the population is

$$g(29) = 4.5(1.015^{29}) \approx 6.93 \text{ billion people.}$$

(b) Twice the population in 2009 is $2(6.93) = 13.86$ billion. We must find the number x such that $g(x) = 13.86$; that is, we must solve the equation

$$4.5(1.015^x) = 13.86.$$

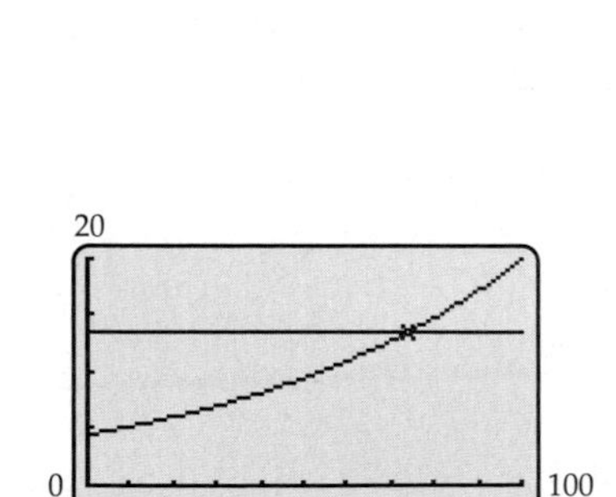

Figure 6–5

This can be done with an equation solver or by graphical means, as in Figure 6–5, which shows the intersection point of $g(x) = 4.5(1.015^x)$ and $y = 13.86$. The solution is $x \approx 75.6$, which corresponds to the year 2055. According to this model, the world population will double in your lifetime. This is what is meant by the population explosion. ■

EXAMPLE 7 Radioactive Decay

One kilogram of plutonium (^{239}Pu) is stored in a nuclear waste facility, where it slowly decays. The amount remaining after x years is approximated by

$$M(x) = .99997^x.$$

(a) Without graphing, describe the shape of the graph of M. Then confirm your analysis by graphing.

(b) How much plutonium will be left after 10,000 years?

SOLUTION

(a) Since M is an exponential function whose base is smaller than, but very close to, 1, its graph falls *very slowly* from left to right.

GRAPHING EXPLORATION

Confirm this statement by graphing $M(x)$ in the window with $0 \le x \le 5000$ and $0 \le y \le 2$.

(b) After 10,000 years, the amount of plutonium remaining is

$$M(10{,}000) = .99997^{10{,}000} \approx .74 \text{ kilogram.}$$

So almost three-fourths of the original amount remains. This is the reason that nuclear waste disposal is such a serious problem. ■

THE NUMBER *e* AND THE NATURAL EXPONENTIAL FUNCTION

There is an irrational number, denoted e, that arises naturally in a variety of phenomena and plays a central role in the mathematical description of the physical universe. Its decimal expansion begins

$$\mathbf{e = 2.718281828459045 \cdots .}$$

Your calculator has an e^x key that can be used to evaluate the **natural exponential function** $f(x) = e^x$. If you key in e^1, the calculator will display the first part of the decimal expansion of e.

The graph of $f(x) = e^x$ has the same shape as the graph of $g(x) = 2^x$ in Figure 6–1 but climbs more steeply.

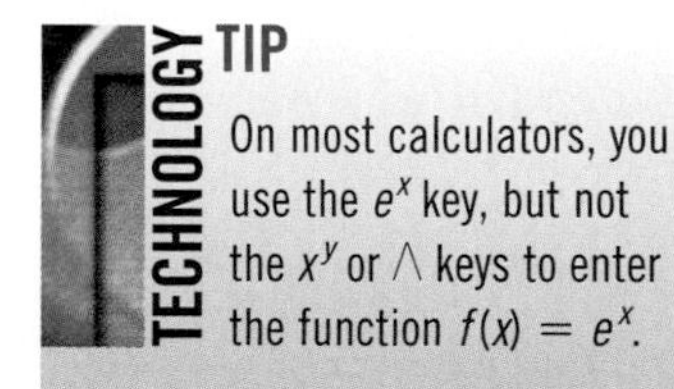

TECHNOLOGY TIP

On most calculators, you use the e^x key, but not the x^y or $\wedge$ keys to enter the function $f(x) = e^x$.

GRAPHING EXPLORATION

Graph $f(x) = e^x$, $g(x) = 2^x$, and $h(x) = 3^x$ on the same screen in a window with $-5 \le x \le 5$. The Technology Tip in the margin may be helpful.

EXAMPLE 8 **Transistor Growth**

For the past 30 years, silicon chip technology has advanced dramatically. The approximate number of transistors that can be placed on a single chip in year x is given by

$$N(x) = 2418e^{.30895x},$$

where $x = 2$ corresponds to 1972.*

*Based on data from Intel.

(a) Estimate the number of transistors per chip in 1995 and 2003.
(b) Assuming that this model remains accurate for the next decade (which is not at all certain), when will it be possible to put 420 million transistors on a chip?

SOLUTION

(a) The year 1995 corresponds to $x = 25$, so the number of transistors was

$$N(25) = 2418e^{.30895(25)} \approx 5{,}468{,}123.$$

In 2003 ($x = 33$), the number was

$$N(33) = 2418e^{.30895(33)} \approx 64{,}750{,}115.$$

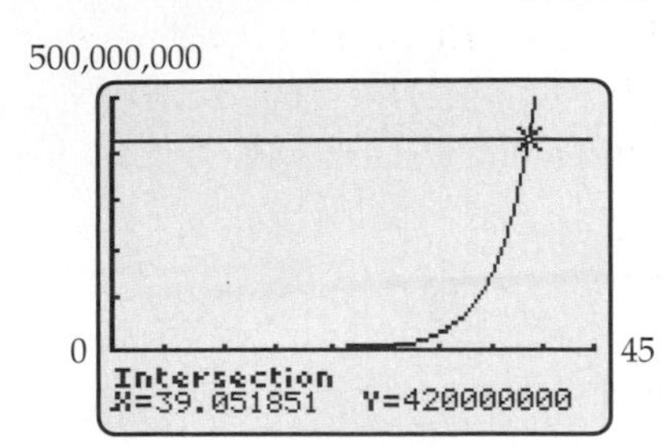

Figure 6–6

(b) We must find the value of x for which $N(x) = 420$ million; that is, we must solve the equation

$$2418e^{.30895x} = 420{,}000{,}000.$$

This can be done with an equation solver or by graphical means. Figure 6–6 shows that the solution is $x \approx 39$, which corresponds to the year 2009. ■

OTHER EXPONENTIAL FUNCTIONS

The population growth models in earlier examples do not take into account factors that may limit population growth in the future (wars, new diseases, etc.). Example 9 illustrates a **logistic function** that is designed to model such situations more accurately.

EXAMPLE 9 Inhibited Population Growth

The world population model in Example 6 is based on data from 1980 to the present and shows a constantly increasing population. Because of decreasing fertility rates and aging populations, however, the population growth rate is expected to slow significantly during this century. The logistic function

$$f(x) = \frac{10}{1 + 232.97e^{-.02944x}}$$

models this situation.* It gives the world population (in billions) in year x, where $x = 0$ corresponds to 1800.

(a) According to this model, what will the world population be in 2009 and in 2055?
(b) If this model is accurate, is there an upper limit on the world's population?

SOLUTION

(a) The year 2009 corresponds to $x = 209$ and

$$f(209) = \frac{10}{1 + 232.97e^{-.02944(209)}} \approx 6.69 \text{ billion.}$$

---

*Based on data from the United Nations Population Division and the U.S. Census Bureau.

This is quite close to the 6.93 billion given by the function in Example 6. For 2055 ($x = 255$), the model gives a population of

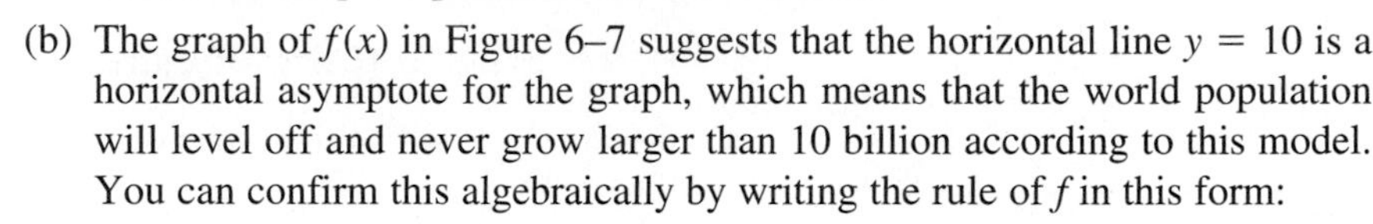

$$f(255) = \frac{10}{1 + 232.97e^{-.02944(255)}} \approx 8.87 \text{ billion},$$

whereas Example 6 predicts close to 14 billion.

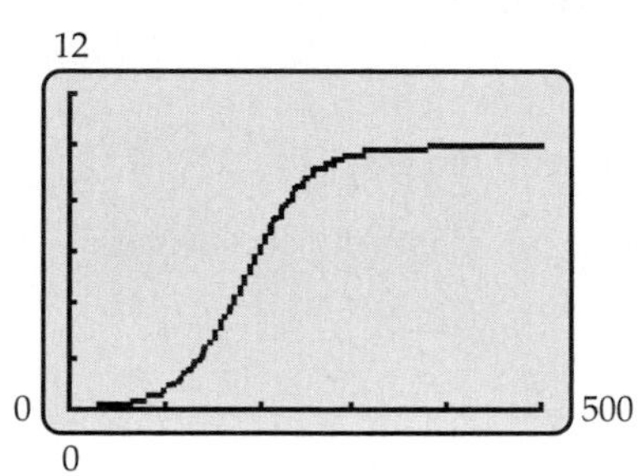

Figure 6–7

(b) The graph of $f(x)$ in Figure 6–7 suggests that the horizontal line $y = 10$ is a horizontal asymptote for the graph, which means that the world population will level off and never grow larger than 10 billion according to this model. You can confirm this algebraically by writing the rule of f in this form:

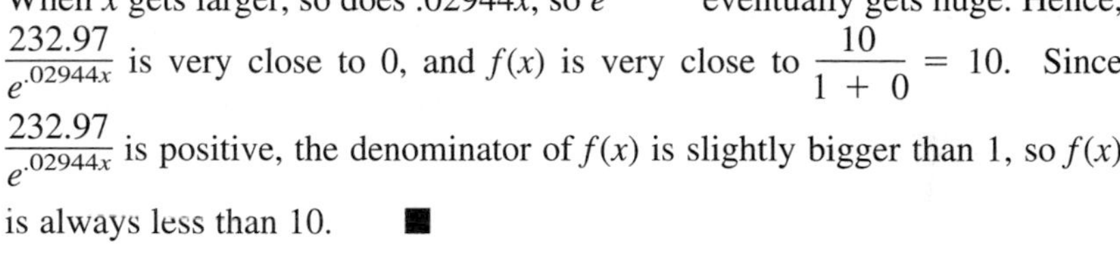

$$f(x) = \frac{10}{1 + 232.97e^{-.02944x}} = \frac{10}{1 + \frac{232.97}{e^{.02944x}}}.$$

When x gets larger, so does $.02944x$, so $e^{.02944x}$ eventually gets huge. Hence, $\frac{232.97}{e^{.02944x}}$ is very close to 0, and $f(x)$ is very close to $\frac{10}{1 + 0} = 10$. Since $\frac{232.97}{e^{.02944x}}$ is positive, the denominator of $f(x)$ is slightly bigger than 1, so $f(x)$ is always less than 10. ■

When a cable, such as a power line, is suspended between towers of equal height, it forms a curve called a **catenary,** which is the graph of a function of the form

$$f(x) = A(e^{kx} + e^{-kx})$$

for suitable constants A and k. The Gateway Arch in St. Louis (Figure 6–8) has the shape of an inverted catenary, which was chosen because it evenly distributes the internal structural forces.

Figure 6–8

GRAPHING EXPLORATION

Graph each of the following functions in the window with $-5 \le x \le 5$ and $-10 \le y \le 80$.

$$y_1 = 10(e^{.4x} + e^{-.4x}), \qquad y_2 = 10(e^{2x} + e^{-2x}),$$

$$y_3 = 10(e^{3x} + e^{-3x}).$$

How does the coefficient of x affect the shape of the graph?

Predict the shape of the graph of $y = -y_1 + 80$. Confirm your answer by graphing.

EXERCISES 6.1

In Exercises 1–6, sketch a complete graph of the function.

1. $f(x) = 4^{-x}$
2. $f(x) = (5/2)^{-x}$
3. $f(x) = 2^{3x}$
4. $g(x) = 3^{x/2}$
5. $f(x) = 2^{x^2}$
6. $g(x) = 2^{-x^2}$

In Exercises 7–12, list the transformations needed to transform the graph of $h(x) = 2^x$ into the graph of the given function. [Section 0.6 may be helpful.]

7. $f(x) = 2^x - 5$
8. $g(x) = -(2^x)$
9. $k(x) = 3(2^x)$
10. $g(x) = 2^{x-1}$
11. $f(x) = 2^{x+2} - 5$
12. $g(x) = -5(2^{x-1}) + 7$

In Exercises 13 and 14, match the functions to the graphs. Assume that $a > 1$ and $c > 1$.

13. $f(x) = a^x$
 $g(x) = a^x + 3$
 $h(x) = a^{x+5}$

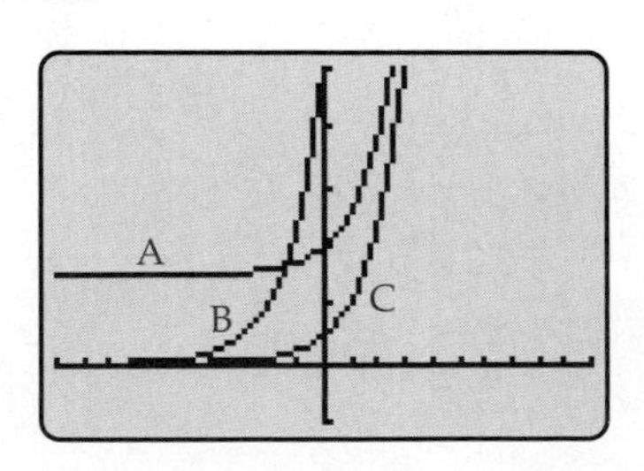

14. $f(x) = c^x$
 $g(x) = -3c^x$
 $h(x) = c^{x+5}$
 $k(x) = -3c^x - 2$

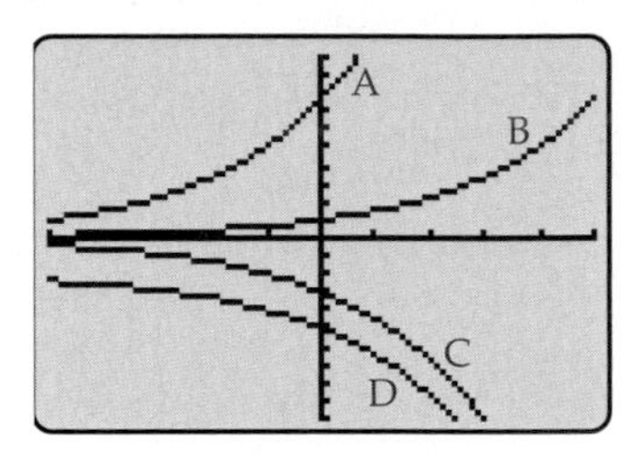

Knowledge of even and odd functions is a prerequisite for Exercises 15–19. Determine whether the function is even, odd, or neither.

15. $f(x) = 10^x$
16. $g(x) = 2^x - x$
17. $f(x) = \dfrac{e^x + e^{-x}}{2}$
18. $f(x) = \dfrac{e^x - e^{-x}}{2}$
19. $f(x) = e^{-x^2}$
20. Explain why $e^x + e^{-x}$ is approximately equal to e^x when x is large.

C *In Exercises 21–24, find the average rate of change of the function.* (See Exercises 21–23 on page 210.)*

21. $f(x) = x2^x$ as x goes from 1 to 3
22. $g(x) = 3^{x^2-x}$ as x goes from -1 to 1
23. $h(x) = 5^{-x^2}$ as x goes from -1 to 0
24. $f(x) = e^x - e^{-x}$ as x goes from -3 to -1

C *In Exercises 25–28, find the difference quotient of the function. (See Exercises 27–28 on page 69.)*

25. $f(x) = 10^x$
26. $g(x) = 5^{x^2}$
27. $f(x) = 2^x + 2^{-x}$
28. $f(x) = e^x - e^{-x}$

* **C** and ○ indicate examples, exercises, and sections that are relevant to calculus.

In Exercises 29–36, find a viewing window (or windows) that shows a complete graph of the function.

29. $k(x) = e^{-x}$

30. $f(x) = e^{-x^2}$

31. $f(x) = \frac{e^x + e^{-x}}{2}$

32. $h(x) = \frac{e^x - e^{-x}}{2}$

33. $g(x) = 2^x - x$

34. $k(x) = \frac{2}{e^x + e^{-x}}$

35. $f(x) = \frac{5}{1 + e^{-x}}$

36. $g(x) = \frac{10}{1 + 9e^{-x/2}}$

In Exercises 37–42, list all asymptotes of the graph of the function and the approximate coordinates of each local maximum and local minimum.

37. $f(x) = x2^x$

38. $g(x) = x2^{-x}$

39. $h(x) = e^{x^2/2}$

40. $k(x) = 2^{x^2-6x+2}$

41. $f(x) = e^{-x^2}$

42. $g(x) = -xe^{x^2/20}$

43. (a) A genetic engineer is growing cells in a fermenter. The cells multiply by splitting in half every 15 minutes. The new cells have the same DNA as the original ones. Complete the following table.

Time (hours)	Number of Cells
0	1
.25	2
.5	4
.75	
1	

(b) Write the rule of the function that gives the number of cells C at time t hours.

44. Do Exercise 43, using the following table.

Time (hours)	Number of Cells
0	300
.25	600
.5	1200
.75	
1	

45. The Gateway Arch (Figure 6–8) is 630 feet high and 630 feet wide at ground level. Suppose it were placed on a coordinate plane with the x-axis at ground level and the y-axis going through the center of the arch. Find a catenary function $g(x) = A(e^{kx} + e^{-kx})$ and a constant C such that the graph of the function $f(x) = g(x) + C$ provides a model of the arch. [*Hint*: Experiment with various values of A, k, C as in the Graphing Exploration on page 432. Many correct answers are possible.]

46. If you deposit $750 at 2.2% interest, compounded annually and paid from the day of deposit to the day of withdrawal, your balance at time t is given by $B(t) = 750(1.022)^t$. How much will you have after two years? After three years and nine months?

47. The number of digital devices (cell phones, PCs, digital cameras, MP3 players, etc.) used worldwide in year x is approximated by

$$g(x) = 1.244 \cdot 1.263^x,$$

where $x = 0$ corresponds to 2000 and $g(x)$ is in billions.*

(a) Estimate the number of digital devices in 2002 and 2005.

(b) When will the number of devices be twice the number in 2005? (Round your answer to the nearest year.)

48. The population of a colony of fruit flies t days from now is given by the function $p(t) = 100 \cdot 3^{t/10}$.

(a) What will the population be in 15 days? In 25 days?

(b) When will the population reach 2500?

49. If the population of the United States continues to grow as it has since 1980, then the approximate population (in millions) in year t is given by

$$P(t) = 227e^{.0093t},$$

where $t = 0$ corresponds to 1980.

(a) Estimate the population in 2005 and 2010.

(b) When will the population reach 400 million?

50. If current rates of deforestation and fossil fuel consumption continue, then the amount of atmospheric carbon dioxide in parts per million (ppm) will be given by $f(x) = 375e^{.00609x}$, where $x = 0$ corresponds to 2000.†

(a) What is the amount of carbon dioxide in 2003? In 2022?

(b) In what year will the amount of carbon dioxide reach 500 ppm?

---

*Based on data and projections from *IDC*.

†Based on projections from the International Panel on Climate Change.

51. The pressure of the atmosphere $p(x)$ (in pounds per square inch) is given by

$$p(x) = ke^{-.0000425x},$$

where x is the height above sea level (in feet) and k is a constant.

(a) Use the fact that the pressure at sea level is 15 pounds per square inch to find k.

(b) What is the pressure at 5000 feet?

(c) If you were in a spaceship at an altitude of 160,000 feet, what would the pressure be?

52. (a) The function $g(t) = 1 - e^{-.0479t}$ gives the percentage of the population (expressed as a decimal) that has seen a new TV show t weeks after it goes on the air. What percentage of people have seen the show after 24 weeks?

(b) Approximately when will 90% of the people have seen it?

53. According to data from the National Center for Health Statistics, the life expectancy at birth for a person born in a year x is approximated by the function

$$D(x) = \frac{79.257}{1 + 9.7135 \times 10^{24} \cdot e^{-.0304x}} \quad (1900 \le x \le 2050).$$

(a) What is the life expectancy of someone born in 1980? In 2000?

(b) In what year was life expectancy at birth 60 years?

54. (a) The beaver population near a certain lake in year t is approximately

$$p(t) = \frac{2000}{1 + 199e^{-.5544t}}.$$

What is the population now ($t = 0$) and what will it be in 5 years?

(b) Approximately when will there be 1000 beavers?

55. The number of subscribers to basic cable TV (in millions) can be approximated by

$$g(x) = \frac{76.7}{1 + 16(.8444^x)},$$

where $x = 0$ corresponds to 1970.*

(a) Estimate the number of subscribers in 1995 and 2005.

(b) When does the number of subscribers reach 70 million?

(c) According to this model, will the number of subscribers ever reach 90 million?

56. An eccentric billionaire offers you a job for the month of September. She says that she will pay you 2¢ on the first day, 4¢ on the second day, 8¢ on the third day, and so on, doubling your pay on each successive day.

(a) Let $P(x)$ denote your salary in *dollars* on day x. Find the rule of the function P.

(b) Would you be better off financially if instead you were paid a flat rate of $10,000 per day? [*Hint:* Consider $P(30)$.]

57. Take an ordinary piece of printer paper and fold it in half; the folded sheet is twice as thick as the single sheet was. Fold it in half again so that it is twice as thick as before. Keep folding it in half as long as you can. Soon the folded paper will be so thick and small that you will be unable to continue, but suppose you could keep folding the paper as many times as you wanted. Assume that the paper is .002 inches thick.

(a) Make a table showing the thickness of the folded paper for the first four folds (with fold 0 being the thickness of the original unfolded paper).

(b) Find a function of the form $f(x) = Pa^x$ that describes the thickness of the folded paper after x folds.

(c) How thick would the paper be after 20 folds?

(d) How many folds would it take to reach the moon (which is 243,000 miles from the earth)? [*Hint:* One mile is 5280 feet.]

58. The figure is the graph of an exponential growth function $f(x) = Pa^x$.

(a) In this case, what is P? [*Hint:* What is $f(0)$?]

(b) Find the rule of the function f by finding a. [*Hint:* What is $f(2)$?]

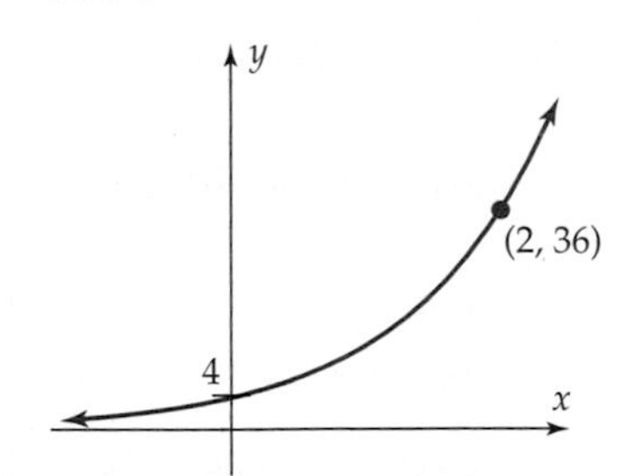

Thinkers

59. Find a function $f(x)$ with the property $f(r + s) = f(r)f(s)$ for all real numbers r and s. [*Hint:* Think exponential.]

60. Find a function $g(x)$ with the property $g(2x) = (g(x))^2$ for every real number x.

*Based on data from *The Cable TV Financial Datebook* and *The Pay TV Newsletter*.

61. (a) Using the viewing window with $-4 \le x \le 4$ and $-1 \le y \le 8$, graph $f(x) = \left(\frac{1}{2}\right)^x$ and $g(x) = 2^x$ on the same screen. If you think of the y-axis as a mirror, how would you describe the relationship between the two graphs?

(b) Without graphing, explain how the graphs of $g(x) = 2^x$ and $k(x) = 2^{-x}$ are related.

62. Approximating exponential functions by polynomials. For each positive integer n, let f_n be the polynomial function whose rule is

$$f_n(x) = 1 + x + \frac{x^2}{2!} + \frac{x^3}{3!} + \frac{x^4}{4!} + \cdots + \frac{x^n}{n!},$$

where $k!$ is the product $1 \cdot 2 \cdot 3 \cdots k$.

(a) Using the viewing window with $-4 \le x \le 4$ and $-5 \le y \le 55$, graph $g(x) = e^x$ and $f_4(x)$ on the same screen. Do the graphs appear to coincide?

(b) Replace the graph of $f_4(x)$ by that of $f_5(x)$, then by $f_6(x)$, $f_7(x)$, and so on until you find a polynomial $f_n(x)$ whose graph appears to coincide with the graph of $g(x) = e^x$ in this viewing window. Use the trace feature to move from graph to graph at the same value of x to see how accurate this approximation is.

(c) Change the viewing window so that $-6 \le x \le 6$ and $-10 \le y \le 400$. Is the polynomial you found in part (b) a good approximation for $g(x)$ in this viewing window? What polynomial is?

6.2 Applications of Exponential Functions

In the previous section, we encountered several exponential functions that modeled growth and decay. In this section, we learn how to construct such exponential models in a variety of real-life situations.

COMPOUND INTEREST

We begin with a subject that interests almost everyone: money and how it grows.

EXAMPLE 1

If you deposit \$5000 in a savings account that pays 3% interest, compounded annually, how much money is in the account after nine years?

SOLUTION After one year, the account balance is

$$5000 + 3\% \text{ of } 5000 = 5000 + (.03)5000 = 5000(1 + .03)$$
$$= 5000(1.03) = \$5150.$$

The initial balance has grown by a factor of 1.03. If you leave the \$5150 in the account, then at the end of the second year, the balance is

$$5150 + 3\% \text{ of } 5150 = 5150 + (.03)5150 = 5150(1 + .03) = 5150(1.03).$$

Once again, the amount at the beginning of the year has grown by a factor of 1.03. The same thing happens every year. A balance of P dollars at the beginning of the year grows to $P(1.03)$. So the balance grows like this:

$$5000 \xrightarrow{\text{Year 1}} 5000(1.03) \xrightarrow{\text{Year 2}} \underbrace{[5000(1.03)](1.03)}_{5000(1.03)^2} \xrightarrow{\text{Year 3}} \underbrace{[5000(1.03)(1.03)](1.03)}_{5000(1.03)^3} \rightarrow \cdots.$$

Consequently, the balance at the end of year x is given by

$$f(x) = 5000 \cdot 1.03^x.$$

The balance at the end of nine years is $f(9) = 5000(1.03^9) = \$6523.87$ (rounded to the nearest penny). ■

The argument used in Example 1 applies in the general case.

Compound Interest Formula

> If P dollars is invested at interest rate r per time period (expressed as a decimal), then the amount A after t periods is
>
> $$A = P(1 + r)^t.$$

In Example 1, for instance, we had $P = 5000$ and $r = .03$ (so $1 + r = 1 + .03 = 1.03$); the number of periods (years) was $t = 9$.

EXAMPLE 2

Suppose you borrow \$50 from your friendly neighborhood loan shark, who charges 18% interest per week. How much do you owe after one year (assuming that he lets you wait that long to pay)?

SOLUTION You use the compound interest formula with $P = 50$, $r = .18$, and $t = 52$ (because interest is compounded weekly and there are 52 weeks in a year). So you figure that you owe

$$A = P(1 + r)^t = 50(1 + .18)^{52} = 50 \cdot 1.18^{52} = \$273{,}422.58.^*$$

```
50*1.18^(365/7)
         279964.6785
```

Figure 6–9

When you try to pay the loan shark this amount, however, he points out that a 365-day year has more than 52 weeks, namely, $\frac{365}{7} = 52\frac{1}{7}$ weeks. So you recalculate with $t = 365/7$ (and careful use of parentheses, as shown in Figure 6–9) and find that you actually owe

$$A = P(1 + r)^t = 50(1 + .18)^{365/7} = 50 \cdot 1.18^{365/7} = \$279{,}964.68.$$

Ouch! ■

As Example 2 illustrates, the compound interest formula can be used even when the number of periods t is not an integer. You must also learn how to read "financial language" to apply the formula correctly, as shown in the following example.

*Here and below, all financial answers are rounded to the nearest penny.

EXAMPLE 3

Determine the amount a \$3500 investment is worth after three and a half years at the following interest rates:

(a) 6.4% compounded annually;

(b) 6.4% compounded quarterly;

(c) 6.4% compounded monthly.

SOLUTION

(a) Using the compound interest formula with $P = 3500$, $r = .064$, and $t = 3.5$, we have

$$A = 3500(1 + .064)^{3.5} = \$4348.74.$$

(b) "6.4% interest, compounded quarterly" means that the interest period is one-fourth of a year and the interest rate per period is $.064/4 = .016$. Since there are four interest periods per year, the number of periods in 3.5 years is $4(3.5) = 14$, so

$$A = 3500\left(1 + \frac{.064}{4}\right)^{14} = 3500(1 + .016)^{14} = \$4370.99.$$

(c) Similarly, "6.4% compounded monthly" means that the interest period is one month (1/12 of a year) and the interest rate per period is $.064/12$. The number of periods (months) in 3.5 years is 42, so

$$A = 3500\left(1 + \frac{.064}{12}\right)^{42} = \$4376.14.$$

Note that the more often interest is compounded, the larger is the final amount. ■

EXAMPLE 4

If \$5000 is invested at 6.5% annual interest, compounded monthly, how long will it take for the investment to double?

SOLUTION

The compound interest formula (with $P = 5000$ and $r = .065/12$) shows that the amount in the account after t months is $5000\left(1 + \frac{.065}{12}\right)^t$. We must find the value of t such that

$$5000\left(1 + \frac{.065}{12}\right)^t = 10{,}000.$$

Algebraic methods for solving this equation will be considered in Section 6.5. For now, we use technology.

GRAPHING EXPLORATION

Solve the equation, either by using an equation solver or by graphical means as follows. Graph

$$y = 5000\left(1 + \frac{.065}{12}\right)^t - 10{,}000$$

in a viewing window with $0 \le t \le 240$ (that's 20 years) and find the t-intercept.

The exploration shows that it will take 128.3 months (approximately 10.7 years) for the investment to double. ■

EXAMPLE 5

What interest rate, compounded annually, is needed for a $16,000 investment to grow to $50,000 in 18 years?

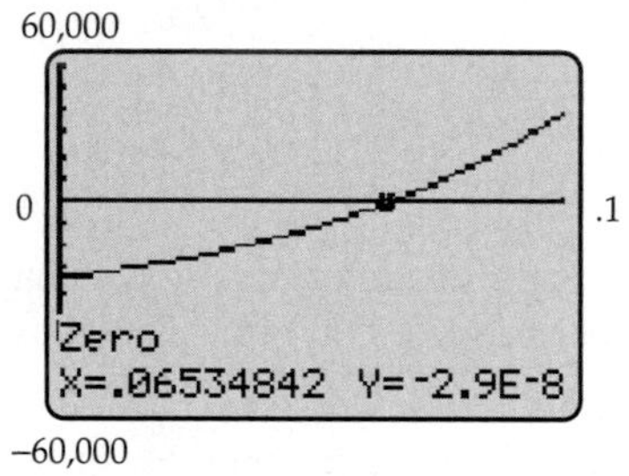

SOLUTION In the compound interest formula, we have $A = 50{,}000$, $P = 16{,}000$ and $t = 18$. We must find r in the equation

$$16{,}000(1 + r)^{18} = 50{,}000.$$

The equation can be solved numerically with an equation solver or by one of the following methods.

60,000

0 .1

Zero
X=.06534842 Y=-2.9E-8

–60,000

Figure 6–10

Graphical: Rewrite the equation as $16{,}000(1 + r)^{18} - 50{,}000 = 0$. Then the solution is the r-intercept of the graph of $y = 16{,}000(1 + r)^{18} - 50{,}000$, as shown in Figure 6–10.

Algebraic:

$$16{,}000(1 + r)^{18} = 50{,}000$$

Divide both sides by 16,000: $$(1 + r)^{18} = \frac{50{,}000}{16{,}000} = 3.125$$

Take 18th roots on both sides: $$\sqrt[18]{(1 + r)^{18}} = \sqrt[18]{3.125}$$

$$1 + r = \sqrt[18]{3.125}$$

$$r = \sqrt[18]{3.125} - 1 \approx .06535.$$

So the necessary interest rate is about 6.535%. ■

CONTINUOUS COMPOUNDING AND THE NUMBER *e*

As a general rule, the more often interest is compounded, the better off you are, as we saw in Example 3. But there is, alas, a limit.

EXAMPLE 6 The Number *e*

You have $1 to invest for 1 year. The Exponential Bank offers to pay 100% annual interest, compounded n times per year and rounded to the nearest penny.

You may pick any value you want for n. Can you choose n so large that your $1 will grow to some huge amount?

SOLUTION Since interest is compounded n times per year and the annual rate is 100% (= 1.00), the interest rate per period is $r = 1/n$ and the number of periods in 1 year is n. According to the formula, the amount at the end of the year will be $A = \left(1 + \frac{1}{n}\right)^n$. Here's what happens for various values of n:

Interest Is Compounded	$n =$	$\left(1 + \frac{1}{n}\right)^n =$
Annually	1	$\left(1 + \frac{1}{1}\right)^1 = 2$
Semiannually	2	$\left(1 + \frac{1}{2}\right)^2 = 2.25$
Quarterly	4	$\left(1 + \frac{1}{4}\right)^4 \approx 2.4414$
Monthly	12	$\left(1 + \frac{1}{12}\right)^{12} \approx 2.6130$
Daily	365	$\left(1 + \frac{1}{365}\right)^{365} \approx 2.71457$
Hourly	8760	$\left(1 + \frac{1}{8760}\right)^{8760} \approx 2.718127$
Every minute	525,600	$\left(1 + \frac{1}{525{,}600}\right)^{525{,}600} \approx 2.7182792$
Every second	31,536,000	$\left(1 + \frac{1}{31{,}536{,}000}\right)^{31{,}536{,}000} \approx 2.7182818$

Since interest is rounded to the nearest penny, your dollar will grow no larger than $2.72, no matter how big n is. ■

The last entry in the preceding table, 2.7182818, is the number e to seven decimal places. This is just one example of how e arises naturally in real-world situations. In calculus, it is proved that e is the *limit* of $\left(1 + \frac{1}{n}\right)^n$, meaning that as n gets larger and larger, $\left(1 + \frac{1}{n}\right)^n$ gets closer and closer to e.

GRAPHING EXPLORATION

Confirm this fact graphically by graphing the function

$$f(x) = \left(1 + \frac{1}{x}\right)^x$$

and the horizontal line $y = e$ in the viewing window with $0 \le x \le 5000$ and $-1 \le y \le 4$ and noting that the two graphs appear to be identical.

When interest is compounded n times per year for larger and larger values of n, as in Example 6, we say that the interest is **continuously compounded.** In this terminology, Example 6 says that $1 will grow to $2.72 in 1 year at an interest rate of 100% compounded continuously. A similar argument with more realistic interest rates (see Exercise 42) produces the following result (Example 6 is the case when $P = 1$, $r = 1$, and $t = 1$).

Continuous Compounding

If P dollars is invested at interest rate r, compounded continuously, then the amount A after t years is

$$A = Pe^{rt}.$$

EXAMPLE 7 

If \$3800 is invested at 9.2% interest, compounded continuously, find:

(a) The amount in the account after seven and a half years.

(b) The number of years for the account balance to reach \$5000.

SOLUTION

(a) Apply the continuous compounding formula with $P = 3800$, $r = .092$, and $t = 7.5$.

$$A = 3800e^{(.092)7.5} = 3800e^{.69} \approx \$7576.12.$$

(b) We must solve the equation

$$3800e^{.092t} = 5000, \qquad \text{or, equivalently,} \qquad 3800e^{.092t} - 5000 = 0.$$

GRAPHING EXPLORATION

Solve the equation graphically and verify that it will take just a few days less than three years for the investment to be worth \$5000. ■

EXPONENTIAL GROWTH

The rules of exponential growth functions are found in the same way that the formula for compound interest was developed.

EXAMPLE 8

The world population in 1980 was about 4.5 billion people and has been increasing at approximately 1.5% per year. Find the rule of the function that gives the world population in year x, where $x = 0$ corresponds to 1980.

SOLUTION

From one year to the next, the population increases by 1.5%. If P is the population at the start of a year, then the population at the end of that year is

$$P + 1.5\% \text{ of } P = P + .015P = P(1 + .015) = P(1.015).$$

So the population increases by a factor of 1.015 every year.

$$4.5 \to \underbrace{4.5 \cdot 1.015}_{\text{Year 1}} \to \underbrace{[4.5 \cdot 1.015]1.015}_{\text{Year 2: } 4.5 \cdot 1.015^2} \to \underbrace{[4.5 \cdot 1.015 \cdot 1.015]1.015}_{\text{Year 3: } 4.5 \cdot 1.015^3} \to \cdots.$$

Consequently, the population (in billions) in year x is given by

$$g(x) = 4.5 \cdot 1.015^x.$$

This function was used in Example 6 of Section 6.1. ■

Note the form of the function in Example 8: The constant 4.5 is the initial population (when $x = 0$), and the number 1.015 is the factor by which the population changes in one year when population grows at a rate of 1.5% per year. This function and the compound interest formula are special cases of the following.

Exponential Growth

Exponential growth can be described by a function of the form

$$f(x) = Pa^x,$$

where $f(x)$ is the quantity at time x, P is the initial quantity (when $x = 0$) and $a > 1$ is the factor by which the quantity changes when x increases by 1.

If the quantity is growing at the rate r per time period, then $a = 1 + r$, and

$$f(x) = Pa^x = P(1 + r)^x.$$

EXAMPLE 9

At the beginning of an experiment, a culture contains 1000 bacteria. Five hours later, there are 7600 bacteria. Assuming that the bacteria grow exponentially, how many will there be after 24 hours?

SOLUTION The bacterial population is given by $f(x) = Pa^x$, where P is the initial population, a is the change factor, and x is the time in hours. We are given that $P = 1000$, so $f(x) = 1000a^x$. The next step is to determine a. Since there are 7600 bacteria when $x = 5$, we have

$$7600 = f(5) = 1000a^5,$$

so

$$\begin{aligned} 1000\,a^5 &= 7600 \\ a^5 &= 7.6 \\ a &= \sqrt[5]{7.6} = (7.6)^{1/5} = (7.6)^{.2}. \end{aligned}$$

Therefore, the population function is $f(x) = 1000(7.6^{.2})^x = 1000 \cdot (7.6)^{.2x}$. After 24 hours, the bacteria population will be

$$f(24) = 1000(7.6)^{.2(24)} \approx 16{,}900{,}721. \quad ■$$

EXPONENTIAL DECAY

In some situations, a quantity *decreases* by a fixed factor as time goes on.

EXAMPLE 10

When tap water is filtered through a layer of charcoal and other purifying agents, 30% of the chemical impurities in the water are removed, and 70% remain. If the water is filtered through a second purifying layer, then the amount of impurities remaining is 70% of 70%, that is, $(.7)(.7) = .7^2 = .49$ or 49%. A third layer results in $.7^3$ of the impurities remaining. Thus, the function

$$f(x) = .7^x$$

gives the percentage of impurities remaining in the water after it passes through x layers of purifying material. How many layers are needed to ensure that 95% of the impurities are removed from the water?

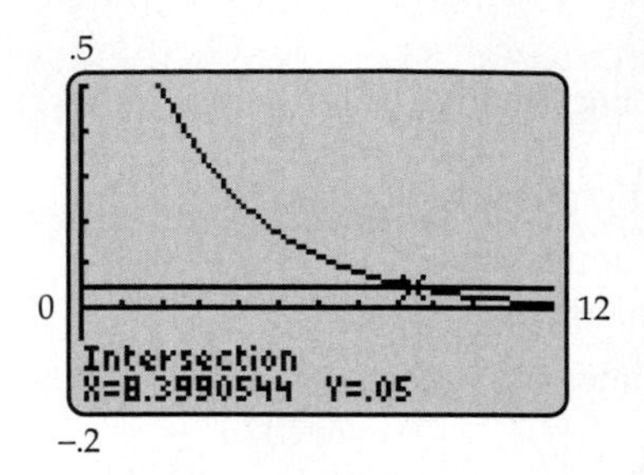

Figure 6–11

SOLUTION If 95% of the impurities are removed, then 5% will remain. Hence, we must find x such that $f(x) = .05$, that is, we must solve the equation $.7^x = .05$. This can be done numerically or graphically. Figure 6–11 shows that the solution is $x \approx 8.4$, so 8.4 layers of material are needed.

GRAPHING EXPLORATION

How many layers are needed to ensure that 99% of the impurities are removed? ■

Example 10 illustrates **exponential decay.** Note that the impurities were removed at a rate of $30\% = .3$ and that the amount of impurities remaining in the water was changing by a factor of $1 - .30 = .7$. The same thing is true in the general case.

Exponential Decay

Exponential decay can be described by a function of the form

$$f(x) = Pa^x,$$

where $f(x)$ is the quantity at time x, P is the initial quantity (when $x = 0$) and $0 < a < 1$. Here, a is the factor by which the quantity changes when x increases by 1.

If the quantity is decaying at the rate r per time period, then $a = 1 - r$, and

$$f(x) = Pa^x = P(1 - r)^x.$$

One of the important uses of exponential functions is to describe radioactive decay. The **half-life** of a radioactive element is the time it takes a given quantity to decay to one-half of its original mass. The half-life depends only on the substance and not on the size of the sample. Exercise 61 proves the following result.

Radioactive Decay

The mass $M(x)$ of a radioactive element at time x is given by

$$M(x) = c(.5^{x/h}),$$

where c is the original mass and h is the half-life of the element.

EXAMPLE 11

When a living organism dies, its carbon-14 decays exponentially. An archeologist determines that the skeleton of a mastodon has lost 64% of its carbon-14.* Use the fact that the half-life of carbon-14 is 5730 years to estimate how long ago the mastodon died.

SOLUTION The amount of carbon-14 in the skeleton at time x is given by $f(x) = P(.5^{x/5730})$, where $x = 0$ corresponds to the death of the mastodon. Since the skeleton has lost 64% of its carbon-14, we know that the amount in the skeleton now is 36% of P, that is, $.36P$. So we must find the value of x for which

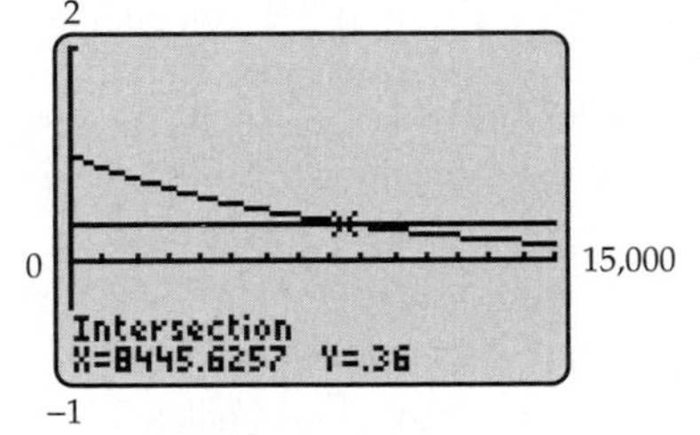

Figure 6–12

$$f(x) = .36P$$

Use the rule of f: $$P(.5^{x/5730}) = .36P$$

Divide both sides by P: $$.5^{x/5730} = .36.$$

To solve this equation, we graph $y_1 = .5^{x/5730}$ and $y_2 = .36$ on the same screen and find the x-coordinate of the intersection point. Figure 6–12 shows that $x \approx 8445.63$. Therefore, the mastodon died about 8446 years ago. ■

EXERCISES 6.2

1. If \$1,000 is invested at 8%, find the value of the investment after 5 years if interest is compounded
   (a) annually. (b) quarterly. (c) monthly.
   (d) weekly.
2. If \$2500 is invested at 11.5%, what is the value of the investment after 10 years if interest is compounded
   (a) annually? (b) monthly? (c) daily?

In Exercises 3–12, determine how much money will be in a savings account if the initial deposit was \$500 and the interest rate is:

3. 2% compounded annually for 8 years.
4. 2% compounded annually for 10 years.
5. 2% compounded quarterly for 10 years.
6. 2.3% compounded monthly for 9 years.
7. 2.9% compounded daily for 8.5 years.
8. 3.5% compounded weekly for 7 years and 7 months.
9. 3% compounded continuously for 4 years.
10. 3.5% compounded continuously for 10 years.
11. 2.45% compounded continuously for 6.2 years.
12. 3.25% compounded continuously for 11.6 years.

A sum of money P that can be deposited today to yield some larger amount A in the future is called the ***present value*** *of A. In Exercises 13–18, find the present value of the given amount A. [Hint: Substitute A, the interest rate per period r, and the number t of periods in the compound interest formula and solve for P.]*

13. \$5000 at 6% compounded annually for 7 years.
14. \$3500 at 5.5% compounded annually for 4 years.
15. \$4800 at 7.2% compounded quarterly for 5 years.
16. \$7400 at 5.9% compounded quarterly for 8 years.
17. \$8900 at 11.3% compounded monthly for 3 years.
18. \$9500 at 9.4% compounded monthly for 6 years.

In Exercises 19–36, use the compound interest formula. In most cases, you will be given three of the quantities A, P, r, t and will have to solve an equation to find the remaining one.

19. A typical credit card company charges 18% annual interest, compounded monthly, on the unpaid balance. If your current balance is \$520 and you don't make any payments for 6 months, how much will you owe (assuming they don't sue you in the meantime)?

---

*The technique involves measuring the ratio of the radioactive isotope of carbon, carbon-14, to ordinary nonradioactive carbon-12 in the skeleton.

20. When his first child was born a father put $3000 in a savings account that pays 4% annual interest, compounded quarterly. How much will be in the account on the child's 18th birthday?

21. Mary Alice has $10,000 to invest for two years. Fund *A* pays 13.2% interest, compounded annually. Fund *B* pays 12.7% interest, compounded quarterly. Fund *C* pays 12.6% interest, compounded monthly. Which fund will return the most money?

22. If you invest $7400 for five years, are you better off with an interest rate of 5% compounded quarterly or 4.8% compounded continuously?

23. If you borrow $1200 at 14% interest, compounded monthly, and pay off the loan (principal and interest) at the end of two years, how much interest will you have paid?

24. A developer borrows $150,000 at 6.5% interest, compounded quarterly, and agrees to pay off the loan in four years. How much interest will she owe?

25. A manufacturer has settled a lawsuit out of court by agreeing to pay $1.5 million four years from now. At this time, how much should the company put in an account paying 6.4% annual interest, compounded monthly, to have $1.5 million in four years? [*Hint:* See Exercises 13–18.]

26. Darlene Amidon-Brent wants to have $30,000 available in five years for a down payment on a house. She has inherited $25,000. How much of the inheritance should be invested at 5.7% annual interest, compounded quarterly, to accumulate the $30,000?

27. You win a contest and have a choice of prizes. You can take $3000 now, or you can receive $4000 in four years. If money can be invested at 6% interest, compounded annually, which prize is more valuable in the long run?

28. If money can be invested at 7% compounded quarterly, which is worth more: $9000 now or $12,500 in five years?

29. If an investment of $1000 grows to $1407.10 in seven years with interest compounded annually, what is the interest rate?

30. If an investment of $2000 grows to $2700 in three and a half years, with an annual interest rate that is compounded quarterly, what is the annual interest rate?

31. If you put $3000 in a savings account today, what interest rate (compounded annually) must you receive to have $4000 after five years?

32. If interest is compounded continuously, what annual rate must you receive if your investment of $1500 is to grow to $2100 in six years?

33. At an interest rate of 8% compounded annually, how long will it take to double an investment of

(a) $100 (b) $500 (c) $1200?

(d) What conclusion about doubling time do parts (a)–(c) suggest?

34. At an interest rate of 6% compounded annually, how long will it take to double an investment of *P* dollars?

35. How long will it take to double an investment of $500 at 7% annual interest, compounded continuously?

36. How long will it take to triple an investment of $5000 at 8% annual interest, compounded continuously?

Exercises 37–40 deal with zero-coupon (noninterest-bearing) bonds.

37. A zero-coupon bond is redeemable for $10,000 in five years. If an investor wants a return of 5% compounded yearly on her money, how much should she offer to pay for the bond? [*Hint:* In other words, what is the present value of the bond under these conditions? See Exercises 13–18.]

38. What price should the investor in Exercise 37 pay for the bond if she wants a return of 6% compounded monthly?

39. A father plans to purchase a $25,000 zero-coupon bond for his daughter as a college graduation present four years from now. (We should all have such nice fathers.) If he wants his investment to earn at least 4% compounded daily, what is the maximum price he should pay for the bond? [Assume a 365-day year.]

40. The father in Exercise 39 actually paid 21,500 for the bond. What was the return on his investment?

41. (a) Suppose *P* dollars is invested for one year at 12% interest compounded quarterly. What interest rate *r* would yield the same amount in one year with annual compounding? *r* is called the **effective rate of interest.** [*Hint:* Solve the equation $P(1 + .12/4)^4 = P(1 + r)$ for *r*. The left side of the equation is the yield after one year at 12% compounded quarterly and the right side is the yield after one year at *r*% compounded annually.]

(b) Fill the blanks in the following table.

12% Compounded	Effective Rate
Annually	12%
Quarterly	
Monthly	
Daily	

42. This exercise provides an illustration of why the continuous compounding formula (page 440) is valid, using a realistic interest rate. We shall determine the value of \$4000 deposited for three years at 5% interest compounded n times per year for larger and larger values of n. In this case, the interest rate per period is $.05/n$ and the number of periods in three years is $3n$, so the amount in the account at the end of three years is

$$A = 4000\left(1 + \frac{.05}{n}\right)^{3n} = 4000\left[\left(1 + \frac{.05}{n}\right)^{n}\right]^{3}.$$

(a) Fill in the missing entries in the following table.

n	$\left(1 + \frac{.05}{n}\right)^n$
1,000	
10,000	
500,000	
1,000,000	
5,000,000	
10,000,000	

(b) Compare the entries in the second column of the table with the number $e^{.05}$ and fill the blank in the following sentence:

As n gets larger and larger, the value of $\left(1 + \frac{.05}{n}\right)^n$ gets closer and closer to the number ______.

(c) Use you answer to part (b) to fill the blank in the following sentence:

As n gets larger and larger, the value of

$$A = 4000\left[\left(1 + \frac{.05}{n}\right)^{n}\right]^{3}$$

gets closer and closer to ______.

(d) Compare your answer in part (c) to the value of the investment given by the continuous compounding formula.

43. A weekly census of the tree-frog population in Frog Hollow State Park produces the following results.

Week	1	2	3	4	5	6
Population	18	54	162	486	1458	4374

(a) Find a function of the form $f(x) = Pa^x$ that describes the frog population at time x weeks.

(b) What is the growth factor in this situation (that is, by what number must this week's population be multiplied to obtain next week's population)?

(c) Each tree frog requires 10 square feet of space and the park has an area of 6.2 square miles. Will the space required by the frog population exceed the size of the park in 12 weeks? In 14 weeks? [Remember: 1 square mile = 5280^2 square feet.]

44. The fruit fly population in a certain laboratory triples every day. Today there are 200 fruit flies.

(a) Make a table showing the number of fruit flies present for the first four days (today is day 0, tomorrow is day 1, etc.).

(b) Find a function of the form $f(x) = Pa^x$ that describes the fruit fly population at time x days.

(c) What is the growth factor here (that is, by what number must each day's population be multiplied to obtain the next day's population)?

(d) How many fruit flies will there be a week from now?

45. The population of Mexico was 100.4 million in 2000 and is expected to grow at the rate of 1.4% per year.

(a) Find the rule of the function f that gives Mexico's population (in millions) in year x, with $x = 0$ corresponding to 2000.

(b) Estimate Mexico's population in 2010.

(c) When will the population reach 125 million people?

46. The United States imported \$15 billion in goods from China in 1990, and imports have been growing at the rate of 19.3% per year.

(a) Find the rule of a function that gives the value of U.S. imports from China (in billions of dollars) in year x, with $x = 0$ corresponding to 1990.

(b) Approximately how much did the United States import from China in 2002?

(c) If this model remains accurate, when will imports reach \$360 billion?

47. Average annual expenditure per pupil in public elementary and secondary schools was \$4966 in 1989–1990 and has been increasing at about 3.9% a year.*

(a) Write the rule of a function that gives the expenditure per pupil in year x, where $x = 0$ corresponds to the 1989–1990 school year.

(b) According to this model, what are the expenditures per pupil in 2003–2004?

(c) In what year do expenditures first exceed \$9000 per pupil?

48. There were four bald eagle nests in Ohio in 1979. Because of a restoration program undertaken by the Ohio Department of Natural Resources, this number has been increasing about 13.7% per year.

(a) Find the rule of a function that gives the approximate number of eagle nests in Ohio in year x, where $x = 0$ corresponds to 1979.

(b) How many nests were there in 2003?

(c) If this growth continues, when will there be 150 nests?

*Based on data from the National Education Association.

49. The U.S. Census Bureau estimates that the Hispanic population in the United States will increase from 32.44 million in 2000 to 98.23 million in 2050.*

(a) Find an exponential function that gives the Hispanic population in year x, with $x = 0$ corresponding to 2000.

(b) What is the projected Hispanic population in 2010 and 2025?

(c) In what year will the Hispanic population reach 55 million?

50. The U.S. Department of Commerce estimated that there were 56.8 million Internet users in the United States in 1997 and 157.6 million in 2002.

(a) Find an exponential function that models the number of Internet users in year x, with $x = 0$ corresponding to 1997.

(b) For how long is this model likely to remain accurate? [*Hint:* The current U.S. population is about 294 million.]

51. At the beginning of an experiment, a culture contains 200 *H. pylori* bacteria. An hour later there are 205 bacteria. Assuming that the *H. pylori* bacteria grow exponentially, how many will there be after 10 hours? After 2 days?

52. The population of India was approximately 1030 million in 2001 and was 865 million a decade earlier. If the population continues to grow exponentially at the same rate, what will it be in 2006?

53. Kerosene is passed through a pipe filled with clay to remove various pollutants. Each foot of pipe removes 25% of the pollutants.

(a) Write the rule of a function that gives the percentage of pollutants remaining in the kerosene after it has passed through x feet of pipe. [See Example 10.]

(b) How many feet of pipe are needed to ensure that 90% of the pollutants have been removed from the kerosene?

54. If inflation runs at a steady 3% per year, then the amount a dollar is worth decreases by 3% each year.

(a) Write the rule of a function that gives the value of a dollar in year x.

(b) How much will the dollar be worth in 5 years? In 10 years?

(c) How many years will it take before today's dollar is worth only a dime?

**Statistical Abstracts of the United States* 2002.

55. Many people make minimum monthly payments on their credit card balances and wonder why they aren't getting ahead. If you owe $800 on your credit card and you make a minimal payment of 2% of the balance each month, how long will it take to reduce your balance to $100? Make the unlikely assumption that there is no interest—if there were, it would take much longer. [*Hint:* If your balance is P dollars, then after your payment, what is it?]

56. As the number of transistors per chip has increased (see Example 8 in Section 6.1), the average price of a transistor has dropped exponentially, from $1 in 1972 to a fraction of a cent in 2002, namely $.0000001. Assuming that this trend continues, estimate the average transistor price in 2006.

57. (a) The half-life of radium is 1620 years. Find the rule of the function that gives the amount remaining from an initial quantity of 100 milligrams of radium after x years.

(b) How much radium is left after 800 years? After 1600 years? After 3200 years?

58. (a) The half-life of polonium-210 is 140 days. Find the rule of the function that gives the amount of polonium-210 remaining from an initial 20 milligrams after t days.

(b) How much polonium-210 is left after 15 weeks? After 52 weeks?

(c) How long will it take for the 20 milligrams to decay to 4 milligrams?

59. How old is a piece of ivory that has lost 58% of its carbon-14? [See Example 11.]

60. How old is a mummy that has lost 49% of its carbon-14?

61. This exercise provides a justification for the claim that the function $M(x) = c(.5)^{x/h}$ gives the mass after x years of a radioactive element with half-life h years. Suppose we have c grams of an element that has a half-life of 50 years. Then after 50 years, we would have $c\left(\frac{1}{2}\right)$ grams. After another 50 years, we would have half of that, namely, $c\left(\frac{1}{2}\right)\left(\frac{1}{2}\right) = c\left(\frac{1}{2}\right)^2$.

(a) How much remains after a third 50-year period? After a fourth 50-year period?

(b) How much remains after t 50-year periods?

(c) If x is the number of years, then $x/50$ is the number of 50-year periods. By replacing the number of periods t in part (b) by $x/50$, you obtain the amount remaining after x years. This gives the function $M(x)$ when $h = 50$. The same argument works in the general case (just replace 50 by h).

6.3 Common and Natural Logarithmic Functions

Roadmap

We begin with common and natural logarithms because they are the most widely used. Instructors who prefer to begin with logarithms to an arbitrary base *b* should cover Special Topics 6.4.A before covering this section.

Instructors who want to introduce logarithmic functions as the inverse functions of exponential functions should first cover the relevant parts of Section 3.4.

From their invention in the 17th century until the development of computers and calculators, logarithms were the only effective tool for numerical computation in the natural sciences and engineering. Although they are no longer needed for computation, logarithmic functions still play an important role in science and engineering. In this section, we examine the two most important types of logarithms: those to base 10 and those to base e.

COMMON LOGARITHMS

The basic idea of logarithms can be seen from an example. We'll use the number 704. The *logarithm* of 704 is denoted log 704 and is defined to be the solution of the equation $10^x = 704$. This number can be approximated in two ways.

Solve the equation $10^x = 704$ graphically (that is, find the intersection of $y = 10^x$ and $y = 704$), as in Figure 6–13,

or

Use the LOG key on your calculator, as in Figure 6–14.

Except for rounding, you get the same answer.

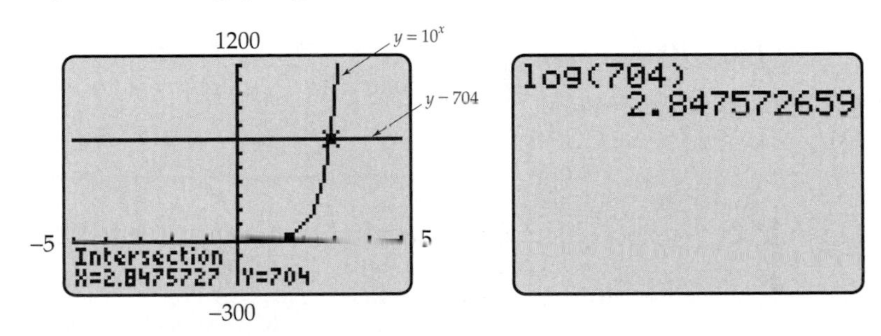

Figure 6–13

Figure 6–14

The fact that the number log 704 is the solution of $10^x = 704$ means that

log 704 is *the exponent to which* 10 *must be raised to produce* 704.

Similarly, the solution of $10^x = 56$ is the number log 56. Figure 6–15 shows that log 56 $\approx$ 1.748188027 and that 10 raised to this exponent is indeed 56.

Figure 6–15

More generally, whenever c is a positive real number, the horizontal line $y = c$ lies above the x-axis and hence intersects the graph of $y = 10^x$, so that the equation $10^x = c$ has a solution. Consequently, we have the following definition.

Definition of Common Logarithms

If $c > 0$, then the **common logarithm** of c, denoted log c, is the solution of the equation $10^x = c$.* In other words,

log c is the exponent to which 10 must be raised to produce c.

*The word "common" is usually omitted except when it is necessary to distinguish these logarithms from other types that are introduced below.

EXAMPLE 1

Without using a calculator, find:

(a) $\log 1000$; (b) $\log 1$: (c) $\log \sqrt{10}$.

SOLUTION

(a) To find $\log 1000$, ask yourself "what power of 10 equals 1000?" The answer is 3 because $10^3 = 1000$. Therefore, $\log 1000 = 3$.

(b) To what power must 10 be raised to produce 1? Since $10^0 = 1$, we conclude that $\log 1 = 0$.

(c) $\text{Log}\sqrt{10} = 1/2$ because $1/2$ is the exponent to which 10 must be raised to produce $\sqrt{10}$, that is, $10^{1/2} = \sqrt{10}$. ■

A calculator is necessary to find most logarithms, but even then you should proceed as in Example 1 to get a rough estimate. For instance, $\log 795$ is the exponent to which 10 must be raised to produce 795. Since $10^2 = 100$ and $10^3 = 1000$, this exponent must be between 2 and 3, that is, $2 < \log 795 < 3$.

Since logarithms are exponents, every statement about logarithms is equivalent to a statement about exponents. For instance, "$\log v = u$" means "u is the exponent to which 10 must be raised to produce v," or in symbols, "$10^u = v$." In other words,

Logarithmic and Exponential Equivalence

Let u and v be real numbers with $v > 0$. Then

$$\log v = u \quad \text{exactly when} \quad 10^u = v.$$

Note for those who have read Section 3.4. If you apply the statement in the preceding box to the functions $g(x) = \log x$ and $f(x) = 10^x$, you obtain

$$\log v = u \quad \text{exactly when} \quad 10^u = v$$

$$g(v) = u \quad \text{exactly when} \quad f(u) = v.$$

Thus, the box says that the common logarithm function $g(x) = \log x$ is the *inverse function* of the exponential function $f(x) = 10^x$.

EXAMPLE 2

Translate each of the following logarithmic statements into an equivalent exponential statement.

$$\log 29 = 1.4624 \qquad \log .47 = -.3279 \qquad \log(k + t) = d.$$

SOLUTION Using the preceding box, we have these translations.

Logarithmic Statement	Equivalent Exponential Statement
$\log 29 = 1.4624^*$	$10^{1.4624} = 29$
$\log .47 = -.3279$	$10^{-.3279} = .47$
$\log (k + t) = d$	$10^d = k + t$ ■

EXAMPLE 3

Translate each of the following exponential statements into an equivalent logarithmic statement:

$$10^{5.5} = 316{,}227.766 \qquad 10^{.66} = 4.5708819 \qquad 10^{rs} = t$$

SOLUTION Translate as follows.

Exponential Statement	Equivalent Logarithmic Statement
$10^{5.5} = 316{,}277.766$	$\log 316{,}227.766 = 5.5$
$10^{.66} = 4.5708819$	$\log 4.5708819 = .66$
$10^{rs} = t$	$\log t = rs$ ■

EXAMPLE 4 

Solve the equation $\log x = 4$.

SOLUTION $\log x = 4$ is equivalent to $10^4 = x$. So the solution is $x = 10{,}000$. ■

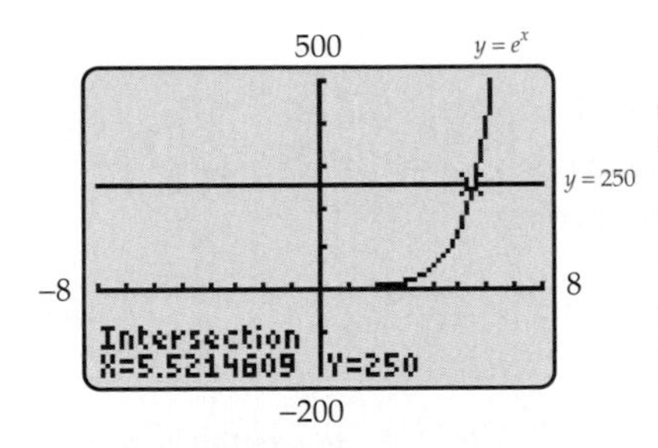

Figure 6–16

Figure 6–17

NATURAL LOGARITHMS

Common logarithms are closely related to the exponential function $f(x) = 10^x$. With the advent of calculus, however, it became clear that the most useful exponential function in science and engineering is $g(x) = e^x$. Consequently, a new type of logarithm, based on the number e instead of 10, was developed. This development is essentially a carbon copy of what was done above, with some minor changes in notation.

For example, the *natural logarithm* of 250 is denoted ln 250 and is defined to be the solution of the equation $e^x = 250$. This number can be approximated in two ways.

Solve the equation $e^x = 250$ graphically (that is, find the intersection of $y = e^x$ and $y = 250$), as in Figure 6–16,

or

Use the LN key on your calculator, as in Figure 6–17.

Except for rounding, you get the same answer.

---

*Here and below, all logarithms are rounded to four decimal places and an equal sign is used rather than the more correct "approximately equal."

The fact that the number ln 250 is the solution of $e^x = 250$ means that

ln 250 is *the exponent to which e must be raised to produce* 250.

The same procedure works for any positive number.

CALCULATOR EXPLORATION

Find ln 87. Then raise e to this exponent. Except possibly for rounding, the answer will be 87.

More generally, we have the following definition.

Definition of Natural Logarithms

If $c > 0$, then the **natural logarithm** of c, denoted $\ln c$, is the solution of the equation $e^x = c$. In other words,

$\ln c$ is the exponent to which e must be raised to produce c.

Since natural logarithms are exponents, every statement about them is equivalent to a statement about exponents to the base e. For instance, "$\ln v = u$" means "u is the exponent to which e must be raised to produce v," or in symbols, "$e^u = v$." In other words, we have the following.

Logarithmic and Exponential Equivalence

Let u and v be real numbers, with $v > 0$. Then

$$\ln v = u \qquad \text{exactly when} \qquad e^u = v.$$

Note for those who have read Section 3.4. The situation here is the same as with common logarithms. If you apply the statement in the preceding box to the functions $g(x) = \ln x$ and $f(x) = e^x$, you obtain

$$\ln v = u \qquad \text{exactly when} \qquad e^u = v$$

$$g(v) = u \qquad \text{exactly when} \qquad f(u) = v.$$

Thus, the box says that the natural logarithm function $g(x) = \ln x$ is the *inverse function* of the exponential function $f(x) = e^x$.

EXAMPLE 5

Translate:

(a) $\ln 14 = 2.6391$ into an equivalent exponential statement.

(b) $e^{5.0626} = 158$ into an equivalent logarithmic statement.

SOLUTION

(a) Using the preceding box, we see that $\ln 14 = 2.6391$ is equivalent to $e^{2.6391} = 14$.

(b) Similarly, $e^{5.0626} = 158$ is equivalent to $\ln 158 = 5.0626$. ■

PROPERTIES OF LOGARITHMS

Since common and natural logarithms have almost identical definitions (just replace 10 by e), it is not surprising that they have the same essential properties. You don't need a calculator to understand these properties. You need only use the definition of logarithms or translate logarithmic statements into equivalent exponential ones (or vice versa).

EXAMPLE 6

What is $\log(-2)$?

Translation: To what power must 10 be raised to produce -2?

Answer: The graph of $f(x) = 10^x$ (see Figure 6–13) shows that every power of 10 is *positive*. So 10^x can *never* be -2 or any negative number or zero; hence, $\log(-2)$ is not defined. Similarly, $\ln(-2)$ is not defined because every power of e is positive. Therefore,

$\ln v$ and $\log v$ are defined only when $v > 0$. ■

EXAMPLE 7

What is $\ln 1$?

Translation: To what power must e be raised to produce 1?

Answer: We know that $e^0 = 1$, which means that $\ln 1 = 0$. Combining this fact with Example 1(b), we have

$$\ln 1 = 0 \qquad \text{and} \qquad \log 1 = 0.$$ ■

EXAMPLE 8

What is $\ln e^9$?

Translation: To what power must e be raised to produce e^9?

Answer: Obviously, the answer is 9. So $\ln e^9 = 9$ and in general

$$\ln e^k = k \qquad \text{for every real number } k.$$

Similarly,

$$\log 10^k = k \qquad \text{for every real number } k$$

because k is the exponent to which 10 must be raised to produce 10^k. In particular, when $k = 1$, we have

$$\ln e = 1 \qquad \text{and} \qquad \log 10 = 1. \qquad \blacksquare$$

EXAMPLE 9

Find $10^{\log 678}$ and $e^{\ln 678}$.

SOLUTION By definition, log 678 is the exponent to which 10 must be raised to produce 678. So if you raise 10 to this exponent, the answer will be 678, that is, $10^{\log 678} = 678$. Similarly, ln 678 is the exponent to which e must be raised to produce 678, so that $e^{\ln 678} = 678$. The same argument works with any positive number v in place of 678:

$$e^{\ln v} = v \qquad \text{and} \qquad 10^{\log v} = v \quad \text{for every } v > 0. \qquad \blacksquare$$

The facts presented in the preceding examples may be summarized as follows.

Properties of Logarithms

Natural Logarithms	Common Logarithms
1. $\ln v$ is defined only when $v > 0$;	$\log v$ is defined only when $v > 0$.
2. $\ln 1 = 0$ and $\ln e = 1$;	$\log 1 = 0$ and $\log 10 = 1$.
3. $\ln e^k = k$ for every real number k;	$\log 10^k = k$ for every real number k.
4. $e^{\ln v} = v$ for every $v > 0$;	$10^{\log v} = v$ for every $v > 0$.

EXAMPLE 10

Applying Property 3 with $k = 2x^2 + 7x + 9$ shows that

$$\ln e^{2x^2+7x+9} = 2x^2 + 7x + 9. \qquad \blacksquare$$

EXAMPLE 11

Solve the equation $\ln (x + 1) = 2$.

SOLUTION Since $\ln (x + 1) = 2$, we have

$$e^{\ln(x+1)} = e^2.$$

Applying Property 4 with $v = x + 1$ shows that

$$x + 1 = e^{\ln(x+1)} = e^2$$

$$x = e^2 - 1 \approx 6.3891. \qquad \blacksquare$$

Property 4 has another interesting consequence. If a is any positive number, then $e^{\ln a} = a$. Hence, the rule of the exponential function $f(x) = a^x$ can be written as

$$f(x) = a^x = (e^{\ln a})^x = e^{(\ln a)x}.$$

For example, $f(x) = 2^x = e^{(\ln 2)x} \approx e^{.6931x}$. Thus, we have this useful result.

Exponential Functions

> Every exponential growth or decay function can be written in the form
>
> $$f(x) = Pe^{kx},$$
>
> where $f(x)$ is the amount at time x, P is the initial quantity, and k is positive for growth and negative for decay.

GRAPHS OF LOGARITHMIC FUNCTIONS

Figure 6–18 shows the graphs of $f(x) = \log x$ and $g(x) = \ln x$. Both are increasing functions with these four properties:

Domain: all positive real numbers; **Range:** all real numbers; **x-intercept:** 1; **Vertical Asymptote:** y-axis.

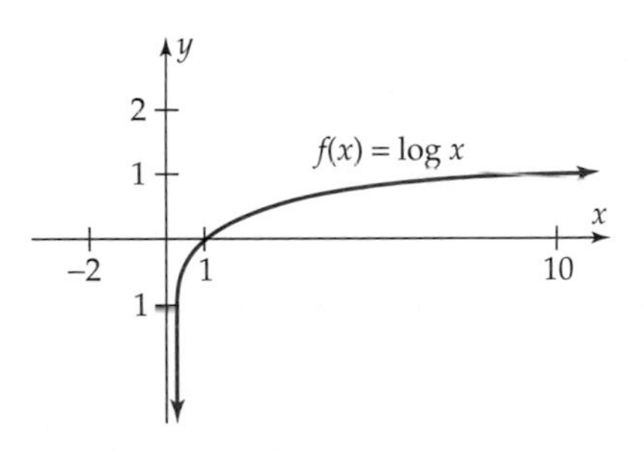

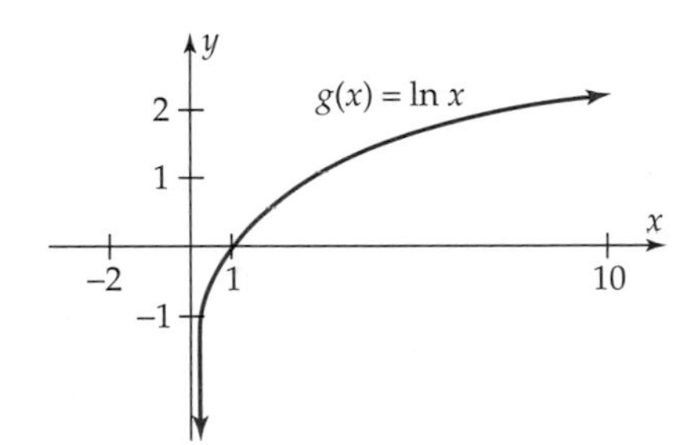

Figure 6–18

Calculators and computers do not accurately show that the y-axis is a vertical asymptote of these graphs. By evaluating the functions at very small numbers (such as $x = 1/10^{500}$), you can see that the graphs go lower and lower as x gets closer to 0. On a calculator, however, the graph will appear to end abruptly near the y-axis (try it!).

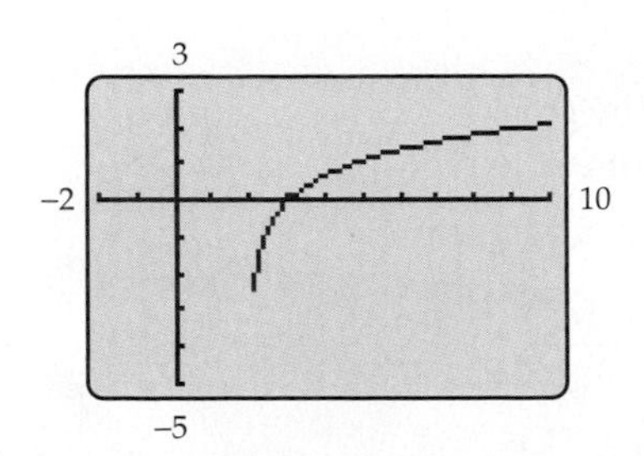

Figure 6–19

EXAMPLE 12

Sketch the graph of $f(x) = \ln(x - 2)$.

SOLUTION Using a calculator to graph $f(x) = \ln(x - 2)$, we obtain Figure 6–19, in which the graph appears to end abruptly near $x = 2$. Fortunately, however, we have read Section 0.6, so we know that this is *not* how the graph looks. From Section 0.6, we know that the graph of $f(x) = \ln(x - 2)$ is the graph

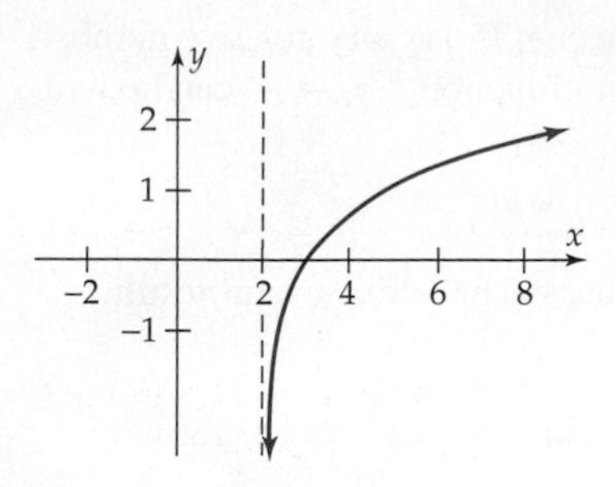

Figure 6–20

of $g(x) = \ln x$ shifted horizontally 2 units to the right, as shown in Figure 6–20. In particular, the graph of f has a vertical asymptote at $x = 2$ and drops sharply downward there. ■

GRAPHING EXPLORATION

Graph $y_1 = \ln(x - 2)$ and $y_2 = -5$ in the same viewing window and verify that the graphs do not appear to intersect, as they should. Nevertheless, try to solve the equation $\ln(x - 2) = -5$ by finding the intersection point of y_1 and y_2. Some calculators will find the intersection point even though it does not show on the screen. Others produce an error message, in which case the SOLVER feature should be used instead of a graphical solution.

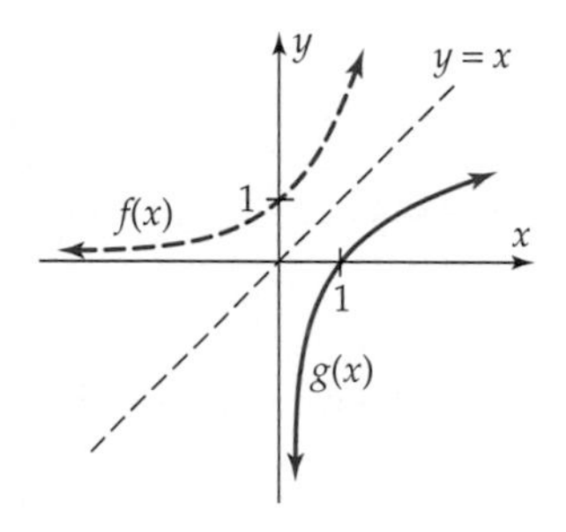

Figure 6–21

EXAMPLE 13

How is the graph of $g(x) = \log x$ related to the graph of $f(x) = 10^x$?

SOLUTION We graph both functions on the same set of axes, together with the line $y = x$, and obtain Figure 6–21.* As the figure suggests, the graph of $g(x) = \log x$ is the mirror image of the graph of $f(x) = 10^x$, with the line $y = x$ being the mirror. In technical terms, the graph of g is the graph of f *reflected in the line* $y = x$.† The graphs of $k(x) = \ln x$ and $h(x) = e^x$ are related similarly: The graph of k is the graph of h reflected in the line $y = x$. ■

Although logarithms are defined only for positive numbers, many logarithmic functions include negative numbers in their domains.

EXAMPLE 14

Find the domain of each of the following functions.

(a) $f(x) = \ln(x + 4)$ (b) $g(x) = \log x^2$

SOLUTION

(a) $f(x) = \ln(x + 4)$ is defined only when $x + 4 > 0$, that is, when $x > -4$. So the domain of f consists of all real numbers greater than -4.

(b) Since $x^2 > 0$ for all nonzero x, the domain of $g(x) = \log x^2$ consists of all real numbers except 0. ■

GRAPHING EXPLORATION

Verify the conclusions of Example 14 by graphing each of the functions. What is the vertical asymptote of each graph?

---

*On a calculator screen, the parts of the graph near the axes may not be visible. Furthermore, to see the situation accurately, you must use a square window.

†This should be no surprise to those who have read Section 3.4 because they know that $g(x) = \log x$ is the inverse function of $f(x) = 10^x$ and that all inverse functions have this reflection property (see page 257).

Logarithms have a variety of applications. For example, the Value Line Geometric Composite Index uses logarithms to compute the average return for a group of 1700 stocks.* Here is another financial application.

EXAMPLE 15

If you invest money at interest rate r (expressed as a decimal and compounded annually), then the time it takes to double your investment is given by

$$D(r) = \frac{\ln 2}{\ln(1 + r)}.$$

(a) How long will it take to double an investment of \$2500 at 6.5%?

(b) What interest rate is needed for the investment in part (a) to double in six years?

SOLUTION

(a) The interest rate is $r = .065$, so the doubling time is

$$D(.065) = \frac{\ln 2}{\ln(1 + .065)} \approx 11 \text{ years.}$$

(b) For the investment to double in six years, we must have $D(r) = 6$. Thus, we must solve the equation

$$\frac{\ln 2}{\ln(1 + r)} = 6.$$

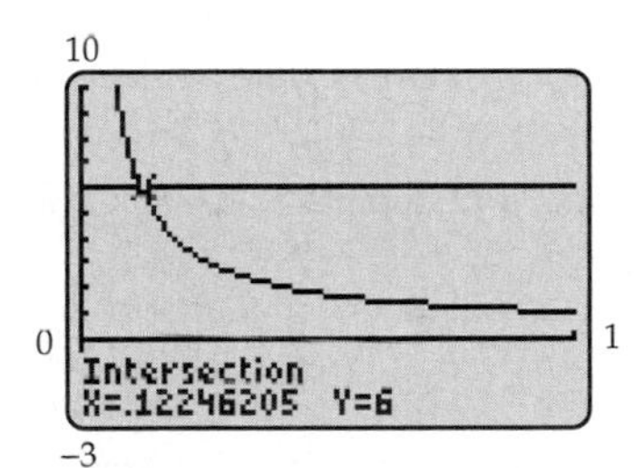

Figure 6–22

Algebraic solution techniques for such equations will be considered in Section 6.5. For now, we solve the equation graphically by finding the intersection of $y_1 = \frac{\ln 2}{\ln(1 + r)}$ and $y_2 = 6$. Figure 6–22 shows that $r \approx .1225$, so an interest rate of about 12.25% is needed to double the investment in six years. ■

EXERCISES 6.3

Unless stated otherwise, all letters represent positive numbers.

In Exercises 1–4, find the logarithm, without using a calculator.

1. $\log 10{,}000$ **2.** $\log .001$

3. $\log \frac{\sqrt{10}}{1000}$ **4.** $\log \sqrt[3]{.01}$

In Exercises 5–14, translate the given logarithmic statement into an equivalent exponential statement.

5. $\log 1000 = 3$ **6.** $\log .001 = -3$

7. $\log 750 = 2.88$ **8.** $\log (.8) = -.097$

9. $\ln 3 = 1.0986$ **10.** $\ln 10 = 2.3026$

11. $\ln .01 = -4.6052$ **12.** $\ln s = r$

13. $\ln (x^2 + 2y) = z + w$ **14.** $\log (a + c) = d$

In Exercises 15–24, translate the given exponential statement into an equivalent logarithmic one.

15. $10^{-2} = .01$ **16.** $10^3 = 1000$

17. $10^{.4771} = 3$ **18.** $10^{7k} = r$

19. $e^{3.25} = 25.79$ **20.** $e^{-4} = .0183$

21. $e^{12/7} = 5.5527$ **22.** $e^k = t$

23. $e^{2/r} = w$ **24.** $e^{4uv} = m$

New York Times, August 31, 2003.

In Exercises 25–36, evaluate the given expression without using a calculator.

25. $\log 10^{\sqrt{43}}$ **26.** $\log 10^{\sqrt{x^2+y^2}}$ **27.** $\ln e^{15}$

28. $\ln e^{3.78}$ **29.** $\ln \sqrt{e}$ **30.** $\ln \sqrt[5]{e}$

31. $e^{\ln 931}$ **32.** $e^{\ln 34.17}$ **33.** $\ln e^{x+y}$

34. $\ln e^{x^2+2y}$ **35.** $e^{\ln x^2}$ **36.** $e^{\ln\sqrt{x+3}}$

In Exercises 37–40, write the rule of the function in the form $f(x) = Pe^{kx}$. (See the discussion and box after Example 11.)

37. $f(x) = 4(25^x)$ **38.** $g(x) = 3.9(1.03^x)$

39. $g(x) = -16(30.5^x)$ **40.** $f(x) = -2.2(.75^x)$

In Exercises 41–44, find the domain of the given function (that is, the largest set of real numbers for which the rule produces well-defined real numbers).

41. $f(x) = \ln(x + 1)$ **42.** $g(x) = \ln(x + 2)$

43. $h(x) = \log(-x)$ **44.** $k(x) = \log(2 - x)$

45. $f(x) = \log x^3$ **46.** $h(x) = \ln x^4$

47. (a) Graph $y = x$ and $y = e^{\ln x}$ in separate viewing windows [or use a split-screen if your calculator has that feature]. For what values of x are the graphs identical?
(b) Use the properties of logarithms to explain your answer in part (a).

48. (a) Graph $y = x$ and $y = \ln(e^x)$ in separate viewing windows [or a split-screen if your calculator has that feature]. For what values of x are the graphs identical?
(b) Use the properties of logarithms to explain your answer in part (a).

In Exercises 49–54, list the transformations that will change the graph of $g(x) = \ln x$ into the graph of the given function. [Section 0.6 may be helpful.]

49. $f(x) = 2 \cdot \ln x$ **50.** $f(x) = \ln x - 7$

51. $h(x) = \ln(x - 4)$ **52.** $k(x) = \ln(x + 2)$

53. $h(x) = \ln(x + 3) - 4$ **54.** $k(x) = \ln(x - 2) + 2$

In Exercises 55–58, sketch the graph of the function.

55. $f(x) = \log(x - 3)$ **56.** $g(x) = 2\ln x + 3$

57. $h(x) = -2\log x$ **58.** $f(x) = \ln(-x) - 3$

In Exercises 59–64, find a viewing window (or windows) that shows a complete graph of the function.

59. $f(x) = \dfrac{x}{\ln x}$ **60.** $g(x) = \dfrac{\ln x}{x}$

61. $h(x) = \dfrac{\ln x^2}{x}$ **62.** $k(x) = e^{2/\ln x}$

63. $f(x) = 10\log x - x$ **64.** $f(x) = \dfrac{\log x}{x}$

C

In Exercises 65–68, find the average rate of change of the function. (See Exercises 21–23 on page 210.)

65. $f(x) = \ln(x - 2)$, as x goes from 3 to 5.

66. $g(x) = x - \ln x$, as x goes from .5 to 1.

67. $g(x) = \log(x^2 + x + 1)$, as x goes from -5 to -3.

68. $f(x) = x\log|x|$, as x goes from 1 to 4.

69. (a) What is the average rate of change of $f(x) = \ln x$, as x goes from 3 to $3 + h$?
(b) What is the value of h when the average rate of change of $f(x) = \ln x$, as x goes from 3 to $3 + h$, is .25?

70. (a) Find the average rate of change of $f(x) = \ln x^2$, as x goes from .5 to 2.
(b) Find the average rate of change of $g(x) = \ln(x - 3)^2$, as x goes from 3.5 to 5.
(c) What is the relationship between your answers in parts (a) and (b) and why is this so?

71. Use the doubling function D of Example 15 for this exercise.
(a) Find the time it takes to double your money at each of these interest rates: 4%, 6%, 8%, 12%, 18%, 24%, 36%.
(b) Round the answers in part (a) to the nearest year and compare them with these numbers:

$$72/4,\ 72/6,\ 72/8,\ 72/12,\ 72/18,\ 72/24,\ 72/36.$$

Use this evidence to state a "rule of thumb" for determining approximate doubling time, without using the function D. This rule of thumb, which has long been used by bankers, is called the **rule of 72.**

72. The concentration of hydrogen ions in a given solution is denoted $[H^+]$ and is measured in moles per liter. For example, $[H^+] = .00008$ for beer and $[H^+] = .0004$ for wine. Chemists define the pH of the solution to be the number $\text{pH} = -\log[H^+]$. The solution is said to be an *acid* if $\text{pH} < 7$ and a *base* if $\text{pH} > 7$.
(a) Is beer an acid or a base? What about wine?
(b) If a solution has a pH of 2, what is its $[H^+]$?
(c) For hominy, $[H^+] = 5 \cdot 10^{-8}$. Is hominy a base?

73. The life expectancy of a woman born in year x is approximated by

$$f(x) = 17.7 + 30.855\log x,$$

where $x = 10$ corresponds to 1910.*
(a) What is the life expectancy of a woman born in 1988?
(b) In what year will the life expectancy of a woman born in that year be 81 years?

---

*National Center for Health Statistics.

74. The number of multiple births (babies born as twins, triplets, etc.) each year is approximated by

$$g(x) = 93{,}201.973 + 10{,}467.499 \log x,$$

where $x = 1$ corresponds to 1989.

(a) Use this model to estimate the number of multiple births in 2005 and 2008.

(b) If this model remains accurate, in what year will there be 108,000 multiple births?

75. The percentage of first-year college students who are male is approximated by

$$M(x) = 53.145 - 6.185 \log x,$$

where $x = 5$ corresponds to 1985.*

(a) Estimate the percentage of male first-year students in 2006.

(b) If this model remains accurate, in what year will the percentage of male first-year students drop below 44%?

76. The height h above sea level (in miles) is related to the barometric pressure p (in inches of mercury) by the equation

$$h = -5 \ln \left(\frac{p}{29.92}\right).$$

If a weather balloon records the following pressures, how high is the balloon?

(a) 29.91 inches

(b) 20.04 inches

(c) 11.89 inches

(d) What is the pressure at a height of 4 miles?

77. Suppose $f(x) = A \ln x + B$, where A and B are constants. If $f(1) = 10$ and $f(e) = 1$, what are A and B?

78. *(Parts of Section 3.4 are prerequisites for this exercise.)* Show that

$g(x) = \ln\left(\dfrac{x}{1-x}\right)$ is the inverse function of

$$f(x) = \frac{1}{1 + e^{-x}}.$$

79. The height h above sea level (in meters) is related to air temperature t (in degrees Celsius), the atmospheric pressure p (in centimeters of mercury at height h), and the atmospheric pressure c at sea level by

$$h = (30t + 8000) \ln (c/p).$$

If the pressure at the top of Mount Rainier is 44 centimeters on a day when sea level pressure is 75.126 centimeters and the temperature is 7°C, what is the height of Mount Rainier?

80. Mount Everest is 8850 meters high. What is the atmospheric pressure at the top of the mountain on a day when the temperature is -25°C and the atmospheric pressure at sea level is 75 centimeters? [See Exercise 79.]

81. Beef consumption in the United States (in billions of pounds) in year x can be approximated by the function

$$f(x) = -154.41 + 39.38 \ln x \qquad (x \geq 90).$$

where $x = 90$ corresponds to 1990.†

(a) How much beef was consumed in 1999 and in 2002?

(b) When will beef consumption reach 30 billion pounds per year?

82. Students in a trigonometry class were given a final exam. Each month thereafter, they took an equivalent exam. The class average on the exam taken after t months is given by

$$F(t) = 82 - 8 \cdot \ln (t + 1).$$

(a) What was the class average after six months?

(b) After a year?

(c) When did the class average drop below 55?

83. One person with a flu virus visited the campus. The number T of days it took for the virus to infect x people was given by:

$$T = -.93 \ln \left[\frac{7000 - x}{6999x}\right].$$

(a) How many days did it take for 6000 people to become infected?

(b) After two weeks, how many people were infected?

84. The population of St. Petersburg, Florida (in thousands) can be approximated by the function

$$g(x) = -127.9 + 81.91 \ln x \qquad (x \geq 70),$$

where $x = 70$ corresponds to 1970.

(a) Estimate the population in 1995 and 2003.

(b) If this model remains accurate, when will the population be 260,000?

85. A bicycle store finds that the number N of bikes sold is related to the number d of dollars spent on advertising by $N = 51 + 100 \cdot \ln (d/100 + 2)$.

(a) How many bikes will be sold if nothing is spent on advertising? If $1000 is spent? If $10,000 is spent?

(b) If the average profit is $25 per bike, is it worthwhile to spend $1000 on advertising? What about $10,000?

(c) What are the answers in part (b) if the average profit per bike is $35?

---

*Based on data from the Higher Education Research Institute at UCLA.

†Based on data from the U.S. Department of Agriculture.

86. **Approximating Logarithmic Functions by Polynomials.** For each positive integer n, let f_n be the polynomial function whose rule is

$$f_n(x) = x - \frac{x^2}{2} + \frac{x^3}{3} - \frac{x^4}{4} + \frac{x^5}{5} - \cdots \pm \frac{x^n}{n},$$

where the sign of the last term is $+$ if n is odd and $-$ if n is even. In the viewing window with $-1 \le x \le 1$ and $-4 \le y \le 1$, graph $g(x) = \ln(1 + x)$ and $f_4(x)$ on the same screen. For what values of x does f_4 appear to be a good approximation of g?

87. Using the viewing window in Exercise 86, find a value of n for which the graph of the function f_n (as defined in Exercise 86) appears to coincide with the graph of $g(x) = \ln(1 + x)$. Use the trace feature to move from graph to graph to see how good this approximation actually is.

6.4 Properties of Logarithms

Logarithms have several important properties in addition to those presented in Section 6.3. These properties, which we shall call *logarithm laws,* arise from the fact that logarithms are exponents. Essentially, they are properties of exponents translated into logarithmic language.

The first law of exponents says that $b^m b^n = b^{m+n}$, or in words,

The exponent of a product is the sum of the exponents of the factors.

Since logarithms are just particular kinds of exponents, this statement translates as follows.

The logarithm of a product is the sum of the logarithms of the factors.

Here is the same statement in symbolic language.

Product Law for Logarithms

For all $v, w > 0$,

$$\ln(vw) = \ln v + \ln w,$$

and

$$\log(vw) = \log v + \log w.$$

Proof According to Property 4 of logarithms (in the box on page 452),

$$e^{\ln v} = v \quad \text{and} \quad e^{\ln w} = w.$$

Therefore, by the first law of exponents (with $m = \ln v$ and $n = \ln w$),

$$vw = e^{\ln v} e^{\ln w} = e^{\ln v + \ln w}.$$

So raising e to the exponent $(\ln v + \ln w)$ produces vw. But the definition of logarithm says that $\ln vw$ is the exponent to which e must be raised to produce vw. Therefore, we must have $\ln vw = \ln v + \ln w$. A similar argument works for common logarithms (just replace e by 10 and "ln" by "log"). ■

EXAMPLE 1

A calculator shows that $\ln 7 = 1.9459$ and $\ln 9 = 2.1972$. Therefore,

$$\ln 63 = \ln(7 \cdot 9) = \ln 7 + \ln 9 = 1.9459 + 2.1972 = 4.1341. \quad \blacksquare$$

CALCULATOR EXPLORATION

We know that $5 \cdot 7 = 35$. Key in LOG(35) ENTER. Then key in LOG(5) + LOG(7) ENTER. The answers are the same by the Product Law. Do you get the same answer if you key in LOG(5) × LOG(7) ENTER?

EXAMPLE 2

Use the Product Law to write

(a) $\log(7xy)$ as a sum of three logarithms.

(b) $\log x^2 + \log y + 1$ as a single logarithm.

SOLUTION

(a) $\log(7xy) = \log 7x + \log y = \log 7 + \log x + \log y$

(b) Note that $\log 10 = 1$ (why?). Hence,

$$\begin{aligned}\log x^2 + \log y + 1 &= \log x^2 + \log y + \log 10 \\ &= \log(x^2 y) + \log 10 \\ &= \log(10x^2 y). \quad \blacksquare\end{aligned}$$

CAUTION

A common error in applying the Product Law for Logarithms is to write the *false* statement

$$\begin{aligned}\ln 7 + \ln 9 &= \ln(7 + 9) \\ &= \ln 16\end{aligned}$$

instead of the correct statement

$$\begin{aligned}\ln 7 + \ln 9 &= \ln(7 \cdot 9) \\ &= \ln 63.\end{aligned}$$

GRAPHING EXPLORATION

Illustrate the Caution in the margin graphically by graphing both

$$f(x) = \ln x + \ln 9 \qquad \text{and} \qquad g(x) = \ln(x + 9)$$

in the standard viewing window and verifying that the graphs are not the same. In particular, the functions have different values at $x = 7$.

The second law of exponents, namely, $b^m/b^n = b^{m-n}$, may be roughly stated in words as follows.

The exponent of the quotient is the difference of exponents.

When the exponents are logarithms, this says

The logarithm of a quotient is the difference of the logarithms.

In other words,

Quotient Law for Logarithms

For all v, $w > 0$,

$$\ln\left(\frac{v}{w}\right) = \ln v - \ln w$$

and

$$\log\left(\frac{v}{w}\right) = \log v - \log w.$$

The proof of the Quotient Law is very similar to the proof of the Product Law (see Exercise 34).

Figure 6–23

EXAMPLE 3

Figure 6–23 illustrates the Quotient Law by showing that

$$\log\left(\frac{297}{39}\right) = \log 297 - \log 39. \quad \blacksquare$$

EXAMPLE 4

For any $w > 0$,

$$\ln\left(\frac{1}{w}\right) = \ln 1 - \ln w = 0 - \ln w = -\ln w,$$

and

$$\log\left(\frac{1}{w}\right) = \log 1 - \log w = 0 - \log w = -\log w. \quad \blacksquare$$

CAUTION

Do not confuse $\ln\left(\frac{v}{w}\right)$ with the quotient $\frac{\ln v}{\ln w}$. They are *different* numbers. For example,

$$\ln\left(\frac{36}{3}\right) = \ln(12) = 2.4849, \qquad \text{but} \qquad \frac{\ln 36}{\ln 3} = \frac{3.5835}{1.0986} = 3.2619.$$

GRAPHING EXPLORATION

Illustrate the preceding Caution graphically by graphing both $f(x) = \ln(x/3)$ and $g(x) = (\ln x)/(\ln 3)$ and verifying that the graphs are not the same at $x = 36$ (or anywhere else, for that matter).

The third law of exponents, namely, $(b^m)^k = b^{mk}$, can also be translated into logarithmic language.

Power Law for Logarithms

For all k and all $v > 0$,

$$\ln (v^k) = k(\ln v),$$

and

$$\log (v^k) = k(\log v).$$

Proof Since $v = 10^{\log v}$ (why?), the third law of exponents (with $b = 10$ and $m = \log v$) shows that

$$v^k = (10^{\log v})^k = 10^{(\log v)k} = 10^{k(\log v)}.$$

So raising 10 to the exponent $k(\log v)$ produces v^k. But the exponent to which 10 must be raised to produce v^k is, by definition, $\log (v^k)$. Therefore, $\log (v^k) = k(\log v)$, and the proof is complete. A similar argument with e in place of 10 and "ln" in place of "log" works for natural logarithms. ■

EXAMPLE 5

Express $\ln \sqrt{19}$ without radicals or exponents.

SOLUTION First write $\sqrt{19}$ in exponent notation, then use the Power Law:

$$\ln \sqrt{19} = \ln 19^{1/2}$$

$$= \frac{1}{2} \ln 19 \quad \text{or} \quad \frac{\ln 19}{2}. \quad ■$$

EXAMPLE 6

Express as a single logarithm: $\dfrac{\log (x^2 + 1)}{3} - \log x$.

SOLUTION

$$\frac{\log (x^2 + 1)}{3} - \log x = \frac{1}{3} \log (x^2 + 1) - \log x$$

$$= \log (x^2 + 1)^{1/3} - \log x \qquad \text{[Power Law]}$$

$$= \log \sqrt[3]{x^2 + 1} - \log x$$

$$= \log \left(\frac{\sqrt[3]{x^2 + 1}}{x} \right) \qquad \text{[Quotient Law]} \quad ■$$

EXAMPLE 7

Express as a single logarithm: $\ln 3x + 4 \ln x - \ln 3xy$.

SOLUTION

$$\begin{aligned}
\ln 3x + 4 \cdot \ln x - \ln 3xy &= \ln 3x + \ln x^4 - \ln 3xy && \text{[Power Law]}\\
&= \ln (3x \cdot x^4) - \ln 3xy && \text{[Product Law]}\\
&= \ln \frac{3x^5}{3xy} && \text{[Quotient Law]}\\
&= \ln \frac{x^4}{y} && \text{[Cancel } 3x\text{]}
\end{aligned}$$

■

EXAMPLE 8

Simplify: $\ln\left(\frac{\sqrt{x}}{x}\right) + \ln \sqrt[4]{ex^2}$.

SOLUTION Begin by changing to exponential notation.

$$\ln\left(\frac{x^{1/2}}{x}\right) + \ln (ex^2)^{1/4} = \ln (x^{-1/2}) + \ln (ex^2)^{1/4}$$

$$\begin{aligned}
&= -\frac{1}{2} \cdot \ln x + \frac{1}{4} \cdot \ln ex^2 && \text{[Power Law]}\\
&= -\frac{1}{2} \cdot \ln x + \frac{1}{4}(\ln e + \ln x^2) && \text{[Product Law]}\\
&= -\frac{1}{2} \cdot \ln x + \frac{1}{4}(\ln e + 2 \cdot \ln x) && \text{[Power Law]}\\
&= -\frac{1}{2} \cdot \ln x + \frac{1}{4} \cdot \ln e + \frac{1}{2} \cdot \ln x\\
&= \frac{1}{4} \cdot \ln e = \frac{1}{4} && [\ln e = 1]
\end{aligned}$$

■

APPLICATIONS

Because logarithmic growth is slow, measurements on a logarithmic scale (that is, on a scale determined by a logarithmic function) can sometimes be deceptive.

EXAMPLE 9 Earthquakes

The magnitude $R(i)$ of an earthquake on the Richter scale is given by $R(i) = \log (i/i_0)$, where i is the amplitude of the ground motion of the earthquake and i_0 is the amplitude of the ground motion of the so-called zero earthquake.* A moderate earthquake might have 1000 times the ground motion of the zero earthquake (that is, $i = 1000i_0$). So its magnitude would be

$$\log (1000i_0/i_0) = \log 1000 = \log 10^3 = 3.$$

*The zero earthquake has ground motion amplitude of less than 1 micron on a standard seismograph 100 kilometers from the epicenter.

An earthquake with 10 times this ground motion (that is, $i = 10 \cdot 1000i_0 = 10{,}000i_0$) would have a magnitude of

$$\log (10{,}000i_0/i_0) = \log 10{,}000 = \log 10^4 = 4.$$

So a *tenfold* increase in ground motion produces only a one-point change on the Richter scale. In general,

Increasing the ground motion by a factor of 10^k increases the Richter magnitude by k units.*

For instance, the 1989 World Series earthquake in San Francisco measured 7.0 on the Richter scale, and the great earthquake of 1906 measured 8.3. The difference of 1.3 points means that the 1906 quake was $10^{1.3} \approx 20$ times more intense than the 1989 one in terms of ground motion. ■

EXERCISES 6.4

In Exercises 1–18, write the given expression as a single logarithm.

1. $\ln x^2 + 3 \ln y$
2. $\ln 2x + 2(\ln x) - \ln 3y$
3. $\log (x^2 - 9) - \log (x + 3)$
4. $\log 3x - 2[\log x - \log (2 + y)]$
5. $2(\ln x) - 3(\ln x^2 + \ln x)$
6. $\ln (e/\sqrt{x}) - \ln \sqrt{ex}$
7. $3 \ln (e^2 - e) - 3$
8. $2 - 2 \log (20)$
9. $\log (10x) + \log (20y) - 1$
10. $\ln (e^2x) + \ln (ey) - 3$
11. $\log (x - 3) + \log (x + 3)$
12. $\log (x - 3) - \log (x + 3)$
13. $2 \log x - 2[\log (x + 1) + \log (x - 2) + \log x]$
14. $\ln (x - 1) - 2[\ln (x + 5) + \ln x]$
15. $\frac{1}{2}[\ln x - \ln (3x + 5)]$
16. $\frac{1}{3} \ln x - \frac{1}{2}[\ln (x^2 + 1) + \ln x]$
17. $\log (\sqrt{x^2 + 2} + \sqrt{2}) + \log (\sqrt{x^2 + 2} - \sqrt{2}) - \log x$
18. $\ln (\sqrt{x + 1} - \sqrt{x}) + \ln (\sqrt{x + 1} + \sqrt{x})$

In Exercises 19–24, let u = ln x and v = ln y. Write the given expression in terms of u and v. For example,

$$\ln x^3y = \ln x^3 + \ln y = 3 \ln x + \ln y = 3u + v.$$

19. $\ln (x^2y^5)$
20. $\ln (x^3y^2)$
21. $\ln (\sqrt{x} \cdot y^2)$
22. $\ln \left(\frac{\sqrt{x}}{y}\right)$
23. $\ln (\sqrt[3]{x^2 \sqrt{y}})$
24. $\ln \left(\frac{\sqrt{x^2y}}{\sqrt[3]{y}}\right)$

In Exercises 25–30, use graphical or algebraic means to determine whether the statement is true or false.

25. $\ln |x| = |\ln x|$?
26. $\ln \left(\frac{1}{x}\right) = \frac{1}{\ln x}$?
27. $\log x^5 = 5(\log x)$?
28. $e^{x \ln x} = x^x \quad (x > 0)$?
29. $\ln x^3 = (\ln x)^3$?
30. $\log \sqrt{x} = \sqrt{\log x}$?

In Exercises 31–32, find values of a and b for which the statement is false.

31. $\dfrac{\log a}{\log b} = \log\left(\dfrac{a}{b}\right)$
32. $\log (a + b) = \log a + \log b$
33. If $\ln b^7 = 7$, what is b?

**Proof:* If one quake has ground motion amplitude i and the other $10^k i$, then

$$\begin{aligned} R(10^k i) &= \log (10^k i/i_0) = \log 10^k + \log (i/i_0) \\ &= k + \log (i/i_0) = k + R(i). \end{aligned}$$

34. Prove the Quotient Law for Logarithms: For $v, w > 0$, $\ln\left(\frac{v}{w}\right) = \ln v - \ln w$. (Use properties of exponents and the fact that $v = e^{\ln v}$ and $w = e^{\ln w}$.)

35. According to the Power Law for Logarithms, $\log x^2 = 2 \log x$. Compare the graphs of $f(x) = \log x^2$ and $g(x) = 2 \log x$. Are they the same? Explain what's going on here.

36. Do Exercise 35 for the functions $f(x) = \log x^3$ and $g(x) = 3 \log x$.

In Exercises 37–40, use the logarithm laws to answer the question. Section 0.6 may be helpful.

37. Why can the graph of $f(x) = \log(5x)$ be obtained by shifting the graph of $g(x) = \log x$ upward? How far up must it be shifted?

38. Why can the graph of $f(x) = \log(x/5)$ be obtained by shifting the graph of $g(x) = \log x$ downward? How far down must it be shifted?

39. Why can the graph of $f(x) = \ln(x^5)$ be obtained by stretching the graph of $g(x) = \ln x$ away from the x-axis? What is the stretching factor?

40. How is the graph of $f(x) = \ln \sqrt{x}$ related to the graph of $g(x) = \ln x$?

41. The *difference quotient* of a function was defined in Exercises 27–28 on page 69. Compute and simplify the difference quotient of $f(x) = \log x$. Your answer should be a single logarithm.

42. If $f(x) = \ln x$, show that $f\left(\frac{1}{x}\right) = -f(x)$.

43. The size of China's automobile market (in millions of vehicles sold per year) is approximated by

$$f(x) = 1.66 + 1.91 \ln x,$$

where $x = 1$ corresponds to 2001.*
(a) How many Chinese vehicles will be sold in 2007?
(b) When will sales of Chinese vehicles reach 6 million?

44. The size of Japan's automobile market (in millions of vehicles sold per year) is approximated by

$$g(x) = \frac{x}{45} + \frac{533}{90},$$

where $x = 1$ corresponds to 2001.* When will the Chinese automobile market be the same size as the Japanese market (see Exercise 43)?

In Exercises 45–48, state the magnitude on the Richter scale of an earthquake that satisfies the given condition.

45. 100 times stronger than the zero quake.

46. $10^{4.7}$ times stronger than the zero quake.

47. 350 times stronger than the zero quake.

48. 2500 times stronger than the zero quake.

Exercises 49–52 deal with the energy intensity i of a sound, which is related to the loudness of the sound by the function $L(i) = 10 \cdot \log(i/i_0)$, where i_0 is the minimum intensity detectable by the human ear and $L(i)$ is measured in decibels. Find the decibel measure of the sound.

49. Ticking watch (intensity is 100 times i_0).

50. Soft music (intensity is 10,000 times i_0).

51. Loud conversation (intensity is 4 million times i_0).

52. Victoria Falls in Africa (intensity is 10 billion times i_0).

53. How much louder is the sound in Exercise 50 than the sound in Exercise 49?

54. The perceived loudness L of a sound of intensity I is given by $L = k \cdot \ln I$, where k is a certain constant. By how much must the intensity be increased to double the loudness? (That is, what must be done to I to produce $2L$?)

Thinkers

55. Compute each of the following pairs of numbers.
(a) $\log 18$ and $\frac{\ln 18}{\ln 10}$
(b) $\log 456$ and $\frac{\ln 456}{\ln 10}$
(c) $\log 8950$ and $\frac{\ln 8950}{\ln 10}$
(d) What do these results suggest?

56. Prove that for any positive number c, $\log c = \frac{\ln c}{\ln 10}$. [*Hint:* We know that $10^{\log c} = c$ (why?). Take natural logarithms on both sides and use a logarithm law to simplify and solve for $\log c$.]

57. Find each of the following logarithms.
(a) $\log 8.753$ (b) $\log 87.53$ (c) $\log 875.3$
(d) $\log 8753$ (e) $\log 87{,}530$
(f) How are the numbers 8.753, 87.53, . . . , 87,530 related to one another? How are their logarithms related? State a general conclusion that this evidence suggests.

58. Prove that for every positive number c, $\log c$ can be written in the form $k + \log b$, where k is an integer and $1 \le b < 10$. [*Hint:* Write c in scientific notation and use logarithm laws to express $\log c$ in the required form.]

*Based on forecasts by J. D. Power and Associates.

6.4.A SPECIAL TOPICS Logarithmic Functions to Other Bases*

The same procedure used in Sections 6.3–6.4 can be carried out with any positive number b in place of 10 and e.

Throughout this section, b is a fixed positive number with $b > 1$.†

The basic idea of logarithms can be seen from an example. The *logarithm of* 150 *to base* 7 is defined to be the solution of the equation $7^x = 150$ and is denoted $\log_7 150$. The solution of $7^x = 150$ (that is, $\log_7 150$) can be approximated graphically by finding the intersection of $y = 7^x$ and $y = 150$, as in Figure 6–24. Subject to rounding, Figure 6–25 shows that 7 raised to this power is 150. In other words, $\log_7 150$ is the exponent to which 7 must be raised to produce 150.

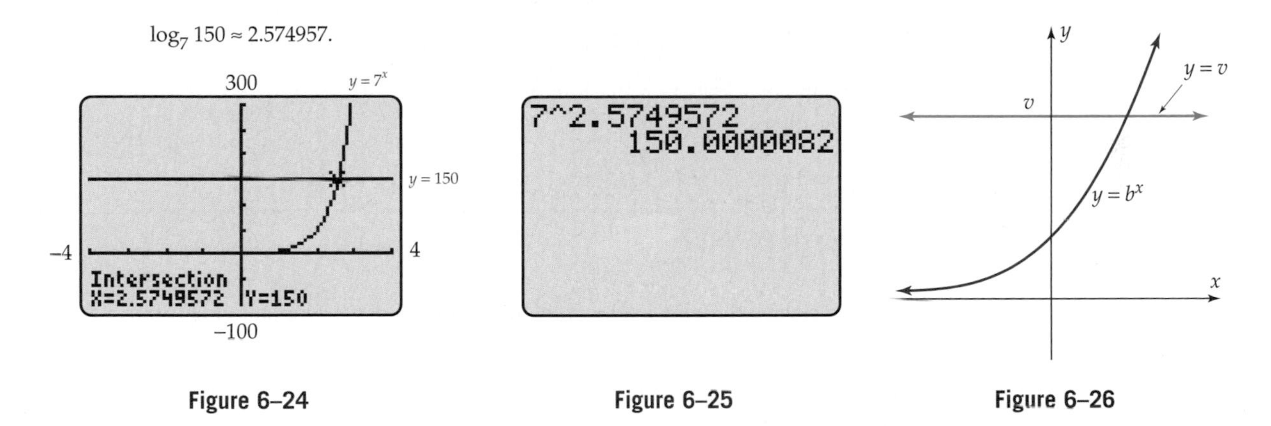

Figure 6–24 Figure 6–25 Figure 6–26

More generally, whenever b and v are positive numbers, then the horizontal line $y = v$ lies above the x-axis and hence intersects the graph of $y = b^x$ (Figure 6–26), so the equation $b^x = v$ has a solution. Consequently, we have this definition.

Definition of Logarithms to Base b

> If b and v are positive numbers, then the *logarithm of v to base b,* denoted $\log_b v$, is the solution of the equation $b^x = v$. In other words,
>
> $\log_b v$ is the exponent to which b must be raised to produce v.

EXAMPLE 1

To find $\log_2 16$, ask yourself, "What power of 2 equals 16?" Since $2^4 = 16$, we see that $\log_2 16 = 4$. Similarly, $\log_2 (1/8) = -3$ because $2^{-3} = 1/8$. ■

*This section replicates the discussion of Sections 6.3 and 6.4 in a more general context. It may be read before Section 6.3, if desired, and is not used in the sequel. Instructors who prefer to introduce logarithms as inverse functions of exponential functions should first cover the relevant parts of Section 3.4.

†The discussion is also valid when $0 < b < 1$, but in that case the graphs have a different shape.

Since logarithms to base b are exponents, every statement about them is equivalent to a statement about exponents to base b. For instance, "$\log_b v = u$" means "u is the exponent to which b must be raised to produce v" or, in symbols, "$b^u = v$." In other words, we have the following.

Alternate Definition of Logarithms

Let u and v be numbers with $v > 0$. Then

$$\log_b v = u \qquad \text{exactly when} \qquad b^u = v.$$

Note for those who have read Section 3.4. If you apply the statement in the preceding box to the functions $g(x) = \log_b x$ and $f(x) = b^x$, you obtain

$$\log_b v = u \qquad \text{exactly when} \qquad b^u = v$$

$$g(v) = u \qquad \text{exactly when} \qquad f(u) = v.$$

Thus, the box says that the logarithm function $g(x) = \log_b x$ is the *inverse function* of the exponential function $f(x) = b^x$.

EXAMPLE 2

Since logarithms are just exponents, every logarithmic statement can be translated into exponential language.

Logarithmic Statement	**Equivalent Exponential Statement**
$\log_3 81 = 4$	$3^4 = 81$
$\log_4 64 = 3$	$4^3 = 64$
$\log_{125} 5 = \frac{1}{3}$	$125^{1/3} = 5$*
$\log_8 \left(\frac{1}{4}\right) = -\frac{2}{3}$	$8^{-2/3} = \frac{1}{4}$ (verify!) ■

EXAMPLE 3

Solve: $\log_5 x = 3$.

SOLUTION The equation $\log_5 x = 3$ is equivalent to the exponential statement $5^3 = x$, so the solution is $x = 125$. ■

*Because $125^{1/3} = \sqrt[3]{125} = 5$.

```
log(.4)
      -.3979400087
log(45.3)
       1.656098202
log(685)
       2.835690571
```

Figure 6–27

EXAMPLE 4

Logarithms to the base 10 are called **common logarithms.** It is customary to write log v instead of $\log_{10} v$. Then

$$\log 100 = 2 \qquad \text{because} \qquad 10^2 = 100;$$

$$\log .001 = -3 \qquad \text{because} \qquad 10^{-3} = \frac{1}{10^3} = \frac{1}{1000} = .001.$$

Calculators have a LOG key for evaluating common logarithms, as shown in Figure 6–27.* ■

```
ln(.5)
      -.6931471806
ln(65)
        4.17438727
ln(158)
       5.062595033
```

Figure 6–28

EXAMPLE 5

The most frequently used base for logarithms in modern applications is the number e ($\approx$ 2.71828 . . .). Logarithms to the base e are called **natural logarithms** and use a different notation: We write ln v instead of $\log_e v$. Calculators also have an LN key for evaluating natural logarithms. For example, see Figure 6–28. ■

You *don't* need a calculator to understand the essential properties of logarithms. You need only translate logarithmic statements into exponential ones (or vice versa).

EXAMPLE 6

What is log (-25)?

Translation: To what power must 10 be raised to produce -25? The graph of $f(x) = 10^x$ lies entirely above the x-axis (use your calculator), which means that *every* power of 10 is *positive.* So 10^x can *never* be -25, or any negative number, or zero. The same argument works for any base b.

$\log_b v$ is defined only when $v > 0$. ■

EXAMPLE 7

What is $\log_5 1$?

Translation: To what power must 5 be raised to produce 1? The answer, of course, is $5^0 = 1$. So $\log_5 1 = 0$. Similarly, $\log_5 5 = 1$ because 1 is the answer to "what power of 5 equals 5?" In general,

$$\log_b 1 = 0 \qquad \text{and} \qquad \log_b b = 1. \quad ■$$

*For convenient reading, logarithms printed in the text are rounded to four decimal places. For example, log .4 = $-.3979$. It is customary to use an equal sign in such cases rather than an "approximately equal" sign ($\approx$).

EXAMPLE 8

What is $\log_2 2^9$?

Translation: To what power must 2 be raised to produce 2^9? Obviously, the answer is 9. So $\log_2 2^9 = 9$, and, in general,

$$\boldsymbol{\log_b b^k = k \qquad \text{for every real number } k.}$$

This property holds even when k is a complicated expression. For instance, if x and y are positive, then

$$\log_6 6^{\sqrt{3x+y}} = \sqrt{3x + y} \qquad (\text{here } k = \sqrt{3x + y}). \qquad ■$$

EXAMPLE 9

What is $10^{\log 439}$? Well, log 439 is the power to which 10 must be raised to produce 439, that is $10^{\log 439} = 439$. Similarly,

$$\boldsymbol{b^{\log_b v} = v \qquad \text{for every } v > 0.} \qquad ■$$

Here is a summary of the facts illustrated in the preceding examples.

Properties of Logarithms

1. $\log_b v$ is defined only when $v > 0$.
2. $\log_b 1 = 0$ and $\log_b b = 1$.
3. $\log_b (b^k) = k$ for every real number k.
4. $b^{\log_b v} = v$ for every $v > 0$.

LOGARITHM LAWS

The first law of exponents states that $b^m b^n = b^{m+n}$, or, in words,

The exponent of a product is the sum of the exponents of the factors.

Since logarithms are just particular kinds of exponents, this statement translates as follows.

The logarithm of a product is the sum of the logarithms of the factors.

The second and third laws of exponents, namely, $b^m/b^n = b^{m-n}$ and $(b^m)^k = b^{mk}$, can also be translated into logarithmic language.

Logarithm Laws

Let b, v, w, k be real numbers, with b, v, w positive and $b \neq 1$.

Product Law: $\log_b (vw) = \log_b v + \log_b w$.

Quotient Law: $\log_b\left(\frac{v}{w}\right) = \log_b v - \log_b w$.

Power Law: $\log_b (v^k) = k(\log_b v)$.

Proof of the Quotient Law According to Property 4 in the first box on page 468,

$$b^{\log_b v} = v \qquad \text{and} \qquad b^{\log_b w} = w.$$

Therefore, by the second law of exponents (with $m = \log_b v$ and $n = \log_b w$), we have

$$\frac{v}{w} = \frac{b^{\log_b v}}{b^{\log_b w}} = b^{\log_b v - \log_b w}.$$

Since $\log_b (v/w)$ is the exponent to which b must be raised to produce v/w, we must have $\log_b (v/w) = \log_b v - \log_b w$. This proves the Quotient Law. The Product and Power Laws are proved in a similar fashion. ■

EXAMPLE 10

Simplify and write as a single logarithm.

(a) $\log_3 (x + 2) + \log_3 y - \log_3 (x^2 - 4)$

(b) $3 - \log_5 (125x)$

SOLUTION

(a) $\log_3 (x + 2) + \log_3 y - \log_3 (x^2 - 4)$

$= \log_3 [(x + 2)y] - \log_3 (x^2 - 4)$ [Product Law]

$= \log_3 \left(\frac{(x + 2)y}{x^2 - 4}\right)$ [Quotient Law]

$= \log_3 \left(\frac{(x + 2)y}{(x + 2)(x - 2)}\right)$ [Factor denominator]

$= \log_3 \left(\frac{y}{x - 2}\right)$ [Cancel common factor]

(b) $3 - \log_5 (125x)$

$= 3 - (\log_5 125 + \log_5 x)$ [Product Law]

$= 3 - \log_5 125 - \log_5 x$

$= 3 - 3 - \log_5 x$ [$\log_5 125 = 3$ because $5^3 = 125$]

$= -\log_5 x$

$= \log_5 x^{-1} = \log_5 \left(\frac{1}{x}\right)$ [Power Law] ■

CAUTION

1. A common error in using the Product Law is to write something like $\log 6 + \log 7 = \log (6 + 7) = \log 13$ instead of the correct statement $\log 6 + \log 7 = \log (6 \cdot 7) = \log 42$.
2. Do not confuse $\log_b\left(\frac{v}{w}\right)$ with the quotient $\frac{\log_b v}{\log_b w}$. They are *different* numbers. For example, when $b = 10$,

$$\log\left(\frac{48}{4}\right) = \log 12 = 1.0792 \qquad \text{but} \qquad \frac{\log 48}{\log 4} = \frac{1.6812}{0.6021} = 2.7922.$$

For graphic illustrations of the errors mentioned in the Caution, see Exercises 86 and 87.

EXAMPLE 11

Given that

$$\log_7 2 = .3562, \qquad \log_7 3 = .5646, \qquad \text{and} \qquad \log_7 5 = .8271,$$

find:

(a) $\log_7 10$; (b) $\log_7 2.5$; (c) $\log_7 48$.

SOLUTION

(a) By the Product Law,

$$\log_7 10 = \log_7 (2 \cdot 5) = \log_7 2 + \log_7 5 = .3562 + .8271 = 1.1833.$$

(b) By the Quotient Law,

$$\log_7 2.5 = \log_7 \left(\frac{5}{2}\right) = \log_7 5 - \log_7 2 = .8271 - .3562 = .4709.$$

(c) By the Product and Power Laws,

$$\begin{aligned} \log_7 48 = \log_7 (3 \cdot 16) = \log_7 3 + \log_7 16 &= \log_7 3 + \log_7 2^4 \\ &= \log_7 3 + 4 \cdot \log_7 2 = .5646 + 4(.3562) \\ &= 1.9894. \end{aligned}$$ ■

Example 11 worked because we were *given* several logarithms to base 7. But there's no $\log_7$ key on the calculator, so how do you find logarithms to base 7 or to any base other than e or 10? *Answer:* Use the LN key on the calculator and the following formula.

Change of Base Formula

> For any positive numbers b and v,
>
> $$\log_b v = \frac{\ln v}{\ln b}.$$

Proof By Property 4 in the box on page 468, $b^{\log_b v} = v$. Take the natural logarithm of each side of this equation.

$$\ln (b^{\log_b v}) = \ln v$$

Apply the Power Law for natural logarithms on the left side.

$$(\log_b v)(\ln b) = \ln v$$

Dividing both sides by $\ln b$ finishes the proof.

$$\log_b v = \frac{\ln v}{\ln b}$$ ■

EXAMPLE 12

To find $\log_7 3$, apply the Change of Base Formula with $b = 7$.

$$\log_7 3 = \frac{\ln 3}{\ln 7} = \frac{1.0986}{1.9459} = .5646 \quad \blacksquare$$

EXAMPLE 13

Environmental scientists study the diversity of species in an ecological community. If there are k different species, with n_1 individuals of species 1, n_2 individuals of species 2, and so on, then the *Shannon index of diversity* H is given by

$$H = \frac{N \log_2 N - [n_1 \log_2 n_1 + n_2 \log_2 n_2 + \cdots + n_k \log_2 n_k]}{N},$$

where $N = n_1 + n_2 + \cdots + n_k$. A study of the prey species of barn owls (that is, the creatures the typical barn owl ate) obtained the following data.

Species	Number
Rats	143
Mice	1405
Birds	452

Find the index of diversity for this prey community.

SOLUTION In this case, $n_1 = 143$, $n_2 = 1405$, $n_3 = 452$, and

$$N = n_1 + n_2 + n_3 = 143 + 1405 + 452 = 2000.$$

So the index of diversity is

$$N = \frac{N \log_2 N - [n_1 \log_2 n_1 + n_2 \log_2 n_2 + n_3 \log_2 n_3]}{N}$$

$$= \frac{2000 \log_2 2000 - [143 \log_2 143 + 1405 \log_2 1405 + 452 \log_2 452]}{2000}.$$

To compute H, we use the Change of Base Formula.

$$H = \frac{2000 \dfrac{\ln 2000}{\ln 2} - \left[143 \dfrac{\ln 143}{\ln 2} + 1405 \dfrac{\ln 1405}{\ln 2} + 452 \dfrac{\ln 452}{\ln 2}\right]}{2000}$$

$$\approx 1.1149. \quad \blacksquare$$

EXERCISES 6.4.A

Note: *Unless stated otherwise, all letters represent positive numbers and* $b \neq 1$.

In Exercises 1–8, fill in the missing entries in each table.

1.

x	0	1	2	4
$f(x) = \log_4 x$				

2.

x	1/25	5	25	$\sqrt{5}$
$g(x) = \log_5 x$				

3.

x		1/6	1	216
$h(x) = \log_6 x$	−2			

4.

x	10/3	4	6	12
$k(x) = \log_3 (x - 3)$				

5.

x	0	1/7	$\sqrt{7}$	49
$f(x) = 2 \log_7 x$				

6.

x			100	1000
$g(x) = 3 \log x$	6	3		

7.

x	−2.75	−1	1	29
$h(x) = 3 \log_2 (x + 3)$				

8.

x	$1/e$	1	e	e^2
$k(x) = 2 \ln x$				

In Exercises 9–18, translate the given exponential statement into an equivalent logarithmic one.

9. $10^{-2} = .01$
10. $10^3 = 1000$
11. $\sqrt[3]{10} = 10^{1/3}$
12. $10^{.4771} \approx 3$
13. $10^{7k} = r$
14. $10^{(a+b)} = c$
15. $7^8 = 5{,}764{,}801$
16. $2^{-3} = 1/8$
17. $3^{-2} = 1/9$
18. $b^{14} = 3379$

In Exercises 19–28, translate the given logarithmic statement into an equivalent exponential one.

19. $\log 10{,}000 = 4$
20. $\log .001 = -3$
21. $\log 750 \approx 2.88$
22. $\log (.8) = -.097$
23. $\log_5 125 = 3$
24. $\log_8 (1/4) = -2/3$
25. $\log_2 (1/4) = -2$
26. $\log_2 \sqrt{2} = 1/2$
27. $\log (x^2 + 2y) = z + w$
28. $\log (a + c) = d$

In Exercises 29–36, evaluate the given expression without using a calculator.

29. $\log 10^{\sqrt{97}}$
30. $\log_{17} (17^{17})$
31. $\log 10^{x^2+y^2}$
32. $\log_{3.5} [3.5^{(x^2-1)}]$
33. $\log_{16} 4$
34. $\log_2 64$
35. $\log_{\sqrt{3}}(27)$
36. $\log_{\sqrt{3}}(1/9)$

In Exercises 37–40, a graph or a table of values for the function $f(x) = \log_b x$ is given. Find b.

37.

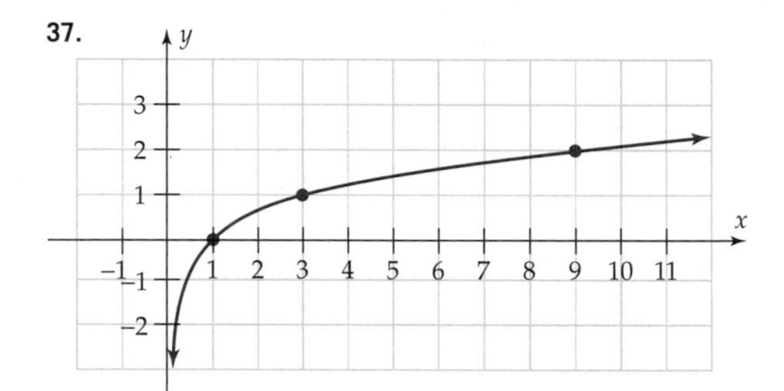

38.

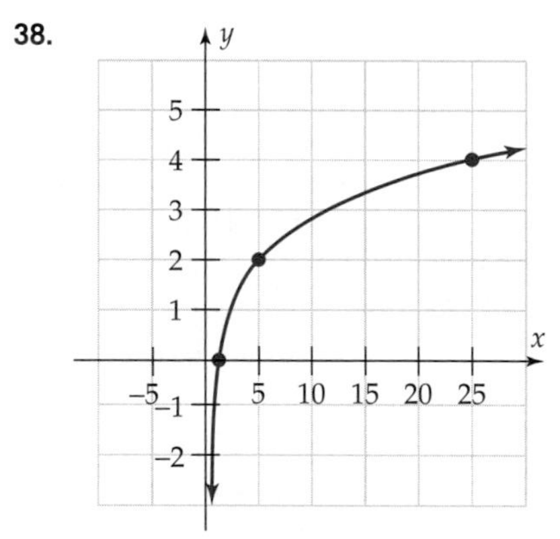

39.

x	.05	1	400	$2\sqrt{5}$
$f(x)$	−1	0	2	1/2

40.

x	1/25	1	5	125
$f(x)$	−4	0	2	6

In Exercises 41–46, find x.

41. $\log_3 243 = x$
42. $\log_{81} 27 = x$
43. $\log_{27} x = 1/3$
44. $\log_5 x = -4$
45. $\log_x 64 = 3$
46. $\log_x (1/9) = -2/3$

In Exercises 47–60, write the given expression as the logarithm of a single quantity, as in Example 10.

47. $2 \log x + 3 \log y - 6 \log z$

48. $5 \log_8 x - 3 \log_8 y + 2 \log_8 z$

49. $\log x - \log (x + 3) + \log (x^2 - 9)$

50. $\log_3 (y + 2) + \log_3 (y - 3) - \log_3 y$

51. $\frac{1}{2} \log_2 (25c^2)$

52. $\frac{1}{3} \log_2 (27b^6)$

53. $-2 \log_4 (7c)$

54. $\frac{1}{3} \log_5 (x + 1)$

55. $2 \ln (x + 1) - \ln (x + 2)$

56. $\ln (z - 3) + 2 \ln (z + 3)$

57. $\log_2 (2x) - 1$

58. $2 - \log_5 (25z)$

59. $2 \ln (e^2 - e) - 2$

60. $4 - 4 \log_5 (20)$

In Exercises 61–68, use a calculator and the change of base formula to find the logarithm.

61. $\log_2 10$

62. $\log_2 22$

63. $\log_7 5$

64. $\log_5 7$

65. $\log_{500} 1000$

66. $\log_{500} 250$

67. $\log_{12} 56$

68. $\log_{12} 725$

Exercises 69–72 deal with the Shannon index of diversity (Example 13). Note that in two communities with the same number of species, a larger index indicates greater diversity.

69. A study of barn owl prey in another area produced the following data.

Species	Number
Rats	662
Mice	907
Birds	531

Find the index of diversity of this community. Is this community more or less diverse than the one in Example 13?

70. An eastern forest is composed of the following trees.

Species	Number
Beech	2754
Birch	689
Hemlock	4428
Maple	629

What is the index of diversity of this community?

71. A community has high divesity when all species have approximately the same number of individuals. It has low diversity when a few species account for most of the total population. Illustrate this fact for the following two communities.

Community 1	Number
Species A	1000
Species B	1000
Species C	1000

Community 2	Number
Species A	2500
Species B	200
Species C	300

72. In a community of k species, the maximum possible index is the number $\log_2 k$ (as is shown in Exercise 90). Use properties of logarithms (*not* a calculator) to show that Community 1 in Exercise 71 has the largest possible index.

In Exercises 73–78, answer true or false and give reasons for your answer.

73. $\log_b (r/5) = \log_b r - \log_b 5$

74. $\dfrac{\log_b a}{\log_b c} = \log_b \left(\dfrac{a}{c}\right)$

75. $(\log_b r)/t = \log_b (r^{1/t})$

76. $\log_b (cd) = \log_b c + \log_b d$

77. $\log_5 (5x) = 5(\log_5 x)$

78. $\log_b (ab)^t = t(\log_b a) + t$

Thinkers

79. Which is larger: 397^{398} or 398^{397}? [*Hint:* $\log 397 \approx 2.5988$ and $\log 398 \approx 2.5999$ and $f(x) = 10^x$ is an increasing function.]

80. If $\log_b 9.21 = 7.4$ and $\log_b 359.62 = 19.61$, then what is $\log_b 359.62/\log_b 9.21$?

In Exercises 81–84, assume that a and b are positive, with $a \neq 1$ and $b \neq 1$.

81. Express $\log_b u$ in terms of logarithms to the base a.

82. Show that $\log_b a = 1/\log_a b$.

83. How are $\log_{10} u$ and $\log_{100} u$ related?

84. Show that $a^{\log b} = b^{\log a}$.

85. If $\log_b x = \frac{1}{2} \log_b v + 3$, show that $x = (b^3)\sqrt{v}$.

86. Graph the functions $f(x) = \log x + \log 7$ and $g(x) = \log (x + 7)$ on the same screen. For what values of x is it true that $f(x) = g(x)$? What do you conclude about the statement

$$\log 6 + \log 7 = \log (6 + 7)?$$

87. Graph the functions $f(x) = \log (x/4)$ and $g(x) = (\log x)/(\log 4)$. Are they the same? What does this say about a statement such as

$$\log\left(\frac{48}{4}\right) = \frac{\log 48}{\log 4}?$$

In Exercises 88–89, sketch a complete graph of the function, labeling any holes, asymptotes, or local maxima or minima.

88. $f(x) = \log_5 x + 2$

89. $h(x) = x \log x^2$

90. Suppose an ecological community with a total of N individuals has maximum diversity: It has k species, each of the same size N/k. Use the properties of logarithms to show that the index of diversity of this community is $\log_2 k$.

6.5 Algebraic Solutions of Exponential and Logarithmic Equations

Most of the exponential and logarithmic equations solved by graphical means earlier in this chapter could also have been solved algebraically. The algebraic techniques for solving such equations are based on the properties of logarithms.

EXPONENTIAL EQUATIONS

The easiest exponential equations to solve are those in which both sides are powers of the same base.

EXAMPLE 1

Solve $8^x = 2^{x+1}$.

SOLUTION Using the fact that $8 = 2^3$, we rewrite the equation as follows.

$$8^x = 2^{x+1}$$
$$(2^3)^x = 2^{x+1}$$
$$2^{3x} = 2^{x+1}$$

Since the powers of 2 are equal, the exponents must be the same, that is,

$$3x = x + 1$$
$$2x = 1$$
$$x = \frac{1}{2}.$$ ■

When different bases are involved in an exponential equation, a different solution technique is needed.

EXAMPLE 2

To solve $5^x = 2$,

Take logarithms on each side:*	$\ln 5^x = \ln 2$
Use the Power Law:	$x(\ln 5) = \ln 2$
Divide both sides by ln 5:	$x = \frac{\ln 2}{\ln 5} \approx \frac{.6931}{1.6094} \approx .4307.$

Remember: $\frac{\ln 2}{\ln 5}$ is *not* $\ln \frac{2}{5}$ or $\ln 2 - \ln 5$. ■

EXAMPLE 3 

To solve $2^{4x-1} = 3^{1-x}$,

Take logarithms of each side:	$\ln 2^{4x-1} = \ln 3^{1-x}$
Use the Power Law:	$(4x - 1)(\ln 2) = (1 - x)(\ln 3)$
Multiply out both sides:	$4x(\ln 2) - \ln 2 = \ln 3 - x(\ln 3)$
Rearrange terms:	$4x(\ln 2) + x(\ln 3) = \ln 2 + \ln 3$
Factor left side:	$(4 \cdot \ln 2 + \ln 3)x = \ln 2 + \ln 3$
Divide both sides by $(4 \cdot \ln 2 + \ln 3)$:	$x = \frac{\ln 2 + \ln 3}{4 \cdot \ln 2 + \ln 3} \approx .4628.$ ■

APPLICATIONS OF EXPONENTIAL EQUATIONS

As we saw on page 442, the mass of a radioactive element at time x is given by

$$M(x) = c(.5^{x/h}),$$

where c is the initial mass and h is the half-life of the element.

EXAMPLE 4

After 43 years, a 20-milligram sample of strontium-90 (^{90}Sr) decays to 6.071 mg. What is the half-life of strontium-90?

SOLUTION The mass of the sample at time x is given by

$$f(x) = 20(.5^{x/h}),$$

where h is the half-life of strontium-90. We know that $f(x) = 6.071$ when $x = 43$,

*We shall use natural logarithms, but using logarithms to other bases would produce the same answer (Exercise 24).

that is, $6.071 = 20(.5^{43/h})$. We must solve this equation for h.

Divide both sides by 20: $$\frac{6.071}{20} = .5^{43/h}$$

Take logarithms on both sides: $$\ln \frac{6.071}{20} = \ln .5^{43/h}$$

Use the Power Law: $$\ln \frac{6.071}{20} = \frac{43}{h} \ln .5$$

Multiply both sides by h: $$h \ln \frac{6.071}{20} = 43 \ln .5$$

Divide both sides by $\ln \frac{6.071}{20}$: $$h = \frac{43 \ln .5}{\ln (6.071/20)} \approx 25$$

Therefore, strontium-90 has a half-life of 25 years. ■

EXAMPLE 5

A certain bacteria is known to grow exponentially, with the population at time t given by a function of the form $g(t) = Pe^{kt}$, where P is the original population and k is the continuous growth rate. A culture shows 1000 bacteria present. Seven hours later, there are 5000.

(a) Find the continuous growth rate k.

(b) Determine when the population will reach one billion.

SOLUTION

(a) The original population is $P = 1000$, so the growth function is $g(t) = 1000e^{kt}$. We know that $g(7) = 5000$, that is,

$$1000e^{k \cdot 7} = 5000.$$

To determine the growth rate, we solve this equation for k.

Divide both sides by 1000: $$e^{7k} = 5$$

Take logarithms of both sides: $$\ln e^{7k} = \ln 5$$

Use the Power Law: $$7k \ln e = \ln 5$$

Since $\ln e = 1$ (why?), this equation becomes

$$7k = \ln 5$$

Divide both sides by 7: $$k = \frac{\ln 5}{7} \approx .22992$$

Therefore, the growth function is $g(t) \approx 1000e^{.22992t}$.

(b) The population will reach one billion when $g(t) = 1{,}000{,}000{,}000$, that is, when

$$1000e^{.22992t} = 1{,}000{,}000{,}000.$$

So we solve this equation for t:

Divide both sides by 1000: $e^{.22992t} = 1{,}000{,}000$

Take logarithms on both sides: $\ln e^{.22992t} = \ln 1{,}000{,}000$

Use the Power Law: $.22992t \ln e = \ln 1{,}000{,}000$

Remember $\ln e = 1$: $.22992t = \ln 1{,}000{,}000$

Divide both sides by .22992: $t = \dfrac{\ln 1{,}000{,}000}{.22992} \approx 60.09.$

Therefore, it will take a bit more than 60 hours for the culture to grow to one billion. ■

EXAMPLE 6 **Inhibited Population Growth**

The population of fish in a lake at time t months is given by the function

$$p(t) = \frac{20{,}000}{1 + 24e^{-t/4}}.$$

How long will it take for the population to reach 15,000?

SOLUTION We must solve this equation for t.

$$15{,}000 = \frac{20{,}000}{1 + 24e^{-t/4}}$$

$$15{,}000(1 + 24e^{-t/4}) = 20{,}000$$

$$1 + 24e^{-t/4} = \frac{20{,}000}{15{,}000} = \frac{4}{3}$$

$$24e^{-t/4} = \frac{1}{3}$$

$$e^{-t/4} = \frac{1}{3} \cdot \frac{1}{24} = \frac{1}{72}$$

$$\ln e^{-t/4} = \ln\left(\frac{1}{72}\right)$$

$$\left(-\frac{t}{4}\right)(\ln e) = \ln 1 - \ln 72$$

$$-\frac{t}{4} = -\ln 72 \qquad [\ln e = 1 \text{ and } \ln 1 = 0]$$

$$t = 4(\ln 72) \approx 17.1067$$

So the population reaches 15,000 in a little over 17 months. ■

LOGARITHMIC EQUATIONS

Equations that involve only logarithmic terms may be solved by using this fact, which is proved in Exercise 23 (and is valid with log replaced by ln).

If $\log u = \log v$, then $u = v$.

EXAMPLE 7

Solve: $\log(3x + 2) + \log(x + 2) = \log(7x + 6)$.

SOLUTION First we write the left side as a single logarithm.

$$\log(3x + 2) + \log(x + 2) = \log(7x + 6)$$

Use the Product Law: $$\log[(3x + 2)(x + 2)] = \log(7x + 6)$$

Multiply out left side: $$\log(3x^2 + 8x + 4) = \log(7x + 6).$$

Since the logarithms are equal, we must have

$$3x^2 + 8x + 4 = 7x + 6$$

Subtract $7x + 6$ from both sides: $$3x^2 + x - 2 = 0$$

Factor: $$(3x - 2)(x + 1) = 0$$

$$3x - 2 = 0 \quad \text{or} \quad x + 1 = 0$$

$$3x = 2 \qquad\qquad x = -1$$

$$x = \frac{2}{3}.$$

```
log(3*2/3+2)+log
(2/3+2)
       1.028028724
log(7*2/3+6)
       1.028028724
```

Figure 6–29

Thus, $x = 2/3$ and $x = -1$ are the *possible* solutions and must be checked in the *original* equation. When $x = 2/3$, both sides of the original equation have the same value, as shown in Figure 6–29. So $2/3$ is a solution. When $x = -1$, however, the right side of the equation is

$$\log(7x + 6) = \log[7(-1) + 6] = \log(-1),$$

which is not defined. So -1 is not a solution. ■

EXAMPLE 8

Solve: $\ln(x + 4) - \ln(x + 2) = \ln x$.

SOLUTION First, write the left side as a single logarithm.

Use the Quotient Law: $$\ln\left(\frac{x + 4}{x + 2}\right) = \ln x$$

Therefore,

$$\frac{x + 4}{x + 2} = x$$

Multiply both sides by $x + 2$: $$x + 4 = x(x + 2)$$

Multiply out right side: $$x + 4 = x^2 + 2x$$

Rearrange terms: $$x^2 + x - 4 = 0$$

This equation does not readily factor, so we use the quadratic formula.

$$x = \frac{-1 \pm \sqrt{1^2 - 4 \cdot 1 \cdot (-4)}}{2 \cdot 1} = \frac{-1 \pm \sqrt{17}}{2}.$$

```
(-1+√(17))/2→A
       1.561552813
ln(A+4)-ln(A+2)
        .445680719
ln(A)
        .445680719
```

Figure 6–30

An easy way to verify that $\frac{-1+\sqrt{17}}{2}$ is a solution is to store this number as A and then evaluate both sides of the original equation at $x = A$, as shown in Figure 6–30. The second possibility, however, is not a solution (because $x = \frac{-1-\sqrt{17}}{2}$ is negative, so $\ln x$ is not defined). ■

Equations that involve both logarithmic and constant terms may be solved by using the basic property of logarithms (see page 452).

(*) $$10^{\log v} = v \quad \textbf{and} \quad e^{\ln v} = v.$$

EXAMPLE 9

Solve $7 + 2 \log 5x = 11$.

SOLUTION We start by getting all the logarithmic terms on one side and the constant on the other.

Subtract 7 from both sides:	$2 \log 5x = 4$
Divide both sides by 2:	$\log 5x = 2$

We know that if two quantities are equal, say $a = b$, then $10^a = 10^b$. We use this fact here, with the two sides of the preceding equation as a and b.

Exponentiate both sides:	$10^{\log 5x} = 10^2$
Use the basic logarithm property (*):	$5x = 100$
Divide both sides by 5:	$x = 20$

Verify that 20 is actually a solution of the original equation. ■

EXAMPLE 10

Solve $\ln(x - 3) = 5 - \ln(x - 3)$.

SOLUTION We proceed as in Example 9, but since the base for these logarithms is e, we use e rather than 10 when we exponentiate.

	$\ln(x-3) = 5 - \ln(x-3)$
Add $\ln(x - 3)$ to both sides:	$2\ln(x-3) = 5$
Divide both sides by 2:	$\ln(x-3) = \frac{5}{2}$
Exponentiate both sides:	$e^{\ln(x-3)} = e^{5/2}$
Use the basic property of logarithms (*):	$x - 3 = e^{5/2}$
Add 3 to both sides:	$x = e^{5/2} + 3 \approx 15.1825$

This is the only possibility for a solution. A calculator shows that it actually is a solution of the original equation. ■

EXAMPLE 11 

Solve $\log(x - 16) = 2 - \log(x - 1)$.

SOLUTION

	$\log(x - 16) = 2 - \log(x - 1)$
Add $\log(x - 1)$ to both sides:	$\log(x - 16) + \log(x - 1) = 2$
Use the Product Law:	$\log[(x - 16)(x - 1)] = 2$
Multiply out left side:	$\log(x^2 - 17x + 16) = 2$
Exponentiate both sides:	$10^{\log(x^2 - 17x + 16)} = 10^2$
Use the basic logarithm property (*):	$x^2 - 17x + 16 = 100$
Subtract 100 from both sides:	$x^2 - 17x - 84 = 0$
Factor:	$(x + 4)(x - 21) = 0$
	$x + 4 = 0$ or $x - 21 = 0$
	$x = -4$ or $x = 21$

You can easily verify that 21 is a solution of the original equation, but -4 is not [when $x = -4$, then $\log(x - 16) = \log(-20)$, which is not defined]. ■

APPLICATIONS OF LOGARITHMIC EQUATIONS

Many phenomena can be modeled by logarithmic functions. In such cases, logarithmic equations are used to answer various questions.

EXAMPLE 12

The number of pounds of fish (in billions) used for human consumption in the United States in year x is approximated by the function

$$f(x) = 10.57 + 1.75 \ln x,$$

where $x = 5$ corresponds to 1995.*

(a) How many pounds of fish were used in 2004?

(b) When will fish consumption reach 16 billion pounds?

SOLUTION

(a) Since 2004 corresponds to $x = 14$, we evaluate $f(x)$ at 14.

$$f(14) = 10.57 + 1.75 \ln 14 \approx 15.19 \text{ billion pounds.}$$

*Based on data from the U.S. National Oceanic and Atmospheric Administration and the National Marine Fisheries Service.

(b) Fish consumption is 16 billion pounds when $f(x) = 16$, so we must solve the equation

$$10.57 + 1.75 \ln x = 16$$

Subtract 10.57 from both sides: $$1.75 \ln x = 5.43$$

Divide both sides by 1.75: $$\ln x = \frac{5.43}{1.75}$$

Exponentiate both sides: $$e^{\ln x} = e^{5.43/1.75}$$

Use the basic property of logarithms (*): $$x \approx 22.26.$$

Since $x = 22$ corresponds to 2012, fish consumption will reach 16 billion pounds in 2012. ■

EXERCISES 6.5

In Exercises 1–8, solve the equation without using logarithms.

1. $3^x = 81$ **2.** $3^x + 3 = 30$ **3.** $3^{x+1} = 9^{5x}$

4. $4^{5x} = 16^{2x-1}$ **5.** $3^{5x}9^{x^2} = 27$

6. $2^{x^2+5x} = 1/16$ **7.** $9^{x^2} = 3^{-5x-2}$

8. $4^{x^2-1} = 8^x$

In Exercises 9–22, solve the equation. First express your answer in terms of natural logarithms (for instance, $x = (2 + \ln 5)/(\ln 3)$). Then use a calculator to find an approximation for the answer.

9. $3^x = 5$ **10.** $5^x = 4$ **11.** $2^x = 3^{x-1}$

12. $4^{x+2} = 2^{x-1}$ **13.** $3^{1-2x} = 5^{x+5}$

14. $4^{3x-1} = 3^{x-2}$ **15.** $2^{1-3x} = 3^{x+1}$

16. $3^{z+3} = 2^z$ **17.** $e^{2x} = 5$

18. $e^{-3x} = 2$ **19.** $6e^{-1.4x} = 21$

20. $3.4e^{-x/3} = 5.6$ **21.** $2.1e^{(x/2)\ln 3} = 5$

22. $7.8e^{(x/3)\ln 5} = 14$

23. Prove that if $\ln u = \ln v$, then $u = v$. [*Hint:* Property (*) on page 479.]

24. (a) Solve $7^x = 3$, using natural logarithms. Leave your answer in logarithmic form; don't approximate with a calculator.

(b) Solve $7^x = 3$, using common (base 10) logarithms. Leave your answer in logarithmic form.

(c) Use the Change of Base Formula in Special Topics 6.4.A to show that your answers in parts (a) and (b) are the same.

In Exercises 25–34, solve the equation as in Examples 7 and 8.

25. $\ln (3x - 5) = \ln 11 + \ln 2$

26. $\log (4x - 1) = \log (x + 1) + \log 2$

27. $\log (3x - 1) + \log 2 = \log 4 + \log (x + 2)$

28. $\ln (x + 6) - \ln 10 = \ln (x - 1) - \ln 2$

29. $2 \ln x = \ln 36$

30. $2 \log x = 3 \log 4$

31. $\ln x + \ln (x + 1) = \ln 3 + \ln 4$

32. $\ln (6x - 1) + \ln x = \frac{1}{2} \ln 4$

33. $\ln x = \ln 3 - \ln (x + 5)$

34. $\ln (2x + 3) + \ln x = \ln e$

In Exercises 35–42, solve the equation, as in Examples 9–11.

35. $\ln (x + 9) - \ln x = 1$

36. $\ln (2x + 1) - 1 = \ln (x - 2)$

37. $\log x + \log (x - 3) = 1$

38. $\log (x - 1) + \log (x + 2) = 1$

39. $\log \sqrt{x^2 - 1} = 2$

40. $\log \sqrt[3]{x^2 + 21x} = 2/3$

41. $\ln (x^2 + 1) - \ln (x - 1) = 1 + \ln (x + 1)$

42. $\dfrac{\ln (x + 1)}{\ln (x - 1)} = 2$

43. Book sales on the Internet (in billions of dollars) in year x are approximated by

$$f(x) = 1.84 + 2.1 \ln x,$$

where $x = 2$ corresponds to 2002.*

(a) How much was spent on Internet book sales in 2004?

(b) When will book sales reach \$6.5 billion?

44. According to data from the U.S. Bureau of Labor Statistics, the consumer price index (CPI) for food is approximated in the year x by

$$g(x) = 10.13 + 52.1 \ln x,$$

where $x = 7$ corresponds to 1987.

(a) What was the CPI for food in 2003?

(b) If this model remains accurate, when will the CPI for food reach 185?

45. U.S. Census Bureau data shows that the number of families in the United States (in millions) in year x is given by

$$h(x) = 51.42 + 15.473 \log x,$$

where $x = 5$ is 1985.

(a) How many families were there in 2002?

(b) When will there be 74 million families?

46. The number of automobiles sold in the United States (in millions of vehicles) in year x is approximated by

$$g(x) = 15.27 + 2.282 \log x,$$

where $x = 3$ corresponds to 2003.†

(a) How many cars were sold in 2004?

(b) If this model remains accurate, when will sales reach 18 million cars?

Exercises 47–56, deal with radioactive decay and the function $M(x) = c(.5^{x/h})$; see Example 4.

47. A sample of 300 grams of uranium decays to 200 grams in .26 billion years. Find the half-life of uranium.

48. It takes 1000 years for a sample of 100 mg of radium-226 to decay to 65 mg. Find the half-life of radium-226.

49. A 3-gram sample of an isotope of sodium decays to 1 gram in 23.7 days. Find the half-life of the isotope of sodium.

50. The half-life of cobalt-60 is 4.945 years. How long will it take for 25 grams to decay to 15 grams?

51. After six days a sample of radon-222 decayed to 33.6% of its original mass. Find the half-life of radon-222. [*Hint:* When $x = 6$, then $M(x) = .336P$.]

52. Krypton-85 loses 6.44% of its mass each year. What is its half-life?

53. A Native American mummy was found recently. If it has lost 26.4% of its carbon-14, approximately how long ago did the Native American die? [*Hint:* Remember that the half-life of carbon-14 is 5730 years and see Example 11 in Section 6.2.]

54. How old is a wooden statue that has only one-third of its original carbon-14?

55. How old is a piece of ivory that has lost 36% of its carbon-14?

56. How old is a mummy that has lost 49% of its carbon-14?

Exercises 57–62, deal with the compound interest formula $A = P(1 + r)^t$, which was discussed in Section 6.2.

57. At what annual rate of interest should \$1000 be invested so that it will double in 10 years if interest is compounded quarterly?

58. How long does it take \$500 to triple if it is invested at 6% compounded: (a) annually, (b) quarterly, (c) daily?

59. (a) How long will it take to triple your money if you invest \$500 at a rate of 5% per year compounded annually?

(b) How long will it take at 5% compounded quarterly?

60. At what rate of interest (compounded annually) should you invest \$500 if you want to have \$1500 in 12 years?

61. How much money should be invested at 5% interest, compounded quarterly, so that 9 years later the investment will be worth \$5000? This amount is called the **present value** of \$5000 at 5% interest.

62. Find a formula that gives the time needed for an investment of P dollars to double, if the interest rate is r% compounded annually. [*Hint:* Solve the compound interest formula for t, when $A = 2P$.]

Exercises 63–70 deal with functions of the form $f(x) = Pe^{kx}$, where k is the continuous exponential growth rate (see Example 5).

63. The present concentration of carbon dioxide in the atmosphere is 364 parts per million (ppm) and is increasing exponentially at a continuous yearly rate of .4% (that is, $k = .004$). How many years will it take for the concentration to reach 500 ppm?

64. The amount P of ozone in the atmosphere is currently decaying exponentially each year at a continuous rate of $\frac{1}{4}$% (that is, $k = -.0025$). How long will it take for half the ozone to disappear (that is, when will the amount be $P/2$)? [Your answer is the half-life of ozone.]

---

*Based on forecasts by Forrester Research.

†Based on data from J. D. Power and Associates.

65. The population of Brazil increased from 151 million in 1990 to 173 million in 2000.*
(a) At what continuous rate was the population growing during this period?
(b) Assuming that Brazil's population continues to increase at this rate, when will it reach 250 million?

66. Outstanding consumer debt increased exponentially from $781.5 billion in 1990 to $1765.5 billion in 2002.†
(a) At what continuous growth rate is consumer debt growing?
(b) Assuming that this rate continues, when will consumer debt reach $2500 billion?

67. The probability P percent of having an accident while driving a car is related to the alcohol level of the driver's blood by the formula $P = e^{kt}$, where k is a constant. Accident statistics show that the probability of an accident is 25% when the blood alcohol level is $t = .15$.
(a) Find k. [Use $P = 25$, not .25.]
(b) At what blood alcohol level is the probability of having an accident 50%?

68. Under normal conditions, the atmospheric pressure (in millibars) at height h feet above sea level is given by $P(h) = 1015e^{-kh}$, where k is a positive constant.
(a) If the pressure at 18,000 feet is half the pressure at sea level, find k.
(b) Using the information from part (a), find the atmospheric pressure at 1000 feet, 5000 feet, and 15,000 feet.

69. One hour after an experiment begins, the number of bacteria in a culture is 100. An hour later, there are 500.
(a) Find the number of bacteria at the beginning of the experiment and the number three hours later.
(b) How long does it take the number of bacteria at any given time to double?

70. If the population at time t is given by $S(t) = ce^{kt}$, find a formula that gives the time it takes for the population to double.

*International Data Base, Bureau of the Census, U.S. Department of Commerce.
†*Federal Reserve Bulletin.*

71. The spread of a flu virus in a community of 45,000 people is given by the function

$$f(t) = \frac{45{,}000}{1 + 224e^{-.899t}},$$

where $f(t)$ is the number of people infected in week t.
(a) How many people had the flu at the outbreak of the epidemic? After three weeks?
(b) When will half the town be infected?

72. The beaver population near a certain lake in year t is approximately

$$p(t) = \frac{2000}{1 + 199e^{-.5544t}}.$$

(a) When will the beaver population reach 1000?
(b) Will the population ever reach 2000? Why?

Thinkers

73. According to one theory of learning, the number of words per minute N that a person can type after t weeks of practice is given by $N = c(1 - e^{-kt})$, where c is an upper limit that N cannot exceed and k is a constant that must be determined experimentally for each person.
(a) If a person can type 50 wpm (words per minute) after four weeks of practice and 70 wpm after eight weeks, find the values of k and c for this person. According to the theory, this person will never type faster than c wpm.
(b) Another person can type 50 wpm after four weeks of practice and 90 wpm after eight weeks. How many weeks must this person practice to be able to type 125 wpm?

74. Hilary Hungerford has been offered two jobs, each with the same starting salary of $32,000 and identical benefits. Assuming satisfactory performance, she will receive a $1600 raise each year at the Great Gizmo Company, whereas the Wonder Widget Company will give her a 4% raise each year.
(a) In what year (after the first year) would her salary be the same at either company? Until then, which company pays better? After that, which company pays better?
(b) Answer the questions in part (a) assuming that the annual raise at Great Gizmo is $2000.

Chapter 6 Review

IMPORTANT

CONCEPTS

IMPORTANT

FACTS & FORMULAS

- *Laws of Exponents:*

$$c^r c^s = c^{r+s} \qquad (cd)^r = c^r d^r$$

$$\frac{c^r}{c^s} = c^{r-s} \qquad \left(\frac{c}{d}\right)^r = \frac{c^r}{d^r}$$

$$(c^r)^s = c^{rs} \qquad c^{-r} = \frac{1}{c^r}$$

- *Exponential Growth Functions:*

$$f(x) = P(1 + r)^x \qquad (r > 0),$$

$$f(x) = Pa^x \qquad (a > 1),$$

$$f(x) = Pe^{kx} \qquad (k > 0)$$

- *Exponential Decay Functions:*

$$f(x) = P(1 - r)^x \qquad (0 < r < 1),$$

$$f(x) = Pa^x \qquad (0 < a < 1),$$

$$f(x) = Pe^{kx} \qquad (k < 0)$$

- *Compound Interest Formula:* $A = P(1 + r)^t$
- *Logarithm Laws:* For all $v, w > 0$ and any k:

$$\ln (vw) = \ln v + \ln w \qquad \log_b (vw) = \log_b v + \log_b w$$

$$\ln \left(\frac{v}{w}\right) = \ln v - \ln w \qquad \log_b \left(\frac{v}{w}\right) = \log_b v - \log_b w$$

$$\ln (v^k) = k(\ln v) \qquad \log_b (v^k) = k(\log_b v)$$

- *Change of Base Formula:* $\log_b v = \dfrac{\ln v}{\ln b}$
- $g(x) = \log x$ is the inverse function of $f(x) = 10^x$, which means that

$$10^{\log v} = v \text{ for all } v > 0 \qquad \text{and} \qquad \log 10^u = u \text{ for all } u.$$

- $g(x) = \ln x$ is the inverse function of $f(x) = e^x$, which means that

$$e^{\ln v} = v \text{ for all } v > 0 \qquad \text{and} \qquad \ln (e^u) = u \text{ for all } u.$$

- $h(x) = \log_b x$ is the inverse function of $k(x) = b^x$, which means that

$$b^{\log_b v} = v \text{ for all } v > 0 \qquad \text{and} \qquad \log_b (b^u) = u \text{ for all } u.$$

REVIEW QUESTIONS 

In Questions 1–8, find a viewing window (or windows) that shows a complete graph of the function.

1. $f(x) = 3.7^x$
2. $g(x) = 2.5^{-x}$
3. $g(x) = 2^x - 1$
4. $f(x) = 2^{x-1}$
5. $f(x) = 2^{x^3-x-2}$
6. $g(x) = \dfrac{850}{1 + 5e^{-.4x}}$
7. $h(x) = \ln (x + 4) - 2$
8. $k(x) = \ln \left(\dfrac{x}{x-2}\right)$

9. A computer software company claims the following function models the "learning curve" for their mathematical software.

$$P(t) = \frac{100}{1 + 48.2e^{-.52t}},$$

where t is measured in months and $P(t)$ is the average percent of the software program's capabilities mastered after t months.
   (a) Initially, what percent of the program is mastered?
   (b) After six months, what percent of the program is mastered?
   (c) Roughly, when can a person expect to "learn the most in the least amount of time"?
   (d) If the company's claim is true, how many months will it take to have completely mastered the program?

10. Compunote has offered you a starting salary of \$60,000 with \$1000 yearly raises. Calcuplay offers you an initial salary of \$30,000 and a guaranteed 6% raise each year.
   (a) Complete the following table for each company.

Year	Compunote
1	\$60,000
2	\$61,000
3	
4	
5	

Year	Calcuplay
1	\$30,000
2	\$31,800
3	
4	
5	

   (b) For each company, write a function that gives your salary in terms of years employed.
   (c) If you plan on staying with the company for only five years, which job should you take to earn the most money?
   (d) If you plan on staying with the company for 20 years, which is your better choice?
   (e) In what year does the salary at Calcuplay exceed the salary at Compunote?

11. Phil borrows \$800 at 9% interest, compounded annually.
   (a) How much does he owe after six years?
   (b) If he pays off the loan at the end of six years, how much interest will he owe?

12. If you invest \$5000 for five years at 9% interest, how much more will you make if interest is compounded continuously than if it is compounded quarterly?

13. Mary Karen invests \$2000 at 5.5% interest, compounded monthly.
   (a) How much is her investment worth in three years?
   (b) When will her investment be worth \$12,000?

14. If a \$2000 investment grows to \$5000 in 14 years, with interest compounded annually, what is the interest rate?

15. Company sales are increasing at 6.5% per year. If sales this year are \$56,000, write the rule of a function that gives the sales in year x ($x = 0$ being the present year).

16. The population of Potterville is decreasing at an annual rate of 1.5%. If the population is now 38,500, what will the population be x years from now?

17. The half-life of carbon-14 is 5730 years. How much carbon-14 remains from an original 16-gram sample after 12,000 years?

18. How long will it take for 4 grams of carbon-14 to decay to 1 gram?

In Questions 19–24, translate the given exponential statement into an equivalent logarithmic one.

19. $e^{6.628} = 756$
20. $e^{5.8972} = 364$
21. $e^{r^2-1} = u + v$
22. $e^{a-b} = c$
23. $10^{2.8785} = 756$
24. $10^{c+d} = t$

In Questions 25–30, translate the given logarithmic statement into an equivalent exponential one.

25. $\ln 1234 = 7.118$
26. $\ln (ax + b) = y$
27. $\ln (rs) = t$
28. $\log 1234 = 3.0913$
29. $\log_5 (cd - k) = u$
30. $\log_d (uv) = w$

In Questions 31–34, evaluate the given expression without using a calculator.

31. $\ln e^3$
32. $\ln \sqrt[3]{e}$
33. $e^{\ln 3/4}$
34. $e^{\ln (x+2y)}$

REVIEW QUESTIONS

35. Simplify: $3 \ln \sqrt{x} + (1/2)\ln x$

36. Simplify: $\ln (e^{4e})^{-1} + 4e$

In Questions 37–39, write the given expression as a single logarithm.

37. $\ln 3x - 3 \ln x + \ln 3y$

38. $\log_7 7x + \log_7 y - 1$

39. $4 \ln x - 2(\ln x^3 + 4 \ln x)$

40. $\log(-.01) = ?$

41. $\log_{20} 400 = ?$

42. You are conducting an experiment about memory. The people who participate agree to take a test at the end of your course and every month thereafter for a period of two years. The average score for the group is given by the model

$$M(t) = 91 - 14 \ln (t + 1) \qquad (0 \le t \le 24)$$

where t is time in months after the first test.

(a) What is the average score on the initial exam?
(b) What is the average score after three months?
(c) When will the average drop below 50%?
(d) Is the magnitude of the rate of memory loss greater in the first month after the course (from $t = 0$ to $t = 1$) or after the first year (from $t = 12$ to $t = 13$)?
(e) Hypothetically, if the model could be extended past $t = 24$ months, would it be possible for the average score to be 0%?

43. Which of the following statements are *true*?

(a) $\ln 10 = (\ln 2)(\ln 5)$
(b) $\ln (e/6) = \ln e + \ln 6$
(c) $\ln (1/7) + \ln 7 = 0$
(d) $\ln(-e) = -1$
(e) None of the above is true.

44. Which of the following statements are *false*?

(a) $10 (\log 5) = \log 50$
(b) $\log 100 + 3 = \log 10^5$
(c) $\log 1 = \ln 1$
(d) $\log 6/\log 3 = \log 2$
(e) All of the above are false.

Use the following six graphs for Questions 45 and 46.

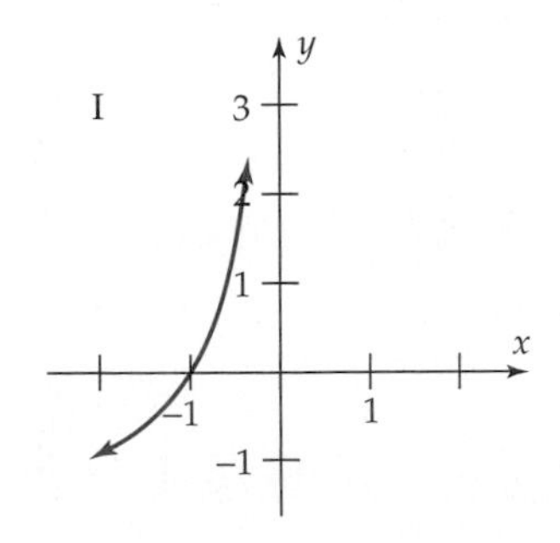

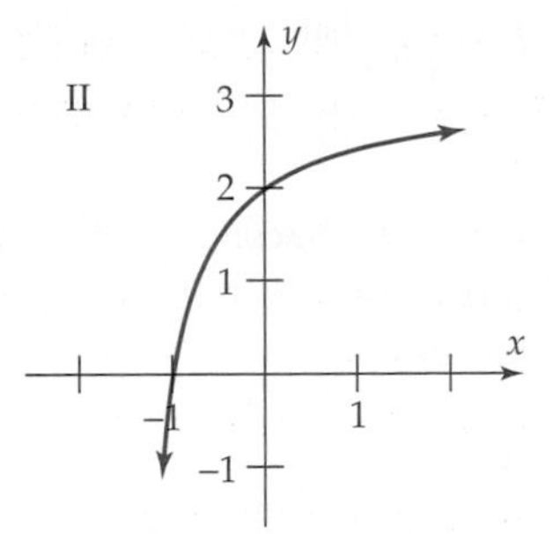

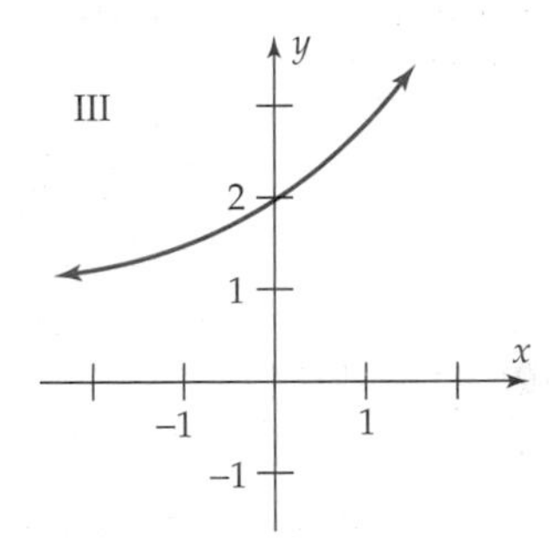

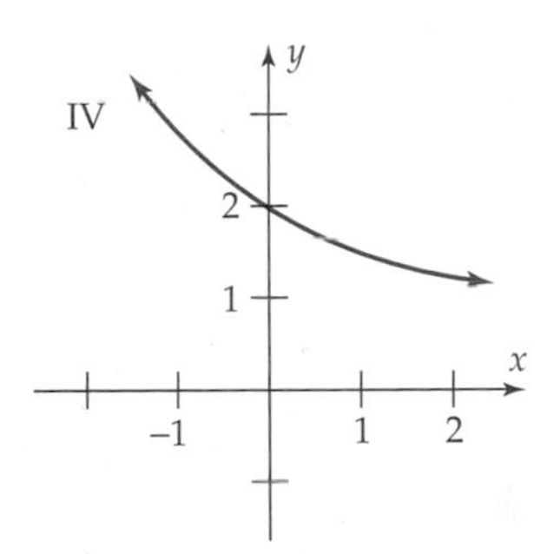

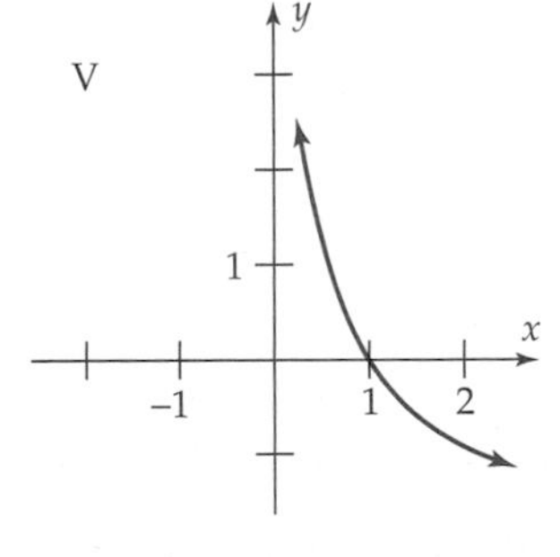

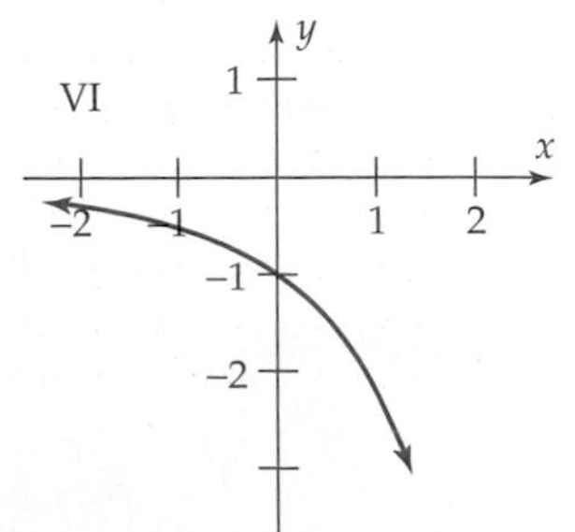

REVIEW QUESTIONS

45. If $b > 1$, then the graph of $f(x) = -\log_b x$ could possibly be:
(a) I (b) IV
(c) V (d) VI
(e) none of these

46. If $0 < b < 1$, then the graph of $g(x) = b^x + 1$ could possibly be:
(a) II (b) III
(c) IV (d) VI
(e) none of these

47. If $\log_3 9^{x^2} = 4$, what is x?

48. What is the domain of the function $f(x) = \ln\left(\frac{x}{x-1}\right)$?

In Questions 49–57, solve the equation for x.

49. $8^x = 4^{x^2-3}$

50. $e^{3x} = 4$

51. $2 \cdot 4^x - 5 = -4$

52. $725e^{-4x} = 1500$

53. $u = c + d \ln x$

54. $2^x = 3^{x+3}$

55. $\ln x + \ln(3x - 5) = \ln 2$

56. $\ln(x + 8) - \ln x = 1$

57. $\log(x^2 - 1) = 2 + \log(x + 1)$

58. At a small community college the spread of a rumor through the population of 500 faculty and students can be modeled by the equation.

$$\ln(n) - \ln(1000 - 2n) = .65t - \ln 998,$$

where n is the number of people who have heard the rumor after t days.
(a) How many people know the rumor initially (at $t = 0$)?
(b) How many people have heard the rumor after four days?
(c) Roughly, in how many weeks will the entire population have heard the rumor?
(d) Use the properties of logarithms to write n as a function of t; in other words solve the model above for n in terms of t.
(e) Enter the function you found in part (d) into your calculator and use the table feature to check your answers to parts (a), (b), and (c). Do they agree?
(f) Now graph the function. Roughly over what time interval does the rumor seem to spread the fastest?

59. The half-life of polonium (^{210}Po) is 140 days. If you start with 10 milligrams, how much will be left at the end of a year?

60. An insect colony grows exponentially from 200 to 2000 in three months' time. How long will it take for the insect population to reach 50,000?

61. Hydrogen-3 decays at a rate of 5.59% per year. Find its half-life.

62. The half-life of radium-88 is 1590 years. How long will it take for 10 grams to decay to 1 gram?

63. How much money should be invested at 8% per year, compounded quarterly, in order to have \$1000 in 10 years?

64. At what annual interest rate should you invest your money if you want to double it in six years?

65. One earthquake measures 4.6 on the Richter scale. A second earthquake is 1000 times more intense than the first. What does it measure on the Richter scale?

DISCOVERY PROJECT 6 Exponential and Logistic Modeling of Diseases

Diseases that are contagious and are transmitted homogeneously through a population often appear to be spreading exponentially. That is, the rate of spread is proportional to the number of people in the population who are already infected. This is a reasonable model as long as the number of infected people is relatively small in comparison with the number of people in the population who can be infected. The standard exponential model looks like this:

$$f(t) = Y_0 e^{rt}.$$

Y_0 is the initial number of infected people (the number on the arbitrarily decided day 0), and r is the rate by which the disease spreads through the population. If the time t is measured in days, then r is the ratio of new infections to current infections each day.

Suppose that in Big City, population 3855, there is an outbreak of dingbat disease. On the first Monday after the outbreak was discovered (day 0), 72 people have dingbat disease. On the following Monday (day 7), 193 people have dingbat disease.

1. Using the exponential model, Y_0 is clearly 72. Calculate the value of r.
2. Using your values of Y_0 and r, predict the number of cases of dingbat disease that will be reported on day 14.

It turns out that eventually, the spread of disease must slow as the number of infected people approaches the number of susceptible people. What happens is that some of the people to whom the disease would spread are already infected. As time goes on, the spread of the disease becomes proportional to the number of susceptible and uninfected people. The disease then follows the logistic model

$$g(t) = \frac{rY_0}{aY_0 + (r - aY_0)e^{-rt}}.$$

Y_0 is still the initial value, and r serves the same function as before, at least at the initial time. The extra parameter a is not so obvious, but it is inversely related to the number of people susceptible to the disease. Unfortunately, the algebra to solve for a is quite complicated. It is much easier to approximate a using the same r from the exponential model.

3. On day 14 in Big City, 481 people have dingbat disease. Using the values of Y_0 and r from Exercise 1 in the rule of the function g, determine the value of a. [*Hint:* $g(14) = 481$.] Does g overestimate or underestimate the number of people with dingbat disease on day 7?
4. Use the function g from Exercise 3 to approximate the number of people in Big City who are susceptible to the disease. Does this model make sense? [Remember, as time goes on, the number of people infected approaches the number of people susceptible.]

DISCOVERY PROJECT 6 Continued

5. In the logistic model, the rate at which the disease spreads tends to fall over time. This means that the value of r you calculated in Exercise 1 is a little low. Raise the value of r and find the new value of a as in Exercise 3. Experiment until you find a value of r for which $g(7) = 193$ (meaning that the model g matches the data on day 7).
6. Using the function g from Exercise 5, repeat Exercise 4.

Appendix GEOMETRY REVIEW

An **angle** consists of two half-lines that begin at the same point P, as in Figure 1. The point P is called the **vertex** of the angle, and the half-lines are called the **sides** of the angle.

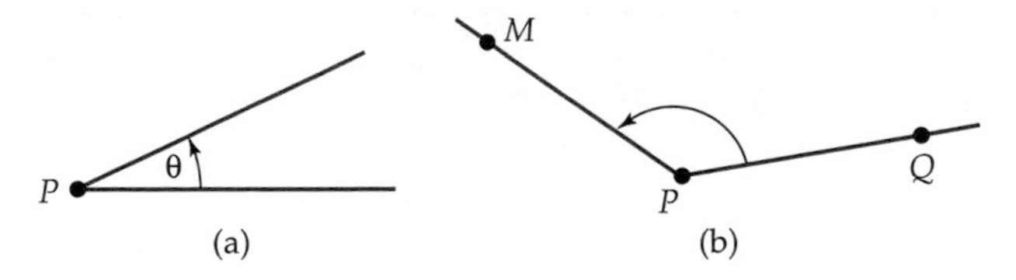

Figure 1

An angle may be labeled by a Greek letter, such as angle θ in Figure 1(a), or by listing three points (a point on one side, the vertex, a point on the other side), such as angle QPM in Figure 1(b).

To measure the size of an angle, we must assign a number to each angle. Here is the classical method for doing this.

1. Construct a circle whose center is the vertex of the angle.
2. Divide the circumference of the circle into 360 equal parts (called **degrees**) by marking 360 points on the circumference, beginning with the point where one side of the angle intersects the circle. Label these points 0°, 1°, 2°, 3°, and so on.
3. The label of the point where the second side of the angle intersects the circle is the degree measure of the angle.

For example, Figure 2 shows an angle θ of measure 25 degrees (in symbols, 25°) and an angle β of measure 135°.

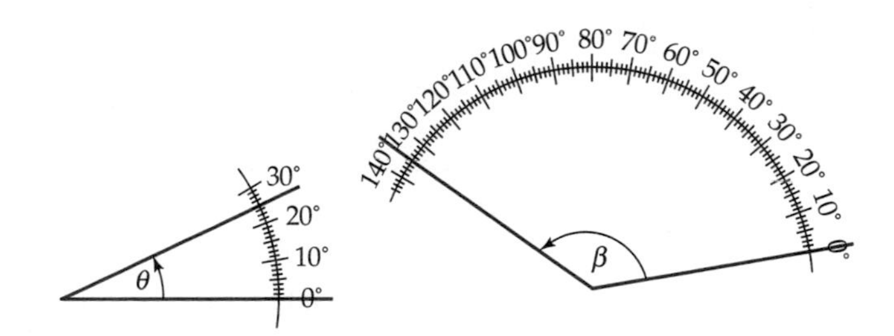

Figure 2

An **acute angle** is an angle whose measure is strictly between 0° and 90°, such as angle θ in Figure 2. A **right angle** is an angle that measures 90°. An **obtuse angle** is an angle whose measure is strictly between 90° and 180°, such as angle β in Figure 2.

TRIANGLES

A **triangle** has three sides (straight line segments) and three angles, formed at the points where the various sides meet. When angles are measured in degrees,

The sum of the measures of all three angles of a triangle is *always* 180°.

For instance, see Figure 3. This fact is proved in Exercise 73.

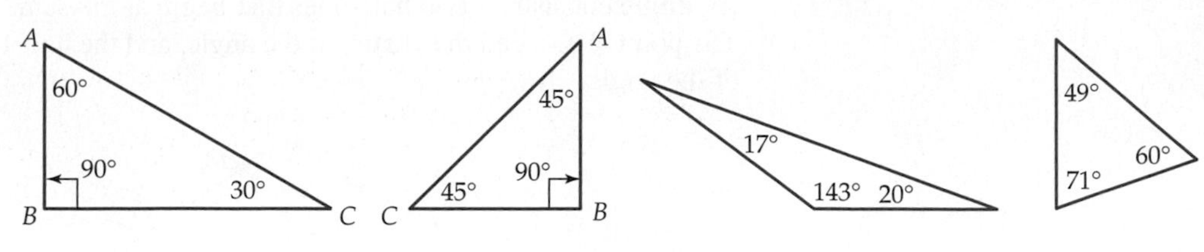

Figure 3

A **right triangle** is a triangle, one of whose angles is a right angle, such as the first two triangles shown in Figure 3. The side of a right triangle that lies opposite the right angle is called the **hypotenuse;** the sides of the right angle are sometimes called the **legs** of the triangle. In each of the right triangles in Figure 3, side AC is the hypotenuse. The legs are sides AB and BC.

Pythagorean Theorem

If the legs of a right triangle have lengths a and b and the hypotenuse has length c, then

$$c^2 = a^2 + b^2.$$

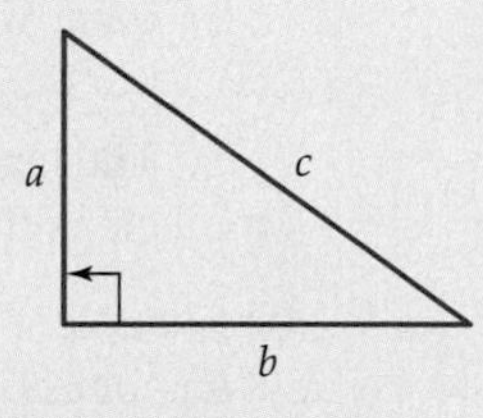

See Exercise 74 for a proof of the Pythagorean Theorem.

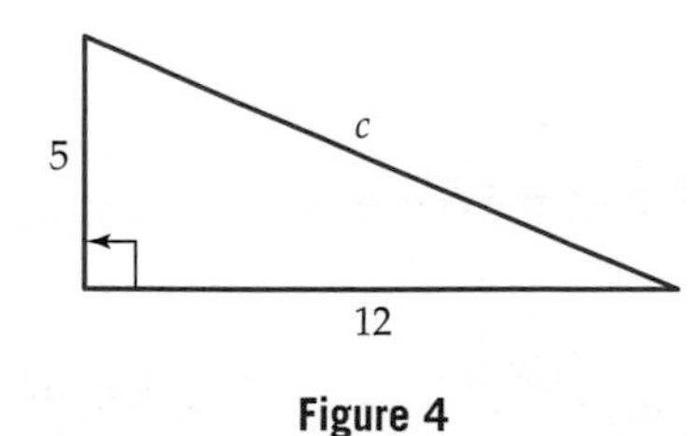

Figure 4

EXAMPLE 1

Consider the right triangle with legs of lengths 5 and 12, as shown in Figure 4.

According to the Pythagorean Theorem, the length c of the hypotenuse satisfies the equation $c^2 = 5^2 + 12^2 = 25 + 144 = 169$. Since $169 = 13^2$, we see that c must be 13. ■

Isosceles Triangle Theorem

If two angles of a triangle are equal, then the two sides opposite these angles have the same length.

Conversely, if two sides of a triangle have the same length, then the two angles opposite these sides are equal.

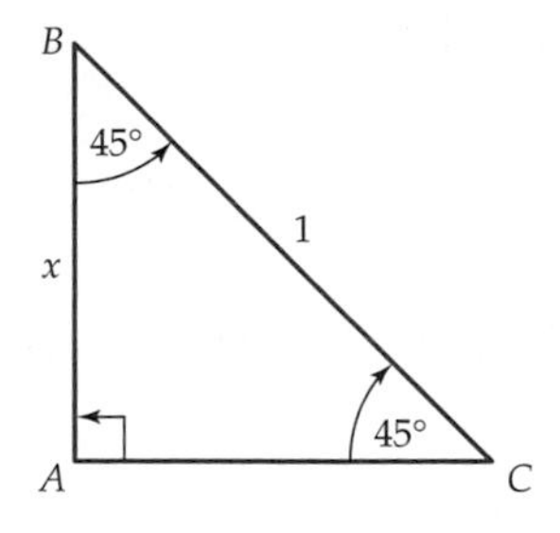

Figure 5

EXAMPLE 2

Suppose the hypotenuse of the right triangle shown in Figure 5 has length 1 and that angles B and C measure 45° each.

Then by the Isosceles Triangle Theorem, sides AB and AC have the same length. If x is the length of side AB, then by the Pythagorean Theorem,

$$x^2 + x^2 = 1^2$$
$$2x^2 = 1$$
$$x^2 = \frac{1}{2}$$
$$x = \sqrt{\frac{1}{2}} = \frac{1}{\sqrt{2}} = \frac{\sqrt{2}}{2}.$$

(We ignore the other solution of this equation, namely, $x = -\sqrt{1/2}$, since x represents a length here and therefore must be nonnegative.) Therefore, the legs of a 90°–45°–45° triangle with hypotenuse 1 are each of length $\sqrt{2}/2$. ■

30-60-90 Triangle Theorem

In a right triangle that has an angle of 30°, the length of the side opposite the 30° angle is one-half the length of the hypotenuse.

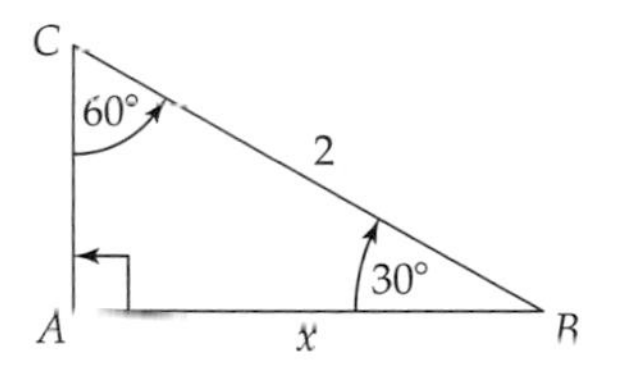

Figure 6

EXAMPLE 3

Suppose that in the right triangle shown in Figure 6 angle B is 30° and the length of hypotenuse BC is 2.

By the 30–60–90 Triangle Theorem, the side opposite the 30° angle, namely, side AC, has length 1. If x denotes the length of side AB, then by the Pythagorean Theorem,

$$1^2 + x^2 = 2^2$$
$$x^2 = 3$$
$$x = \sqrt{3}. \quad ■$$

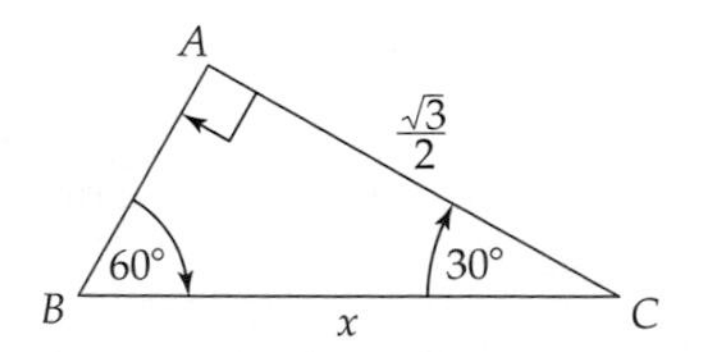

Figure 7

EXAMPLE 4

The right triangle shown in Figure 7 has a 30° angle at C, and side AC has length $\sqrt{3}/2$.

Let x denote the length of the hypotenuse BC. By the 30–60–90 Triangle Theorem, side AB has length $\frac{1}{2}x$. By the Pythagorean Theorem,

$$\left(\frac{1}{2}x\right)^2 + \left(\frac{\sqrt{3}}{2}\right)^2 = x^2$$
$$\frac{x^2}{4} + \frac{3}{4} = x^2$$

$$\frac{3}{4} = \frac{3}{4}x^2$$
$$x^2 = 1$$
$$x = 1.$$

Therefore, the triangle has hypotenuse of length 1 and legs of lengths $1/2$ and $\sqrt{3}/2$. ■

Figures 6 and 7 illustrate properties shared by all triangles.

The longest side of a triangle is always opposite its largest angle.

The shortest side of a triangle is always opposite its smallest angle.

You may find these facts helpful when checking solutions to triangle problems.

CONGRUENT TRIANGLES

Two angles are said to be **congruent** if they have the same measure, and two line segments are **congruent** if they have the same length. We say that two triangles are **congruent** if the three sides and three angles of one are congruent respectively to the corresponding sides and angles of the other, as illustrated in Figure 8.

Congruent Triangles

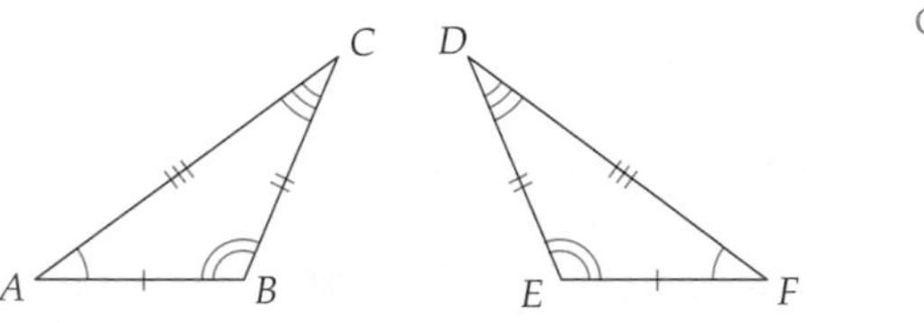

Corresponding Congruent Sides*	Corresponding Congruent Angles
$AB \cong FE$	$\angle A \cong \angle F$
$BC \cong ED$	$\angle B \cong \angle E$
$CA \cong DF$	$\angle C \cong \angle D$

Figure 8

When two triangles are congruent, you can place one on top of the other so that they coincide. (You might have to rotate or flip one of the triangles over to do this, as is the case in Figure 8.) Thus,

Congruent triangles have the same size and same shape.

The key facts about congruent triangles are proved in high school geometry.

Congruent Triangles Theorem

If two triangles are given, then any one of the following conditions guarantees that they are congruent.

1. SAS: Two sides and the angle between them in one triangle are congruent to the corresponding sides and angle in the other triangle.
2. ASA: Two angles and the side between them in one triangle are congruent to the corresponding angles and side in the other triangle.
3. SSS: The three sides of one triangle are congruent to the corresponding sides of the other triangle.

*The symbol ≅ means "is congruent to."

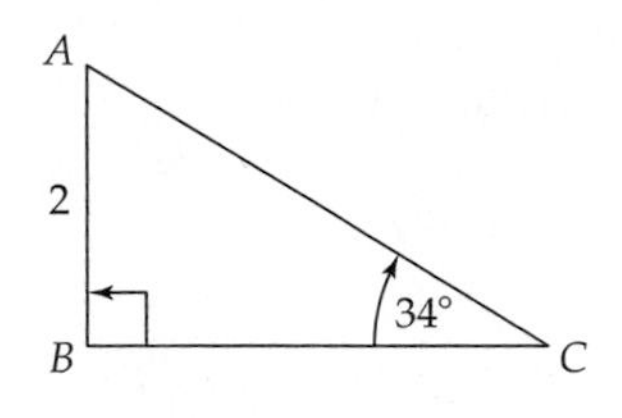

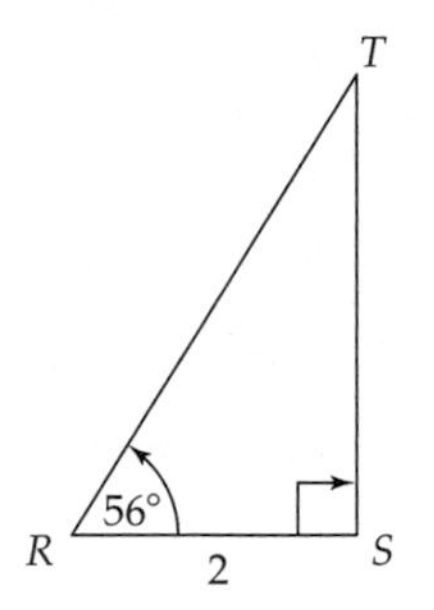

Figure 9

EXAMPLE 5

Show that the right triangles in Figure 9 are congruent.

SOLUTION Triangle ABC has angles of 90° and 34°. Since the sum of all three angles is 180°, angle A must measure 56°. Hence, we have these congruences.

$$\angle A \cong \angle R \quad \text{[Both measure 56°.]}$$

$$\angle B \cong \angle S \quad \text{[Both measure 90°.]}$$

$$AB \cong RS. \quad \text{[Both have length 2.]}$$

Since two angles and the side between them in triangle ABC are congruent to the corresponding angles and side in triangle DEF, the triangles are congruent by ASA. ■

SIMILAR TRIANGLES

Two triangles are said to be **similar** if the three angles of one are congruent respectively to the three angles of the other, as illustrated in Figure 10.

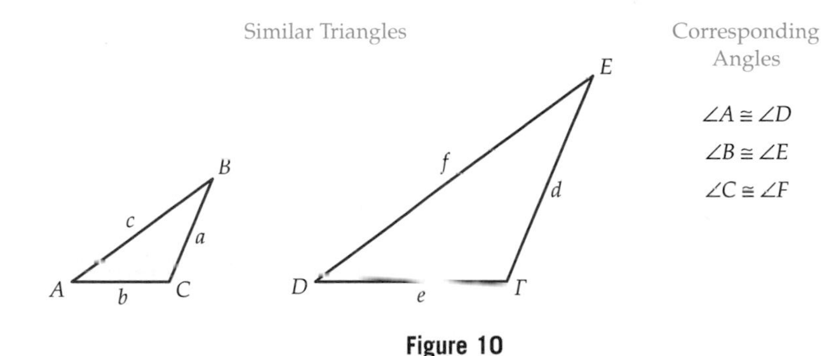

Figure 10

Thus,

Similar triangles have the same shape, but not necessarily the same size.

In Figure 10, for example, the sides of triangle DEF are twice as long as the corresponding sides of triangle ABC:

$$d = 2a, \qquad e = 2b, \qquad f = 2c,$$

which is equivalent to:

$$\frac{d}{a} = 2, \qquad \frac{e}{b} = 2, \qquad \frac{f}{c} = 2.$$

Hence,

$$\frac{d}{a} = \frac{e}{b} = \frac{f}{c}.$$

A similar fact holds in the general case.

Ratios Theorem

Suppose triangle ABC is similar to triangle DEF (with $\angle A \cong \angle D$; $\angle B \cong \angle E$; $\angle C \cong \angle F$).

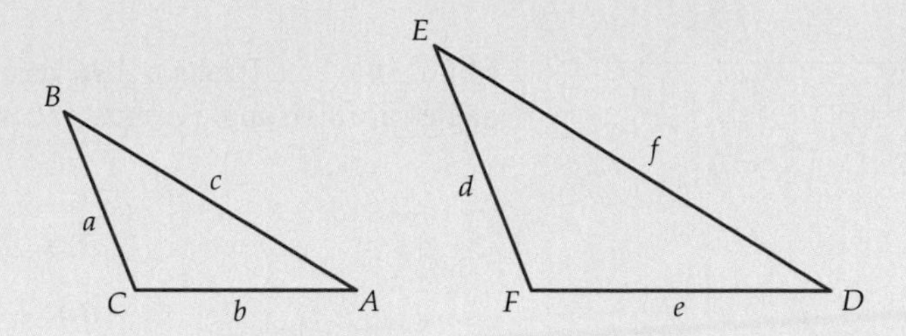

Then the ratios of corresponding sides are equal, that is,

$$\frac{d}{a} = \frac{e}{b} = \frac{f}{c}.$$

NOTE: There are many ways to rewrite the conclusions of the Ratios Theorem. For example,

$$\frac{d}{a} = \frac{f}{c} \quad \text{is equivalent to} \quad \frac{a}{c} = \frac{d}{f}.$$

To see this, multiply the first equation by $\frac{a}{f}$. Similarly, multiplying by $\frac{b}{f}$ shows that

$$\frac{e}{b} = \frac{f}{c} \quad \text{is equivalent to} \quad \frac{b}{c} = \frac{e}{f}.$$

EXAMPLE 6

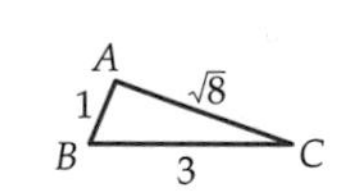

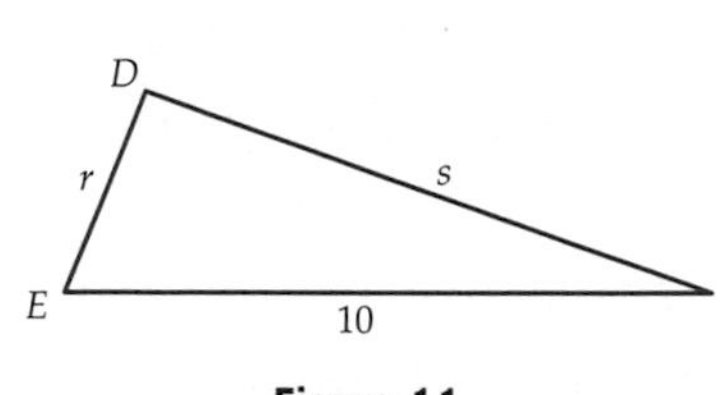

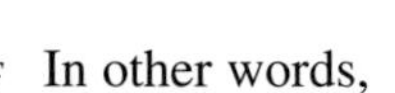

Figure 11

Suppose the triangles in Figure 11 are similar and that the sides have the lengths indicated. Find r and s.

SOLUTION By the Ratios Theorem,

$$\frac{\text{Length } DF}{\text{Length } AC} = \frac{\text{Length } EF}{\text{Length } BC}.$$

In other words,

$$\frac{s}{\sqrt{8}} = \frac{10}{3},$$

so

$$3s = 10\sqrt{8}$$

$$s = \left(\frac{10}{3}\right)\sqrt{8}.$$

Similarly,

$$\frac{\text{Length } DE}{\text{Length } AB} = \frac{\text{Length } EF}{\text{Length } BC},$$

so

$$\frac{r}{1} = \frac{10}{3}$$

$$r = \frac{10}{3}.$$

Therefore, the sides of triangle DEF are of lengths 10, $\frac{10}{3}$, and $\frac{10}{3}\sqrt{8}$. ■

EXAMPLE 7

Tom, who is 6 feet tall, stands in the sunlight near the main library on campus, and Anne determines that his shadow is 7 feet long. At the same time, their friend Mary Alice finds that the library casts a 100-foot-long shadow. How high is the main library?

SOLUTION Both Tom and the library are perpendicular to the ground. So we have the right triangles in Figure 12 (which is not to scale).

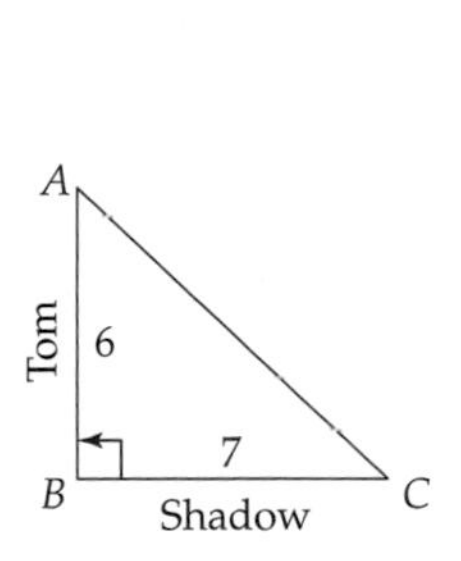

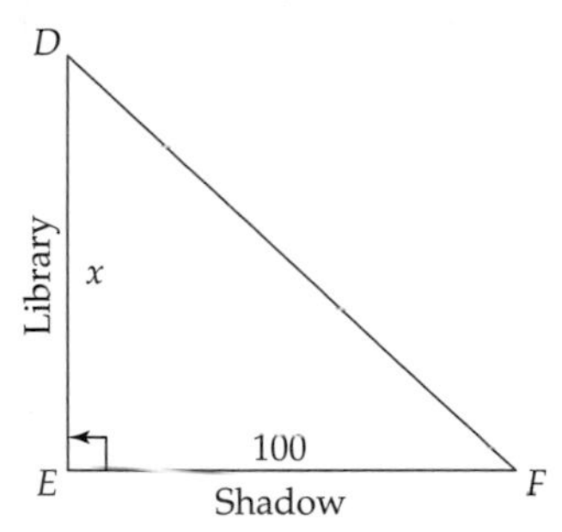

Figure 12

Angles B and E each measure 90°; hence,

$$\angle B \cong \angle E.$$

Because the sun hits both Tom and the library at the same angle,

$$\angle A \cong \angle D.$$

If each of these angles measures $k°$, then angle C measures $180° - 90° - k°$ (because the sum of the angles in triangle ABC must be 180°). Similarly, angle F must also measure $180° - 90° - k°$, so

$$\angle C \cong \angle F.$$

Therefore, triangle ABC is similar to triangle DEF. By the Ratios Theorem,

$$\frac{\text{Length } DE}{\text{Length } AB} = \frac{\text{Length } EF}{\text{Length } BC}$$

$$\frac{x}{6} = \frac{100}{7}$$

$$x = 6\left(\frac{100}{7}\right) \approx 85.7.$$

So the library is about 85.7 feet high. ■

In addition to the definition, there are several other ways to show that two triangles are similar.

Similar Triangles Theorem

Given triangles ABC and DEF,

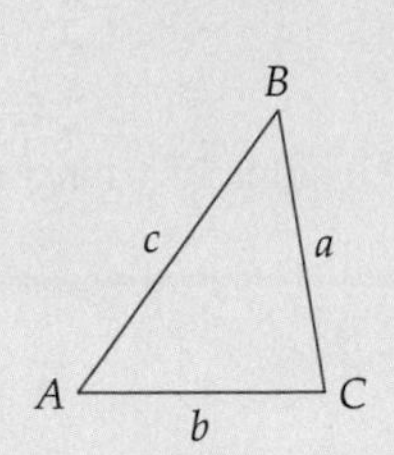

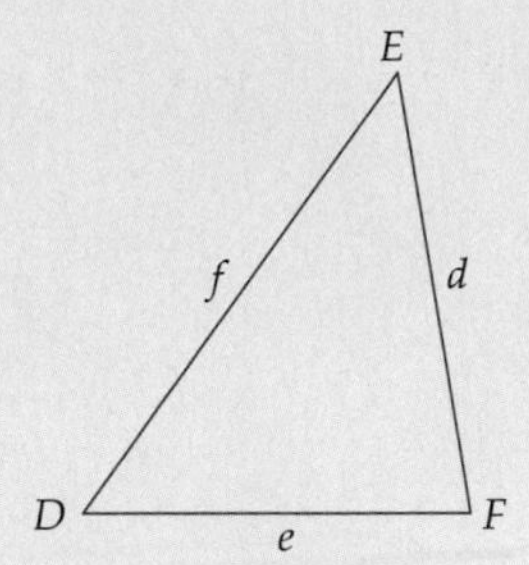

then any one of the following conditions guarantees that they are similar.

1. AA: Two angles of one are congruent to two angles of the other.
2. S/S/S: The ratios of corresponding sides are all equal:

$$\frac{d}{a} = \frac{e}{b} = \frac{f}{c}.$$

3. S/A/S: An angle in one is congruent to an angle in the other, and the ratios of the corresponding sides of these angles are the same:

$$\angle A \cong \angle D \quad \text{or} \quad \angle B \cong \angle E \quad \text{or} \quad \angle C \cong \angle F$$

$$\frac{e}{b} = \frac{f}{c} \qquad \frac{d}{a} = \frac{f}{c} \qquad \frac{d}{a} = \frac{e}{b}.$$

Example 7 shows why the first condition guarantees similarity: If two pairs of angles are congruent, then the third angles are also congruent because the sum of the angles in each triangle is 180°. The proofs of the other conditions will be omitted.

EXAMPLE 8

Show that the triangles in Figure 13 are similar, and find angles C, D, and E and side f.

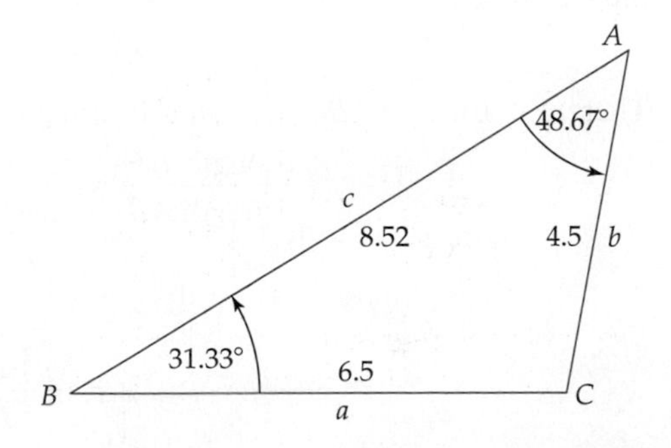

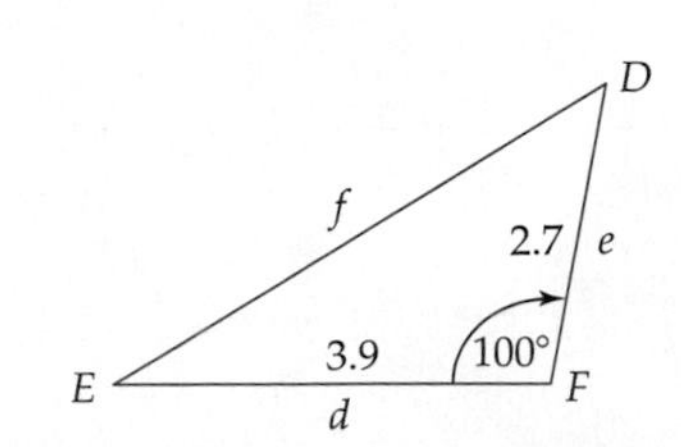

Figure 13

SOLUTION Since the sum of the measures of the angles in triangle ABC is 180°, angle C must measure

$$180° - 31.33° - 48.67° = 100°.$$

Hence, $\angle C \cong \angle F$. The ratios of the corresponding sides of these angles are

$$\frac{d}{a} = \frac{3.9}{6.5} = .6 \quad \text{and} \quad \frac{e}{b} = \frac{2.7}{4.5} = .6$$

Thus,

$$\angle C \cong \angle F \quad \text{and} \quad \frac{d}{a} = \frac{e}{b}.$$

Therefore, triangle ABC is similar to triangle DEF by S/A/S of the Similar Triangles Theorem. Consequently, corresponding angles are congruent. In particular, angle D measures 48.67° and angle E measures 31.33° because $\angle D \cong \angle A$ and $\angle E \cong \angle B$. Since corresponding side ratios are equal,

$$\frac{f}{c} = \frac{d}{a}$$

$$\frac{f}{8.52} = \frac{3.9}{6.5} = .6$$

$$f = .6(8.52) = 5.112. \quad \blacksquare$$

LINES

The most useful tool for dealing with straight lines is the concept of *slope,* which is defined as follows.

Slope of a Line

> If (x_1, y_1) and (x_2, y_2) are points with $x_1 \neq x_2$, then the **slope** of the line through these points is the number
>
> $$\frac{\text{change in } y}{\text{change in } x} = \frac{y_2 - y_1}{x_2 - x_1}.$$

EXAMPLE 9

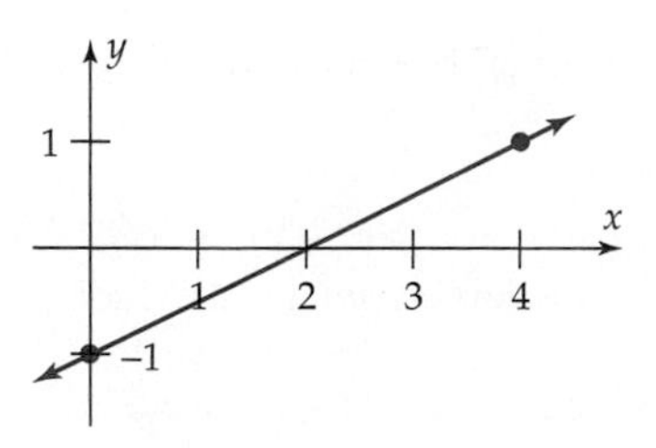

Figure 14

To find the slope of the line through (0, −1) and (4, 1) (see Figure 14), we apply the formula in the previous box with $x_1 = 0$, $y_1 = -1$ and $x_2 = 4$, $y_2 = 1$:

$$\text{Slope} = \frac{y_2 - y_1}{x_2 - x_1} = \frac{1 - (-1)}{4 - 0} = \frac{2}{4} = \frac{1}{2}.$$

The order of the points makes no difference; if you use (4, 1) for (x_1, y_1) and (0, −1) for (x_2, y_2), you obtain the same number:

$$\text{Slope} = \frac{y_2 - y_1}{x_2 - x_1} = \frac{-1 - 1}{0 - 4} = \frac{-2}{-4} = \frac{1}{2}. \quad \blacksquare$$

CAUTION

When finding slopes, you must subtract the y-coordinates and x-coordinates in the same order. With the points (3, 4) and (1, 8), for instance, if you use $8 - 4$ in the numerator, you must use $1 - 3$ in the denominator (*not* $3 - 1$).

EXAMPLE 10

The lines shown in Figure 15 are determined by these points:

L_1: $(-1, -1)$ and $(0, 2)$ L_2: $(0, 2)$ and $(2, 4)$ L_3: $(-6, 2)$ and $(3, 2)$

L_4: $(-3, 5)$ and $(3, -1)$ L_5: $(1, 0)$ and $(2, -2)$.

Their slopes are as follows.

$$L_1: \frac{2-(-1)}{0-(-1)} = \frac{3}{1} = 3$$

$$L_2: \frac{4-2}{2-0} = \frac{2}{2} = 1$$

$$L_3: \frac{2-2}{3-(-6)} = \frac{0}{9} = 0$$

$$L_4: \frac{-1-5}{3-(-3)} = \frac{-6}{6} = -1$$

$$L_5: \frac{-2-0}{2-1} = \frac{-2}{1} = -2$$

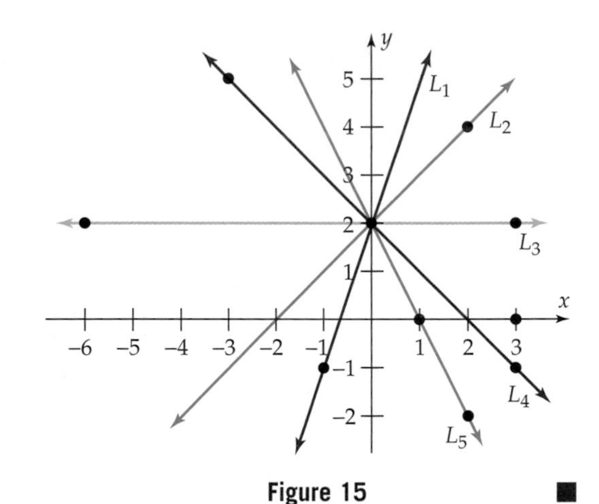

Figure 15 ■

Example 10 illustrates how the slope measures the steepness of the line, as summarized below.

Properties of Slope

The slope of a nonvertical line is a number m that measures how steeply the line rises or falls.

If $m > 0$, the line rises from left to right; the larger m is, the more steeply the line rises. [Lines L_1 and L_2 in Example 10]

If $m = 0$, the line is horizontal. [Line L_3]

If $m < 0$, the line falls from left to right; the larger $|m|$ is, the more steeply the line falls. [Lines L_4 and L_5]

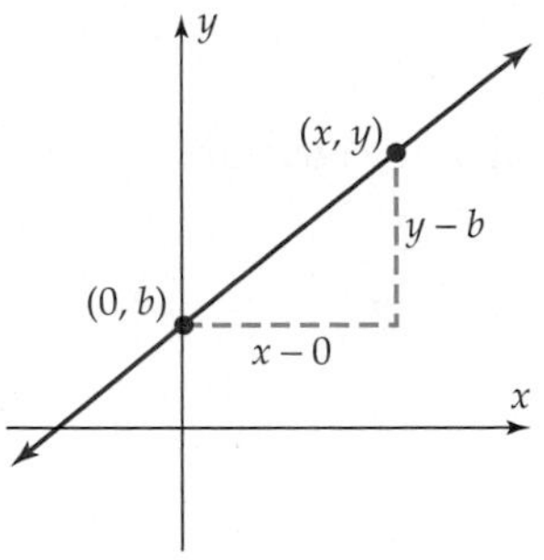

SLOPE-INTERCEPT FORM

Figure 16

Let L be a nonvertical line with slope m and y-intercept b. Then $(0, b)$ is a point on L. Let (x, y) be any other point on L. Using the points $(0, b)$ and (x, y) to compute the slope of L (see Figure 16), we have

$$\text{Slope of } L = \frac{y - b}{x - 0}.$$

Since the slope of L is m, this equation becomes

$$m = \frac{y - b}{x}$$

$$mx = y - b$$

$$y = mx + b.$$

Thus the coordinates of any point on L satisfy the equation $y = mx + b$. So we have the following fact.

Slope-Intercept Form

> The line with slope m and y-intercept b is the graph of the equation
>
> $$y = mx + b.$$

EXAMPLE 11

Show that the graph of $2y - 5x = 2$ is a straight line. Find its slope, and graph the line.

SOLUTION We begin by solving the equation for y:

$$2y = 5x + 2$$

$$y = 2.5x + 1.$$

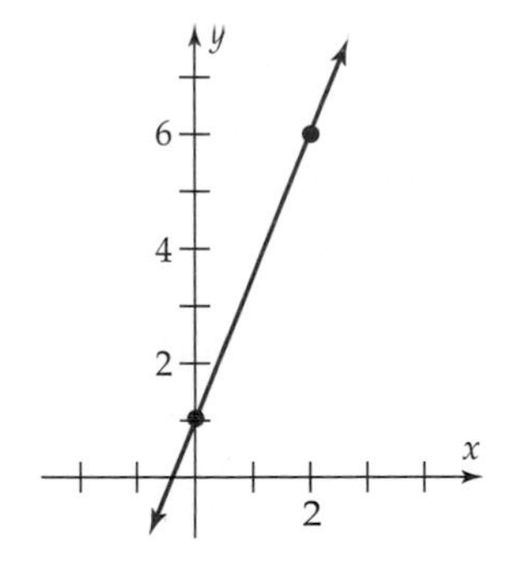

Figure 17

The equation now has the form in the preceding box, with $m = 2.5$ and $b = 1$. Therefore, its graph is the line with slope 2.5 and y-intercept 1.

Since the y-intercept is 1, the point $(0, 1)$ is on the graph. To find another point on the line, choose a value for x, say, $x = 2$, and compute the corresponding value of y:

$$y = 2.5x + 1 = 2.5(2) + 1 = 6.$$

Hence, $(2, 6)$ is on the line. Plotting the line through $(0, 1)$ and $(2, 6)$ produces Figure 17. ■

EXAMPLE 12

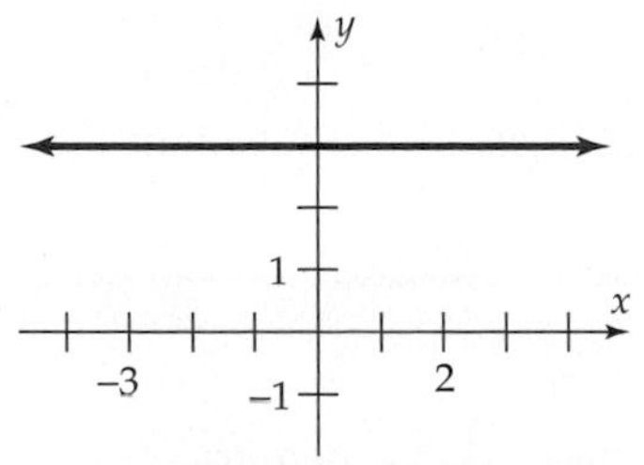

Figure 18

Describe and sketch the graph of the equation $y = 3$.

SOLUTION We can write $y = 3$ as $y = 0x + 3$. So its graph is a line with slope 0, which means that the line is horizontal, and y-intercept 3, which means that the line crosses the y-axis at 3. This is sufficient information to obtain the graph in Figure 18. ■

Example 12 is an illustration of this fact.

Horizontal Lines

> The horizontal line with y-intercept b is the graph of the equation
> $$y = b.$$

POINT-SLOPE FORM

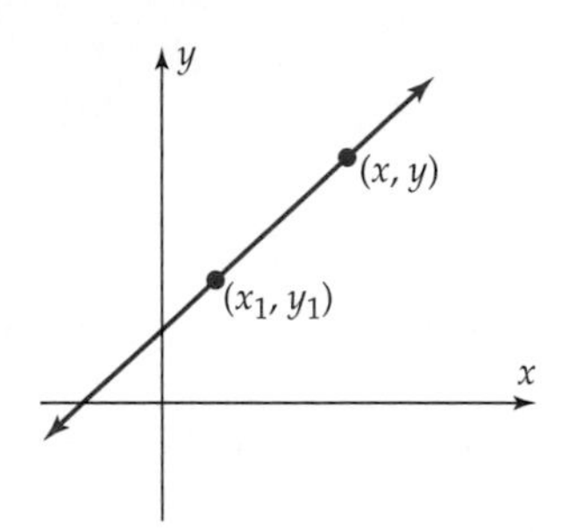

Figure 19

Suppose the line L passes through the point (x_1, y_1) and has slope m. Let (x, y) be any other point on L. Using the points (x_1, y_1) and (x, y) to compute the slope m of L (see Figure 19), we have

$$\frac{y - y_1}{x - x_1} = \text{slope of } L$$

$$\frac{y - y_1}{x - x_1} = m$$

$$y - y_1 = m(x - x_1).$$

Thus the coordinates of every point on L satisfy the equation $y - y_1 = m(x - x_1)$, and we have this fact:

Point-Slope Form

> The line with slope m through the point (x_1, y_1) is the graph of the equation
> $$y - y_1 = m(x - x_1).$$

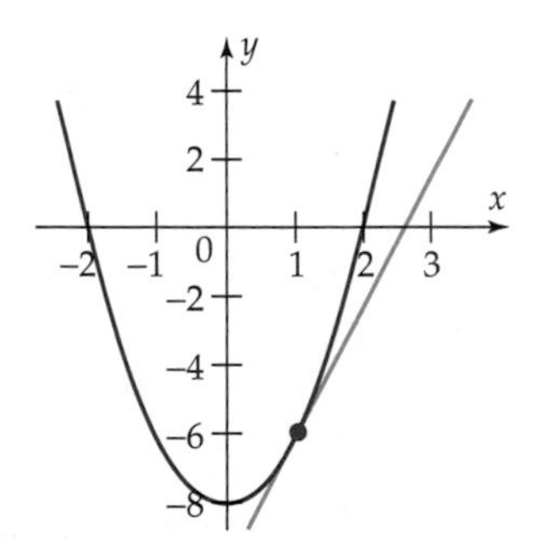

Figure 20

EXAMPLE 13

The tangent line to the graph of $y = 2x^2 - 8$ at the point $(1, -6)$ is shown in Figure 20. In calculus, it is shown that this line passes through $(1, -6)$ and has slope 4. Find the equation of the tangent line.

SOLUTION Substitute 4 for m and $(1, -6)$ for (x_1, y_1) in the point-slope equation:

$$y - y_1 = m(x - x_1)$$

$$y - (-6) = 4(x - 1) \quad \text{[point-slope form]}$$

$$y + 6 = 4x - 4$$

$$y = 4x - 10 \quad \text{[slope-intercept form]}$$ ■

The preceding discussion does not apply to vertical lines, whose equations have a different form from those examined earlier.

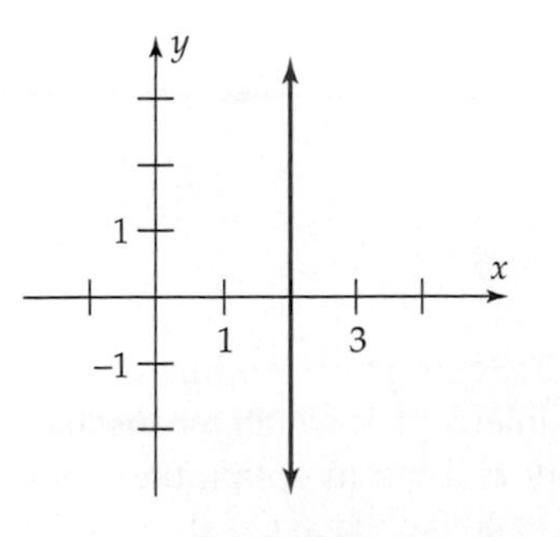

Figure 21

EXAMPLE 14

Every point on the vertical line in Figure 21 has first coordinate 2. Thus every point on the line satisfies $x + 0y = 2$. Thus the line is the graph of the equation

$x = 2$. If you try to compute the slope of this line, say, using (2, 1) and (2, 4), you obtain $\frac{4-1}{2-2} = \frac{4}{0}$, which is not defined. ■

Example 14 illustrates these facts:

Vertical Lines

> The vertical line with x-intercept c is the graph of the equation
> $$x = c.$$
> The slope of this line is undefined.

PARALLEL AND PERPENDICULAR LINES

The slope of a line measures how steeply it rises or falls. Since parallel lines rise or fall equally steeply, the following fact should be plausible.

Parallel Lines

> Two nonvertical lines are parallel exactly when they have the same slope.

EXAMPLE 15

Find the equation of the line L through $(2, -1)$ that is parallel to the line M whose equation is $3x - 2y + 6 = 0$.

SOLUTION First find the slope of M by rewriting its equation in slope-intercept form:

$$3x - 2y + 6 = 0$$
$$-2y = -3x - 6$$
$$y = \frac{3}{2}x + 3.$$

Therefore M has slope $3/2$. The parallel line L must have the same slope, $3/2$. Since $(2, -1)$ is on L, we can use the point-slope form to find its equation:

$$y - y_1 = m(x - x_1)$$
$$y - (-1) = \frac{3}{2}(x - 2) \quad \text{[point-slope form]}$$
$$y + 1 = \frac{3}{2}x - 3$$
$$y = \frac{3}{2}x - 4 \quad \text{[slope-intercept form]}$$ ■

Two lines that meet in a right angle (90° angle) are said to be **perpendicular.** As you might suspect, there is a close relationship between the slopes of two perpendicular lines.

Perpendicular Lines

> Two nonvertical lines are perpendicular exactly when the product of their slopes is -1.

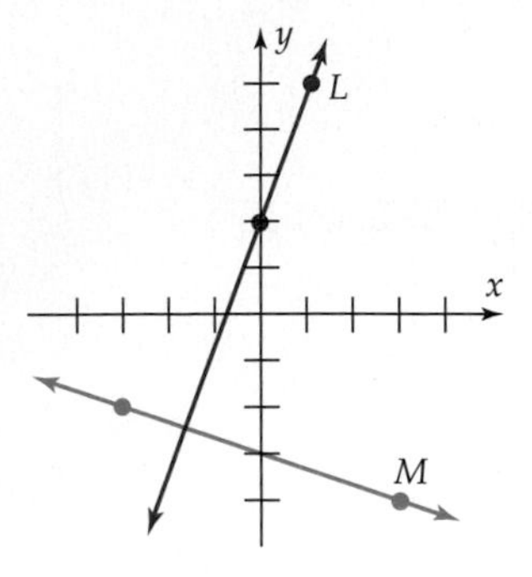

Figure 22

EXAMPLE 16

In Figure 22, the line L through (0, 2) and (1, 5) appears to be perpendicular to the line M through $(-3, -2)$ and $(3, -4)$. We can confirm this fact by computing the slopes of these lines:

$$\text{Slope } L = \frac{5-2}{1-0} = 3 \qquad \text{and} \qquad \text{slope } M = \frac{-4-(-2)}{3-(-3)} = \frac{-2}{6} = -\frac{1}{3}.$$

Since $3(-1/3) = -1$, the lines L and M are perpendicular. ■

EXERCISES

In Exercises 1–8, find the missing side(s) of the triangle.

1.

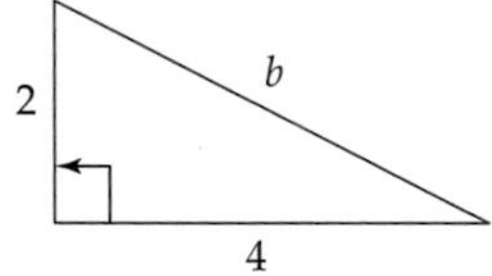

2.

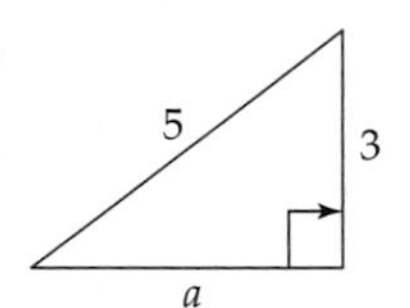

3.

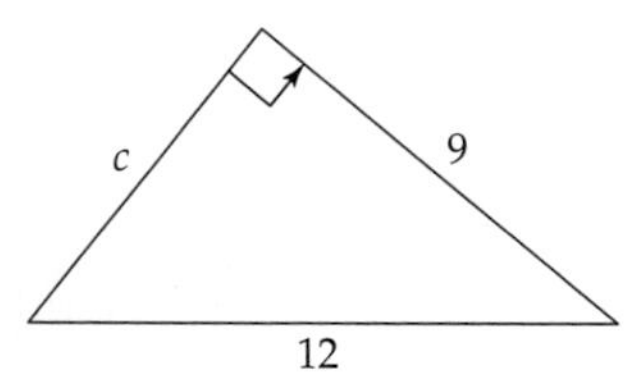

4.

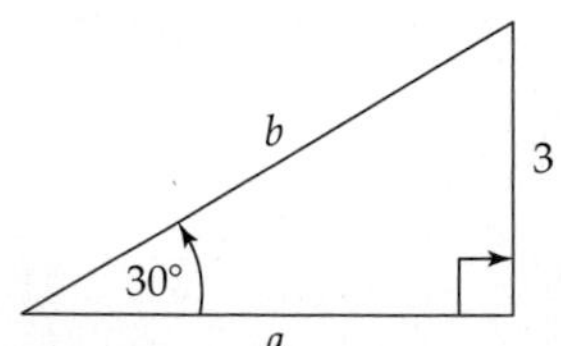

5.

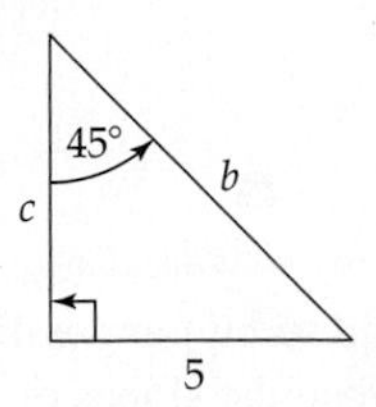

6.

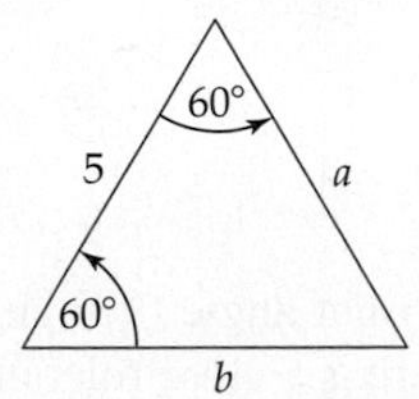

7.

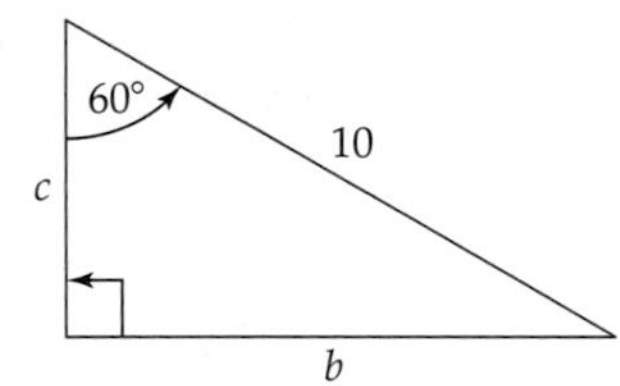

8.

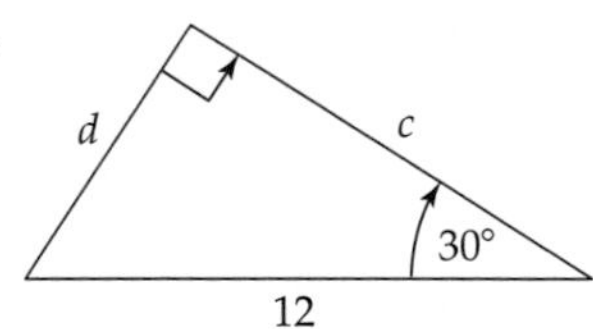

In Exercises 9 and 10, two congruent triangles are given. Find the missing sides and angles.

9.

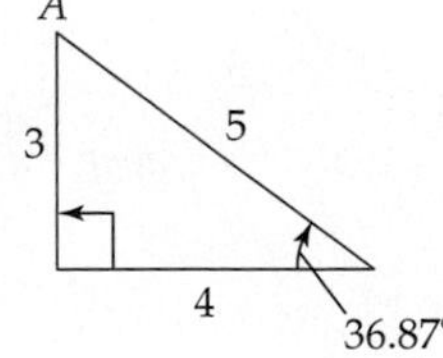

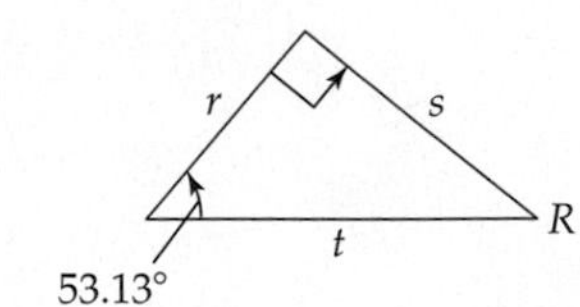

10.

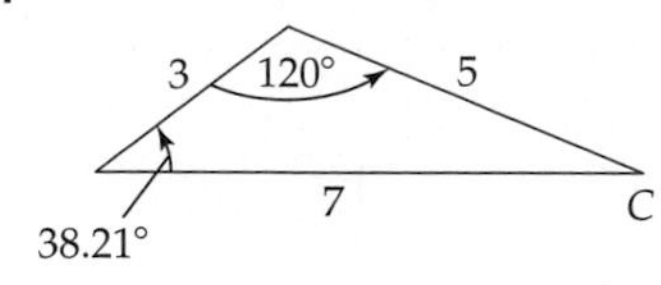

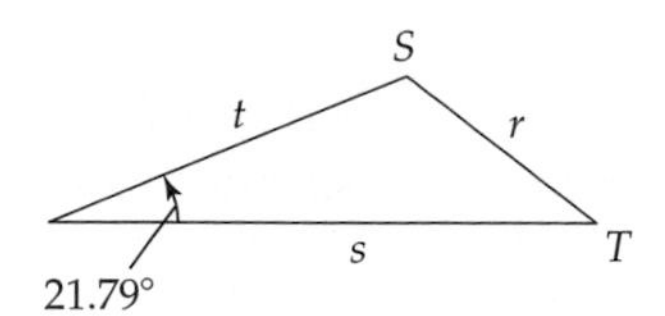

In Exercises 11 and 12, prove that the two triangles are congruent.

11.

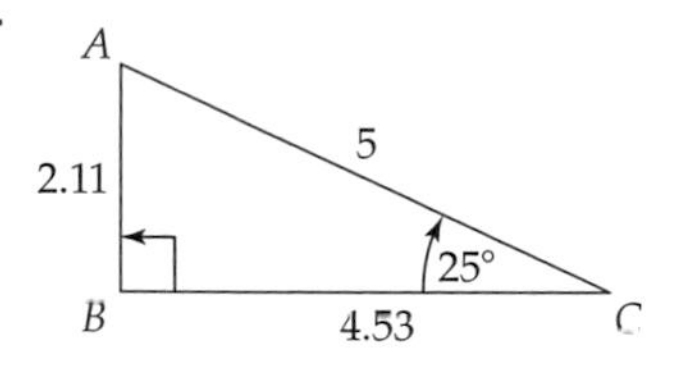

12.

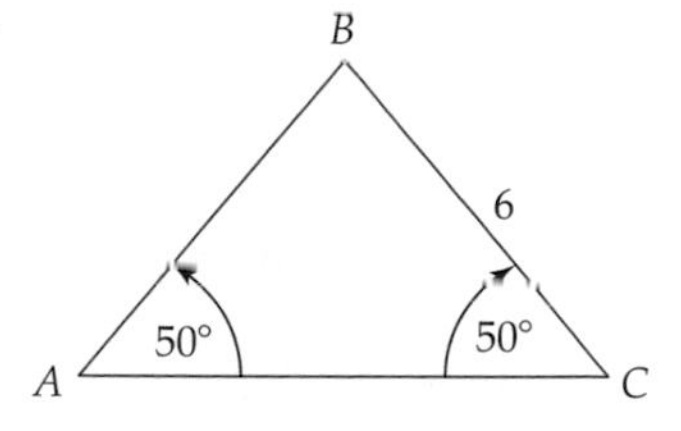

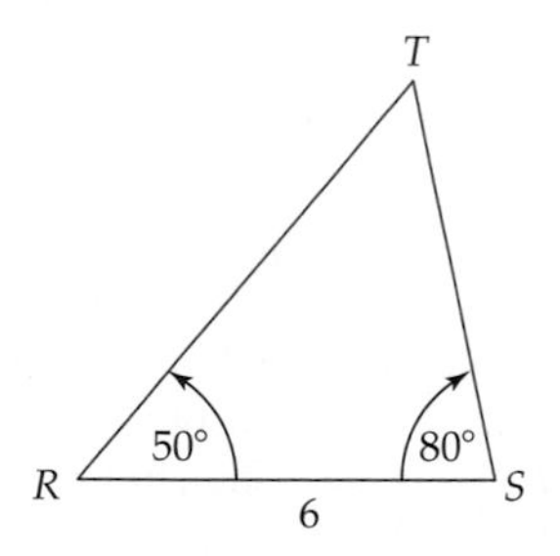

13. If $\angle u \cong \angle v$ and $OP \cong OQ$, show that $PD \cong DQ$ and that PQ is perpendicular to OD. [*Hint:* Prove that triangles POD and DOQ are congruent.]

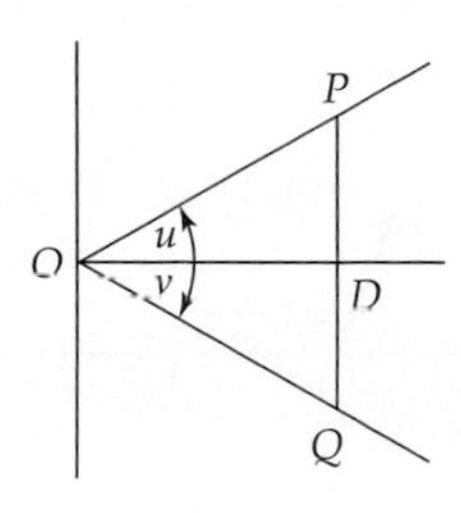

14. If $AB \cong AD$, $BC \cong DC$, $\angle x \cong \angle y$, and $\angle u \cong \angle v$, prove that triangles ADC and ABC are congruent.

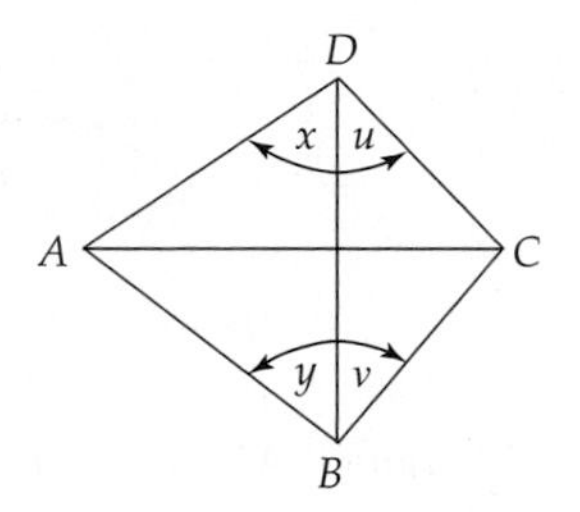

15. If $AB \cong DC$ and $\angle u \cong \angle v$, prove that triangles ABC and ACD are congruent.

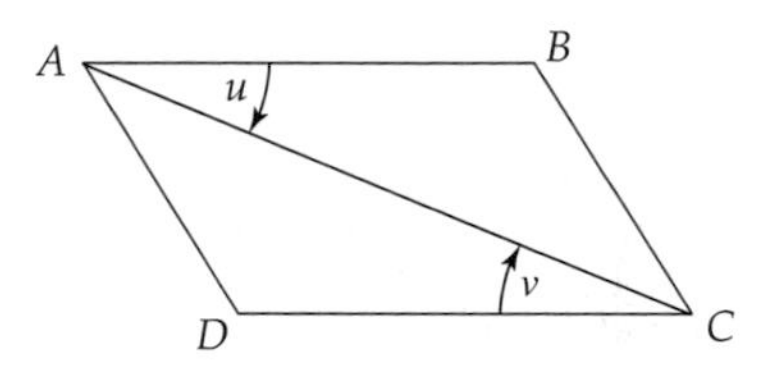

16. If $AB \cong CB$ and $\angle u \cong \angle v$, prove that triangles ABD and BCD are congruent.

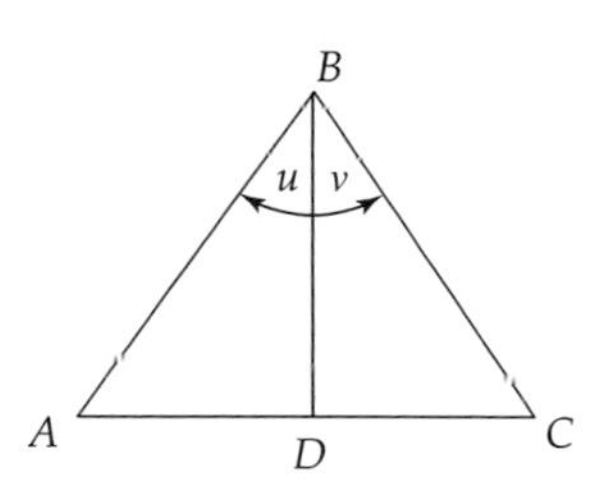

17. If $\angle x \cong \angle y$ and $\angle u \cong \angle v$, prove that triangles ABC and ADC are congruent.

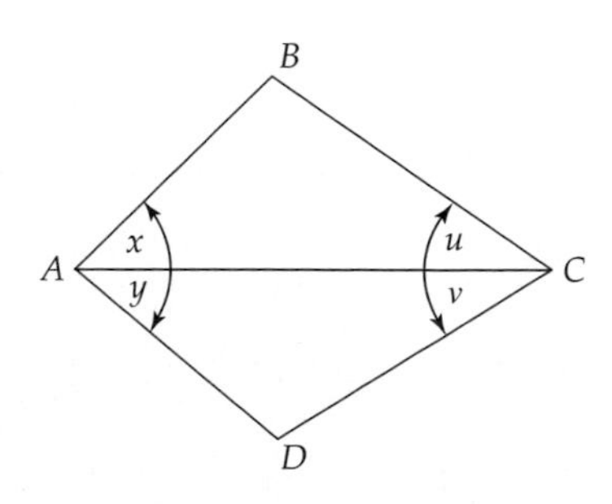

18. If $\angle x \cong \angle y$ and $\angle u \cong \angle v$, prove that triangles ABD and DBC are congruent.

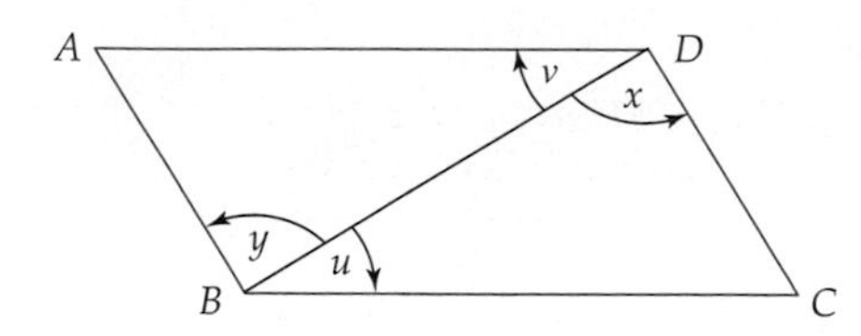

In Exercises 19–22, triangle ABC is similar to triangle RST. The following notation is used: The side opposite angle A is denoted a, the side opposite angle B is denoted b, the side opposite angle R is denoted r, and so on.

19. If $\angle B = 37°$, $\angle C = 90°$, $a = 6$, $b = 4.5$ and $\angle R = 53°$, $\angle T = 90°$, $s = 21$, find c, r, t.

20. If $\angle A = 90°$, $c = 220$ and $\angle T = 90°$, $\angle S \cong \angle B$, $r = 100$, $s = 50$, find a, b, t.

21. If $a = 7$, $b = 5$, $c = 6$, and $\angle A \cong \angle T$, $\angle C \cong \angle R$, $s = 2.5$, find r and t.

22. If $a = 12$, $b = 9$, $c = 4$, and $\angle B \cong \angle T$, $\angle C \cong \angle R$, $t = 6.3$, find r and s.

In Exercises 23–26, prove that the triangles are similar, and find the missing sides and angles.

23.

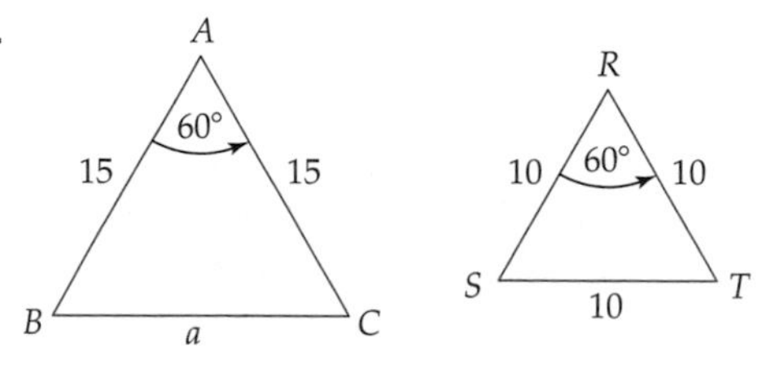

24.

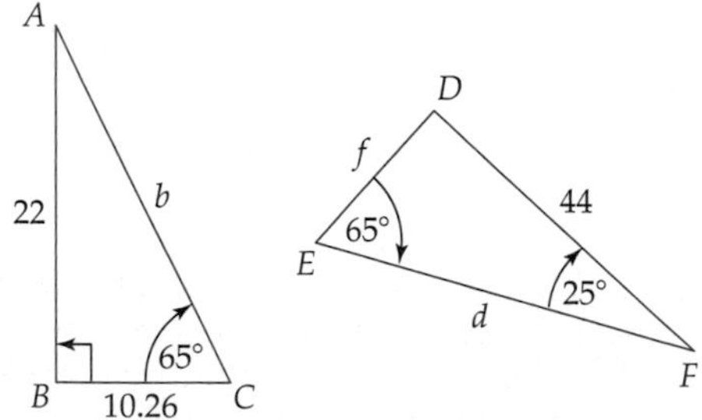

25.

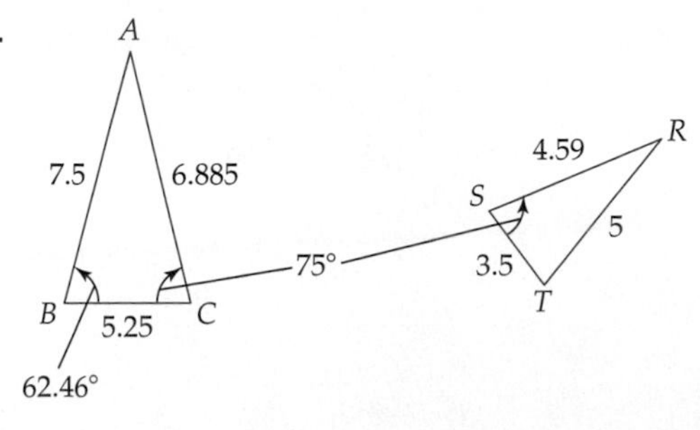

26.

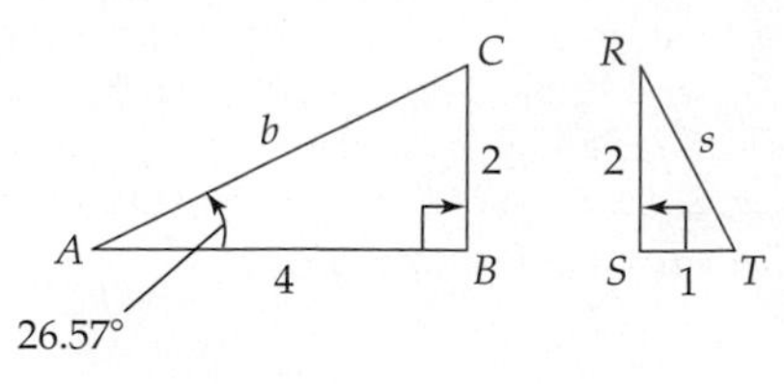

In Exercises 27–30, prove that triangles ABC and RST are similar, and find the missing sides and angles. The notation is the same as in Exercises 19–22.

27. $\angle C = 90°$, $a = 5$, $b = 5$ and $\angle T = 90°$, $r = 17.5$, $s = 17.5$.

28. $\angle B = (160/3)°$, $\angle C = 90°$, $a = 3.5$, $b = 4.7$ and $\angle R = (110/3)°$, $\angle T = 90°$, $r = 7$, $s = 9.4$

29. $\angle A = 105°$, $\angle B = 33.5°$, $a = 7$, $b = 4$, $c = 4.8$ and $\angle R = 105°$, $\angle T = 33.5°$, $r = 3.5$, $t = 2$.

30. $\angle B = 25°$, $\angle C = 142.8°$, $a = 5$, $b = 10$, $c = 14.3$, and $\angle S = 25°$, $\angle T = 12.2°$, $t = 10$

31. A flagpole casts a 99-foot shadow at the same time as a 5-foot-long stick (perpendicular to the ground) casts a 9-foot shadow. How high is the flagpole?

32. If a 6-foot-tall person casts a 10-foot shadow, how long is the shadow cast by a 28.3-foot-high tree that is next to the person?

33. A scale model of a pyramid is $1/20$ the size of the original. If the model pyramid is 4 feet tall, how tall is the original pyramid?

34. A scale model of a ship is 8 inches long, and its mast is 4.5 inches high. If the real ship is 65 feet long, how high is the real mast?

35. On a map of Ohio, Columbus is 9.2 centimeters from Cincinnati and 8.5 centimeters from Marietta. On the map, Cincinnati is 15.75 centimeters from Marietta. If the actual distance from Columbus to Cincinnati is 100 miles, how far is Marietta from Columbus and from Cincinnati?

36. A small mirror is placed on level ground 4 feet from the base of a flagpole. A person whose eyes are 66 inches above the ground sees the top of the pole in the mirror when he is 2 feet from the mirror. The base of the pole, the mirror, and the person are aligned on a straight line. How high is the flagpole?

37. For which of the line segments in the figure is the slope
(a) largest? (b) smallest?
(c) largest in absolute value? (d) closest to zero?

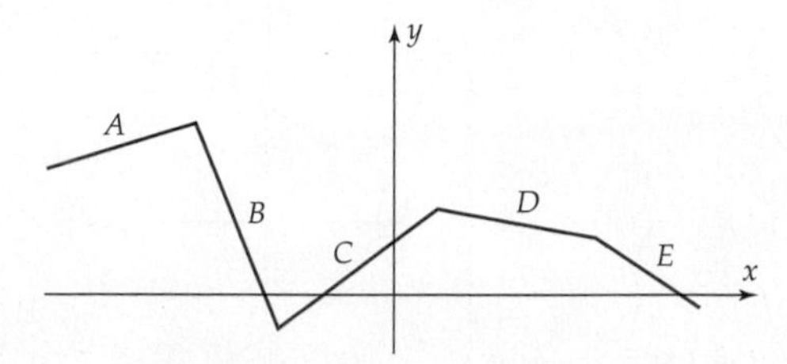

38. The doorsill of a campus building is 5 feet above ground level. To allow wheelchair access, the steps in front of the door are to be replaced by a straight ramp with constant slope $1/12$, as shown in the figure. How long must the ramp be? [The answer is *not* 60 feet.]

In Exercises 39–42, find the slope of the line through the given points.

39. $(1, 2); (3, 7)$
40. $(-1, -2); (2, -1)$
41. $(1/4, 0); (3/4, 2)$
42. $(\sqrt{2}, -1); (2, -9)$

In Exercises 43–46, find a number t such that the line passing through the two given points has slope −2.

43. $(0, t); (9, 4)$
44. $(1, t); (-3, 5)$
45. $(t + 1, 5); (6, -3t + 7)$
46. $(t, t); (5, 9)$

In Exercises 47–50, find the slope and y-intercept of the line whose equation is given.

47. $2x - y + 5 = 0$
48. $3x + 4y = 7$
49. $3(x - 2) + y = 7 - 6(y + 4)$
50. $2(y - 3) + (x - 6) = 4(x + 1) - 2$

In Exercises 51–54, find the equation of the line with slope m that passes through the given point.

51. $m = 1; (3, 5)$
52. $m = 2; (-2, 1)$
53. $m = -1; (6, 2)$
54. $m = 0; (-4, -5)$

In Exercises 55–58, find the equation of the line through the given points. [Hint: *First, find the slope.*]

55. $(0, -5)$ and $(-3, -2)$
56. $(4, 3)$ and $(2, -1)$
57. $(4/3, 2/3)$ and $(1/3, 3)$
58. $(6, 7)$ and $(6, 15)$

In Exercises 59–62, determine whether the line through P and Q is parallel or perpendicular to the line through R and S or neither.

59. $P = (2, 5), Q = (-1, -1)$ and $R = (4, 2), S = (6, 1)$.
60. $P = (0, 3/2), Q = (1, 1)$ and $R = (2, 7), S = (3, 9)$.
61. $P = (-3, 1/3), Q = (1, -1)$ and $R = (2, 0)$, $S = (4, -2/3)$.
62. $P = (3, 3), Q = (-3, -1)$ and $R = (2, -2)$, $S = (4, -5)$.

In Exercises 63–64, determine whether the lines whose equations are given are parallel, perpendicular, or neither.

63. $2x + y - 2 = 0$ and $4x + 2y + 18 = 0$.
64. $3x + y - 3 = 0$ and $6x + 2y + 17 = 0$.

In Exercises 65–72, find an equation for the line satisfying the given conditions.

65. Through $(-2, 1)$ with slope 3.
66. y-intercept -7 and slope 1.
67. Through $(2, 3)$ and parallel to $3x - 2y = 5$.
68. Through $(1, -2)$ and perpendicular to $y = 2x - 3$.
69. x-intercept 5 and y-intercept -5.
70. Through $(-5, 2)$ and parallel to the line through $(1, 2)$ and $(4, 3)$.
71. Through $(-1, 3)$ and perpendicular to the line through $(0, 1)$ and $(2, 3)$.
72. y-intercept 3 and perpendicular to $2x - y + 6 = 0$.

Thinkers

73. If ABC is a triangle, prove that the sum of its angles is 180°. [*Hint:* In the figure below, use congruent triangles to show that $\angle u \cong \angle x$ and $\angle v \cong \angle y$. You may assume that opposite sides of a rectangle have the same length.]

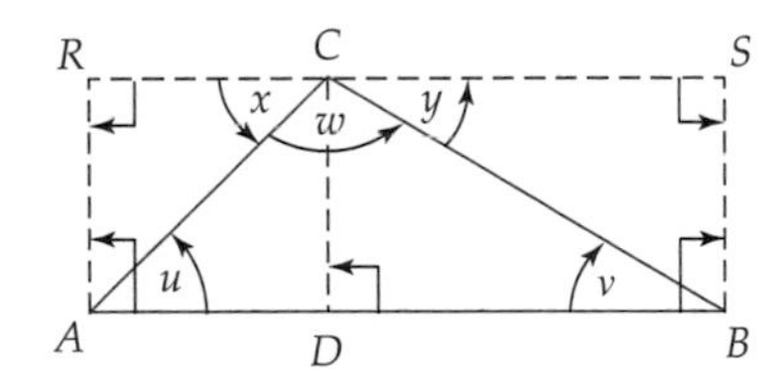

74. Prove the Pythagorean Theorem. [*Hint:* Use the figure below, in which triangle ABC has a right angle at C. Find three similar triangles in the figure, and use the properties of similar triangles to show that $a^2 + b^2 = c^2$.]

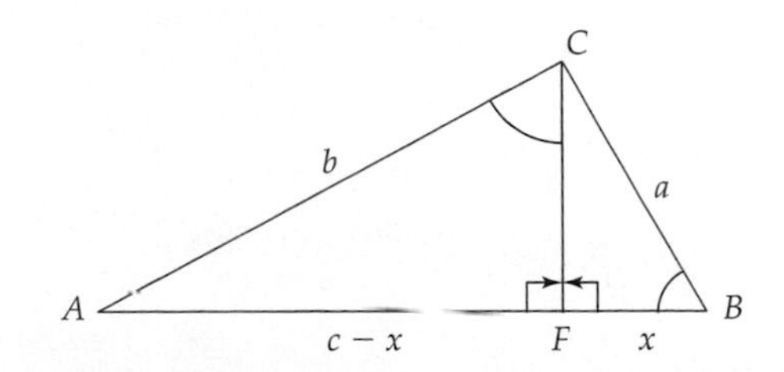

ANSWERS TO ODD-NUMBERED EXERCISES

CHAPTER 0

Section 0.1, page 11

1.

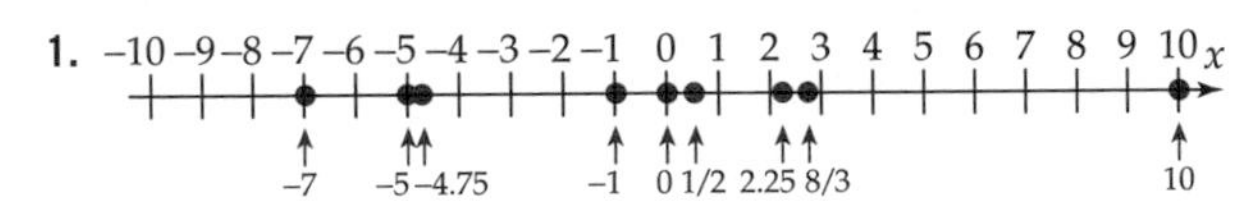

3. $12 > 9$ **5.** $-4 > -8$ **7.** $\pi < 100$

9. $y \leq 7.5$ **11.** $t > 0$ **13.** $c \leq 3$

15. $-6 < -2$ **17.** $3/4 = .75$ **19.** $1/3 > .33$

21. 5 **23.** 6 **25.** 0

27. $\frac{2040}{523}, \frac{189}{37}, \sqrt{27}, \frac{4587}{691}, 6.735, \sqrt{47}$

29. $a = b + c$

31. a lies to the right of b.

33. $b - a > 0$ *or* $a < b$.

35. 0 **37.** 10 **39.** 169

41. $\pi - \sqrt{2}$

43. $|-2| < |-5|$ **45.** $|3| > -|4|$

47. $-7 < |-1|$ **49.** 7

51. $29/2$ **53.** $\pi - 3$

55. $|316.3 - 408.6|$

57. For 5 ft 8 in., $|x - 143| \leq 21$; for 6 ft 0 in., $|x - 163| \leq 26$.

59. $|21 - (-12)| = 33$ **61.** 2000, 2003, 2004

63. 3 or 5 **65.** -2 or 7

67. -3 or $1/2$ **69.** -2 or $-1/4$

71. $-4/3$ or 1 **73.** $-4/3$ or $1/4$

75. $2 \pm \sqrt{3}$ **77.** $-3 \pm \sqrt{2}$

79. no real number solutions

81. $\frac{1}{2} \pm \sqrt{2}$ **83.** $1 \pm \frac{\sqrt{3}}{2}$

85. -6 or -3 **87.** $\frac{-1 \pm \sqrt{2}}{2}$ **89.** $-3/2$ or 5

91. no real number solutions **93.** no real number solutions

95. approximately 1.8240 or .4701

97. (a) 2116.1 pounds per square foot, 670.5229 pounds per square foot
(b) 14,410 feet

99. 1960

101. (a) About 27.4 miles
(b) About 46.7 miles
(c) About 416.663 ft

103. about 1.753 ft **105.** 2 m

107. about 6.32 seconds. The time in Example 7 is less because in that example the ball is thrown down.

109. (a) about 4.38 seconds
(b) after 50 seconds

111. If all of a, b, and c were zero, then by property 1, $|a| = 0$, $|b| = 0$, and $|c| = 0$, so $|a| + |b| + |c|$ would be zero.

Section 0.2, page 21

1. $A(-3, 3)$; $B(-1.5, 3)$; $C(-2.5, 0)$; $D(-1.5, -3)$; $E(0, 2)$; $F(0, 0)$; $G(2, 0)$; $H(3, 1)$; $I(3, -1)$

3. $(-6, 3)$ **5.** $(4, 2)$

7. (a)

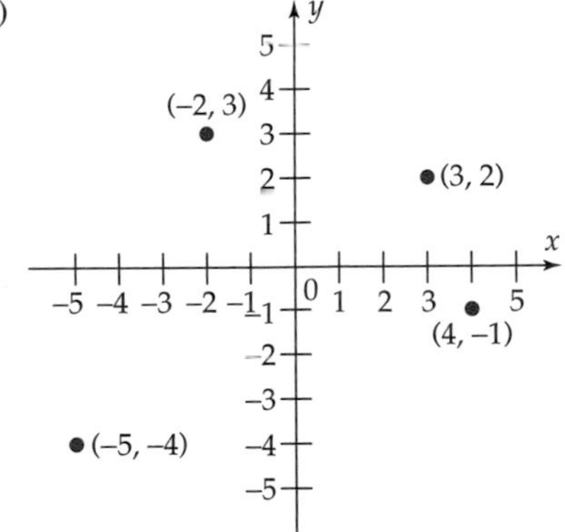

(b)

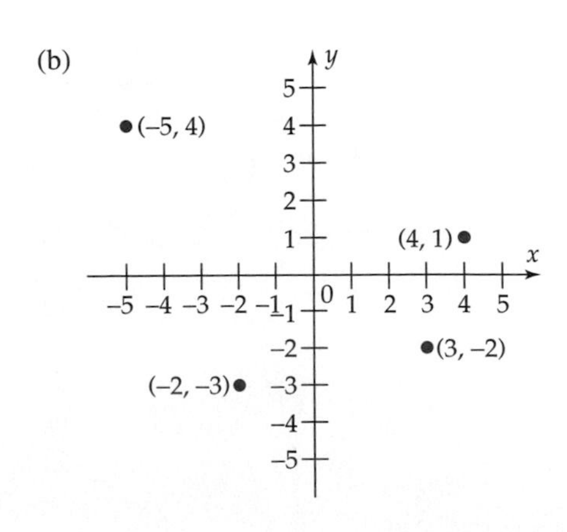

(c) The points (a, b) and $(a, -b)$ are on opposite sides of, and the same distance from, the x-axis.

9. $13; (-1/2, -1)$ **11.** $\sqrt{17}; (3/2, -3)$

13. $\sqrt{6 - 2\sqrt{6}}; \left(\frac{\sqrt{2}+\sqrt{3}}{2}, \frac{3}{2}\right)$

15. $|a - b|\sqrt{2}; \left(\frac{a+b}{2}, \frac{a+b}{2}\right)$

17. $13 + 2\sqrt{2} + \sqrt{13} \approx 19.434$

19. The distances are $(0, 0)$ to $(1, 1)$: $\sqrt{2}$; $(0, 0)$ to $(2, -2)$: $2\sqrt{2}$; $(1, 1)$ to $(2, -2)$: $\sqrt{10}$. But $(\sqrt{2})^2 + (2\sqrt{2})^2 = 2 + 8 = 10$ and $(\sqrt{10})^2 = 10$, so the triangle is a right triangle with hypotenuse $\sqrt{10}$.

21. (a)

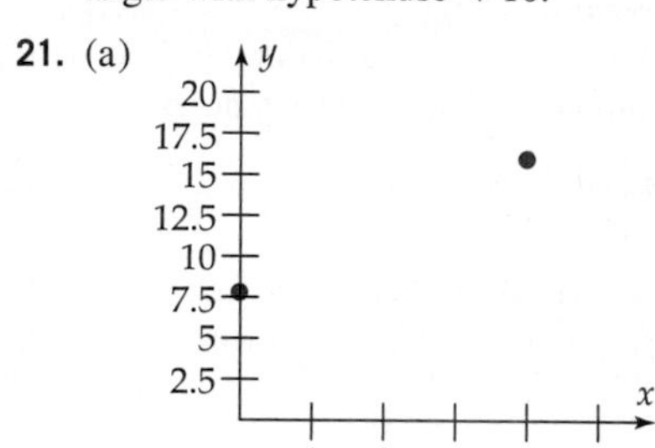

Year 2000 is represented by 0

(b) $(2, 11.9)$

(c) The midpoint might be interpreted to project sales of organically grown food in the United States to be \$11.9 billion in 2002. This requires the assumption of linear growth.

23. (a) QB$(20, 10)$; R$\left(48\frac{1}{3}, 45\right)$; distance ≈ 45.03 yd
(b) $\left(34\frac{1}{6}, 27\frac{1}{2}\right)$

25. yes **27.** yes **29.** no

31. $(x + 3)^2 + (y - 4)^2 = 4$ **33.** $x^2 + y^2 = 2$

35. $(x - 5)^2 + (y + 2)^2 = 5$

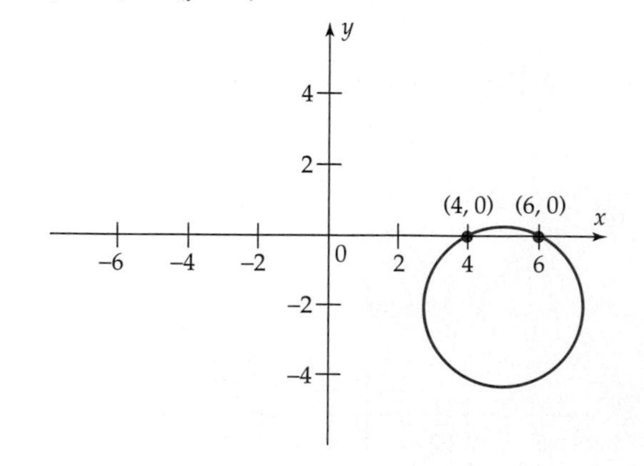

37. $(x + 1)^2 + (y - 3)^2 = 9$

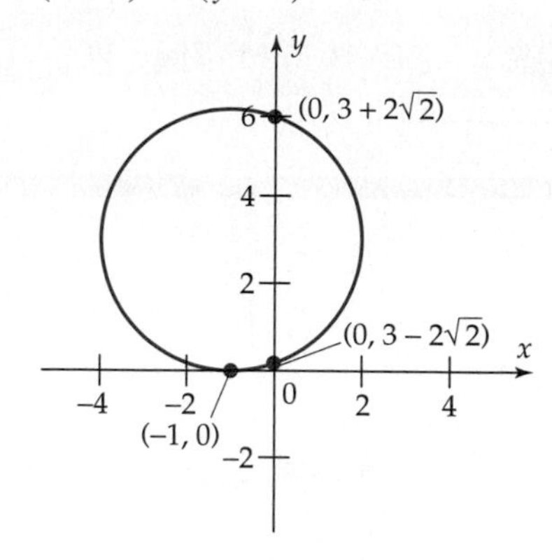

39. $(x - 2)^2 + (y - 2)^2 = 8$

41. $(x - 1)^2 + (y - 2)^2 = 8$

43. $(x + 5)^2 + (y - 4)^2 = 16$

45. $(x - 2)^2 + (y - 1)^2 = 5$

47. $(-3, -4)$ and $(2, 1)$

49. $x = 6$

51. (a) about 7,250,000
(b) about 2007 and 7,750,000
(c) before 2003 and after 2011

53. (a) under 56; (b) age 56;
(c) retiring at age 60: about \$30,000; retiring at age 65: about \$35,000.

55. B **57.** C

59. Using the coordinates suggested by the hint: The coordinates of M are $\left(\frac{0+s}{2}, \frac{r+0}{2}\right) = \left(\frac{s}{2}, \frac{r}{2}\right)$. The distances from M to the two endpoints are $\sqrt{\left(0 - \frac{s}{2}\right)^2 + \left(r - \frac{r}{2}\right)^2} = \frac{\sqrt{r^2+s^2}}{2}$ and $\sqrt{\left(\frac{s}{2} - s\right)^2 + \left(\frac{r}{2} - 0\right)^2} = \frac{\sqrt{r^2+s^2}}{2}$.
The distance from M to the origin is $\sqrt{\left(0 - \frac{s}{2}\right)^2 + \left(0 - \frac{r}{2}\right)^2} = \frac{\sqrt{r^2+s^2}}{2}$. The distances are identical.

61. The circles are all those that are tangent to the y-axis and have center on the x-axis.

63. The midpoint of the segment joining (c, d) and $(-c, -d)$ is the point $\left(\frac{c+(-c)}{2}, \frac{d+(-d)}{2}\right)$, which is $(0, 0)$, the origin. The points (c, d) and $(-c, -d)$ thus lie on a straight line through the origin and equidistant from the origin. Since the x- and y-coordinates both have opposite signs, the points are on opposite sides of the origin.

Section 0.3, page 37

1. $P(3, 5)$, $Q(-6, 2)$

3. $P(-12, 8)$, $Q(-3, -9)$

5.

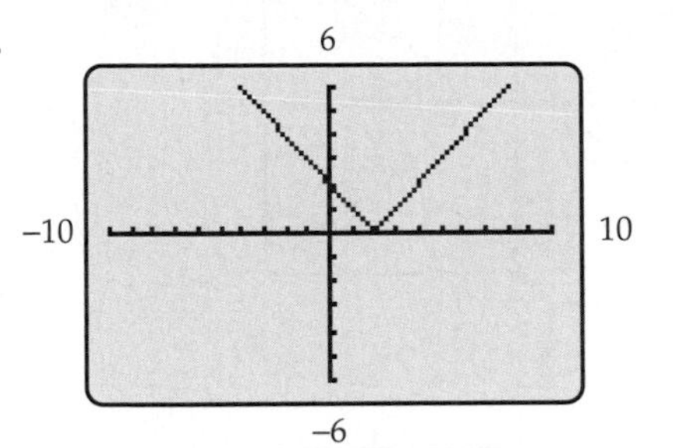

7.

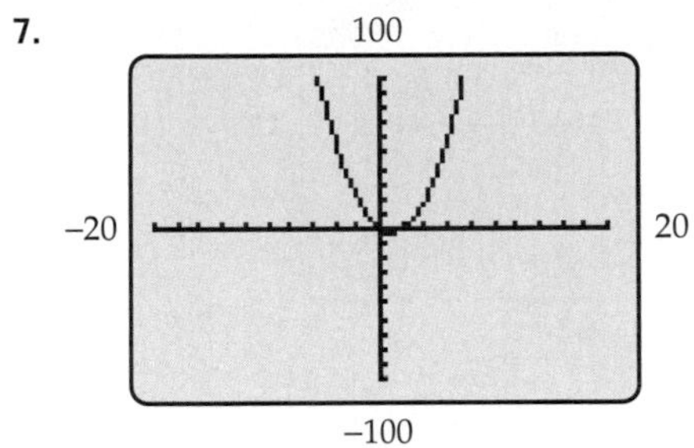

9.

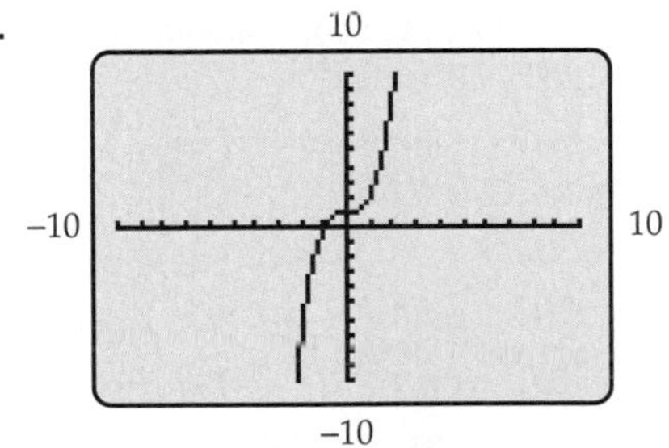

11.

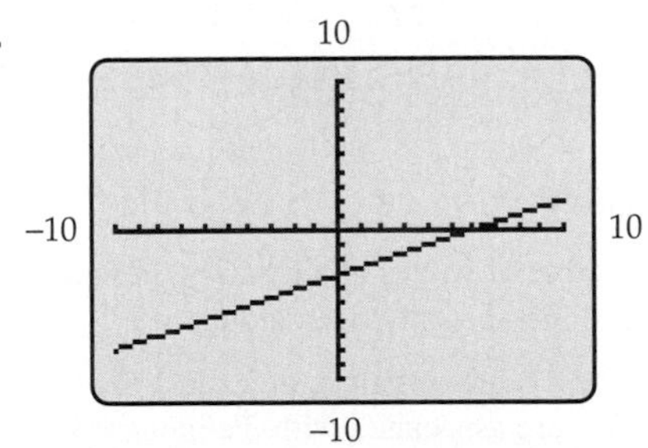

13.

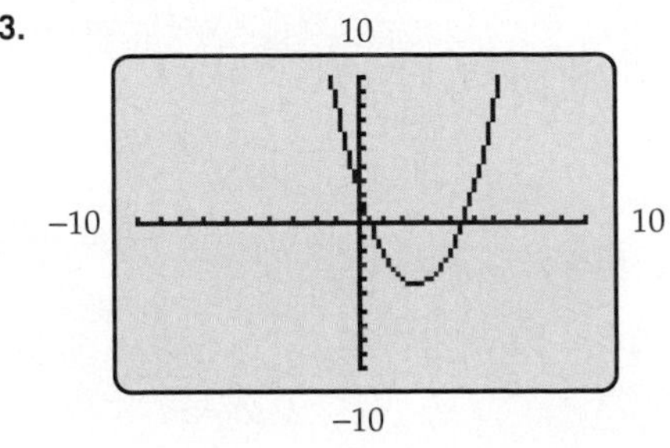

15.

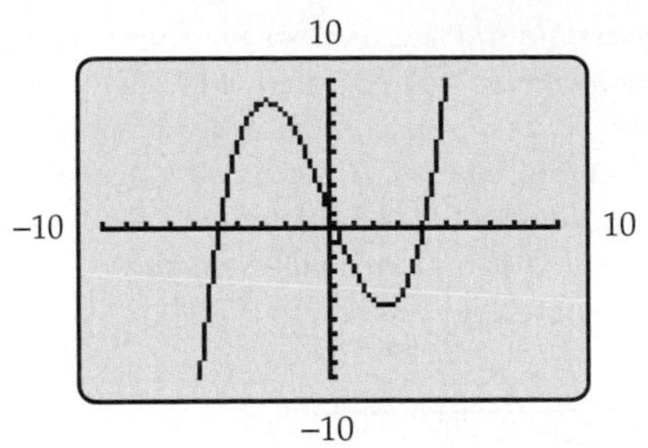

17. c

19. d

21. e

23. (c)

25. (a)

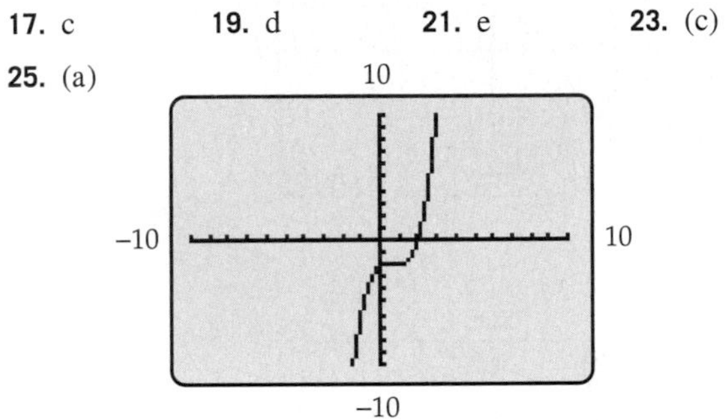

(b)

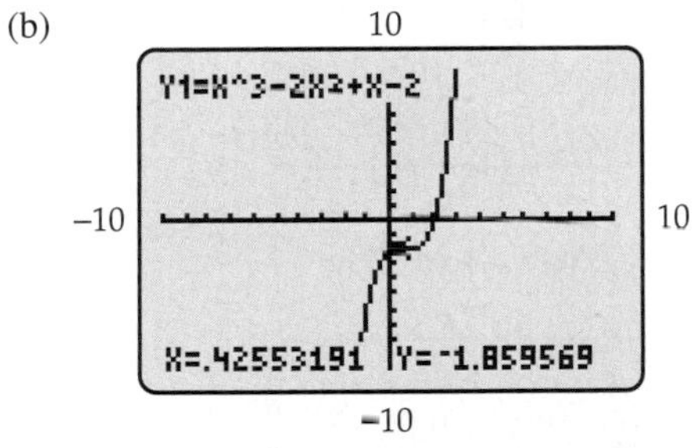

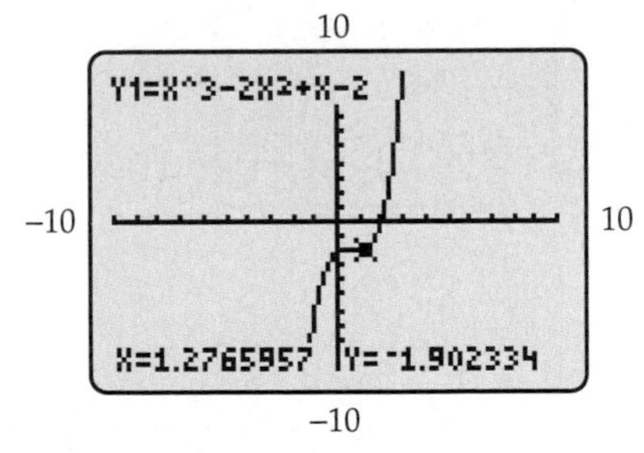

(c)

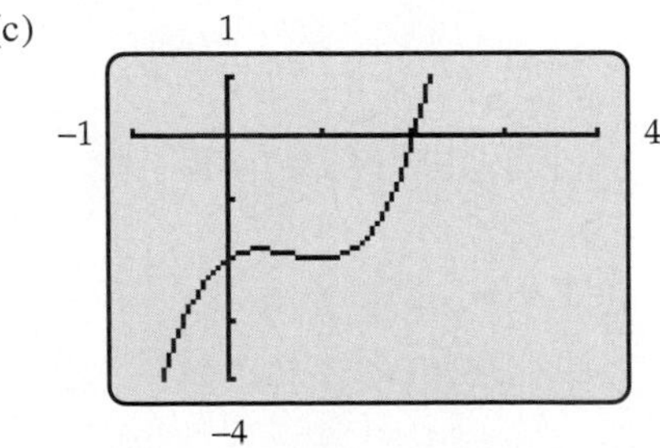

There are many possible square windows for Exercises 27–36; these were made on a TI-83+. To have a square window with the same y-axis shown here on wider screen calculators (such as TI-86/89), the x-axis should be longer than the one shown here. Because of the calculator's limited resolution, some graphs that should be connected (such as circles) may appear to have breaks in them.

27.

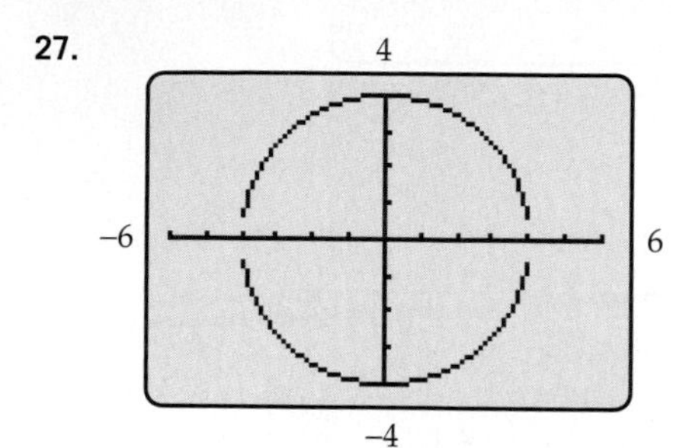

29.

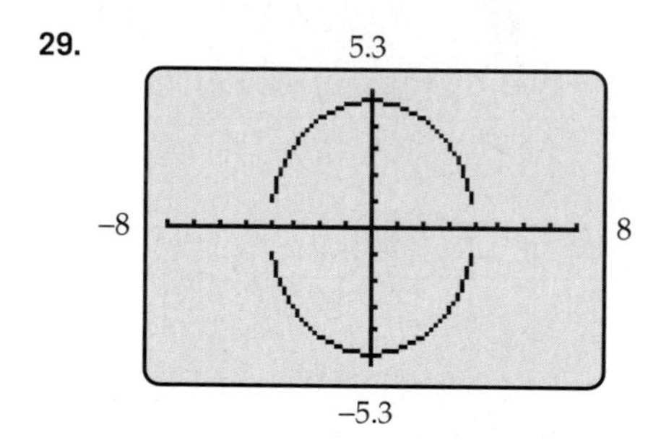

31.

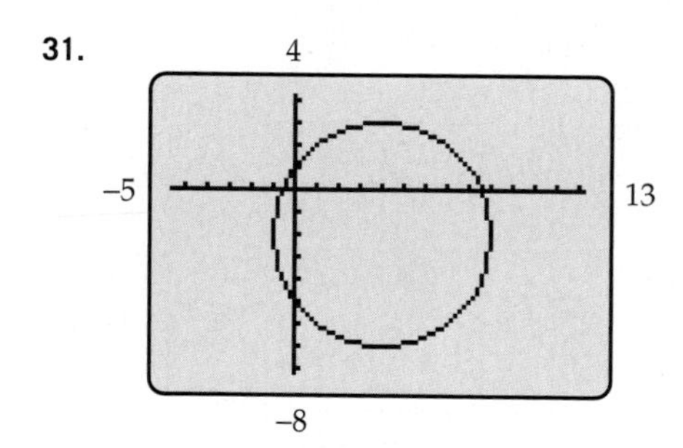

33.

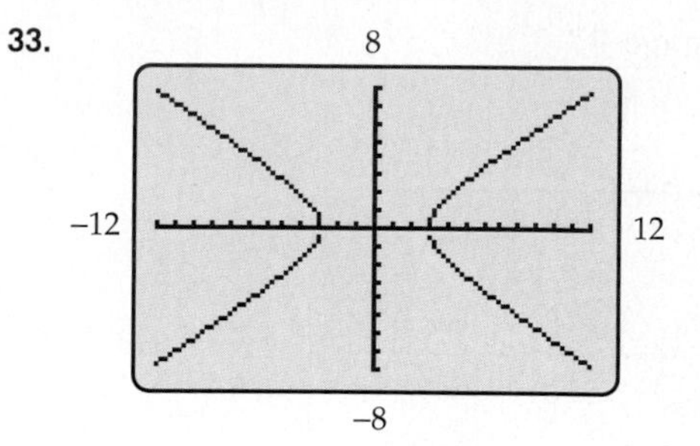

35.

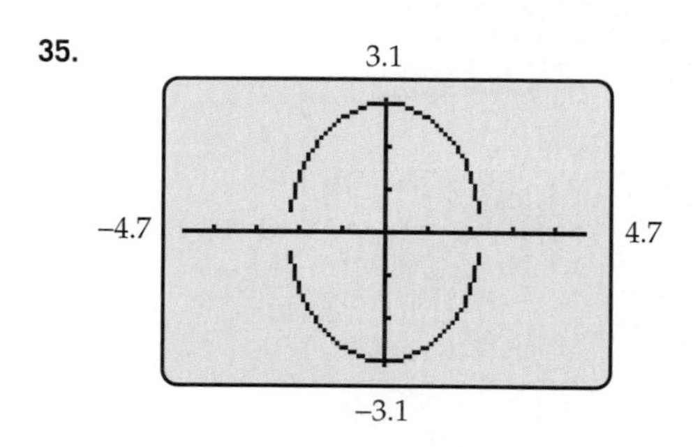

37. Maximum at about (.7922, 4.4849);
Minimum at about (4.2078, -3.4849)

39. The model suggests a maximum population of 613,230 in the year 1961 or 1962.

41. The maximum number of subscribers, about 22.5 million, occurred in 2004.

43. 3 **45.** 3 **47.** 2

49. -2.4265 **51.** 1.1640 **53.** 1.2372

55. -1.60051 **57.** 2.1017 **59.** -1.7521

61. .9505 **63.** 0, 2.2074

65. 2.3901 **67.** $-.6514$, 1.1514

69. 7.0334 **71.** 2006

Section 0.4, page 48

1. This could be a table of values of a function, because each input determines one and only one output.

3. This could not be a table of values of a function, because two output values are associated with the input -5.

5. This defines y as a function of x.

7. This defines x as a function of y.

9. This defines both y as a function of x and x as a function of y.

11. Neither

13.

X	Y1
-2	-2
-1.5	-3.25
-1	-4
-.5	-4.25
0	-4
.5	-3.25
1	-2

X= -2

X	Y1
1.5	-.25
2	2
2.5	4.75
3	8
3.5	11.75
4	16
4.5	20.75

X=4.5

15.

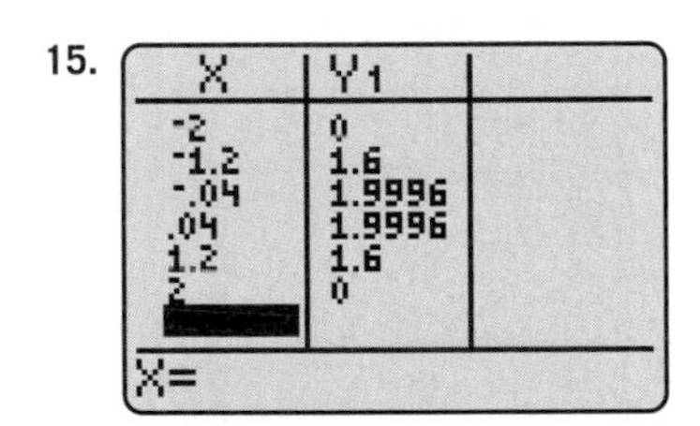

17. Postage is a function of weight, but weight is not a function of postage; for example, all letters less than one ounce use the same postage amount.

19. (a) Average man:

Drinks in 1 hour	2	3	4	5
Blood alcohol content	.03	.05	.07	.10

Average woman:

Drinks in 1 hour	2	3
Blood alcohol content	.05	.08

(b) These tables define functions. For the average man, the domain consists of all x such that $2 \le x \le 5$, and the range of all y such that $.03 \le y \le .10$. For the average woman, the domain consists of all x such that $2 \le x \le 3$, and the range of all y such that $.05 \le y \le .08$.

21. (a) $A = \pi r^2$ (b) $A = \dfrac{\pi d^2}{4}$

23. $V = 4x^3$

25. (a) $-.8$ (b) $-.75$ (c) $-.4$ (d) $-.125$ (e) 0

27. $\sqrt{3} + 1$

29. $\sqrt{\sqrt{2} + 3} - \sqrt{2} + 1$

31. 4

33. 34/3

35. 59/12

37. $(a + k)^2 + \dfrac{1}{a + k} + 2$

39. $6 - 4x + x^2 + \dfrac{1}{2 - x}$

41. 8

43. -1

45. $s^2 + 2s$

47. $t^2 - 1$

49. -2

51. 2

53. (a) $f(a) = a^2, f(b) = b^2, f(a + b) = (a + b)^2 = a^2 + 2ab + b^2$, so $(a + b)^2 \neq a^2 + b^2$, that is, $f(a + b) \neq f(a) + f(b)$
(b) $f(a) = 3a, f(b) = 3b, f(a + b) = 3(a + b) = 3a + 3b = f(a) + f(b)$.

55. all real numbers

57. all real numbers

59. $x \ge 0$

61. all real numbers except 0

63. all real numbers

65. all real numbers except 3 and -2

67. $6 \le x \le 12$

69. Many examples including $f(x) = x^2$ and $g(x) = x^4$

71. $f(x) = \sqrt{2x - 5}$

73. $f(x) = \dfrac{x^3 + 6}{5}$

75. (a) The first model appears to be the best for the given data.
(b) 56.6 million

77. (a) $f(x) = 200 + 5x$ $g(x) = 10x$
(b)

x	20	35	50
$f(x)$	300	375	450
$g(x)$	200	350	500

(c) 45 suits

79. (a) $y = 108 - 4x$
(b) $V(x) = 108x^2 - 4x^3$

81. $x^3 - 3x + 2$; $-x^3 - 3x + 2$; $x^3 + 3x - 2$

83. $x^2 + 2x - 5 + \frac{1}{x}$; $\frac{1}{x} - x^2 - 2x + 5$; $x^2 + 2x - 5 - \frac{1}{x}$

85. $-3x^4 + 2x^3$; $\frac{-3x + 2}{x^3}$; $\frac{x^3}{-3x + 2}$

87. $(x^2 - 3)\sqrt{x - 3}$; $\frac{x^2 - 3}{\sqrt{x - 3}}$; $\frac{\sqrt{x - 3}}{x^2 - 3}$

89. all real numbers except 0; all real numbers except 0

91. $-4/3 \le x \le 2$; $-4/3 < x \le 2$

93. 0

95. 30

Section 0.5, page 57

1. yes, 0

3. no

5.

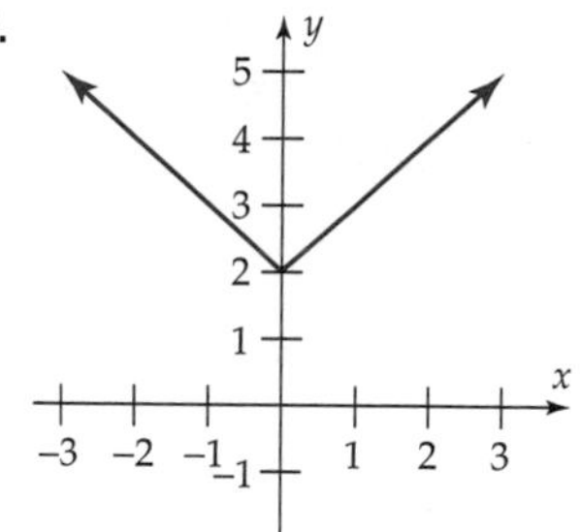

7.

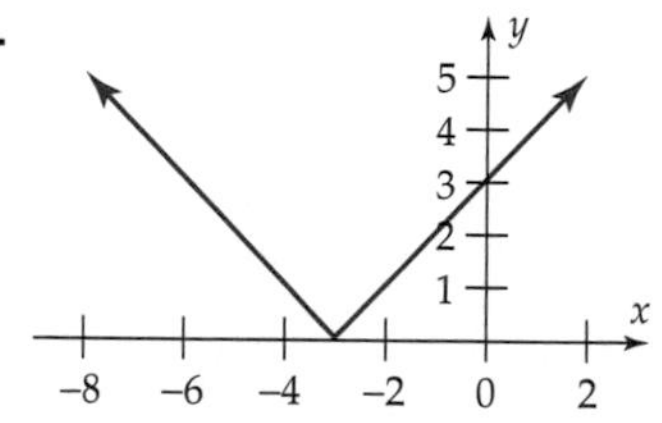

9. maximum at $(-.57735, .3849)$, minimum at $(.57735, -.3849)$

11. maximum at $(1, .5)$, minimum at $(-1, -.5)$

13. maximum at $(.4367, 2.1767)$, minimum at $(.7633, 2.1593)$

15. increasing when $-2.5 < x < 0$ and when $1.7 < x < 4$; decreasing when $-6 < x < -2.5$ and when $0 < x < 1.7$

17. constant when $x < -1$ and when $x > 1$; decreasing when $-1 < x < 1$

19. decreasing when $x < -5.7936$ and when $x > .4603$; increasing when $-5.7936 < x < .4603$

21. decreasing when $x < 0$ and when $.867 < x < 2.883$; increasing when $0 < x < .867$ and when $x > 2.883$

23. One of many possible graphs is shown.

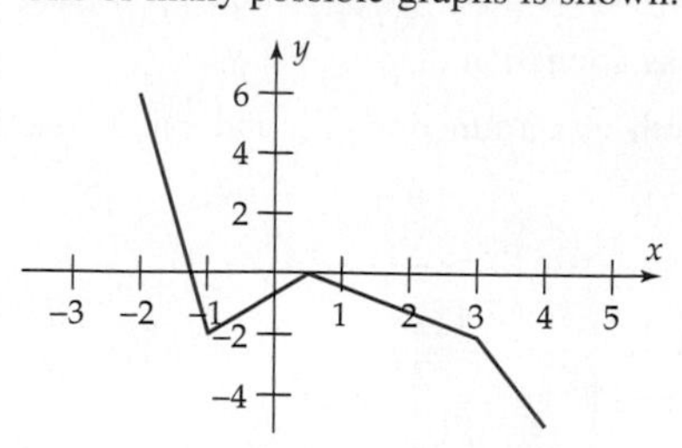

25. (a) $2x + 2z = 100$
(b) $A(x) = x(50 - x)$
(c)

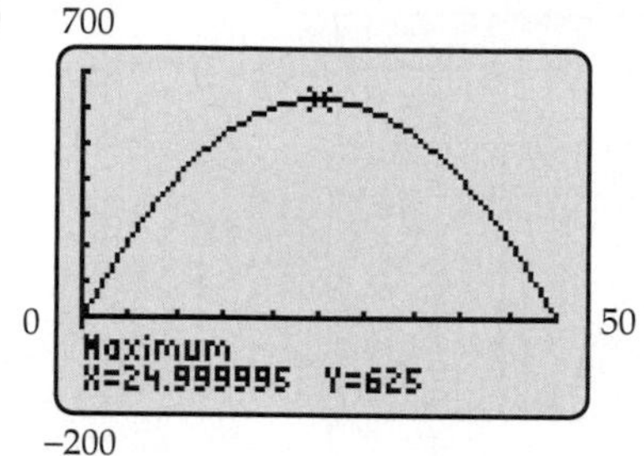

$x = 25$ inches, $z = 25$ inches

27. (a) $S = 2x^2 + 4xh$
(b) $x^2h = 867$
(c) $S = 2x^2 + \frac{3468}{x}$
(d)

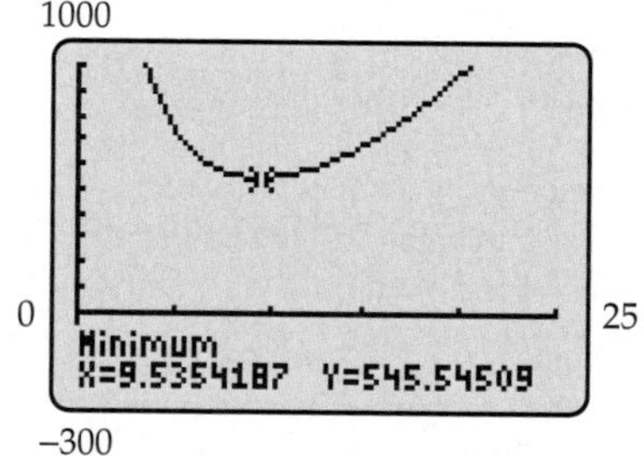

$x = 9.5354$ inches, $h = 9.5354$ inches

29. (a) (iv) (b) (i) (c) (v) (d) (iii) (e) (ii)

In Exercises 30–32, many correct answers are possible, one of which is given here.

31.

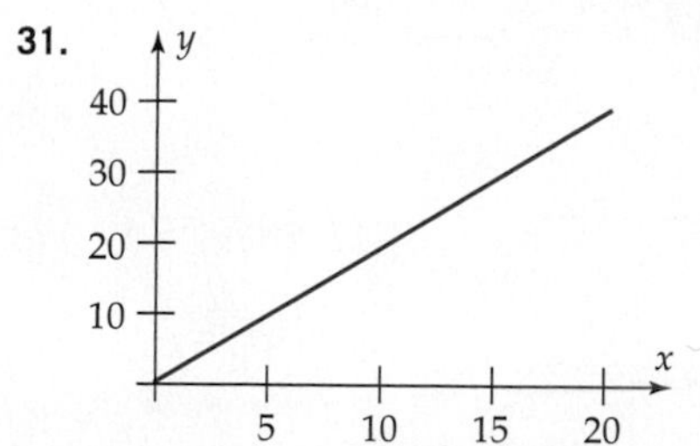

Domain is $x \ge 0$, range is $y \ge 0$.

33.

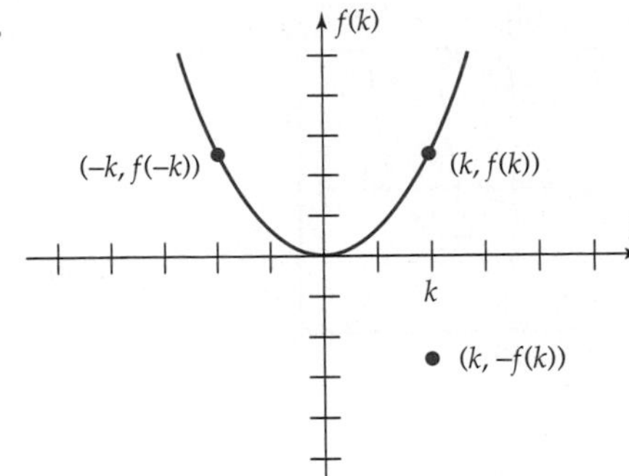

35. (a) 11% (b) 8% (c) 6%
(d) lowest: (late) 1993; highest: 1990
(e) 1980–1982, graph has steepest slope

37. $-3 \leq x \leq 5$ **39.** 4 **41.** 3.5

43. 4.5 **45.** 1 and 5

47. 39, 75 **49.** $12 \leq x \leq 20$

51. $\sqrt{4 - x^2}$

53. $-10 \leq x \leq 30$ and $-8 \leq y \leq 3$ $(-10 \leq t \leq 10)$

55. $-10 \leq x \leq 10$ and $-10 \leq y \leq 10$ $(-10 \leq t \leq 10)$

57. $-30 \leq x \leq 0$ and $0 \leq y \leq 10$ $(0 \leq t \leq 10)$

59. $-10 \leq x \leq 45$ and $-20 \leq y \leq 20$

61. $-2 \leq x \leq 32$ and $-10 \leq y \leq 75$; near y-axis: $-2 \leq x \leq 5$ and $-10 \leq y \leq 10$

63. $-15 \leq x \leq 2$ and $-25 \leq y \leq 75$; near y-axis: $-2 \leq x \leq 2$ and $-4 \leq y \leq 2$

65. During the next five minutes, she rested, then continued her run at her original pace for ten minutes. She then ran home at this pace, a 25-minute run.

67. seven times

Section 0.6, page 68

1. H **3.** F **5.** K **7.** C

9.

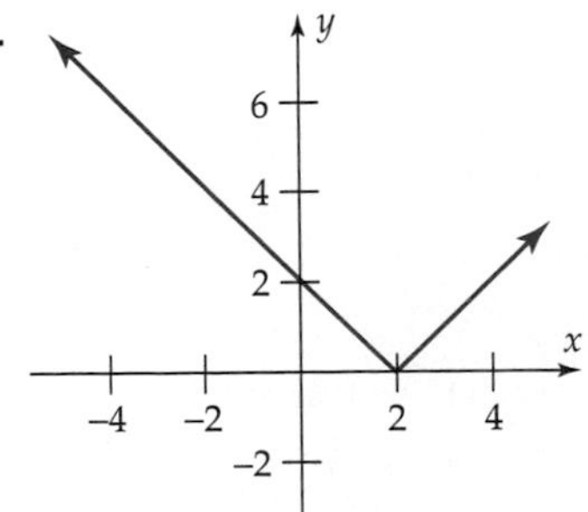

11.

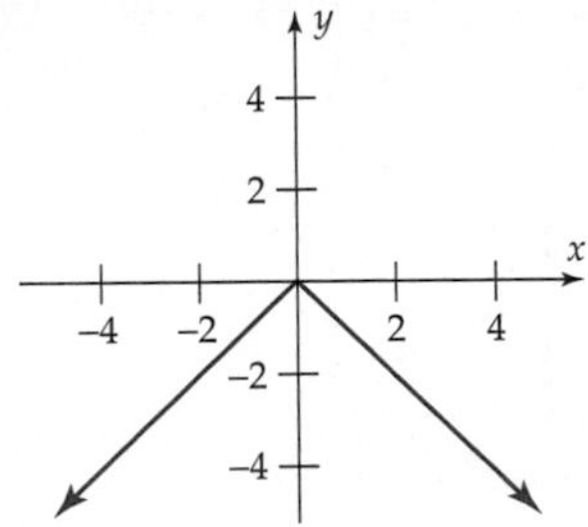

13. $-10 \leq x \leq 10$ and $-40 \leq y \leq 45$

15. $-14 \leq x \leq 12$ and $0 \leq y \leq 15$

17.

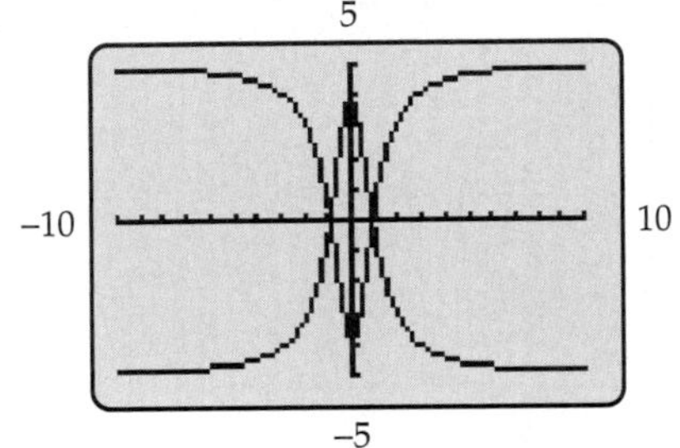

19. Shift 3 units to the right then 2 units up.

21. Reflect across the x-axis, then shrink vertically by a factor of $1/2$, and then shift 6 units down.

23. $g(x) = (x + 5)^2 + 6$ **25.** $g(x) = 2\sqrt{x - 6} - 3$

27. (a) $g(x) = x^2 + 3x + 2$
(b) both difference quotients are $2x + h + 3$

29.

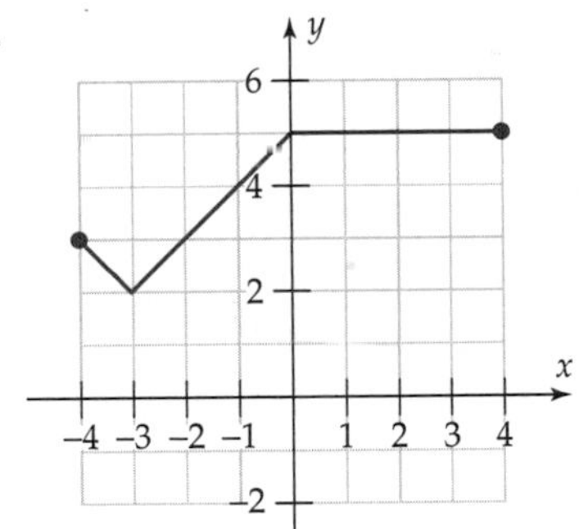

31.

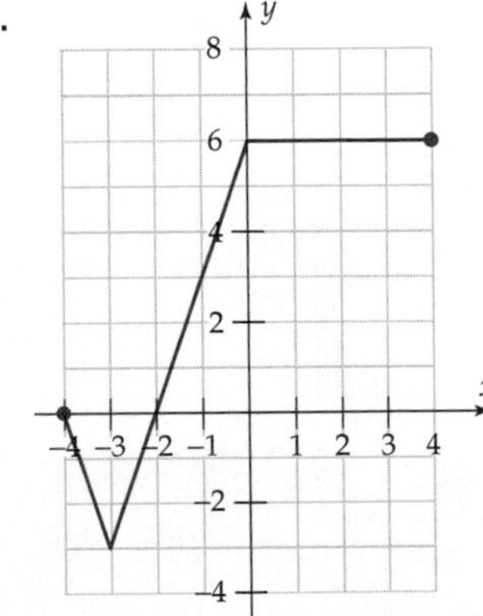

33. 

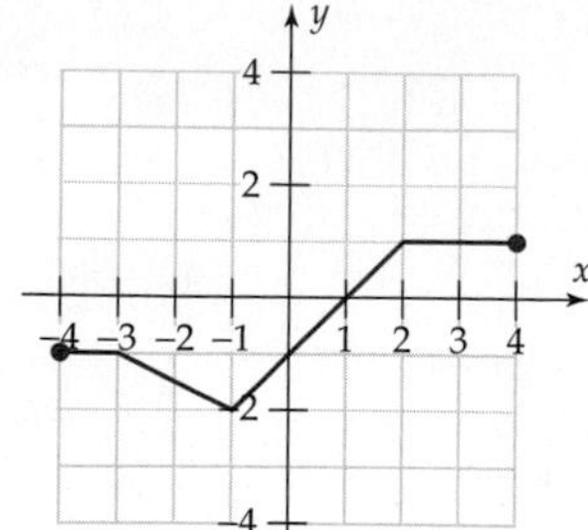

35.

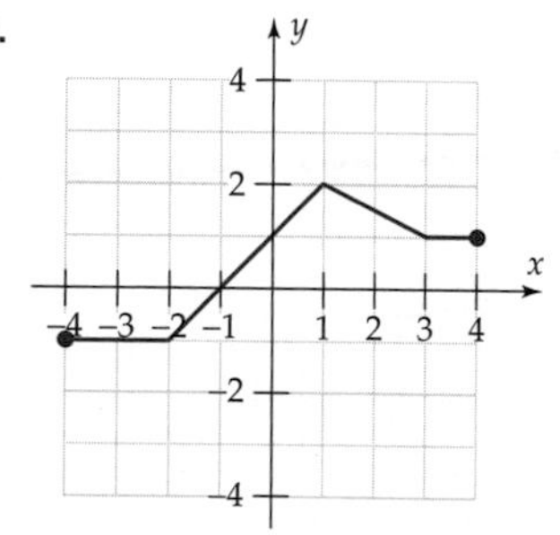

37.

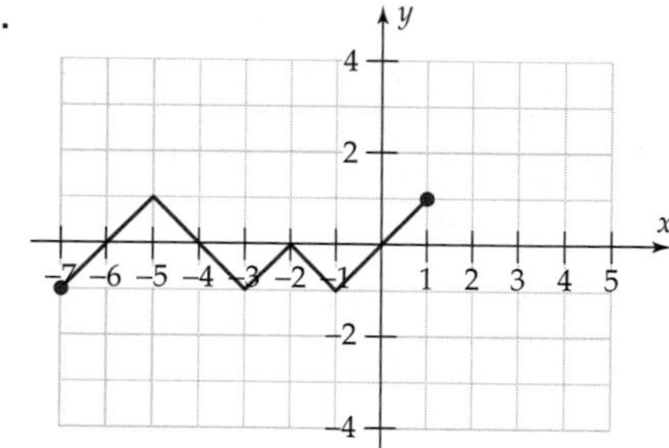

39.

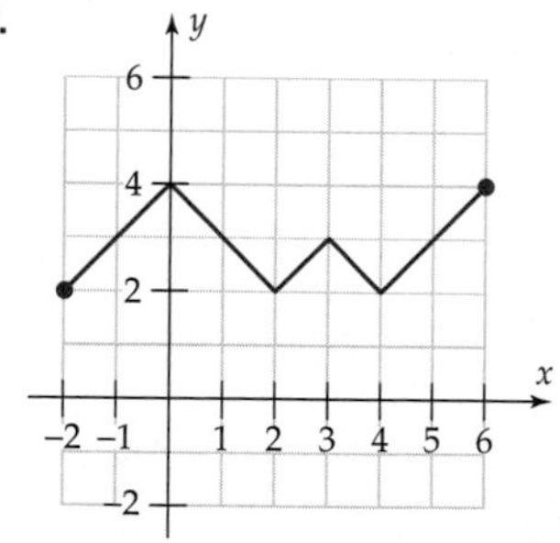

41.

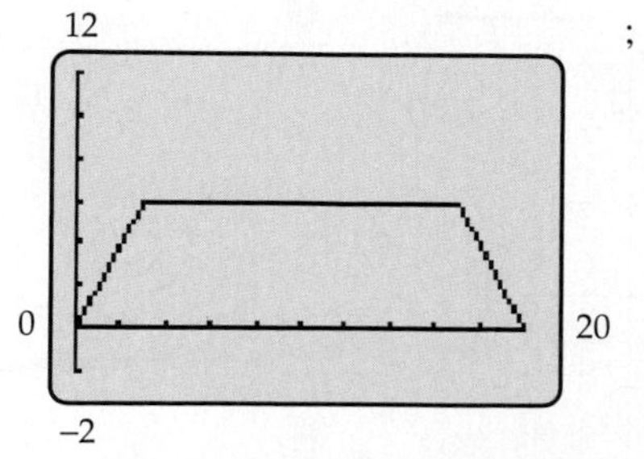

$g(x) = |x - 3| + |x - 17| - 8$

43.

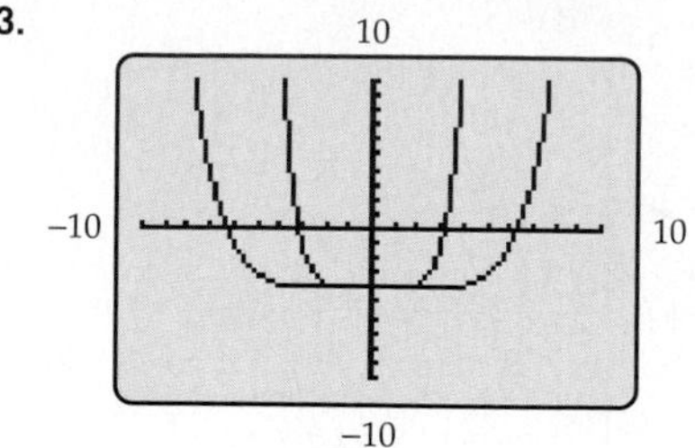

45.

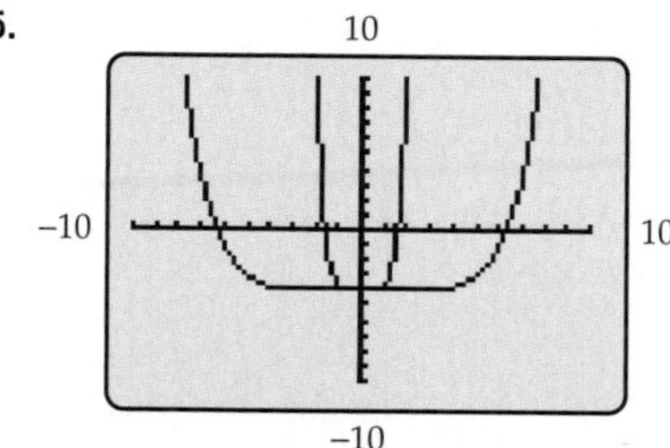

47.

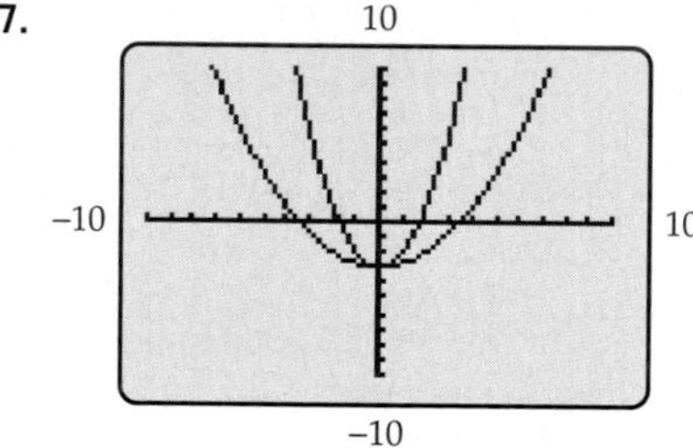

49.

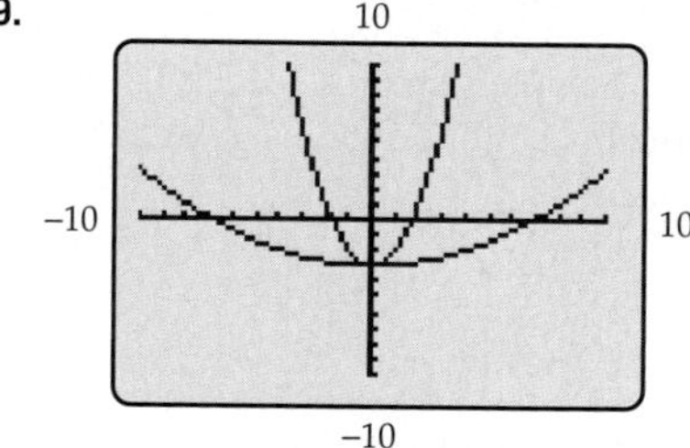

Chapter 0 Review, page 72

1. (a) $>$ (b) $<$ (c) $<$ (d) $>$ (e) $=$.

3. (a) $-10 < y < 0$ (b) $0 \le x \le 10$

5. (a) $|x + 7| < 3$ (b) $|y| > |x - 3|$

7. (a) $7 - \pi$ (b) $\sqrt{23} - \sqrt{3}$

9. (c)

11. no real solution

13. $5/3$ or -1

15. $\sqrt{58}$

17. $\sqrt{c^2 + d^2}$

19. $\left(d, d + \frac{c}{2}\right)$

21. (a) $\sqrt{17}$ (b) $(x - 2)^2 + (y + 3)^2 = 17$

23.

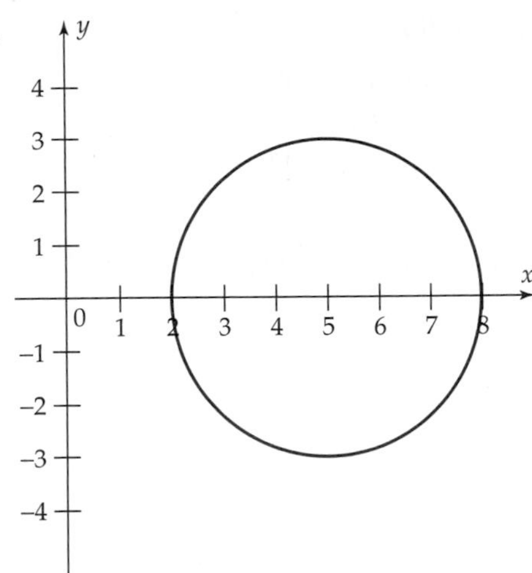

25. (b) and (d)

27. (a) a, d
(b) b and c do not show the peaks and valleys; e shows only a vertical segment.
(c) $-4 \le x \le 7$, $-6 \le y \le 8$

29. (a) None of the windows shows a complete graph though c comes close.
(b) a, b and d do not show any peaks or valleys; c does not show the valleys; no graph appears in window e.
(c) $-8 \le x \le 12$, $-1000 \le y \le 500$

31. (a) b, c
(b) a does not show peaks or valleys; d shows what appears to be several vertical lines; no graph appears in window e.
(c) $-9 \le x \le 9$, $-150 \le y \le 150$

33. (a) 1999 (b) 1992 (c) 2003

35. $x = 3.2678$

37. $x = -3.2843$

39. $x = 1.6511$

41. 1999

43. $\pm\sqrt{5}$

45. $1, -2$

47. $\dfrac{5 - \sqrt{5}}{2}$

49. no solution

51.

x	0	1	2	-4	t	k	$b - 1$	$1 - b$	$6 - 2u$
$f(x)$	7	5	3	15	$7 - 2t$	$7 - 2k$	$9 - 2b$	$5 + 2b$	$4u - 5$

53. Many possible answers, including these.
(a) If $f(x) = x + 1$, $a = 1$, $b = 2$, then $f(a + b) = f(1 + 2) = f(3) = 4$, but $f(a) + f(b) = f(1) + f(2) = 2 + 3 = 5$
(b) If $f(x) = x + 1$, $a = 1$, $b = 2$, then $f(ab) = f(1 \cdot 2) = f(2) = 3$, but $f(a) \cdot f(b) = f(1) \cdot f(2) = 2 \cdot 3 = 6$.

55. $r \ge 4$

57. $t^2 + t - 2$

59. (a) $f(t) = 50\sqrt{t}$
(b) $g(t) = 2500\pi t$
(c) radius is 150 meters, area is $22{,}500\pi$ square meters
(d) after approximately 12.7 hours

61. No local maximum, local minimum at $-.5$, increasing when $x > -.5$, decreasing when $x < -.5$

63. Local maximum at -5.0737, local minimum at $-.263$, increasing when $x < -5.073$ and when $x > -.263$, decreasing when $-5.0737 < x < -.263$.

65.

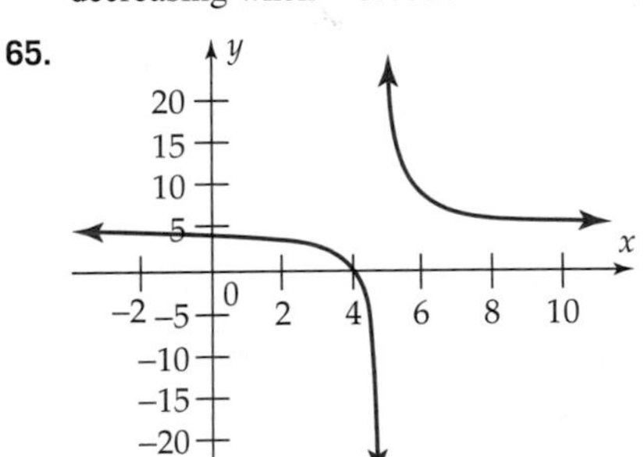

67. 3

69. false

71.

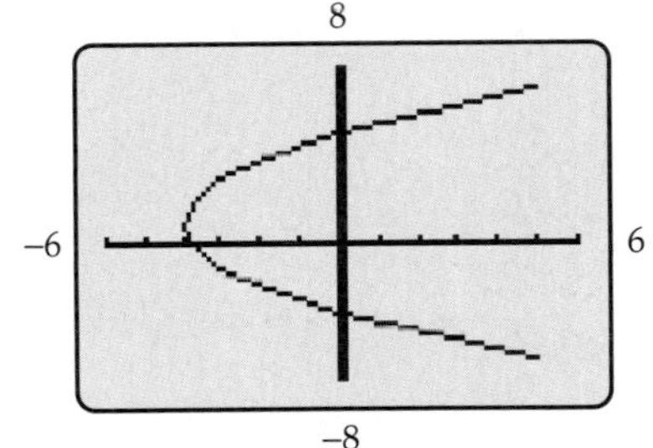

73. Shift 2 units right.

75. Shift 4 units left, then reflect in the x-axis, and then shift 5 units down.

77.

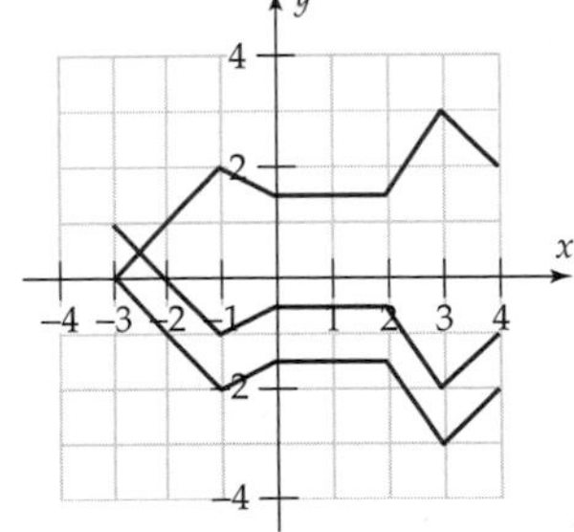

79. (a) -1 (b) -1 (c) 2

CHAPTER 1

Section 1.1, page 87

1. $\sin\theta = \frac{\sqrt{2}}{\sqrt{11}}$
   $\cos\theta = \frac{3}{\sqrt{11}}$
   $\tan\theta = \frac{\sqrt{2}}{3}$

3. $\sin\theta = \frac{\sqrt{3}}{\sqrt{7}}$
   $\cos\theta = \frac{2}{\sqrt{7}}$
   $\tan\theta = \frac{\sqrt{3}}{2}$

5. $\sin\theta = \frac{h}{m}$
   $\cos\theta = \frac{d}{m}$
   $\tan\theta = \frac{h}{d}$

7. $c = 36$ 9. $c = 36$ 11. $c = 8.4$

13. $h = 25\sqrt{2}/2$ 15. $h = 300$ 17. $h = 50\sqrt{3}$

19. $c = 4\sqrt{3}/3$ 21. $a = 10\sqrt{3}/3$

23. $a = 6.4$, $c = 7.7$ 25. $b = 24.8$, $c = 24.1$

27. $a = 10.7$, $b = 11.8$ 29. $\theta \approx 48.6°$

31. $\theta \approx 48.2°$ 33. $A = 33.7°$, $C = 56.3°$

35. $A = 44.4°$, $C = 45.6°$ 37. $A = 48.2°$, $C = 41.8°$

39. (a) $A = \frac{1}{2}ah$
   (b) $\sin\theta = \frac{h}{b}$
   (c) $h = b\sin\theta$ follows immediately from (b).
   (d) $A = \frac{1}{2}ah = \frac{1}{2}ab\sin\theta$

41. $A = 29.6$ 43. $A = 220$

45. (a) $c = \sqrt{a^2 + b^2}$
   (b) $\sin\theta = \frac{a}{c}$, $\cos\theta = \frac{b}{c}$
   (c) $(\sin\theta)^2 + (\cos\theta)^2 = \left(\frac{a}{c}\right)^2 + \left(\frac{b}{c}\right)^2 = \frac{a^2 + b^2}{c^2} = \frac{c^2}{c^2} = 1$

47. $\cos\theta = \frac{4}{5}$, $\tan\theta = \frac{3}{4}$

49. $\sin\theta = \frac{\sqrt{7}}{4}$, $\tan\theta = \frac{\sqrt{7}}{3}$

51. (a) $\sin\theta = a/c = \cos\alpha$
   (b) Since $\theta + \alpha + 90° = 180°$, $\theta + \alpha = 90°$.
   (c) Since $\theta + \alpha = 90°$, $\alpha = 90° - \theta$; thus, $\sin\theta = \cos\alpha = \cos(90° - \theta)$

Section 1.2, page 98

1. $\angle A = 74°$, $c = 29.0$, $a = 27.9$

3. $\angle A = 36.9°$, $\angle B = 53.1°$

5. $\angle C = 60°$, $b = 10$, $c = 8.7$

7. $\angle C = 18°$, $a = 3.3$, $c = 1.1$

9. $b = 2.9$, $\angle A = 60.8°$, $\angle C = 29.2°$

11. $c = 8$, $\angle A = 36.9°$, $\angle C = 53.1°$

13. $w = 100.7$ ft 15. About 68° 17. 23.0 m

19. About 15.3 feet 21. 276.3 feet

23. About 3402.8 feet

25. 85.9 feet

27. 27.5 feet. It is not safe for him to jump.

29. No

31. 8598.3 feet (1.6 miles)

33. 30° 35. 1.53 miles 37. 25.4 m

39. 3.52 m 41. 52.5 mph

43. 19.25 ft, $\theta \approx 54.0°$

45. 205.7 feet 47. 173.2 miles

49. (a) 56.7 feet (b) 9.7 feet

51. (a) Area $= 200\sin t\cos t$
   (b) $t = 45°$; approximately 14.14 by 7.07 feet

53. 1401.3 m

Section 1.3, page 112

1. $\sin\theta = \frac{8}{\sqrt{89}}$, $\cos\theta = \frac{5}{\sqrt{89}}$, $\tan\theta = \frac{8}{5}$

3. $\sin\theta = \frac{\sqrt{5}}{3}$, $\cos\theta = -\frac{2}{3}$, $\tan\theta = -\frac{\sqrt{5}}{2}$

5. $\sin\theta = \frac{7}{\sqrt{53}}$, $\cos\theta = \frac{2}{\sqrt{53}}$, $\tan\theta = \frac{7}{2}$

7. $\sin\theta = \frac{6}{\sqrt{61}}$, $\cos\theta = \frac{-5}{\sqrt{61}}$, $\tan\theta = -\frac{6}{5}$

9. $\sin\theta = \frac{\sqrt{15}}{5}$, $\cos\theta = \frac{\sqrt{10}}{5}$, $\tan\theta = \frac{\sqrt{6}}{2}$

11. .5446 13. -2.7475

15. $\sin\theta = \frac{2}{\sqrt{5}}$, $\cos\theta = \frac{1}{\sqrt{5}}$, $\tan\theta = 2$

17. $\sin\theta = \frac{2}{\sqrt{29}}$, $\cos\theta = \frac{-5}{\sqrt{29}}$, $\tan\theta = -\frac{2}{5}$

19. $\sin 150° = \frac{1}{2}$, $\cos 150° = -\frac{\sqrt{3}}{2}$, $\tan 150° = -\frac{\sqrt{3}}{3}$

21. $a = 4.2$; $B = 125.0°$; $C = 35.0°$

23. $c = 13.9$; $A = 22.5°$; $B = 39.5°$

25. $a = 24.4$; $B = 18.4°$; $C = 21.6°$

27. $c = 21.5$; $A = 33.5°$; $B = 67.9°$

29. $A = 120°$; $B = 21.8°$; $C = 38.2°$

31. $A = 24.1°$; $B = 30.8°$; $C = 125.1°$

33. $A = 38.8°$; $B = 34.5°$; $C = 106.7°$

35. $A = 34.1°$; $B = 50.5°$; $C = 95.4°$

37. $A = 77.5°$; $B = 48.4°$; $C = 54.1°$

39. 334.9 kilometers **41.** 54.7° **43.** 154.5 feet

45. 63.7 feet **47.** 84.9°

49. 26,205 miles

51. For the left-hand cable, $A = 34.1°$; for the right-hand cable, $B = 25.6°$.

53. 8.4 km **55.** 7.0 miles

57. 425.7 feet and 469.4 feet **59.** 33.4°

61. 978 miles

63. (a) Since the triangles are similar,

$$\frac{v}{\sqrt{u^2 + v^2}} = \frac{y}{\sqrt{x^2 + y^2}}$$

These are the two computations for sin θ, and they are equal.

(b) Again, the triangles are similar; hence,

$$\frac{u}{\sqrt{u^2 + v^2}} = \frac{x}{\sqrt{x^2 + y^2}}$$

These are the two computations for cos θ, and they are equal.

(c) From similarity,

$$\frac{v}{u} = \frac{y}{x}$$

These are the two computations for tan θ, and they are equal.

65. $a = 19.5$; $b = 21.23$; $c = 24.27$.
$A = 50.2°$; $B = 56.8°$; $C = 73.0°$.

Section 1.4, page 123

1. $C = 110°$, $b = 2.5$, $c = 6.3$

3. $B = 14°$, $b = 2.2$, $c = 6.8$

5. $A = 88°$, $a = 17.3$, $c = 12.8$

7. $C = 41.5°$, $b = 9.7$, $c = 10.9$

9. $\sin C > 1$, hence no such triangle exists.

11. Case 1: $A = 55.2°$, $C = 104.8°$, $c = 14.1$
Case 2: $A = 124.8°$, $C = 35.2°$, $c = 8.4$

13. $\sin C > 1$ hence no such triangle exists.

15. Case 1: $B = 65.8°$, $A = 58.2°$, $a = 10.3$
Case 2: $B = 114.2°$, $A = 9.8°$, $a = 2.1$

17. $b = 14.7$, $c = 15.2$, $C = 72°$

19. $a = 9.8$, $B = 23.3°$; $C = 81.7°$

21. $A = 18.6°$, $B = 39.6°$, $C = 121.8°$

23. $c = 13.9$, $A = 60.1°$, $B = 72.9°$

25. $A = 77.7°$, $C = 39.8°$, $a = 18.9$

27. $\sin B > 1$; hence, no such triangle exists.

29. $a = 18.5$, $B = 36.7°$, $C = 78.3°$

31. $C = 126°$, $b = 193.8$, $c = 273.3$

33. 135.5 meters

35. 9.99 meters **37.** 30.1 kilometers

39. $\alpha = 5.4°$ **41.** 5.0 feet **43.** 5.3°

45. About 9641 feet

47. (a) Solve triangle ABC to find $\angle BAC$, the angle at A. $180° - \angle BAC = \angle EAB$. Solve triangle ABD to find $\angle ABD$, the angle at B. $180° - \angle ABD = \angle EBA$. Now solve triangle EAB using the two angles and included side to find EA.
(b) About 94 feet

49. 13.36 meters

Special Topics 1.4.A, page 128

1. 7.3 square units **3.** 32.5 square units

5. 82.3 square units **7.** 31.4 square units

9. 6.5 square units **11.** 7691 square units

13. 235.9 square feet **15.** 5.8 gallons

17. 11.18 square units

19. (a) 50 (b) $6\sqrt{3}$

21. Since $12 + 20 < 36$, this violates the Triangle Inequality, which states that the sum of two sides of a triangle is greater than the third side. This is not a triangle, and the area is undefined.

Chapter 1 Review, page 131

1. $\sin\theta = \frac{3}{\sqrt{34}}$; $\cos\theta = \frac{5}{\sqrt{34}}$; $\tan\theta = \frac{3}{5}$

3. (d) **5.** (e)

7. $A = 42.7°$; $C = 47.3°$, $b = 17.7$

9. $b = 14.65$; $c = 8.40$; $A = 55°$

11. 225.9 feet **13.** 1.52°

15. $\sin\theta = \frac{1}{\sqrt{2}}$, $\cos\theta = -\frac{1}{\sqrt{2}}$, $\tan\theta = -1$

17. $\sin\theta = \frac{5}{\sqrt{34}}$, $\cos\theta = \frac{3}{\sqrt{34}}$, $\tan\theta = \frac{5}{3}$

19. $A = 52.9°$; $B = 41.6°$; $C = 85.5°$

21. $b = 21.8$; $A = 20.6°$; $C = 29.4°$

23. 301 miles

25. $A = 25°$, $a = 2.9$, $b = 5.6$

27. $A = 52.0°$, $B = 66.0°$, $b = 86.9$

29. Case 1: $B = 81.8°$, $C = 38.2°$, $c = 2.5$
Case 2: $B = 98.2°$, $C = 21.8°$, $c = 1.5$

31. 147.4 square units **33.** 13.4 km

35. $C = 75°$, $a = 41.6$, $c = 54.1$

37. $b = 8.3$; $A = 35.5°$; $C = 68.5°$

39. Joe is 217.9 meters from the flagpole. Alice is 240 meters from the flagpole.

41. (a) 3618 feet (b) 4019 feet (c) 3642 feet

43. 71.89°

45. 10 square units **47.** 37.95 square units

CHAPTER 2

Section 2.1, page 142

1. $\frac{2\pi}{9}$ **3.** $\frac{\pi}{9}$ **5.** $\frac{\pi}{18}$

7. $9\pi/4, 17\pi/4, -7\pi/4, -15\pi/4$

9. $11\pi/6, 23\pi/6, -13\pi/6, -25\pi/6$

11. $5\pi/3$ **13.** $3\pi/4$ **15.** $3\pi/5$

17. $7 - 2\pi$ **19.** $\pi/30$ **21.** $-\pi/15$

23. $5\pi/12$ **25.** $3\pi/4$ **27.** $-5\pi/4$

29. $31\pi/6$ **31.** 36° **33.** −18°

35. 135° **37.** 4° **39.** −75°

41. 972° **43.** $4\pi/3$ **45.** $7\pi/6$

47. $41\pi/6$ **49.** 7π radians **51.** 4π radians

53. 42.5π radians **55.** $2\pi k$ radians

Special Topics 2.1.A, page 146

1. $8\pi \approx 25.13$ cm **3.** 4.25 radians

5. 50/9 radians **7.** 2000 radians

9. 5 **11.** 8.75

13. About 3491 miles **15.** About 942.5 miles

17. 3 radians

19. (a) 400π radians per minute
(b) $\frac{200\pi}{3} \approx 2513.3$ inches per minute; $800\pi \approx 209.4$ feet per minute

21. 24.37 radians per second

23. (a) 5π radians per second
(b) 6.69 miles per hour

25. About 15.9 feet

Section 2.2, page 154

1. $\sin t = \frac{7}{\sqrt{53}}$, $\cos t = \frac{2}{\sqrt{53}}$, $\tan t = \frac{7}{2}$

3. $\sin t = \frac{-6}{\sqrt{61}}$, $\cos t = \frac{-5}{\sqrt{61}}$, $\tan t = \frac{6}{5}$

5. $\sin t = \frac{-10}{\sqrt{103}}$, $\cos t = \sqrt{\frac{3}{103}}$, $\tan t = -\frac{10}{\sqrt{3}}$

7. −1 **9.** 0 **11.** 0

13. 0 **15.** 0

17. $\sin t = \frac{1}{\sqrt{5}}$, $\cos t = -\frac{2}{\sqrt{5}}$, $\tan t = -\frac{1}{2}$

19. $\sin t = -\frac{4}{5}$, $\cos t = -\frac{3}{5}$, $\tan t = \frac{4}{3}$

21. $\sin\frac{\pi}{3} = \frac{\sqrt{3}}{2}$, $\cos\frac{\pi}{3} = \frac{1}{2}$, $\tan\frac{\pi}{3} = \sqrt{3}$

23. $\sin\frac{\pi}{4} = \frac{\sqrt{2}}{2}$, $\cos\frac{\pi}{4} = \frac{\sqrt{2}}{2}$, $\tan\frac{\pi}{4} = 1$

25. $\sin\frac{3\pi}{4} = \frac{\sqrt{2}}{2}$, $\cos\frac{3\pi}{4} = -\frac{\sqrt{2}}{2}$, $\tan\frac{3\pi}{4} = -1$

27. $\sin\frac{5\pi}{6} = \frac{1}{2}$, $\cos\frac{5\pi}{6} = -\frac{\sqrt{3}}{2}$, $\tan\frac{5\pi}{6} = -\frac{\sqrt{3}}{3}$

29. $\sin\left(-\frac{23\pi}{6}\right) = \frac{1}{2}$, $\cos\left(-\frac{23\pi}{6}\right) = \frac{\sqrt{3}}{2}$, $\tan\left(-\frac{23\pi}{6}\right) = \frac{\sqrt{3}}{3}$

31. $\sin\left(-\frac{19\pi}{3}\right) = -\frac{\sqrt{3}}{2}$, $\cos\left(-\frac{19\pi}{3}\right) = \frac{1}{2}$, $\tan\left(-\frac{19\pi}{3}\right) = -\sqrt{3}$

33. $\sin\left(-\frac{15\pi}{4}\right) = \frac{\sqrt{2}}{2}$, $\cos\left(-\frac{15\pi}{4}\right) = \frac{\sqrt{2}}{2}$, $\tan\left(-\frac{15\pi}{4}\right) = 1$

35. $\sin\left(-\frac{17\pi}{2}\right) = -1$, $\cos\left(-\frac{17\pi}{2}\right) = 0$, $\tan\left(-\frac{17\pi}{2}\right)$ is undefined.

37. $-\frac{\sqrt{3}}{2}$ **39.** $-\frac{\sqrt{2}}{2}$ **41.** $\frac{-\sqrt{6} + \sqrt{2}}{4}$

43. (a) 467 mph (b) 1458 mph

45. (a)

Date	Jan 1	Mar 1	May 1
Average Temperature	31.7	43.8	68.3

Date	July 1	Sept 1	Nov 1
Average Temperature	80.9	69.0	44.5

(b)

Date	June 1	June 4	June 7	June 10
Average Temperature	77.4	78.0	78.6	79.1

Date	June 13	June 16	June 19	June 22
Average Temperature	79.6	80.0	80.3	80.5

Date	June 25	June 28
Average Temperature	80.7	80.8

47. $\sin t = \dfrac{-3}{\sqrt{10}}$, $\cos t = \dfrac{1}{\sqrt{10}}$, $\tan t = -3$

49. $\sin t = -\dfrac{5}{\sqrt{34}}$, $\cos t = \dfrac{3}{\sqrt{34}}$, $\tan t = -\dfrac{5}{3}$

51. $\sin t = \dfrac{1}{\sqrt{5}}$, $\cos t = \dfrac{-2}{\sqrt{5}}$, $\tan t = -\dfrac{1}{2}$

53.

Quadrant II

sin t	+
cos t	−
tan t	−

Quadrant I

sin t	+
cos t	+
tan t	+

Quadrant III

sin t	−
cos t	−
tan t	+

Quadrant IV

sin t	−
cos t	+
tan t	−

55. Many correct answers, including: Since $0 < 1 < \pi/2$, the terminal side of an angle of 1 radian lies in the first quadrant, so $y = \sin 1$ is positive.

57. Since $\pi/2 < 3 < \pi$, the terminal side of an angle of 3 radians lies in the second quadrant, so $y/x = \tan 3$ is negative.

59. Since $0 < 1.5 < \pi/2$, the terminal side of an angle of 1.5 radians lies in the first quadrant, so $y/x = \tan 1.5$ is positive.

61. $t = \dfrac{\pi}{2} + 2\pi n$, n any integer

63. $t = \pi n$, n any integer

65. $t = \dfrac{\pi}{2} + \pi n$, n any integer

67. $\sin(\cos 0) < \cos(\sin 0)$

69. (a) Horse A will reach the position occupied by horse B at a time t after rotation of $\pi/4$ radians, which is an eighth of a complete circle. This takes $1/8$ of a minute. Thus, $B(t) = A(t + 1/8)$.
(b) Horse A will reach the position occupied by horse C at a time t after rotation of $1/3$ of a complete circle. This takes $1/3$ of a minute. Thus, $C(t) = A(t + 1/3)$.
(c) $E(t) = D(t + 1/8)$; $F(t) = D(t + 1/3)$
(d) P, A, and D are always collinear, and $A(t)$ and $D(t)$ are defined as the vertical distances from each horse to the x-axis; this is illustrated by Figure S. Thus, $D(t)/A(t) = 5/1$ so that $D(t) = 5A(t)$.
(e) $E(t) = D(t + 1/8) = 5A(t + 1/8)$; $F(t) = D(t + 1/3) = 5A(t + 1/3)$
(f) In t minutes, the merry-go-round rotates $2\pi t$ radians. Therefore, $A(t) = \sin 2\pi t$.
(g) $B(t) = \sin 2\pi(t + 1/8) = \sin(2\pi t + \pi/4)$; $C(t) = \sin 2\pi(t + 1/3) = \sin(2\pi t + 2\pi/3)$
(h) $D(t) = 5 \sin 2\pi t$; $E(t) = 5 \sin(2\pi t + \pi/4)$; $F(t) = 5 \sin(2\pi t + 2\pi/3)$

Section 2.3, page 170

1. $(fg)(t) = 3 \sin^2 t + 6 \sin t \cos t$

3. $(fg)(t) = 3 \sin^3 t + 3 \sin^2 t \tan t$

5. $(\cos t + 2)(\cos t - 2)$

7. $(\sin t + \cos t)(\sin t - \cos t)$

9. $(\tan t + 3)^2$

11. $(3 \sin t + 1)(2 \sin t - 1)$

13. $(\cos^2 t + 5)(\cos t + 1)(\cos t - 1)$

15. $(f \circ g)(t) = \cos(2t + 4)$; $(g \circ f)(t) = 2 \cos t + 4$

17. $(f \circ g)(t) = \tan(t^2 + 2)$; $(g \circ f)(t) = \tan^2(t + 3) - 1$

19. Yes **21.** No **23.** No

25. $\sin t = -\sqrt{3}/2$

27. $\sin t = \sqrt{3}/2$

29. $\sin(-t) = -3/5$

31. $\sin(2\pi - t) = -3/5$

33. $\tan t = 3/4$

35. $\tan(2\pi - t) = -3/4$ (using Exercise 33)

37. $\sin t = -\sqrt{21}/5$

39. $\cos(2\pi - t) = -2/5$

41. $\sin(4\pi + t) = -\sqrt{21}/5$

43. $\cos\dfrac{\pi}{8} = \dfrac{\sqrt{2 + \sqrt{2}}}{2}$

45. $\sin\dfrac{17\pi}{8} = \dfrac{\sqrt{2 - \sqrt{2}}}{2}$

47. $\sin^2 t - \cos^2 t$

49. $\sin t$

51. $|\sin t \cos t|\sqrt{\sin t}$

53. $\dfrac{1}{4}$

55. $\cos t + 2$

57. $\cos t$

59. Use $k = \pi$. $f(t + k) = \sin 2(t + \pi) = \sin(2t + 2\pi) = \sin 2t = f(t)$ for all t.

61. Use $k = \pi$. $f(t + k) = \sin 4(t + \pi) = \sin(4t + 4\pi) = \sin 4t = f(t)$ for all t. [$k = \pi/2$ also works.]

63. Use $k = \pi/2$. $f(t + k) = \tan 2(t + \pi/2) = \tan(2t + \pi) = \tan 2t = f(t)$ for all t in the domain.

65. (a) If $0 < k < 2\pi$, there is no number such that $\cos k = 1$.
(b) If $\cos(t + k) = \cos t$ for every number t, then use $t = 0$.
$$\cos(0 + k) = \cos 0$$
$$\cos k = 1$$
(c) If there were such a number k, then by part (b), $\cos k = 1$, which is impossible by part (a). Therefore, there is no number k with $0 < k < 2\pi$ such that $\cos(t + k) = \cos t$ for every number t, and the period of cosine is 2π.

Section 2.4, page 180

1. $t = n\pi$ where n is any integer

3. $t = \dfrac{\pi}{2} + 2n\pi$ where n is any integer

5. $t = \pi + 2n\pi$ where n is any integer

7. $\tan t = 11$

9. $\tan t = 1.4$

11. Shift 3 units vertically up.

13. Reflect the graph in the horizontal axis.

15. Shift 5 units vertically up.

17. Stretch by a factor of 3 away from the horizontal axis.

19. Stretch by a factor of 3 away from the horizontal axis, then shift 2 units vertically up.

21. Shift 2 units horizontally to the right.

23. Two solutions

25. Two solutions

27. Two solutions

29. Two solutions

31. Possibly an identity

33. Possibly an identity

35. Possibly an identity

37. Not an identity

39. Possibly an identity

41. Not an identity

43. (a) The graphs will appear identical, but the functions are not the same.
(b) The calculator plots 95–159 points, each of which has an x-coordinate that is a multiple of 2π. Since all y-coordinates are 1, the graph looks like the horizontal line $y = 1$.

45. (a) There should be 80 full waves in an interval of length $80 \cdot 2\pi$.
(b) The calculator graph can't show 80 waves, since it plots only 95–159 points. So you obtain something like the graph below (or something worse) that does not have 80 waves.

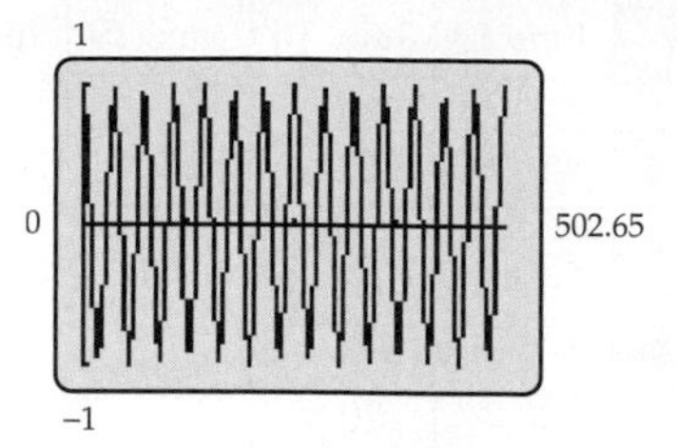

47. (a) $-\pi \leq t \leq \pi$
(b) $n = 15$. Agreement is good over the entire interval for f_{15}. In fact, $\sin(2) = .9092974268$, $f_{15}(2) = .9092974265$.

49. $\dfrac{r(t)}{s(t)}$, where $r(t) = f_{15}(t)$ in Exercise 47 and $s(t) = f_{16}(t)$ in Exercise 48.

51. The y-coordinate of the point on the graph of the cosine function is the same as the x-coordinate of the point on the unit circle, because of the definition of the cosine function as the x-coordinate of a point on the unit circle.

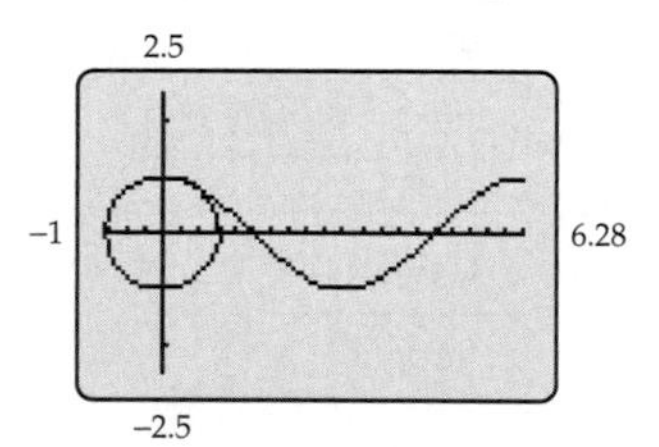

Section 2.5, page 192

1. Amplitude = 3
Period = π
Phase shift = $\pi/2$

3. Amplitude = 7
Period = $2\pi/7$
Phase shift = $-1/49$

5. Amplitude = 1
Period = 1
Phase shift = 0

7. Amplitude = 6
Period = 2/3
Phase shift = $-1/3\pi$

9. $f(t) = 3 \sin(8t - 8\pi/5)$ or $f(t) = 3 \cos(8t - 8\pi/5)$

11. $f(t) = (2/3)\sin 2\pi t$ or $f(t) = (2/3)\cos 2\pi t$

13. $f(t) = 7 \sin\left(\dfrac{6\pi}{5}t + \dfrac{3\pi^2}{5}\right)$ or $f(t) = 7 \cos\left(\dfrac{6\pi}{5}t + \dfrac{3\pi^2}{5}\right)$

15. $f(t) = 2 \sin 4t$

17. $f(t) = 1.5 \cos(t/2)$

19. (a) Period $= 2\pi/300 = \pi/150$.
(b) The graph should have 300 complete waves between 0 and 2π.
(c) $0 \leq x \leq 2\pi/75$, $-2 \leq y \leq 2$.

21. (a) Period $= 2\pi/900 = \pi/450$.
(b) The graph should have 900 complete waves between 0 and 2π.
(c) $0 \le x \le 2\pi/225$, $-2 \le y \le 2$.

23. (a) $f(t) = 12\sin(10t - \pi/2)$
(b) $g(t) = -12\cos(10t)$

25. (a) $f(t) = -\sin(2t)$
(b) $g(t) = \cos(2t + \pi/2)$

27.

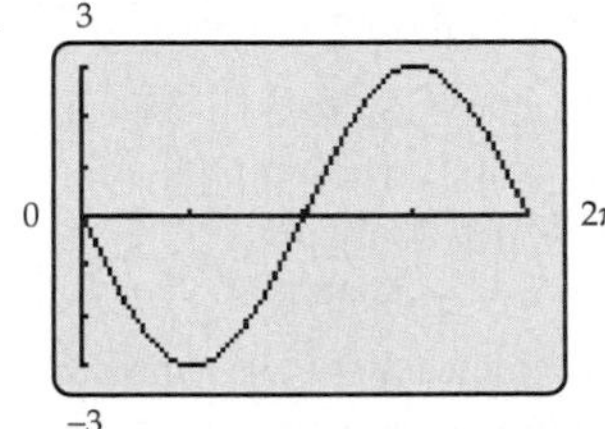

29.

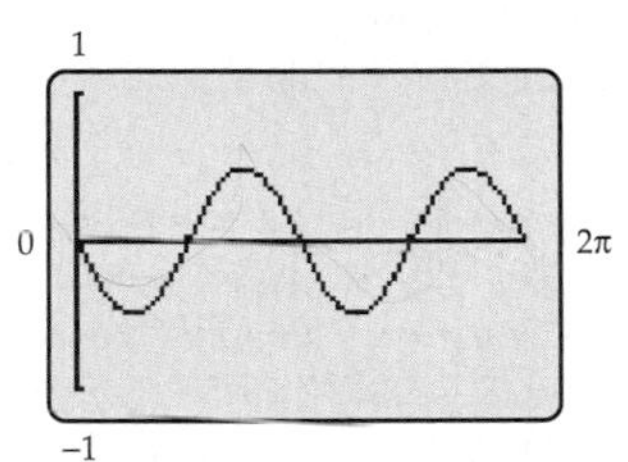

31.

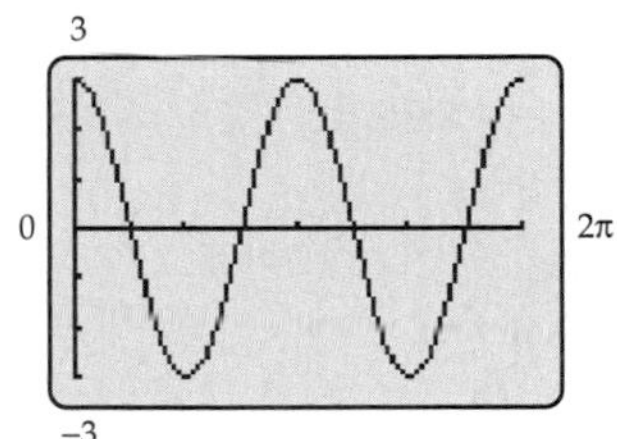

33. Local maximum at $t = \frac{5\pi}{6}$; local minimum $t = \frac{11\pi}{6}$.

35. Local maxima at $t = \frac{\pi}{6}, \frac{5\pi}{6}, \frac{3\pi}{2}$; local minima at $t = \frac{\pi}{2}, \frac{7\pi}{6}, \frac{11\pi}{6}$.

37. $f(t) = 5.3852\sin(t + 1.1903)$

39. $f(t) = 3.8332\sin(4t + 1.4572)$

41. The waves in the graph of g all have the same height, which is not the case with the graph of f.

43. (a) The person's blood pressure is 134/92.
(b) The pulse rate is 75.

45. (a) Maximum is 79.5 cubic inches; minimum is 30.5 cubic inches.
(b) every 4 seconds
(c) 15

47. Period is $\frac{1}{980{,}000}$ seconds. Frequency is 980,000 cycles per second (known as 980 kilohertz).

49. $h(t) = 6\sin(\pi t/2)$.

51. $h(t) = 6\cos(\pi t/2)$

53. $f(t) = 125\sin\left(\frac{\pi}{5}t\right)$

55. x-coordinate of P is $\cos(20\pi t) + \sqrt{16 - \sin^2(20\pi t)}$.

57. $d(t) = 10\sin(\pi t/2)$.

59. (a) At least four (starting point, high point, low point, ending point)
(b) 301
(c) Every calculator is different; the TI-83+ plots 95 points; others plot as many as 239.
(d) Obviously, 239 points (or fewer) are not enough when the absolute minimum is 301.

61. (a)

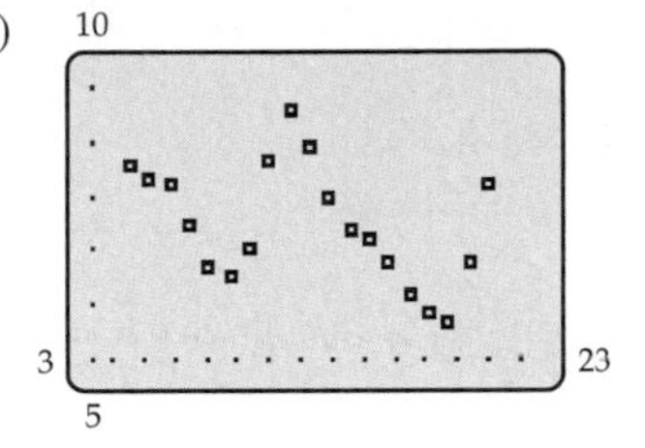

(b) The SinReg feature of the TI-83+ yields

Other calculators may give somewhat different answers.

(c) Graphing the model found in part (b) over the scatter plot (see figure below) shows that the model is a poor fit, and hence is unlikely to be of much use for making predictions.

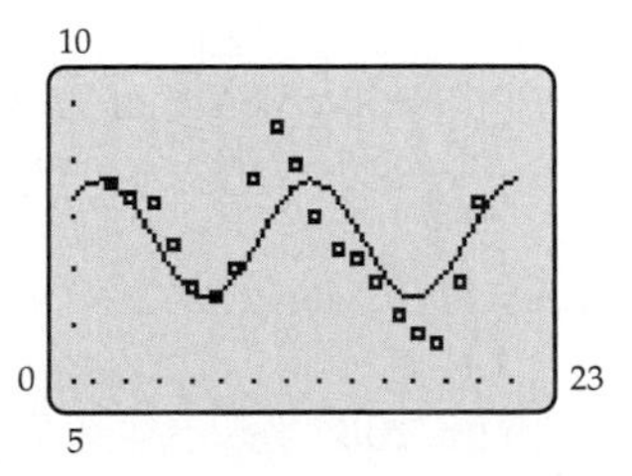

63. (a) The SinReg feature of the TI-83+ yields

(b) Using this value of b (other calculators may give somewhat different results) yields a period of 15.7 months. One would expect a period of 12 months.

(c) Plotting another year of the same data versus the model yields

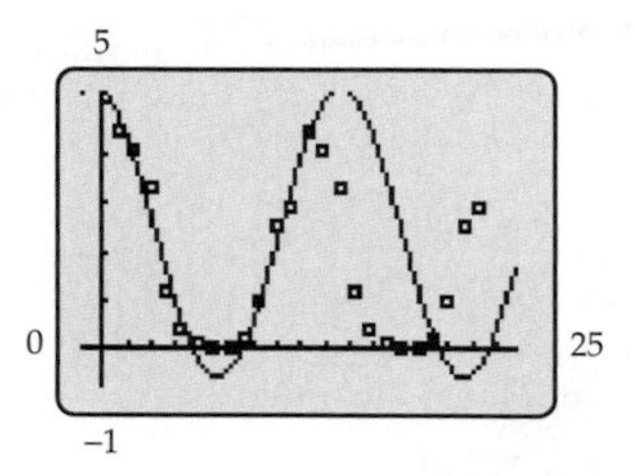

Clearly the 15.7-month period produces a very poor fit.

(d) Now we obtain the following model and the following graph:

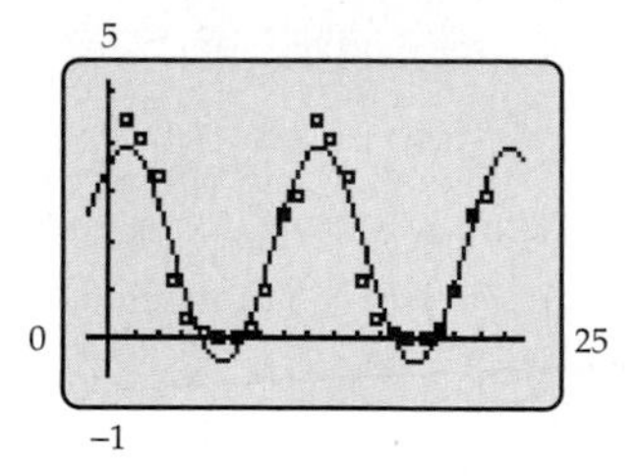

(e) Using this value of b yields a period of 12.2 months. This seems more reasonable, and the model fits the data better.

65. (a) $k = 9.8/\pi^2$ meter (b) about 397.43 seconds

Special Topics 2.5.A, page 203

1. $f(t) = 2.2361 \sin(t + 1.1071)$

3. $f(t) = 5.3852 \sin(4t - 1.1903)$

5. $f(t) = 5.1164 \sin(3t - .7442)$

7. $0 \le x \le 2\pi$, $-5 \le y \le 5$

9. $-10 \le x \le 10$, $-10 \le y \le 10$

11. $0 \le x \le .02\pi$, $-2 \le y \le 2$

13. $0 \le x \le .04$, $-7 \le y \le 7$

15. $0 \le x \le 10$, $-10 \le y \le 10$

17. To the left of the y-axis, the graph lies above the t-axis, which is a horizontal asymptote of the graph. To the right of the y-axis, the graph makes waves of amplitude 1, of shorter and shorter period as you move away from the origin. Window: $-3 \le x \le 3.2$ and $-2 \le y \le 2$

19. The graph is symmetric with respect to the y-axis and consists of waves along the t-axis, whose amplitude slowly increases as you move away from the origin in either direction. Window: $-30 \le x \le 30$ and $-6 \le y \le 6$

21. The graph, which is not defined at $t = 0$, is symmetric with respect to the y-axis and consists of waves along the t-axis, whose amplitude rapidly decreases as you move away from the origin in either direction. Window: $-30 \le x \le 30$ and $-.3 \le y \le 1$

23. The function has period π (why?). The graph lies on or below the t-axis because $|\cos t|$ is always between 0 and 1, so its natural logarithm is negative. The graph has vertical asymptotes when $t = \pm\pi/2, \pm 3\pi/2, \pm 5\pi/2, \pm 7\pi/2, \ldots$ ($\cos t = 0$ at these points and ln 0 is not defined). Window: $-2\pi \le x \le 2\pi$ and $-3 \le y \le 1$ (four periods)

25. (a)

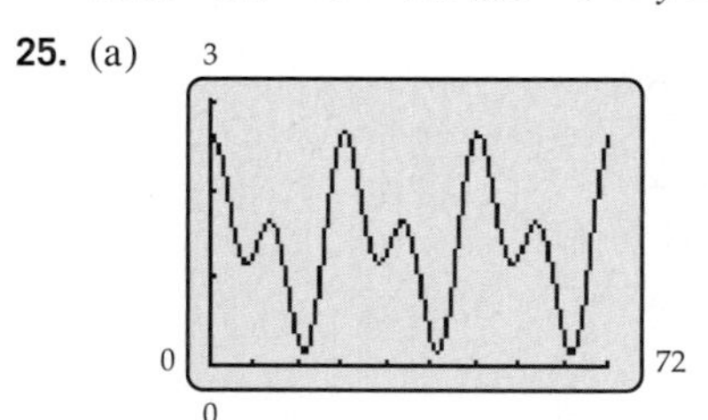

(b) Maximum at $t = .2$, $t = 24.4$, and $t = 48.6$, that is, shortly after midnight. Minimum at $t = 17.1$, $t = 41.3$, and $t = 65.5$, that is, around 5:30 P.M.

(c) 24.2 hours

Section 2.6, page 209

1. IV **3.** II **5.** IV

7. $\sin t = 4/5$ $\cos t = 3/5$ $\tan t = 4/3$
$\cot t = 3/4$ $\sec t = 5/3$ $\csc t = 5/4$

9. $\sin t = 12/13$ $\cos t = -5/13$ $\tan t = -12/5$
$\cot t = -5/12$ $\sec t = -13/5$ $\csc t = 13/12$

11. $\sin t = \dfrac{5}{\sqrt{26}}$ $\cos t = -\dfrac{1}{\sqrt{26}}$ $\tan t = -5$
$\cot t = -\dfrac{1}{5}$ $\sec t = -\sqrt{26}$ $\csc t = \dfrac{\sqrt{26}}{5}$

13. $\sin t = \dfrac{\sqrt{3}}{\sqrt{5}}$ $\cos t = \dfrac{\sqrt{2}}{\sqrt{5}}$ $\tan t = \dfrac{\sqrt{3}}{\sqrt{2}}$

$\cot t = \dfrac{\sqrt{2}}{\sqrt{3}}$ $\sec t = \dfrac{\sqrt{5}}{\sqrt{2}}$ $\csc t = \dfrac{\sqrt{5}}{\sqrt{3}}$

15. $\sin t = \dfrac{3}{\sqrt{12+2\sqrt{2}}}$ $\cot t = \dfrac{1+\sqrt{2}}{3}$

$\cos t = \dfrac{1+\sqrt{2}}{\sqrt{12+2\sqrt{2}}}$ $\sec t = \dfrac{\sqrt{12+2\sqrt{2}}}{1+\sqrt{2}}$

$\tan t = \dfrac{3}{1+\sqrt{2}}$ $\csc t = \dfrac{\sqrt{12+2\sqrt{2}}}{3}$

17. $\sin\dfrac{4\pi}{3} = -\dfrac{\sqrt{3}}{2}$ $\cos\dfrac{4\pi}{3} = -\dfrac{1}{2}$ $\tan\dfrac{4\pi}{3} = \sqrt{3}$

$\cot\dfrac{4\pi}{3} = \dfrac{1}{\sqrt{3}}$ $\sec\dfrac{4\pi}{3} = -2$ $\csc\dfrac{4\pi}{3} = -\dfrac{2}{\sqrt{3}}$

19. $\sin\dfrac{7\pi}{4} = -\dfrac{\sqrt{2}}{2}$ $\cos\dfrac{7\pi}{4} = \dfrac{\sqrt{2}}{2}$ $\tan\dfrac{7\pi}{4} = -1$

$\cot\dfrac{7\pi}{4} = -1$ $\sec\dfrac{7\pi}{4} = \sqrt{2}$ $\csc\dfrac{7\pi}{4} = -\sqrt{2}$

21. -3.8287

23. (a) $h = .01$, rate $= 5.6511$; $h = .001$, rate $= 5.7618$; $h = .0001$, rate $= 5.7731$; $h = .00001$, rate $= 5.7743$
(b) $(\sec 2)^2 \approx 5.7744$.

25. $\cos t + \sin t$

27. $1 - 2\sec t + \sec^2 t$

29. $\cot^3 t - \tan^3 t$

31. $\csc t(\sec t - \csc t)$

33. $-\tan^2 t - \sec^2 t$

35. $(\cos t - \sec t)(\cos^2 t + 1 + \sec^2 t)$

37. $\cot t$

39. $\dfrac{2\tan t + 1}{3\sin t + 1}$

41. $4 - \tan t$

43. $\tan t = \dfrac{\sin t}{\cos t} = \dfrac{1}{\cos t/\sin t} = \dfrac{1}{\cot t}$

45. $1 + \cot^2 t = 1 + \dfrac{\cos^2 t}{\sin^2 t} = \dfrac{\sin^2 t + \cos^2 t}{\sin^2 t} = \dfrac{1}{\sin^2 t} = \csc^2 t$

47. $\sec(-t) = \dfrac{1}{\cos(-t)} = \dfrac{1}{\cos t} = \sec t$

49. $\sin t = \dfrac{\sqrt{3}}{2}$
$\cos t = -1/2$
$\tan t = -\sqrt{3}$
$\cot t = -\dfrac{1}{\sqrt{3}}$
$\sec t = -2$
$\csc t = \dfrac{2}{\sqrt{3}}$

51. $\sin t = 1$
$\cos t = 0$
$\tan t$ is undefined
$\cot t = 0$
$\sec t$ is undefined
$\csc t = 1$

53. $\tan t = -12/5$
$\cos t = -5/13$
$\sin t = 12/13$
$\cot t = -5/12$
$\sec t = -13/5$
$\csc t = 13/12$

55. Possibly an identity

57. Not an identity

59. Look at the graph of $y = \sec t$ in Figure 2–80. If you draw in the line $y = t$, it will pass through $(-\pi/2, -\pi/2)$ and $(\pi/2, \pi/2)$ and obviously will not intersect the graph of $y = \sec t$ when $-\pi/2 \le t \le -\pi/2$. But it will intersect each part of the graph that lies above the t-axis to the right of $t = \pi/2$; it will also intersect those parts that lie below the t-axis to the left of $-\pi/2$. The first coordinate of these infinitely many intersection points will be a solution of $\sec t = t$.

61. (a) area $OCA = \dfrac{\sin\theta\cos\theta}{2}$
(b) area $ODB = \tan\theta/2$
(c) area $OCB = \theta/2$

Chapter 2 Review, page 213

1. $\pi/3$ **3.** 324° **5.** $11\pi/9$ radians

7. $-495°$ **9.** $-1/3$ **11.** 0

13. .809 **15.** $-\sqrt{3}/2$ **17.** $\sqrt{3}/2$

19.

t	0	$\pi/6$	$\pi/4$	$\pi/3$	$\pi/2$
$\sin t$	0	$1/2$	$\sqrt{2}/2$	$\sqrt{3}/2$	1
$\cos t$	1	$\sqrt{3}/2$	$\sqrt{2}/2$	$1/2$	0

21. $9/4$ **23.** 0 **25.** $-\sqrt{\dfrac{2}{3}}$

27. $-3/5$ **29.** $-12/13$ **31.** $-1/2$ **33.** $\sqrt{3}/2$

35. (c) **37.** $\sin t = 7/\sqrt{58}$

39. $\tan t = -7/3$ **41.** $y = -\sqrt{3}x$

43. $-3/2$ **45.** $2/\sqrt{13}$

47. See Figure 2–81 on page 209.

49. (d) **51.** $-3/2$ **53.** 0 **55.** (d)

57. $-1/\sqrt{3}$ **59.** (a) $3/2$ (b) $\pi/5$

61.

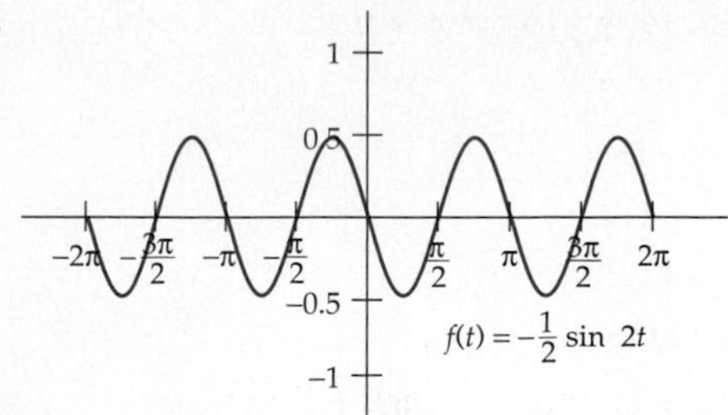

63. Not an identity. **65.** Possibly an identity.

67. 1/2

69. (a) Minimum: 8.9 hours; maximum: 15.1 hours
(b) January 1–March 2 and October 9–December 31

71. $g(t) = 2\cos(5t/2)$

73. $f(t) = 8\sin\left(\frac{2\pi}{5}t - \frac{28\pi}{5}\right)$ (Other answers are possible.)

75. $f(t) \approx 10.5588\sin(4t + .4580)$

77. $0 \le x \le \pi/50$ and $-5 \le y \le 5$

79. (a)

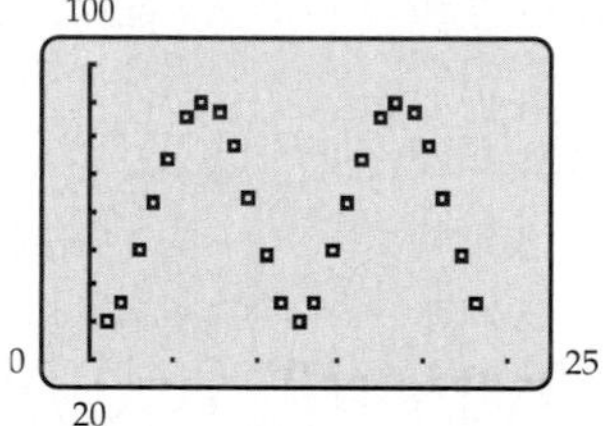

(b)

(c) The period is 12.05 months. The graph of the function, superimposed on the data points, suggests a good fit.

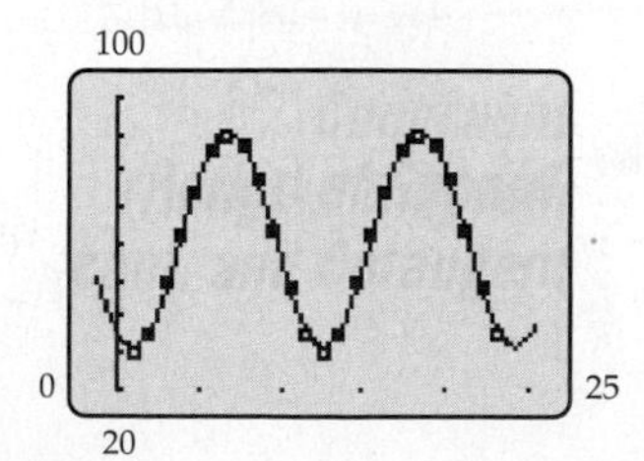

CHAPTER 3

Section 3.1, page 228

1. Possibly an identity **3.** Possibly an identity

5. B **7.** E

9. $\tan x\cos x = \frac{\sin x}{\cos x}\cdot\cos x = \sin x$

11. $\cos x\sec x = \cos x\cdot\frac{1}{\cos x} = 1$

13. $\tan x\csc x = \frac{\sin x}{\cos x}\cdot\frac{1}{\sin x} = \frac{1}{\cos x} = \sec x$

15. $\frac{\tan x}{\sec x} = \frac{\sin x}{\cos x} \div \frac{1}{\cos x} = \frac{\sin x}{\cos x}\cdot\frac{\cos x}{1} = \sin x$

17. $(1 + \cos x)(1 - \cos x) = 1 - \cos^2 x = \sin^2 x$

19. This is not an identity.

21. $\frac{\sin(-x)}{\cos(-x)} = \frac{-\sin x}{\cos x} = -\tan x$

23. $\cot(-x) = \frac{\cos(-x)}{\sin(-x)} = \frac{\cos x}{-\sin x} = -\cot x$

25. This is not an identity.

27. $\sec^2 x - \csc^2 x = (\tan^2 x + 1) - (\cot^2 x + 1) = \tan^2 x + 1 - \cot^2 x - 1 = \tan^2 x - \cot^2 x$

29. $\sin^2 x(\cot x + 1)^2 = [\sin x(\cot x + 1)]^2 = (\cos x + \sin x)^2 = \left(\cos x\left(1 + \frac{\sin x}{\cos x}\right)\right)^2 = [\cos x(1 + \tan x)]^2 = \cos^2 x(\tan x + 1)^2$

31. $\sin^2 x - \tan^2 x = \sin^2 x - \frac{\sin^2 x}{\cos^2 x} = \frac{\sin^2 x\cos^2 x - \sin^2 x}{\cos^2 x} = \frac{\sin^2 x(\cos^2 x - 1)}{\cos^2 x} = \tan^2 x(\cos^2 x - 1) = \tan^2 x(-\sin^2 x) = -\sin^2 x\tan^2 x$

33. $(\cos^2 x - 1)(\tan^2 x + 1) = (\cos^2 x - 1)\sec^2 x = \cos^2 x\sec^2 x - \sec^2 x = \cos^2 x\frac{1}{\cos^2 x} - \sec^2 x = 1 - \sec^2 x = -\tan^2 x$

35. $\frac{\sec x}{\csc x} = \frac{1/\cos x}{1/\sin x} = \frac{1}{\cos x}\cdot\frac{\sin x}{1} = \frac{\sin x}{\cos x} = \tan x$

37. $\cos^4 x - \sin^4 x = (\cos^2 x)^2 - (\sin^2 x)^2 = (\cos^2 x + \sin^2 x)(\cos^2 x - \sin^2 x) = 1(\cos^2 x - \sin^2 x) = \cos^2 x - \sin^2 x$

39. This is not an identity.

41. $\frac{\sec x}{\csc x} + \frac{\sin x}{\cos x} = \frac{1/\cos x}{1/\sin x} + \frac{\sin x}{\cos x} = \tan x + \frac{\sin x}{\cos x} = \tan x + \tan x = 2\tan x$

43. $\frac{\sec x + \csc x}{1 + \tan x} = \frac{1/\cos x + 1/\sin x}{1 + \sin x/\cos x} = \left(\frac{1}{\cos x} + \frac{1}{\sin x}\right) \div \left(1 + \frac{\sin x}{\cos x}\right) = \left(\frac{\sin x + \cos x}{\sin x\cos x}\right) \div \left(\frac{\cos x + \sin x}{\cos x}\right) = \frac{\sin x + \cos x}{\sin x\cos x}\cdot\frac{\cos x}{\cos x + \sin x} = \frac{1}{\sin x} = \csc x$

45. $\dfrac{1}{\csc x - \sin x} = \dfrac{1}{\dfrac{1}{\sin x} - \sin x} = \dfrac{1}{\dfrac{1 - \sin^2 x}{\sin x}} =$
$\dfrac{\sin x}{1 - \sin^2 x} = \dfrac{\sin x}{\cos^2 x} = \dfrac{1}{\cos x} \cdot \dfrac{\sin x}{\cos x} = \sec x \tan x$

47. This is not an identity.

49.

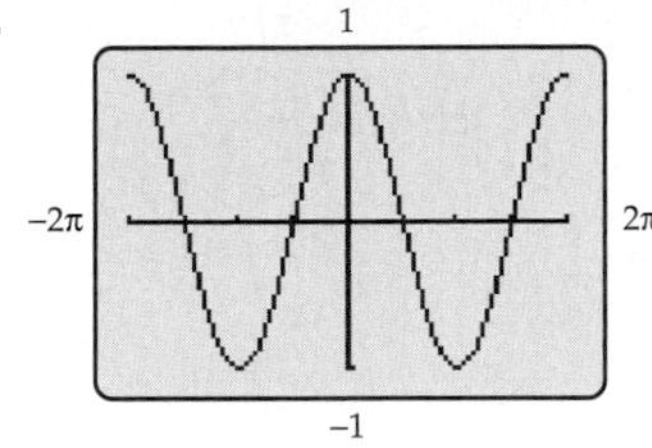

Conjecture: Right side $= \cos x$

Proof: $1 - \dfrac{\sin^2 x}{1 + \cos x} = \dfrac{1 + \cos x}{1 + \cos x} - \dfrac{\sin^2 x}{1 + \cos x} =$
$\dfrac{1 + \cos x - \sin^2 x}{1 + \cos x} = \dfrac{\cos^2 x + \sin^2 x + \cos x - \sin^2 x}{1 + \cos x}$
$= \dfrac{\cos^2 x + \cos x}{1 + \cos x} = \dfrac{\cos x(\cos x + 1)}{1 + \cos x} = \cos x$

51.

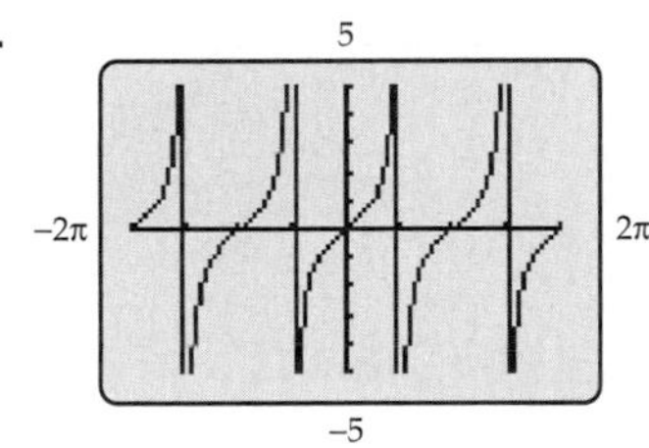

Conjecture: Right side $= \tan x$

Proof: $(\sin x + \cos x)(\sec x + \csc x) - \cot x - 2 =$
$(\sin x + \cos x)\left(\dfrac{1}{\cos x} + \dfrac{1}{\sin x}\right) - \cot x - 2 =$
$\tan x + 1 + 1 + \cot x - \cot x - 2 = \tan x$

53. $\dfrac{\cos^3 x}{1 + \sin x} = \dfrac{\cos^3 x(1 - \sin x)}{(1 + \sin x)(1 - \sin x)} = \dfrac{\cos^3 x(1 - \sin x)}{1 - \sin^2 x}$
$= \dfrac{\cos^3 x(1 - \sin x)}{\cos^2 x} = \cos x(1 - \sin x) = \dfrac{1 - \sin x}{\sec x}$

55. $\dfrac{\cos x}{1 - \sin x} = \dfrac{\cos x(1 + \sin x)}{(1 - \sin x)(1 + \sin x)} = \dfrac{\cos x(1 + \sin x)}{1 - \sin^2 x}$
$= \dfrac{\cos x(1 + \sin x)}{\cos^2 x} = \dfrac{1 + \sin x}{\cos x} = \dfrac{1}{\cos x} + \dfrac{\sin x}{\cos x}$
$= \sec x + \tan x$

57. Use Strategy 4: Prove that $(\cos x \cot x)(\cos x \cot x) = (\cot x + \cos x)(\cot x - \cos x)$.
$(\cot x + \cos x)(\cot x - \cos x) = \cot^2 x - \cos^2 x =$
$\dfrac{\cos^2 x}{\sin^2 x} - \cos^2 x = \dfrac{\cos^2 x - \cos^2 x \sin^2 x}{\sin^2 x} =$
$\dfrac{\cos^2 x(1 - \sin^2 x)}{\sin^2 x} = \dfrac{\cos^2 x}{\sin^2 x}(1 - \sin^2 x) = \cot^2 x \cos^2 x$
$= (\cos x \cot x)(\cos x \cot x)$
Therefore, $\dfrac{\cos x \cot x}{\cot x - \cos x} = \dfrac{\cot x + \cos x}{\cos x \cot x}$

59. (a)

(b) Since $\cos t = x - 2$ and $\sin t = y - 4$, the Pythagorean identity yields

$$\cos^2 t + \sin^2 t = 1$$
$$(x - 2)^2 + (y - 4)^2 = 1$$

(c) The graph is a circle with radius 1 and center at (2, 4).

61. Since $x = \dfrac{2}{\cos t}$, $\dfrac{x^2}{4} = \dfrac{4}{\cos^2 t} \div 4 = \dfrac{1}{\cos^2 t} = \sec^2 t$. Also, since $y = 4 \tan t$, $\dfrac{y^2}{16} = \dfrac{16 \tan^2 t}{16} = \tan^2 t$. Therefore, $\dfrac{x^2}{4} - \dfrac{y^2}{16} = \sec^2 t - \tan^2 t = 1$. The equation $\dfrac{x^2}{4} - \dfrac{y^2}{16} = 1$ is satisfied by every point on the curve.

Section 3.2, page 236

1. $\dfrac{\sqrt{6} - \sqrt{2}}{4}$ **3.** $2 - \sqrt{3}$ **5.** $2 - \sqrt{3}$

7. $-2 - \sqrt{3}$ **9.** $-2 - \sqrt{3}$ **11.** $\dfrac{\sqrt{6} + \sqrt{2}}{4}$

13. $\cos x$ **15.** $-\sin x$ **17.** $-1/\cos x$

19. $-\sin 2$ **21.** $\cos x$ **23.** $-2 \sin x \sin y$

25. $\dfrac{4 + \sqrt{2}}{6}$ **27.** $\dfrac{-\sqrt{3} + 2\sqrt{6}}{10}$

29. $\dfrac{4 + 3\sqrt{3}}{10}$ **31.** $\dfrac{4 - 3\sqrt{3}}{10}$

33. From the figure, $\sin t = \frac{4}{5}$ and $\cos t = \frac{3}{5}$. So,
$5\sin(x + t) = 5(\sin x \cos t + \cos x \sin t) =$
$\sin x \cdot 5\cos t + \cos x \cdot 5\sin t = \sin x \cdot 5\left(\frac{3}{5}\right) +$
$\cos x \cdot 5\left(\frac{4}{5}\right) = 3\sin x + 4\cos x$

35. $\frac{f(x+h) - f(x)}{h} = \frac{\cos(x+h) - \cos x}{h} =$
$\frac{\cos x \cos h - \sin x \sin h - \cos x}{h} =$
$\frac{\cos x(\cos h - 1) - \sin x \sin h}{h} =$
$\cos x\left(\frac{\cos h - 1}{h}\right) - \sin x \frac{\sin h}{h}$

37. $\sin(x + y) = -44/125$; $\tan(x + y) = 44/117$; $x + y$ lies in quadrant III.

39. $\cos(x + y) = -\frac{56}{65}$; $\tan(x + y) = \frac{33}{56}$; $x + y$ lies in quadrant III.

41. $\sin u \cos v \cos w + \cos u \sin v \cos w +$
$\cos u \cos v \sin w - \sin u \sin v \sin w$

43. If $x + y = \frac{\pi}{2}$, then $y = \frac{\pi}{2} - x$ and $\sin y = \sin\left(\frac{\pi}{2} - x\right)$
$= \cos x$. Thus, $\sin^2 x + \sin^2 y = \sin^2 x + \cos^2 x = 1$.

45. $\sin(x - \pi) = \sin x \cos \pi - \cos x \sin \pi = \sin x(-1) -$
$\cos x(0) = -\sin x$

47. $\cos(\pi - x) = \cos \pi \cos x + \sin \pi \sin x = (-1)\cos x +$
$(0)\sin x = -\cos x$

49. $\sin(x + \pi) = \sin x \cos \pi + \cos x \sin \pi = \sin x \cdot (-1) +$
$\cos x \cdot (0) = -\sin x$

51. $\tan(x + \pi) = \frac{\tan x + \tan \pi}{1 - \tan x \tan \pi} = \frac{\tan x + 0}{1 - \tan x \cdot 0} = \tan x$

53. $\sin(x + y)\sin(x - y) =$
$(\sin x \cos y + \cos x \sin y)(\sin x \cos y - \cos x \sin y) =$
$\sin^2 x \cos^2 y - \cos^2 x \sin^2 y$

55. $\frac{1}{2}[\cos(x - y) - \cos(x + y)] =$
$\frac{1}{2}[\cos x \cos y + \sin x \sin y - (\cos x \cos y - \sin x \sin y)]$
$= \frac{1}{2} \cdot 2\sin x \sin y = \sin x \sin y$

57. $\frac{\cos(x - y)}{\sin x \cos y} = \frac{\cos x \cos y + \sin x \sin y}{\sin x \cos y} = \frac{\cos x}{\sin x} + \frac{\sin y}{\cos y}$
$= \cot x + \tan y$

59. This is not an identity.

Special Topics 3.2.A, page 240

1. 3/4 **3.** $-4/5$ **5.** undefined

7. 1.3734 **9.** $\pi/4$

11. 1.3909 **13.** .1419

Section 3.3, page 247

1. $\sin 2x = \frac{120}{169}$, $\cos 2x = \frac{119}{169}$, $\tan 2x = \frac{120}{119}$

3. $\sin 2x = \frac{24}{25}$, $\cos 2x = -\frac{7}{25}$, $\tan 2x = -\frac{24}{7}$

5. $\sin 2x = \frac{24}{25}$, $\cos 2x = \frac{7}{25}$, $\tan 2x = \frac{24}{7}$

7. $\sin 2x = \frac{\sqrt{15}}{8}$, $\cos 2x = 7/8$, $\tan 2x = \frac{\sqrt{15}}{7}$

9. (a) Length $= 2x$, height $= y$, area $= 2xy$
(b) $x = 3\cos t$, $y = 3\sin t$
(c)
$$A = 2xy$$
$$A = 2(3\cos t)(3\sin t)$$
$$A = 9(2\sin t\cos t)$$
$$A = 9\sin 2t$$

11. $\frac{\sqrt{2 + \sqrt{2}}}{2}$ **13.** $\frac{\sqrt{2 + \sqrt{2}}}{2}$ **15.** $2 - \sqrt{3}$

17. $\frac{\sqrt{2 + \sqrt{3}}}{2}$ **19.** $\frac{\sqrt{2 - \sqrt{2}}}{2}$ **21.** $1 - \sqrt{2}$

23. $\sin\frac{x}{2} = \frac{\sqrt{30}}{10}$, $\cos\frac{x}{2} = \frac{\sqrt{70}}{10}$, $\tan\frac{x}{2} = \sqrt{\frac{3}{7}}$

25. $\sin\frac{x}{2} = \frac{1}{\sqrt{10}}$, $\cos\frac{x}{2} = -\frac{3}{\sqrt{10}}$, $\tan\frac{x}{2} = -\frac{1}{3}$

27. $\sin\frac{x}{2} = \frac{\sqrt{50 + 20\sqrt{5}}}{10}$, $\cos\frac{x}{2} = -\frac{\sqrt{50 - 20\sqrt{5}}}{10}$,
$\tan\frac{x}{2} = -(\sqrt{5} + 2)$

29. $\frac{1}{2}\sin 10x - \frac{1}{2}\sin 2x$ **31.** $\frac{1}{2}\cos 6x + \frac{1}{2}\cos 2x$

33. $\frac{1}{2}\cos 20x - \frac{1}{2}\cos 14x$ **35.** $2\sin 4x\cos x$

37. $2\cos 7x\sin 2x$ **39.** .96

41. .28 **43.** .316

45. $4\cos^3 x - 3\cos x$

47. $\cos x$ **49.** $\sin 4y$ **51.** 1

53. Use the third product to sum identity with $\frac{1}{2}(x + y)$ in place of x and $\frac{1}{2}(x - y)$ in place of y.

$$2\cos\left[\frac{1}{2}(x + y)\right]\cos\left[\frac{1}{2}(x - y)\right] =$$
$$2 \cdot \frac{1}{2}\left[\cos\left(\frac{1}{2}(x + y) + \frac{1}{2}(x - y)\right) + \cos\left(\frac{1}{2}(x + y) - \frac{1}{2}(x - y)\right)\right] = \cos x + \cos y$$

55. $\sin 16x = \sin 2(8x) = 2\sin 8x \cos 8x$

57. $\cos^4 x - \sin^4 x = (\cos^2 x + \sin^2 x)(\cos^2 x - \sin^2 x) = 1(\cos^2 x - \sin^2 x) = \cos 2x$

59. This is not an identity.

61. $\dfrac{1 + \cos 2x}{\sin 2x} = \dfrac{1 + 2\cos^2 x - 1}{2\sin x \cos x} = \dfrac{2\cos x \cos x}{2\sin x \cos x} = \dfrac{\cos x}{\sin x} = \cot x$

63. $\sin 3x = \sin(2x + x) = \sin 2x \cos x + \cos 2x \sin x = 2\sin x \cos x \cos x + (1 - 2\sin^2 x)\sin x = 2\sin x \cos^2 x + \sin x - 2\sin^3 x = 2\sin x(1 - \sin^2 x) + \sin x - 2\sin^3 x = \sin x(2 - 2\sin^2 x + 1 - 2\sin^2 x) = \sin x(3 - 4\sin^2 x)$

65. This is not an identity.

67. $\csc^2\left(\dfrac{x}{2}\right) = \dfrac{1}{\sin^2\left(\dfrac{x}{2}\right)} = \dfrac{1}{\left(\pm\sqrt{\dfrac{1 - \cos x}{2}}\right)^2} = \dfrac{1}{\dfrac{1 - \cos x}{2}} = \dfrac{2}{1 - \cos x}$

69. $\dfrac{\sin x - \sin 3x}{\cos x + \cos 3x} = \dfrac{2\cos\left(\dfrac{x + 3x}{2}\right)\sin\left(\dfrac{x - 3x}{2}\right)}{2\cos\left(\dfrac{x + 3x}{2}\right)\cos\left(\dfrac{x - 3x}{2}\right)} = \dfrac{\sin(-x)}{\cos(-x)} = \dfrac{-\sin x}{\cos x} = -\tan x$

71. (a) Use Strategy 4: Prove that $(1 - \cos x)(1 + \cos x) = \sin x \sin x$.
$(1 - \cos x)(1 + \cos x) = 1 - \cos^2 x = \sin^2 x = \sin x \sin x$

Therefore, $\dfrac{1 - \cos x}{\sin x} = \dfrac{\sin x}{1 + \cos x}$.

(b) On page 245, we proved that $\tan\left(\dfrac{x}{2}\right) = \dfrac{1 - \cos x}{\sin x}$.

Hence, by part (a) $\tan\left(\dfrac{x}{2}\right) = \dfrac{1 - \cos x}{\sin x} = \dfrac{\sin x}{1 + \cos x}$.

Section 3.4, page 258

1. $g(f(x)) = g(x + 1) = x + 1 - 1 = x$
$f(g(x)) = f(x - 1) = x - 1 + 1 = x$

3. $g(f(x)) = g\left(\dfrac{1}{x + 1}\right) = \dfrac{1 - \dfrac{1}{x + 1}}{\dfrac{1}{x + 1}} = \dfrac{x + 1 - 1}{1} = x$

$f(g(x)) = f\left(\dfrac{1 - x}{x}\right) = \dfrac{1}{\dfrac{1 - x}{x} + 1} = \dfrac{x}{1 - x + x} = x$

5. $\pi/2$ **7.** $-\pi/4$ **9.** 0
11. $\pi/6$ **13.** $-\pi/4$ **15.** $-\pi/3$
17. $2\pi/3$ **19.** .3576 **21.** -1.2728
23. .7168 **25.** $-.8584$ **27.** 2.2168
29. $\cos u = 1/2$; $\tan u = -\sqrt{3}$
31. $\pi/2$ **33.** $5\pi/6$ **35.** $-\pi/3$
37. $\pi/3$ **39.** $\pi/6$ **41.** 4/5
43. 4/5 **45.** 5/12

47. $\cos(\sin^{-1} v) = \sqrt{1 - v^2} \quad (-1 \le v \le 1)$

49. $\tan(\sin^{-1} v) = \dfrac{v}{\sqrt{1 - v^2}} \quad (-1 < v < 1)$

51.

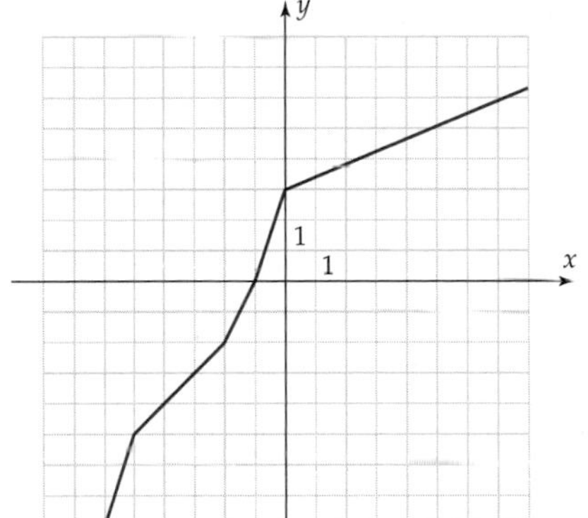

53.

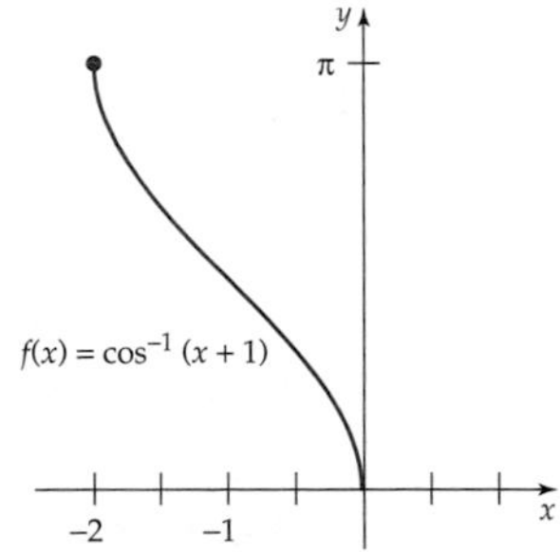

55.

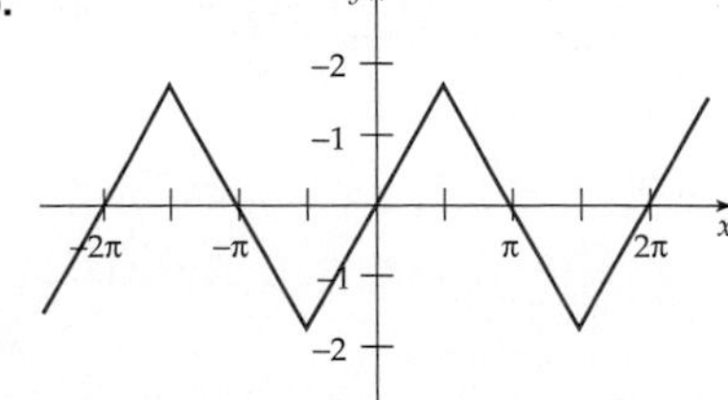

57. (a) $t = \dfrac{1}{2\pi f}\sin^{-1}\left(\dfrac{V}{V_{\max}}\right) + \dfrac{n}{f} \quad (n = 0, \pm 1, \pm 2, \ldots)$
(b) $t = 5.8219 \times 10^{-4}$ sec

59. (a) $\theta = \tan^{-1}\left(\dfrac{25}{x}\right) - \tan^{-1}\left(\dfrac{10}{x}\right)$ (b) 15.8 ft

61. Verify that the restricted secant function is one-to-one (see Figure 2–80). Hence, the restricted secant function has an inverse function.

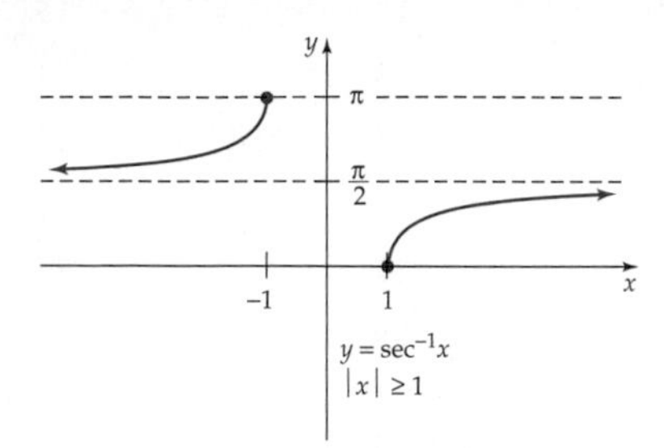

63. Verify that the restricted cotangent function is one-to-one (see Figure 2–81). Hence, the restricted cotangent function has an inverse function.

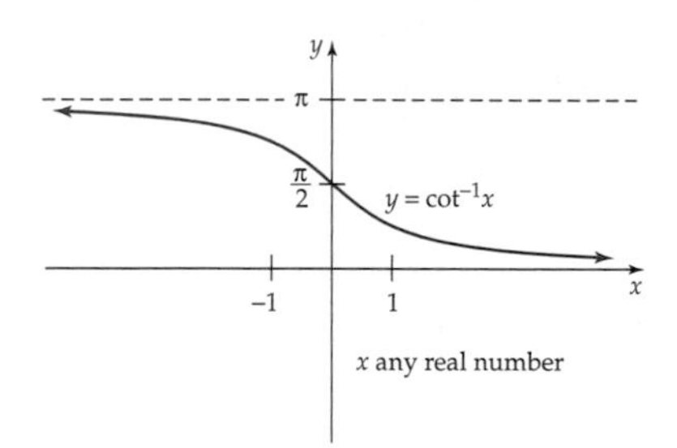

65. Let $u = \sin^{-1}(-x)$. Then

$$\begin{aligned} -x &= \sin u \quad (-\pi/2 \le u \le \pi/2) \\ x &= -\sin u \\ x &= \sin(-u) \\ \sin^{-1} x &= -u \quad (-\pi/2 \le -u \le \pi/2) \\ \sin^{-1} x &= -\sin^{-1}(-x) \\ \sin^{-1}(-x) &= -\sin^{-1} x \end{aligned}$$

67. Let $u = \cos^{-1}(-x)$. Then $-x = \cos u$. Note that $0 \le u \le \pi$; hence, $-\pi \le -u \le 0$, which implies that $0 \le \pi - u \le \pi$. Hence,

$$\begin{aligned} x &= -\cos u \\ x &= \cos(\pi - u) \quad \text{(from the stated identity)} \\ \pi - u &= \cos^{-1} x \\ u &= \pi - \cos^{-1} x \\ \cos^{-1}(-x) &= \pi - \cos^{-1} x \end{aligned}$$

69. Let $u = \cot x$. Then

$$\tan(\pi/2 - x) = \cot x = u \quad (0 < u < \pi, \text{ hence, } -\pi/2 < \pi/2 - u < \pi/2)$$

$$\pi/2 - x = \tan^{-1}(u) = \tan^{-1}(\cot x)$$

71. Let $u = \sin^{-1} x$. Then $\sin u = x$, $(-\pi/2 < u < \pi/2)$

$$\tan u = \frac{\sin u}{\cos u} = \frac{\sin u}{\sqrt{1 - \sin^2 u}} = \frac{x}{\sqrt{1 - x^2}}$$

(positive square root, since u lies in quadrant I or IV)

$$u = \tan^{-1}\left(\frac{x}{\sqrt{1 - x^2}}\right)$$

$$\sin^{-1} x = \tan^{-1}\left(\frac{x}{\sqrt{1 - x^2}}\right)$$

73. The statement is false. For example, let $x = 1/2$.

$$\tan^{-1}\left(\frac{1}{2}\right) \approx .4636, \text{ but } \frac{\sin^{-1}(1/2)}{\cos^{-1}(1/2)} = \frac{\pi/6}{\pi/3} = \frac{1}{2}$$

75. (a) Slope $= (a - b)/(b - a) = -1$
(b) The slopes are 1 and -1; their product is -1, so the lines are perpendicular.
(c) Length of $PR = \sqrt{(c - a)^2 + (c - b)^2}$, length of $RQ = \sqrt{(b - c)^2 + (a - c)^2}$. Clearly, these are equal, since $(b - c)^2 = (c - b)^2$ and $(c - a)^2 = (a - c)^2$. Thus, the line $y = x$ is perpendicular to PQ and bisects PQ, so the line is the perpendicular bisector of PQ, and thus P and Q are symmetric with respect to the line $y = x$.

77. (a) Suppose $a \ne b$. If $f(a) = f(b)$, then $g(f(a)) = g(f(b))$. But $a = g(f(a))$ by (1) and $b = g(f(b))$. Hence, $a = b$, contrary to our hypothesis. Therefore, it cannot happen that $f(a) = f(b)$, that is, $f(a) \ne f(b)$. Hence, f is one-to-one
(b) If $g(y) = x$, then $f(g(y)) = f(x)$. But $f(g(y)) = y$ by (2). Hence, $y = f(g(y)) = f(x)$.
(c) If $f(x) = y$, then $g(f(x)) = g(y)$. But $g(f(x)) = x$ by (1). Hence, $x = g(f(x)) = g(y)$.

Section 3.5, page 266

In the answers for this section, $k = 0, \pm 1, \pm 2, \pm 3, \ldots$.

1. $x = -.4836 + 2k\pi$ or $3.6252 + 2k\pi$

3. $x = 2.1700 + 2k\pi$ or $-2.1700 + 2k\pi$

5. $x = -.2327 + k\pi$

7. $x = .4101 + k\pi$

9. $x = \pm 1.9577 + 2k\pi$

11. $x = .1193$ or 3.0223

13. $x = 1.3734$ or 4.5150

15. $x = \dfrac{\pi}{3} + 2k\pi$ or $\dfrac{2\pi}{3} + 2k\pi$

17. $x = -\dfrac{\pi}{3} + k\pi$

19. $x = 5\pi/6 + 2k\pi$ or $-5\pi/6 + 2k\pi$

21. $x = -\frac{\pi}{6} + 2k\pi$ or $\frac{7\pi}{6} + 2k\pi$

23. $x = \frac{\pi}{6} + 2k\pi$ or $\frac{5\pi}{6} + 2k\pi$

25. $x = .5275 + k\pi$ or $x = 1.6868 + k\pi$

27. $x = .4959, 1.2538, 1.5708, 1.8877, 2.6457$, or 4.7124. Since the function is periodic with period 2π, all solutions are given by each of these six roots plus $2k\pi$.

29. $x = .1671, 1.8256, 2.8867$, or 4.5453. Since the function is periodic with period 2π, all solutions are given by each of these four roots plus $2k\pi$.

31. $x = 1.2161 + 2k\pi$ or $x = 5.0671 + 2k\pi$

33. $x = 2.4620 + 2k\pi$ or $x = 3.8212 + 2k\pi$

35. $x = .5166 + 2k\pi$ or $x = 5.6766 + 2k\pi$

37. $\theta = 30°$ or $150°$ **39.** $\theta = 30°$ or $330°$

41. $\theta = 82.8°$ or $262.8°$ **43.** $\theta = 114.8°$ or $245.2°$

45. $\theta = 45°$ or $315°$

47. $t = 1.8546$; $A = 11.18$

49. $t = 1.6961$; $A = .35$

51. $t = 1.9346$, $L = 16.47$ **53.** $t = 1.6236$, $L = 11.61$

55. (a) $A = 2x(2\cos 2x) = 4x\cos 2x$
(b) $x = .3050$ and $x = .5490$

57. $\alpha = 65.4°$ **59.** $\alpha = 30°$

61. $\theta = 27.6°$ **63.** $\theta = 14.2°$

65. (a) days 60 and 282 (March 2 and October 10)
(b) day 171 (June 21)

67. $\theta = .5475$ or 1.0233 **69.** $13.25°$ or $76.75°$

71. The basic equations $\sin x = c$ and $\cos x = c$ have no solutions if $c > 1$ or $c < -1$.

Special Topics 3.5.A, page 271

In the answers for this section $k = 0, \pm1, \pm2, \pm3, \ldots$.

1. $x = -\frac{\pi}{6} + k\pi$ or $\frac{2\pi}{3} + k\pi$

3. $x = \pm\frac{\pi}{2} + 4k\pi$ **5.** $x = -\frac{\pi}{9} + \frac{k\pi}{3}$

7. $x = \pm.7381 + \frac{2k\pi}{3}$ **9.** $x = 2.2143 + 2k\pi$

11. $x = 5.9433$ and 3.4814

13. $2.1588, 5.3004, 3\pi/4$, and $7\pi/4$

15. $x = \pi/2, 3\pi/2, \pi/4, 5\pi/4$

17. $x = \pi/2, 3\pi/2, \pi/6, 5\pi/6$

19. $x = .8481, 2.2935, 1.7682, 4.9098$

21. $x = .8213, 2.3203$.

23. $x = 1.2059, 4.3475, .3649$, and 3.5065

25. $x = 1.0591, 4.2007, 2.8679$, and 6.0095

27. No solution

29. $x = \pi/2, 3\pi/2, 7\pi/6, 11\pi/6$

31. $x = \pi/6, 5\pi/6$

33. $x = 7\pi/6, 11\pi/6, \pi/2$

35. $x = 0, \pi/3, 5\pi/3, 2\pi$

37. The solutions $x = 0, \pi$ are missed owing to dividing by $\sin x$.

Chapter 3 Review, page 274

1. $1/3 + \cot t$ **3.** $\sin^4 x$

5. $\sin^4 t - \cos^4 t = (\sin^2 t + \cos^2 t)(\sin^2 t - \cos^2 t)$
$= 1(\sin^2 t - (1 - \sin^2 t)) = \sin^2 t - 1 + \sin^2 t = 2\sin^2 t - 1$

7. Use Strategy 4. Prove that $\sin t \sin t = (1 - \cos t)(1 + \cos t)$ is an identity.
$\sin t \sin t = \sin^2 t = 1 - \cos^2 t = (1 - \cos t)(1 + \cos t)$
Therefore, $\frac{\sin t}{1 - \cos t} = \frac{1 + \cos t}{\sin t}$.

9. This is not an identity.

11. $(\sin x + \cos x)^2 - \sin 2x = \sin^2 x + 2\sin x\cos x + \cos^2 x - \sin 2x = \sin^2 x + \cos^2 + 2\sin x\cos x - 2\sin x\cos x = 1 + 0 = 1$

13. $\frac{\tan x - \sin x}{2\tan x} = \frac{\sin x/\cos x - \sin x}{2(\sin x/\cos x)} =$

$\frac{\frac{\sin x - \sin x\cos x}{\cos x}}{\frac{2\sin x}{\cos x}} = \frac{\sin x - \sin x\cos x}{2\sin x} =$

$\frac{\sin x(1 - \cos x)}{2\sin x} = \frac{1 - \cos x}{2} = \sin^2(x/2)$

15. $\cos(x + y)\cos(x - y) =$
$(\cos x\cos y - \sin x\sin y)(\cos x\cos y + \sin x\sin y) =$
$\cos x^2\cos^2 y - \sin^2 x\sin^2 y =$
$\cos^2 x(1 - \sin^2 y) - (1 - \cos^2 x)\sin^2 y =$
$\cos^2 x - \cos^2 x\sin^2 y - \sin^2 y + \cos^2 x\sin^2 y =$
$\cos^2 x - \sin^2 y$

17. Use Strategy 4. Prove that $(\sec x + 1)(\sec x - 1) = \tan x\tan x$ is an identity.
$(\sec x + 1)(\sec x - 1) = \sec^2 x - 1 = \tan^2 x = \tan x\tan x$
Therefore, $\frac{\sec x + 1}{\tan x} = \frac{\tan x}{\sec x - 1}$ is an identity.

19. $\frac{1 + \tan^2 x}{\tan^2 x} = \frac{1}{\tan^2 x} + \frac{\tan^2 x}{\tan^2 x} = \cot^2 x + 1 = \csc^2 x$

21. $\tan^2 x - \sec^2 x = \sec^2 x - 1 - \sec^2 x = -1 = \csc^2 x - 1 - \csc^2 x = \cot^2 x - \csc^2 x$

23. 120/169

25. $-\dfrac{56}{65}$

27. $\dfrac{\sqrt{45}+1}{8}$ or $\dfrac{3\sqrt{5}+1}{8}$

29. Yes. If $\sin x = 0$, $\sin 2x = 2\sin x\cos x = 2\cdot 0\cdot\cos x = 0$.

31. Using the half-angle identity,

$$\cos\frac{\pi}{12} = \cos\left[\frac{1}{2}\left(\frac{\pi}{6}\right)\right] = \sqrt{\frac{1+\cos(\pi/6)}{2}} = \sqrt{\frac{1+\sqrt{3}/2}{2}} = \sqrt{\frac{2+\sqrt{3}}{4}} = \frac{\sqrt{2+\sqrt{3}}}{2}$$

Using the subtraction identity for cosine,

$$\cos\frac{\pi}{12} = \cos\left(\frac{\pi}{3}-\frac{\pi}{4}\right) = \cos\frac{\pi}{3}\cos\frac{\pi}{4} + \sin\frac{\pi}{3}\sin\frac{\pi}{4} = \frac{1}{2}\cdot\frac{\sqrt{2}}{2} + \frac{\sqrt{3}}{2}\cdot\frac{\sqrt{2}}{2} = \frac{\sqrt{2}+\sqrt{6}}{4}$$

Hence, $\dfrac{\sqrt{2+\sqrt{3}}}{2} = \dfrac{\sqrt{2}+\sqrt{6}}{4}$ and $\sqrt{2+\sqrt{3}} = \dfrac{\sqrt{2}+\sqrt{6}}{2}$.

33. $\dfrac{\sqrt{6}+\sqrt{2}}{4} = \dfrac{\sqrt{2+\sqrt{3}}}{2}$

35. (a)

37. .96

39. $\pi/4$ radians

41. $\pi/4$

43. $\pi/3$

45. $5\pi/6$

47. .75

49. $\pi/3$

51.

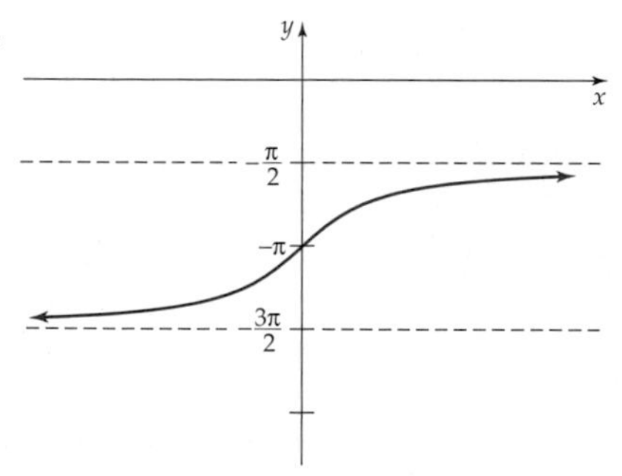

53. $\sqrt{15}/4$

In the following answers, $k = 0, \pm1, \pm2, \pm3, \ldots.$

55. $x = \dfrac{\pi}{6} + 2k\pi$ or $\dfrac{5\pi}{6} + 2k\pi$

57. $x = -\pi/4 + k\pi$

59. $x = .7754 + 2k\pi$ or $2.3662 + 2k\pi$

61. $x = 1.4940 + k\pi$

63. $x = \pi/6 + 2k\pi$ or $5\pi/6 + 2k\pi$

65. $x = -\dfrac{\pi}{6} + 2k\pi$ or $\dfrac{7\pi}{6} + 2k\pi$

67. $x = \pm\pi/3 + k\pi$

69. $x = .8959 + 2k\pi$ or $2.2457 + 2k\pi$

71. $x = k\pi$

73. $x = .8419, 2.2997, 4.1784$, or 5.2463. Since the function is periodic with period 2π, all solutions are given by $x = .8419 + 2k\pi$, $x = 2.2997 + 2k\pi$, and so on.

75. $\theta = 314.5°$ and $225.5°$

77. $\theta = 9.06°$ or $80.94°$

CHAPTER 4

Section 4.1, page 284

1. $8 + 2i$

3. $-2 - 10i$

5. $-1/2 - 2i$

7. $\dfrac{\sqrt{2}-\sqrt{3}}{2} + 2i$

9. $1 + 13i$

11. $-10 + 11i$

13. $-21 - 20i$

15. 4

17. $-i$

19. i

21. i

23. $\dfrac{5}{29} + \dfrac{2}{29}i$

25. $-\dfrac{1}{3}i$

27. $\dfrac{12}{41} - \dfrac{15}{41}i$

29. $-\dfrac{5}{41} - \dfrac{4}{41}i$

31. $\dfrac{10}{17} - \dfrac{11}{17}i$

33. $\dfrac{7}{10} + \dfrac{11}{10}i$

35. $-\dfrac{113}{170} + \dfrac{41}{170}i$

37. $6i$

39. $\sqrt{14}i$

41. $-4i$

43. $11i$

45. $(\sqrt{15} + 3\sqrt{2})i$

47. 2/3

49. $-41 - i$

51. $(2 + 5\sqrt{2}) + (\sqrt{5} - 2\sqrt{10})i$

53. $\dfrac{1}{3} - \dfrac{\sqrt{2}}{3}i$

55. $x = 2, y = -2$

57. $x = -3/4, y = 3/2$

59. $\dfrac{1}{3} + \dfrac{\sqrt{14}}{3}i, \dfrac{1}{3} - \dfrac{\sqrt{14}}{3}i$

61. $-\dfrac{1}{2} + \dfrac{\sqrt{7}}{2}i, -\dfrac{1}{2} - \dfrac{\sqrt{7}}{2}i$

63. $\dfrac{1}{4} + \dfrac{\sqrt{31}}{4}i, \dfrac{1}{4} - \dfrac{\sqrt{31}}{4}i$

65. $\dfrac{3+\sqrt{3}}{2} + 0i, \dfrac{3-\sqrt{3}}{2} + 0i$

67. $2 + 0i, -1 + \sqrt{3}i, -1 - \sqrt{3}i$

69. $1 + 0i, -1 + 0i, 0 + i, 0 - i$

71. -1

73. $\bar{\bar{z}} = \overline{\overline{a+bi}} = \overline{a-bi} = a + bi = z$

75. $\dfrac{1}{z} = \dfrac{a}{a^2+b^2} - \dfrac{b}{a^2+b^2}i$

Section 4.2, page 290

1–7.

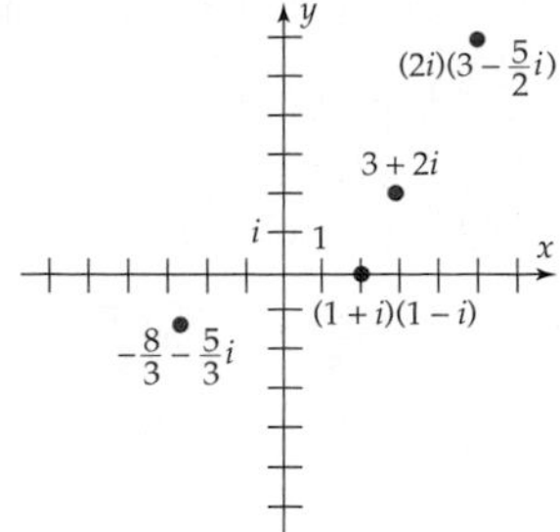

9. 13 **11.** $\sqrt{3}$ **13.** 12

15. Many correct answers, including $z = 1$, $w = i$

17.

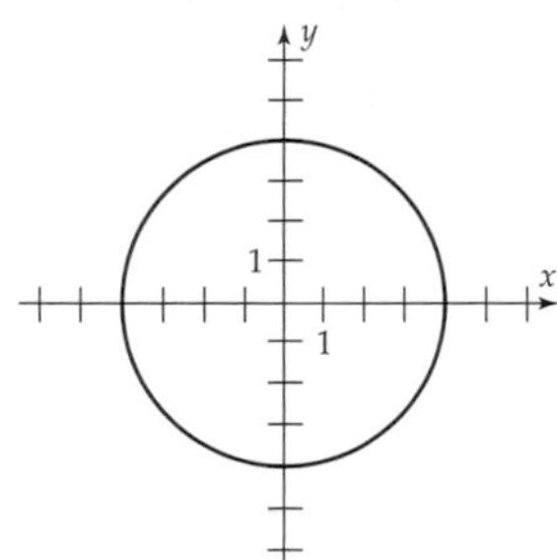

19.

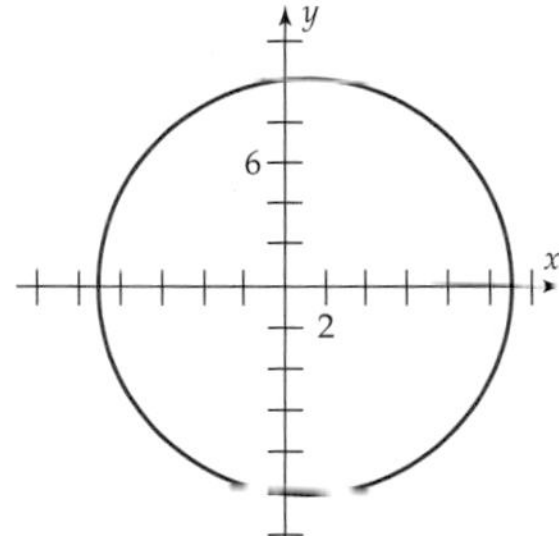

21.

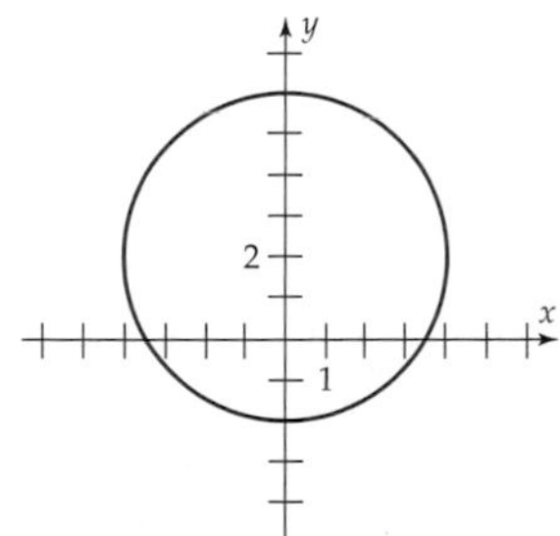

23.

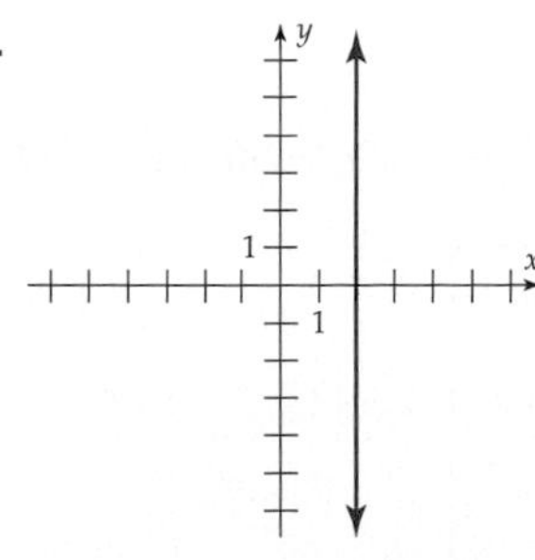

25. $\sqrt{2} + i\sqrt{2}$

27. $0 + i = i$

29. $-\frac{5}{2} + i\frac{5\sqrt{3}}{2}$

31. $\frac{3\sqrt{3}}{4} + \frac{3}{4}i$

33. $-1.3073 - 1.5136i$

35. $-1.6646 + 3.6372i$

37. $3\sqrt{2}\left(\cos\frac{\pi}{4} + i\sin\frac{\pi}{4}\right)$

39. $4\left(\cos\frac{\pi}{3} + i\sin\frac{\pi}{3}\right)$

41. $6\left(\cos\frac{11\pi}{6} + i\sin\frac{11\pi}{6}\right)$

43. $\sqrt{6}\left(\cos\frac{5\pi}{4} + i\sin\frac{5\pi}{4}\right)$

45. $5(\cos(.9273) + i\sin(.9273))$

47. $13(\cos(5.1072) + i\sin(5.1072))$

49. $\sqrt{5}(\cos(1.1071) + i\sin(1.1071))$

51. $\frac{\sqrt{74}}{2}(\cos(2.191) + i\sin(2.191))$

53. $\cos\frac{3\pi}{2} + i\sin\frac{3\pi}{2} = 0 + i(-1) = -i$

55. $12\left(\cos\frac{\pi}{3} + i\sin\frac{\pi}{3}\right) = 6 + 6\sqrt{3}i$

57. $36\left(\cos\frac{\pi}{2} + i\sin\frac{\pi}{2}\right) = 36i$

59. $\cos\frac{\pi}{3} + i\sin\frac{\pi}{3} = \frac{1}{2} + \frac{\sqrt{3}}{2}i$

61. $2\left(\cos\frac{\pi}{6} + i\sin\frac{\pi}{6}\right) = \sqrt{3} + i$

63. $\frac{3}{2}\left(\cos\frac{\pi}{4} + i\sin\frac{\pi}{4}\right) = \frac{3\sqrt{2}}{4} + \frac{3\sqrt{2}}{4}i$

65. $2\sqrt{2}\left(\cos\frac{7\pi}{12} + i\sin\frac{7\pi}{12}\right)$

67. $\cos\frac{\pi}{2} + i\sin\frac{\pi}{2}$

69. $12\left(\cos\frac{2\pi}{3} + i\sin\frac{2\pi}{3}\right)$

71. $2\sqrt{2}\left(\cos\frac{19\pi}{12} + i\sin\frac{19\pi}{12}\right)$

73. Since $z = r(\cos\theta + i\sin\theta)$ and $i = \cos\frac{\pi}{2} + i\sin\frac{\pi}{2}$ we see that $iz = r\left(\cos\left(\theta + \frac{\pi}{2}\right) + i\sin\left(\theta + \frac{\pi}{2}\right)\right)$. The argument of iz is 90° more than that of z; hence, iz is equivalent to rotating z through 90° in the complex plane.

75. (a) $\frac{b}{a}$

(b) $\frac{d}{c}$

(c) $y - b = \frac{d}{c}(x - a)$

(d) $y - d = \frac{b}{a}(x - c)$

(e)

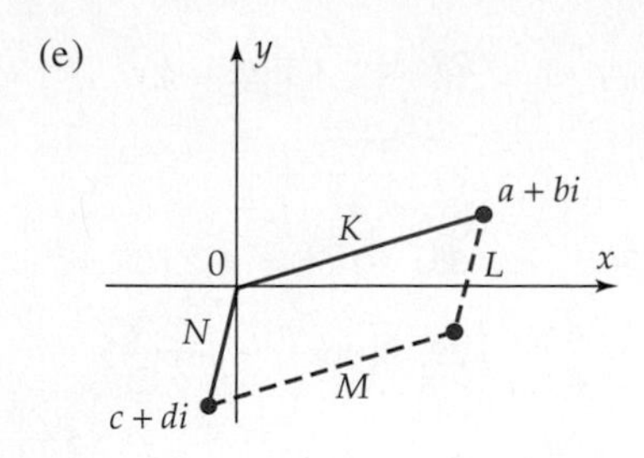

(f) Substituting $(a + c, b + d)$ in the equation of line L yields

$$y - b = \frac{d}{c}(x - a)$$
$$b + d - b = \frac{d}{c}(a + c - a)$$
$$d = d.$$

Similarly, substituting $(a + c, b + d)$ in the equation of line M yields

$$y - d = \frac{b}{a}(x - c)$$
$$b + d - d = \frac{b}{a}(a + c - c)$$
$$b = b.$$

Thus, $(a + c, b + d)$ lies on both lines L and M.

77. (a) $r_2(\cos\theta_2 + i\sin\theta_2)(\cos\theta_2 - i\sin\theta_2) = r_2(\cos^2\theta_2 + \sin^2\theta_2) = r_2 \cdot 1 = r_2$

(b) $r_1(\cos\theta_1 + i\sin\theta_1)(\cos\theta_2 - i\sin\theta_2)$

$= r_1(\cos\theta_1\cos\theta_2 - i\cos\theta_1\sin\theta_2 + i\sin\theta_1\cos\theta_2 - i^2\sin\theta_1\sin\theta_2)$

$= r_1(\cos\theta_1\cos\theta_2 + \sin\theta_1\sin\theta_2 + i(\sin\theta_1\cos\theta_2 - \cos\theta_1\sin\theta_2))$

$= r_1(\cos(\theta_1 - \theta_2) + i\sin(\theta_1 - \theta_2))$

Section 4.3, page 299

1. $0 + 1i = i$

3. $128 + 128i\sqrt{3}$

5. $-\frac{243\sqrt{3}}{2} - \frac{243}{2}i$

7. $-1 + 0i = -1$

9. $-64 + 0i = -64$

11. $\frac{1}{2} - \frac{\sqrt{3}}{2}i$

13. $0 + i1 = i$

15. $k = 0$: 1; $k = 1$: i; $k = 2$: -1; $k = 3$: $-i$

17. $k = 0$: $6\left(\cos\frac{\pi}{6} + i\sin\frac{\pi}{6}\right)$; $k = 1$: $6\left(\cos\frac{7\pi}{6} + i\sin\frac{7\pi}{6}\right)$

19. $k = 0$: $4\left(\cos\frac{\pi}{15} + i\sin\frac{\pi}{15}\right)$; $k = 1$: $4\left(\cos\frac{11\pi}{15} + i\sin\frac{11\pi}{15}\right)$; $k = 2$: $4\left(\cos\frac{7\pi}{5} + i\sin\frac{7\pi}{5}\right)$

21. $k = 0$: $3\left(\cos\frac{\pi}{48} + i\sin\frac{\pi}{48}\right)$; $k = 1$: $3\left(\cos\frac{25\pi}{48} + i\sin\frac{25\pi}{48}\right)$; $k = 2$: $3\left(\cos\frac{49\pi}{48} + i\sin\frac{49\pi}{48}\right)$; $k = 3$: $3\left(\cos\frac{73\pi}{48} + i\sin\frac{73\pi}{48}\right)$

23. $k = 0$: $\cos\frac{\pi}{5} + i\sin\frac{\pi}{5}$; $k = 1$: $\cos\frac{3\pi}{5} + i\sin\frac{3\pi}{5}$; $k = 2$: $\cos\pi + i\sin\pi = -1$; $k = 3$: $\cos\frac{7\pi}{5} + i\sin\frac{7\pi}{5}$; $k = 4$: $\cos\frac{9\pi}{5} + i\sin\frac{9\pi}{5}$

25. $k = 0$: $\cos\frac{\pi}{10} + i\sin\frac{\pi}{10}$; $k = 1$: $\cos\frac{\pi}{2} + i\sin\frac{\pi}{2}$; $k = 2$: $\cos\frac{9\pi}{10} + i\sin\frac{9\pi}{10} = i$; $k = 3$: $\cos\frac{13\pi}{10} + i\sin\frac{13\pi}{10}$; $k = 4$: $\cos\frac{17\pi}{10} + i\sin\frac{17\pi}{10}$

27. $k = 0$: $\sqrt[4]{2}\left(\cos\frac{\pi}{8} + i\sin\frac{\pi}{8}\right)$; $k = 1$: $\sqrt[4]{2}\left(\cos\frac{9\pi}{8} + i\sin\frac{9\pi}{8}\right)$

29. $k = 0$: $2\left(\cos\frac{\pi}{24} + i\sin\frac{\pi}{24}\right)$; $k = 1$: $2\left(\cos\frac{13\pi}{24} + i\sin\frac{13\pi}{24}\right)$; $k = 2$: $2\left(\cos\frac{25\pi}{24} + i\sin\frac{25\pi}{24}\right)$; $k = 3$: $2\left(\cos\frac{37\pi}{24} + i\sin\frac{37\pi}{24}\right)$

31. $\cos\frac{\pi}{6} + i\sin\frac{\pi}{6} = \frac{\sqrt{3}}{2} + \frac{1}{2}i$, $\cos\frac{7\pi}{6} + i\sin\frac{7\pi}{6} = -\frac{\sqrt{3}}{2} - \frac{1}{2}i$, $\cos\frac{\pi}{2} + i\sin\frac{\pi}{2} = i$
$\cos\frac{3\pi}{2} + i\sin\frac{3\pi}{2} = -i$, $\cos\frac{5\pi}{6} + i\sin\frac{5\pi}{6} = -\frac{\sqrt{3}}{2} + \frac{1}{2}i$,
$\cos\frac{11\pi}{6} + i\sin\frac{11\pi}{6} = \frac{\sqrt{3}}{2} - \frac{1}{2}i$

33. $\cos\frac{\pi}{6} + i\sin\frac{\pi}{6} = \frac{\sqrt{3}}{2} + \frac{1}{2}i$, $\cos\frac{5\pi}{6} + i\sin\frac{5\pi}{6} = -\frac{\sqrt{3}}{2} + \frac{1}{2}i$, $\cos\frac{3\pi}{2} + i\sin\frac{3\pi}{2} = -i$

35. $3i$, $-\frac{3\sqrt{3}}{2} - \frac{3i}{2}$, $\frac{3\sqrt{3}}{2} - \frac{3i}{2}$

37. $3\left[\cos\left(\frac{\pi + 4k\pi}{10}\right) + i\sin\left(\frac{\pi + 4k\pi}{10}\right)\right]$, $k = 0, 1, 2, 3, 4$

39. $\sqrt[4]{2}\left(\frac{\sqrt{3}}{2} + \frac{1}{2}i\right)$, $\sqrt[4]{2}\left(-\frac{1}{2} + \frac{\sqrt{3}}{2}i\right)$, $\sqrt[4]{2}\left(-\frac{\sqrt{3}}{2} - \frac{1}{2}i\right)$, $\sqrt[4]{2}\left(\frac{1}{2} - \frac{\sqrt{3}}{2}i\right)$

41. 1, $.6235 \pm .7818i$, $-.2225 \pm .9749i$, $-.9010 \pm .4339i$

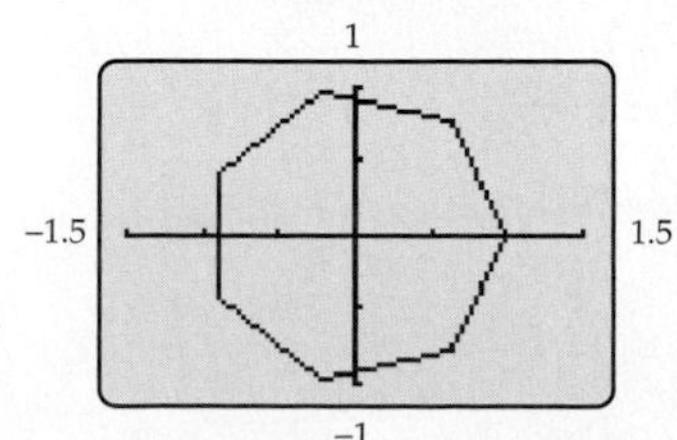

43. $\pm 1, \pm i, .7071 \pm .7071i, -.7071 \pm 7071i$

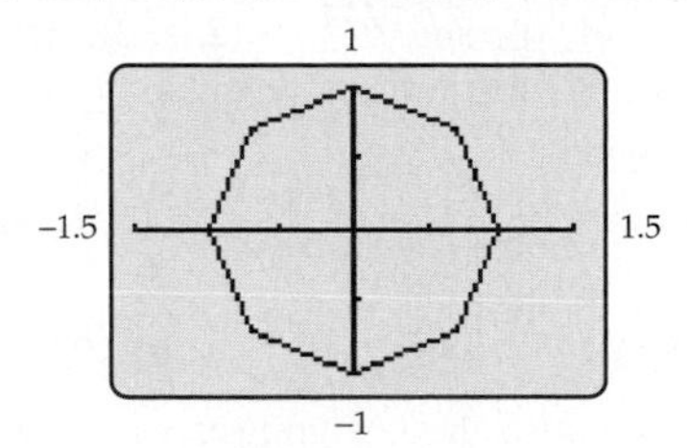

45. $1, .7660 \pm .6428i, .1736 \pm .9848i, -.5 \pm .8660i,$
$-.9397 \pm .3420i$

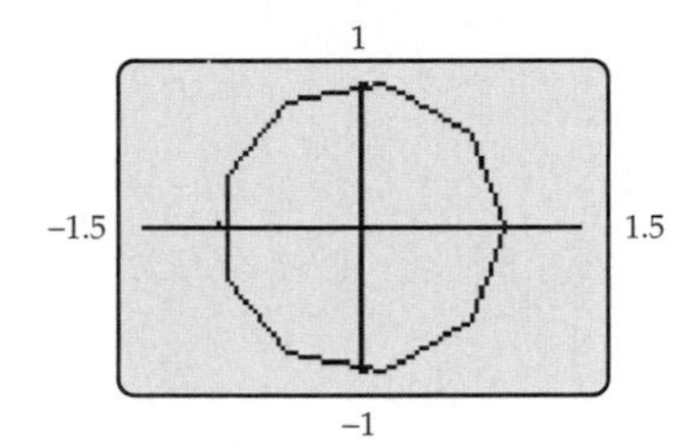

47. $-1, i, -i$

49. $(-1 \pm \sqrt{3}i)/2, (1 \pm \sqrt{3}i)/2$, and -1

51. 12 trips around the circle

53. Since the u_i are distinct, clearly, the vu_i are distinct. Since v is a solution of $z^n = r(\cos\theta + i\sin\theta)$, we can write $v^n = r(\cos\theta + i\sin\theta)$, and similarly $u_i^n = 1$. Then $(vu_i)^n = v^n u_i^n = r(\cos\theta + i\sin\theta)\cdot 1 = r(\cos\theta + i\sin\theta)$. Thus, the vu_i are the solutions of the equation.

Section 4.4, page 312

1. $3\sqrt{5}$ **3.** $\sqrt{34}$

5. $\langle 6, 6\rangle$ **7.** $\langle -6, 10\rangle$

9. $\langle 13/5, -2/5\rangle$

11. $\mathbf{u}+\mathbf{v} = \langle 4, 5\rangle$; $\mathbf{u}-\mathbf{v} = \langle -8, 3\rangle$; $3\mathbf{u}-2\mathbf{v} = \langle -18, 10\rangle$

13. $\mathbf{u}+\mathbf{v} = \langle 3+4\sqrt{2}, 3\sqrt{2}+1\rangle$;
$\mathbf{u}-\mathbf{v} = \langle 3-4\sqrt{2}, 3\sqrt{2}-1\rangle$;
$3\mathbf{u}-2\mathbf{v} = \langle 9-8\sqrt{2}, 9\sqrt{2}-2\rangle$

15. $\mathbf{u}+\mathbf{v} = \langle -23/4, 13\rangle$; $\mathbf{u}-\mathbf{v} = \langle -9/4, 7\rangle$;
$3\mathbf{u}-2\mathbf{v} = \langle -17/2, 24\rangle$

17. $\mathbf{u}+\mathbf{v} = 14\mathbf{i}-4\mathbf{j}$; $\mathbf{u}-\mathbf{v} = 2\mathbf{i}+4\mathbf{j}$; $3\mathbf{u}-2\mathbf{v} = 12\mathbf{i}+8\mathbf{j}$

19. $\mathbf{u}+\mathbf{v} = -\frac{5}{4}\mathbf{i} - \frac{3}{2}\mathbf{j}$; $\mathbf{u}-\mathbf{v} = -\frac{11}{4}\mathbf{i} - \frac{3}{2}\mathbf{j}$;
$3\mathbf{u}-2\mathbf{v} = -\frac{15}{2}\mathbf{i} - \frac{9}{2}\mathbf{j}$

21. $\langle -7, 0\rangle$ **23.** $\langle -2, 1/2\rangle$

25. $\langle 6, -13/4\rangle$ **27.** $\langle 4, 0\rangle$

29. $\langle -5\sqrt{2}, -5\sqrt{2}\rangle$ **31.** $\langle 4.5963, 3.8567\rangle$

33. $\langle -.1710, -.4698\rangle$ **35.** $\|\mathbf{v}\| = 4\sqrt{2}, \theta = 45°$

37. $\|\mathbf{v}\| = 8, \theta = 180°$ **39.** $\|6\mathbf{j}\| = 6, \theta = 90°$

41. $\|-2\mathbf{i}+8\mathbf{j}\| = 2\sqrt{17}, \theta = 104.04°$

43. $\left\langle \frac{4}{\sqrt{41}}, -\frac{5}{\sqrt{41}}\right\rangle$ **45.** $\left\langle \frac{1}{\sqrt{5}}, \frac{2}{\sqrt{5}}\right\rangle$

47. $\|\mathbf{u}+\mathbf{v}\| = 108.2$ pounds, $\theta = 46.1°$

49. $\|\mathbf{u}+\mathbf{v}\| = 17.4356$ newtons, $\theta = 213.4132°$

51. $\mathbf{v} = \langle 8, 2\rangle$

53. $\mathbf{v}+\mathbf{0} = \langle c, d\rangle + \langle 0, 0\rangle = \langle c, d\rangle = \mathbf{v}$ and
$\mathbf{0}+\mathbf{v} = \langle 0, 0\rangle + \langle c, d\rangle = \langle c, d\rangle = \mathbf{v}$

55. $r(\mathbf{u}+\mathbf{v}) = r(\langle a, b\rangle + \langle c, d\rangle) = r\langle a+c, b+d\rangle =$
$\langle ra+rc, rb+rd\rangle = \langle ra, rb\rangle + \langle rc, rd\rangle =$
$r\langle a, b\rangle + r\langle c, d\rangle = r\mathbf{u} + r\mathbf{v}$

57. $(rs)\mathbf{v} = rs\langle c, d\rangle = \langle rsc, rsd\rangle = r\langle sc, sd\rangle = r(s\mathbf{v})$ and
$\langle rsc, rsd\rangle = \langle src, srd\rangle = s\langle rc, rd\rangle = s(r\mathbf{v})$

59. 48.575 pounds

61. 32.1 pounds parallel to plane; 38.3 pounds perpendicular to plane

63. 66.4°

65. ground speed: 253.2 mph; course: 69.1°

67. ground speed: 304.1 mph; course: 309.5°

69. air speed: 424.3 mph; direction: 62.4°

71. 69.08°

73. 341.77 lb on **v**; 170.32 lb on **u**

75. 517.55 lb on the 28° rope; 579.90 lb on the 38° rope

77. (a) $\mathbf{v} = \langle x_2 - x_1, y_2 - y_1\rangle$; $k\mathbf{v} = \langle kx_2 - kx_1, ky_2 - ky_1\rangle$

(b) $\|\mathbf{v}\| = \sqrt{(x_2-x_1)^2 + (y_2-y_1)^2}$
$\|k\mathbf{v}\| = \sqrt{(kx_2 - kx_1)^2 + (ky_2 - ky_1)^2}$

(c) $\|k\mathbf{v}\| = \sqrt{k^2(x_2-x_1)^2 + k^2(y_2-y_1)^2}$
$= \sqrt{k^2}\sqrt{(x_2-x_1)^2 + (y_2-y_1)^2}$
$= |k|\sqrt{(x_2-x_1)^2 + (y_2-y_1)^2} = |k|\,\|\mathbf{v}\|$

(d) $\tan\theta = \frac{y_2-y_1}{x_2-x_1} = \frac{k}{k}\frac{y_2-y_1}{x_2-x_1} = \frac{ky_2-ky_1}{kx_2-kx_1} = \tan\beta$
Since the angles have the same tangent, they can differ only by a multiple of π and so are parallel. They have either the same or the opposite direction.

(e) If $k > 0$, the signs of the components of $k\mathbf{v}$ are the same as those of **v**, so the two vectors lie in the same quadrant. Therefore, they must have the same direction. If $k < 0$, then the components of $k\mathbf{v}$ and the components of **v** must have opposite signs, and the two vectors do not lie in the same quadrant. Therefore, they do not have the same direction and must have opposite directions.

79. (a) Since $\mathbf{u} - \mathbf{v} = \langle a - c, b - d\rangle$, $\|\mathbf{u} - \mathbf{v}\| = \sqrt{(a - c)^2 + (b - d)^2}$. The magnitude of $\mathbf{w}$ is given by the distance between the points (a, b) and (c, d). Hence, $\|\mathbf{u} - \mathbf{v}\| = \|\mathbf{w}\|$.

(b) $\mathbf{u} - \mathbf{v}$ lies on the straight line through $(0, 0)$ and $(a - c, b - d)$, which has slope $\frac{(b - d) - 0}{(a - c) - 0} = \frac{b - d}{a - c}$. $\mathbf{w}$ lies on the line joining (a, b) and (c, d). This also has slope $\frac{b - d}{a - c}$. Since the slopes are the same, the vectors, $\mathbf{u} - \mathbf{v}$ and $\mathbf{w}$ are parallel. Therefore, they either point in the same direction or in opposite directions. We can see that they have the same direction by considering the signs of the components of $\mathbf{u} - \mathbf{v}$. If $a - c > 0$ and $b - d > 0$, then $a > c$, $b > d$ and $\mathbf{u} - \mathbf{v}$ and $\mathbf{w}$ both point up and right. If $a - c > 0$ and $b - d < 0$, both vectors will point left and up. If $a - c < 0$ and $b - d < 0$, both vectors will point left and down. In any case, they point in the same direction.

Section 4.5, page 323

1. $\mathbf{u} \cdot \mathbf{v} = -7$, $\mathbf{u} \cdot \mathbf{u} = 25$, $\mathbf{v} \cdot \mathbf{v} = 29$

3. $\mathbf{u} \cdot \mathbf{v} = 6$, $\mathbf{u} \cdot \mathbf{u} = 5$, $\mathbf{v} \cdot \mathbf{v} = 9$

5. $\mathbf{u} \cdot \mathbf{v} = 12$, $\mathbf{u} \cdot \mathbf{u} = 13$, $\mathbf{v} \cdot \mathbf{v} = 13$

7. 6 **9.** 20 **11.** -28

13. 1.75065 radians **15.** 2.1588 radians

17. $\pi/2$ radians **19.** Orthogonal

21. Parallel **23.** Neither

25. $k = 2$ **27.** $k = \sqrt{2}$

29. $\text{proj}_{\mathbf{u}}\mathbf{v} = \langle 12/17, -20/17\rangle$; $\text{proj}_{\mathbf{v}}\mathbf{u} = \langle 6/5, 2/5\rangle$

31. $\text{proj}_{\mathbf{u}}\mathbf{v} = \langle 0, 0\rangle$; $\text{proj}_{\mathbf{v}}\mathbf{u} = \langle 0, 0\rangle$

33. $\text{comp}_{\mathbf{v}}\mathbf{u} = 22/\sqrt{13}$ **35.** $\text{comp}_{\mathbf{v}}\mathbf{u} = 3/\sqrt{10}$

37. $\mathbf{u} \cdot (\mathbf{v} + \mathbf{w}) = \langle a, b\rangle \cdot (\langle c, d\rangle + \langle r, s\rangle)$
$= \langle a, b\rangle \cdot \langle c + r, d + s\rangle = a(c + r) + b(d + s)$
$= ac + ar + bd + bs$
$\mathbf{u} \cdot \mathbf{v} + \mathbf{u} \cdot \mathbf{w} = \langle a, b\rangle \cdot \langle c, d\rangle + \langle a, b\rangle \cdot \langle r, s\rangle$
$= (ac + bd) + (ar + bs) = ac + ar + bd + bs$

39. $\mathbf{0} \cdot \mathbf{u} = \langle 0, 0\rangle \cdot \langle a, b\rangle = 0a + 0b = 0$

41. If $\theta = 0$ or π, then $\mathbf{u}$ and $\mathbf{v}$ are parallel, so $\mathbf{v} = k\mathbf{u}$ for some real number k. We know that $\|\mathbf{v}\| = |k|\ \|\mathbf{u}\|$ (Exercise 77, Section 4.4). If $\theta = 0$, then $\cos\theta = 1$ and $k > 0$. Since $k > 0$, $|k| = k$ and so $\|\mathbf{v}\| = k\|\mathbf{u}\|$. Therefore, $\mathbf{u} \cdot \mathbf{v} = \mathbf{u} \cdot k\mathbf{u} = k\mathbf{u} \cdot \mathbf{u} = k\|\mathbf{u}\|^2 = \|\mathbf{u}\|(k\|\mathbf{u}\|) = \|\mathbf{u}\|\|\mathbf{v}\| = \|\mathbf{u}\|\|\mathbf{v}\| \cos\theta$. On the other hand, if $\theta = \pi$, then $\cos\theta = -1$ and $k < 0$. Since $k < 0$, $|k| = -k$ and so $\|\mathbf{v}\| = -k\|\mathbf{u}\|$. Then $\mathbf{u} \cdot \mathbf{v} = \mathbf{u} \cdot k\mathbf{u}, = k\mathbf{u} \cdot \mathbf{u} = k\|\mathbf{u}\|^2 = \|\mathbf{u}\|(-k\|\mathbf{u}\|) = -\|\mathbf{u}\|\|\mathbf{v}\| = \|\mathbf{u}\|\|\mathbf{v}\| \cos\theta$. In both cases we have shown $\mathbf{u} \cdot \mathbf{v} = \|\mathbf{u}\|\|\mathbf{v}\| \cos\theta$.

43. If $A = (1, 2)$, $B = (3, 4)$, and $C = (5, 2)$, then the vector $\overrightarrow{AB} = \langle 2, 2\rangle$, $\overrightarrow{AC} = \langle 4, 0\rangle$, and $\overrightarrow{BC} = \langle 2, -2\rangle$. Since $\overrightarrow{AB} \cdot \overrightarrow{BC} = 0$, $\overrightarrow{AB}$ and $\overrightarrow{BC}$ are perpendicular, so the angle at vertex B is a right angle.

45. Many possible answers: One is $\mathbf{u} = \langle 1, 0\rangle$, $\mathbf{v} = \langle 1, 1\rangle$, and $\mathbf{w} = \langle 1, -1\rangle$.

47. 300 lb ($= 600 \sin 30°$) **49.** 13 **51.** 24

53. The force in the direction of the lawnmower's motion is $30 \cos 60° = 15$ lb. Thus, the work done is

$$15(75) = 1125 \text{ ft-lb}.$$

55. 1368 ft-lb

Chapter 4 Review, page 326

1. $(\sqrt{2} + \sqrt{3}) - i$ **3.** $-i$

5. $x = 2 \pm i$

7. $\sqrt{20} + \sqrt{10}$

9. The graph is a circle of radius 2 centered at the origin.

11. $2\left(\cos\frac{\pi}{3} + i \sin\frac{\pi}{3}\right)$

13. $4\sqrt{2} + i\,4\sqrt{2}$ **15.** $2\sqrt{3} + 2i$

17. $\frac{81}{2} - \frac{81\sqrt{3}}{2}i$

19. $\cos 0 + i \sin 0$, $\cos\frac{\pi}{3} + i \sin\frac{\pi}{3}$, $\cos\frac{2\pi}{3} + i \sin\frac{2\pi}{3}$, $\cos\pi + i \sin\pi$, $\cos\frac{4\pi}{3} + i \sin\frac{4\pi}{3}$, $\cos\frac{5\pi}{3} + i \sin\frac{5\pi}{3}$

21. $\cos\frac{\pi}{8} + i \sin\frac{\pi}{8}$, $\cos\frac{5\pi}{8} + i \sin\frac{5\pi}{8}$, $\cos\frac{9\pi}{8} + i \sin\frac{9\pi}{8}$, $\cos\frac{13\pi}{8} + i \sin\frac{13\pi}{8}$

23. $\langle 11, -1\rangle$ **25.** $2\sqrt{29}$ **27.** $-11\mathbf{i} + 8\mathbf{j}$

29. $\sqrt{10}$ **31.** $\left\langle \frac{5\sqrt{2}}{2}, \frac{5\sqrt{2}}{2}\right\rangle$

33. $-\frac{1}{\sqrt{5}}\mathbf{i} + \frac{2}{\sqrt{5}}\mathbf{j}$

35. Course: 126.18°; ground speed: 321.87 mph

37. -26 **39.** 3

41. .70 radians **43.** $= 2\mathbf{i} + \mathbf{j}$

45. $(\mathbf{u} + \mathbf{v}) \cdot (\mathbf{u} - \mathbf{v}) = \mathbf{u} \cdot \mathbf{u} - \mathbf{u} \cdot \mathbf{v} + \mathbf{v} \cdot \mathbf{u} - \mathbf{v} \cdot \mathbf{v} = \|\mathbf{u}\|^2 - \mathbf{u} \cdot \mathbf{v} + \mathbf{v} \cdot \mathbf{u} - \|\mathbf{v}\|^2 = \|\mathbf{u}\|^2 - \|\mathbf{v}\|^2 = 0$ (because $\|\mathbf{u}\| = \|\mathbf{v}\|$)

47. 1750 lb

CHAPTER 5

Section 5.1, page 345

1. $x^2 + (y-3)^2 = 4$

3. $x^2 + 6y^2 = 18$

5. $(x-4)^2 + (y-3)^2 = 4$

7. Center $(-4, 3)$, radius $2\sqrt{10}$

9. Center $(-3, 2)$, radius $2\sqrt{7}$

11. Center $(-5/2, -5)$, radius $\dfrac{\sqrt{677}}{2} \approx 13.01$

13. Ellipse

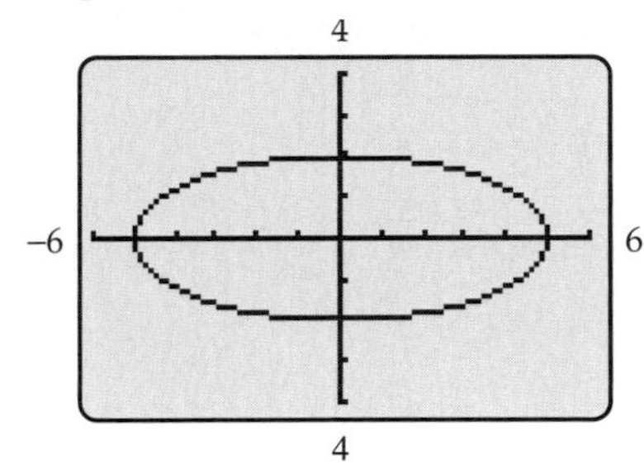

15. Ellipse

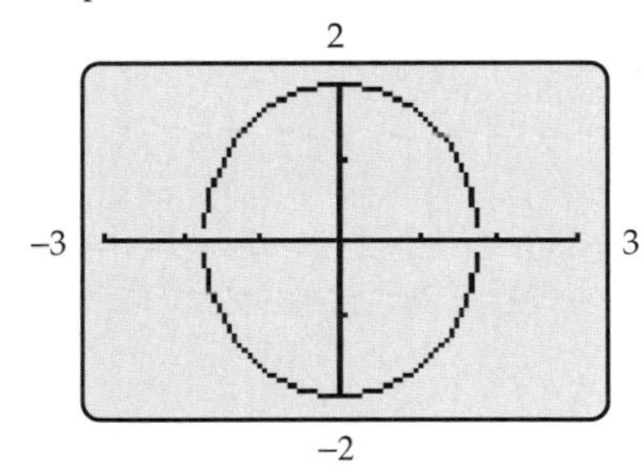

17. Ellipse

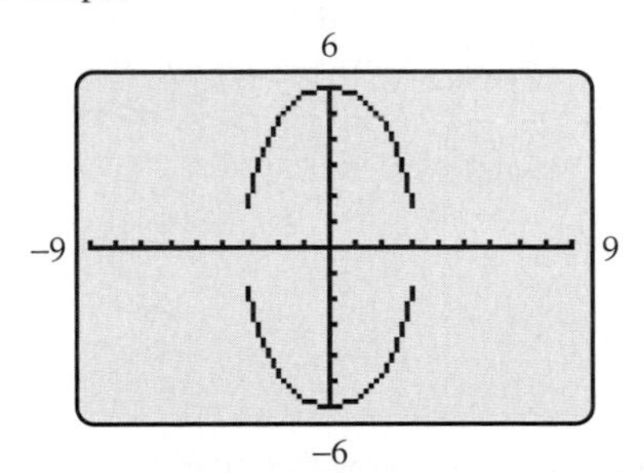

19. Circle

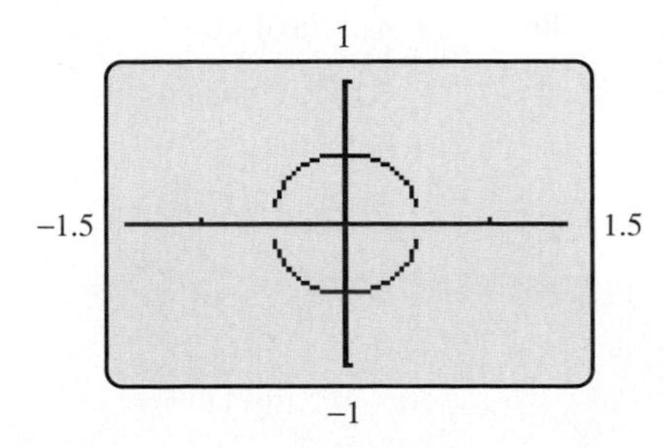

21. $\dfrac{x^2}{49} + \dfrac{y^2}{4} = 1$

23. $\dfrac{x^2}{36} + \dfrac{y^2}{16} = 1$

25. $\dfrac{x^2}{9} + \dfrac{y^2}{49} = 1$

27. 8π

29. $2\pi\sqrt{3}$

31. $7\pi/\sqrt{3}$

33. Ellipse, center $(1, 5)$

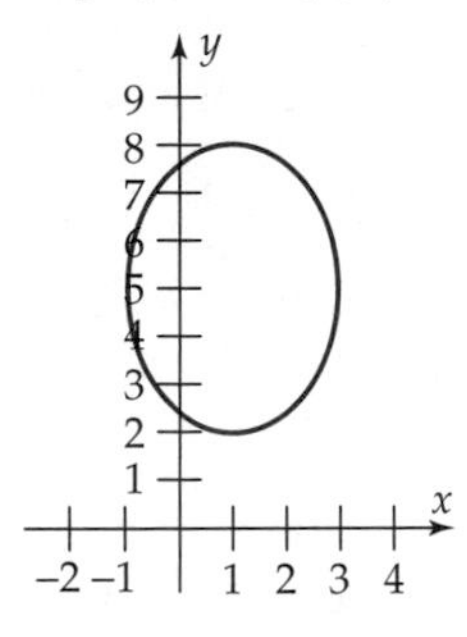

35. Ellipse, center $(-1, 4)$

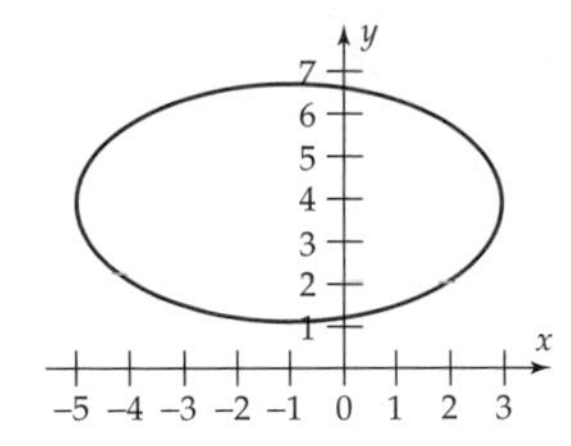

37. Ellipse, center $(-3, 1)$

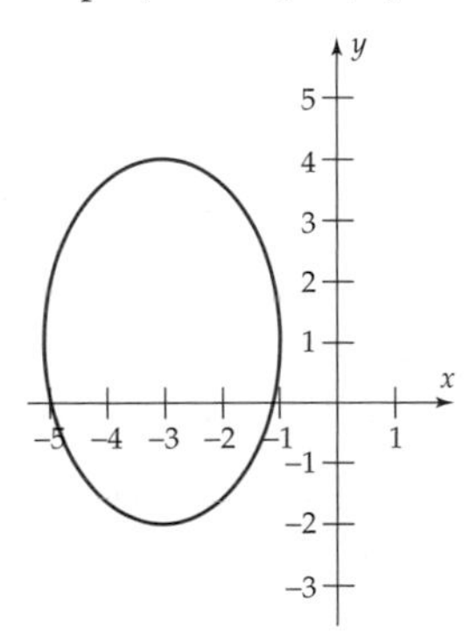

39. Circle, center $(-3, 4)$, radius $\sqrt{20}$

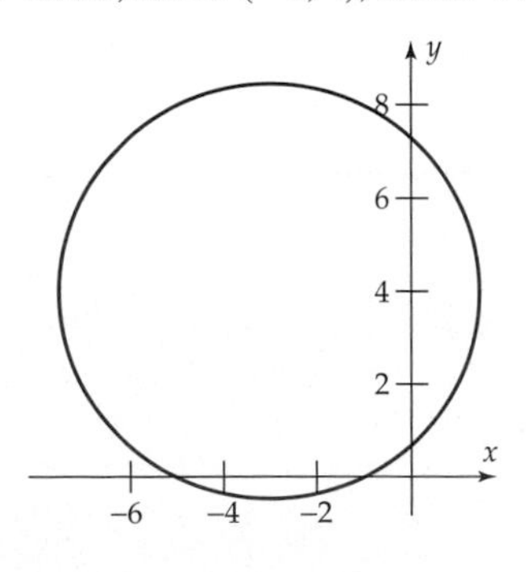

41. Ellipse, center $(-3, 2)$

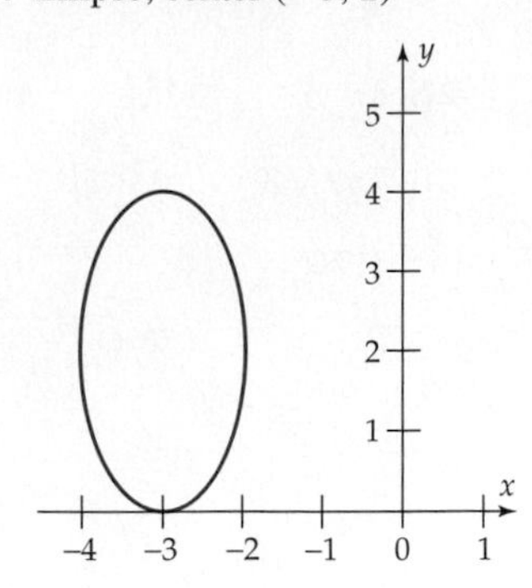

43. $x = \sqrt{10} \cos t,\ y = 6 \sin t \quad (0 \le t \le 2\pi)$

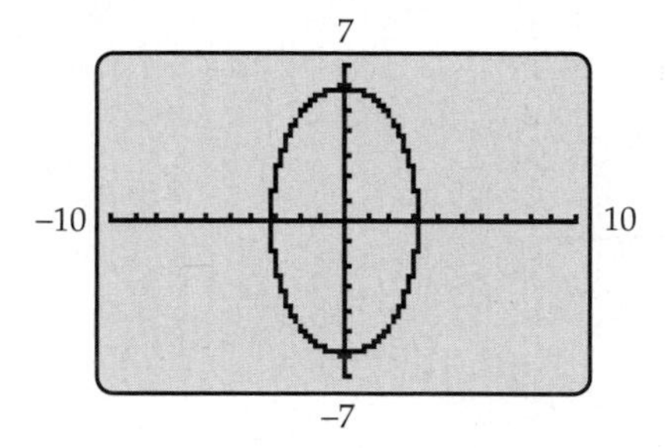

45. $x = \frac{1}{2} \cos t,\ y = \frac{1}{2} \sin t \quad (0 \le t \le 2\pi)$

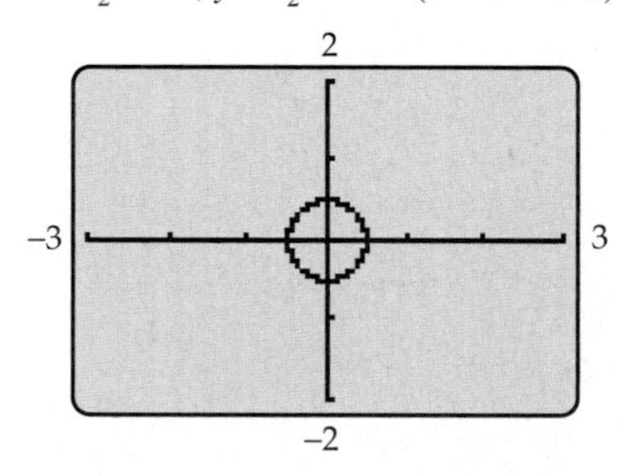

47. $x = 2 \cos t + 1,\ y = 3 \sin t + 5 \quad (0 \le t \le 2\pi)$

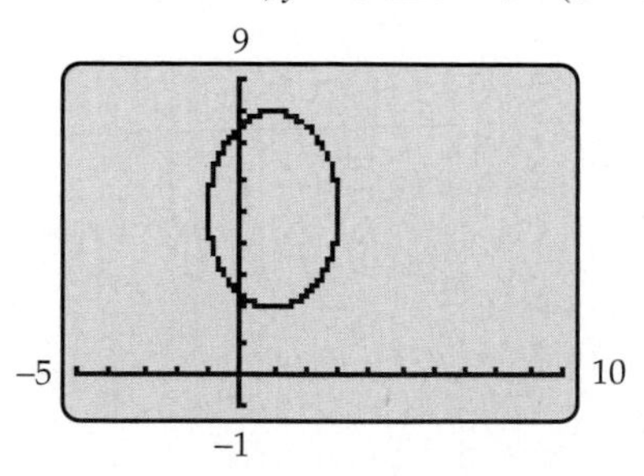

49. $x = 4 \cos t - 1,\ y = \sqrt{8} \sin t + 4 \quad (0 \le t \le 2\pi)$

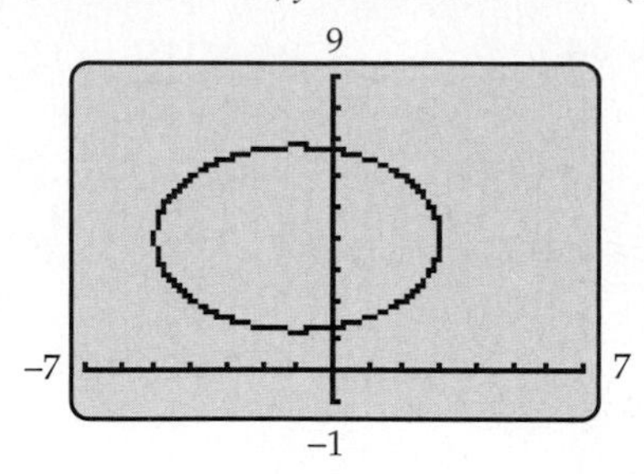

51. $\dfrac{(x-2)^2}{4} + \dfrac{(y-3)^2}{16} = 1$ **53.** $\dfrac{(x-7)^2}{25/4} + \dfrac{(y+4)^2}{36} = 1$

55. $\dfrac{(x-3)^2}{36} + \dfrac{(y+2)^2}{16} = 1$

57. $\dfrac{(x+5)^2}{49} + \dfrac{(y-3)^2}{16} = 1$ or $\dfrac{(x+5)^2}{16} + \dfrac{(y-3)^2}{49} = 1$

59. $2x^2 + 2y^2 - 8 = 0$

61. $2x^2 + y^2 - 8x - 6y + 9 = 0$

63. $\dfrac{(x+3)^2}{4} + \dfrac{(y+3)^2}{8} = 1$

65. The ellipse has y-intercepts $(0, \pm 4)$. Therefore, the circle, which has center at the origin, has radius 4. The equation of the circle must be $x^2 + y^2 = 16$.

67. The minimum distance is 226,335 miles. The maximum distance is 251,401 miles.

69. 80 feet **71.** 17.3 feet

73. Eccentricity $= .1$

75. Eccentricity $= \dfrac{\sqrt{3}}{2} \approx .87$

77. The closer the eccentricity is to zero, the more the ellipse resembles a circle. The closer the eccentricity is to 1, the more elongated the ellipse becomes.

79. Eccentricity $\approx .38$.

81. If $a = b$, the equation becomes

$$\frac{x^2}{a^2} + \frac{y^2}{a^2} = 1,$$
$$x^2 + y^2 = a^2.$$

This is the equation of a circle with center at the origin, radius a.

83. The fence is an ellipse with major axis 100 ft, minor axis $50\sqrt{3} \approx 86.6$ ft.

Section 5.2, page 357

1. $x^2 + 4y^2 = 1$
3. $2x^2 - y^2 = 8$
5. $6x^2 + 2y^2 = 18$
7. Hyperbola

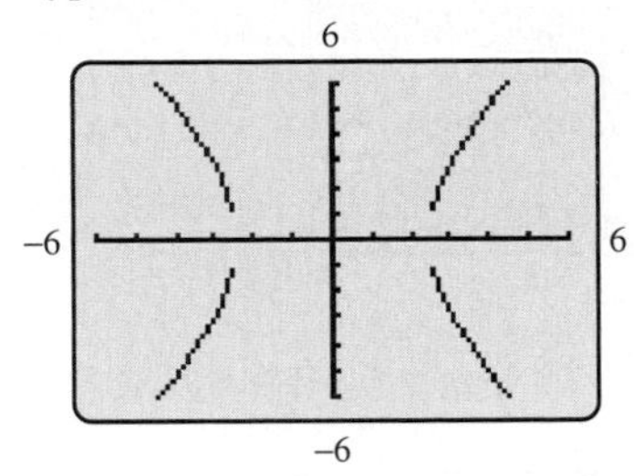

Note: Because of limited resolution, this calculator graph does not show that the top and bottom halves of the graph are connected.

9. Hyperbola (See Note in Exercise 7 answer.)

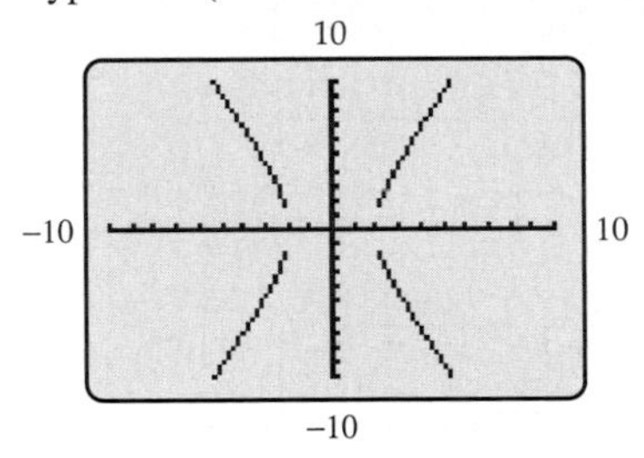

11. Hyperbola

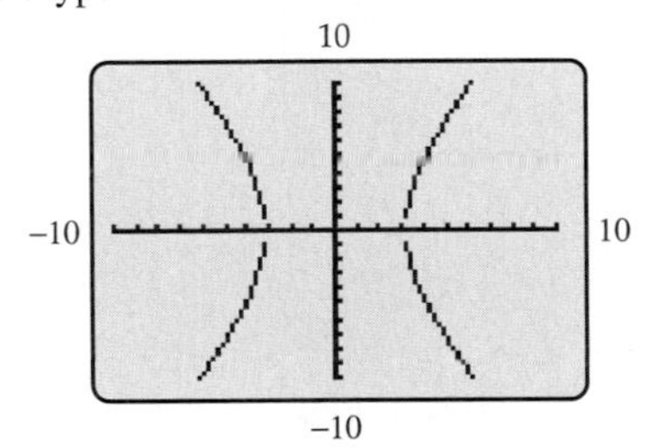

13. Hyperbola

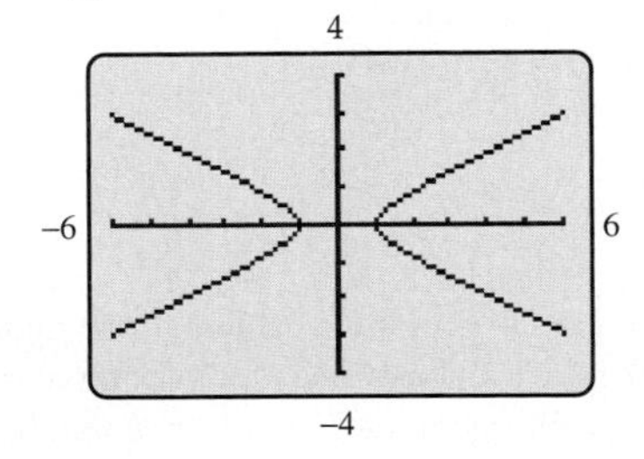

15. $\dfrac{x^2}{9} - \dfrac{y^2}{36} = 1$
17. $\dfrac{x^2}{4} - \dfrac{y^2}{1} = 1$
19. Hyperbola, center $(-1, -3)$

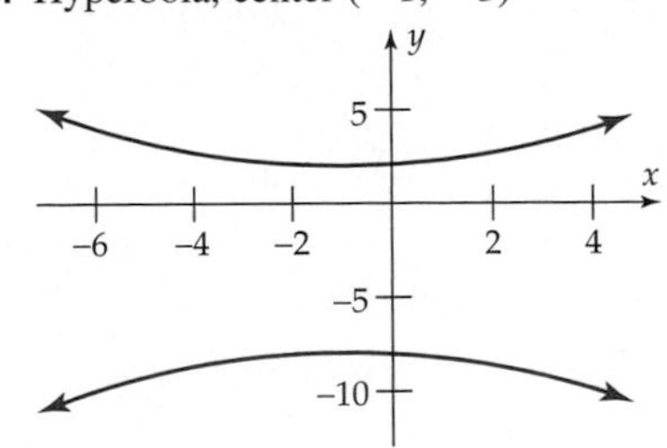

21. Hyperbola, center $(-3, 2)$

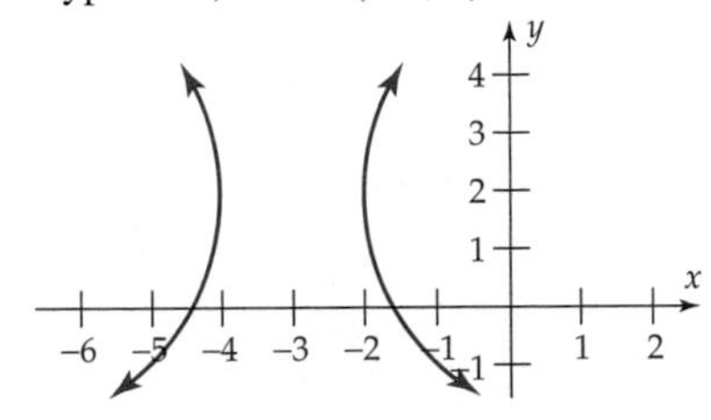

23. Hyperbola, center $(1, -4)$

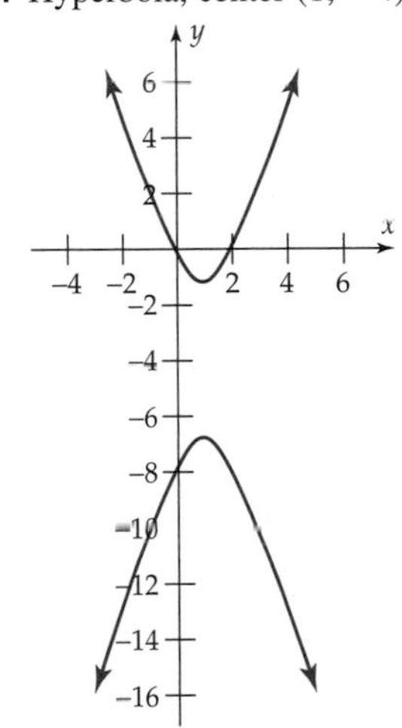

25. Hyperbola, center $(3, 3)$

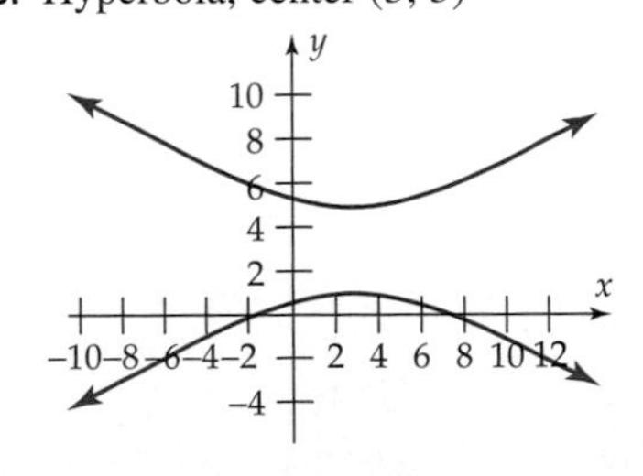

27. Circle, center (3, 4)

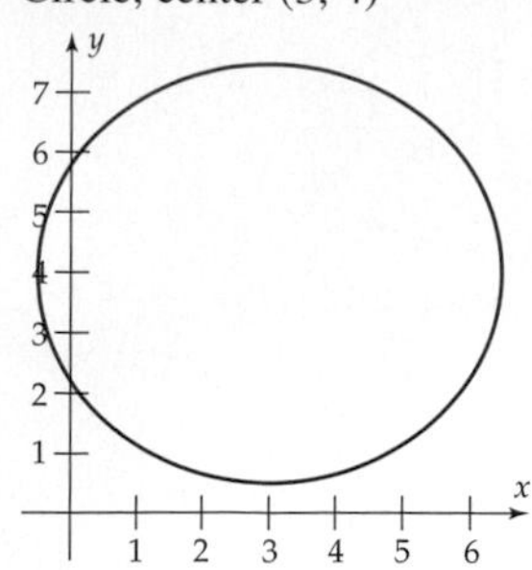

29. Ellipse, center (3, 4)

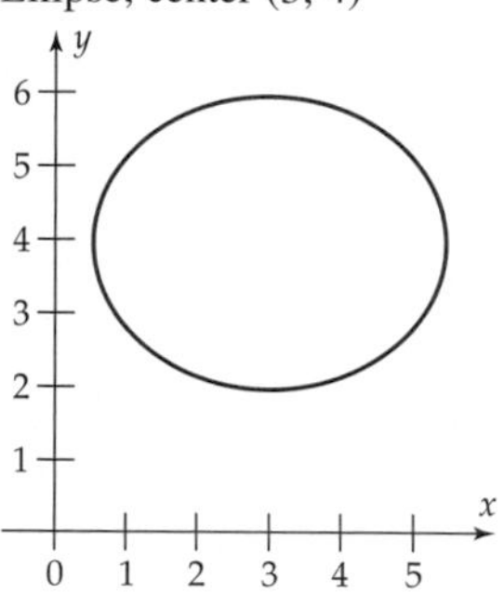

31. Hyperbola, center $(-2, 2)$

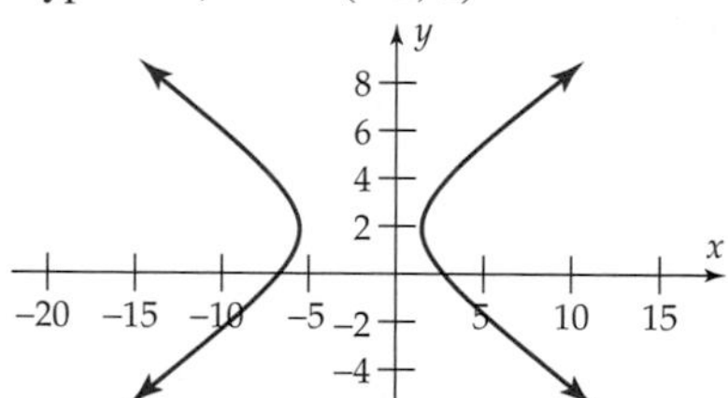

33. $\dfrac{(y-3)^2}{4} - \dfrac{(x+2)^2}{6} = 1$

35. $\dfrac{(x-4)^2}{9} - \dfrac{(y-2)^2}{16} = 1$

37. $y^2 - 2x^2 = 6$

39. $\dfrac{(y-2)^2}{4} - \dfrac{(x-3)^2}{9} = 1$

41. $\dfrac{(x+3)^2}{3} - \dfrac{(y+3)^2}{4} = 1$

43. $x = \sqrt{10}/\cos t,\ y = 6 \tan t \quad (0 \le t \le 2\pi)$

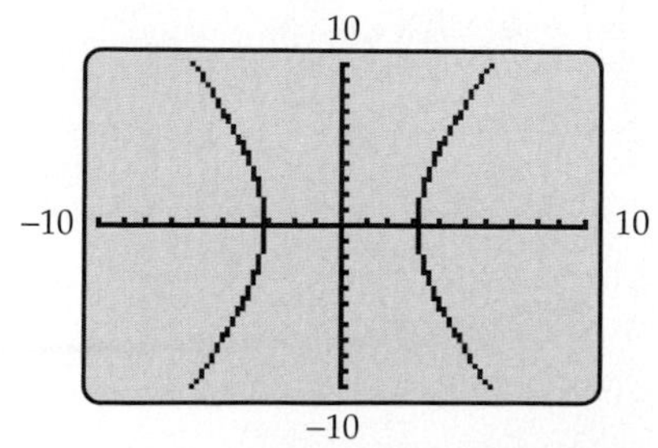

45. $x = 1/\cos t,\ y = \frac{1}{2} \tan t \quad (0 \le t \le 2\pi)$

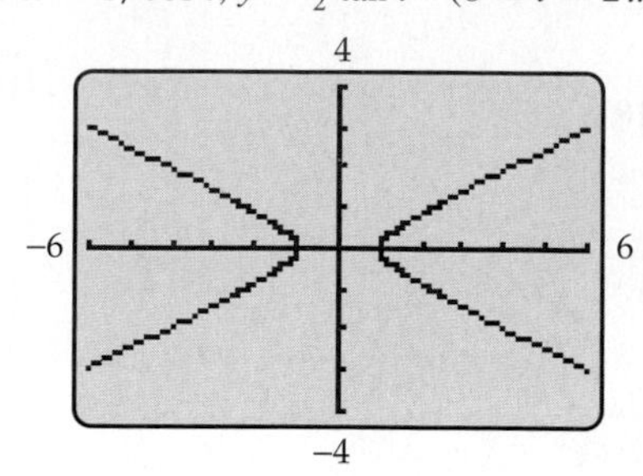

47. $x = 4 \tan t - 1,\ y = 5/\cos t - 3 \quad (0 \le t \le 2\pi)$

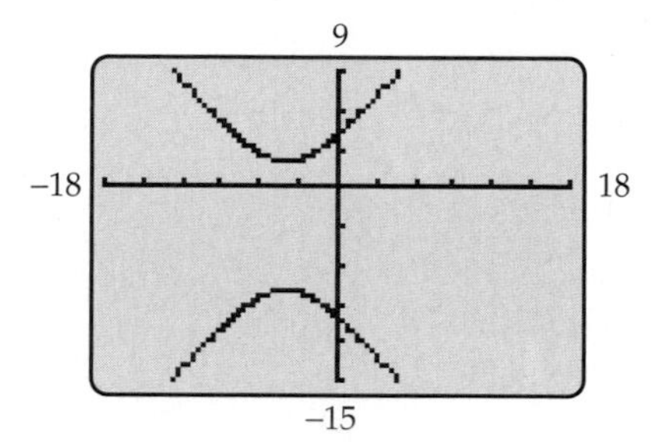

49. $x = 1/\cos t - 3,\ y = 2 \tan t + 2 \quad (0 \le t \le 2\pi)$

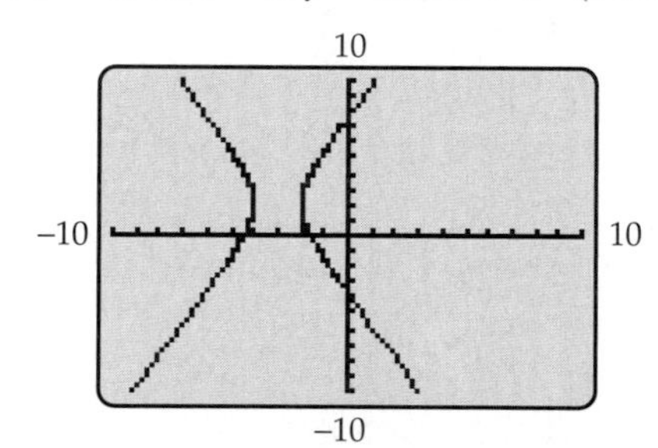

51. The graphs are shown below on one set of coordinate axes.

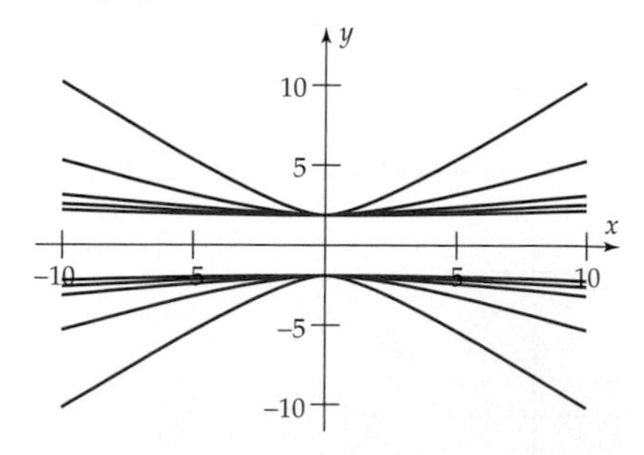

As b increases, the graphs get flatter. Although the hyperbola may look like two horizontal lines for very large b, it is still a hyperbola, with asymptotes $y = \pm 2x/b$ that have nonzero slopes.

53. The asymptotes of $\dfrac{x^2}{a^2} - \dfrac{y^2}{a^2} = 1$ are $y = \pm\dfrac{a}{a}x$ or $y = \pm x$. Since they have slopes 1 and -1 and $1(-1) = -1$, they are perpendicular.

55. The equation is $\frac{x^2}{1{,}210{,}000} - \frac{y^2}{5{,}759{,}600} = 1$ (measurement in feet). The exact location cannot be determined from only the given information.

57. $\sqrt{5} \approx 2.24$ **59.** $\sqrt{3}$ **61.** $3/2$

Section 5.3, page 367

1. $6x = y^2$ **3.** $2x^2 - y^2 = 8$

5. $x^2 + 6y^2 = 18$

7. Focus: $(0, 1/12)$, directrix $y = -1/12$

9. Focus: $(0, 1)$, directrix $y = -1$

11. Opens to the right.
Vertex: $(2, 0)$, focus: $(9/4, 0)$, directrix: $x = 7/4$

13. Opens to the left.
Vertex: $(2, -1)$, focus: $(7/4, -1)$, directrix: $x = 9/4$

15. Opens to the right.
Vertex: $(2/3, -3)$, focus: $(17/12, -3)$, directrix: $x = -1/12$

17. Opens to right.
Vertex: $(-81/4, 9/2)$, focus: $(-20, 9/2)$, directrix: $x = -41/2$

19. Opens upward.
Vertex: $(-1/6, -49/12)$, focus: $(-1/6, -4)$, directrix: $y = -25/6$

21. Opens downward.
Vertex: $(2/3, 19/3)$, focus: $(2/3, 25/4)$, directrix: $y = 77/12$

23.

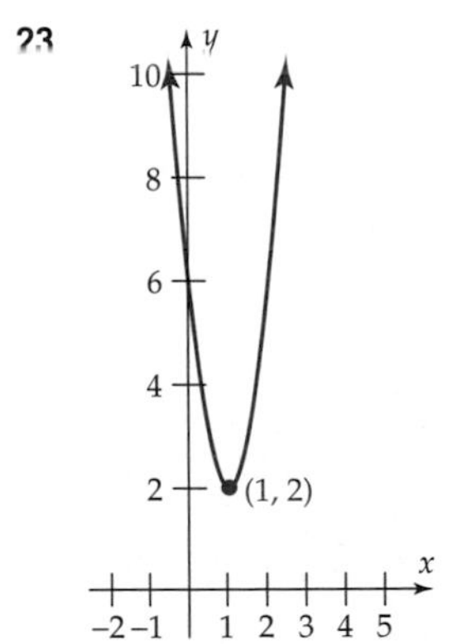

25.

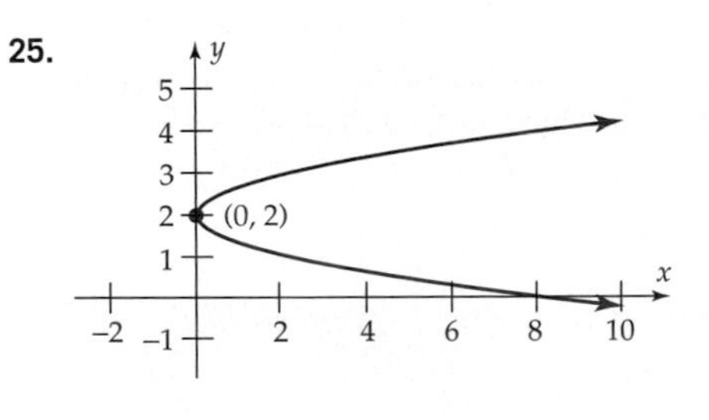

27.

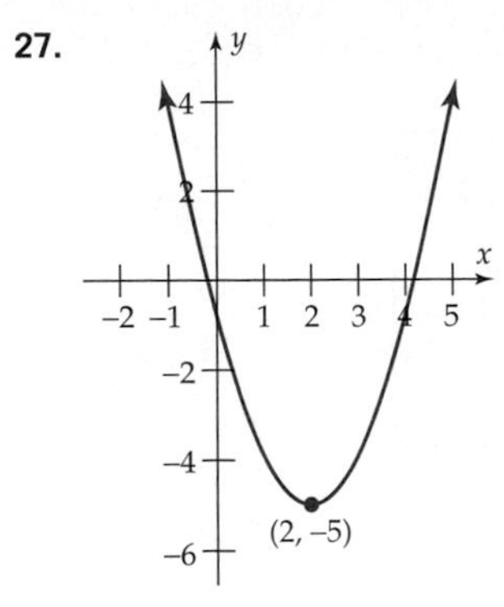

29.

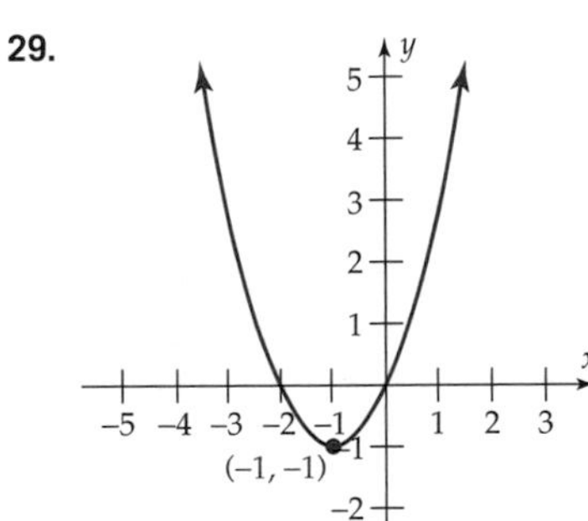

31. $y = 3x^2$ **33.** $y = 13(x - 1)^2$

35. $x - 2 = 3(y - 1)^2$ **37.** $x + 3 = 4(y + 2)^2$

39. $y - 1 = 2(x - 1)^2$ **41.** $(y - 3)^2 = x + 1$

43. $x = t^2/4$, $y = t$

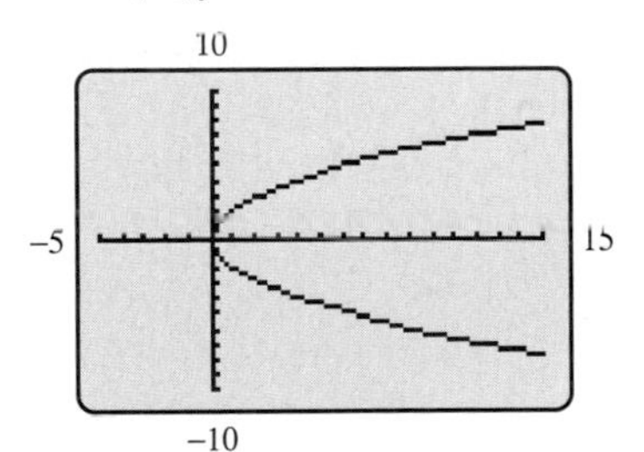

45. $x = t$, $y = 4(t - 1)^2 + 2$

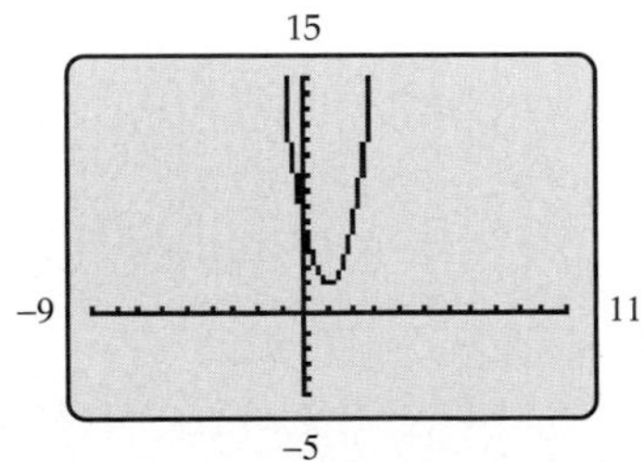

47. $x = 2(t-2)^2, y = t$

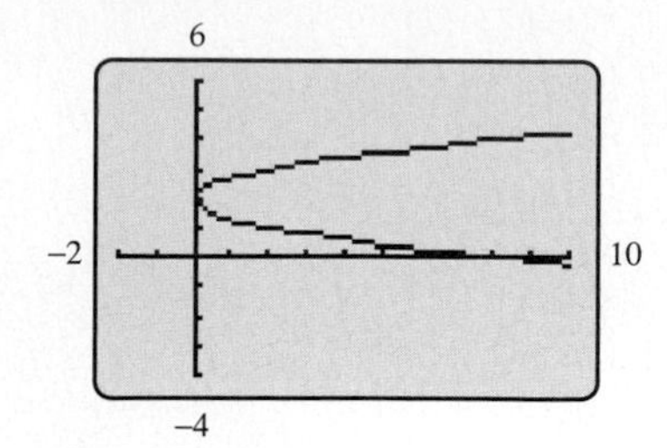

49. Parabola, vertex (3, 4)

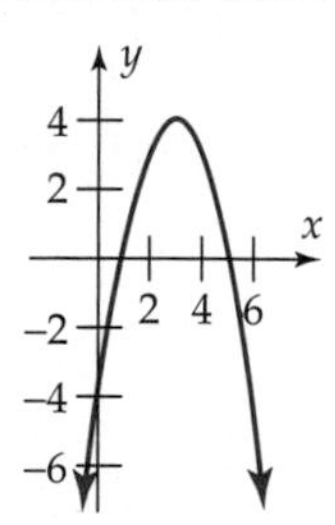

51. Parabola, vertex (2/3, 1/3)

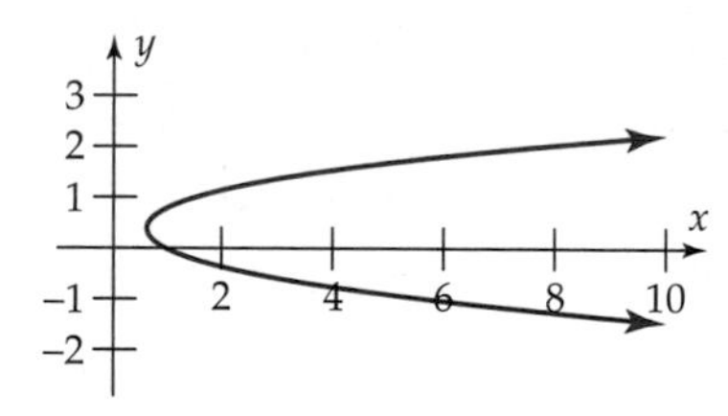

53. Circle, center (1, 2)

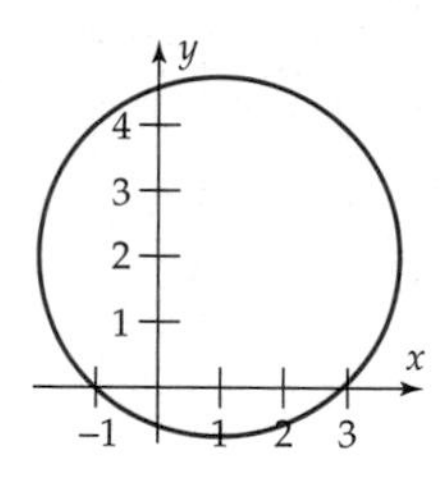

55. Hyperbola, center $(-4, 2)$, vertices $\left(-4 \pm \sqrt{2}, 2\right)$

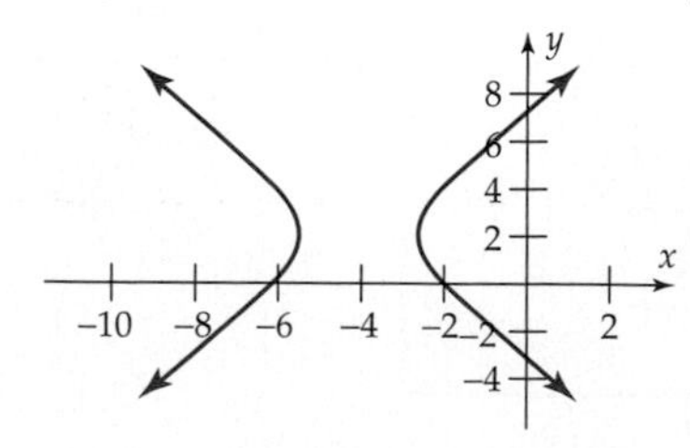

57. $y = (x-4)^2 + 2$ **59.** $y = -x^2 - 8x - 18$

61. $y = (x+5)^2 - 3$ **63.** $b = 0$

65. $\left(9, \dfrac{-1 \pm \sqrt{34}}{2}\right)$

67. The receiver should be placed at the focus, $\frac{2}{3}$ feet or 8 inches from the vertex.

69. 43.3 feet **71.** 5.072 meters

73. 27.44 feet

Section 5.4, page 373

1. Ellipse **3.** Parabola **5.** Hyperbola

7. Ellipse. Window $-6 \le x \le 3$ and $-2 \le y \le 4$.

9. Hyperbola. Window $-7 \le x \le 13$ and $-3 \le y \le 9$.

11. Parabola. Window $-1 \le x \le 8$ and $-3 \le y \le 3$.

13. Ellipse. Window $-1.5 \le x \le 1.5$ and $-1 \le y \le 1$.

15. Hyperbola. Window $-15 \le x \le 15$ and $-10 \le y \le 10$.

17. Parabola. Window $-19 \le x \le 2$ and $-1 \le y \le 13$.

19. Hyperbola. Window $-15 \le x \le 15$ and $-15 \le y \le 15$.

21. Ellipse. Window $-6 \le x \le 6$ and $-4 \le y \le 4$.

23. Parabola. Window $-9 \le x \le 4$ and $-2 \le y \le 10$.

Special Topics 5.4.A, page 378

1. $\left(\dfrac{5\sqrt{2}}{2}, -\dfrac{\sqrt{2}}{2}\right)$ **3.** $\left(\dfrac{\sqrt{3}}{2}, -\dfrac{1}{2}\right)$

5. $\dfrac{u^2}{2} - \dfrac{v^2}{2} = 1$ **7.** $\dfrac{u^2}{4} + v^2 = 1$

9. $\theta \approx 53.13°; x = \dfrac{3}{5}u - \dfrac{4}{5}v; y = \dfrac{4}{5}u + \dfrac{3}{5}v$

11. $\theta \approx 36.87°; x = \dfrac{4}{5}u - \dfrac{3}{5}v; y = \dfrac{3}{5}u + \dfrac{4}{5}v$

13. (a) $(A\cos^2\theta + B\cos\theta\sin\theta + C\sin^2\theta)u^2 + (B\cos^2\theta - 2A\cos\theta\sin\theta + 2C\cos\theta\sin\theta - B\sin^2\theta)uv + (C\cos^2\theta - B\cos\theta\sin\theta + A\sin^2\theta)v^2 + (D\cos\theta + E\sin\theta)u + (E\cos\theta - D\sin\theta)v + F = 0$

(b) $B' = B\cos^2\theta - 2A\cos\theta\sin\theta + 2C\cos\theta\sin\theta - B\sin^2\theta = 2(C-A)\sin\theta\cos\theta + B(\cos^2\theta - \sin^2\theta)$ since B' is the coefficient of uv

(c) Since $\sin 2\theta = 2\sin\theta\cos\theta$ and $\cos 2\theta = \cos^2\theta - \sin^2\theta$, $B' = (C-A)\sin 2\theta + B\cos 2\theta$.

(d) If $\cot 2\theta = \dfrac{A-C}{B}$, then $\dfrac{\cos 2\theta}{\sin 2\theta} = \dfrac{A-C}{B}$. Hence, $(A-C)\sin 2\theta = B\cos 2\theta$. So $-(A-C)\sin 2\theta + B\cos 2\theta = 0$. Since $-(A-C) = (C-A)$, we have $B' = (C-A)\sin 2\theta + B\cos 2\theta = 0$.

15. (a) From Exercise 13(a) we have $(B')^2 - 4A'C' =$ $(B\cos^2\theta - 2A\cos\theta\sin\theta + 2C\cos\theta\sin\theta -$ $B\sin^2\theta)^2 - 4(A\cos^2\theta + B\cos\theta\sin\theta +$ $C\sin^2\theta)(C\cos^2\theta - B\cos\theta\sin\theta + A\sin^2\theta) =$ $[B(\cos^2\theta - \sin^2\theta) + 2(C - A)\cos\theta\sin\theta]^2 -$ $4(A\cos^2\theta + B\cos\theta\sin\theta + C\sin^2\theta)(C\cos^2\theta -$ $B\cos\theta\sin\theta + A\sin^2\theta) = B^2(\cos^2\theta - \sin^2\theta)^2 +$ $4(C - A)^2\cos^2\theta\sin^2\theta + 4B(C - A)(\cos^2\theta -$ $\sin^2\theta)\cos\theta\sin\theta - [4AC(\cos^4\theta + \sin^4\theta) +$ $4(A^2 + C^2 - B^2)\cos^2\theta\sin^2\theta -$ $4AB(\cos^3\theta\sin\theta - \cos\theta\sin^3\theta) +$ $4BC(\cos^3\theta\sin\theta - \cos\theta\sin^3\theta)] = B^2(\cos^4\theta -$ $2\cos^2\theta\sin^2\theta + \sin^4\theta + 4\cos^2\theta\sin^2\theta) -$ $4AC(2\cos^2\theta\sin^2\theta + \cos^4\theta + \sin^4\theta)$ (everything else cancels) $=$ $(B^2 - 4AC)(\cos^4\theta + 2\cos^2\theta\sin^2\theta + \sin^4\theta) =$ $(B^2 - 4AC)(\cos^2\theta + \sin^2\theta)^2 = B^2 - 4AC$

(b) If $B^2 - 4AC < 0$, then also $(B')^2 - 4A'C' < 0$. Since $B' = 0$, $-4A'C' < 0$ and so $A'C' > 0$. By Exercise 14, the graph is an ellipse. The other two cases are proved in the same way.

Section 5.5, page 388

1. $-5 \le x \le 6$ and $-2 \le y \le 2$

3. $-3 \le x \le 4$ and $-2 \le y \le 3$

5. $0 \le x \le 14$ and $-15 \le y \le 0$

7. $-2 \le x < 20$ and $-11 \le y \le 11$

9. $-12 \le x \le 12$ and $-12 \le y \le 12$

11. $-2 \le x \le 20$ and $-20 \le y \le 4$

13. $-25 \le x \le 22$ and $-25 \le y \le 26$

15. $y = 2x + 7$ **17.** $y = 2x + 5$

19. $x = e^y$ (or $y = \ln x$) **21.** $x^2 + y^2 = 9$

23. $x = 7 + 6\cos t$ and $y = -4 + 6\sin t$ $(0 \le t \le 2\pi)$

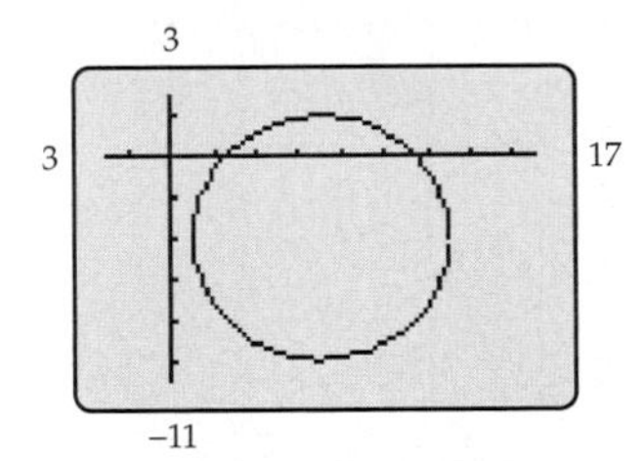

25. Same answer as Exercise 23.

27. (a) $\dfrac{d - b}{c - a}$ (b) $y - b = \dfrac{d - b}{c - a}(x - a)$

(c) Solving both equations for t, we obtain $t = \dfrac{x - a}{c - a}$ and $t = \dfrac{y - b}{d - b}$.

Hence, $\dfrac{y - b}{d - b} = \dfrac{x - a}{c - a}$; that is,

$y - b = \dfrac{d - b}{c - a}(x - a)$.

29. $x = 18t - 6$ and $y = 12 - 22t$ $(0 \le t \le 1)$

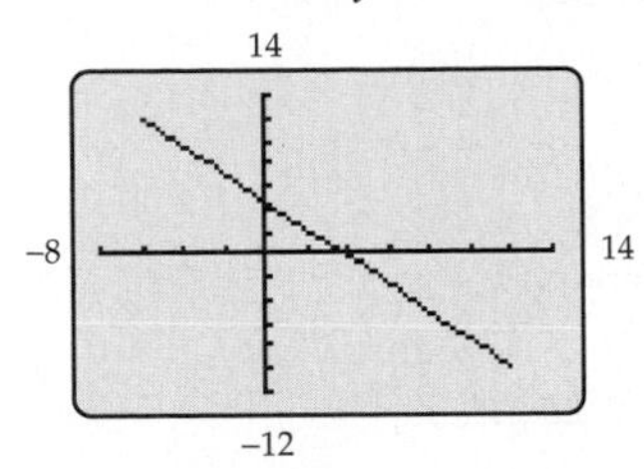

31. $x = 18 - 34t$ and $y = 10t + 4$ $(0 \le t \le 1)$

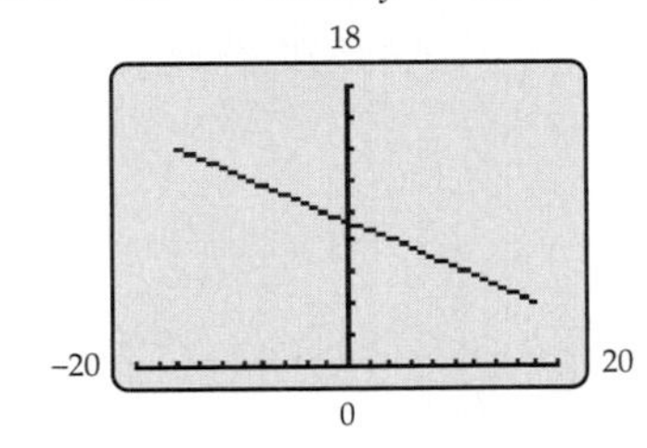

33. (a) $k = 1$

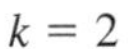
$k = 2$

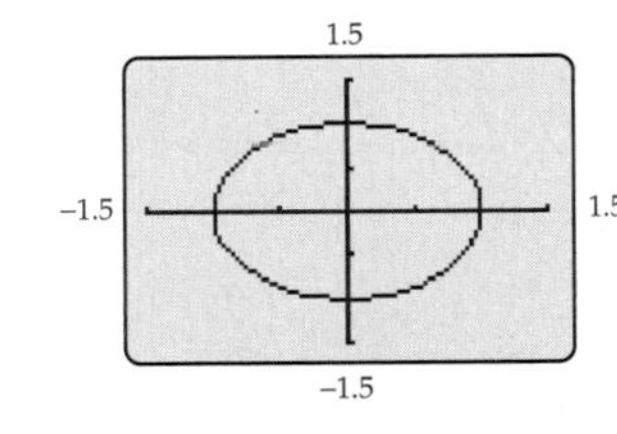

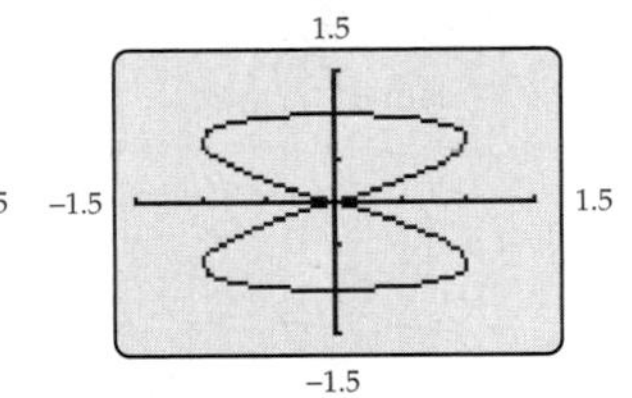

$k - 3$ $k = 4$

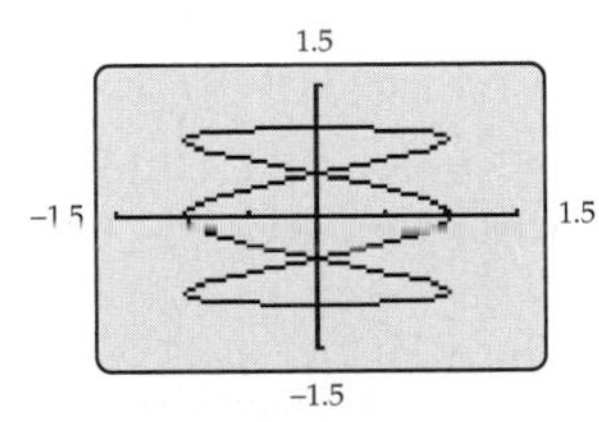

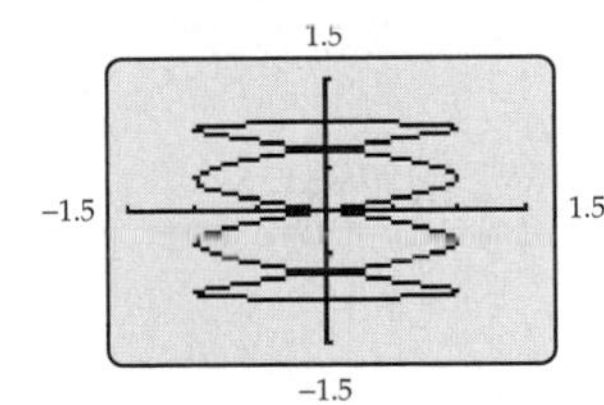

(b) $k = 5$

$k = 6$

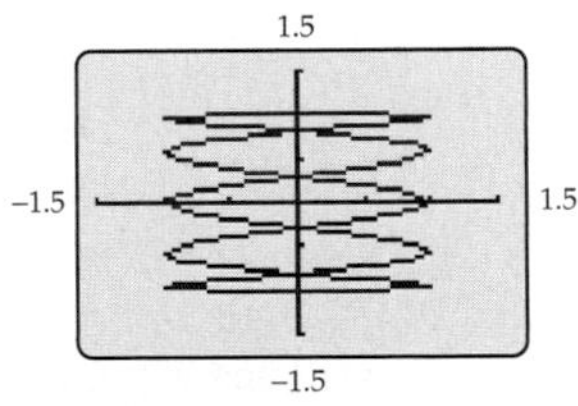

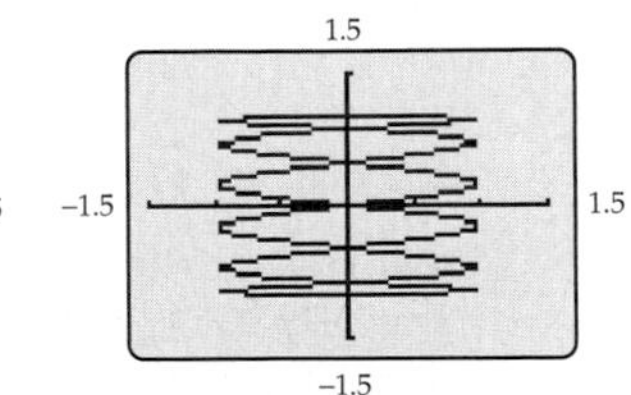

35. (a) Calculator is in degree mode for this graph.

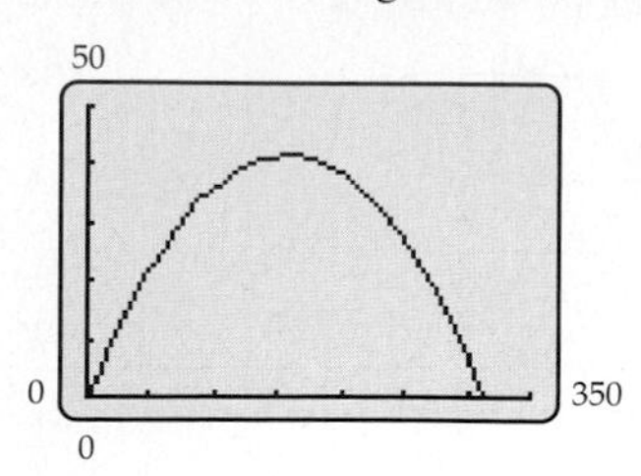

(b) $t = 3.23$ seconds

(c) $y = 41.67$ feet

37. (a)

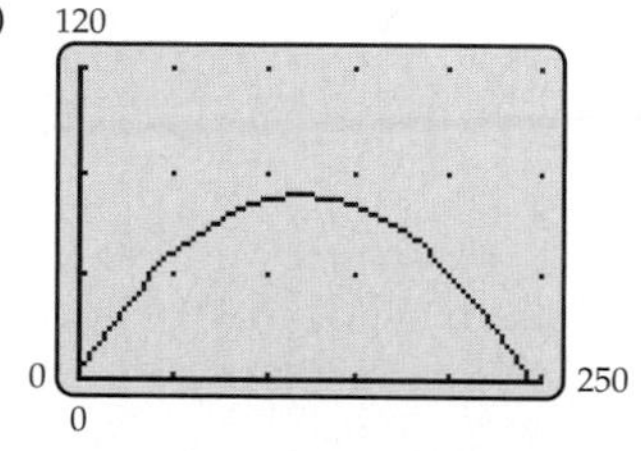

(b) The arrow will go over the wall.

39. $v \approx 105.29$ ft/sec.

41. (a)

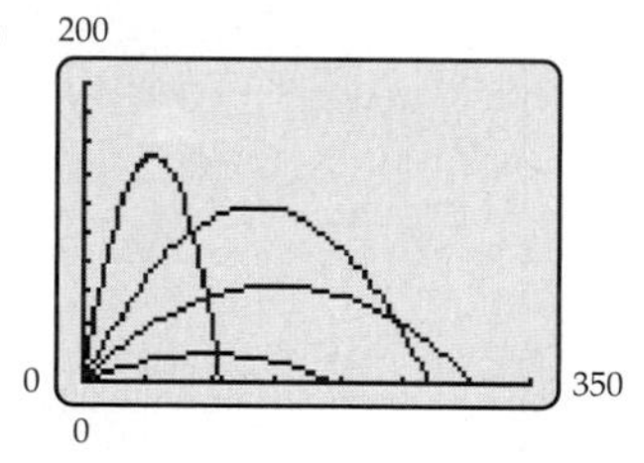

(b) 40°

(c) It appears that the ball would go farthest for a 45° angle.

43. Both graphs give a straight-line segment between $P = (-4, 7)$ and $Q = (2, -5)$. However, the graph in (a) moves from P to Q, while the graph in (b) moves from Q to P.

45. (a) The diagram shows that $\frac{\pi}{2} + \theta = t$, $\theta = t - \frac{\pi}{2}$. In right triangle PQC, $\cos \theta = CQ/3$, so $CQ = 3 \cos \theta$. Similarly, $\sin \theta = PQ/3$, so $PQ = 3 \sin \theta$. Hence, $x = OT - CQ = 3t - 3\cos(t - \pi/2)$ and $y = CT + PQ = 3 + 3\sin(t - \pi/2)$.

(b) $\cos(t - \pi/2) = \cos t \cos \pi/2 + \sin t \sin \pi/2 = \sin t$. Similarly, $\sin(t - \pi/2) = \sin t \cos \pi/2 - \cos t \sin \pi/2 = -\cos t$. Hence, $x = 3t - 3 \sin t = 3(t - \sin t)$ and $y = 3 - 3 \cos t = 3(1 - \cos t)$.

47. (a) The diagram shows that $\frac{3\pi}{2} + \theta = t$, $\theta = t - \frac{3\pi}{2}$. In right triangle PQC, $\cos \theta = CQ/3$, so $CQ = 3 \cos \theta$. Similarly, $\sin \theta = PQ/3$, so $PQ = 3 \sin \theta$. Hence, $x = OT + CQ = 3t + 3 \cos(t - 3\pi/2)$ and $y = CT - PQ = 3 - 3 \sin(t - 3\pi/2)$.

(b) $\cos(t - 3\pi/2) = \cos t \cos(3\pi/2) + \sin t \sin(3\pi/2) = -\sin t$. Similarly, $\sin(t - 3\pi/2) = \sin t \cos(3\pi/2) - \cos t \sin(3\pi/2) = \cos t$. Hence, $x = 3t - 3 \sin t = 3(t - \sin t)$ and $y = 3 - 3 \cos t = 3(1 - \cos t)$.

49. (a) Particles A and B do not collide.

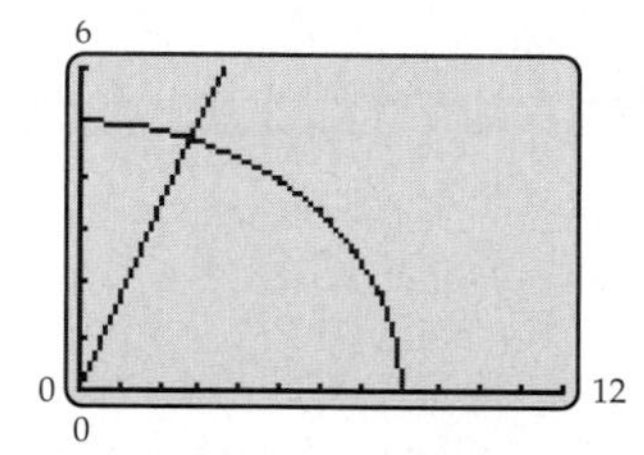

(b) $t \approx 1.10$.

(c) A and C very nearly collide near $t = 1.13$.

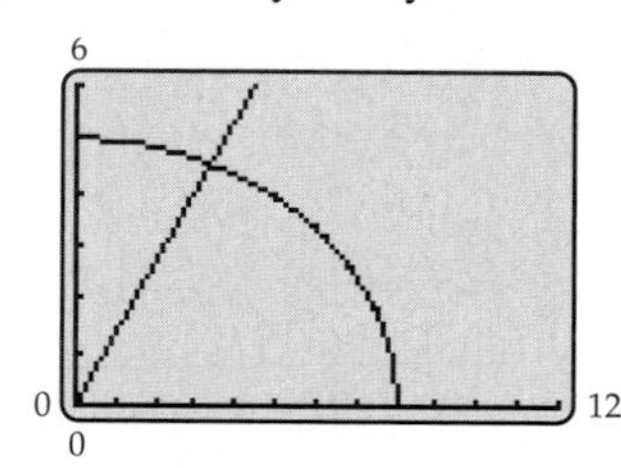

(d) The given function is obtained by applying the distance formula to the points $(8 \cos t, 5 \sin t) = A$ and $(3t, 4t) = C$. $y = \sqrt{(8 \cos t - 3t)^2 + (5 \sin t - 4t)^2}$. There is a minimum at $t = 1.1322$; however, the minimum is not zero.

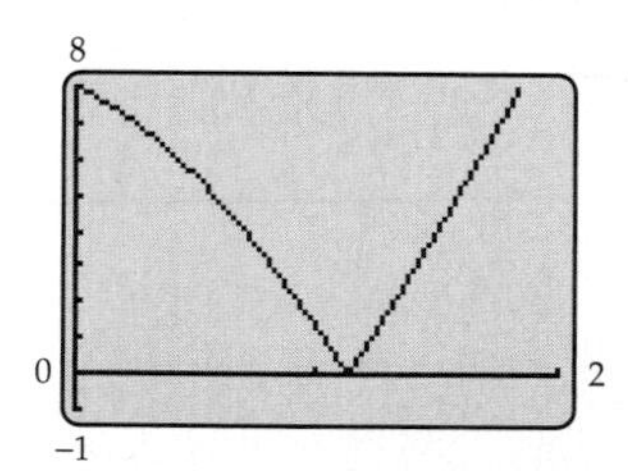

51. (a) As shown in diagram below, the center C of the small circle is always at distance $a - b$ from the origin O. Suppose t is the angle that OC makes with the x-axis. Then the coordinates of C are $x = (a - b)\cos t$, $y = (a - b)\sin t$. Examining the smaller circle in detail, we see that the change in x-coordinate from C to P is $b \cos u$, where u is the angle that PC makes with the positive x-axis. Likewise, the change in y-coordinate from C to P is $-b \sin u$. Therefore, the coordinates of P are

$$x = (a - b)\cos t + b\cos u$$
$$y = (a - b)\sin t - b\sin u.$$

The angles t and u are related by the fact that the inner circle must roll without "slipping." This means the arc length that P has moved around the inner circle must equal the arc length that the inner circle has moved along the circumference of the larger circle. In other words, the arc length from P to W must equal the arc length from S to V. Since the length of a circular arc is the radius times the angle, this means

$$bu = (a - b)t, \text{ or } u = (a - b)t/b.$$

Substituting this for u in the above equations will give the desired parametric equations.

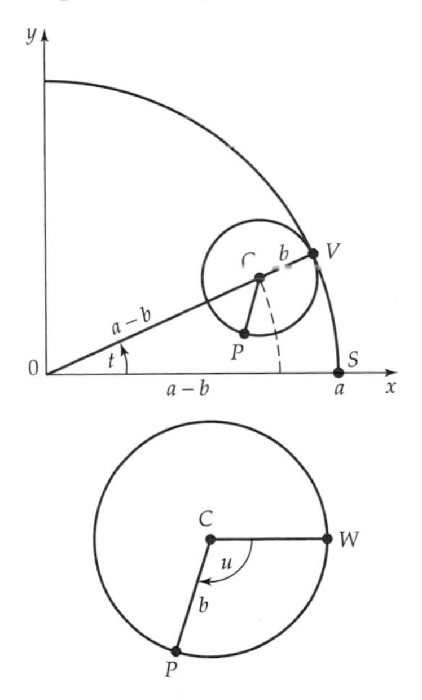

(b)

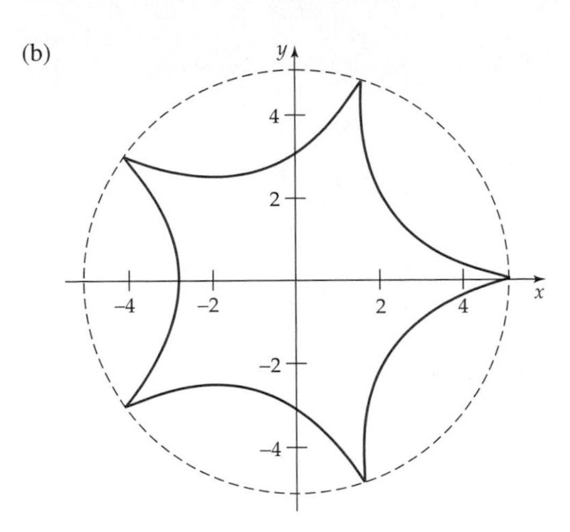

(c)

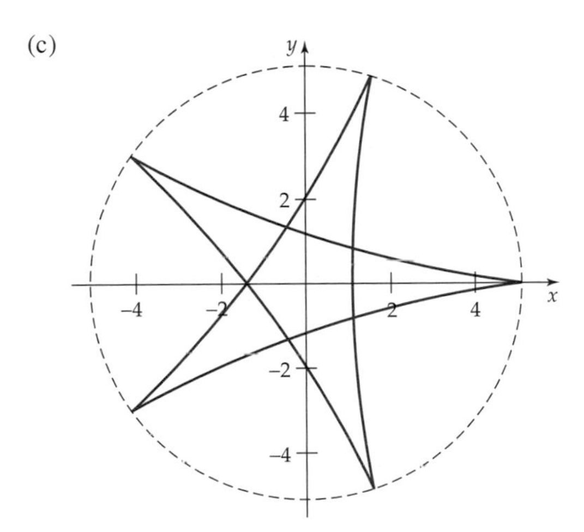

Section 5.6, page 403

1. $P\!:\left(2, \frac{\pi}{4}\right)$, $Q\!:\left(3, \frac{2\pi}{3}\right)$, $R\!: (5, \pi)$, $S\!:\left(7, \frac{7\pi}{6}\right)$, $T\!:\left(4, \frac{3\pi}{2}\right)$, $U\!:\left(6, -\frac{\pi}{3}\right)$, $V\!: (7, 0)$

3.

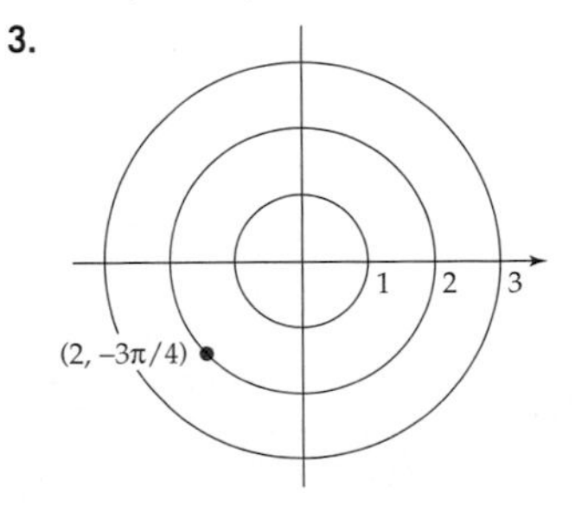

5.

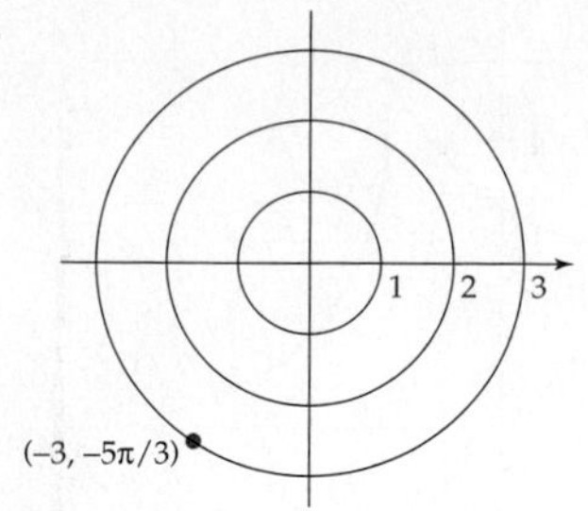

Many answers are possible in Exercises 7–11 in addition to those given here.

7. $\left(3, \frac{7\pi}{3}\right), \left(3, -\frac{5\pi}{3}\right), \left(-3, \frac{4\pi}{3}\right), \left(-3, -\frac{2\pi}{3}\right)$

9. $\left(2, \frac{4\pi}{3}\right), \left(2, -\frac{8\pi}{3}\right), \left(-2, \frac{\pi}{3}\right), \left(-2, -\frac{5\pi}{3}\right)$

11. $\left(\sqrt{3}, \frac{11\pi}{4}\right), \left(\sqrt{3}, -\frac{5\pi}{4}\right), \left(-\sqrt{3}, \frac{7\pi}{4}\right), \left(-\sqrt{3}, -\frac{\pi}{4}\right)$

13. $\left(\frac{3}{2}, \frac{3\sqrt{3}}{2}\right)$

15. $\left(\frac{\sqrt{3}}{2}, -\frac{1}{2}\right)$

17. $(.4255, -1.438)$

19. $(-3.604, 1.736)$

21. $\left(6, -\frac{\pi}{6}\right)$

23. $\left(\sqrt{2}, \frac{\pi}{4}\right)$

25. $\left(6, \frac{\pi}{3}\right)$

27. $(2\sqrt{5}, 1.107)$

29. $(5.59, 2.6679)$

31. $\left(2, \frac{3\pi}{2}\right)$

33. $(4.47, 2.0344)$

35. $r = 5$

37. $r = 12 \sec \theta$

39. $r = \frac{1}{\sin \theta - 2 \cos \theta}$

41. $x^2 + y^2 = 9$

43. $y = \frac{x}{\sqrt{3}}$

45. $x = 1$

47. $x^3 + xy^2 = y$

49. $x^2 + (y - 1)^2 = 1$

51. $x^2 = 16 - 8y$

53.

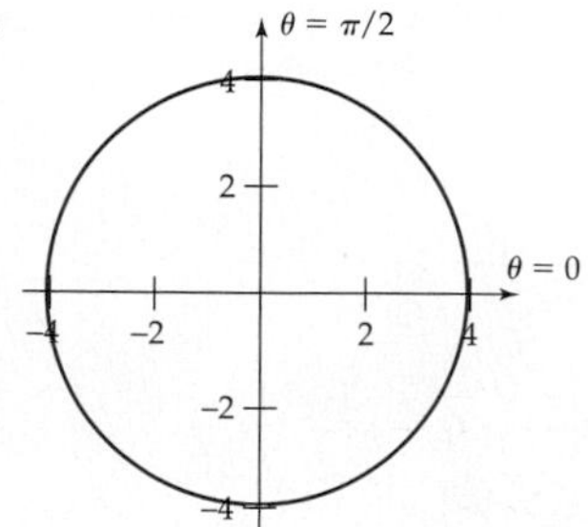

55.

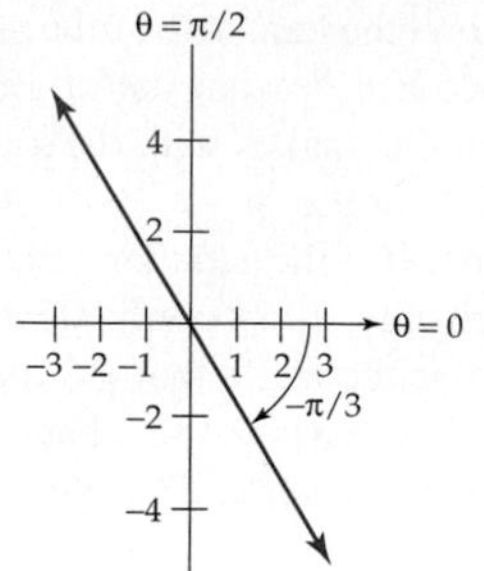

57.

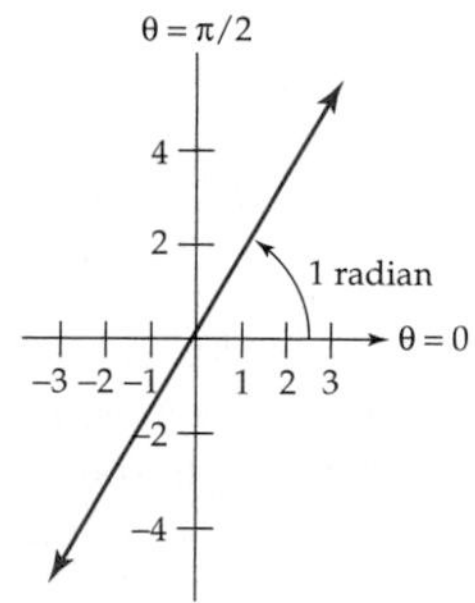

59.

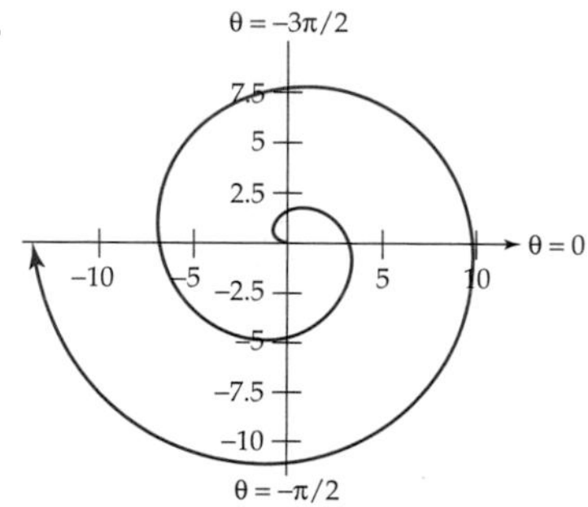

61.

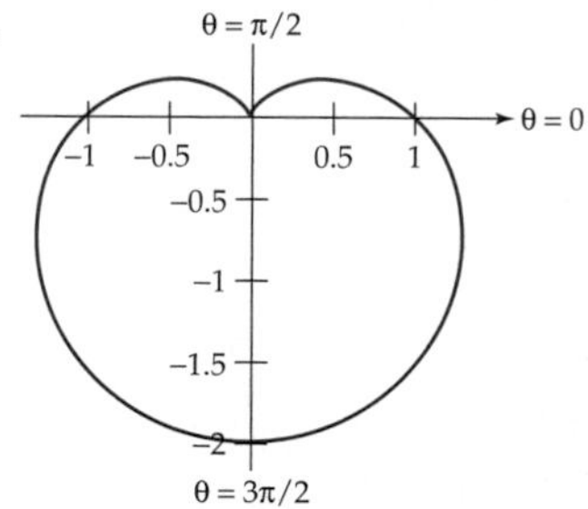

63.

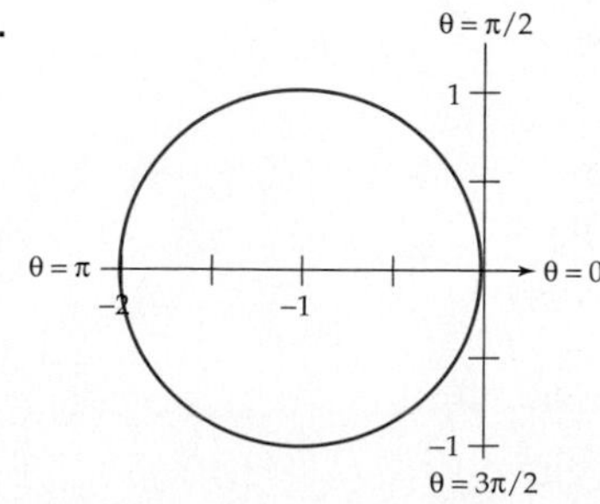

65.

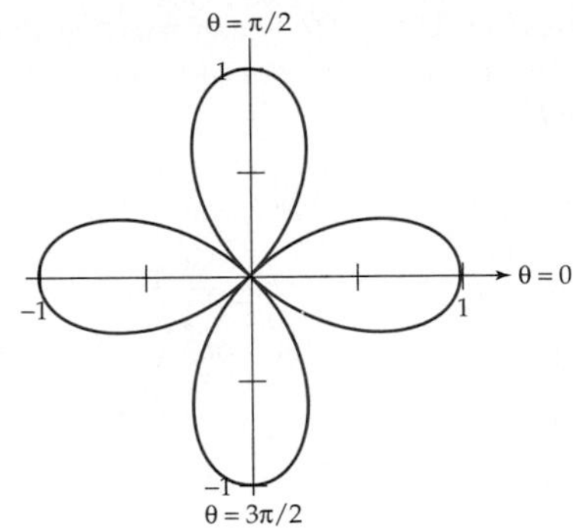

67.

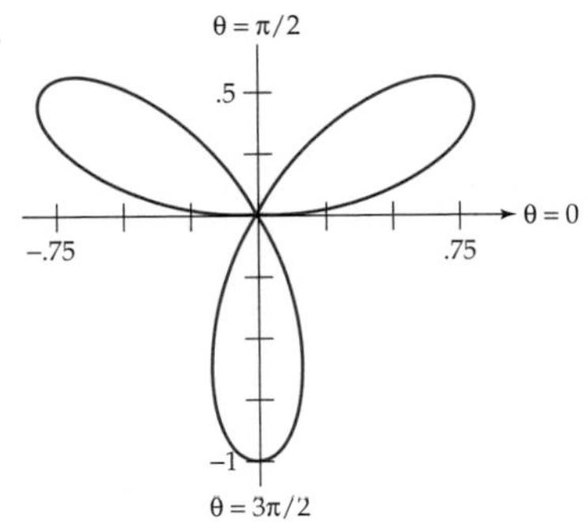

69.

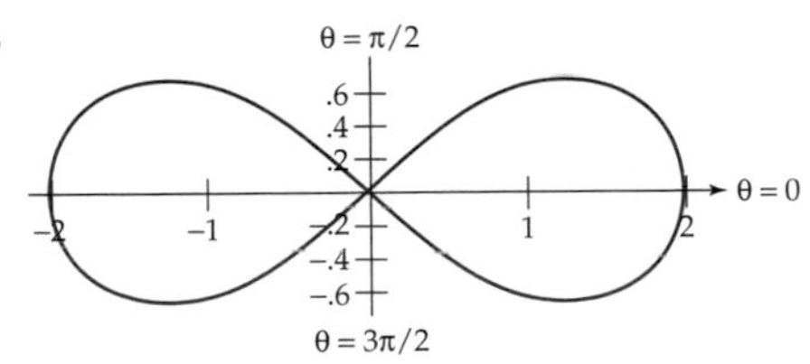

71.

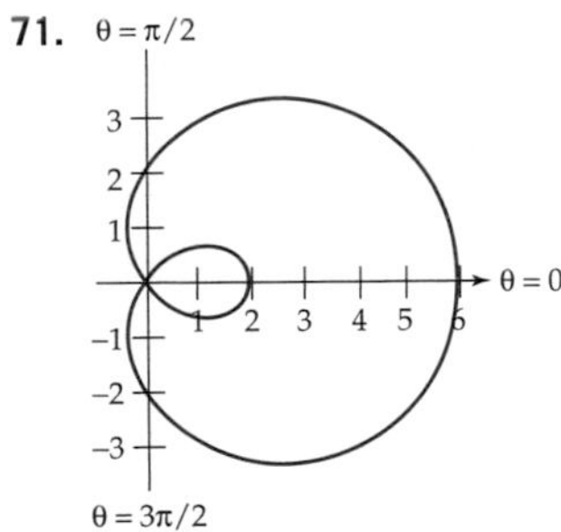

73.

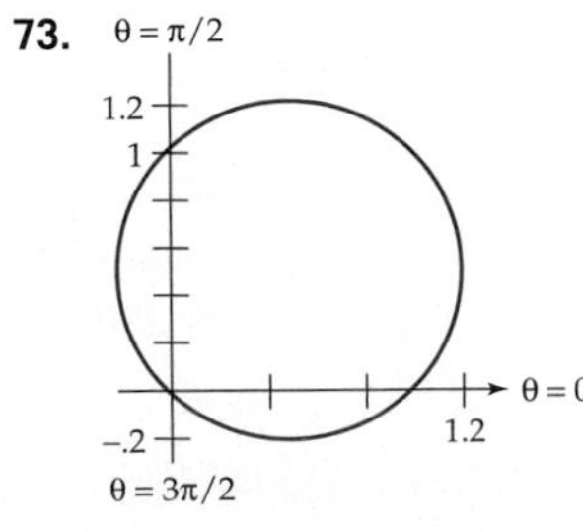

75.

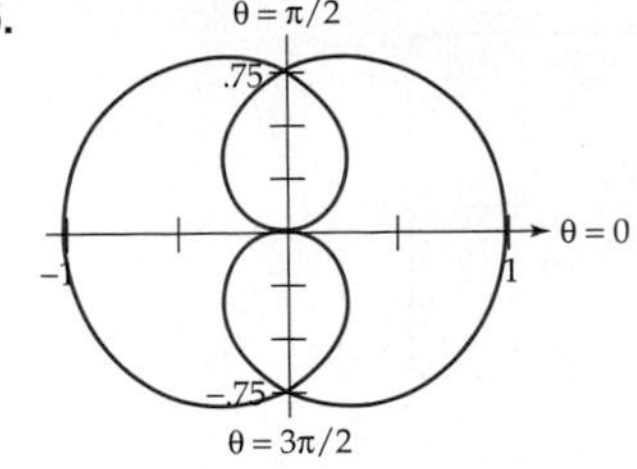

77.

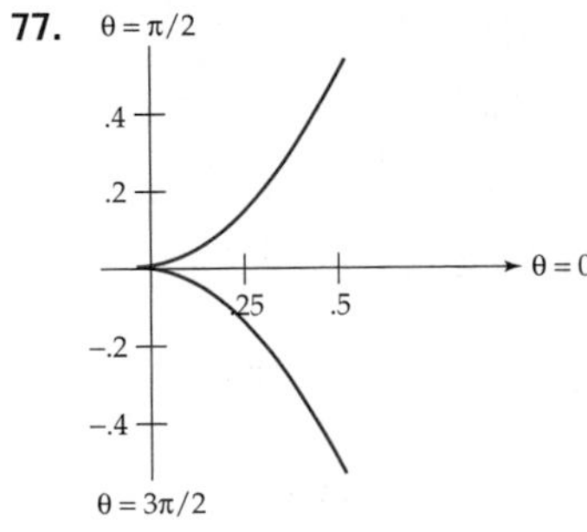

79.

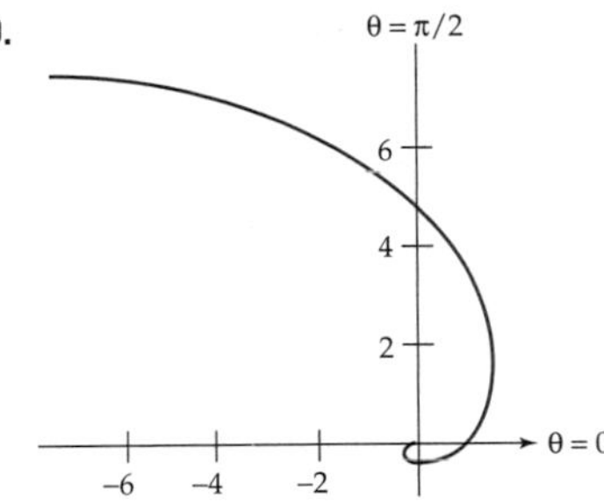

81.

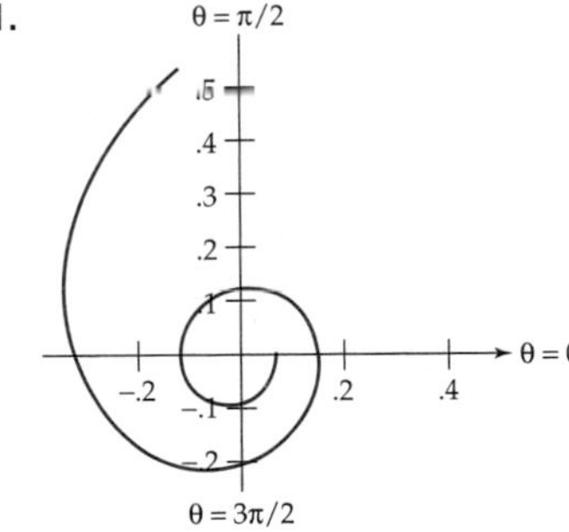

83. (a)

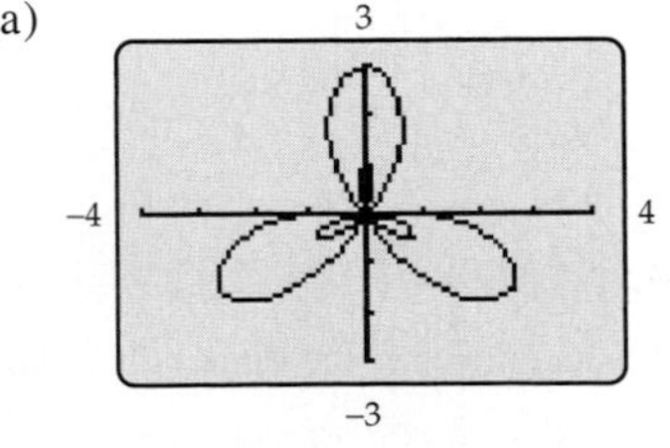

(b)

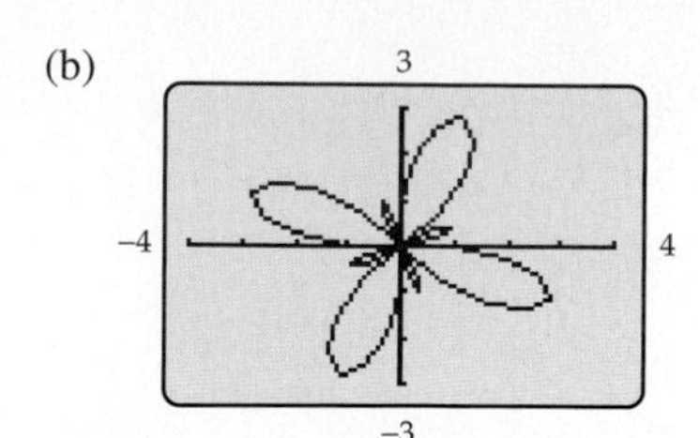

(c)

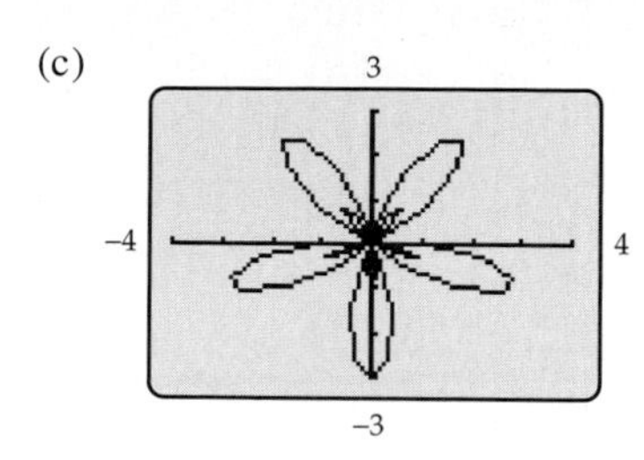

85. $r = a \sin \theta + b \cos \theta$

$$r^2 = ar \sin \theta + br \cos \theta$$
$$x^2 + y^2 = ay + bx$$
$$x^2 - bx + \frac{b^2}{4} + y^2 - ay + \frac{a^2}{4} = \frac{a^2 + b^2}{4}$$
$$\left(x - \frac{b}{2}\right)^2 + \left(y - \frac{a}{2}\right)^2 = \frac{a^2 + b^2}{4}$$

This is the equation of a circle.

87. The distance from (r, θ) to (s, β) is given by the Law of Cosines.

$$d^2 = r^2 + s^2 - 2rs \cos(\theta - \beta)$$
$$d = \sqrt{r^2 + s^2 - 2rs \cos(\theta - \beta)}$$

Section 5.7, page 413

1. (d) **3.** (c) **5.** (a)

7. Hyperbola, $e = 4/3$ **9.** Parabola, $e = 1$

11. Ellipse, $e = 2/3$ **13.** .1

15. $\sqrt{5}$ **17.** $5/4$

19. (a)

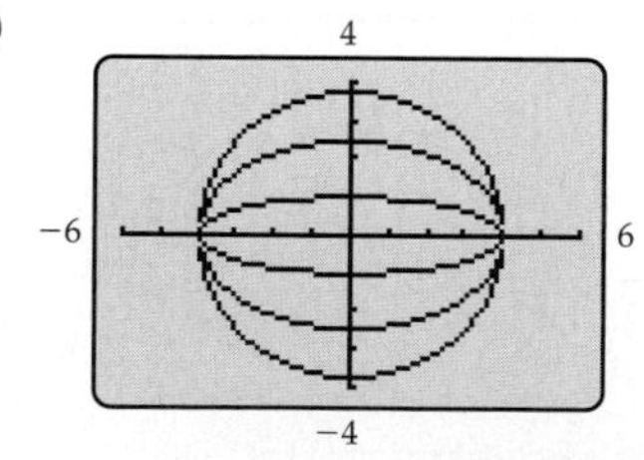

(b) $\sqrt{15}/4, \sqrt{10}/4, \sqrt{2}/4$

(c) The smaller the eccentricity, the closer the shape is to circular.

In the graphs for Exercises 21–31, the x- and y-axes with scales are given for convenience, but coordinates of points are in polar *coordinates.*

21.

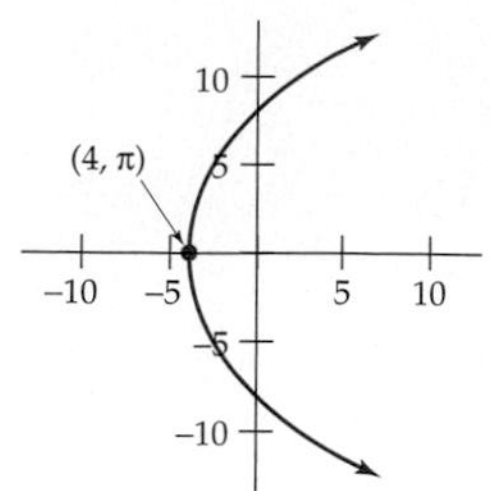

23.

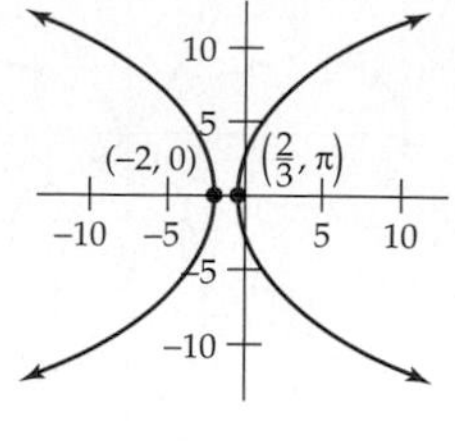

25.

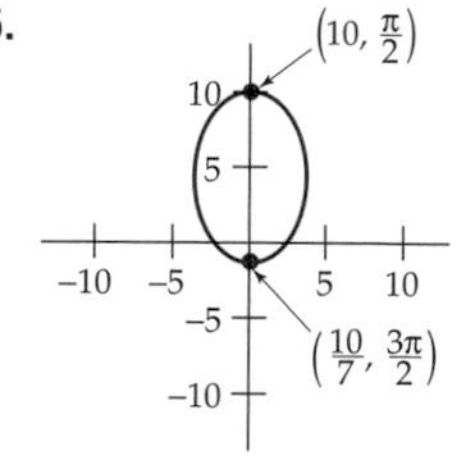

27.

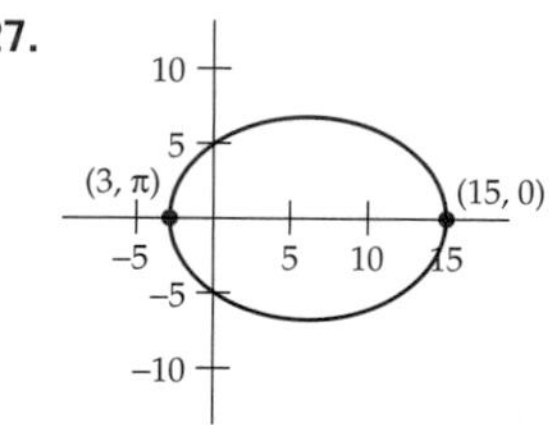

29.

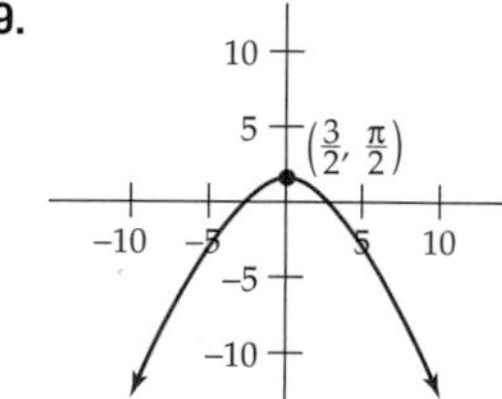

31.

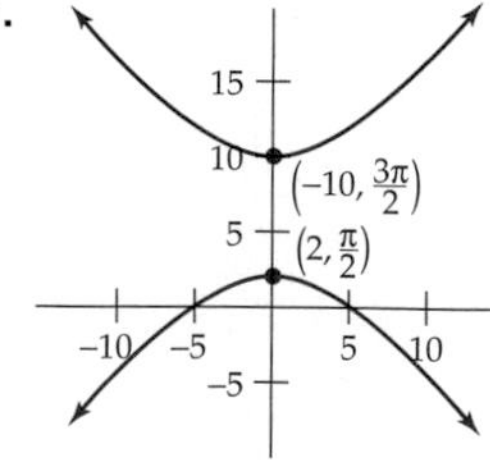

33. $r = \dfrac{6}{1 - \cos \theta}$ **35.** $r = \dfrac{16}{5 + 3 \sin \theta}$

37. $r = \dfrac{3}{1 + 2 \cos \theta}$ **39.** $r = \dfrac{8}{1 - 4 \cos \theta}$

41. $r = \dfrac{3}{1 - \sin \theta}$ **43.** $r = \dfrac{2}{2 + \cos \theta}$

45. $r = \dfrac{2}{1 - 2 \cos \theta}$

47. Since $0 < e < 1$, $0 < 1 - e^2 < 1$ as well. The formulas for a^2 and b^2 show that $a^2 = b^2/(1 - e^2)$, so $a^2 > b^2$. Since a and b are both positive, $a > b$.

49. $r = \dfrac{3 \cdot 10^7}{1 - \cos\theta}$

Chapter 5 Review, page 417

1. Ellipse, vertices $\left(0, \pm 2\sqrt{5}\right)$, foci $(0, \pm 2)$

3. Ellipse, vertices $(1, -1)$ and $(1, 7)$, foci $(1, 0)$ and $(1, 6)$

5. Focus: $(0, 5/14)$, directrix: $y = -5/14$

7. 

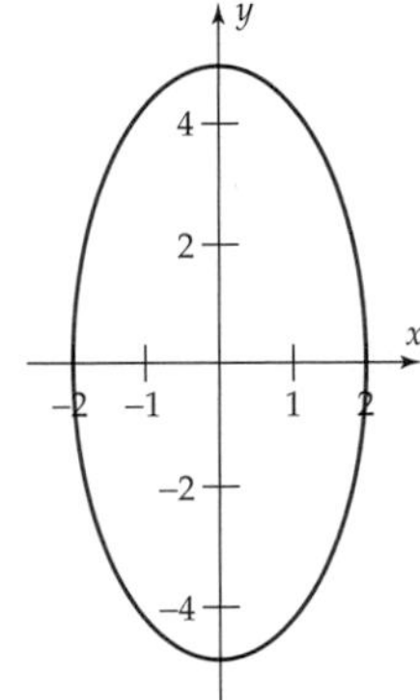

9.

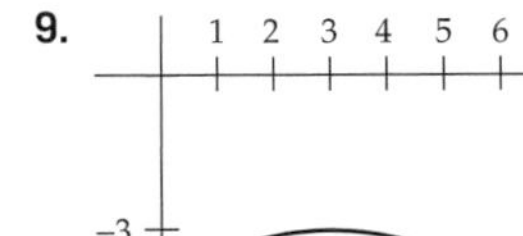

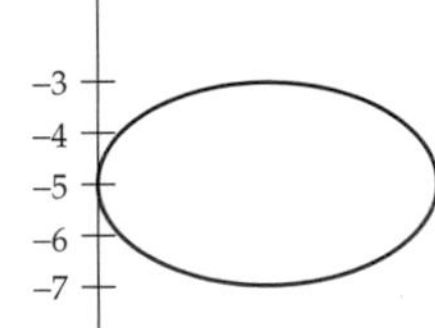

11. Asymptotes: $y + 4 = \pm\dfrac{5}{2}(x - 1)$

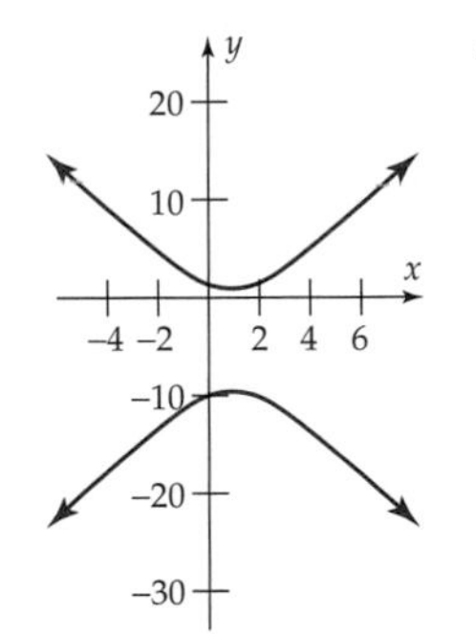

13.

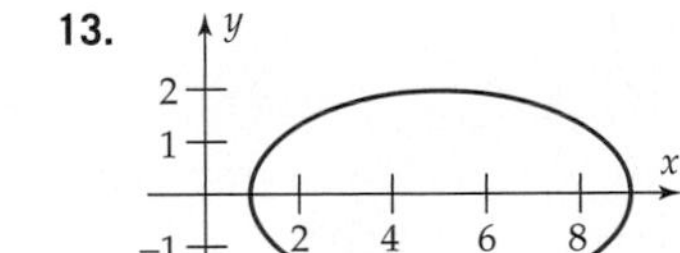

15.

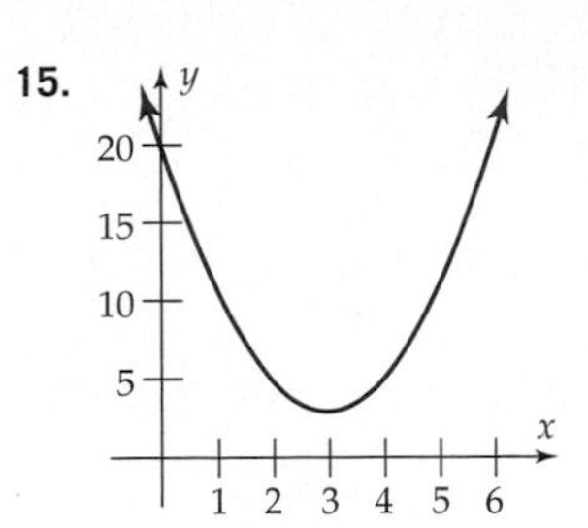

17.

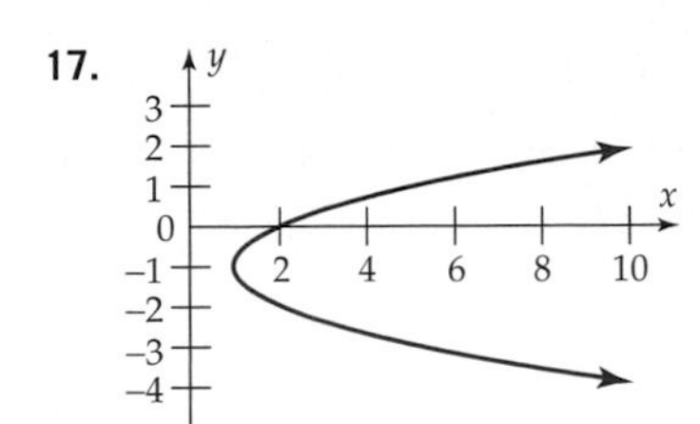

19.

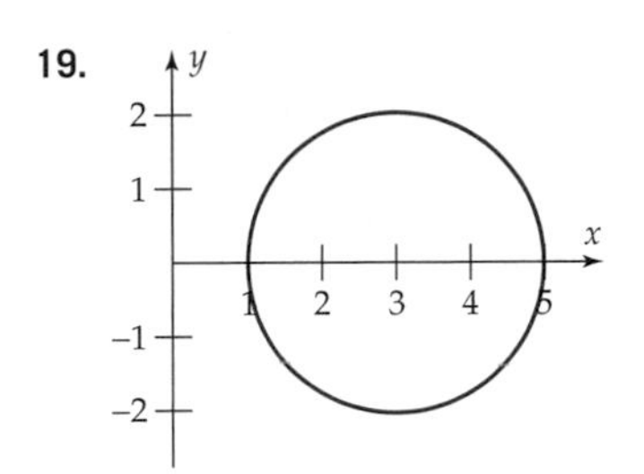

21. Center: $(-4, -5)$, radius: $2\sqrt{2}$

23. $(4, -6)$

25. $\dfrac{(x-3)^2}{4} + \dfrac{(y-1)^2}{2} = 1$

27. $\dfrac{y^2}{4} - \dfrac{(x-3)^2}{16} = 1$

29. $\left(y + \dfrac{1}{2}\right)^2 = -\dfrac{1}{2}\left(x - \dfrac{3}{2}\right)$

31.

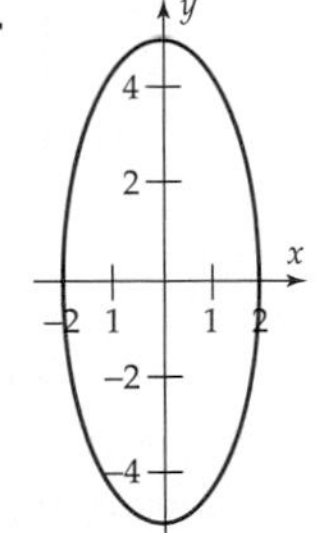

33.

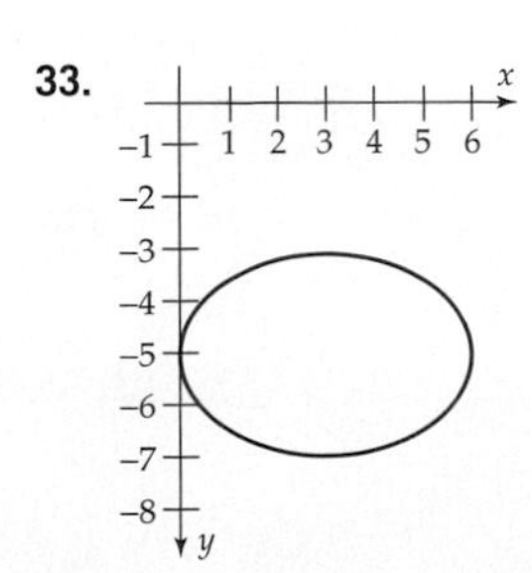

35.

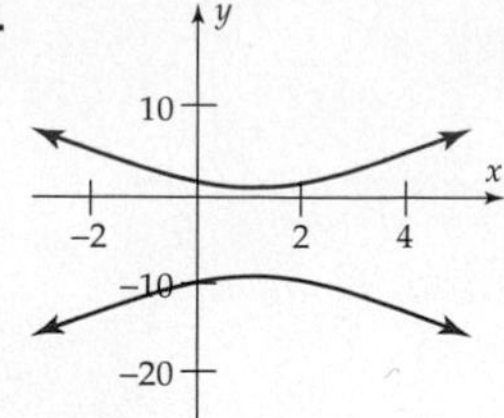

37. 6 feet

39. The receiver should be placed at the focus, $\frac{9}{16}$ feet or 6.75 inches from the vertex.

41. Ellipse **43.** Hyperbola

45. $-6 \le x \le 6$ and $-4 \le y \le 4$

47. $-9 \le x \le 9$ and $-6 \le y \le 6$

49. $-15 \le x \le 10$ and $-10 \le y \le 20$

51. $x = \frac{u}{2} - \frac{\sqrt{3}}{2}v$; $y = \frac{\sqrt{3}}{2}u + \frac{v}{2}$

53. 45°

55. $-15 \le x \le 15$ and $-10 \le y \le 10$

57. $-35 \le x \le 32$ and $-2 \le y \le 16$

59. $x = 3 - 2y$

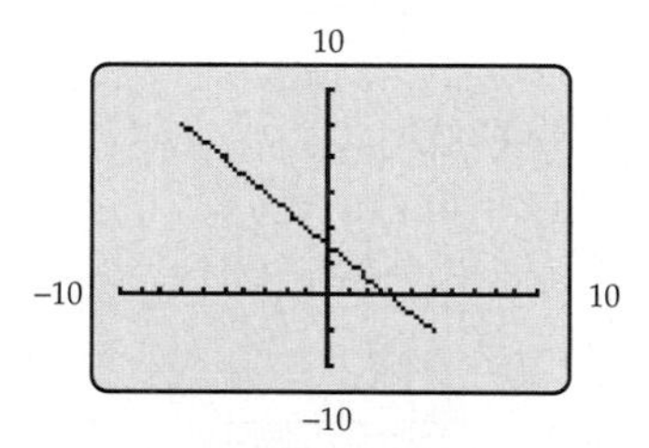

61. $y = 2 - 2x^2$

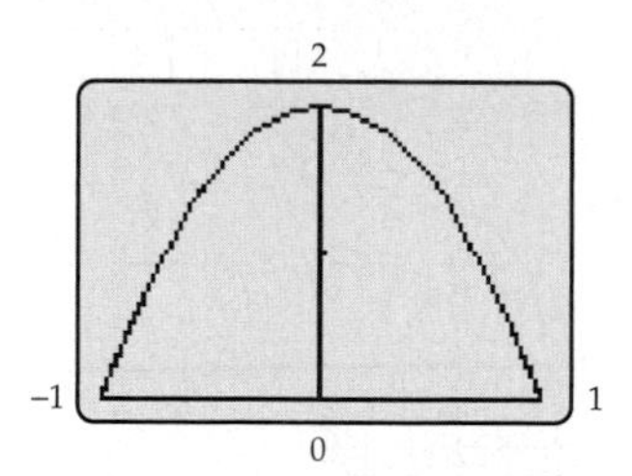

63. (b) and (c)

65.

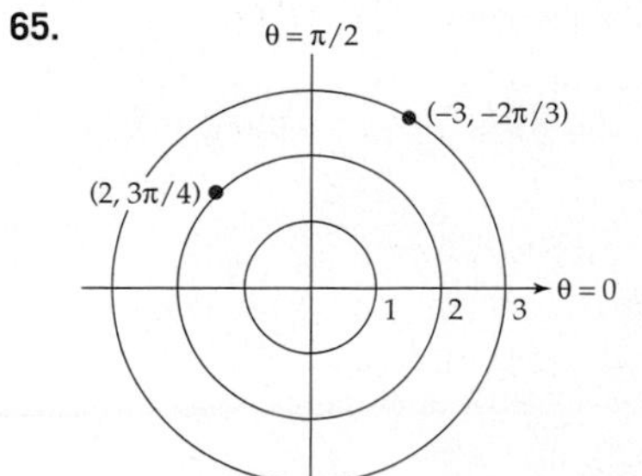

67.

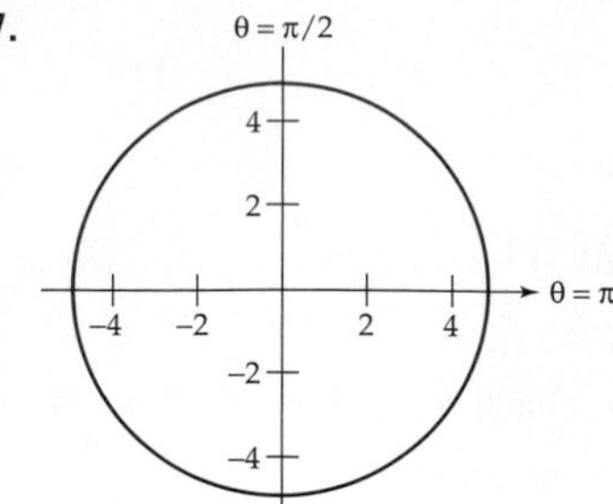

69.

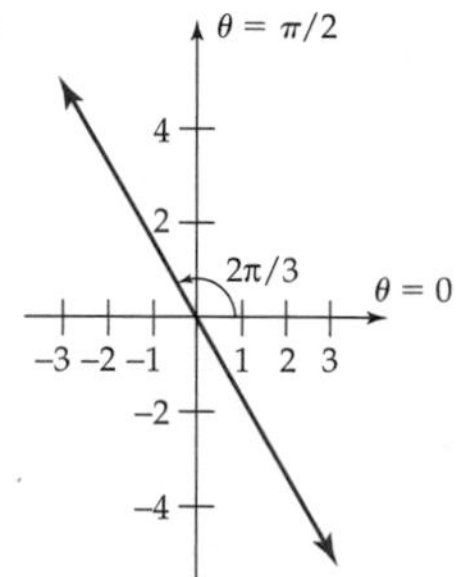

71.

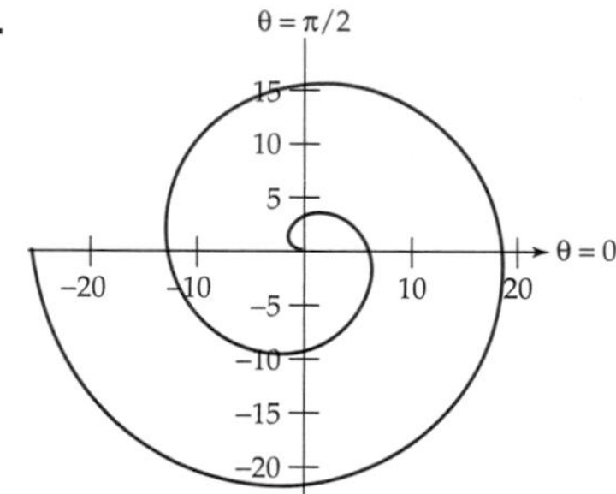

73.

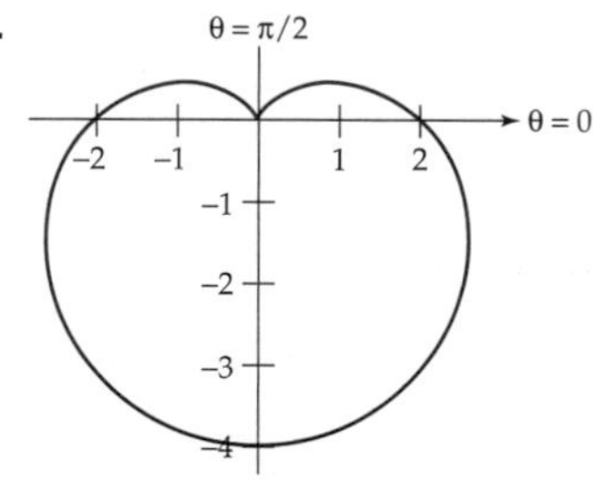

75.

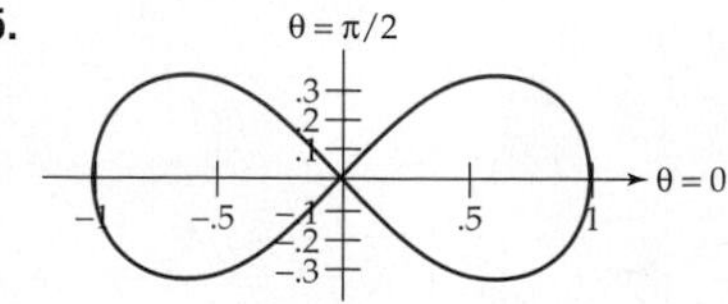

77. $\left(-\frac{3}{2}, -\frac{3\sqrt{3}}{2}\right)$

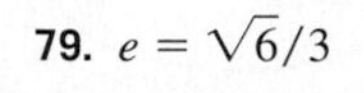

79. $e = \sqrt{6}/3$

81. Ellipse

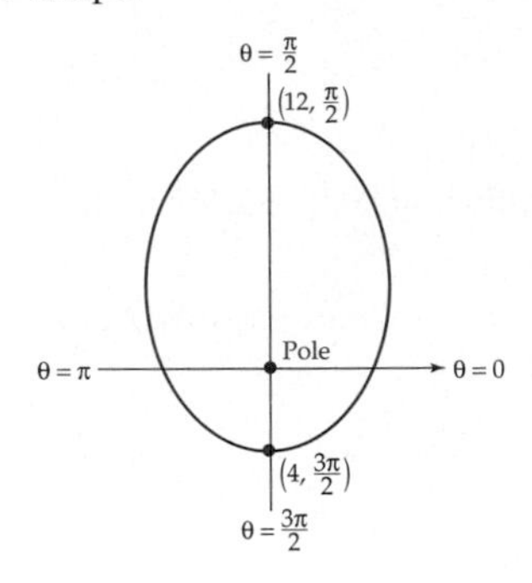

83. Hyperbola

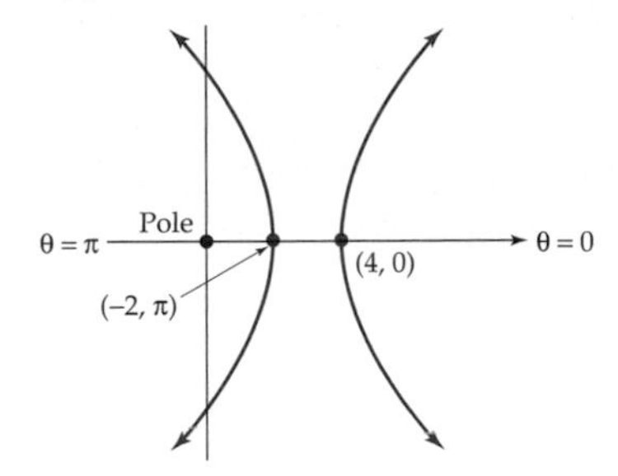

85. $r = \dfrac{24}{5 + \cos\theta}$

87. $r = \dfrac{2}{1 + \cos\theta}$

CHAPTER 6

Section 6.1, page 432

1.

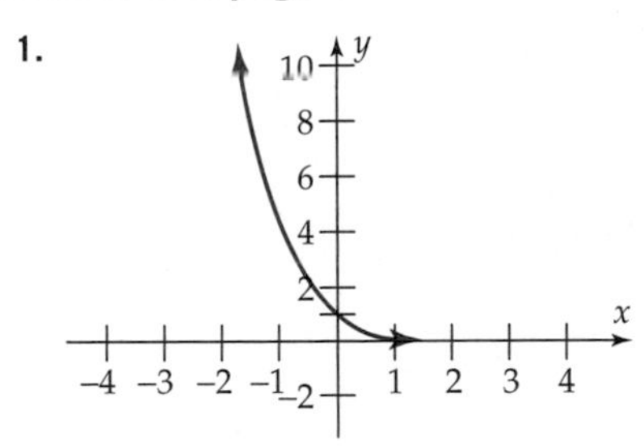

3.

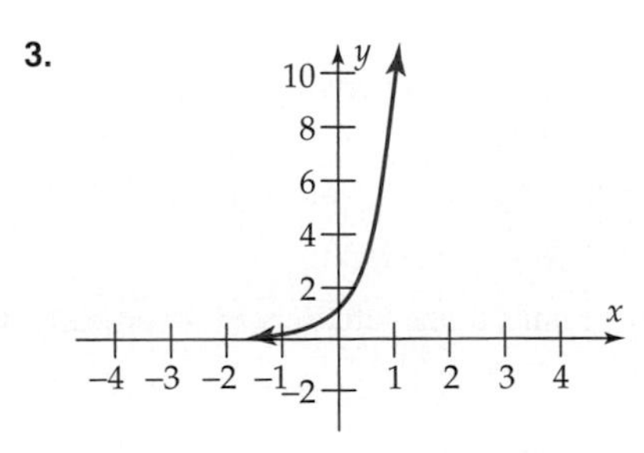

5.

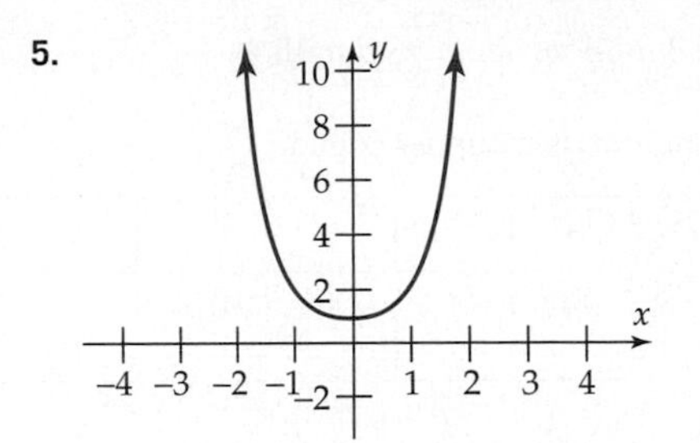

7. Shift vertically 5 units down.

9. Stretch away from the x-axis by a factor of 3.

11. Shift horizontally 2 units left, then shift vertically 5 units down.

13. $f(x)$ has graph C, $g(x)$ has graph A, $h(x)$ has graph B.

15. Neither **17.** Even **19.** Even

21. 11 **23.** 4/5 **25.** $\dfrac{10^{x+h} - 10^x}{h}$

27. $\dfrac{2^{x+h} + 2^{-x-h} - 2^x - 2^{-x}}{h}$

29. $-4 \le x \le 4,\ -1 \le y \le 10$

31. $-4 \le x \le 4,\ -1 \le y \le 10$

33. $-10 \le x \le 10,\ -10 \le y \le 10$

35. $-5 \le x \le 10,\ -1 \le y \le 6$

37. The negative x-axis is an asymptote; $(-1.443, -.531)$ is a local minimum.

39. No asymptote; (0, 1) is a local minimum.

41. The x-axis is an asymptote; (0, 1) is a local maximum.

43. (a)

Time (hours)	Number of Cells
0	1
.25	2
.5	4
.75	8
1	16

(b) $C(t) = 2^{4t}$

45. $g(x) = -315(e^{.00418x} + e^{-.00418x})$; $C = 1260$

47. (a) About 1.984 billion, about 3.998 billion.
(b) 2008

49. (a) About 286.4 million, about 300.1 million
(b) About 2041

51. (a) $k = 15$ (b) About 12.13 psi
(c) About .0167 psi

53. (a) About 74.1 years, about 76.3 years
(b) 1930

55. (a) About 62.2 million, about 73.5 million.
(b) 2000
(c) No. 76.7 million is an upper bound.

57. (a)

Fold	0	1	2	3	4
Thickness (inches)	.002	.004	.008	.016	.032

(b) $f(x) = .002(2^x)$
(c) 2097.152 inches
(d) 43 folds

59. $f(x) = 2^x$ is one example.

61. (a)

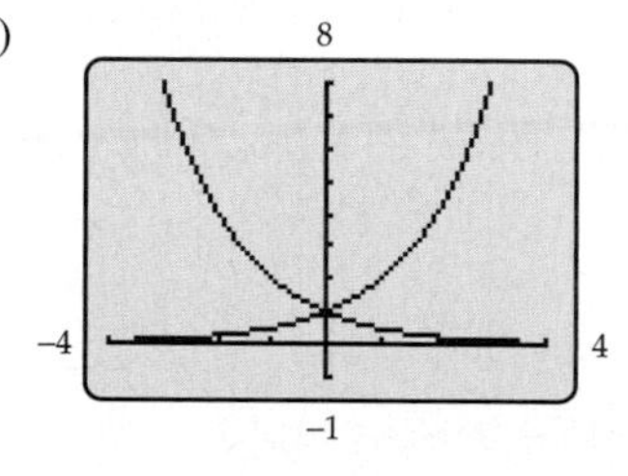

Each graph is the reflection of the other in the y-axis.
(b) Each graph is the reflection of the other in the y-axis.

Section 6.2, page 443

1. (a) \$1469.33
(b) \$1485.95
(c) \$1489.85
(d) \$1491.37

3. \$585.83 **5.** \$610.40 **7.** \$639.76

9. \$563.75 **11.** \$582.02 **13.** \$3325.29

15. \$3359.59 **17.** \$6351.16 **19.** \$568.59

21. Fund C **23.** \$385.18

25. \$1,162,003.14 **27.** \$4000 after 4 years

29. About 5% **31.** About 5.92%

33. (a) 9 years
(b) 9 years
(c) 9 years
(d) Doubling time is independent of investment amount.

35. About 9.9 years **37.** \$7835.26

39. \$21,303.78

41. (a) About 12.55%
(b) Quarterly: 12.55%, monthly: 12.6825%, daily: 12.7475%.

43. (a) $f(x) = 6(3^x)$
(b) 3
(c) In 12 weeks no, in 14 weeks yes.

45. (a) $f(x) = 100.4(1.014^x)$
(b) About 115.375 million
(c) In 2015

47. (a) $f(x) = 4966\,(1.039^x)$
(b) \$8484.46
(c) 2005–2006

49. (a) $f(x) = 32.44(1.0224^x)$
(b) In 2010: 40.485 million, in 2025: 56.442 million
(c) 2023

51. 256 bacteria after 10 hours, 654 bacteria after 2 days

53. (a) $f(x) = .75^x$
(b) 8 feet

55. 103 months

57. (a) $f(x) = 100(.5^{x/1620})$
(b) After 800 years: 71.0 milligrams, after 1600 years: 50.4 milligrams, after 3200 years: 25.4 milligrams.

59. About 7171 years

61. (a) After the third 50-year period: $c\left(\frac{1}{2}\right)^3$, after the fourth 50-year period: $c\left(\frac{1}{2}\right)^4$.
(b) $c\left(\frac{1}{2}\right)^t$
(c) $c\left(\frac{1}{2}\right)^{x/50}$ becomes $c\left(\frac{1}{2}\right)^{x/h}$.

Section 6.3, page 455

1. 4 **3.** -2.5

5. $10^3 = 1000$ **7.** $10^{2.88} = 750$

9. $e^{1.0986} = 3$ **11.** $e^{-4.6052} = .01$

13. $e^{z+w} = x^2 + 2y$ **15.** $\log .01 = -2$

17. $\log 3 = .4771$ **19.** $\ln 25.79 = 3.25$

21. $\ln 5.5527 = 12/7$ **23.** $\ln w = 2/r$

25. $\sqrt{43}$ **27.** 15 **29.** .5

31. 931 **33.** $x + y$ **35.** x^2

37. $f(x) = 4e^{3.2188x}$ **39.** $g(x) = -16e^{3.4177x}$

41. $x > -1$ **43.** $x < 0$ **45.** $x > 0$

47. (a) For all $x > 0$.
(b) According to the fourth property of logarithms, $e^{\ln x} = x$ for every $x > 0$.

49. Stretch vertically by a factor of 2 away from the x-axis.

51. Shift horizontally 4 units to the right.

53. Shift horizontally 3 units to the left, then shift vertically 4 units down.

55.

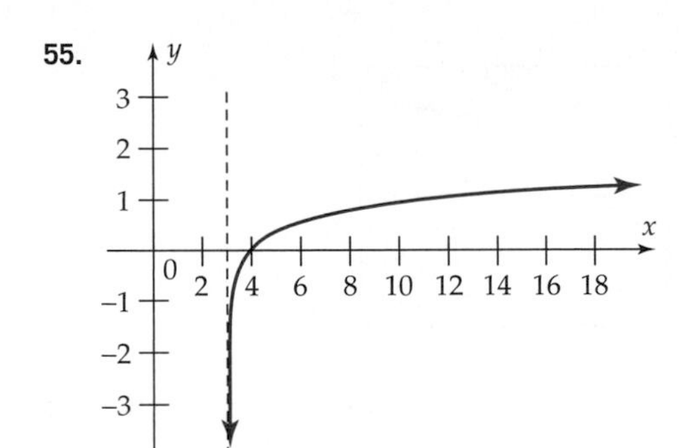

57.

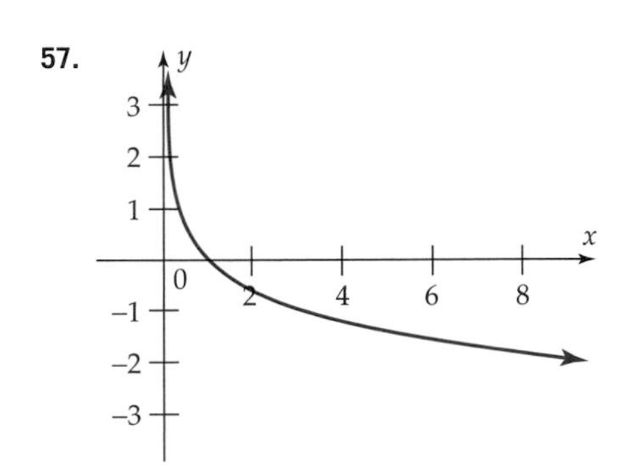

59. $0 \le x \le 20$ and $-10 \le y \le 10$; then $0 \le x \le 2$ and $-20 \le y \le 20$

61. $-10 \le x \le 10$ and $-3 \le y \le 3$

63. $0 \le x \le 20$ and $-6 \le x \le 3$

65. .5493 **67.** −.2386

69. $\dfrac{\ln(3+h) - \ln 3}{h}$
(b) $h = 2.2008$

71. (a) 4%: 17.7 years, 6%: 11.9 years, 8%: 9 years, 12%: 6.1 years, 18%: 4.2 years, 24%: 3.2 years, 36%: 2.3 years
(b) 18, 12, 9, 6, 4, 3, 2; they are the same. The doubling time for an investment at $p\%$ is approximately $72/p$ years.

73. (a) 77.7
(b) 2012

75. (a) 44.4%
(b) 2010

77. $A = -9$, $B = 10$

79. About 4392 meters

81. (a) In 1999: about 26.55 billion pounds, in 2002: about 27.72 billion pounds.
(b) 2008

83. (a) About 9.9 days
(b) 6986 people

85. (a) No ads: about 120 bikes, $1000: about 299 bikes, $10,000: about 513 bikes.
(b) $1000: yes, $10,000: no.
(c) $1000: yes, $10,000: yes.

87. About $n = 30$ on TI-83.

Section 6.4, page 463

1. $\ln(x^2y^3)$ **3.** $\log(x-3)$ **5.** $\ln\left(\dfrac{1}{x^7}\right)$

7. $\ln(e-1)^3$ **9.** $\log(20xy)$ **11.** $\log(x^2-9)$

13. $\log\left(\dfrac{1}{(x+1)(x-2)}\right)^2$ **15.** $\ln\sqrt{\dfrac{x}{3x+5}}$

17. $\log x$ **19.** $2u + 5v$

21. $\frac{1}{2}u + 2v$ **23.** $\dfrac{2}{3}u + \dfrac{1}{6}v$

25. False **27.** True **29.** False

31. $a = 100$, $b = 10$, among many examples

33. $b = e$

35. The graphs are not the same. Negative real numbers are in the domain of $v = \log x^2$ but not in the domain of $y = 2\log x$.

37. The graph of $f(x) = \log(5x) = \log 5 + \log x$ is the graph of $g(x) = \log x$ shifted upward by $\log 5$ units.

39. The graph of $f(x) = \ln(x^5) = 5\ln x$ is the graph of $g(x) = \ln x$ stretched away from the x-axis by a factor of 5.

41. $\log\left(\dfrac{x+h}{x}\right)^{\frac{1}{h}}$

43. (a) About 5.38 million vehicles
(b) 2009

45. 2 **47.** 2.5 **49.** 20 **51.** 66

53. Twice as loud

55. (a) Both are approximately 1.255.
(b) Both are approximately 2.659.
(c) Both are approximately 3.952.
(d) These results suggest that $\log c = \dfrac{\ln c}{\ln 10}$.

57. (a) .9422
(b) 1.9422
(c) 2.9422
(d) 3.9422
(e) 4.9422
(f) The numbers 8.753, 87.53, . . ., 87,530 are related in that each is 10 times the previous number. Their logarithms are related in that each is 1 more than the previous logarithm. The numbers a and b are related as follows: If $b = 10a$, then $\log b = 1 + \log a$.

Special Topics 6.4.A, page 472

1.

x	0	1	2	4
$f(x) = \log_4 x$	Not defined	0	.5	1

3.

x	1/36	1/6	1	216
$h(x) = \log_6 x$	−2	−1	0	3

5.

x	0	1/7	$\sqrt{7}$	49
$f(x) = 2\log_7 x$	Not defined	−2	1	4

7.

x	−2.75	−1	1	29
$h(x) = 3\log_2(x+3)$	−6	3	6	15

9. $\log .01 = -2$ **11.** $\log \sqrt[3]{10} = \frac{1}{3}$

13. $\log r = 7k$ **15.** $\log_7 5{,}764{,}801 = 8$

17. $\log_3 (1/9) = -2$ **19.** $10^4 = 10{,}000$

21. $10^{2.88} \approx 750$ **23.** $5^3 = 125$

25. $2^{-2} = 1/4$ **27.** $10^{z+w} = x^2 + 2y$

29. $\sqrt{97}$ **31.** $x^2 + y^2$ **33.** .5

35. 6 **37.** $b = 3$ **39.** $b = 20$

41. 5 **43.** 3 **45.** 4

47. $\log\left(\frac{x^2y^3}{z^6}\right)$ **49.** $\log[x(x-3)]$

51. $\log_2[5|c|]$ **53.** $\log_4\left(\frac{1}{49c^2}\right)$

55. $\ln\left(\frac{(x+1)^2}{x+2}\right)$ **57.** $\log_2 x$

59. $\ln(e-1)^2$ **61.** 3.3219 **63.** .8271

65. 1.1115 **67.** 1.6199

69. 1.5497, more diverse

71. For Community 1, $H = 1.5850$; for Community 2, $H = .8118$.

73. True by the Quotient Law.

75. True by the Power Law.

77. False. By the Product Law, the logarithm of $5x$ is the logarithm of 5 plus the logarithm of x, not 5 times the logarithm of x.

79. 397^{398} is larger. **81.** $\log_b u = \frac{\log_a u}{\log_a b}$

83. $\log_{10} u = 2\log_{100} u$

85. $\log_b x = \frac{1}{2}\log_b v + 3 = \frac{1}{2}\log_b v + 3\log_b b = \log_b v^{1/2} + \log_b b^3 = \log_b\left(b^3\sqrt{v}\right)$, so $x = b^{\log_b x} = b^{\log_b (b^3\sqrt{v})} = b^3\sqrt{v}$.

87.

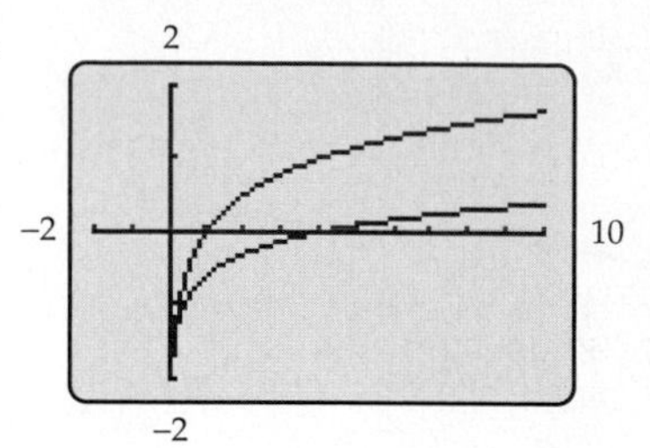

The graphs are not the same; they intersect only when $x = 10^{\log 4/(1-1/\log 4)}$, or approximately 0.1228. Statements such as the one given are generally false.

89.

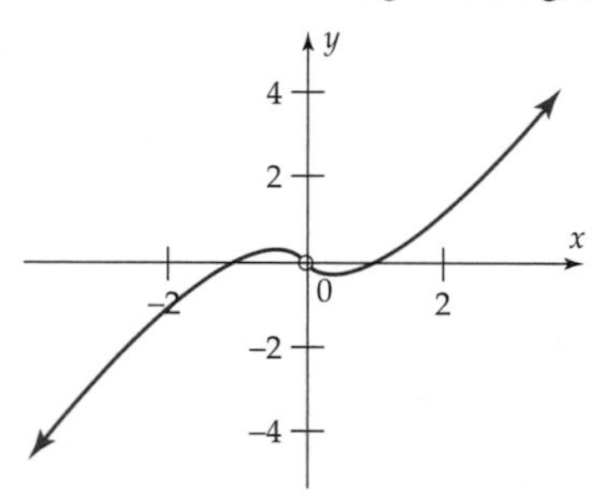

Local minimum at approximately (.3679, −.3196). Local maximum at approximately (−.3679, .3196). No asymptotes. There is a hole at $x = 0$ because $h(0)$ is not defined.

Section 6.5, page 481

1. $x = 4$ **3.** $x = 1/9$

5. $x = 1/2$ or -3 **7.** $x = -2$ or $-1/2$

9. $x = \ln 5/\ln 3 \approx 1.4650$

11. $x = \ln 3/\ln 1.5 \approx 2.7095$

13. $x = \frac{\ln 3 - 5\ln 5}{\ln 5 + 2\ln 3} \approx -1.8254$

15. $x = \frac{\ln 2 - \ln 3}{\ln 3 + 3\ln 2} \approx -.1276$

17. $x = (\ln 5)/2 \approx .8047$

19. $x = (\ln 3.5)/(-1.4) \approx -.8948$

21. $x = \frac{2\ln(5/2.1)}{\ln 3} \approx 1.5793$

23. If $\ln u = \ln v$, then $u = e^{\ln u} = e^{\ln v} = v$.

25. $x = 9$ **27.** $x = 5$

29. $x = 6$ **31.** $x = 3$

33. $x = \frac{-5 + \sqrt{37}}{2}$ **35.** $x = 9/(e-1)$

37. $x = 5$ **39.** $x = \pm\sqrt{10{,}001}$

41. $x = \sqrt{\frac{e+1}{e-1}}$

43. (a) About $4.75 billion
(b) 2009

45. (a) About 72.2 million families
(b) In 2008

47. About .444 billion years

49. About 14.95 days

51. About 3.813 days

53. About 2534 years ago

55. About 3689 years

57. 6.99%

59. (a) About 22.52 years
(b) About 22.11 years

61. \$3197.05

63. About 79.4 years

65. (a) 1.36%
(b) 2027

67. (a) $k = 21.459$
(b) .182

69. (a) At the beginning, 20 bacteria; three hours later, 2500 bacteria
(b) About .43 hours

71. (a) 200 people; 2795 people
(b) After 6 weeks

73. (a) $k \approx .2290$, $c \approx 83.33$
(b) About 12.4 weeks

Chapter 6 Review, page 486

1. $-3 \le x \le 3$ and $0 \le y \le 10$.

3. $-3 \le x \le 3$ and $-2 \le y \le 8$.

5. $-3 \le x \le 3$ and $0 \le y \le 2$.

7. $-5 \le x \le 10$ and $-5 \le y \le 2$.

9. (a) About 2.03%
(b) About 31.97%
(c) From the sixth to the tenth month
(d) Never. The line $y = 100$ is a horizontal asymptote, so $P(t)$ never reaches 100; but after 17 months, P is over 99%.

11. (a) \$1341.68
(b) \$541.68

13. (a) \$2357.90
(b) After 392 months (32 years, 8 months)

15. $S(x) = 56{,}000(1.065)^x$

17. About 3.747 grams

19. $\ln 756 = 6.628$

21. $\ln(u + v) = r^2 - 1$

23. $\log 756 = 2.8785$

25. $e^{7.118} = 1234$

27. $e^t = rs$

29. $5^u = cd - k$

31. 3

33. $3/4$

35. $2 \ln x$ or $\ln x^2$

37. $\ln\left(\dfrac{9y}{x^2}\right)$

39. $\ln x^{-10}$

41. 2

43. (c)

45. (c)

47. $\pm\sqrt{2}$

49. $\dfrac{3 \pm \sqrt{57}}{4}$

51. $-1/2$

53. $e^{(u-c)/d}$

55. 2

57. 101

59. About 1.64 milligrams

61. About 12.05 years

63. \$452.89

65. 7.6

APPENDIX, page 504

1. $b = \sqrt{20} = 2\sqrt{5}$

3. $c = \sqrt{63} = 3\sqrt{7}$

5. $b = \sqrt{50} = 5\sqrt{2}$, $c = 5$

7. $b = \sqrt{75} = 5\sqrt{3}$, $c = 5$

9. Angle $A = 53.13°$
Angle $R = 36.87°$
$r = 3$, $s = 4$, $t = 5$

11. Angle $A = 65°$ and Angle $R = 25°$.
Thus, $\angle A \cong \angle T$, $\angle C \cong \angle R$, $AC \cong TR$.
By ASA, the triangles are congruent.

13. $\angle u \cong \angle v$, that is, $\angle POD \cong \angle QOD$. $\angle PDO \cong \angle QDO$, since all right angles are congruent. $OD \cong OD$. By ASA, $PDO \cong QDO$. Hence, corresponding parts are congruent and $PD \cong QD$.

15. $\angle u \cong \angle v$, that is, $\angle BAC \cong \angle DCA$. $AB \cong DC$ and $AC \cong AC$. By SAS, the triangles ABC and ACD are congruent.

17. $\angle x \cong \angle y$, that is, $\angle BAC \cong \angle DAC$. $\angle u \cong \angle v$, that is, $\angle BCA \cong \angle DCA$. $AC \cong AC$. By ASA, the triangles ABC and ADC are congruent.

19. $c = 7.5$, $r = 28$, $t = 35$

21. $r = 3$, $t = 3.5$

23. Since $\dfrac{AB}{RS} = \dfrac{15}{10} = \dfrac{AC}{RT}$ and $\angle A \cong \angle R$, the triangles are similar by S/A/S. $a = 15$, $\angle S \cong \angle T$, and both have measure 60°, since the triangles are equilateral.

25. Since $\dfrac{AC}{RS} = \dfrac{6.885}{4.59} = \dfrac{5.25}{3.5} = \dfrac{BC}{TS}$ and $\angle C \cong \angle S$, the triangles are similar by S/A/S. (S/S/S can also be used.) $\angle T \cong \angle B$, and both have measure 62.46°. $\angle A \cong \angle R$, and both have measure 42.54°.

27. Since $\dfrac{a}{r} = \dfrac{5}{17.5} = \dfrac{b}{s}$ and $\angle C \cong \angle T$, the triangles are similar by S/A/S. Since the triangles are isosceles right triangles, $\angle A$, $\angle B$, $\angle R$, $\angle S$ all have measure 45°. $c = 5\sqrt{2}$ and $t = 17.5\sqrt{2}$.

29. Since $\angle A \cong \angle R$ and $\angle B \cong \angle T$, the triangles are similar by AA. $\angle C \cong \angle S$ and both have measure 41.5°; $s = 2.4$.

31. 55 ft

33. 80 ft

35. Distance from Marietta to Columbus: 92.4 mi., distance from Marietta to Cincinnati: 171.2 mi.

37. (a) C (b) B (c) B (d) D

39. $\frac{5}{2}$

41. 4

43. $t = 22$

45. $t = \frac{12}{5}$

47. Slope 2; y-intercept 5

49. Slope, $\frac{-3}{7}$; y-intercept $\frac{-11}{7}$

51. $y = x + 2$

53. $y = -x + 8$

55. $y = -x - 5$

57. $y = \frac{-7}{3}x + \frac{34}{9}$

59. Perpendicular

61. Parallel

63. Parallel

65. $y = 3x + 7$

67. $y = \frac{3}{2}x$

69. $y = x - 5$

71. $y = -x + 2$

73. In the figure, $AD \cong CR$, $CD \cong AR$, $AC \cong CA$. Hence, by SSS, triangle $ADC \cong$ triangle CRA, and so $\angle x \cong \angle u$. Similarly, $\angle v \cong \angle y$. Since $\angle x + \angle w + \angle y = 180°$, it follows that $\angle u + \angle w + \angle y = 180°$.

INDEX *of Applications*

Astronomy

Athletics

Biology and Life Sciences

Business and Manufacturing

Chemistry and Physics

Construction

Consumer Affairs

Education

Financial

Miscellaneous

Population

Time and Distance

Weather

SUBJECT INDEX

A